Geography
An Integrated Approach

Third edition

David Waugh

Academic research by
Dr Elizabeth Clutton

Additional case studies by
Sheila Morris

Examination questions by
Pete Murray & John Smith

Contents

Case Studies

Places

Frameworks

Decision Making Exercises

Focus on key skills

Index

Introduction for the reader

Geography: An Integrated Approach (more easily referred to as *GAIA*) has been written as much for those students who have an interest in Geography, an enquiring mind and a concern for the future of the planet upon which they live, as for those specialising in the subject. The text has been written as concisely as seemed practical in order to minimise the time needed for reading and note-taking, and to maximise the time available for discussion, individual enquiry and wider reading. Full colour has been used throughout for photographs, sketches and maps, which aim to illustrate the wide range of natural and human-created environments. Annotated diagrams are included to show inter-relationships and to help explain the more difficult concepts and theories. A wide range of graphical skills has been used to portray geo-graphical data – data which are as recent as possible at the time of writing and which you can continue to update for yourself by referring to relevant web sites on the Internet.

It is because Geography is concerned with interrelationships that this book has included, and aims to integrate, several fields of study. These involve physical environments (atmos-phere, lithosphere and hydrosphere) and the living world (biosphere); economic develop-ment (or lack of it); the frequent misuse of the environment, the long-overdue concern over the resultant consequences, and the need for careful management and sustainable develop-ment; together with the application, where appropriate, of a modern scientific approach using statistical methods in investigations. It is intended that this single book will:

■ satisfy the requirements of the new Advanced Subsidiary (AS) and Advanced GCE Geography specifications, and

■ allow you to read more widely in Geography than just to be limited to the core and option modules in your examination specifications.

By coincidence, the initial letters of the title of this book form the word *GAIA*. In Ancient Greece, Gaia was the goddess of the Earth. Today the term has been reintroduced to mean 'a new look at life on Earth', an approach that looks at the Earth in its entirety as a living organism. Hopefully, this book reflects aspects of this approach.

As the reader, you should be aware that there is no rigid, prescribed sequence in the order either of the chapters themselves or in their structure. Each is open to several routes of enquiry. Terminology can be a major problem, as geographers may use several terms, some borrowed from other disciplines, to describe the same phenomenon. When a term is introduced for the first time it is shown by emboldened print, has a list of alternatives (one of which is subsequently retained for consistency), and is defined. Alternative terms and specific examples often appear in brackets in order to save space. The detailed index, to allow you to cross-reference, has the key page(s) for each entry displayed in bold print.

The book sets out to provide a background store of information which will help you to understand basic processes and concepts, to enter discussions and to develop your own informed, rather than prejudiced, values and attitudes. Theory is, whenever possible, supported by specific examples, which have been highlighted in the text as **Places**. Although there are over one hundred Places, due to limited space many tend to be shorter than I would have wished. Nevertheless they should enable you either to build upon your earlier knowledge or to stimu-late you into reading more widely. At the end of each chapter is a more detailed **Case Study**. These include natural hazards, problems created by population growth, and by the mis-use of the natural environment, and the attempts – or lack of – to manage the environment and the Earth's resources.

References given at the conclusion of each chapter are those to which the author has himself referred (they are not intended to be a comprehensive list) and, in this edition, they include suggested reliable and helpful **Websites**. As the reader, it is essential you appreciate that Geography is a dynamic subject with data, views, policies and terms which change constantly. Consequently, your own research must not be limited to textbooks, which in any case are out of date even before their publication, but should be widened to include the use of the Internet, CD-ROMs, newspapers, journals, television, radio and many 'non-academic' media.

GAIA also includes 19 **Frameworks** whose function is to stimulate discussion on method-ological and theoretical issues. They illustrate some of the skills required, and the problems involved, in geographical enquiry, e.g. the uses, limitations and reliability of models; quantitative techniques; the collection of data, including using the Internet; making classifications; and the dangers of stereotyping and of making broad generalisations. Geography is also concerned with the development of graphical skills. The media show an increasing amount of data in a graphical form, and this is likely to grow as geographical information systems develop. It is assumed that the reader already understands those skills covered by current GCSE and Standard Grade examination specifications and therefore only new skills are explained in this book. Quantitative and statistical techniques are incorporated at appropriate points, although each may be relevant elsewhere in many of the physical and human/economic chapters. Following an explanation of each technique, there is a worked example.

New to this edition are the **specimen AS and A2 examination questions** at the end of each chapter, written by two experienced AS and Advanced GCE Examiners. Mark schemes for all the questions are provided on the accompanying *GAIA* Website. The *GAIA* site also contains hotlinks to the Websites listed in the References in each chapter and to other useful sources of information on the Internet. Go to: **www.gaia.nelson.co.uk**

Also included in this new edition are two extended **Decision Making Exercises** and three **Focus on Key Skills** pages which cover the main key skills of Application of Number, Communication, and Information Technology.

Finally, I hope that the book, as well as helping you earn a good grade, will give you some of the enjoyment which Geography has given me, and that it will encourage you to take every opportunity to travel. The best way to learn about the world and to try to understand the way of life of its different people, is through first-hand experience.

Best wishes with your studies,

David Waugh

Author's Acknowledgements

To help me write this third edition of *Geography: An Integrated Approach*, several leading geographers were asked to comment on the current accuracy and relevance of the second edition, and to indicate recent academic changes in terminology, concepts and approach. I am, therefore, most grateful to the following:
Dr David Chester (University of Liverpool) for 'Plate tectonics, earthquakes and volcanoes'
Professor Andrew Goudie (University of Oxford) for 'Weathering and slopes', 'Periglaciation', 'Deserts' and 'Rock types and landforms'
Professor Malcolm Newson (University of Newcastle) for 'Drainage basins and rivers'
Dr David Evans (University of Glasgow) for 'Glaciation'
Professor John Pethick (University of Newcastle) for 'Coasts'
Dr Grant Bigg (University of East Anglia) for 'Weather and climate'
Dr Steven Trudgill (University of Cambridge) for 'Soils'
Dr Antoinette Mannion (University of Reading) for 'Biogeography'
Michael John Alexander (University of Durham) for 'Soils' and 'World climate, soils and vegetation'
Professor David Phillips (Lingnan University, Hong Kong) for 'Population' and 'World development'
Dr Nicholas Ford (University of Exeter) for 'Population'
Dr Robert Gant (Kingston University) for 'Settlement' and 'Urbanisation'
Dr Ian Bowler (University of Leicester) for 'Farming and food supply' and 'Rural land use'
Dr Nick Middleton (University of Oxford) for 'Energy resources'
Dr Louise Crewe (University of Nottingham) for 'Manufacturing industries'
Dr Paul Bull (Birkbeck College) for 'Manufacturing industries'
Dr Richard Knowles (University of Salford) for 'Transport and interdependence'
Dr Alisdair Rogers (University of Oxford) for 'World development'

My thanks also to the following contributors:
Pete Murray for the AS/A2 questions in Chapters 1–12 and the Focus on Key Skills pages in Chapters 3, 9 and 13
John Smith for the AS/A2 questions in Chapters 13–22 and the Decision Making Exercise in Chapter 11
Sheila Morris for Frameworks 1 and 18, Places 43, Case Studies 3B, 6, 11, 12B, 15B, 16 and 17, and the Decision Making Exercise in Chapter 15
And finally thanks to all the important back-up team of Dr Elizabeth Clutton, Katherine James and my wife Judith.

Chapter 1

Plate tectonics, earthquakes and volcanoes

'. . . how does a supercontinent begin to rift and how do the pieces move apart? What effects do such movements have on the shaping of the continental landscapes, on hot climates and ice ages, on the evolution of life in general and on humanity's relationship with the upper crust of the Earth in particular?'

The Making of a Continent, Ron Redfern, 1983

The history of the Earth

It is estimated that the Earth was formed about 4 600 000 000 years ago. Even if this figure is simplified to 4600 million years, it still presents a timescale far beyond our understanding. Nigel Calder, in his book *The Restless Earth*, made a more comprehensible analogy by reducing the timespan to 46 years. He ignored the eight noughts and compared the 46 years with a human lifetime (Places 1).

Figure 1.1

The geological timescale

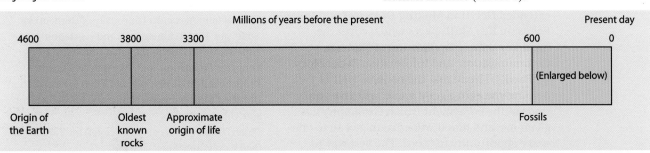

Era		Geological period	Millions of years before present	Conditions and rocks in Britain	Major world events
CENOZOIC	**Quaternary**	Holocene	0.01	Post ice age. Alluvium deposited, peat formed	Early civilisations
		Pleistocene	2	Ice age, with warm periods	Emergence of the human
	Tertiary	Pliocene	7	Warm climate: Crag rocks in East Anglia	
		Miocene	26	No deposits in Britain	Formation of the Alps
		Oligocene	38	Warm shallow seas in south of England	Rockies and Himalayas begin to form
		Eocene	54	Nearly tropical: London clay	Volcanic activity in Scotland
MESOZOIC		Cretaceous	136	Chalk deposited: Atlantic ridge opens	End of the dinosaurs/Age of the dinosaurs
		Jurassic	195	Oxford clays and limestones: warm	Pangaea breaks up
		Triassic	225	Desert: sandstones	First mammals
PALAEOZOIC		Permian	280	Desert: New Red Sandstones, limestones	Formation of Pangaea
		Carboniferous	345	Tropical coast with swamps: coal	First amphibians and insects
		Devonian	395	Warm desert coastline: sandstones	First land animals
		Silurian	440	Warm seas with coral: limestones	First land plants
		Ordovician	500	Warm seas: volcanoes (Snowdonia) sandstones, shales	First vertebrates
		Cambrian	570	Cold at times: sea conditions	Abundant fossils begin
		Pre-Cambrian		Igneous and sedimentary rocks	

'…Or we can depict Mother Earth as a lady of 46, if her "years" are megacenturies. The first seven of those years are wholly lost to the biographer, but the deeds of her later childhood are to be seen in old rocks in Greenland and South Africa. Like the human memory, the surface of our planet distorts the record, emphasising more recent events and letting the rest pass into vagueness – or at least into unimpressive joints in worn down mountain chains.

Most of what we recognise on Earth, including all substantial animal life, is the product of the past six years of the lady's life. She flowered, literally, in her middle age. Her continents were quite bare of life until she was getting on for 42 and flowering plants did not appear until she was 45 – just one year ago. At that time, the great reptiles, including the dinosaurs, were her pets and the break-up of the last supercontinent was in progress.

The dinosaurs passed away eight months ago and the upstart mammals replaced them. In the middle of last week, in Africa, some man-like apes turned into ape-like men and, at the weekend, Mother Earth began shivering with the latest series of ice ages. Just over four hours have elapsed since a new species calling itself *Homo sapiens* started chasing the other animals and in the last hour it has invented agriculture and settled down. A quarter of an hour ago, Moses led his people to safety across a crack in the Earth's shell, and about five minutes later Jesus was preaching on a hill farther along the fault line. Just one minute has passed, out of Mother Earth's 46 "years", since man began his industrial revolution, three human lifetimes ago. During that time he has multiplied his numbers and skills prodigiously and ransacked the planet for metal and fuel.'

N. Calder, The Restless Earth (1972)

Geologists have been able to study rocks and fossils formed during the last 600 million years, equivalent to the last 'six years of the lady's life', and have produced a time chart, or **geological timescale**. Not only have they been able to add dates with increasing confidence, but they have made progress in describing and accounting for the major changes in the Earth's surface, e.g. sea-level fluctuations and landform development, and in its climate. The timescale, shown in Figure 1.1, should be a useful reference for later parts of this book.

Earthquakes

Even the earliest civilisations were aware that the crust of the Earth is not rigid and immobile. The first major European civilisation, the Minoan, based in Crete, constructed buildings such as the Royal Palace at Knossos which withstood a succession of earthquakes. However, this civilisation may have been destroyed by the effects of a huge volcanic eruption on the nearby island of Thera (Santorini). Later, inhabitants of places as far apart as Lisbon (1755), Tokyo (1923), San Francisco (1906 – Places 2, page 11), Mexico City (1985), Los Angeles (1994 – Case Study 15A) and Kobe (1995) were to suffer from the effects of major earth movements.

It was by studying earthquakes that geologists were first able to determine the structure of the Earth (Figure 1.2). At the **Mohorovičič** or '**Moho' discontinuity**, it was found that shock waves begin to travel faster, indicating a change of structure – in this case, the junction of the Earth's **crust** and **mantle** (Figure 1.2). The 'Moho' discontinuity is the junction between the Earth's crust and the mantle where seismic waves are modified. The Moho is at about 35–40 km beneath continents (reaching 70 km under mountain chains) and at 6–10 km below the oceans.

Earthquakes result from a slow build-up of pressure within crustal rocks. If this pressure is suddenly released then parts of the surface may experience a jerking movement. Within the crust, the point at which the release in pressure occurs is known as the **focus**. Above this, on the surface and usually receiving the worst of the shock or **seismic waves**, is the **epicentre**. Unfortunately, it is not only the immediate or primary effects of the earthquake that may cause loss of life and property; often the secondary or after-effects are even more serious. These (Places 2) may include fires from broken gas pipes, disruption of transport and other services, exposure caused by a lack of shelter, a shortage of food, clean water and medical equipment, and disease caused by polluted water supplies. These problems may be exacerbated by after-shocks which often follow the main earthquake.

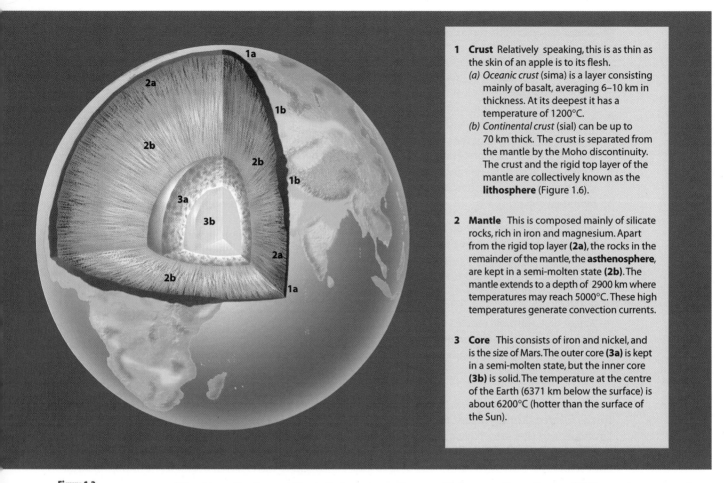

1 **Crust** Relatively speaking, this is as thin as the skin of an apple is to its flesh.
  (a) *Oceanic crust* (sima) is a layer consisting mainly of basalt, averaging 6–10 km in thickness. At its deepest it has a temperature of 1200°C.
  (b) *Continental crust* (sial) can be up to 70 km thick. The crust is separated from the mantle by the Moho discontinuity. The crust and the rigid top layer of the mantle are collectively known as the **lithosphere** (Figure 1.6).

2 **Mantle** This is composed mainly of silicate rocks, rich in iron and magnesium. Apart from the rigid top layer (**2a**), the rocks in the remainder of the mantle, the **asthenosphere**, are kept in a semi-molten state (**2b**). The mantle extends to a depth of 2900 km where temperatures may reach 5000°C. These high temperatures generate convection currents.

3 **Core** This consists of iron and nickel, and is the size of Mars. The outer core (**3a**) is kept in a semi-molten state, but the inner core (**3b**) is solid. The temperature at the centre of the Earth (6371 km below the surface) is about 6200°C (hotter than the surface of the Sun).

Figure 1.2

The internal structure of the Earth

The strength of an earthquake is measured on the Richter scale (Figure 1.3). To cover the huge range of earthquakes, the magnitude of the scale is logarithmic, each unit representing a tenfold increase in strength and a 30-fold increase in energy. This means that the 1755 Lisbon earthquake was 10 times stronger and released 30 times more energy than the 1985 Mexico City earthquake, and was nearly 100 times stronger and released almost 900 times more energy than the 1989 San Francisco earthquake (Figure 1.3).

Figure 1.3

The Richter scale

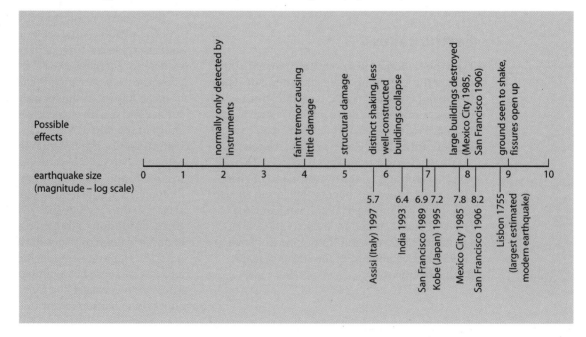

1906

At 0512 hours on the morning of 18 April, the ground began to shake. There were three tremors, each one increasingly more severe. The ground moved by over 6 m in an earthquake which measured 8.2 on the Richter scale. Many apartment buildings collapsed, bridges were destroyed – the Golden Gate had not then been built – and water pipes fractured. The worst damage was 'downtown' where the housing density was greatest. Although many people were trapped within collapsed buildings there were relatively few deaths.

Then came the fire! It started in numerous places resulting from overturned stoves or sparked by electricity or the ignition of gas escaping from the broken mains. As the water pipes had been fractured, it hardly mattered that there were only 38 horse-drawn fire engines to cope with 52 fires. As the fire spread, houses were blown up with dynamite to try to create gaps to thwart the flames, but the explosions only caused further fires. It took over three days to put out the fires, by which time over 450 people (mainly those previously trapped) had died, 28 000 buildings in 500 blocks had been destroyed, and an area six times greater than that destroyed by the Great Fire of London had been ravaged. Fortunately, the threat of disease and starvation was averted.

1989

During the early evening rush-hour on 17 October 1989, an earthquake measuring 6.9 on the Richter scale shook the city for 15 seconds. The early-warning system had given no clues. Skyscrapers swayed 3 m, fractured gas pipes caused fires in one residential area and parts of a downtown shopping centre collapsed. The greatest loss of life occurred when 1.5 km of the upper section of the two-tier Interstate Highway 880 collapsed onto the lower portion, killing people in their vehicles. The final casualty figures of 67 dead and 2000 homeless were, however, low compared with an earthquake of similar magnitude in Armenia 11 months earlier which killed 55 000 people. In San Francisco, money was available and precautions had been taken to reduce the effects of an earthquake. San Francisco also has fully equipped and trained emergency services. Armenia has none of these things – which is why the death toll there was so much greater.

Earthquakes, volcanoes and young fold mountains

These do not occur at random over the Earth's surface but have a clearly identifiable pattern. This can be seen by working through the following activities.

1   On an outline map of the world, **mark by a dot** (there is no need to name the places) the location of the following earthquakes:

1920	Taiwan, Fiji
1921	Peru
1922	Chile
1923	Japan
1924	Philippines
1925	California
1926	Rhodes
1927	Japan
1928	Chile
1929	Aleutians, Japan
1931	New Zealand
1932	Mexico
1933	California
1935	Sumatra
1938	Java
1939	Chile, Turkey

1940	Burma, Peru
1941	Ecuador, Guatemala
1943	Philippines, Java
1944	Japan
1946	West Indies, Japan
1949	Alaska
1950	Japan, Assam
1953	Turkey, Japan
1956	California
1957	Mexico
1958	Alaska
1960	Chile, Morocco
1962	Iran
1963	Yugoslavia
1964	Alaska, Turkey, Mexico, Japan, Taiwan
1965	El Salvador, Greece
1966	Chile, Peru, Turkey
1967	Colombia, Yugoslavia, Java, Japan
1968	Iran
1970	Peru
1971	New Guinea, California
1972	Nicaragua
1976	Guatemala, Italy, China, Philippines, Turkey
1978	Japan
1980	Italy
1985	Mexico, Colombia
1988	Armenia
1989	San Francisco, Iran
1993	Java, Japan, India, Egypt
1994	Los Angeles

1995	Japan, Greece
1996	China, Indonesia
1997	Afghanistan, Italy, Iran
1998	Iraq, Afghanistan
1999	Turkey, Taiwan

2   On a tracing overlay, mark and **name** the following **volcanoes**: Aconcagua, Chimborazo, Cotopaxi, Nevado del Ruiz, Paricutín, Popocatepetl, Mount St Helens, Fuji, Mount Pinatubo, Mayon, Krakatoa, Merapi, Ruapehu, Erebus, Helgafell, Surtsey, Azores, Ascension, St Helena, Tristan da Cunha, Vesuvius, Etna, Pelée, Montserrat, Mauna Loa, Kilauea.

3   On a second overlay, **mark and name** the following **fold mountains**: Andes, Rockies, Atlas, Pyrenees, Alps, Caucasus, Hindu Kush, Himalayas, Southern Alps.

4   Use the Internet (see Framework 1, page 22) to find the names of more earthquakes and volcanic eruptions, after 1999.

Plate tectonics

As early as 1620 Francis Bacon noted the jigsaw-like fit between the east coast of South America and the west coast of Africa. Others were later to point out similarities between the shapes of coastlines of several adjacent continents.

In 1912, a German meteorologist, Alfred Wegener, published his theory that all the continents were once joined together in one large supercontinent which he named **Pangaea**. Later, this landmass somehow split up and the various continents, as we know them, drifted apart. Wegener collated evidence from several sciences:

- **Biology** Mesosaurus was a small reptile living in Permian times (Figure 1.1); its remains have been found only in South Africa and Brazil. A plant which existed when coal was being formed has only been located in India and Antarctica.
- **Geology** Rocks of similar type, age, formation and structure occur in south-east Brazil and South Africa, and the Appalachian Mountains of the eastern USA correspond geologically with mountains in north-west Europe.
- **Climatology** Coal, formed under warm, wet conditions, is found beneath the Antarctic ice-cap, and evidence of glaciation had been noted in tropical Brazil and central India. Coal, sandstone and limestone could not have formed in Britain with its present climate.

Wegener's theory of **continental drift** combined information from several subject areas, but his ideas were rejected by specialists in those disciplines, partly because he was not regarded as an expert himself but perhaps mainly because he could not explain how solid continents had changed their positions. He was unable to suggest a mechanism for drift.

Figure 1.4a shows Wegener's Pangaea and how it began to divide up into two large continents, which he named **Laurasia** and **Gondwanaland**; it also suggests how the world may look in the future if the continents continue to drift.

Figure 1.4

The wandering continents

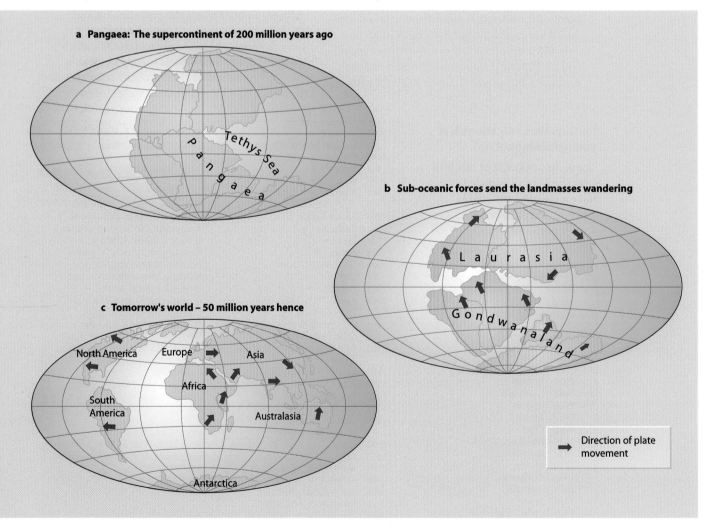

a **Pangaea: The supercontinent of 200 million years ago**

b **Sub-oceanic forces send the landmasses wandering**

c **Tomorrow's world – 50 million years hence**

Direction of plate movement

Plate tectonics, earthquakes and volcanoes

Since Wegener first put forward this theory, three groups of new evidence have become available to support his ideas.

1 **The discovery and study of the Mid-Atlantic Ridge** While investigating islands in the Atlantic in 1948, Maurice Ewing noted the presence of a continuous mountain range extending the whole length of the ocean bed. This mountain range, named the Mid-Atlantic Ridge, is about 1000 km wide and rises to 2500 m in height. Ewing also noted that the rocks of this range were volcanic and recent in origin – not ancient as previously assumed was the case in mid-oceans. Later investigations show similar ranges on other ocean floors, the one in the eastern Pacific extending for nearly 5000 km (Figure 1.8).

2 **Studies of palaeomagnetism in the 1950s** During underwater volcanic eruptions, basaltic lava is intruded into the crust and cools (Figure 1.31). During the cooling process, individual minerals, especially iron, align themselves along the Earth's magnetic field, i.e. in the direction of the magnetic pole. Recent refinements in dating techniques enable the time at which rocks were formed to be accurately calculated. It was known before the 1950s that the Earth's magnetic pole varied a little from year to year, but only then was it discovered that the magnetic field reverses periodically, i.e. the magnetic pole is in the south for a period of time and then in the north for a further period, and so on. It is claimed that there have been 171 reversals over 76 million years. If formed when the magnetic pole was in the north, new basalt would be aligned to the north. After a reversal in the magnetic poles, newer lava would be oriented to the south. After a further reversal, the alignment would again be to the north. Subsequent investigations have shown that these alternations in alignment are almost symmetrical in rocks on either side of the Mid-Atlantic Ridge (Figure 1.5).

3 **Sea floor spreading** In 1962, Harry Hess studied the age of rocks from the middle of the Atlantic outwards to the coast of North America. He confirmed that the newest rocks were in the centre of the ocean, and were still being formed in Iceland, and that the oldest rocks were those nearest to the USA and the Caribbean. He also suggested that the Atlantic could be widening by up to 5 cm a year.

One major difficulty resulting from this concept of sea floor spreading was the implication that the Earth must be increasing in size. Since this is not so, evidence was needed to show that else-where parts of the crust were being destroyed. Such areas were found to correspond to the fringes of the Pacific Ocean – the region where you plotted some major earthquakes and volcanic eruptions (page 11). These discoveries led to the development of the theory of **plate tectonics** which is now virtually universally accepted, but which may still be modified following further investigation and study.

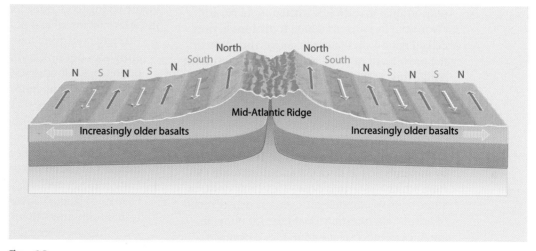

Figure 1.5

The repeated reversal of the Earth's magnetic field

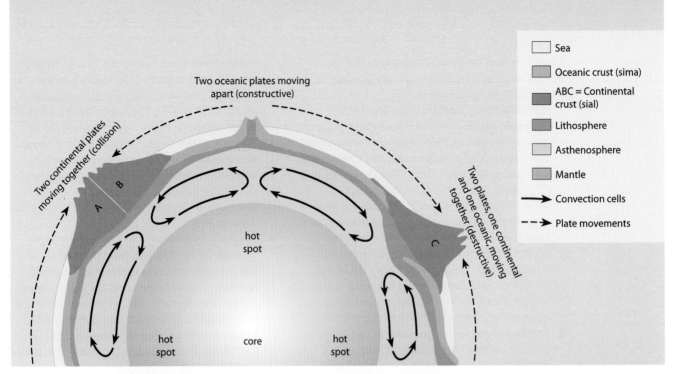

Figure 1.6

How plates move

Figure 1.7

Differences between continental and oceanic crust

	Continental crust (sial)	Oceanic crust (sima)
Thickness	35–40 km on average, reaching 60–70 km under mountain chains	6–10 km on average
Age of rocks	very old, mainly over 1500 million years	very young, mainly under 200 million years
Weight of rocks	lighter, with an average density of 2.6	heavier, with an average density of 3.0
Nature of rocks	light in colour; many contain silica and aluminium; numerous types, granite is the most common	dark in colour; many contain silica and magnesium; few types, mainly basalt

The theory of plate tectonics

The **lithosphere** (the Earth's crust and the rigid upper part of the mantle) is divided into seven large and several smaller **plates**. The plates, which are rigid, float like rafts on the underlying semi-molten mantle (the **asthenosphere**) and are moved by currents which form **convection cells** (Figure 1.6). Plate tectonics is the study of the movement of these plates and their resultant landforms. There are two types of plate: **continental** and **oceanic**. However, these terms do not refer to actual continents and oceans but to different types of crust or rock.

Continental crust, or **sial**, is composed of older, lighter rock of granitic type. It is dominated by minerals rich in silica (Si) and aluminium (Al), from which the term is derived. Oceanic crust, or **sima**, consists of much younger and denser rock of basaltic composition. Its dominant minerals are silica (Si) and magnesium (Ma). The major differences between the two types of crust are summarised in Figure 1.7.

Plate movement

As a result of the convection cells generated by heat from the centre of the Earth, plates may move towards, away from or sideways along adjacent plates. It is at plate boundaries that most of the world's major landforms occur, and where earthquake, volcanic and mountain-building zones are located (Figure 1.8). However, before trying to account for the formation of these landforms several points should be noted.

1 Due to its relatively low density, continental crust does not sink and so is permanent; being denser, oceanic crust can sink. Oceanic crust is being formed and destroyed continuously.
2 Continental plates, such as the Eurasian Plate, may consist of both continental and oceanic crust.
3 Continental crust may extend far beyond the margins of the landmass.
4 Plates cannot overlap. This means that either they must be pushed upwards on impact to form mountains (AB on Figure 1.6) or one plate must be forced downwards into the mantle and destroyed (C on Figure 1.6).
5 No 'gaps' may occur on the Earth's surface so, if two plates are moving apart, new oceanic crust originating from the mantle must be being formed.

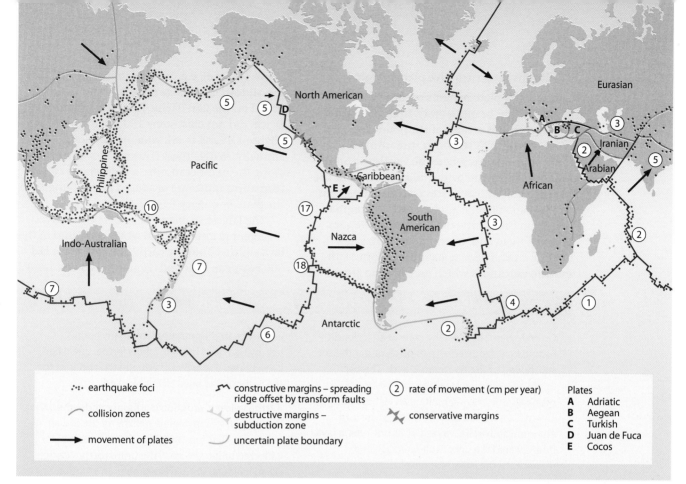

Legend:

- ∴ earthquake foci
- ⌒ collision zones
- → movement of plates
- ∿ constructive margins – spreading ridge offset by transform faults
- destructive margins – subduction zone
- uncertain plate boundary
- ② rate of movement (cm per year)
- ✕ conservative margins

Plates
A Adriatic
B Aegean
C Turkish
D Juan de Fuca
E Cocos

6 The Earth is neither expanding nor shrinking in size. Thus when new oceanic crust is being formed in one place, older oceanic crust must be being destroyed in another.

7 Plate movement is slow (though not in geological terms) and is usually continuous. Sudden movements are detected as earthquakes.

8 Most significant landforms (fold mountains, volcanoes, island arcs, deep-sea trenches, and batholith intrusions) are found at plate boundaries. Very little change occurs in plate centres (shield lands). Figure 1.9 summarises the major landforms resulting from different types of plate movement.

Type of plate boundary	Description of changes	Examples
A Constructive margins (spreading or divergent plates)	two plates move away from each other; new oceanic crust appears forming mid-ocean ridges with volcanoes	Mid-Atlantic Ridge (Americas moving away from Eurasian and African Plates) East Pacific Rise (Nazca and Pacific Plates moving apart)
B Destructive margins (subduction zones)	oceanic crust moves towards continental crust but, being heavier, sinks and is destroyed forming deep-sea trenches and island arcs with volcanoes	Nazca sinks under South American Plate (Andes) Juan de Fuca sinks under North American Plate (Rockies) Island arcs of the West Indies and Aleutians
Collision zones	two continental crusts collide and, as neither can sink, are forced up into fold mountains	Indian Plate collided with Eurasian Plate, forming Himalayas African Plate collided with Eurasian Plate, forming Alps
C Conservative or passive margins (transform faults)	two plates move sideways past each other – land is neither formed nor destroyed	San Andreas Fault in California
Note: centres of plates are rigid...	rigid plate centres form	
	a shields lands (cratons) of ancient worn-down rocks	Canadian (Laurentian) Shield, Brazilian Shield
	b depressions on edges of the shield which develop into large river basins	Mississippi – Missouri, Amazon
...with one main exception	Africa dividing to form a rift valley and possibly a new sea	African Rift Valley and the Red Sea

Landforms at constructive plate margins

Constructive plate margins occur where two plates diverge, or move away, from each other and new crust is created at the boundary. This process, known as **sea-floor spreading**, occurs in the mid-Atlantic where the North and South American Plates are being pulled apart from the Eurasian and African Plates by convection cells. As the plates diverge, molten rock or **magma** rises from the mantle to fill any possible gaps between them and, in doing so, creates new oceanic crust. The magma initially forms **submarine volcanoes** which may in time grow above sea-level, e.g. Surtsey, south of Iceland on the Mid-Atlantic Ridge (Places 3) and Easter Island on the East Pacific Rise. The Atlantic Ocean did not exist some 150 million years ago (Figure 1.4) and is still widening by some 2–5 cm annually. Where there is lateral movement along the mid-ocean ridges, large cracks called **transform faults** are produced at right-angles to the plate boundary (Figure 1.8).

The largest visible product of constructive divergent plates is Iceland where one-third of the lava emitted onto the Earth's surface in the last 500 years can be found (Figures 1.10b and 1.26).

Places 3   Iceland: a constructive plate margin

On 14 November 1963, the crew of an Icelandic fishing boat reported an explosion under the sea south-west of the Westman Islands. This was followed by smoke, steam and emissions of pumice stone. Having built up an ash cone of 130 m from the seabed, the island of Surtsey emerged above the waves. On 4 April 1964, a lava flow covered the unconsolidated ash and guaranteed the island's survival.

Just before 0200 hours on 23 January 1973, an earth tremor stopped the clock in the main street of Heimaey, Iceland's main fishing port. Once again the North American and Eurasian Plates were moving apart (Figure 1.10b). Fishermen at sea witnessed the crust of the Earth break open and lava and ash pour out of a fissure 2 km in length (page 25). Eventually the activity became concentrated on the volcanic cone of Helgafell and the inhabitants of Heimaey were evacuated to safety. By the time volcanic activity ceased six months later, many homes nearby had been burned; others farther afield had been buried under 5 m of ash; and the entrance to the harbour had been all but blocked.

A large volcanic eruption in a fissure under the Vatnajökull icecap melted 3000 m³ of the glacier above it in October 1996. The resultant meltwater collected under the ice in the Grimsvotn volcanic crater (caldera) until, in November, an eruption spewed a 4270 m high column of ash into the air and released the trapped water. The subsequent torrent, which contained house-sized blocks of ice and black sulphurous water, demolished three of Iceland's largest bridges and several kilometres of the south coast ring road (Figure 1.25). A further event in December 1998 resulted in five craters within the caldera becoming active along a 1300 m long fissure and the creation of an eruption plume 10 km in height.

Figure 1.10

A constructive plate margin: Iceland

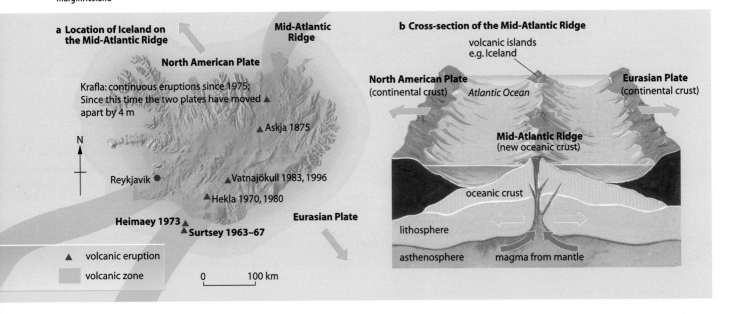

a **Location of Iceland on the Mid-Atlantic Ridge**

Mid-Atlantic Ridge

North American Plate

Krafla: continuous eruptions since 1975; Since this time the two plates have moved apart by 4 m

▲ Askja 1875

N

Reykjavik ●

▲ Vatnajökull 1983, 1996

▲ Hekla 1970, 1980

Heimaey 1973 ▲

▲ Surtsey 1963–67

Eurasian Plate

▲ volcanic eruption

☐ volcanic zone

0   100 km

b **Cross-section of the Mid-Atlantic Ridge**

volcanic islands e.g. Iceland

North American Plate (continental crust)   *Atlantic Ocean*   **Eurasian Plate** (continental crust)

Mid-Atlantic Ridge (new oceanic crust)

oceanic crust

lithosphere

asthenosphere   magma from mantle

Figure 1.11

The African Rift Valley

a Location

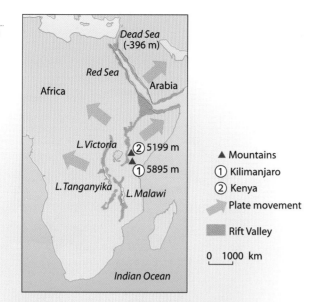

b Idealised cross-section

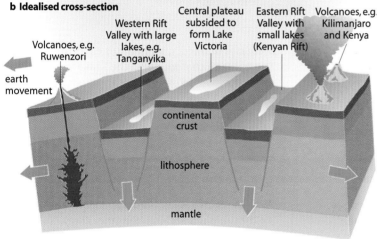

Volcanoes, e.g. Ruwenzori

earth movement

Western Rift Valley with large lakes, e.g. Tanganyika

Central plateau subsided to form Lake Victoria

Eastern Rift Valley with small lakes (Kenyan Rift)

Volcanoes, e.g. Kilimanjaro and Kenya

continental crust

lithosphere

mantle

Figure 1.12

A destructive plate margin – the Nazca and South American Plate boundary

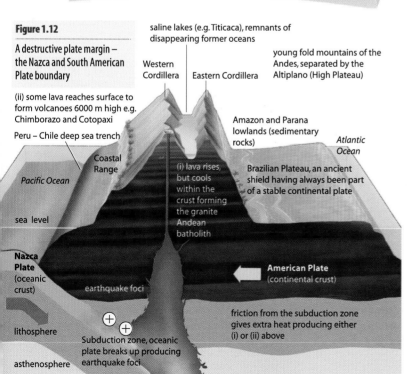

saline lakes (e.g. Titicaca), remnants of disappearing former oceans

young fold mountains of the Andes, separated by the Altiplano (High Plateau)

Western Cordillera

Eastern Cordillera

(ii) some lava reaches surface to form volcanoes 6000 m high e.g. Chimborazo and Cotopaxi

Peru – Chile deep sea trench

Coastal Range

Pacific Ocean

sea level

(i) lava rises, but cools within the crust forming the granite Andean batholith

Amazon and Parana lowlands (sedimentary rocks)

Atlantic Ocean

Brazilian Plateau, an ancient shield having always been part of a stable continental plate

Nazca Plate (oceanic crust)

earthquake foci

American Plate (continental crust)

lithosphere

Subduction zone, oceanic plate breaks up producing earthquake foci

asthenosphere

friction from the subduction zone gives extra heat producing either (i) or (ii) above

The Atlantic Ocean was formed as the continent of Laurasia split into two, a process that may be repeating itself today in East Africa. Here the brittle crust has fractured and, as sections moved apart, the central portion dropped to form the Great African Rift Valley (Figure 1.11) with its associated volcanic activity. In Africa the rift valley extends for 4000 km from Mozambique to the Red Sea. In places its sides are over 600 m in height while its width varies between 10 and 50 km. Where the land has been pulled apart and dropped sufficiently, it has been invaded by the sea. It has been suggested that the Red Sea is a newly forming ocean. Looking 50 million years into the future (Figure 1.4c), it is possible that the east of Africa may detach itself from the remainder of the continent.

Landforms at destructive plate margins

Destructive margins occur where continental and oceanic plates converge. The Pacific Ocean, which extends over five oceanic plates, is surrounded by several continental plates (Figure 1.8). The Pacific Plate, the largest of the oceanic plates, and the Philippines Plate move north-west to collide with eastern Asia. In contrast, the smaller Nazca, Cocos and Juan de Fuca Plates travel eastwards towards South America, Central America and North America respectively. Figure 1.12 shows how the Nazca Plate, made of oceanic crust which cannot override continental crust, is forced to dip downwards at an angle to form a **subduction zone** with its associated **deep-sea trench**. As oceanic lithosphere descends, the increase in pressure can trigger off major earthquakes, while the heat produced by friction helps to convert the disappearing crust back into magma. Being less dense than the mantle, the newly formed magma will try to rise to the Earth's surface. Where the magma does reach the surface (Figure 1.13 and Places 4) volcanoes will occur. These volcanoes are likely to form either a long chain of **fold mountains** (e.g. the Andes) or, if the eruptions take place offshore, an **island arc** (e.g. Japan, West Indies). Estimates claim that 80 per cent of the world's present active volcanoes are located above subduction zones. As the rising magma at destructive margins is more acidic than the lava of constructive margins (page 24), it is more viscous and flows less easily. It may solidify within the mountain mass to form large **intrusive** features called **batholiths** (Figure 1.31).

Plate tectonics, earthquakes and volcanoes

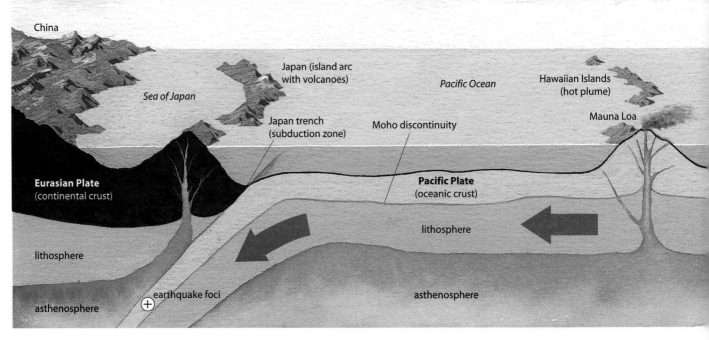

Figure 1.13

Landforms resulting from plate tectonics in the Pacific Ocean

Tsunamis are giant tidal waves generated, usually at destructive margins, either by volcanic eruptions, as with Krakatoa in 1883, or by submarine earthquakes. They can travel across oceans at speeds of up to 600 km/hr. As a tsunami approaches the land, its height increases markedly, sometimes exceeding 15 m. The tsunami that followed the Krakatoa eruption drowned over 36 000 people in coastal villages of Java and Sumatra before travelling three times around the world (Place 35, page 289).

Places 4 New Zealand: a destructive plate margin

New Zealand lies on a destructive plate margin where the oceanic crust of the Pacific Plate is subducted beneath the continental crust of the Indo-Australian Plate (Figure 1.14). In the North Island, this subduction has created the offshore Tonga–Kermadec deep-sea trench, formed numerous volcanoes and causes frequent, and occasionally life-threatening, earthquakes. In the South Island, the subduction has pushed the edges of the continental plate upwards to form the Southern Alps. Earthquake activity becomes less frequent and less severe towards the south. Most of New Zealand's volcanoes and geothermal features are located in the 20–40 km wide Taupo Volcanic Zone which stretches almost 300 km from Mount Ruapehu to White Island and which forms part of the Pacific 'ring of fire'. Seismic evidence suggests that the Pacific Plate descends swiftly at this point causing the oceanic crust to melt and the resultant magma to rise to the surface. Since 1886, five volcanoes have erupted in this zone, the latest, apart from the permanently active White Island, being Ruapehu which erupted in 1995 and again in 1996. The Ruapehu eruption ejected steam, ash and car-sized

rocks and caused several **lahars** (mudflows – Figure 1.47). At various times local roads, the railway, airspace and a nearby HEP station were forced to close, although the biggest loss was to the tourist industry as two skiing seasons were lost.

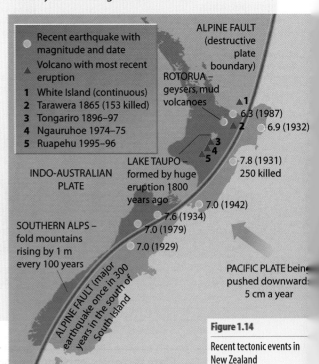

- ○ Recent earthquake with magnitude and date
- ▲ Volcano with most recent eruption
- **1** White Island (continuous)
- **2** Tarawera 1865 (153 killed)
- **3** Tongariro 1896–97
- **4** Ngauruhoe 1974–75
- **5** Ruapehu 1995–96

ALPINE FAULT (destructive plate boundary)

ROTORUA – geysers, mud volcanoes

▲1
6.3 (1987)
▲2   6.9 (1932)

▲3
▲4
▲5

LAKE TAUPO – formed by huge eruption 1800 years ago

INDO-AUSTRALIAN PLATE

7.8 (1931) 250 killed

7.0 (1942)

7.6 (1934)

7.0 (1979)

7.0 (1929)

SOUTHERN ALPS – fold mountains rising by 1 m every 100 years

ALPINE FAULT (major earthquake once in 300 years in the south of South Island)

PACIFIC PLATE being pushed downward: 5 cm a year

Figure 1.14

Recent tectonic events in New Zealand

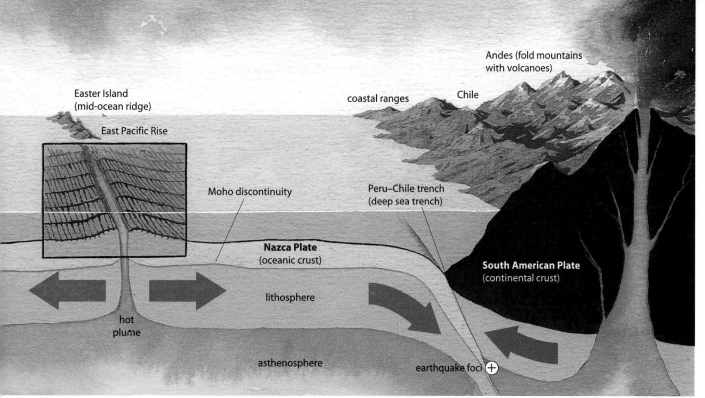

Easter Island
(mid-ocean ridge)

East Pacific Rise

Andes (fold mountains
with volcanoes)

coastal ranges   Chile

Moho discontinuity

Peru–Chile trench
(deep sea trench)

Nazca Plate
(oceanic crust)

South American Plate
(continental crust)

lithosphere

hot
plume

asthenosphere

earthquake foci ⊕

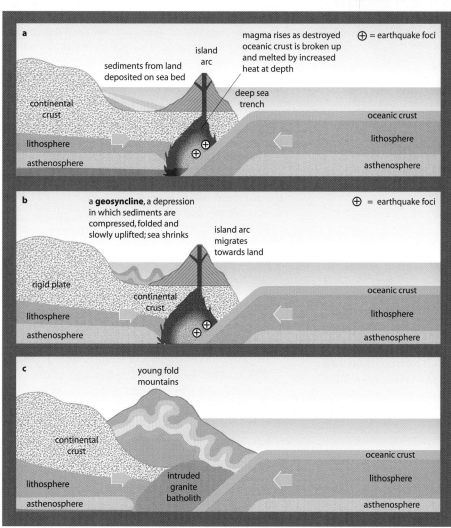

a

⊕ = earthquake foci

sediments from land
deposited on sea bed

island
arc

magma rises as destroyed
oceanic crust is broken up
and melted by increased
heat at depth

deep sea
trench

continental
crust

oceanic crust

lithosphere

lithosphere

asthenosphere

asthenosphere

⊕

b

⊕ = earthquake foci

a **geosyncline**, a depression
in which sediments are
compressed, folded and
slowly uplifted; sea shrinks

island arc
migrates
towards land

rigid plate

continental
crust

oceanic crust

lithosphere

lithosphere

asthenosphere

asthenosphere

⊕

c

young fold
mountains

continental
crust

intruded
granite
batholith

oceanic crust

lithosphere

lithosphere

asthenosphere

asthenosphere

Landforms at collision plate margins

The formation of fold mountains is often extremely complex. As has already been explained in the context of the Pacific, fold mountains often occur where oceanic crust is subducted by continental crust (Figure 1.15). A second, though less frequent, occurrence is when two plates composed of continental crust move together. In Places 5 the Indian subcontinent, forming part of the Indo-Australian Plate, is shown to have moved north-eastwards and to have collided with the Eurasian Plate. Because continental crust cannot sink, the subsequent collision caused the intervening sediments, which contained seashells, to be pushed upwards to form the Himalayas – an uplift that is still continuing. It is where these continental collisions occur that fold mountains form and the Earth's crust is at its thickest (Figures 1.6 and 1.7).

Figure 1.15

A collision plate margin – the formation of fold mountains (orogenesis)

Measurements of current convergence rates suggest that the Indo-Australian Plate is moving towards the Eurasian Plate at a rate of 5.8 cm/year. Although the convergence of two plates of continental crust has pushed up the Himalayas and caused the formation of the Tibetan Plateau, in parts the Indian Plate is being pushed under Tibet to form the mountain roots shown on Figure 1.16. This movement causes great stresses which are released by periodic earthquakes, e.g. in Central India in 1993 when an earthquake caused the deaths of 22 000 people. Scientists also believe that this plate movement is causing Mount Everest to rise by 2.5 cm/year (Figure 1.17).

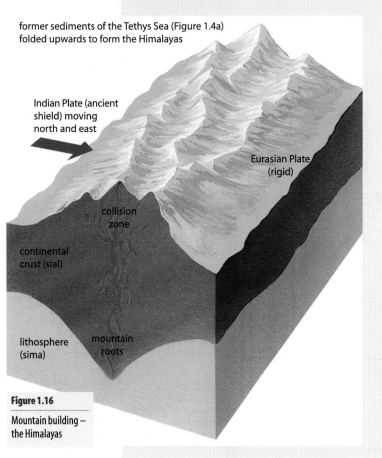

former sediments of the Tethys Sea (Figure 1.4a) folded upwards to form the Himalayas

Indian Plate (ancient shield) moving north and east

Eurasian Plate (rigid)

collision zone

continental crust (sial)

lithosphere (sima)

mountain roots

Figure 1.16

Mountain building – the Himalayas

Figure 1.17

Is Mount Everest still rising?

MOUNT EVEREST could be growing taller by up to an inch [2.5 cm] a year, according to scientists who hope to measure its precise height for the first time.

Although the mountain is usually referred to as being 29,002 ft [8848 m] high, experts admit they are not sure exactly how high it is because the summit is covered in a 20 ft [6 m] layer of snow and ice.

A new expedition, which should reach the summit in May, will carry an ice-coring kit able to drill through this layer and find the rock below. The team will then use the global positioning system (GPS) to work out the peak's precise height.

The mountain is believed to be growing slightly each year, pushed up as two of the Earth's tectonic plates collide beneath it. Researchers hope that by monitoring the position of the summit, they will be able to predict when earthquakes could hit the region.

According to Bradford Washburn, who is planning this summer's expedition, the movement has never been monitored.

'We know that Everest is on a fault line that is moving. As these plates come together, Everest is actually being pushed upwards. However, until we find out exactly where the mountain ends and the ice begins at the summit, this is impossible to monitor. But we think it could be moving as much as 1.2 inches [3 cm] a year.'

From *The Sunday Times*, 1 February 1998

NB The team later measured Everest to be 29 035 ft (8850 m).

Landforms at conservative plate margins

Conservative margins occur where two plates move parallel or nearly parallel to each other. Although frequent small earth tremors and occasional severe earthquakes may occur as a consequence of the plates trying to slide past each other, the margin between the plates is said to be **conservative** because crustal rocks are being neither created nor destroyed here. The boundary between the two plates is characterised by pronounced transform faults (Figure 1.18a). The San Andreas Fault is the most notorious of several hundred known transform faults in California (Places 6 and Case Study 15A).

The San Andreas Fault forms a junction between the North American and Pacific Plates. Although both plates are moving north-west, the Pacific Plate moves faster giving the illusion that they are moving in opposite directions. The Pacific Plate moves about 6 cm a year, but sometimes it sticks (like a machine without oil) until pressure builds up enabling it to jerk forwards. The last major movement resulted in the San Francisco earthquake of 1989 (Places 2, page 11). Should these plates continue to slide past each other, it is likely that Los Angeles will eventually be on an island off the Canadian coast.

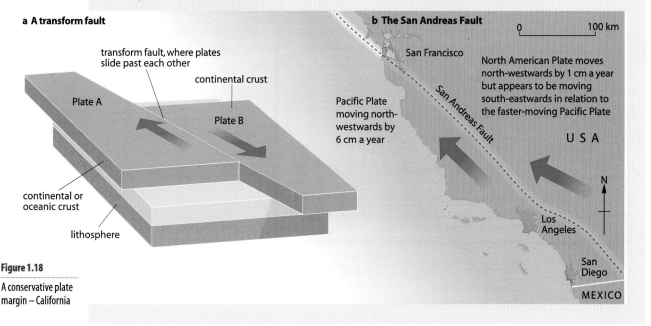

a A transform fault

transform fault, where plates slide past each other

continental crust

Plate A

Plate B

continental or oceanic crust

lithosphere

b The San Andreas Fault

0   100 km

San Francisco

Pacific Plate moving north-westwards by 6 cm a year

San Andreas Fault

North American Plate moves north-westwards by 1 cm a year but appears to be moving south-eastwards in relation to the faster-moving Pacific Plate

USA

N

Los Angeles

San Diego

MEXICO

Figure 1.18

A conservative plate margin – California

Plate tectonics and the British Isles

During the Cambrian period (Figure 1.1), northern Scotland lay on the American Plate while the rest of Britain was on the Eurasian Plate, as it is today. Both plates are thought then to have been in the latitude of present-day South Africa. In the Ordovician and Silurian periods, the two plates began to converge causing volcanic activity and the formation of mountains in Snowdonia and the Lake District (a collision zone). Being continental crust, sediment between the plates was pushed up to form the Caledonian Mountains which linked Scotland to the rest of Britain. During the Devonian period, the locked plates drifted northwards through a desert environment (the present Kalahari Desert) when the Old Red Sandstones were deposited (page 201). This northward movement continued in Carboniferous times, accompanied by a sinking of the land which allowed the limestones of that period to form in warm, clear seas (page 196).

As the land began to emerge from these seas, millstone grit was formed from sediments in a shallow sea, and then coal measures were laid down under the hot, wet, swampy conditions usually associated with equatorial areas. It was during the Permian and Triassic periods that the continents collided to form Pangaea (Figure 1.4a). Africa moved towards Europe, and Britain's New Red Sandstones (page 201) were laid down under dry, hot desert conditions (in the position of the present-day Sahara Desert). A further submergence during Jurassic/Cretaceous times enabled the Cotswold limestones and then the chalk of the Downs to form – again in warm, clear seas (page 196).

During the Tertiary era, the North American and Eurasian Plates split apart forming a constructive boundary and the volcanoes of north-western Scotland (page 29). At the same time, the African Plate moved further north pushing up the Alps and the hills of southern England. Subsequently, although Britain has been located away from the volcanoes and severe earthquakes associated with various plate margins, its landscape has been modified both during and since the ice ages. These, however, have been due to climatic change rather than plate movement.

What is the Internet?

The Internet is a collection of linked worldwide computer networks by which information can be exchanged rapidly. You can use this World Wide Web at college, at school or at home, if you are connected to an Internet Service Provider (ISP) such as AOL, Compuserve and Freeserve.

Browser   software such as Internet Explorer and Netscape Navigator which allows you to visit different Web pages and linked sites.

Search engine   a huge database which uses search terms (key words) to locate relevant Websites on the database.

Directory   a smaller database containing a range of selected Websites on a variety of topics.

URL   (Universal Resource Location) Website address, e.g. http://www.nelsonthornes.com/gaia

Hotlinks   cross-reference links created in a Website allowing you to transfer to another site directly.

Surfing   following links to find information on a variety of sites.

Bookmarking   storing a site address on a list of favourites, enabling a quick return without further 'surfing'.

Why use the Web?

The major advantage of the Web is the immediacy of the information available. For example, during Hurricane Mitch in October 1998, students were able to follow the track of the hurricane almost as it happened, and to obtain accounts of its impact immediately, long before the newspaper reports arrived. It was also possible to see the damage and to read first-hand accounts from people in affected areas. There are hundreds of similar examples – see Figure 1.19 for instance.

Finding material on the Web

Once 'logged on' to the Web it is worth thinking of it as a vast library with information published by a wide variety of sources. If you require up-to-date information, e.g. on hurricanes, there are two main ways of searching for this.

1 **Use a search engine** (using key words). Type in your key words, and the search engine will provide a list of sites featuring your key words. Be as specific as possible because the list can be extensive. For example, using AltaVista (http://www.altavista.com), the key word

Figure 1.19

The Montserrat Government Website continues to provide regular information on the Soufrière Hills Volcano

The Government of Montserrat and the Montserrat Volcano Observatory

Official Releases Concerning the Situation at Soufrière Hills Volcano, Montserrat, West Indies

http://www.geo.mtu.edu/volcanoes/west.indies/soufriere/govt/

'hurricanes' brings up more than 100 000 sites. Many of these will be irrelevant to your needs and it may take a lot of time and money to find the information you want, but you can discover some very useful sites.

2 **Use a directory** (using an index). To find hurricanes, go to Yahoo, for example (http://www.yahoo.co.uk). Select Society and Culture/Environment from the subject index. Follow the relevant links until you find the selection of sites for Hurricanes, Typhoons and Tropical Cyclones, which has links to 93 relevant sites at this stage (Figure 1.20). This may not offer the wide variety of sites that a search engine can provide but the selection is more logical and often more up to date.

However, the most effective means of finding information is to identify a likely source and find out its URL before logging on. For example, the official UK government site is the most likely to provide up-to-date UK population data. Type in the URL (http://www.open.gov.uk) and follow the index links to locate the relevant statistics:

> http://www.open.gov.uk ⇒ Organisation index ⇒ Office for National Statistics ⇒ United Kingdom in figures ⇒ The People: Population and Vital Statistics

A note of warning: it can be interesting to follow links and 'surf' for information, but it is also time-consuming and expensive. Once a useful site is found you should 'bookmark' it for speedy future reference.

Figure 1.20

Search route through the Yahoo! directory

YAHOO!®

Home > Society & Culture > Environment and Nature > Meteorology > Weather Phenomena > Hurricanes, Typhoons, and Tropical Cyclones (93)

Click on the subject index links until you find the 93 sites listed for Hurricanes, Typhoons and Tropical Cyclones.

What geographical material is available?

- There is a wealth of material from around the world on most geography topics. There is comprehensive coverage of Disasters on the Web. Researching for information on Hazards can be easier than for other topics because it is often very newsworthy material.

- Using a URL is one of the fastest ways of obtaining information. A selection of URLs is given in the References at the end of each chapter in this book. Others may be found in geography magazines and newspaper supplements such as 'Guardian On-Line', and 'Connected' (*Daily Telegraph*).

- A number of sites on the Web provide geographical reference lists and hotlinks, e.g. the GAIA or Geography: An Integrated Approach site (http://www.nelsonthornes.com/gaia). The University of Texas (http://asnic.utexas.edu) has an extensive set of links on economic and social development. But again it takes time to search for the most relevant sites.

- Up-to-date news material can be easily accessed from newspapers 'on line', e.g. *The Times* (http://www.the-times.co.uk), *Guardian* (http://www.guardian.co.uk) and *Daily Telegraph* (http://www.telegraph.co.uk). It can be interesting to get a different slant on an event by looking at a 'local' paper, such as the *Los Angeles Times* (http://www.latimes.com) or the *Daily Nation* in Kenya (http://www.nationaudio.com).

Note that there is no overall control over the Web – anyone may set up a site, and so the reliability of the information depends on the source. You must remember too that the Web is used by many organisations to publicise their own attitudes and values on important topics. Users have to attempt to analyse the materials for bias, and where possible to obtain another viewpoint in order to achieve a balance.

There are now organisations such as Encyclopaedia Britannica and the Virtual Teaching Centre which evaluate sites. Britannica uses a star rating system for this purpose (http://www.britannica.com/bcom/internet_guide_display_page). The Herodot Project offers advice on evaluating sites for yourself (http://www.hope.ac.uk/ebs/herodot/h'dot/wwweval.htm).

Words of advice

- At present much of the material on the Web comes from American sources. You need to be aware of this, although material from other parts of the world is increasing. Remember too that there is material in all languages on the Web, and alternative sites in other languages may be useful.

- The Web is not a geography textbook, but it needs to be recognised as an **additional tool** which can help towards a greater understanding of geographical and environmental issues. The drawbacks include bias of material and difficulty in finding an appropriate level of language and detail. Research will find pages of information on most topics, so 'skim and scan' reading in order to get the main facts is important. Remember also to check the date when each site was last updated.

- Downloading material to your own site or to a printer is common practice. This allows study and analysis. However, inserting material directly into essays without further analysis is of little use, and also contravenes copyright. All material taken from the Web *must be acknowledged*, in the same way that you should acknowledge the origin of material from textbooks when using it in individual studies and coursework.

- Personal contact with people on other Websites can provide a different perspective. For example, scientists in the British Antarctic Survey are in regular contact with schools studying Wilderness regions. You can also link with colleges in other countries, which has practical advantages for research. Individual students can link via e-mail with students elsewhere to collect and share material for their studies.

- On a practical note, remember that for users in the UK it is much cheaper and quicker to access the Web early in the day, before the North American continent comes on-line. Wherever you are in the world, bear this in mind.

Advantages of the Web	Disadvantages of the Web
Immediacy of much information	Serious bias of some information
Detail of material on academic sites	Source of material needs to be checked for accuracy and bias
Extensive links to other sites	Sites can be quickly outdated
Lively interactive presentations on some sites	Information can take a long time to locate – needs good browsing skills
Surfing can be fun	Surfing can be expensive, both in time and money

Volcanology

The term **volcanology** includes all the processes by which solid, liquid or gaseous materials are forced into the Earth's crust or are ejected onto the surface. Although material in the mantle has a high temperature, it is kept in a semi-solid state because of the great pressure exerted upon it. However, if this pressure is released locally by folding, faulting or other movements at plate boundaries, some of the semi-solid material becomes molten and rises, forcing its way into weaknesses in the crust, or onto the surface, where it cools, crystallises and solidifies.

The molten rock is called **magma** when it is below the surface and **lava** when on the surface. When lava and other materials reach the surface they are called **extrusive**. The resulting landforms vary in size from tiny cones to widespread lava flows. Materials injected into the crust are referred to as **intrusive**. These may later be exposed at the surface by erosion of the overlying rocks. Both extrusive and intrusive materials cooled from magma are known as **igneous rocks**.

Extrusive landforms

There are several types of extrusive landform whose nature depends on how gaseous and/or viscous the lava is when it reaches the Earth's surface (Figure 1.21).

- Lava produced by the upward movement of material from the mantle is **basaltic** and tends to be located along mid-ocean ridges, over hot spots and alongside rift valleys.

- Lava that results from the process of subduction is described as **andesitic** (after the Andes) and occurs as island arcs or at destructive plate boundaries where oceanic crust is being destroyed.

- **Pyroclastic material** (meaning 'fire broken') is material ejected by volcanoes in a fragmented form. Tephra, fragments of different sizes, include ash, lapilli (small stones) and bombs (larger material) which are thrown into the air before falling back to earth. Pyroclastic flows are materials that move down the side of a volcano (Figure 1.46). The fragments may occur as mudflows (or lahars), e.g. Mt Pinatubo (Case Study 1) and Nevado del Ruiz (Figure 2.27 in Case Study 2B), or as a fast-moving cloud known as a nuée ardente, e.g. Mt Pelée. The deposited material is **ignimbrite**.

How can volcanoes be classified?

Because of the large number of volcanoes and wide variety of eruptions, it is convenient to group together those with similar characteristics (Framework 7, page 167). Unfortunately, there is no universally accepted method of classification. One of the two most quoted groupings is according to the **shape** of the volcano and its vent which, because it describes landforms, is arguably of more value to the geographer (page 25). The other is the nature of the **eruption**, which has traditionally been the method used by volcanologists (page 28).

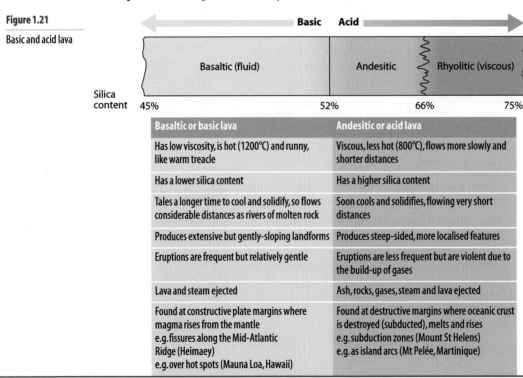

Figure 1.21

Basic and acid lava

Basaltic or basic lava	Andesitic or acid lava
Has low viscosity, is hot (1200°C) and runny, like warm treacle	Viscous, less hot (800°C), flows more slowly and shorter distances
Has a lower silica content	Has a higher silica content
Tales a longer time to cool and solidify, so flows considerable distances as rivers of molten rock	Soon cools and solidifies, flowing very short distances
Produces extensive but gently-sloping landforms	Produces steep-sided, more localised features
Eruptions are frequent but relatively gentle	Eruptions are less frequent but are violent due to the build-up of gases
Lava and steam ejected	Ash, rocks, gases, steam and lava ejected
Found at constructive plate margins where magma rises from the mantle e.g. fissures along the Mid-Atlantic Ridge (Heimaey) e.g. over hot spots (Mauna Loa, Hawaii)	Found at destructive margins where oceanic crust is destroyed (subducted), melts and rises e.g. subduction zones (Mount St Helens) e.g. as island arcs (Mt Pelée, Martinique)

The shape of the volcano and its vent

1 **Fissure eruptions**   When two plates move apart, lava may be ejected through fissures rather than via a central vent (Figure 1.22a). The Heimaey eruption of 1973 (Places 3, page 16) began with a fissure 2 km in length. This was small in comparison with that at Laki, also in Iceland, where in 1783 a fissure exceeding 30 km opened up. The basalt may form large plateaus, filling in hollows rather than building up into the more typical cone-shaped volcanic peak. The remains of one such lava flow, formed when the Eurasian and North American Plates began to move apart, can be seen in Northern Ireland, north-west Scotland, Iceland and Greenland. The columnar jointing produced by the slow cooling of the lava provides tourist attractions at the Giant's Causeway in Northern Ireland (Figure 1.27) and Fingal's Cave on the Isle of Staffa.

2 **Basic or shield volcanoes**   In volcanoes such as Mauna Loa on Hawaii, lava flows out of a central vent and can spread over wide areas before solidifying. The result is a 'cone' with long, gentle sides made up of many layers of lava from repeated flows (Figure 1.22b).

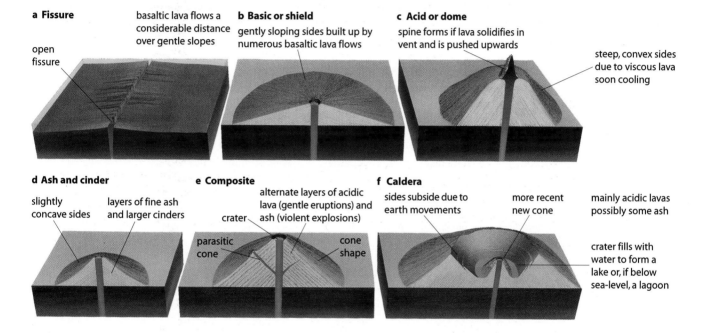

a Fissure
open fissure
basaltic lava flows a considerable distance over gentle slopes

b Basic or shield
gently sloping sides built up by numerous basaltic lava flows

c Acid or dome
spine forms if lava solidifies in vent and is pushed upwards
steep, convex sides due to viscous lava soon cooling

d Ash and cinder
slightly concave sides
layers of fine ash and larger cinders

e Composite
alternate layers of acidic lava (gentle eruptions) and ash (violent explosions)
crater
parasitic cone
cone shape

f Caldera
sides subside due to earth movements
more recent new cone
mainly acidic lavas possibly some ash
crater fills with water to form a lake or, if below sea-level, a lagoon

Figure 1.22

Classification of volcanoes based on their shape (not to scale)

3 **Acid or dome volcanoes**   Acid lava quickly solidifies on exposure to the air. This produces a steep-sided, convex cone as in most cases the lava solidifies near to the crater (Figure 1.22c). In one extreme instance, that of Mt Pelée, the lava actually solidified as it came up the vent and produced a spine rather than flowing down the sides.

4 **Ash and cinder cones**   (Figure 1.22d) Paricutín, for example, was formed in the 1940s by ash and cinders building up into a symmetrical cone.

5 **Composite cones**   Many of the larger, classically shaped volcanoes result from alternating types of eruption in which first ash and then lava (usually acidic) are ejected, e.g. Mt Etna and Fujiyama (Figure 1.22e).

6 **Calderas**   When the build-up of gases becomes extreme, huge explosions may clear the magma chamber beneath the volcano and remove the summit of the cone. This causes the sides of the crater to subside, thus widening the opening to several kilometres in diameter. In the cases of both Thera (Santorini) and Krakatoa, the enlarged craters or calderas have been flooded by the sea and later eruptions have formed smaller cones within the resultant lagoons (Figures 1.22f and 1.29).

Minor extrusive features

These are often associated with, but are not exclusive to, areas of declining volcanic activity. They include solfatara, fumaroles, geysers and mud volcanoes (Figure 1.23, Places 7 and Figure 17.16).

Figure 1.23

Minor extrusive landforms

a Mud volcano: hot water mixes with mud and surface deposits

b Solfatara: created when gases, mainly sulphurous, escape onto the surface

c Geyser: water in the lower crust is heated by rocks and turns to steam; pressure increases and the steam and water explode onto the surface

d Fumaroles: superheated water turns to steam as its pressure drops when it emerges from the ground

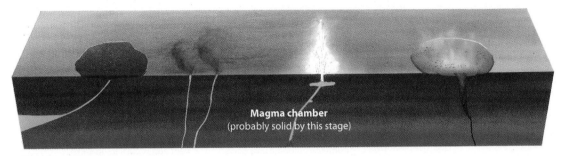

Magma chamber
(probably solid by this stage)

Places 7   Solfatara, Italy: an area of declining volcanic activity

Solfatara is a small volcano on the outskirts of Naples. Its crater is 2 km in diameter, making it larger than that of nearby Vesuvius, but there is no volcanic cone. Solfatara takes its name from the gases which escape to the surface; they are mainly sulphurous and can be smelled from a considerable distance. Many rocks are coated with sulphur. **Solfatara** has given its name to all similar features of this type. **Fumaroles**, resulting from superheated water being turned to steam as it cools on its ejection through the thin crust, are numerous in the area (Figure 1.24). Evidence of the thinness of the crust (magma is only 3 m below the surface) is provided by a guide who throws a boulder onto the surface and makes groups of tourists jump in harmony to hear the hollowness of the ground. The guide, who is needed to keep visitors safely away from bubbling mud volcanoes and areas too hot to walk upon, also shows volcanic activity by lighting twigs and stirring loose material to cause a miniature eruption.

The only minor feature missing is the **geyser**, an intermittent fountain of hot water (e.g. Old Faithful, Yellowstone National Park, USA, Figure 17.16).

During the mid-1980s the temperature (160°C), pressure and surface of Solfatara all rose, giving rise to fears of a new eruption – the last was in 1198. Despite the appearance of a small fissure near to the observatory, which led to its abandonment, activity appears to have stabilised.

Figure 1.24

Inside the Solfatara crater, near Naples, Italy

Figure 1.25

Results of the 1996 Grimsvötn eruption, Iceland

Figure 1.26

The boundary between the North American and Eurasian Plates in Iceland, showing the split and a volcano along the boundary margin

Figure 1.27

The Giant's Causeway, Northern Ireland

Figure 1.28

Vesuvius: notice the new cone within the old crater of Monte Somma

Figure 1.29

Krakatoa with Anak (meaning 'child of') Krakatoa

Plate tectonics, earthquakes and volcanoes

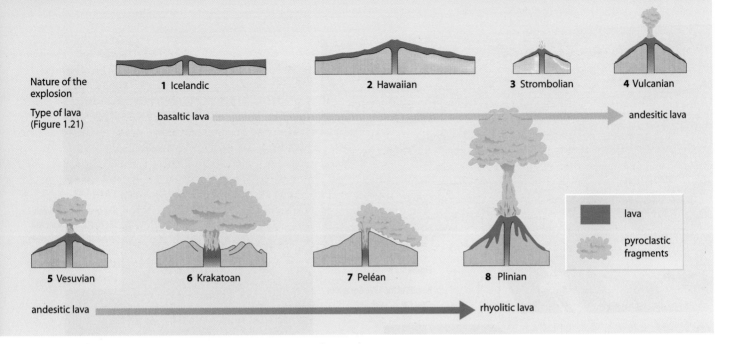

Figure 1.30

Classification of volcanoes according to the nature of the explosion

The nature of the eruption

This classification of volcanoes is based on the degree of violence of the explosion which is a consequence of the pressure and amount of gas in the magma (Figure 1.30). Its categories may be summarised as follows:

- Icelandic, where lava flows gently from a fissure
- Hawaiian, where lava is emitted gently but from a vent
- Strombolian, where small but very frequent eruptions occur
- Vulcanian, which is more violent and less frequent (this is becoming a dated term)
- Vesuvian, which has a violent explosion after a long period of inactivity (Figure 1.28)
- Krakatoan, which has an exceptionally violent explosion that may remove much of the original cone (Figure 1.29)
- Peléan, where a violent eruption is accompanied by pyroclastic flows that may include a nuée ardente ('glowing cloud')

- Plinian, where large amounts of lava and pyroclastic material are ejected.

Hydrovolcanism has been a major area of research since the 1980s. This is where hot magma interacts with water at or near the Earth's surface, e.g. in crater lakes (Ruapehu – Places 4, page 18) or with sea water (Surtsey – Places 3, page 16, and Hawaii), ice (Iceland – Places 3, page 16) and groundwater. Landforms resulting from hydro-volcanism include:

- on ocean floors – basaltic pillow lava flows together with, in less deep water, fragmented material
- on land – falls (material ejected into the air and which soon falls back to the ground), surges (including violent blasts which eject material in all directions, e.g. Mount St Helens) and lahars (fragmented material transported by water, e.g. Mt Pinatubo, Case Study 1 and Nevado del Ruiz, Case Study 2B).

Figure 1.31

Diagrammatic model showing intrusive landforms: batholith, dyke and sills.

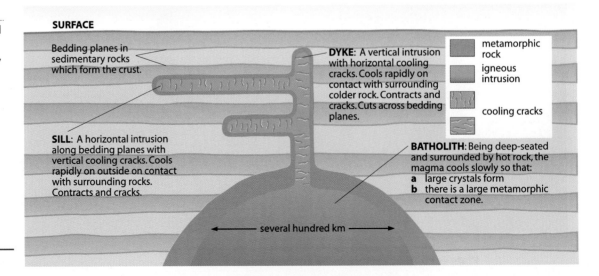

Intrusive landforms

Usually, only a relatively small amount of magma actually reaches the surface as most is intruded into the crust, where it solidifies. Such intrusions may initially have little impact upon the surface **geomorphology**, but if the overlying rocks are later worn away, distinctive landforms may then develop (Figure 1.32).

During the Tertiary era, an upthrust of magma was **intruded** into the sedimentary rocks of Arran to form the Northern Granite. As the magma slowly cooled, it formed large crystals (unlike on the surface where rapid cooling forms fine crys-

tals), contracted and cracked resulting in a series of joints. The magma also produced a large, deep-seated, dome-shaped **batholith** as it solidified.

Surrounding the batholith is a **metamorphic aureole** where the original sedimentary rocks have been changed (metamorphosed) by the heat and pressure of the intrusion from sandstones into **schists**. Since then, the overlying rocks have been removed by water, ice and even the sea to leave the granite batholith with its jointing exposed (Figure 1.32). These joints have been widened by chemical weathering (pages 42–44) to form the large granite slabs and tors surrounding Goatfell (Figure 8.14).

Figure 1.32

Idealised transect through northern Arran

Figure 1.33

Fieldsketch of a dyke at Kildonan, Arran

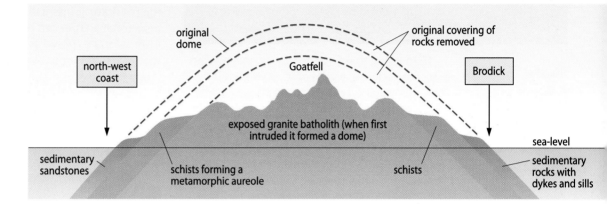

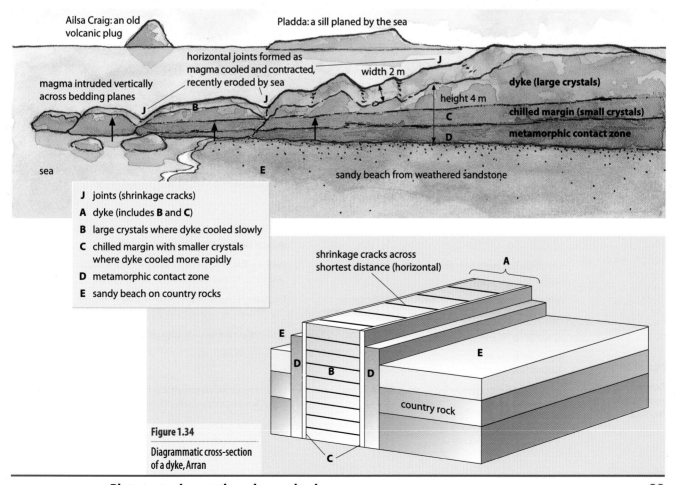

J joints (shrinkage cracks)

A dyke (includes **B** and **C**)

B large crystals where dyke cooled slowly

C chilled margin with smaller crystals where dyke cooled more rapidly

D metamorphic contact zone

E sandy beach on country rocks

Figure 1.34

Diagrammatic cross-section of a dyke, Arran

Figure 1.35

Dyke at Kildonan, Arran

Figure 1.36

Fieldsketch of a sill exposed at Drumadoon, Arran

although those parts that come into contact with the surrounding rock will cool more rapidly to produce a chilled margin (Figure 1.34). Most dykes on Arran were formed after, and radiate from, the batholith intrusion; they are so numerous that they have been termed a 'dyke swarm'. Most of the dykes are more resistant to erosion than the surrounding sandstones and so where they cross the island's beaches they stand up like groynes (Figure 1.35). Although averaging 3 m, these dykes vary from 1 to 15 m in width.

A **sill** is formed when the igneous rock is intruded along the bedding planes between the existing sedimentary rocks (Figure 1.31). The magma cools and contracts but this time the resultant joints will be vertical and their hexagonal shapes can be seen when the land-form is later exposed as on headlands such as that at Drumadoon on the west coast of Arran (Figures 1.36 and 1.37) and The Giant's Causeway in Northern Ireland (Figure 1.27). The sill at Drumadoon is 50 m thick.

If, in trying to rise to the surface, magma cuts across the bedding planes of the sedimentary rock, it is called a **dyke** (Figures 1.31 and 1.33). The material which forms the dyke cools slowly

original covering of sandstone removed

vertical joints (columnar jointing) formed as magma cooled and contracted

magma intruded horizontally between bedding planes

50 m

metamorphic contact zone under a chilled margin

30 m

Talus (scree) covering sandstone (11 m)

raised beach

sea

Figure 1.37

Sill at Drumadoon, Arran

Benefits	Hazards
Lava and ash weather rapidly. Basic lava may produce fertile soils ideal for farming (the region surrounding Mt Etna) if it is carefully managed. The fertility of acid lava is much lower.	Earthquakes destroy buildings and result in loss of life.
Igneous rock contains minerals such as gold, copper, lead and silver.	Violent eruptions with blast waves and gas may destroy life and property (Mt Pelée, Mount St Helens).
Extinct volcanoes may provide defensive settlement sites (Edinburgh).	Mudflows/lahars may be caused by heavy rain and melting snow (Armero in Colombia and Pinatubo in the Philippines).
Igneous rock is used for building purposes (Naples, Aberdeen).	Tidal waves/tsunamis (following the eruption of Krakatoa).
Geothermal power is being developed (Iceland, New Zealand).	Ejection of ash and lava ruins crops and kills animals.
Geysers and volcanoes are tourist attractions (Yellowstone National Park), generating revenue for local communities.	Interrupts communications.
Volcanic eruptions may produce spectacular sunsets (Krakatoa).	Short-term climatic changes occur as volcanic dust absorbs solar energy, lowering temperatures and increasing rainfall.

Figure 1.38

Benefits and hazards resulting from tectonic processes

What are natural hazards?

Natural hazards, which include earthquakes, volcanic eruptions, floods, drought and storms, result from natural processes within the environment (Figure 1.39). They are, therefore, different from environmental disasters, such as desertification, ozone depletion and acid rain, which are caused by human activity and the mis-management of the environment. It is important, however, to stress the difference between a **natural hazard** and a **hazard event**. Natural hazards have the potential to affect people and the environment; it is the hazard event that causes the damage. An event only becomes a hazard if it affects, or threatens, people and property. For example, the submarine volcanic eruption which created the new island of Surtsey (Places 3, page 16) was hardly a hazard event as it posed no threat to people or property. In contrast, the eruption of Mount St Helens was a hazard event as it killed 61 people, destroyed roads and property and interrupted human activities. The impact of a hazard event may be felt over a wide area; the effects may be long-term as well as immediate; and the event can be costly to property and dangerous to people.

Figure 1.39

Types of natural hazard

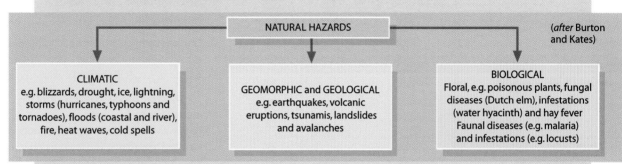

NATURAL HAZARDS

(*after* Burton and Kates)

CLIMATIC
e.g. blizzards, drought, ice, lightning, storms (hurricanes, typhoons and tornadoes), floods (coastal and river), fire, heat waves, cold spells

GEOMORPHIC and GEOLOGICAL
e.g. earthquakes, volcanic eruptions, tsunamis, landslides and avalanches

BIOLOGICAL
Floral, e.g. poisonous plants, fungal diseases (Dutch elm), infestations (water hyacinth) and hay fever
Faunal diseases (e.g. malaria) and infestations (e.g. locusts)

The International Decade for Natural Disaster Reduction (IDNDR)

Figure 1.40

The problems of defining natural disasters

The United Nations resolved that, during the 10-year period 1990–99, they would try to reduce the loss of life, property damage and social and economic destruction caused by natural disasters, especially in economically less developed countries. However, as Keith Smith claimed in *Teaching Geography*, September 1996, there is no general agreement on what constitutes a 'natural disaster' (Figure 1.40).

(1) Many natural disasters arise from a combination of individual geophysical events. In these cases it may be impossible to allocate all the loss to a single cause. For example, are the recorded deaths due primarily to *flood* (as opposed to a *storm* with intense rainfall) or to *landslide* (as opposed to an *earthquake* creating mass movements)? Such questions raise doubts about the classification of events under particular and usually competing headings, and can lead to some double-counting of extreme physical events. On the other hand, drought, which is an important disaster type in many developing countries, is frequently excluded from analysis. This is because of the difficulties involved in distinguishing its role in causing famine from that of ongoing malnutrition in the poorest countries. As a result, drought effects can go unrecorded.

(2) Even direct deaths and damage, suffered immediately after the event, may be difficult to quantify accurately in the case of a major Third World disaster. When reports can speak only of 'hundreds killed' or of 'damage estimated in millions of dollars' it is because many developing countries lack reliable census data or population registers. There is probably significant under-reporting of people recorded as 'missing' after an event, to say nothing of those who die perhaps weeks later from disease or malnutrition created by the disaster. In all cases, little attempt is usually made to assess the more indirect losses such as those resulting from injury, loss of employment, or forced relocation after a disaster.

From an article by Keith Smith in *Teaching Geography*, September 1996

Natural disasters, some increasingly caused by human mismanagement of the environment, threaten sustainable development (Framework 16, page 499). Data provided by the IDNDR suggests that about 60 per cent of natural disasters and up to 90 per cent of deaths caused by those disasters, occur in developing countries, especially those in the Asia/Pacific region (Figures 1.41 and 1.42). Developing countries are less likely to have the equipment to predict the occurrence of a hazard and less money either to reduce the impact of an event or to organise rapid and effective response to that event. Even so, mortality rates do appear to have declined since 1990, mainly due to improvements in early storm- and flood-warning systems, e.g. hurricane prediction in the West Indies and flood warning schemes in Bangladesh. The large loss of life in Turkey and Taiwan caused by the 1999 earthquakes may prove to be an exception.

Figure 1.41

Significant natural disasters, 1963–92

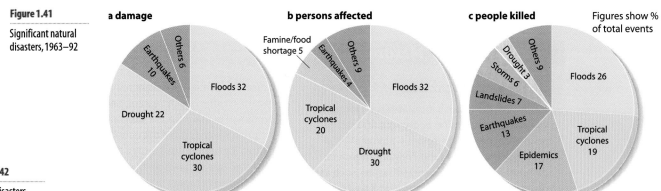

a damage

Others 6
Earthquakes 10
Drought 22
Floods 32
Tropical cyclones 30

b persons affected

Famine/food shortage 5
Earthquakes 4
Others 9
Floods 32
Tropical cyclones 20
Drought 30

c people killed

Others 9
Drought 3
Storms 6
Landslides 7
Earthquakes 13
Epidemics 17
Floods 26
Tropical cyclones 19

Figures show % of total events

Figure 1.42

Natural disasters, 1967–91

Type	Number of events	Number killed	References: Case Studies	References: Places	References: Other pages
Geological events					
Earthquake	758	646 300	2A, 15A	2,4,6	9,15
Volcanic eruption	102	2800	1,2A	3,4,35	15,24
Tsunami	20	6400	15	–	18
Weather events					
Tropical cyclone (hurricane/typhoon)	894	896 100	2A	19,30,31	235
Flood – river	1358	304 900	3A,3C,12,15A	12	61,76
Flood – coastal			6	19	147,169
Storm (depression/tornado/lightning)	819	55 000	–	19,29	230
Cold wave. Blizzard. Heatwave	133	4900	–	32	–
Drought	430	1 333 700	7,15A	61	223,502
Weather-related events					
Avalanche	29	1300	2A,4	–	–
Landslide	238	42 000	2A,6,15A	8	48
Fire (bush)	729	82 000	15A	77	293
Insect infestation/locusts	68	0	–	75	–
Famine/food shortage	37	606 000	–	75	502

How may people react to natural hazards?

The study of natural hazards, involving the relationship and interaction between physical systems and human systems (i.e. people and the environment), falls into the domain of geographers – although not exclusively so.

The following questions need to be asked when studying either the risk of a potential natural hazard or a specific hazard event.

1 What are people's perception of the natural hazard?

Perception is how individuals or groups of people view the hazard risk, and it often depends upon their knowledge and experience of the event. The inhabitants of Pompeii, many of whom were killed during the eruption of Vesuvius in AD 79, did not realise until that event that the mountain was indeed a volcano. Since then, Vesuvius has erupted on numerous occasions. So why do people continue to rebuild and live in this, and other, hazardous areas? It may be because they:

- perceive the area as providing the best of opportunities to earn a living
- are too concerned with day-to-day problems to consider the hazard risk
- have the capital and technology to cope with the hazard event.

2 What are the immediate and long-term effects of the event?

3 How do people respond to the event (Figure 1.43)?

4 How might people adjust to and plan for a future event?

It has been suggested that people have six options. They may try to: prevent the event; modify the hazard; lessen the possible amount of damage; spread the losses caused by the event; claim for losses; or do nothing but pray that the event will not occur again (at least not in their own lifetime).

5 Can a future event be predicted?

This involves predicting where the next event will take place, when it is likely to occur and how big it is likely to be.

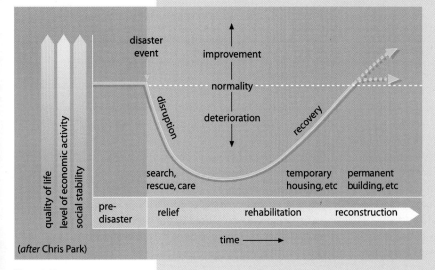

Figure 1.43

The responses to a hazard event

disaster event
improvement
normality
deterioration
disruption
recovery
search, rescue, care
temporary housing, etc
permanent building, etc
quality of life
level of economic activity
social stability
pre-disaster
relief
rehabilitation
reconstruction
time
(*after* Chris Park)

Plate tectonics, earthquakes and volcanoes

Volcanic eruptions – Mount Pinatubo

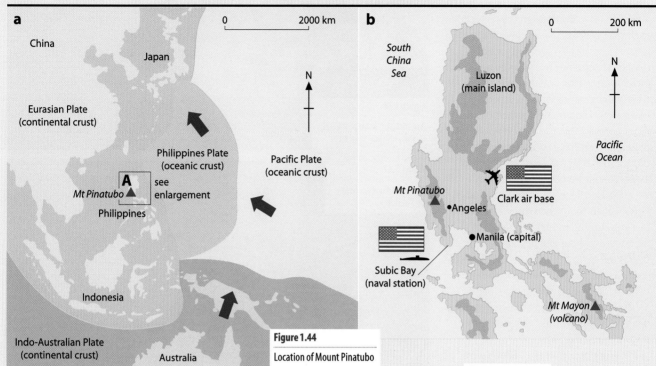

a

China

Japan

Eurasian Plate
(continental crust)

Philippines Plate
(oceanic crust)

Pacific Plate
(oceanic crust)

0   2000 km

N

A see
   enlargement

Mt Pinatubo ▲

Philippines

Indonesia

Indo-Australian Plate
(continental crust)

Australia

b

South
China
Sea

Luzon
(main island)

0   200 km

N

Mt Pinatubo ▲

Clark air base

●Angeles

Pacific
Ocean

●Manila (capital)

Subic Bay
(naval station)

Mt Mayon ▲
(volcano)

Figure 1.44

Location of Mount Pinatubo

Why is Mount Pinatubo in a hazard risk area?

Mount Pinatubo is located in the Philippines (Figure 1.44). The Philippines lie on a destructive plate margin where the Philippines Plate, composed of oceanic crust, moves towards and is subducted by the Eurasian Plate, which consists of continental crust. As the oceanic plate is subducted, it is converted into magma which rises to the surface and forms volcanoes. The Philippines owe their existence to the frequent ejection of lava over a period of several million years. Even before Pinatubo erupted in 1991, there were over 30 active volcanoes in the Philippines.

Why did people live in this hazard risk area?

As Mount Pinatubo had not erupted since 1380, people living in the area no longer considered it to be a hazard. During that time, ash and lava from earlier eruptions had weathered into a fertile soil, ideal for rice growing. By 1991, people no longer perceived Pinatubo to be a danger. On the lower slopes of the mountain, the Aeta, recognised as the aboriginal inhabitants of the islands,

practised subsistence farming (slash and burn agriculture, Places 66, page 480). Near the foothills was the rapidly growing city of Angeles, together with an American air base and a naval station (Figure 1.44).

What were the nature, effects and consequences of the eruption?

1 Immediate effects

The volcano began to show signs of erupting in early June, 1991. Fortunately, there were several advance-warning signs which allowed time for the evacuation of thousands of people from Angeles and the 15 000 personnel from the American air base. The number and size of eruptions increased after 9 June. On 12 June, an explosion sent a cloud of steam and ash 30 km into the atmosphere – the third-largest eruption experienced anywhere in the world this century (Figure 1.45). Up to 50 cm of ash fell nearby, and over 10 cm within a 600 km radius. The eruptions were, characteristically, accompanied by earthquakes and torrential rain – except that the rain, combining with the ash, fell as thick mud. The ash destroyed all crops on adjacent farmland and its weight caused buildings to

Figure 1.45

Eye-witness account
of the eruption

Seismologists said a mixture of searing gas, ash and molten rock quickly raced down the mountain's west and northern flanks and into the Marella, Maraunot and O'Donnel rivers [Figure 1.46]. Ash also rained down on seven towns in the region and traces of ash were detected near the Subic Bay naval base, 50 miles [80 km] to the south-west. Pumice fragments measuring up to 1.2 inches [3 cm] long fell on villages south-west of the volcano. At a refugee centre in Olongapo, about 35 miles [56 km] south-west of the volcano, survivors said they saw the sky grow dark and then heard a tremendous explosion followed by a rain of ash and stones as big as a man's head.

Other reporters described panic as people fled with their belongings and livestock over roads made slippery by the falling ash. Refugees wore cardboard boxes with air and peepholes to protect themselves from the ash. The ash was so thick in the air that at noon motorists were driving with their headlights on and wipers operating to clear the debris.

The Independent, 13 June 1991

collapse, including 200 000 homes, a local hospital and many factories. Power supplies were cut off for three weeks and water supplies became contaminated. Relief operations were hindered as many roads became impassable and bridges were destroyed.

2 Longer-term effects

The thick fall of ash not only ruined the harvest of 1991, but made planting impossible for 1992. Over one million farm animals died, many through starvation due to the lack of grass. Several thousand farmers and their families had to take refuge in large cities. The majority were forced to seek food and shelter in shanty-type refugee camps. Disease, especially malaria, chickenpox and diarrhoea, spread rapidly and doctors had to treat hundreds of people for respiratory and stomach disorders. Soon after the event, and again in 1993, typhoons brought heavy rainfall which caused flooding and lahars (mudflows). Lahars form when surface water picks up large amounts of volcanic ash in mountainous areas and deposits it as mud over lower-lying areas (Figure 1.47). The ash that was ejected into the atmosphere is believed to have caused changes in the Earth's climate, including the lowering of world temperatures and ozone depletion (Figure 1.48).

The eruption and its after-effects were blamed for about 700 deaths. Of these, only six were believed to have been a direct result of the eruption itself. Over 600 people were to die from disease and a further 70 from suffocation by lahars.

Figure 1.46

A pyroclastic cloud produced by the eruption, June 1991

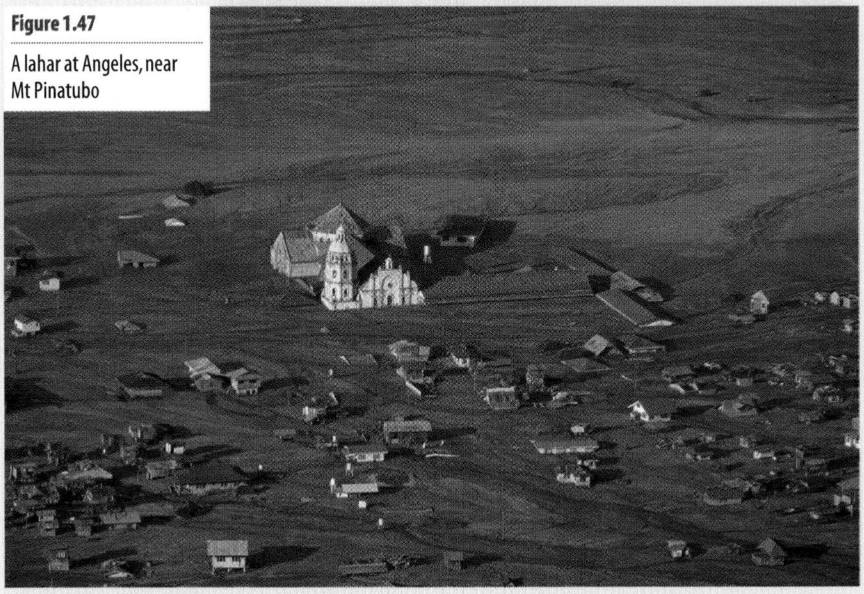

Figure 1.47

A lahar at Angeles, near Mt Pinatubo

IT HAS been described as the world's greatest climatic experiment, but unlike most scientific endeavours it was unplanned. When the tropical tranquillity of the Philippines was shattered last June by a volcanic explosion, Mount Pinatubo was a relatively obscure volcano, known in the scientific community only to a handful of geologists. Having sent more than 20 million tonnes of dust and ash into the atmosphere, altering its heat balance and accelerating ozone depletion over a large part of the globe, Pinatubo has become the focus of several far-reaching studies.

Climatologists now use the term 'Pinatubo effect' to describe how volcanic ash and debris, if sent high enough into the atmosphere, can influence temperature and weather for several years afterwards. The dust from Pinatubo was ejected as high as 20 miles [32 km] above the Earth. From the haven of Earth orbit, satellites observed the plume of volcanic ash as it girdled the globe at speeds approaching 75 miles [120 km] per hour. A month after the eruption which killed 350 people, a 3000 mile [4800 km] cloud of ash and sulphur compounds circled the Earth.

Satellite temperature measurements confirmed that the dust had effectively shaded the surface of the Earth from the sun's rays, resulting in a lowering of the average global temperature. A NASA team at the Goddard Institute for Space Studies in New York, led by James Hansen, tried to assess what effect the cooling caused by the dust of Mount Pinatubo would have on global warming caused by man-made emissions of carbon dioxide. They concluded that Pinatubo would in effect delay global warming by several years. While global warming experts argue about the effect of Pinatubo's eruption on average temperatures, ozone specialists are interested in the effect the volcano has had and will have on the ozone layer. The volcano has spewed out huge quantities of sulphate aerosols, particles containing sulphur that remain suspended in the atmosphere for several years. These sulphate particles are important in the chemistry of ozone destruction for two reasons: first, they act as sites where ozone-destroying reactions take place; and secondly, they mop up nitrogen-containing compounds that help to prevent ozone destruction. This winter American and European scientists undertook the most intensive investigation of ozone depletion over the northern hemisphere, including Europe and North America. More than 300 scientists from 17 countries were involved and their work has shown that ozone levels fell by 10 to 20 per cent more than expected. 'The eruption of Mount Pinatubo has increased the abundance of natural sulphate particles, potentially enhancing ozone losses due to chemical reactions that occur on particle surfaces,' the NASA ozone monitoring team said earlier this month.

Figure 1.48

The climatic effects of the eruption

The Independent on Sunday, 10 May 1992

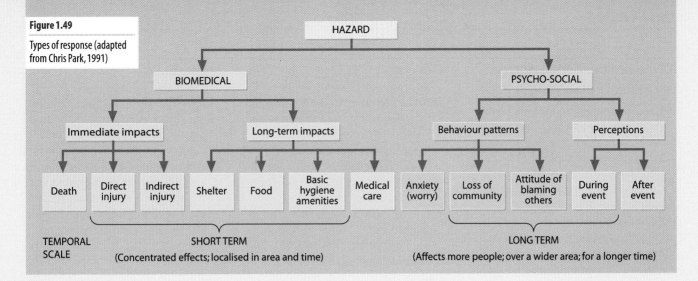

Figure 1.49

Types of response (adapted from Chris Park, 1991)

HAZARD

BIOMEDICAL — PSYCHO-SOCIAL

BIOMEDICAL: Immediate impacts — Long-term impacts

PSYCHO-SOCIAL: Behaviour patterns — Perceptions

Immediate impacts: Death | Direct injury | Indirect injury

Long-term impacts: Shelter | Food | Basic hygiene amenities | Medical care

Behaviour patterns: Anxiety (worry) | Loss of community | Attitude of blaming others

Perceptions: During event | After event

TEMPORAL SCALE

SHORT TERM (Concentrated effects; localised in area and time)

LONG TERM (Affects more people; over a wider area; for a longer time)

How did people respond to the hazard event?

Chris Park has divided human responses during and after any hazard event into two categories (Figure 1.49).

Within a few weeks of the major Pinatubo eruption, groups of evacuees from the affected area began to consider their future options and their next move. Their range of responses included the following:

1 Some members of the Aeta tribe (Figure 1.50) decided not to return to their former homes. As a spokesperson explained: 'Everything we have planted has been destroyed. There is no point in going back. The government will have to put us somewhere else.'

2 In contrast, the majority of the Aeta tribe decided to return. To them, the mountain slopes, although vastly changed, were still their home and the hard way of life in the hills was preferable to the foreign habits of the lowlanders and to living in urban areas.

3 Most of the people who fled from the City of Angeles have, so far, opted against returning home. To them, life in the shanty refugee camps is safer than returning to an area where eruptions and earthquakes are still occurring and where the heavy rain is likely to cause lahars for several years until the regrowth of vegetation stabilises the slopes.

Can future eruptions be predicted?

It is now possible to modify some natural hazards (e.g. cloud seeding to reduce drought) and to forecast and control other actual hazard events (e.g. implementing flood control schemes, installing early-warning hurricane systems and constructing buildings designed to withstand earthquakes). However, at present, while it may be possible to predict fairly accurately where volcanic eruptions are likely to occur (i.e. at constructive and destructive plate margins, Figure 1.8), there seems little prospect of people being able to predict either the time or the scale of a specific event. Also, in places where a volcanic eruption has not occured for several centuries, as in the case of Pinatubo, the human perception of the hazard risk decreases and the event, should it take place, catches more people and organisations unprepared.

Figure 1.50

Members of the Aeta tribe

UN recommendations for the surveillance of volcanoes

1 Regional networks of seismographic stations should be set up in risk areas that are populated.
2 Portable seismographs and tilt-meters should be available for instant installation on any volcano showing signs of activity.
3 Trained volcanologists should be made available to interpret events.
4 Detailed histories of volcanoes should be produced. Some, but by no means all, volcanoes have a cycle of eruptions; if this cycle is known, it can aid prediction of future events.
5 Detailed maps should be produced to show deposits of lava and pyroclastic materials from previous eruptions. These maps can show areas most likely to be at risk from the hazard (such a map would not have predicted the effects on Mount St Helens).
6 Periodic measurements should be taken to show possible changes in temperature and pressure within the volcano. Chemical analyses of fumaroles and hot springs might indicate variations from the norm.
7 Emergency procedures, including evacuation procedures, should be established for all communities.
8 Aerial magnetic and infrared photo-graphic surveys should be made at regular intervals. It has been shown that, before an eruption, there are local variations in the Earth's magnetic field, while infrared photos could show a build-up in temperature and any possible swellings (bulging) of the cone.

These procedures are easier to adopt in an economically more developed country with its greater wealth and technology. UN procedures are also more likely to be followed where the expected frequency of the event is high, and in risk areas having high population densities.

Predicting and planning for earthquakes

Scientists can use sensitive instruments to measure increases in earth movements and a build-up of pressure. They can also map the epicentres and frequency of previous earthquakes to see if there is either a repeat location or a time-interval pattern. In Kanto, the region surrounding Tokyo, there has been a severe earthquake, on average, every 70–80 years for the last five centuries. As the last event was in 1923, with an estimated 14 000 deaths, then an equally severe earthquake might be expected to occur early in the 21st century. Even so, such methods can predict neither the precise timing nor the exact location of the earthquake. A less scientific method, but successfully used in China, has been the observation of unusual animal behaviour shortly before a major earth movement, e.g. mice have fled houses, dogs have howled, fish have jumped out of water and the giant panda has moaned.

In earthquake-prone areas, especially in more wealthy countries, buildings can be constructed to withstand earthquakes. They are built with steel (which can sway during earth movement) and fire-resistant materials – never with bricks or reinforced concrete blocks. Foundations are sunk deep into bedrock and are separated from the superstructure by shock-absorbers. Open spaces should be provided for people to assemble, and roads made sufficiently wide to allow rapid access by emergency services. The emergency services themselves need to be trained and well-equipped, while local residents need to be made aware as to how they should respond both during and after the event.

References

Calder, N. (1972) *The Restless Earth*, BBC Publications.

Chester, D. (1993) *Volcanoes and Society*, Edward Arnold.

Decker, D. W. and Decker, B. (1991) *Mountains of Fire*, Cambridge University Press.

Donert, K. (1997) *A Geographer's Guide to the Internet*, Geographical Association.

Earthquakes (1983) Geological Museum Publications.

Francis, P. (1984) *Volcanoes*, Penguin Books.

Goudie, A. (1993) *The Nature of the Environment*, Basil Blackwell.

Kennedy, A.J. (2000) *The Internet: The Rough Guide*, Rough Guides.

Lamb, S. and Sington, D. (1998) *Earth Story*, BBC Publications.

Park, C. (1991) *Environmental Hazards*, Thomas Nelson.

Prosser, R. (1992) *Natural Systems and Human Responses*, Thomas Nelson.

Websites
Plate tectonics:
http://vulcan.wr.usgs.gov/Glossary/Plate Tectonics/framework.html

Earthquake information:
http://quake.wr.usgs.gov/

Home page of Volcano World:
http://volcano.und.nodak.edu/vw.html

List of volcano and earth-science-oriented web servers:
http://vulcan.wr.usgs.gov/Servers/volcservers.html

See also for more links:
http://www.nelsonthornes.com/gaia

1 a Study Figure 1.51 and identify the internal structure of the Earth by naming A, B, C and D **(4 marks)**

b Identify the **two** types of crust of the Earth and describe the differences between them. **(4 marks)**

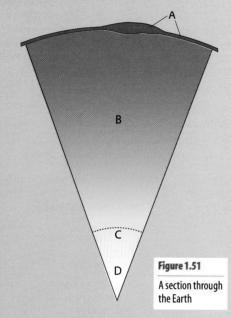

Figure 1.51

A section through the Earth

c Why, and in what ways, do crustal plates move?**(8 marks)**

d For any **one** type of plate margin, identify its type of movement and explain how this creates distinctive landforms. **(9 marks)**

2 a With the aid of a diagram, describe the characteristic features of the materials which make up the different kinds of Earth's **crust**. **(4 marks)**

b Explain the theory of plate tectonics. **(3 marks)**

c Study Figure 1.8 (page 15).
Choose any **one** plate on the Earth's surface.
i Name the plate and describe its characteristics, extent and location. **(5 marks)**
ii Identify the kinds of margins which outline it. **(4 marks)**

d For **either** a named destructive plate margin **or** a named constructive plate margin:
i Draw a diagram to show the relationships between the active Earth and the landforms being created.
ii Describe the relationships you have identified. (This may be by annotations on your diagram or by written material.) **(9 marks)**

3 a Making use of an annotated diagram, identify the relationship between the Earth's crust and the interior structure of the Earth. **(5 marks)**

b Explain why crustal plates move. **(8 marks)**

c For any **one** plate margin, explain the active processes and their relationship to resulting landforms. **(12 marks)**

4 Describe the theory of plate tectonics and explain **three** pieces of evidence which provide support for the theory. **(25 marks)**

5 a In areas where there are volcanic eruptions, earthquakes also occur. Suggest how volcanoes and earthquakes are linked to each other. **(8 marks)**

b Earthquakes occur in areas where there is no evidence of volcanic eruption. For **one** area where there are earthquakes but no volcanoes, explain the causes of earthquake activity. **(10 marks)**

c Choose **one** feature of the landscape of an area where earthquakes occur which suggests that such events have occurred in the past. Describe the landscape feature and explain how it suggests past earthquake activity. **(7 marks)**

6 a i Draw an annotated diagram and describe the features which may be found associated with a **constructive plate margin**. **(8 marks)**

ii For **one** of these features, explain the processes that have led to its formation. **(8 marks)**

b i Explain **one** way in which areas close to a constructive plate margin may be of economic value.
ii Suggest how people can exploit the economic resource you have identified. **(9 marks)**

7 a i Draw an annotated diagram and describe the features associated with a **destructive plate margin**. **(8 marks)**

ii For **one** of these features, explain the processes that have led to its formation. **(8 marks)**

b i Explain **one** way in which an area close to a destructive plate margin may be of economic value.
ii Suggest how people can exploit the economic resource you have identified. **(9 marks)**

8 a Why does the crust of the Earth move? **(8 marks)**

b Identify and locate **two** pieces of evidence that the **continental crust** of the Earth has moved in the past. For **each** of these pieces of evidence, explain how it suggests that this movement has occurred. **(10 marks)**

c Explain how the **oceanic crust** of the Earth can be shown to have been formed over a period of time. **(7 marks)**

9 Study Figure 1.52 and answer the following questions.
  a i Name an example of **each** of the following from the map: shield lands (cratons); fold mountains; deep-sea trenches. **(3 marks)**
     ii Explain the meaning of **each** of these terms: shield lands (cratons); fold mountains; deep-sea trenches. **(6 marks)**
  b i Identify the **compass direction** for the movement of the Earth's crust at each of Ascension Island and Easter Island. **(2 marks)**
     ii For **each** of these places, explain why you think the crust moves in that direction. **(4 marks)**
  c i For any **one** volcano on the map, explain the movement of plates which cause it to erupt. **(4 marks)**
     ii Why do earthquakes occur in areas where there is a danger of volcanic eruptions? **(6 marks)**

10 a i What is 'lava'? **(2 marks)**
        ii What happens to lava when it is exposed on the ground surface? **(1 mark)**
        iii Why does some lava flow quickly and some flow more slowly? **(4 marks)**
     b Making use of annotated diagrams, describe **two** different kinds of volcano. Name an example of **each** of your kinds of volcano. **(6 marks)**
     c Choose **two** different examples of landforms which are formed *after* a volcano becomes dormant. For each one: name a place where it may be found and explain how it has been formed from a volcanic landscape. **(8 marks)**
     d Why do people continue to live close to active volcanoes? **(4 marks)**

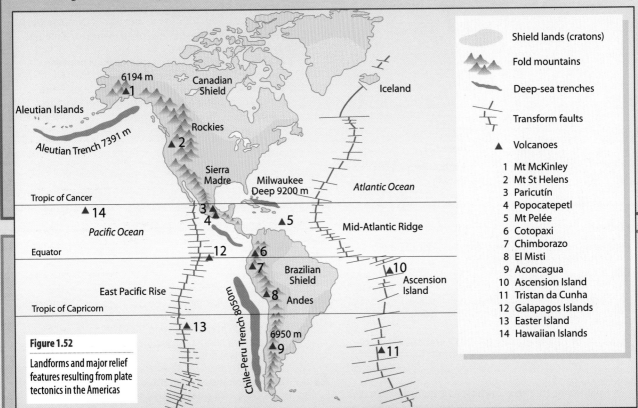

Figure 1.52
Landforms and major relief features resulting from plate tectonics in the Americas

11 Look at Figure 1.52 and make use of Figure 1.8 (page 15).
  a Describe the distribution of **cratons** (shield areas) and of young fold mountain ranges in the Americas. **(8 marks)**
  b Explain, with the use of diagrams, the origins of:
     i Ascension Island (number 10 on Figure 1.52) **(7 marks)**
     ii the Chile–Peru trench and the volcanic mountains numbered 7, 8 and 9 on Figure 1.52. **(10 marks)**

12 Why do earthquakes in Japan have a range of focuses from a shallow focus to a deep focus, while those in Iceland all have a shallow focus? **(25 marks)**

13 Is it possible for people in any area to manage volcanic eruptions successfully? **(25 marks)**

14 a Identify **two different ways** in which volcanoes may be classified. For **one** of the ways you have identified, explain how **one** type of volcano fits into the classification. **(8 marks)**
     b Why do people continue to live close to active volcanoes? **(7 marks)**
     c Using an example of a real upland area, explain what happens to a volcanic area once volcanic activity ceases. **(10 marks)**

15 a i Using an annotated diagram **only**, illustrate the features of a **composite volcano**. **(5 marks)**
        ii Choose **one** feature of **intrusive** volcanic activity, describe it and explain how the feature is created and eventually becomes a **surface** landform. **(10 marks)**
     b Making good use of examples, explain the economic importance of areas of volcanic activity. **(10 marks)**

16 a Draw a labelled diagram to show the features of a **composite volcano**. (**4 marks**)

  b What is the difference between an **intrusive** and an **extrusive** volcanic landform? (**4 marks**)

  c Using Figure 1.24 (page 26) (the Solfatara crater, Italy):

    i What is a 'solfatara'? (**2 marks**)

    ii Describe the features of this landform. *Credit will be given for direct reference to the photograph.* (**4 marks**)

    iii Suggest why these features are present. (**4 marks**)

  d Choose and name **one** intrusive landform. Suggest how it has been formed and describe its shape within the Earth's crust. (**7 marks**)

17 Study Figure 1.3 (page 10).

  a What is an earthquake? (**3 marks**)

  b Why is an earthquake which measures 7.0 on the Richter scale 100 times more severe than one that measures 5.0? (**3 marks**)

  c How severe was the earthquake in San Francisco in 1989? (**1 mark**)

  d How much bigger was the earthquake in San Francisco in 1906 than the one in Kobe in 1972? (**3 marks**)

  e Describe one way in which buildings may be made 'earthquake proof'. (**4 marks**)

  f List **two** rules which you would need to follow if your home was in an earthquake area. Explain why they would be important. (**4 marks**)

  g How do local and national authorities try to prepare for earthquakes in areas where they may occur? (**7 marks**)

Heading	Description from case study	
Location	Identify where the disaster occurred	(2 marks)
Pre-disaster potential	Description of geology of the area to identify the reason for an event to occur	(3 marks)
Disaster event	Timing, size and nature of the disaster	(3 marks)
Disruption	Details of immediate damage	(3 marks)
Relief	Types of immediate relief needed	(3 marks)
Recovery	Nature of the required recovery programme	(4 marks)
Time	Timescale of the continuing impact of the event	(3 marks)
Reconstruction	Type and amount of long-term aid required	(4 marks)

18 Create a table using the headings in the left column of the table above. Use it to provide details of a volcanic or earthquake event you have studied.

19 For **either** a volcanic eruption **or** an earthquake you have studied:

  a Draw a sketch map to show the location of the area where it occurred. (**3 marks**)

  b Describe the hazard event. (**3 marks**)

  c Explain, with the aid of a diagram, the causes of the event. (**4 marks**)

  d How big was the event? (**2 marks**)

  e How frequently do such events occur in this area? (**2 marks**)

  f How large an area was affected by the event? (**3 marks**)

  g Describe the effects of the event on the area. (**4 marks**)

  h What lessons for the future were learned from this event? (**4 marks**)

AS

20 Study Figure 1.53.

  a For **one** geological factor: identify the factor and explain the reason for the difference in hazard assessment. (**8 marks**)

  b Explain **two** ways in which a volcanic eruption could affect an urban area **outside** the zone of direct lava and pyroclastic outfall. (**9 marks**)

  c Use any case study information you may have available to describe the ways a more economically developed area may prepare for the danger of a volcanic eruption nearby. (**8 marks**)

21 a i What is a **natural hazard**?

    ii Under what circumstances can a volcanic eruption be described as a hazard event? (**8 marks**)

  b For any **one** volcanic event you have studied:

    i Identify the causes and timetable of the event. (**8 marks**)

    ii What were the longer-term effects of the event? (**9 marks**)

22 a i Identify the area where this hazard may occur. Describe the causes of the volcanic hazard. (**5 marks**)

    ii Why do people continue to live in the area? (**5 marks**)

    iii For any one volcanic event which has occurred in the area, describe the volcanic activity involved and the reaction of people and organisations to the event. (**8 marks**)

  b How can future volcanic hazards in the area be predicted? (**7 marks**)

23 For any **one** area that experiences volcanic and / or earthquake hazards which you have studied, explain how people perceive and manage the hazard. (**25 marks**)

Figure 1.53

Range of factors affecting volcanic hazards

Danger factor	Assessment of danger
Geological factor	
Plate margin type	There will be more explosive activity on a destructive margin than on a constructive margin.
Volcano type	A shield volcano will be less explosive than a stratovolcano.
Extruded material	A lava eruption is less dangerous than a pyroclastic eruption.
Silica content	Silica-rich magmas produce more explosive eruptions than silica-poor magmas.
Dormancy period	Volcanoes with longer periods of dormancy tend to be more explosive than those with shorter dormancy periods.
Environmental and topographical factors	
Wind direction	Pyroclastic flows are thicker downwind from an active vent.
Topography	Valleys funnel pyroclastic and other flows. Ridges across the route of flows can shelter areas within a blast zone.
Social and economic factors	
Settlement density	More densely settled areas will be at greater risk of immediate damage.
Economic status	Total cost will be greater in more economically developed areas but response will be faster and more effective. Loss of life will be lower. In less developed areas, loss of life will be greater and economic damage will be greater in proportion to total.

24 'The perceptions of and response to earthquake events depend more upon the wealth of the people in the country concerned than on any other factor.' Discuss this statement with reference to earthquake events you have studied. (**25 marks**)

A2

Chapter 2

Weathering and slopes

*'Every valley shall be exalted, and every mountain
and hill shall be made low: and the crooked shall be
made straight, and the rough places plain.'*

The Bible, Isaiah, 40:4

Weathering

The majority of rocks have been formed at high
temperatures (igneous and many metamorphic
rocks) and/or under great pressure (igneous,
metamorphic and sedimentary rocks), but in the
absence of oxygen and water. If, later, these
rocks become exposed on the Earth's surface,
they will experience a release of pressure, be
subjected to fluctuating temperatures, and be
exposed to oxygen in the air and to water. They
are therefore vulnerable to **weathering**, which is
the disintegration and decomposition of rock *in
situ* – i.e. in its original position. Weathering is,
therefore, the natural breakdown of rock and
can be distinguished from erosion because it
need not involve any movement of material.
Weathering is the first stage in the **denudation**
or wearing down of the landscape; it loosens
material which can subsequently be transported
by such agents of erosion as running water
(Chapter 3), ice (Chapter 4), the sea (Chapter 6)
and the wind (Chapter 7). The degree of weathering depends upon the structure and mineral
composition of the rocks, local climate and
vegetation, and the length of time during
which the weathering processes operate.

There are two main types of weathering:
1 **Mechanical** (or **physical**) **weathering**
is the disintegration of rock into smaller particles by mechanical processes but without
any change in the chemical composition of
that rock. It is more likely to occur in areas
devoid of vegetation, such as deserts, high
mountains and arctic regions. Physical
weathering usually produces sands.
2 **Chemical weathering** is the decomposition
of rock resulting from a chemical change. It
produces changed substances and solubles,
and usually forms clays. Chemical weath-

ering is more likely to take place in warmer,
more moist climates where there is an associated vegetation cover.

It should be appreciated that although in any
given area either mechanical or chemical weathering may be locally dominant, both processes
usually operate together rather than in isolation.

Mechanical weathering

Frost shattering

This is the most widespread form of mechanical
weathering. It occurs in rocks that contain
crevices and joints (e.g. joints formed in granite
as it cooled, bedding planes found in sedimentary rocks, and pore spaces in porous rocks),
where there is limited vegetation cover and
where temperatures fluctuate around 0°C. In the
daytime, when it is warmer, water enters the
joints, but during cold nights it freezes. Frost
leads to mechanical breakdown in two ways:
1 As ice occupies 9 per cent more volume than
water, it exerts pressure within the joints.
2 When water freezes within the rock it attracts
small particles of water, creating increasingly
large ice crystals.

In each case the alternating **freeze–thaw
process**, or **frost shattering**, slowly widens the
joints and, in time, causes pieces of rock to
shatter from the main body. Where this **block
disintegration** occurs on steep slopes, large
angular rocks collect at the foot of the slope as
scree or **talus** (Figure 2.1); if the slopes are
gentle, however, large **blockfields** (felsenmeer)
tend to develop. Frost shattering is more
common in upland regions of Britain where
temperatures fluctuate around freezing point for
several months in winter, than in polar areas
where temperatures rarely rise above 0°C.

Salt crystallisation

If water entering the pore spaces in rocks is
slightly saline then, as it evaporates, salt crystals
are likely to form. As the crystals become larger,
they exert stresses upon the rock, causing it to disintegrate. This process occurs in hot deserts where

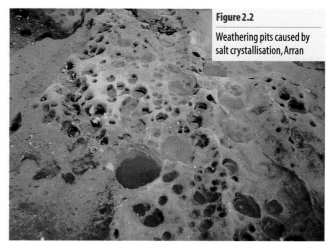

Figure 2.2

Weathering pits caused by salt crystallisation, Arran

Figure 2.1

The formation of screes resulting from frost shattering: Moraine Lake, Banff National Park, Canada

Figure 2.3

An exfoliation dome: Sugar Loaf Mountain in Rio de Janeiro, Brazil

capillary action draws water to the surface and where the rock is sandstone. Individual grains of sand are broken off by **granular disintegration**. Salt crystallisation also occurs on coasts where the constant supply of salt can lead to the development of weathering pits (Figure 2.2).

Pressure release

As stated earlier, many rocks, especially intrusive jointed granites, have developed under considerable pressure. The confining pressure increases the strength of the rocks. If these rocks, at a later date, are exposed to the atmosphere, then there will be a substantial release of pressure. (If you had 10 m of bedrock sitting on top of you, you would be considerably relieved were it to be removed!) The release of pressure weakens the rock allowing other agents to enter it and other processes to develop. Where cracks develop parallel to the surface, a process called **sheeting** causes the outer layers of rock to peel away. This process is now believed to be responsible for the formation of large, rounded rocks called **exfoliation domes** (Figures 2.3 and 7.6) and, in part, for the granite tors of Dartmoor and the Isle of Arran (Figures 8.14 and 8.15). Jointing, caused by pressure release, has also accentuated the characteristic shapes of glacial cirques and troughs (Figures 2.4, 4.14 and 4.15).

Thermal expansion or insolation weathering

Like all solids, rocks expand when heated and contract when cooled. In deserts, where cloud and vegetation cover are minimal, the diurnal range of temperature can exceed 50°C. It was believed that, because the outer layers of rock warm up faster and cool more rapidly than the inner ones, stresses were set up that would cause the outer thickness to peel off like the layers of an onion – the process of **exfoliation** (page 181). Initially, it was thought that it was this expansion–contraction process which produced exfoliation domes. Changes in temperature will also cause different minerals within a rock to expand and contract at different rates. It has been suggested that this causes granular disintegration in rocks composed of several minerals (e.g. granite which consists of quartz, feldspar and mica), whereas in homogeneous rocks it is more likely to cause block disintegration.

Laboratory experiments (e.g. by Griggs in 1936 and Goudie in 1974) have, however, cast doubt upon the effectiveness of insolation weathering (page 181).

Biological weathering

Tree roots may grow along bedding planes or extend into joints, widening them until blocks of rock become detached (Figure 2.5). It is also claimed that burrowing creatures, such as worms and rabbits, may play a minor role in the excavation of partially weathered rocks.

Figure 2.4

The process of pressure release tends to perpetuate landforms: as new surfaces are exposed, the reduction in pressure causes further jointing parallel to the surface

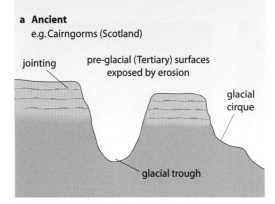

a Ancient
e.g. Cairngorms (Scotland)

jointing   pre-glacial (Tertiary) surfaces exposed by erosion

glacial cirque

glacial trough

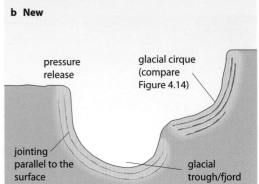

b New

pressure release

glacial cirque (compare Figure 4.14)

jointing parallel to the surface

glacial trough/fjord

Chemical weathering

Chemical weathering tends to:

- attack certain minerals selectively
- occur in zones of alternate wetting and drying, e.g. where the level of the water table fluctuates

Figure 2.5

Mechanical (biological) weathering caused by expanding tree roots in Geltsdale, Cumbria

Figure 2.6

Oxidation in Geltsdale, Cumbria

- occur mostly at the base of slopes where it is likely to be wetter and warmer.

This type of weathering involves a number of specific processes which may operate in isolation but which are more likely to be found in conjunction with one another. Formulae for the various chemical reactions are listed at the end of the chapter.

Oxidation

This occurs when rocks are exposed to oxygen in the air or water. The simplest and most easily recognised example is when iron in a **ferrous** state is changed by the addition of oxygen into a **ferric** state. The rock or soil, which may have been blue or grey in colour (characteristic of a lack of oxygen), is discoloured into a reddish-brown – a process better known as **rusting** (Figure 2.6). Oxidation causes rocks to crumble more easily.

In waterlogged areas, oxidation may operate in reverse and is known as **reduction**. Here, the amount of oxygen is reduced and the soils take on a blue/green/grey tinge (see **gleying**, page 272).

Hydration

Certain rocks, especially those containing salt minerals, are capable of absorbing water into their structure, causing them to swell and to become vulnerable to future breakdown. For example, gypsum is the result of water having been added to anhydrite ($CaSO_4$). This process appears to be most active following successive periods of wet and dry weather and is important in forming clay particles. Hydration is in fact a physio-chemical process as the rocks may swell and exert pressure as well as changing their chemical structure.

Hydrolysis

This is possibly the most significant chemical process in the decomposition of rocks and formation of clays. Hydrogen in water reacts with minerals in the rock or, more specifically, there is a combination of the H+ and OH– ions in the water and the ions of the mineral (i.e. the water combines with the mineral rather than dissolving it).

The rate of hydrolysis depends upon the amount of H+ ions, which in turn depends upon the composition of air and water in the soil (Figure 10.4), the activity of organisms (page 268), the presence of organic acids (page 271) and the cation exchange (page 269). An example of hydrolysis is the breakdown of feldspar (Figure 2.7), a mineral found in igneous rocks such as granite, into a residual clay deposit known as kaolinite (china clay). Granite consists of three minerals – quartz, mica and feldspar (Figure 8.2c) – and, as the table below shows, each reacts at a different rate with water.

Quartz	Mica	Feldspar
Not affected by water, remains unchanged as sand (Figure 2.7)	May be affected by water under more acid conditions releasing aluminium and iron	Readily attracts water producing a chemical change which turns the feldspar into clay (kaolin or china clay)

Figure 2.7

Decomposition of granite by hydrolysis on Goatfell, Arran

Figure 2.8

Carbonation of limestone near Ingleton, North Yorkshire

Carbonation

Rainwater contains carbon dioxide in solution which produces carbonic acid (H_2CO_3). This weak acid reacts with rocks which are composed of calcium carbonate, such as limestone. The limestone dissolves and is removed in solution (calcium bicarbonate) by running water. Carboniferous limestone is well-jointed and bedded (Chapter 8), which results in the development of a distinctive group of landforms (Figure 2.8).

Solution

Some minerals, e.g. rock salt, are soluble in water and simply dissolve *in situ*. The rate of solution can be affected by acidity since many minerals become more soluble as the pH of the solvent increases (page 269).

Organic weathering

Humic acid, derived from the decomposition of vegetation (humus), contains important elements such as calcium, magnesium and iron. These are released by a process known as **chelation** (page 271). The action of bacteria and the respiration of plant roots tends to increase carbon dioxide levels which helps accelerate solution processes, especially carbonation. Lichen can also extract iron from certain rocks through the process of reduction. Recent research suggests that lichen and blue-green algae, which form the pioneer community in the development of a lithosere (page 288), play a far greater weathering role than was previously thought. However, it should be remembered that the presence of a vegetation cover dramatically reduces the extent of mechanical weathering.

Acid rain

Human economic activities (such as power generation and transport) release increasingly more carbon dioxide, sulphur dioxide and nitrogen oxide into the atmosphere. These gases then form acids in solution in rainwater (page 222). Acid rain readily attacks limestones and, to a lesser extent, sandstones, as shown by crumbling buildings and statues (Figure 2.9). The increased level of acidity in water passing through the soil tends to release more hydrogen and so speeds up the process of hydrolysis. An indirect consequence of acid rain is the release from certain rocks of toxic metals, such as aluminium, cadmium, copper and zinc, which can be harmful to plants and soil biota (page 268).

Some authorities, including Andrew Goudie, prefer to divide weathering into three categories rather than the two described here. Their alternative classification includes, as a third category, **biological weathering**. Instead of including 'biological' under mechanical weathering and 'organic' under chemical weathering, they would group these two types together under the heading 'biological weathering'.

Climatic controls on weathering

Mechanical weathering

Frost shattering is important if temperatures fluctuate around 0°C, but will not operate if the climate is too cold (permanently frozen), too warm (no freezing), too dry (no moisture to freeze), or too wet (covered by vegetation). Mechanical weathering will not take place at **X** on Figure 2.10a where it is too warm and there is insufficient moisture, while at **Y**, the high temperature and heavy rainfall will give a thick protective vegetation cover against insolation.

Chemical weathering

This increases as temperatures and rainfall totals increase. It has been claimed that the rate of chemical weathering doubles with every 10°C temperature increase. Recent theories suggest that, in humid tropical areas, direct removal by solution may be the major factor in the lowering of the landscape, due to the continuous flow of water through the soil. Chemical weathering will be rapid at **S** (Figure 2.10b) due to humic acid from the vegetation. It will be limited at **P**, because temperatures are low, and at **R**, where there is insufficient moisture for the chemical decomposition of rocks. Carbon dioxide is an exception in that, being more soluble at lower as opposed to higher temperatures, it can accelerate rates of solution in cold climates.

Weathering regions

Peltier, an American physicist and climatologist, attempted to predict the type and rate of weathering at any given place in the world from its mean annual temperature and mean annual rainfall (Figure 2.10c). It should be realised that mechanical and chemical weathering usually operate together at the same time and at the same place, but it is likely that in each situation one type or the other will be the more significant.

Figure 2.9

Acid rain damage to stone statues, Exeter Cathedral

Figure 2.10

Climatic controls on weathering (*after* Peltier)

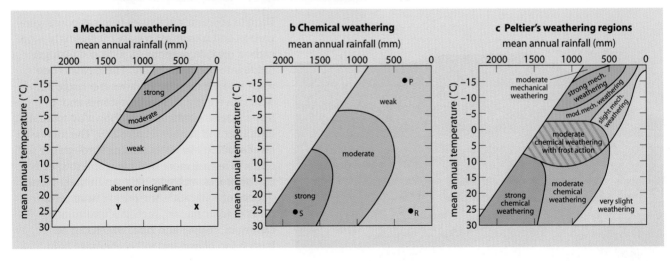

One type of model (Framework 12, page 352) widely adopted by geographers to help explain phenomena is the **system**. The system is a method of analysing relationships within a unit and consists of a number of components between which there are linkages. The model is usually illustrated schematically as a flow diagram.

Systems may be described in three ways:

- **Isolated:** in which there is no input or output of energy or matter. Some suggest the universe is the sole example of this type; others claim the idea is not applicable in geography.

- **Closed:** there is input, transfer and output of energy but not of matter (or mass).

- **Open:** most environmental systems are open and there are inputs and outputs of both energy and matter.

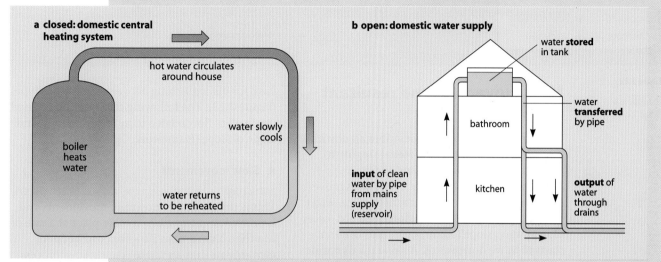

a closed: domestic central heating system

hot water circulates around house

water slowly cools

boiler heats water

water returns to be reheated

b open: domestic water supply

water **stored** in tank

water **transferred** by pipe

bathroom

input of clean water by pipe from mains supply (reservoir)

kitchen

output of water through drains

Figure 2.11

Closed and open systems in the house

Examples of the systems approach used and referred to in this book (chapter number is given in brackets):

Geomorphological	Climate, soils and vegetation	Human and economic
Slopes (2)	Atmosphere energy budget (9)	Population change (13)
Drainage basins (3)	Hydrological cycle (9)	Farming (16)
Glaciers (4)	Soils (10)	Industry (19)
	Ecosystems (11)	Transport (21)
	Nutrient cycle (12)	

When opposing forces, or inputs and outputs, are balanced, the system is said to be in a state of **dynamic equilibrium**. If one element in the system changes because of some outside influence, then it upsets this equilibrium and affects the other components. For example, equilibrium is upset when:

- prolonged heavy rainfall causes an increase in the discharge and velocity of a river or a lowering of base level (page 81), both of which lead to an increase in the rate of erosion

- an increase in carbon dioxide into the atmosphere causes global temperatures to rise (global warming, Case Study 9)

- drought affects the carrying capacity of animals (or people) grazing (living) in an area as the water shortage reduces the availability of grass (food supplies) (page 378)

- an increase in the number of tourists to places of scenic attraction harms the environment (especially where it is fragile) that was the original source of the attraction (page 591).

Weathering and slopes

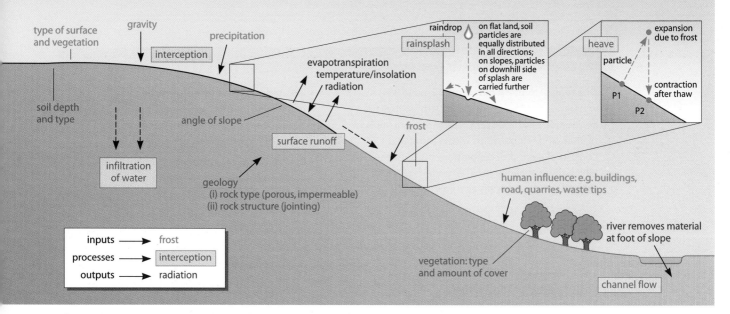

Mass movement and resultant landforms

The term **mass movement** describes all downhill movements of weathered material (**regolith**), including soil, loose stones and rocks, in response to gravity. However, it excludes movements where the material is carried by ice, water or wind. When gravitational forces exceed forces of resistance, slope failure occurs and material starts to move downwards. A slope is a **dynamic open system** (Framework 3) affected by biotic, climatic, gravitational, groundwater and tectonic inputs which vary in scale and time. The amount, rate and type of movement depend upon the degree of slope failure (Figure 2.12).

Although by definition mass movement refers only to the movement downhill of material under the force of gravity, in reality water is usually present and assists the process. When Carson and Kirkby (1972) attempted to group mass movements, they used the speed of movement and the amount of moisture present as a

basis for distinguishing between the various types (Figure 2.13). The following classification is based on speed of flow related to moisture content and angle of slope (Framework 7, page 167).

A Slow movements

Soil creep

This is the slowest of downhill movements and is difficult to measure as it takes place at a rate of less than 1 cm a year. However, unlike faster movements, it is an almost continuous process. Soil creep occurs mainly in humid climates where there is a vegetation cover. There are two major causes of creep, both resulting from repeated expansion and contraction.

- **Wet–dry periods**   During times of heavy rainfall, moisture increases the volume and weight of the soil, causing expansion and allowing the regolith to move downhill under gravity. In a subsequent dry period, the soil will dry out and then contract, especially if it is clay. An extreme case of contraction in clays occurred in south-east England during the 1976 drought when buildings sited on almost imperceptible slopes suffered major structural damage.
- **Freeze–thaw**   When the regolith freezes, the presence of ice crystals increases the volume of the soil by 9 per cent. As the soil expands, particles are lifted at right-angles to the slope in a process called **heave** (Figure 2.12 and page 132). When the ground later thaws and the regolith contracts, these particles fall back vertically under the influence of gravity and so move downslope.

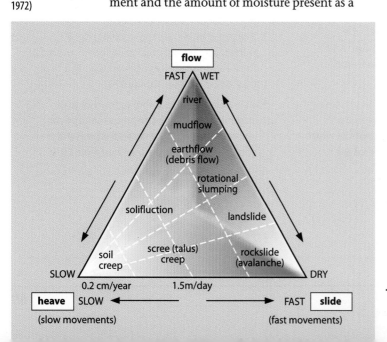

Figure 2.14

Terracettes in Wharfedale, Yorkshire Dales

vegetation cover is limited. During the winter season, both the bedrock and regolith are frozen. In summer, the surface layer thaws but the underlying layer remains frozen and acts like impermeable rock. Because surface meltwater cannot infiltrate downwards and temperatures are too low for much effective evaporation, any topsoil will soon become saturated and will flow as an **active layer** over the frozen subsoil and rock (page 131). This process produces **solifluction sheet** or **lobes** (Figure 5.12), rounded, tongue-like features reaching up to 50 m in width, and **head**, a mixture of sand and clay formed in valleys and at the foot of sea cliffs (Figure 5.13). Solifluction was widespread in southern Britain during the Pleistocene ice age; covered most of Britain following the Pleistocene; and continues to take place in the Scottish Highlands today.

B Flow movements

Earthflows

When the regolith on slopes of 5–15° becomes saturated with water, it begins to flow downhill at a rate varying between 1 and 15 km per year. The movement of material may produce short **flow tracks** and small bulging lobes or tongues, yet may not be fast enough to break the vegetation.

Mudflows

These are more rapid movements, occurring on steeper slopes, and exceeding 1 km/hr. When Nevado del Ruiz erupted in Colombia in 1985, the resultant mudflow reached the town of Armero at an estimated speed of over 40 km/hr (Case Study 2A). Mudflows are most likely to occur following periods of intensive rainfall, when both volume and weight are added to the soil giving it a higher water content than an earthflow. Mudflows may result from a combination of several factors (Figure 2.16).

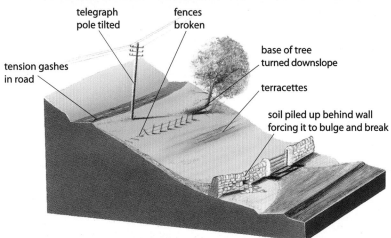

Figure 2.15

The effects of soil creep

Soil creep usually occurs on slopes of about 5° and produces **terracettes** (Figure 2.14). These are step-like features, often 20–50 cm in height, which develop as the vegetation is stretched and torn: they are often used and accentuated by grazing animals, especially sheep. The effects of soil creep are shown in Figure 2.15.

Solifluction

This process, meaning 'soil flow', is a slightly faster movement usually averaging between 5 cm and 1 m a year. It often takes place under periglacial conditions (Chapter 5) where

Figure 2.16

Fieldsketch showing the causes of a mudflow, Glen Rosa, Arran

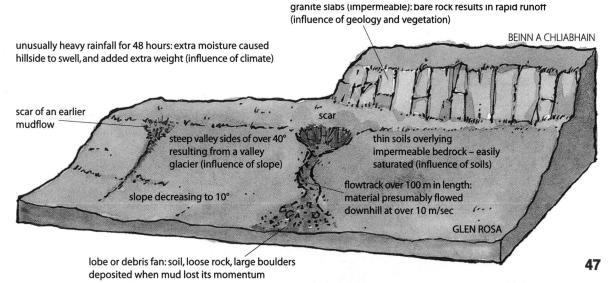

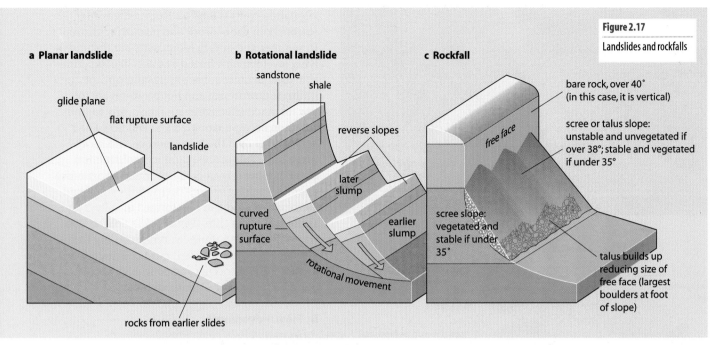

Figure 2.17

Landslides and rockfalls

a Planar landslide

glide plane

flat rupture surface

landslide

rocks from earlier slides

b Rotational landslide

sandstone

shale

reverse slopes

later slump

earlier slump

curved rupture surface

rotational movement

c Rockfall

bare rock, over 40° (in this case, it is vertical)

scree or talus slope: unstable and unvegetated if over 38°; stable and vegetated if under 35°

free face

scree slope: vegetated and stable if under 35°

talus builds up reducing size of free face (largest boulders at foot of slope)

Figure 2.18

Landslides on the Norfolk coast

Figure 2.19

Rockfalls in the crater of Vesuvius, Italy

C Rapid movements

Slides

The fundamental difference between slides and flows is that flows suffer internal derangement whilst, in contrast, slides move 'en masse' and are not affected by internal derangement. Rocks that are jointed or have bedding planes roughly parallel to the angle of slope are particularly susceptible to landslides. Slides may be planar or rotational (Figure 2.17a and b). In a planar slide, the weathered rock moves downhill leaving behind it a flat rupture surface (Figure 2.17a). Where rotational movement occurs, a process sometimes referred to as **slumping**, a curved rupture surface is produced (Figure 2.17b). Rotational movement can occur in areas of homogeneous rock, but is more likely where softer materials (clay or sands) overlie more resistant or impermeable rock (limestone or granite). Slides are common in many coastal areas of southern and eastern England. In Figure 2.18, the cliffs, composed of glacial deposits, are retreating rapidly due to frequent slides. The slumped material can be seen at the foot of the cliff.

Very rapid movements

Rockfalls

These are spontaneous, though relatively rare, debris movements on slopes that exceed 40°. They may result from extreme physical or chemical weathering in mountains, pressure release, storm-wave action on sea cliffs, or earthquakes. Material, once broken from the surface, will either bounce or fall vertically to form scree, or talus, at the foot of a slope (Figures 2.17c and 2.19).

On 8 May 1998, Italy declared a state of emergency in the flood-stricken area between Naples and Salerno, two cities 60 km apart. Two days after mudslides had swept through several small towns and villages, the authorities admitted that they were unsure as to how many people were missing, while the headlines of Rome's *Il Messeggero* claimed that 'under the mud lies the Pompeii of the Year 2000'.

The mudslides had occurred on 6 May following 48 hours of torrential rain. Rivers of mud, water and debris, arriving as mini-tidal waves up to 3 m in height, burst into town centres, tearing down houses and bridges, swallowing cars and causing residents in the densely populated rural area near to Mount Vesuvius, to flee for safety. Early estimates suggested that over 1000 people had been made homeless, many having to spend the first night on rooftops or on the highest floors of apartment blocks until they could be rescued the next day by helicopters. Over 3000 firemen, soldiers and policemen, together with sniffer dogs trained to find avalanche victims, spent several days searching through mud for up to 300 missing people.

Worst hit was Sarno, a small town with a population of 2000 (Figure 2.20). Of 143 bodies eventually recovered from the flood-affected area, over 100 came from this one town. Amongst buildings affected was the local hospital where, the day after the event, the bodies of two doctors, a nurse and several patients were pulled from under the rubble and mud. The authorities admitted that they had no idea as to how many people remained missing as many houses – as elsewhere in the region – had been built illegally, and there was no official data as to the number of occupants.

With a realisation that there was neither national nor local emergency planning for such natural disasters, concern was to be expressed that up to 4000 villages and small towns in Italy were built in potentially hazardous areas and, therefore, at risk from flash floods, landslides, earthquakes and avalanches. This lack of planning includes the area around Vesuvius up whose slopes settlement, especially the suburbs of Naples, is spreading. The lessons of Pompeii and Herculaneum seem long since forgotten (Figure 2.21).

Figure 2.20

The Sarno mudslide

Figure 2.21

A natural or human disaster?

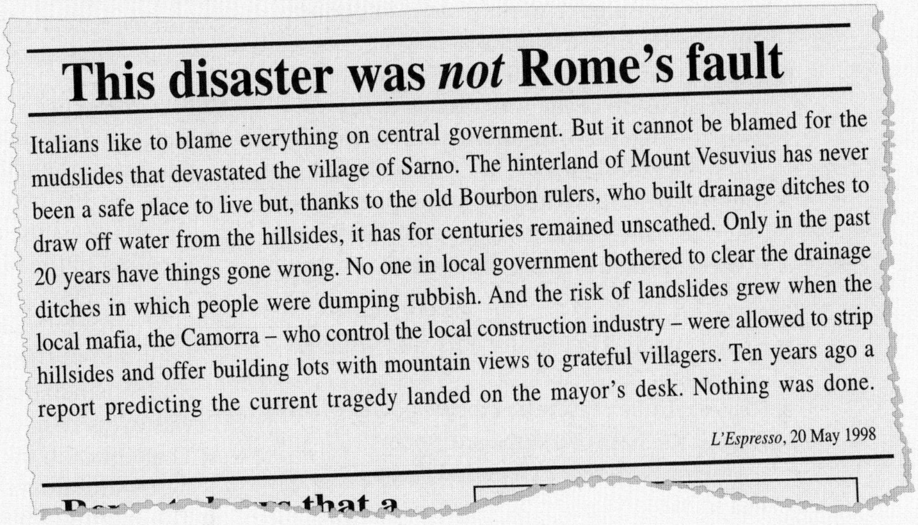

This disaster was *not* Rome's fault

Italians like to blame everything on central government. But it cannot be blamed for the mudslides that devastated the village of Sarno. The hinterland of Mount Vesuvius has never been a safe place to live but, thanks to the old Bourbon rulers, who built drainage ditches to draw off water from the hillsides, it has for centuries remained unscathed. Only in the past 20 years have things gone wrong. No one in local government bothered to clear the drainage ditches in which people were dumping rubbish. And the risk of landslides grew when the local mafia, the Camorra – who control the local construction industry – were allowed to strip hillsides and offer building lots with mountain views to grateful villagers. Ten years ago a report predicting the current tragedy landed on the mayor's desk. Nothing was done.

L'Espresso, 20 May 1998

Development of slopes

Slope development is the result of the interaction of several factors. Rock structure and lithology, soil, climate, vegetation and human activity are probably the most significant. All are influenced by the time over which the processes operate. Slopes are an integral part of the drainage basin system (Chapter 3) as they provide water and sediment for the river channel.

The effects of rock structure and lithology

- Areas of bare rock are vulnerable to mechanical weathering (e.g. frost shattering) and some chemical weathering processes.
- Areas of alternating harder/more resistant rocks and softer/less resistant rocks are more likely to experience movement, e.g. clays on limestones (Vaiont Dam, Case Study 2B).

- An impervious underlying rock will cause the topsoil to become saturated more quickly, e.g. glacial deposits overlying granite.
- Steep gradients are more likely to suffer slope failure than gentler ones. In Britain, especially in lowland areas, most slopes are under 5° and few are over 40°.
- Failure is also likely on slopes where the equilibrium (balance) of the system (Framework 3, page 45), has been disturbed, e.g. a glaciated valley.
- The presence of joints, cracks and bedding planes can allow increased water content and so lead to sliding (Vaiont Dam, Case Study 2B).
- Earthquakes (Mt Huascaran in Peru) and volcanic eruptions (Nevado del Ruiz in Colombia) can cause extreme slope movements (Case Study 2A).

Figure 2.22

The effect of pore-water pressure and capillary action on soil movement

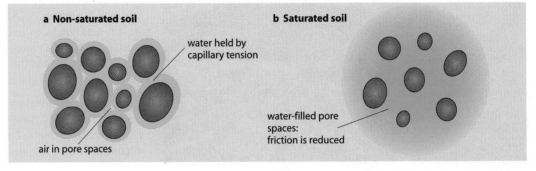

Soil

- Thin soils tend to be more unstable. As they can support only limited vegetation, there are fewer roots to bind the soil together.
- Unconsolidated sands have lower internal cohesion than clays.
- A porous soil, e.g. sand, is less likely to become saturated than one that is impermeable, e.g. clay.
- In a non-saturated soil (Figure 2.22a), the surface tension of the water tends to draw particles together. This increases cohesion and reduces soil movement. In a saturated soil (Figure 2.22b), the pore water pressure (page 267) forces the particles apart, reducing friction and causing soil movement.

Climate

- Heavy rain and meltwater both add volume and weight to the soil.
- Heavy rain increases the erosive power of any river at the base of a slope and so, by removing material, makes that slope less stable.
- Areas with freeze–thaw or wet–dry periods are subjected to alternating expansion–contraction of the soil.

- Heavy snowfall adds weight and is thus conducive to rapid movements, e.g. avalanches, Case Study 4.

Vegetation

- A lack of vegetation means that there are fewer roots to bind the soil together.
- Sparse vegetation cover will encourage surface runoff as precipitation is not intercepted (page 59).

Human influence

- Deforestation increases (afforestation decreases) the rate of slope movement.
- Road construction or quarrying at the foot of slopes upsets the equilibrium, e.g. during the building of the M5 in the Bristol area.
- Slope development processes may be accentuated either by building on steep slopes (Hong Kong and Rio de Janeiro, Case Study 2B) or by using them to deposit industrial or mining waste (Aberfan, Case Study 2B).
- The vibration caused by heavy traffic can destabilise slopes (Mam Tor, Derbyshire).
- The grazing of animals and ploughing help loosen soil and remove the protective vegetation cover.

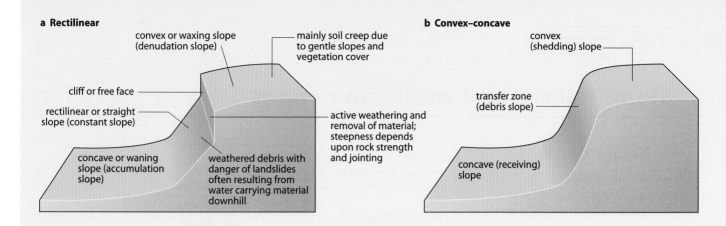

a Rectilinear

convex or waxing slope (denudation slope)

mainly soil creep due to gentle slopes and vegetation cover

cliff or free face

rectilinear or straight slope (constant slope)

concave or waning slope (accumulation slope)

weathered debris with danger of landslides often resulting from water carrying material downhill

active weathering and removal of material; steepness depends upon rock strength and jointing

b Convex–concave

convex (shedding) slope

transfer zone (debris slope)

concave (receiving) slope

Figure 2.23

Slope element models

Slope elements

Two models try to show the shape and form of a typical slope. The first, Figure 2.23a, is more widely used than the second (Figure 2.23b) – although, in this author's view, the first is less easily seen in the British landscape. Regardless of which model is used, confusion unfortunately arises because of the variation in nomenclature used to describe the different facets of the slope.

In reality, few slopes are likely to match up perfectly with either model, and each individual slope is likely to show more elements than those in Figure 2.23.

Figure 2.24

Slope development theories

Slope development through time

How slopes have developed over time is one of the more controversial topics in geomorphology.

This is partly due to the time needed for slopes to evolve and partly due to the variety of combinations of processes acting upon slopes in various parts of the world. Slope development in different environments has led to three divergent theories being proposed: **slope decline**, **slope replacement** and **parallel retreat**. Figure 2.24 is a summary of these theories.

None of the theories of slope development can be universally accepted, although each may have local relevance in the context of the climate and geology (structure) of a specific area. At the same time, two different climates or processes may produce the same type of slope, e.g. cliff retreat due to sea action in a humid climate or to weathering in a semi-arid climate.

	Slope decline (W. M. Davis, 1899)	Slope replacement (W. Penck, 1924)	Parallel retreat (L. C. King, 1948, 1957)
Region of study	Theory based on slopes in what was to Davis a normal climate, north-west Europe and north-east USA.	Conclusions drawn from evidence of slopes in the Alps and Andes.	Based on slopes in South Africa.
Climate	Humid climates.	Tectonic areas.	Semi-arid landscapes. Sea cliffs with wave-cut platforms.
Description of slope	Steepest slopes at beginning of process with a progressively decreasing angle in time to give a convex upper slope and a concave lower slope.	The maximum angle decreases as the gentler lower slopes erode back to replace the steeper ones giving a concave central portion to the slope.	The maximum angle remains constant as do all slope facets apart from the lower one which increases in concavity.
	slope decline — stage 4, stage 3, stage 2, stage 1, concave curve, watershed worn down, convex curve, peneplain. By stage 4 land has been worn down into a convex-concave slope	**slope replacement** — stage 3, stage 2, stage 1, A A A, C B B C C B. talus-scree slope B will replace slope A; slope C will eventually replace slope B	**slope retreat** — stage 4, stage 3, stage 2, stage 1, convex, free face. concave debris slope pediment (can be removed by flash floods)
Changes over time	Assumed a rapid uplift of land with an immediate onset of denudation. The uplifted land would undergo a cycle of erosion where slopes were initially made steeper by vertical erosion by rivers but later became less steep (slope decline) until the land was almost flat (peneplain).	Assumed landscape started with a straight rock slope with equal weathering overall. As scree (talus) collected at the foot of the cliff it gave a gentler slope which, as the scree grew, replaced the original one.	Assumed that slopes had two facets – a gently concave lower slope or pediment and a steeper upper slope (scarp). Weathering caused the parallel retreat of the scarp slope allowing the pediment to extend in size.

Slope failure and mass movement

A Natural causes

All slopes are affected by gravity and, consequently, by one or more of the several mass movement processes by which weathered material is transported downhill (pages 46–48). Where slopes are gentle (about 5°), the movement of material is slow and has relatively little effect upon property, life or human activity. As slope angles increase, however, so too do the rate and frequency of slope movement and the risk of sudden slope failure. Slope failure, occurring in the form of either mudflows or landslides, is a natural event. When this failure occurs in densely populated areas, it becomes a potentially dangerous natural hazard (Framework 2, page 31). Three examples of how slope failure caused by natural events can cause serious loss of property and life (Figure 2.25) are:

  (i) earthquakes
 (ii) volcanic activity
(iii) excessive rainfall.

(i) Earthquakes – avalanches and rockfalls (Peru 1970)

In 1970 an offshore earthquake measuring 7.7 on the Richter scale shook parts of Peru to the north of its capital, Lima. The shock waves loosened a mass of unstable ice and snow near the summit of Huascaran, the country's highest peak (6768 m). The falling ice and snow formed a huge avalanche which rushed downhill, falling 3000 m into the Rio Santo Valley, collecting rocks and boulders en route. In its path stood the town of Yungay with a population of 20 000.

Estimates suggest that the avalanche was travelling at a speed of 480 km/hr when it hit the settlement. It took rescue workers three days to reach the town. Once there, they found very few survivors and only the tops of several 30 m palm trees, which marked the location of the former town square (Figures 2.25 and 2.26).

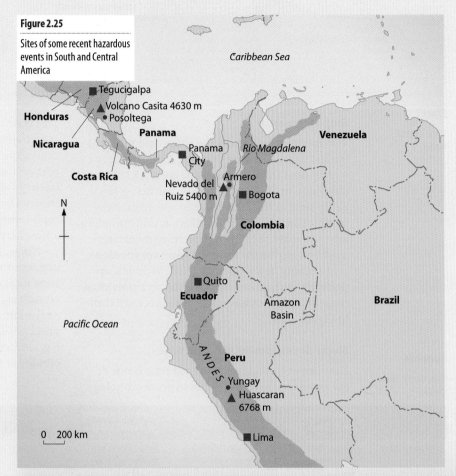

Figure 2.25

Sites of some recent hazardous events in South and Central America

Figure 2.26

The site of Yungay after the avalanche

(ii) Volcanic eruptions – mudflows (Colombia 1985)

The Colombian volcano of Nevado del Ruiz had not erupted since 1595 until, in November 1985, it showed signs of activity by emitting gas and steam. As an increasing amount of magma welled upwards towards the crater, the whole peak must have become warmer, as was made evident by the increased melting of ice and snow around its summit. A mudflow, 20 m in height, which travelled 27 km down the Lagunillas Valley, proved an advance warning that went unheeded. Ice and snow continued to melt until, on 13 November, there was a major eruption. Although this eruption was small in comparison with other eruptions such as Mount St Helens, the lava, ash and hot rocks ejected were sufficient to melt the remaining ice and snow, releasing a tremendous volume of meltwater. This meltwater, swelled by torrential rain (often associated with volcanic eruptions), raced down the Lagunillas Valley collecting with it large amounts of ash deposited from previous eruptions. The resultant mud tidal wave (a lahar), estimated to have been 30 m in height, travelled down the valley at over 80 km/hr.

Some 50 km from the crater, the mudflow emerged onto more open ground on which was situated the town of Armero. The time was 2300 hours when the mudflow struck, and most of the 22 000 inhabitants had already gone to bed. The few survivors claimed that the first onrush of muddy water was ice-cold, but became increasingly warmer. By morning a layer of mud, up to 8 m deep, covered Armero and the surrounding area (Figures 2.25 and 2.27). The death toll was put at 21 000, making this the worst single natural disaster ever to have affected people in the western hemisphere.

(iii) Heavy rainfall – Hurricane Mitch (Honduras and Nicaragua 1998)

Hurricane Mitch swept across Central America early in the week of 27 October 1998 before becoming stationary off the Caribbean coast of Honduras (Places 31, page 238 and Figure 9.51). In the next few days torrential rains caused flooding and landslides which, by affecting 3 million people, made the event the worst natural

Figure 2.27

Armero, Colombia

disaster in Central America's history. In Nicaragua, heavy rain falling onto Volcano Casita caused devastating mudflows which killed 1900 people and virtually destroyed the town of Posoltega. Worse affected was Tegucigalpa, the capital of Honduras (Figures 2.25 and 2.28). Here, despite flood warnings issued by the government, few people believed the river would rise to their doorsteps, and certainly they never dreamed that it would rise over their rooftops, reaching the height of eight-storey buildings. A week after the event, reports from Tegucigalpa (which was still inaccessible by road) stated that debris had blocked the arches of the remaining battered bridges, turning them into dykes which were holding back muddy water, water that was stagnant and almost a solid mass of sludge, dead animals and human bodies

Survivors expected to see, when the river returned to its normal level, a dead town with thousands of bodies trapped against bridges and in ruined houses. The authorities, fearing outbreaks of cholera, malaria and dengue fever, prohibited people from removing dead bodies, unless they were state forensic doctors who had to go about their dreadful task dressed in spacemen-like suits, masks and gloves.

Figure 2.28

Aftermath of Hurricane Mitch, Tegucigalpa, Honduras

B Human mismanagement

The probability of slope failure in populated areas is often increased by thoughtless planning, or a total lack of it, or where human activity exerts too much pressure upon the land available. Three examples of how slope instability and the risk of slope failure may be increased by human activity are when land is used for:

(i) building dams to create reservoirs

(ii) the extraction of a natural resource or the dumping of waste material

(iii) rapid urbanisation.

(i) Building dams to create reservoirs (Italy 1963)

The Vaiont Dam, built in the Italian Alps, was completed in 1960. The dam, the third highest in the world at that time, was built in a narrow valley with steep sides consisting of alternate layers of clays and limestone (Figure 2.29), and where landslides were not uncommon. Down the valley were several hamlets and the small town of Longarone.

Heavy rain in October 1963 saturated the clay. Just before midnight on 9 October, a landslide of rocks, clay, mud and vegetation slid over the harder beds of limestone and into the reservoir. The dam itself stood, but a wave of water spilled over the lip creating a towering wall of water which swept down the valley. Longarone was virtually destroyed. The final death toll was put at almost 1900, although several bodies were never recovered. Debris from the landslide filled in almost two-thirds of the lake. A court of enquiry concluded that the site was geologically unstable and that even during construction many smaller landslides had occurred. The dam was closed.

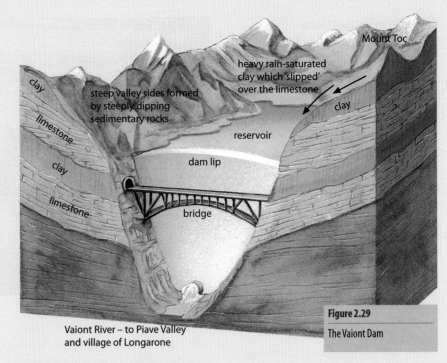

clay

limestone

clay

limestone

steep valley sides formed by steeply dipping sedimentary rocks

Mount Toc

heavy rain-saturated clay which 'slipped' over the limestone

clay

reservoir

dam lip

bridge

Vaiont River – to Piave Valley and village of Longarone

Figure 2.29

The Vaiont Dam

Figure 2.30

Aberfan immediately after the mudflow

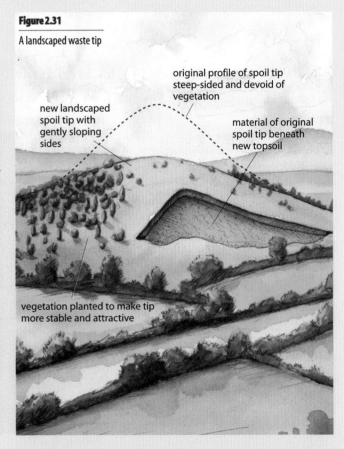

Figure 2.31

A landscaped waste tip

original profile of spoil tip steep-sided and devoid of vegetation

new landscaped spoil tip with gently sloping sides

material of original spoil tip beneath new topsoil

vegetation planted to make tip more stable and attractive

(ii) Dumping waste material (Aberfan 1966)

Aberfan, like many other settlements in the South Wales valleys, grew up around its colliery. However, the valley floors were rarely wide enough to store the coal waste and so it became common practice to tip it high above the towns on the steep valley sides. At Aberfan, the spoil tips were on slopes of 25°, over 200 m above the town and, unknowingly, on a line of springs. Water from these springs added weight to the waste heaps, which reduced their internal cohesion. Following a wet October in 1966 and a night of heavy rain, slope failure resulted in the waste material suddenly and rapidly moving downhill. The resultant mudflow, estimated to contain over 100 000 m³ of material, engulfed part of the town which included the local junior school (Figure 2.30). The time was just after 0900 hours on 21 October, soon after lessons in the school had begun. Of the 147 deaths in Aberfan that morning, 116 were children and five their teachers.

Since then, the colliery has closed and, as elsewhere in the former coal-mining valleys, the potentially dangerous waste tips have been lowered, regraded and landscaped to try to prevent any occurrence of a similar event (Figure 2.31).

(iii) Urbanisation (Hong Kong 1948 to 1998)

Many parts of the world, especially in economically less developed countries, are experiencing rapid urbanisation (page 418). As most of the best sites for residential development have long since been used, it means that newcomers to a city are forced to live on land previously considered unusable (e.g. flood-prone valleys in Nairobi – Places 58, page 444), or unsafe (e.g. steep hillsides in Caracas 1999, and Rio de Janeiro – Places 57, page 443).

In Hong Kong, landslips have been responsible for 480 deaths since 1948 (Figure 2.32). Most landslips during this time have been attributed to two factors: the inadequacies of hillside construction works in the last 50 years, and deficiencies in maintaining slopes once they are utilised (Figure 2.33). In 1966, torrential rainstorms triggered massive landslides which killed 64 people, made 2500 homeless and

caused 8000 to be evacuated. In 1976, a major landslide led to 22 deaths. The consequence of this was the setting up, in 1977, of the Geotechnical Engineering Office (GEO). GEO's main functions were:

- to investigate slopes for potential risk and to take preventive measures
- to control geotechnic aspects of new buildings and roads
- to promote slope maintenance by owners
- to undertake landslide warning and emergency services
- to advise on land-use plans to minimise public risk.

In 1997, most of Hong Kong experienced over 300 mm of rain in 24 hours. At the centre of the storm, 110 mm fell in one hour and 800 mm in the day. Resultant landslips caused the death of one person, injuries to eight people, the disruption of the Kowloon–Guangzhou railway (Case Study 21) and the closure of a six-lane highway for several hours. These losses and disruptions were, however, relatively minor, because the community had learned to cope better with the landslip hazard. Indeed the Hong Kong authorities now collaborate with their counterparts in other cities in Asia and South America with similar climatic and topographic characteristics, and where economic and social development is creating an unacceptable level of landslip risk.

Figure 2.32

Number of landslip fatalities in Hong Kong, 1948–98

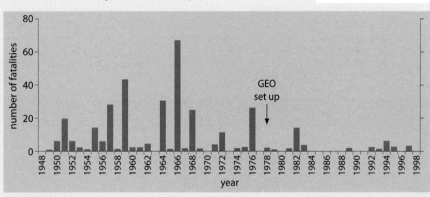

Figure 2.33

Consequences of a landslip in Hong Kong

Formulae for chemical weathering processes referred to in text:

Oxidation $4\,FeO + O_2 \rightarrow 2Fe_2O_3$
(ferrous oxide + oxygen → ferric oxide)

Hydration $CaSO_4 + 2H_2O \rightarrow CaSO_4 2H_2O$
(anhydrite + water → gypsum)

Hydrolosis The formula for hydrolosis varies depending on the rock type involved in the reaction, so no single formula is applicable. The following, for the hydrolosis of feldspar/granite to kaolin, is a common example:

$K_2O, Al_2O_3, 6SiO_2 + H_2O \rightarrow$
$Al_2O_3, 2SiO_2, 2H_2O$
(feldspar + water → kaolin)

Carbonation This process is in two stages:
$H_2O + CO_2 \rightarrow H_2CO_3$
(water + carbon dioxide → carbonic acid)
$CaCO_3 + H_2CO_3 \rightarrow Ca(HCO_3)_2$
(calcium carbonate + carbonic acid → calcium bicarbonate)

Acid rain $2SO_2 + O_2 + 2H_2O \rightarrow 2H_2SO_4$
(sulphur dioxide + oxygen + water → weak sulphuric acid)

References

Goudie, A. (1993) *The Nature of the Environment*, Basil Blackwell.

Ollier, C. (1991) *Weathering and Landforms*, Thomas Nelson.

Small, J. (1999) *A Modern Dictionary of Geography* (4th edition), Hodder and Stoughton.

Trudgill, S. T. (1983) *Weathering and Erosion*, Butterworths.

White, I., Mottershead, D. and Harrison, S. (1992) *Environmental Systems: An Introductory Text*, Chapman & Hall.

Websites

Geoweb, Landslides:
http://www.pacificnet.net/~gimills/slides.htm

See also for more links:
http://www.nelsonthornes.com/gaia

Q Questions

1 Study the slope model in Figure 2.34.
   a i What is meant by the term 'slope element'?
      ii State where on such a slope you could expect to find each of the following landforms:
        scree; terracettes; lobe.   (5 marks)
   b i What is the meaning of 'mass movement' when applied to slope processes?   (4 marks)
      ii Choose **two** of the features marked 1–5 on Figure 2.34. Describe the shape of each of the features you have chosen.   (4 + 4 marks)
      iii For **each** feature, explain the role of mass movement in its formation.   (4+4 marks)

2 a Define the term 'weathering'.   (4 marks)
   b Choose **one** type of **mechanical weathering**.
      i Making good use of diagrams, explain the processes involved in the type of weathering.
      ii Describe the landscape features which result from the weathering type you have chosen.   (8 marks)
   c Choose any **one** climatic region and identify the type of **chemical weathering** which will dominate the area. Explain why this type of chemical weathering will be dominant.   (6 marks)
   d Human activity can influence the rate of weathering that occurs in an area. With the aid of specific examples, explain how human activity influences the rate of weathering.   (7 marks)

AS

3 a i Study Figure 2.34. On a copy of the graph (a), label each of the following types of slope movement to identify their speed of flow: earth / mudflow; solifluction; rockfall; slide; soil creep.   (5 marks)
      ii For any **two** of the flow movements above, explain how the process occurs and describe the landform shape which results.   (10 marks)

   b Use examples of two types of rural land use you have studied to explain how people in rural areas try to manage slopes to reduce the down-slope movement of soil.   (10 marks)

4 'The shape of a slope is a function of structure, process and time.' Discuss this statement with reference to slope shapes in two contrasting rock types and two contrasting environments.   (25 marks)

A2

Figure 2.34
Mass movements

a speed of movement

Extremely slow	Very slow	Slow	Moderate	Rapid	Very rapid	Extremely rapid
1 cm/year	1 m/year	1 km/year	1 km/month	1 km/hour	25 km/hour	10 m/sec

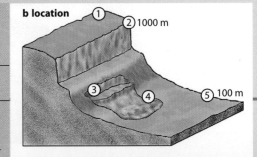

b location

5 Study Figure 2.35 and answer the following questions.
- a i Explain the meaning of each of the following slope movement terms:
  earth flow; mud flow; slide; rock fall. **(6 marks)**
  - ii Name **two** types of slope movement it is possible to see in the photograph. State where they can be found. **(4 marks)**
  - iii Identify **two** ways in which people have tried to protect slopes in this photograph. For each one suggest how it is intended to work. **(6 marks)**
- b Had the slope movement finished when this photograph was taken? Suggest reasons for your answer. **(4 marks)**
- c Should cliffs, such as the one in the photograph, be protected? Give reasons for your answer. **(5 marks)**

6 Use Case Study 2B(iii) on Hong Kong (page 55) to answer the following question.
- a Describe the **physical features** of the hillside shown in the photograph. **(3 marks)**
- b Why have people settled on this hillside? **(3 marks)**
- c Why is a hillside, such as the one in the photograph, in danger of rapid mass movement even without human activities? **(7 marks)**
- d Give **two** examples of human activities which increase the danger of rapid mass movement on such slopes. Explain how they increase the danger. **(6 marks)**
- e The heavy rainfall in 1997 was an extreme climatic event but it created relatively little damage. Explain **one** way in which authorities such as those in Hong Kong are trying to manage the problems caused by the physical environment in which they operate. **(6 marks)**

Figure 2.35

Holbeck Hall Hotel, Scarborough

7 a Study at the photograph of Holbeck Hall Hotel (Figure 2.35).
  - i Draw an annotated diagram or sketch map **only** to illustrate the landscape features of the slopes. **(8 marks)**
  - ii Explain what has happened to these slopes and suggest why it has occurred. **(8 marks)**
- b Making good use of examples, explain how human activities can increase the stability of some slopes and destabilise other slopes. **(9 marks)**

8 'A range of processes, which differ in contrasting environments, affect slope shapes.' Discuss this statement with reference to slopes you have studied. In your answer you should refer to:
- ■ the variation of slope elements in different environments
- ■ the variation in importance of types of weathering process in different environments
- ■ the interaction of factors within environments to create slopes. **(25 marks)**

9 Making good use of examples from your case studies in a range of environments, explain how an understanding of natural slope processes can be used in planning urban developments. **(25 marks)**

10 Choose a drainage basin in the UK that you have studied.
- a Describe and suggest reasons for the variation in slope types that exist within the drainage basin. **(10 marks)**
- b For any one slope, identify and explain changes which are likely to affect the slope in the future. **(8 marks)**
- c Suggest how human activity can influence the rate of change and shape of slopes. **(7 marks)**

11 a Suggest and explain **one** hypothesis you could test through fieldwork to explore the variation in slope angle on a rural valley side. **(5 marks)**
- b How would you use fieldwork techniques to measure the variations in slope angle? In your answer you should refer to equipment you would use, and the methods and recording techniques involved. **(10 marks)**
- c Once fieldwork is completed you will need to manipulate and present your results. Explain **two** techniques you would use to prepare your results for use. **(10 marks)**

Drainage basins and rivers

'All the rivers run into the sea; yet the sea is not full; unto the place from whence the rivers come, thither they return again.'

The Bible, Ecclesiastes, 1:7

A **drainage basin** is an area of land drained by a river and its tributaries. Its boundary is marked by a ridge of high land beyond which any precipitation will drain into adjacent basins. This boundary is called a **watershed**.

A drainage basin may be described as an **open system** and it forms part of the hydrological or water cycle. If a drainage basin is viewed as a system (Framework 3, page 45) then its characteristics are:

- **inputs** in the form of precipitation (rain and snow)
- **outputs** where the water is lost from the system either by the river carrying it to the sea or through **evapotranspiration** (the loss of water directly from the ground, water surfaces and vegetation).

Within this system, some of the water:

- is **stored** in lakes and/or in the soil, or
- passes through a series of **transfers** or flows, e.g. infiltration, percolation, throughflow.

Elements of the drainage basin system

Figure 3.1

The drainage basin as an open system

Figure 3.1 shows the drainage basin system as it is likely to operate in a temperate humid region such as the British Isles.

Precipitation

This forms the major input into the system, though amounts vary over time and space. As a rule, the greater the intensity of a storm, the shorter its duration. Convectional thunderstorms are short, heavy and may be confined to small areas, whereas the passing of a warm front of a depression (page 231) will give a longer period of more steady rainfall extending over all the basin.

Evapotranspiration

The two components of evapotranspiration contribute to form an output from the system. **Evaporation** is the physical process by which moisture is lost directly into the atmosphere from water surfaces, including vegetation, and the soil due to the effects of air movement and the sun's heat. **Transpiration** is a biological process by which water is lost from a plant through the minute pores (stomata) in its leaves. Evaporation rates are affected by temperature, wind speed, humidity, hours of sunshine and other climatic factors. Transpiration rates depend on the time of year, the type and amount of vegetation, the availability of moisture and the length of the growing season. It is also possible to distinguish between the potential and the actual evapotranspiration of an area. For example, in deserts there is a high **potential evapotranspiration** because the amount of moisture that could be lost is greater

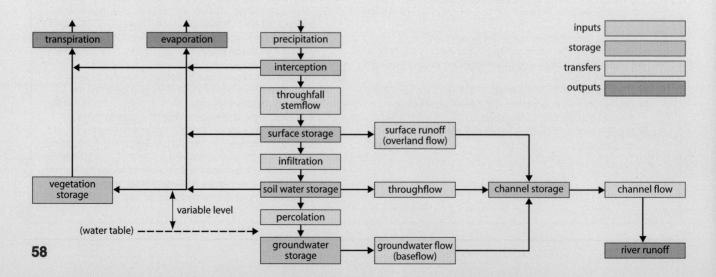

than the amount of water actually available. On the other hand, in Britain the amount of water available for evapotranspiration nearly always exceeds the amount which actually takes place, hence the term **actual evapotranspiration**. In other words, transpiration is limited by the availability of water in the soil.

Interception

The first raindrops of a rainfall event will fall on vegetation which shelters the underlying ground. This is called **interception storage**. It is greater in a woodland area or where tree crops are grown than on grass or arable land. If the precipitation is light and of short duration, much of the water may never reach the ground and it may be quickly lost from the system through evaporation. Estimates suggest that in a woodland area up to 30 per cent of the precipitation may be lost through interception, which helps to explain why soil erosion is limited in forests. According to Newson (1975), 'Interception is a dynamic process of filling and emptying a shallow store (about 2 mm in most UK trees). The emptying occurs because evaporation is very efficient for small raindrops held on tree surfaces.' In an area of deciduous trees, both interception and evapotranspiration rates will be higher in summer, although the two processes do not occur simultaneously.

If a rainfall event persists, then water begins to reach the ground by three possible routes: dropping off the leaves, or **throughfall**; flowing down the trunk, or **stemflow**; and by undergoing **secondary interception** by undergrowth. Following a warm, dry spell in summer, the ground may be hard; at the start of a rainfall event water will then lie on the surface (**surface storage**) until the upper layers become sufficiently moistened to allow it to soak slowly downwards. If precipita-

tion is very heavy initially, or if the soil becomes saturated, then excess water will flow away over the surface, a transfer known as **surface runoff** (or, in Horton's term, **overland flow**) (Figure 3.2).

Infiltration

In most environments, overland flow is relatively rare except in urban areas – which have impermeable coverings of tarmac and concrete – or during exceptionally heavy storms. Soil will gradually admit water from the surface, if the supply rate is moderate, allowing it slowly to infiltrate vertically through the pores in the soil. The maximum rate at which water can pass through the soil is called its **infiltration capacity** and is expressed in mm/hr. The rate of infiltration depends upon the amount of water already in the soil (**antecedent precipitation**), the **porosity** (Figure 8.2) and structure of the soil, the nature of the soil surface (e.g. crusted, cracked, ploughed), and the type, amount and seasonal changes in vegetation cover. Some of the water will flow laterally as **throughflow**. During drier periods, some water may be drawn up towards the surface by **capillary action**.

Percolation

As water reaches the underlying soil or rock layers, which tend to be more compact, its progress is slowed. This constant movement, called percolation, creates **groundwater storage**. Water eventually collects above an impermeable rock layer, or it may fill all pore spaces, creating a **zone of saturation**. The upper boundary of the saturated material, i.e. the upper surface of the groundwater layer, is known as the **water table**. Water may then be slowly transferred laterally as **groundwater flow** or **baseflow**. Except in areas of Carboniferous limestone, groundwater levels usually respond slowly to surface storms or short periods of drought (Figure 3.5). During a lengthy dry period, some of the groundwater store will be utilised as river levels fall. In a subsequent wetter period, groundwater must be replaced before the level of the river can rise appreciably (Figure 3.3). If the water table reaches the surface, it means that the ground is saturated; excess water will then form a marsh where the land is flat, or will become surface runoff if the ground is sloping.

Channel flow

Although some rain does fall directly into the channel of a river (**channel precipitation**), most water reaches it by a combination of three transfer processes: surface runoff (overland flow), throughflow, or groundwater flow (baseflow). Once in the river, as **channel storage**, water flows towards the sea and is lost from the drainage basin system.

Figure 3.2

Surface runoff (overland flow) Blyford, Suffolk

Drainage basins and rivers

The water balance

This shows the state of equilibrium in the drainage basin between the inputs and outputs. It can be expressed as:

$P = Q + E \pm$ change in storage

where:

$P =$ precipitation (measured using rain gauges)

$Q =$ runoff (measured by discharge flumes in the river channel), and

$E =$ evapotranspiration. (This is far more difficult to measure – how can you measure accurately transpiration from a forest?)

In Britain, the annual precipitation nearly always, in most years and in most places, exceeds evapotranspiration. As, therefore, precipitation input exceeds evapotranspiration loss, then there is **positive water balance** (or water budget). However in some years, e.g 1974 and 1975, and 1995 and 1996, the long, dry summers, especially in the south and east of the country, resulted in evapotranspiration exceeding precipitation to give a temporary **negative water balance**. Changes in **storage** in the water balance reflect the amount of moisture in the soil. The **soil moisture budget** is, according to Newson, a sub-system of the catchment water balance.

Figure 3.3 is a graph showing the soil moisture balance for an area in south-east England. During winter, precipitation exceeds evapotranspiration creating a **soil moisture surplus** which results in considerable surface runoff and a rise in river levels. In summer, evapotranspiration exceeds precipitation and so plants and humans have to **utilise** water from the soil store leaving it depleted and causing river levels to fall. By autumn, when precipitation again exceeds evapotranspiration, the first of the surplus water has to be used to **recharge** the soil until it reaches its **field capacity** (page 267). At no time in Figure 3.3 was the utilisation of water sufficient to create a **soil moisture deficit** (as in Figure 3.4b).

Figure 3.3

A model illustrating soil moisture budget

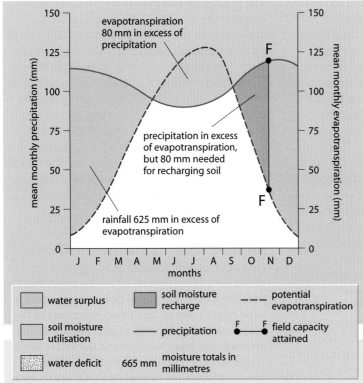

Figure 3.4

Soil moisture budget for two towns in the USA

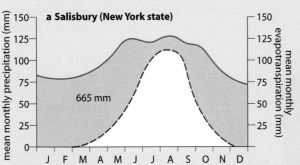

As precipitation is above potential evapotranspiration throughout the year then there is, in an average year, neither a water shortage nor a need to utilise moisture from the soil.

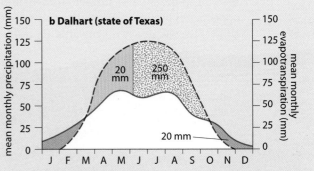

By spring, potential evapotranspiration is greater than precipitation. As there is no water surplus, then plants have to utilise moisture from the soil. By midsummer, water in the soil has been used up and there is a water deficit – meaning that plants can only survive if they are either drought-resistant or if they can obtain water through irrigation. When precipitation does exceed potential evapotranspiration, in winter, the rain is needed to replace (recharge) that taken from the soil earlier in the year, and amounts are insufficient to give a water surplus.

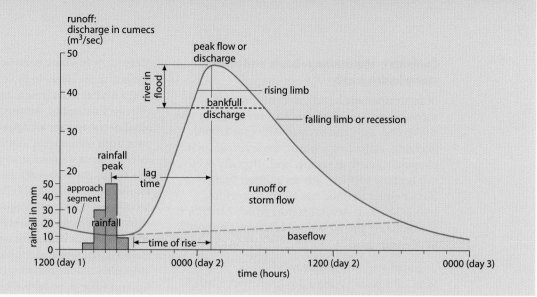

Figure 3.5

The storm hydrograph

The storm hydrograph

An important aspect of **hydrology** (the study of water, **precipitation**, **runoff** and **evaporation/ transpiration processes**) is how a drainage basin reacts to a period of rain. This is important because it can be used in predicting the flood risk and in making the necessary precautions to avoid damage to property and loss of life. The response of a river can be studied by using the **storm** or **flood hydrograph**. The hydrograph is a means of showing the discharge of a river at a given point over a short period of time.

Discharge is the amount of water originating as precipitation which reaches the channel by surface runoff, throughflow and baseflow. Discharge is therefore the water *not* stored in the drainage basin by interception, as surface storage, soil moisture storage or groundwater storage or lost through evapotranspiration (Figure 3.1). The model of a storm hydrograph, Figure 3.5, shows how the discharge of a river responds to an individual rainfall event.

Measuring discharge

Discharge is the velocity (speed) of the river, measured in metres (m) per second, multiplied by the cross-sectional area of the river, measured in m^2. This gives the volume in m^3/sec or **cumecs**. It can be expressed as:

$Q = A \times V$

where:

Q = discharge

A = cross-sectional area

V = velocity.

Interpreting the hydrograph

Refer to the hydrograph in Figure 3.5. The graph includes the **approach segment** which shows the discharge of the river before the storm (the antecedent flow rate). When the storm begins, the river's response is negligible for although some of the rain does fall directly into the channel, most falls elsewhere in the basin and takes time to reach the channel. However, when the initial surface runoff and, later, the throughflow eventually reach the river there is a rapid increase in discharge as indicated by the **rising limb**. The steeper the rising limb, the faster the response to rainfall – i.e. water reaches the channel more quickly. The **peak discharge** (peak flow) occurs when the river reaches its highest level. The period between maximum precipitation and peak discharge is referred to as the **lag time**. The lag time varies according to conditions within the drainage basin, e.g. soil and rock type, slope and size of the basin, drainage density, type and amount of vegetation and water already in storage. Rivers with a short lag time tend to experience a higher peak discharge and are more prone to flooding than rivers with a long lag time. The **falling** or **recession limb** is the segment of the graph where discharge is decreasing and river levels are falling. This segment is usually less steep than the rising limb because throughflow is being released relatively slowly into the channel. By the time all the water from the storm has passed through the channel at a given location, the river will have returned to its baseflow level – unless there has been another storm within the basin. **Stormflow** is the discharge, both surface and subsurface flow, attributed to a single storm. **Baseflow** is very slow to respond to a storm, but by continually releasing groundwater it maintains the river's flow during periods of low precipitation. Indeed, baseflow is more significant over a longer period of time than an individual storm and reflects seasonal changes in precipitation, snowmelt, vegetation and evapotranspiration. Finally, on the graph, **bankfull discharge** occurs when a river's water level reaches the top of its channel; any further increase in discharge will result in flooding of the surrounding land. This happens, on average, once every year or two.

Controls in the drainage basin and on the storm hydrograph

In some drainage basins, river discharge increases very quickly after a storm and may give rise to frequent, and occasionally catastrophic, flooding. Following a storm, the levels of such rivers fall almost as rapidly and, after dry spells, can become very low. Rivers in other basins seem neither to flood nor to fall to very low levels. There are several factors which contribute to regulating the ways in which a river responds to precipitation.

1 Basin size, shape and relief

Size   If a basin is small it is likely that rainfall will reach the main channel more rapidly than in a larger basin where the water has much further to travel. Lag time will therefore be shorter in the smaller basin.

Shape   It has long been accepted that a circular basin is more likely to have a shorter lag time and a higher peak flow than an elongated basin (Figure 3.6a and b). All the points on the watershed of the former are approximately equidistant from the gauging station, whereas in the latter it takes longer for water from the extremities of the basin to reach the gauging station. However, Newson (1994) has pointed out that studies made in many regions of the world have shown that basin shape is less reliable as a flood indicator than basin size and slope.

Relief   The slope of the basin and its valley sides also affect the hydrograph. In steep-sided upland valleys, water is likely to reach the river more quickly than in gently sloping lowland areas (Figure 3.6c).

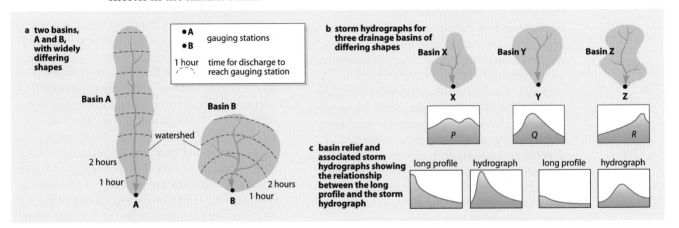

Figure 3.6

Drainage basin shape

2 Types of precipitation

Prolonged rainfall   Flooding most frequently occurs following a long period of heavy rainfall when the ground has become saturated and infiltration has been replaced by surface runoff (overland flow).

Intense storms   (e.g. convectional thunderstorms)   When heavy rain occurs, the rainfall intensity may be greater than the infiltration capacity of the soil (e.g. in summer in Britain, when the ground may be harder). The resulting surface runoff is likely to produce a rapid rise in river levels (flash floods).

Snowfall   Heavy snowfall means that water is held in storage and river levels drop. When temperatures rise rapidly (in Britain, this may be with the passage of a warm front and its associated rainfall, page 231), meltwater soon reaches the main river. It is possible that the ground will remain frozen for some time, in which case infiltration will be impeded.

3 Temperature

Extremes of temperature can restrict infiltration (very cold in winter, very hot and dry in summer) and so increase surface runoff. If evapotranspiration rates are high, then there will be less water available to flow into the main river.

4 Land use

Vegetation   Vegetation may help to prevent flooding by intercepting rainfall (storing moisture on its leaves before it evaporates back into the atmosphere – page 59). Estimates suggest that tropical rainforests intercept up to 80 per cent of rainfall (30 per cent of which may later evaporate) whereas arable land may intercept only 10 per cent. Interception is less during the winter in Britain when deciduous trees have shed their leaves and crops have been harvested to expose bare earth. Plant roots, especially those of trees, reduce throughflow by taking up water from the soil.

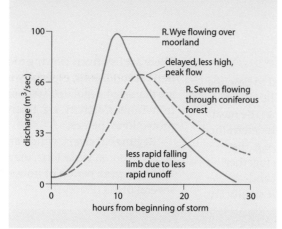

Figure 3.7

The effect of vegetation on the storm hydrographs of the Rivers Wye and Severn (geology and precipitation are the same in both basins)

Flooding is more likely to occur in deforested areas, e.g. the increasingly frequent and serious flooding in Bangladesh is attributed to the removal of trees in Nepal and other Himalayan areas. In areas of afforestation, flooding may initially increase as the land is cleared of old vegetation and drained, but later decrease as the planted trees mature. Newson (1994) points out that, after 20 years of data collecting, the evidence suggests that the canopy has more effect on medium flows than on high flows, as the main ditches remain active.

Figure 3.7 contrasts the storm hydrographs of two rivers. Although they rise very close together, the River Wye flows over moors and grassland, whereas the River Severn flows through an area of coniferous forest.

Urbanisation Urbanisation has increased flood risk. Water cannot infiltrate through tarmac and concrete, and gutters and drains carry water more quickly to the nearest river. Small streams may be either canalised so that

(with friction reduced) the water flows away more quickly, or culverted, which allows only a limited amount of water to pass through at one time (Figure 3.8).

5 Rock type (geology)
Rocks that allow water to pass through them are said to be **permeable**. There are two types of permeable rock:

- **Porous**, e.g. sandstone and chalk, which contain numerous pores able to fill with and store water (Figure 8.2).
- **Pervious**, e.g. Carboniferous limestone, which allow water to flow along bedding planes and down joints within the rock, although the rock itself is impervious (Figure 8.1).

As both types permit rapid infiltration, there is little surface runoff and only a limited number of surface streams. In contrast **impermeable rocks**, such as granite, do not allow water to pass through them and so they produce more surface runoff and a greater number of streams.

6 Soil type
This controls the rate and volume of infiltration, the amount of soil moisture storage and the rate of throughflow (page 265). Sandy soils, with large pore spaces, allow rapid infiltration and do not encourage flooding. Clays have much smaller pore spaces and they are less well connected; this reduces infiltration and throughflow, but encourages surface runoff and increases the risk of flooding.

7 Drainage density
This refers to the number of surface streams in a given area (page 67). The density is higher on impermeable rocks and clays, and lower on permeable rocks and sands. The higher the density, the greater is the probability of flash floods. A **flash flood** is a sudden rise of water in a river, shown on the hydrograph as a shorter lag time and a higher peak flow in relation to normal discharge.

8 Tides and storm surges
High spring tides tend to prevent river floodwater from escaping into the sea. Floodwater therefore builds up in the lower part of the valley. If high tides coincide with gale-force winds blowing onshore and a narrowing estuary, the result may be a **storm surge** (Places 19, page 148). This happened in south-east England and in the Netherlands in 1953 and prompted the construction of the Thames Barrier and the implementation of the Dutch Delta Plan.

Figure 3.8

An urban river

River regimes

The regime of a river is the term used to describe the annual variation in discharge. The average regime, which can be shown by either the mean daily or the mean monthly figures, is determined primarily by the climate of the area, e.g. the amount and distribution of rainfall, together with the rates of evapotranspiration and snowmelt. Local geology may also be significant.

There are few rivers flowing today under wholly natural conditions, especially in Britain. Most are managed, regulated systems which result from human activity, e.g. reservoirs and flood protection schemes.

Regimes of rivers, which are used to demonstrate seasonal variations, may be either simple, with one peak period of flow, or complex with several peaks (Places 9).

Places 9
River Don, Yorkshire and River Torridge, Devon: river discharge

Figure 3.9 shows the rainfall and runoff figures for the River Don (South Yorkshire) for one year. Discharge is usually at its highest in winter when Britain receives most of its depressions and when evapotranspiration is limited due to the low temperatures. Early spring may also show a peak if the source of the river is in an upland area liable to heavy winter snowfalls – in this case, the Pennines. It is possible for runoff to exceed precipitation, e.g. when heavy snowfall at the end of a month melts during a milder, drier period at the beginning of the next month. In contrast, river levels are lowest in summer when most of Britain receives less rainfall and when evapotranspiration rates are at their highest. There is often a correlation, or relationship, between the two variables of rainfall and runoff. This relationship can be shown by means of a scattergraph (Framework 19, page 635). Rainfall is plotted along the base (the x axis) because it is the independent variable, i.e. it does not depend on the amount of runoff. Runoff is plotted on the vertical or y axis because it is the dependent variable, i.e. runoff does depend upon the amount of rainfall.

The Environment Agency (EA) also produce hydrographs covering longer periods of time than for a single storm (Figure 3.5) but with far greater, and more useful, data than that given for the annual regime of a river (Figure 3.9).

Figure 3.10 gives rainfall and discharge for a wet month in late 1992 for the River Torridge in Devon. It shows that:

a as most of the peak discharges occur within a day of peak rainfall then the river must respond quickly to rainfall and, therefore, is likely to pose a flood risk

b the highest discharge (on the 30th) came after several very wet days during which river levels had no time to drop, rather than after a very wet day (the 17th) which followed a relatively dry spell of weather.

Figure 3.9

Rainfall and runoff for the River Don, Yorkshire

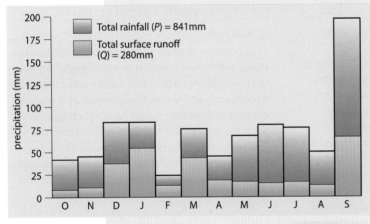

Figure 3.10

Hydrograph for the River Torridge at Torrington, Devon, late 1992

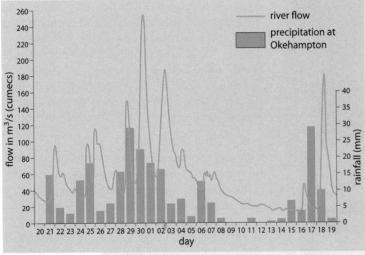

Morphometry of drainage basins

Morphometry means 'the measurement of shape or form'. The development of morphometric techniques was a major advance in the quantitative (as opposed to the qualitative) description of drainage basins (Framework 4). Instead of studies being purely subjective, it became possible to compare and contrast different basins with precision. Much of the early work in this field was by R.E. Horton. In the mid-1940s he devised the 'Laws of drainage composition' which established a hierarchy of streams ranked according to 'order'. One of these laws, the **law of stream number**, states that within a drainage basin a constant geometric relationship exists between stream order and stream number (Figure 3.12a).

Figure 3.11 shows how one of Horton's successors, A.N. Strahler, defined streams of different order. All the initial, unbranched source tributaries he called **first order** streams. When two first order streams join they form a **second order**; when two second order streams merge they form a **third order**; and so on. Notice that it needs two stream segments of equal order to join to produce a segment of a higher order, while the order remains unchanged if a lower order segment joins a higher order segment. For example, a second order plus a second order gives a third order but if a second order stream joins a third order, the resultant stream remains as a third order. A basin may therefore be described in terms of the highest order stream within it, e.g. a 'third order basin' or a 'fourth order basin'.

If the number of segments in a stream order is plotted on a semi-log graph against the stream order, then the resultant best-fit-line will be straight (Figure 3.12a). On a semi-log graph, the vertical scale, showing the dependent variable (Framework 19, page 635), is divided into cycles, each of which begins and ends ten times greater than the previous cycle, e.g. a range of 1 to 10, 10 to 100, 100 to 1000, and so on. (If the horizontal scale, showing the independent variable, had also been divided into cycles instead of having an arithmetic scale, then Figure 3.12 would have been referred to as a log-log graph (Figure 18.25).) Logarithmic graphs are valuable when:

- the rate of change is of more interest than the amount of change: the steeper the line the greater the rate of change
- there is a greater range in the data than there is space to express on an arithmetic scale (a log scale compresses values)
- there are considerably more data at one end of the range than the other.

Figure 3.12a shows a perfect negative correlation (Figure 22.7): as the independent variable (in this case the stream order) increases, then the dependent variable (the number of streams) decreases. Studies of stream ordering for most rivers in the world produce a similar straight-line relationship. For any exceptions to Horton's law of stream ordering, further studies can be made to determine which local factors alter the relationship. Relationships also exist between stream order and the mean length of streams (Figure 3.12b), and stream order and mean drainage basin area (Figure 3.12c).

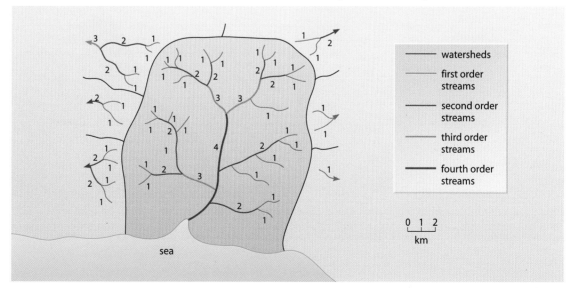

Figure 3.11

Strahler's method of stream ordering

Figure 3.12

Relationships
between stream order
and other variables

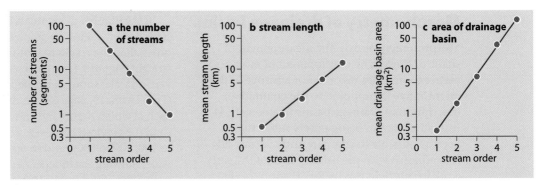

Comparing drainage basins

Horton's work has made it possible to compare different drainage basins scientifically (quantitatively) rather than relying on subjective (qualitative) descriptions by individuals. It allows people studying drainage basin morphometry in different parts of the world to use the same standards, measurements and 'language'.

Figure 3.13 shows two imaginary and adjacent basins. These can be compared in several different ways, including:

- the bifurcation ratio, and
- drainage density.

Figure 3.13

A comparison
between two adjacent
drainage basins on
clays and sands

The bifurcation ratio
This is the relationship between the number of streams of one order and those of the next highest order. It is obtained by dividing the number of streams in one order by the number in the next highest order, e.g. for basin **A** in Figure 3.13:

$$\frac{N1}{N2} \quad \frac{\text{(number of first order streams)}}{\text{(number of second order streams)}} = \frac{26}{6} = 4.33$$

and then finding the mean of all the ratios in the basin being studied, i.e.

$$\frac{4.33 + 3.00 + 2.00}{3} = 3.11 = \text{bifurcation ratio for basin } \mathbf{A}$$

The human significance of the bifurcation ratio is that as the ratio is reduced so the risk of flooding within the basin increases. It also indicates the flood risk for parts, rather than all, of the basin. Most British rivers have a bifurcation ratio of between 3 and 5.

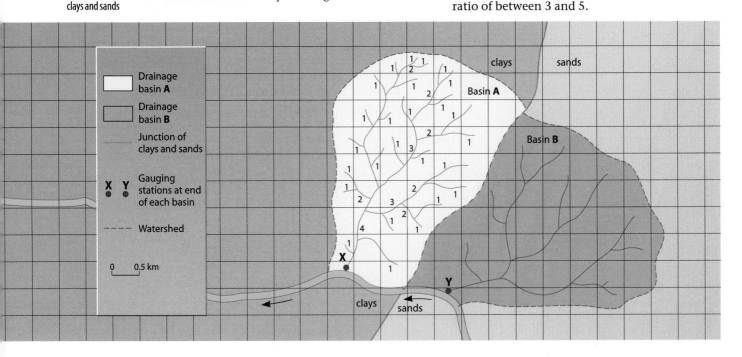

Drainage density

This is found by measuring the total length of all the streams within the basin (*L*) and dividing by the area of the whole basin (*A*). It is therefore the average length of stream within each unit area. For basin **A** in Figure 3.13, this will be:

$$\frac{L}{A} = \frac{22.65}{12.50} = 1.81 \text{ km per km}^2$$

In Britain most drainage densities lie between 2 and 4 km per km^2 but this varies considerably according to local conditions. A number of factors influence drainage density. It tends to be highest in areas where the land surface is impermeable, where slopes are steep, where rainfall is heavy and prolonged, and where vegetation cover is lacking.

a   **Geology and soils**   On very permeable rocks or soils (e.g. chalk, sands) drainage densities may be under 1 km per km^2, whereas this increases to over 5 km per km^2 on highly impermeable surfaces (e.g. granite, clays).

In Figure 3.13 with two adjacent drainage basins of approximately equal size, shape and probably rainfall, the difference in drainage density is likely to be due to basin **A** being on clays and basin **B** on sands.

b   **Land use**   The drainage density, especially of first order streams, is much greater in areas with little vegetation cover. The density decreases, as does the number of first order streams, if the area becomes afforested. Deserts tend to have the highest densities of first order channels, even if the channels are dry for most of the time.

c   **Time**   As a river pattern develops over a period of time, the number of tributaries will decrease, as will the drainage density.

d   **Precipitation**   Densities are usually highest in areas where rainfall totals and intensity are also high.

e   **Relief**   Density is usually greater on steeper slopes than on more gentle slopes.

Framework 4
Quantitative techniques and statistical methods of data interpretation

As geography adjusted to a more scientific approach in the 1960s, a series of statistical techniques were adopted which could be used to quantify field data and add objectivity to the testing of hypotheses and theories. This period is often referred to as the 'Quantitative Revolution'.

At first it seemed to many, the author included, that mathematics had taken over the subject, but it is now accepted that these techniques are a useful aid provided they are not seen as an end in themselves. They provide a tool which, if carefully handled and understood, gives greater precision to arguments, helps in the identification of patterns and may contribute to the discovery of relationships and possible cause–effect links. In short, by providing greater accuracy in handling data they reduce the reliance upon subjective conclusions.

It is essential to select the most appropriate techniques for the data and for the job in hand. Therefore some understanding of the statistical methods involved is important.

Statistical methods may be profitably employed in these areas.

   1 **Sampling**   (Framework 6, page 159) Rapid collection of the data is made possible.

   2 **Correlation and regression**   (Framework 19, page 635) This not only shows possible relationships between two variables but quantifies or measures the strength of those relationships.

   3 **Spatial distributions**   (Framework 19, page 635) Not only may this approach be used to identify patterns, but it may also demonstrate how likely it is that the resultant distributions occurred by chance.

When these new techniques first appeared in schools in the 1970s, they appeared extremely daunting until it was realised that often the difficulty of the worked examples detracted from the usefulness of the technique itself. Where such techniques appear in this book, the mathematics have been simplified to show more clearly how methods may be used and to what effect. With the wider availability of calculators and computers it has become easier to take advantage of more complex calculations to test geographical hypotheses (Framework 10, page 299). Much of the 'number crunching' has now been removed by the increasing availability of statistical packages for micro-computers.

Figure 3.14

Turbulence in a river: the confluence of the Rio Amazon (red with silt from the Andes) and the Rio Negro (black with plant acids)

zontal movement of water so rarely experienced in rivers that it is usually discounted. Such a form of flow, if it existed, would travel over sediment on the river bed without disturbing it. Turbulent flow, the dominant method, consists of a series of erratic eddies, both vertical and horizontal, in a downstream direction (Figures 3.14 and 3.15b). Turbulence varies with the velocity of the river which, in turn, depends upon the amount of energy available after friction has been overcome. It is estimated that under 'normal' conditions about 95 per cent of a river's energy is expended in order to overcome friction.

Influence of velocity on turbulence

■ If the velocity is high, the amount of energy still available after friction has been overcome will be greater and so turbulence increases. This results in sediment on the bed being disturbed and carried downstream. The faster the flow of the river, the larger the quantity and size of particles which can be transported. The transported material is referred to as the river's **load**.

■ When the velocity is low, there is less energy to overcome friction. Turbulence decreases and may not be visible to the human eye. Sediment on the river bed remains undisturbed. Indeed, as turbulence maintains the transport of the load, a reduction in turbulence may lead to deposition of sediment.

The velocity of a river is influenced by three main factors:
1 channel shape in cross-section
2 roughness of the channel's bed and banks, and
3 channel slope.

River form and velocity

A river will try to adopt a channel shape that best fulfils its two main functions: transporting water and sediment. It is important to understand the significance of channel shape in order to identify the controls on the flow of a river.

Types of flow

As water flows downhill under gravity, it seeks the path of least resistance – i.e. a river possesses potential energy and follows a route that will maximise the rate of flow (velocity) and minimise the loss of this energy caused by friction. Most friction occurs along the banks and bed of the river, but the internal friction of the water and air resistance on the surface are also significant.

There are two patterns of flow, **laminar** and **turbulent**. Laminar flow (Figure 3.15a) is a hori-

Figure 3.15

Types of flow in a river

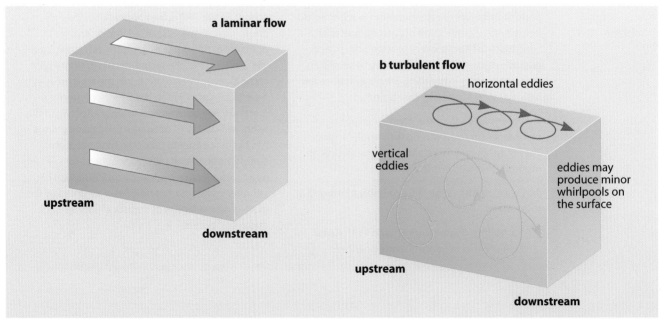

a laminar flow

upstream

downstream

b turbulent flow

horizontal eddies

vertical eddies

eddies may produce minor whirlpools on the surface

upstream

downstream

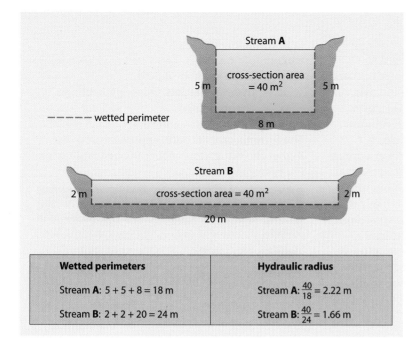

Figure 3.16

The wetted perimeter, hydraulic radius and efficiency of two different-shaped channels with equal area

Wetted perimeters	Hydraulic radius
Stream **A**: 5 + 5 + 8 = 18 m	Stream **A**: $\frac{40}{18}$ = 2.22 m
Stream **B**: 2 + 2 + 20 = 24 m	Stream **B**: $\frac{40}{24}$ = 1.66 m

depth of the channel. The **wetted perimeter** is the total length of the bed and bank sides in contact with the water in the channel. Figure 3.16 shows two channels with the same cross-section area but with different shapes and hydraulic radii.

Stream **A** has a larger hydraulic radius, meaning that it has a smaller amount of water in its cross-section in contact with the wetted perimeter. This creates less friction which in turn reduces energy loss and allows greater velocity. Stream **A** is said to be the more efficient of the two rivers.

Stream **B** has a smaller hydraulic radius, meaning that a larger amount of water is in contact with the wetted perimeter. This results in greater friction, more energy loss and reduced velocity. Stream **B** is less efficient than stream **A**.

The shape of the cross-section controls the point of maximum velocity in a river's channel. The point of maximum velocity is different in a river with a straight course where the channel is likely to be approximately symmetrical (Figure 3.17a) compared with a meandering channel where the shape is asymmetrical (Figure 3.17b).

1 Channel shape

This is best described by the term **hydraulic radius**, i.e. the ratio between the area of the cross-section of a river channel and the length of its wetted perimeter. The cross-section area is obtained by measuring the width and the mean

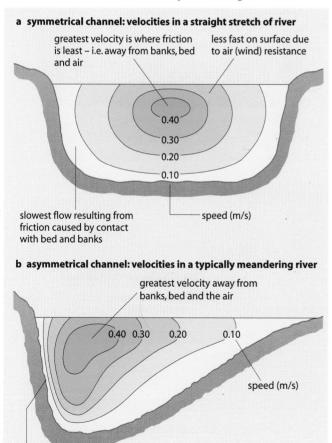

Figure 3.17

Cross-sections of a symmetrical and an asymmetrical stream channel

Figure 3.18

Tiger Leaping Gorge on the River Yangtze, China

2 Roughness of channel bed and banks

A river flowing between banks composed of coarse material with numerous protrusions and over a bed of large, angular rocks (Figure 3.18) meets with more resistance than a river with cohesive clays and silts forming its bed and banks.

Figure 3.19 shows why the velocity of a mountain stream is less than that of a lowland river. As bank and bed roughness increase, so does turbulence. Therefore a mountain stream is likely to pick up loose material and carry it downstream.

Roughness is difficult to measure, but Manning, an engineer, calculated a **roughness coefficient** by which he interrelated the three factors affecting the velocity of a river. In his formula, known as 'Manning's N':

$$v = \frac{R^{0.67} \, S^{0.5}}{n}$$

where:

- v = mean velocity of flow
- R = hydraulic radius
- S = channel slope
- n = boundary roughness.

The formula gives a useful approximation: the higher the value, the rougher the bed and banks. For example:

Bed profile	Sand and gravel	Coarse gravel	Boulders
Uniform	0.02	0.03	0.05
Undulating	0.05	0.06	0.07
Highly irregular	0.08	0.09	0.10

Figure 3.19

Why a river increases in velocity towards its mouth

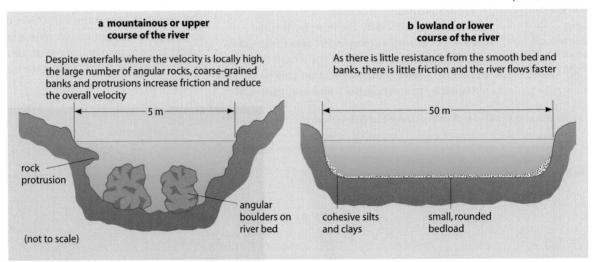

a mountainous or upper course of the river

Despite waterfalls where the velocity is locally high, the large number of angular rocks, coarse-grained banks and protrusions increase friction and reduce the overall velocity

5 m

rock protrusion

angular boulders on river bed

(not to scale)

b lowland or lower course of the river

As there is little resistance from the smooth bed and banks, there is little friction and the river flows faster

50 m

cohesive silts and clays

small, rounded bedload

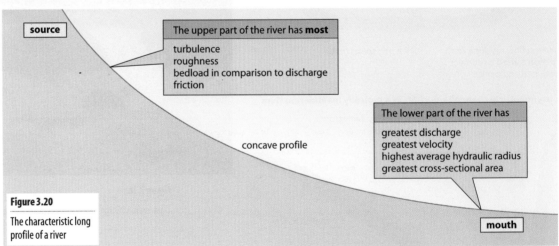

source

The upper part of the river has **most**

turbulence
roughness
bedload in comparison to discharge
friction

concave profile

The lower part of the river has

greatest discharge
greatest velocity
highest average hydraulic radius
greatest cross-sectional area

mouth

Figure 3.20

The characteristic long profile of a river

3 Channel slope

As more tributaries and water from surface runoff, throughflow and groundwater flow join the main river, the discharge, the channel cross-section area and the hydraulic radius will all increase. At the same time, less energy will be lost through friction and the erosive power of bedload material will decrease. As a result, the river flows over a gradually decreasing gradient – the characteristic concave **long profile (thalweg)** as shown in Figure 3.20.

In summarising this section it should be noted that:

- a river in a deep, broad channel, often with a gentle gradient and a small bedload, will have a greater velocity than a river in a shallow, narrow, rock-filled channel – even if the gradient of the latter is steeper
- the velocity of a river increases as it nears the sea – unless, like the Colorado (see Case Study 3C) and the Nile (Places 73, page 490), it flows through deserts where water is lost through evaporation or by human extraction for water supply
- the velocity increases as the depth, width and discharge of a river all increase
- as roughness increases, so too does turbulence and the ability of the river to pick up and transport sediment.

Transportation

Any energy remaining after the river has overcome friction can be used to transport sediment. The amount of energy available increases rapidly as the discharge, velocity and turbulence increase, until the river reaches flood levels. A river in flood has a large wetted perimeter and the extra friction is likely to cause deposition on the floodplain. A river at bankfull stage can move large quantities of soil and rock – its load – along its channel. In Britain, most material carried by a river is either sediment being redistributed from its banks, or material reaching the river from mass movement on its valley sides.

The load is transported by three main processes: **suspension**, **solution** and as **bedload** (Figure 3.21 and Places 10, page 73).

Suspended load
Very fine particles of clay and silt are dislodged and carried by turbulence in a fast-flowing river. The greater the turbulence and velocity, the larger the quantity and size of particles which can be picked up. The material held in suspension usually forms the greatest part of the total load; it increases in amount towards the river's mouth, giving the water its brown or black colour.

Dissolved or solution load
Water flowing within a river channel contains acids (e.g. carbonic acid from precipitation). If the bedrock is readily soluble, like limestone, it is constantly dissolved in the running water and removed in solution. Except in limestone areas, the material in solution forms only a relatively small proportion of the total load.

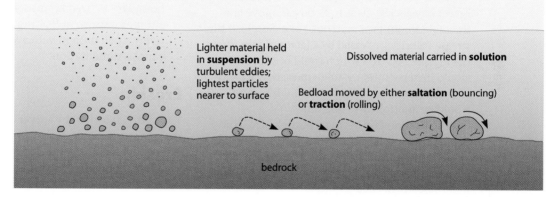

Figure 3.21

Transportation processes in a river or stream

Bedload
Larger particles which cannot be picked up by the current may be moved along the bed of the river in one of two ways. **Saltation** occurs when pebbles, sand and gravel are temporarily lifted up by the current and bounced along the bed in a hopping motion (compare saltation in deserts, page 183). **Traction** occurs when the largest cobbles and boulders roll or slide along the bed. The largest of these may only be moved during times of extreme flood.

It is much more difficult to measure the bedload than the suspended or dissolved load. Its contribution to the total load may be small unless the river is in flood. It has been suggested that the proportion of material carried in one year by the River Tyne is 57 per cent in suspension, 35 per cent in solution and 8 per cent as bedload. This is the equivalent of a 10-tonne lorry tipping its load into the river every 20 minutes throughout the year. In comparison, the Amazon's load is equivalent to four such lorries tipping every minute of the year!

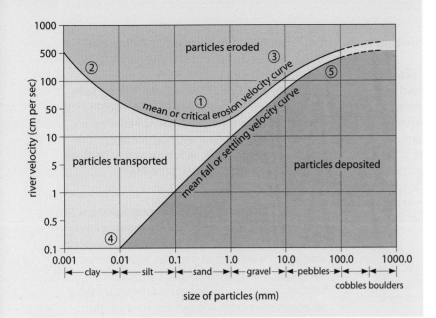

Figure 3.22

The Hjulström graph, showing the relationship between velocity and particle size. It shows the velocities necessary ('critical') for the initiation of movement (erosion); for deposition (sedimentation); and the area where transportation will continue to occur once movement has been initiated

Competence and capacity

Two further terms should be noted at this point: the competence and capacity of a river. **Competence** is the maximum size of material which a river is capable of transporting. **Capacity** is the total load actually transported. When the velocity is low, only small particles such as clay, silt and fine sand can be picked up (Figure 3.22). As the velocity increases, larger material can be moved. Because the maximum particle mass which can be moved increases with the sixth power of velocity, rivers in flood can move considerable amounts of material. For example, if the stream velocity increased by a factor of four, then the mass of boulders which could be moved would increase by 4^6 or 4096 times; if by a factor of five, the maximum mass it could transport would be multiplied 15 625 times.

The relationship between particle size (competence) and water velocity is shown on the Hjulström graph (Figure 3.22). The **mean**, or **critical**, **erosion velocity** curve gives the approximate velocity needed to pick up and transport, in suspension, particles of various sizes. The material carried by the river (capacity) is responsible for most of the subsequent erosion. The **mean fall** or **settling velocity** curve shows the velocities at which particles of a given size become too heavy to be transported and so will fall out of suspension and be deposited.

The graph shows two important points:

1 Sand can be transported at lower velocities than either finer or coarser particles. Particles of about 0.2 mm diameter can be picked up by a velocity of 20 cm per second (labelled **1** on the graph) whereas finer clay particles (**2**), because of their cohesive properties, need a velocity similar to that of pebbles (**3**) to be dislodged. During times of high discharge and velocity, the size and amount of the river's load will increase considerably, causing increased erosion within the channel.

2 The velocity required to maintain particles in suspension is less than the velocity needed to pick them up. Indeed, for very fine clays (**4**) the velocity required to maintain them is virtually nil – at which point the river must almost have stopped flowing! This means that material picked up by turbulent tributaries and lower order streams can be kept in suspension by a less turbulent, higher order main river. For coarser particles (**5**), the boundary between transportation and deposition is narrow, indicating that only a relatively small drop in velocity is needed to cause sedimentation.

Erosion

The material carried by a river can contribute to the wearing away of its banks and, to a lesser extent and mainly in the upper course, its bed. there are four main processes of erosion.

Corrasion

Corrasion occurs when the river picks up material and rubs it along its bed and banks, wearing them away by **abrasion**, rather like sandpaper. This process is most effective during times of flood and is the major method by which the river erodes both vertically and horizontally. If there are hollows in the river bed, pebbles are likely to become trapped. Turbulent eddies in the current can swirl pebbles around to form **potholes** (Figure 3.23).

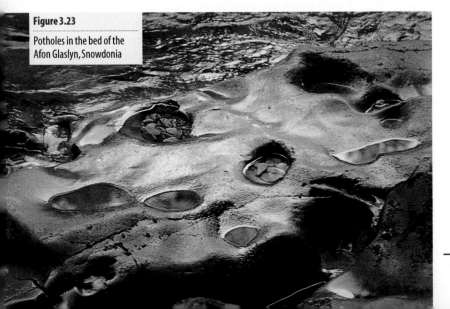

Figure 3.23

Potholes in the bed of the Afon Glaslyn, Snowdonia

Attrition

As the bedload is moved downstream, boulders collide with other material and the impact may break the rock into smaller pieces. In time, angular rocks become increasingly rounded in appearance.

Hydraulic action

The sheer force of the water as the turbulent current hits river banks (on the outside of a meander), pushes water into cracks. The air in the cracks is compressed, pressure is increased and, in time, the bank will collapse. **Cavitation** is a form of hydraulic action caused by bubbles of air collapsing. The resultant shock waves hit and slowly weaken the banks. This is the slowest and least effective erosion process.

Solution, or corrosion

This occurs continuously and is independent of river discharge or velocity. It is related to the chemical composition of the water, e.g. the concentration of carbonic acid and humic acid.

Deposition

When the velocity of a river begins to fall, it has less energy and so no longer has the competence or capacity to carry all its load. So, starting with the largest particles, material begins to be deposited (Figure 3.22). Deposition occurs when:

- discharge is reduced following a period of low precipitation
- velocity is lessened on entering the sea or a lake (resulting in a delta)
- shallower water occurs on the inside of a meander (Figure 3.25)
- the load is suddenly increased (caused by debris from a landslide)
- the river overflows its banks so that the velocity outside the channel is reduced (resulting in a floodplain).

As the river loses energy, the following changes are likely:

- The heaviest or bedload material is deposited first. It is for this reason that the channels of mountain streams are often filled with large boulders (Figures 3.18 and 3.27). Large boulders increase the size of the wetted perimeter.
- Gravel, sand and silt – transported either as bedload or in suspension – will be carried further, to be deposited over the floodplain (Figure 3.31) or in the channel of the river as it nears its mouth (Figure 3.32).
- The finest particles of silt and clay, which are carried in suspension, may be deposited where the river meets the sea – either to infill an estuary or to form a delta (Figure 3.33).
- The dissolved load will not be deposited, but will be carried out to sea where it will help to maintain the saltiness of the oceans.

Places 10   Afon Glaslyn, North Wales: river processes

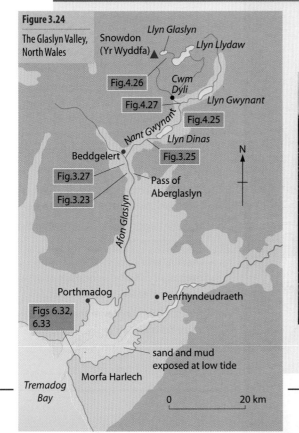

Figure 3.24

The Glaslyn Valley, North Wales

The Afon Glaslyn rises near the centre of the Snowdon massif and flows in a general southerly direction towards Tremadog Bay (Figure 3.24).

Figure 3.25

Erosion and deposition in the middle Afon Glaslyn

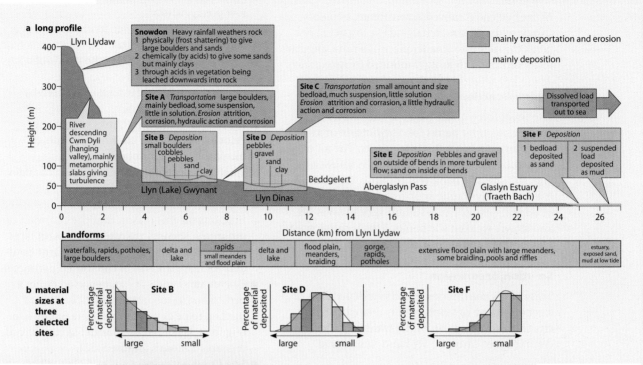

a long profile

Snowdon Heavy rainfall weathers rock
1 physically (frost shattering) to give large boulders and sands
2 chemically (by acids) to give some sands but mainly clays
3 through acids in vegetation being leached downwards into rock

Llyn Llydaw

River descending Cwm Dyli (hanging valley), mainly metamorphic slabs giving turbulence

Site A *Transportation* large boulders, mainly bedload, some suspension, little in solution. *Erosion* attrition, corrasion, hydraulic action and corrosion

Site B *Deposition* small boulders cobbles pebbles sand clay

Llyn (Lake) Gwynant

Site C *Transportation* small amount and size bedload, much suspension, little solution *Erosion* attrition and corrasion, a little hydraulic action and corrosion

Site D *Deposition* pebbles gravel sand clay

Llyn Dinas

Beddgelert

Site E *Deposition* Pebbles and gravel on outside of bends in more turbulent flow; sand on inside of bends

Aberglaslyn Pass

Glaslyn Estuary (Traeth Bach)

Dissolved load transported out to sea

Site F *Deposition*
1 bedload deposited as sand
2 suspended load deposited as mud

mainly transportation and erosion
mainly deposition

Distance (km) from Llyn Llydaw

Landforms

| waterfalls, rapids, potholes, large boulders | delta and lake | rapids / small meanders and flood plain | delta and lake | flood plain, meanders, braiding | gorge, rapids, potholes | extensive flood plain with large meanders, some braiding, pools and riffles | estuary, exposed sand, mud at low tide |

b material sizes at three selected sites

Site B — Percentage of material deposited — large — small

Site D — Percentage of material deposited — large — small

Site F — Percentage of material deposited — large — small

Figure 3.27

The boulder-strewn river bed of the upper Afon Glaslyn

Figure 3.26

The Afon Glaslyn, showing processes and landforms at selected sites

The long profile of the Glaslyn, as shown in Figure 3.26, does not, however, match the smooth curve of the model shown in Figure 3.20. This is partly because of:

- the effect of glaciation in the upper course (Figure 4.25) and
- differences in rock structure in the middle course (the Aberglaslyn Pass in Figure 3.27).

Figure 3.26 (a summary of an Open University programme) shows the relationships between the processes of fluvial transportation, erosion and deposition. By studying this diagram, do you agree with the following hypotheses (Framework 10, page 299):

- that as the competence of the river decreases, material is likely to be carried greater distances
- that the largest material, carried as the bedload, will be deposited first
- that material carried in suspension will be deposited over the floodplain or in the channel of the river as it nears its mouth
- that the finest material and the dissolved load will be carried out to sea?

Figure 3.28

V-shaped valley with inter-locking spurs, small rapids and no floodplain: Peak District National Park

Fluvial landforms

As the velocity of a river increases, surplus energy becomes available which may be har-. nessed to transport material and cause erosion. Where the velocity decreases, an energy deficit is likely to result in depositional features.

Effects of fluvial erosion

V-shaped valleys and interlocking spurs
As was seen in Figure 3.27, the channel of a river in its upper course is often choked with large, angular boulders. This bedload produces a large wetted perimeter which uses up much of the river's energy. Erosion is minimal because little energy is left to pick up and transport material. However, following periods of heavy rainfall or after rapid snowmelt, the discharge of a river may rise rapidly. As the water flows between boulders, turbulence increases and may result either in the bedload being taken up into suspension or, as is more usual because of its size, in its being rolled or bounced along the river bed. The result is intensive **vertical erosion** which enables the river to create a steep-sided valley with a charac-teristic V shape (Figure 3.28). The steepness of the valley sides depends upon several factors.

- **Climate**   Valleys are steeper where there is sufficient rainfall:
  - a to instigate mass movement on the valley sides and
  - b to create sufficient discharge to allow the river to create enough energy to move its bedload and, therefore, to erode vertically, or
  - c for rivers to cross desert areas which have no rain to wash down the valley sides, e.g. the Grand Canyon (Figure 7.19).
- **Rock structure**   Resistant, permeable rocks like Carboniferous limestone (Figure 8.5) often produce almost vertical sides in con-trast to less resistant, impermeable rocks such as clay which are likely to produce more gentle slopes.
- **Vegetation**   Vegetation may help to bind the soil together and thus keep the hillslope more stable.

Interlocking spurs form because the river is forced to follow a winding course around the protrusions of the surrounding highland. As the resultant spurs interlock, the view up or down the valley is restricted (Figure 3.28).

A process characteristic at the source of a river is **headward erosion**, or **spring sapping**. Here, where throughflow reaches the surface, the river may erode back towards its watershed as it undercuts the rock, soil and vegetation.

Waterfalls
A waterfall forms when a river, after flowing over relatively hard rock, meets a band of less resistant rock or, as is common in South America and Africa, where it flows over the edge of a plateau. As the water approaches the brink of the falls, velocity increases because the water in front of it loses contact with its bed and so is unhampered by friction (Figure 3.29). The underlying softer rock is worn away as water falls onto it. In time, the harder rock may become undercut and unstable and may eventually collapse. Some of this collapsed rock may be swirled around at the foot of the falls by turbulence, usually at times of high discharge, to create a deep **plunge pool**. As this process is repeated, the waterfall retreats upstream leaving a deep, steep-sided **gorge** (Places 11). At Niagara, where a hard band of limestone overlies softer shales and sandstone, the Niagara River plunges 50 m causing the falls to retreat by 1 m a year and so creating the Niagara Gorge.

Rapids
Rapids develop where the gradient of the river bed increases without a sudden break of slope (as in a waterfall) or where the stream flows over a series of gently dipping bands of harder rock. Rapids increase the turbulence of a river and hence its erosive power (Figure 3.27).

Drainage basins and rivers

The Iguaçu River, a tributary of the Parana, forms part of the border between Brazil and Argentina. At one point along its course, the Iguaçu plunges 80 m over a 3 km wide, crescent-shaped precipice (Figure 3.30). The Iguaçu Falls occur where the river leaves the resistant basaltic lava which forms the southern edge of the Brazilian plateau and flows onto less resistant rock, while their crescent shape results from the retreat of the falls upstream (Figure 3.29).

By the end of the rainy season (January/February) up to 4 million litres of water a day can pour over the individual cascades – numbering up to 275 – which combine to form the falls. The main attraction is the Devil's Throat where 14 separate falls unite to create a deafening noise, volumes of spray, foaming water and a large rainbow. In contrast, by the end of the dry season (June/July), river levels may be very low – indeed, for one month in 1978 it actually dried up.

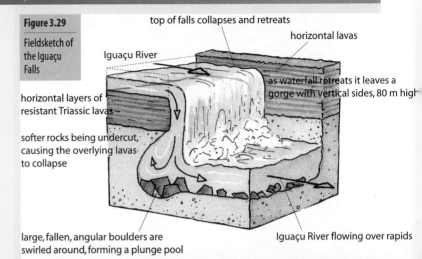

Figure 3.29

Fieldsketch of the Iguaçu Falls

top of falls collapses and retreats

horizontal lavas

Iguaçu River

as waterfall retreats it leaves a gorge with vertical sides, 80 m high

horizontal layers of resistant Triassic lavas

softer rocks being undercut, causing the overlying lavas to collapse

large, fallen, angular boulders are swirled around, forming a plunge pool

Iguaçu River flowing over rapids

Figure 3.30

The Iguaçu Falls

Effects of fluvial deposition

Deposition of sediment takes place when there is a decrease in energy or an increase in capacity which makes the river less competent to transport its load. This can occur anywhere from the upper course, where large boulders may be left, to the mouth, where fine clays may be deposited.

Floodplains

Rivers have most energy when at their bankfull stage. Should the river continue to rise, then the water will cover any adjacent flat land. The land susceptible to flooding in this way is known as the **floodplain** (Figure 3.31 and Places 12, page 80). As the river spreads over its floodplain, there will be a sudden increase in both the wetted perimeter and the hydraulic radius. This results in an increase in friction, a corresponding decrease in velocity and the deposition of material previously held in suspension. The thin veneer of silt, deposited by each flood, increases the fertility of the land, while the successive flooding causes the floodplain to build up in height (as yet it has proved impossible to bore down to bedrock in the lower Nile valley). The floodplain may also be made up of material deposited as point bars on the inside of meanders (Figure 3.38) and can be widened by the **lateral erosion** of the meanders. The edge of the floodplain is often marked by a prominent slope known as the **bluff line** (Figure 3.31).

Levées

When a river overflows its banks, the increase in friction produced by the contact with the floodplain causes material to be deposited. The coarsest material is dropped first to form a small, natural embankment (or levée) alongside the channel (Figure 3.31). During subsequent periods of low discharge, further deposition will occur within the main channel causing the bed of the river to rise and the risk of flooding to increase. To try to contain the river, the embankments are sometimes artificially strengthened and heightened (the levée protecting St Louis from the Mississippi is 15.8 m higher than the floodplain which it is meant to protect). Some rivers, such as the Mississippi and Yangtze, flow above the level of their floodplains which means that if the levées collapse there can be serious damage to property, and loss of life (Case Studies 3A and 3B).

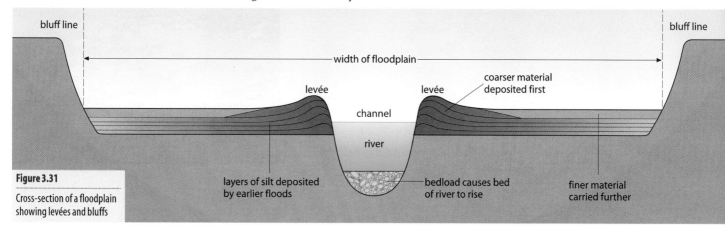

Figure 3.31

Cross-section of a floodplain showing levées and bluffs

(labels: bluff line — bluff line — width of floodplain — levée — levée — coarser material deposited first — channel — river — layers of silt deposited by earlier floods — bedload causes bed of river to rise — finer material carried further)

Braiding

For short periods of the year, some rivers carry a very high load in relation to their velocity, e.g. during snowmelt periods in Alpine or Arctic areas. When a river's level falls rapidly, competence and capacity are reduced, and the channel may become choked with material, causing the river to braid – that is, to divide into a series of diverging and converging segments (Figures 3.32 and 5.16).

Figure 3.32

A braided river, South Island, New Zealand

Deltas

A delta is usually composed of fine sediment which is deposited when a river loses energy and competence as it flows into an area of slow-moving water such as a lake (Figure 4.22) or the sea. When rivers like the Mississippi or the Nile reach the sea, the meeting of fresh and salt water produces an electric charge which causes clay particles to coagulate and to settle on the seabed, a process called **flocculation**.

Deltas are so called because it was thought that their shape resembled that of delta, the fourth letter of the Greek alphabet (Δ). In fact, deltas vary greatly in shape but geomorphologists have grouped them into three basic forms:

■ **arcuate:** having a rounded, convex outer margin, e.g. the Nile
■ **cuspate:** where the material brought down by a river is spread out evenly on either side of its channel, e.g. the Tiber
■ **bird's foot:** where the river has many distributaries bounded by sediment and which extend out to sea like the claws of a bird's foot, e.g. the Mississippi (Figure 3.33).

Although deltas provide some of the world's most fertile land, their flatness makes them high flood-risk areas, while the shallow and frequently changing river channels hinder navigation.

77

Effects of combined erosion and deposition

Pools, riffles and meanders

Rivers rarely flow in a straight line. Indeed, testing under laboratory conditions suggests that a straight course is abnormal and unstable. How meanders begin to form is uncertain, but they appear to have their origins during times of flood and in relatively straight sections where pools and riffles develop (Figure 3.34). The usual spacing between **pools**, areas of deeper water, and

Figure 3.33

The Mississippi delta

Figure 3.34

A possible sequence in the development of a meander

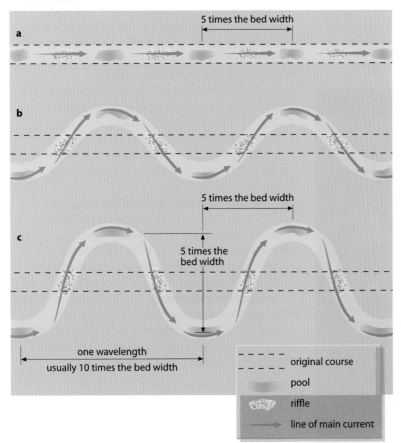

Figure 3.35

A pool and riffles in the River Gelt, Cumbria

riffles, areas of shallower water, is usually very regular, being five to six times that of the bed width. The pool is an area of greater erosion where the available energy in the river builds up due to a reduction in friction. Energy is dissipated across the riffle area. As a higher proportion of the total energy is then needed to overcome friction, the erosive capacity is decreased and, except at times of high discharge, material is deposited (Figure 3.35). The regular spacings of pools and riffles, spacings which are almost perfect in an alluvial stretch of river, are believed to result from a series of secondary flows which exist within the main flow. Secondary flows include **helicoidal flow**, a corkscrew movement, as shown in Figure 3.15b, and a series of converging and diverging lateral rotations. Helicoidal flow is believed to be responsible for moving material from the outside of one meander bend and then depositing much of it on the inside of the next bend. It is thought, therefore, that it is the secondary flows that increase the sinuosity (the curving nature) of the meander (Figure 3.36), producing a regular meander wavelength which is about ten times that of the bed width. Sinuosity is described as:

$$\frac{\text{actual channel length}}{\text{straight-line distance}}$$

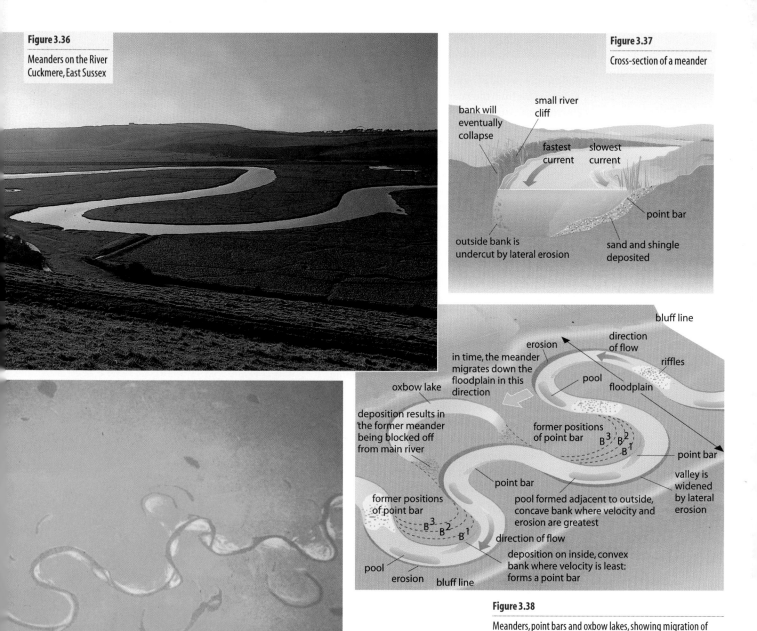

Figure 3.36

Meanders on the River Cuckmere, East Sussex

Figure 3.37

Cross-section of a meander

bank will eventually collapse

small river cliff

fastest current

slowest current

outside bank is undercut by lateral erosion

point bar

sand and shingle deposited

bluff line

erosion

direction of flow

riffles

in time, the meander migrates down the floodplain in this direction

pool

floodplain

oxbow lake

deposition results in the former meander being blocked off from main river

former positions of point bar

B^3 B^2 B^1

point bar

valley is widened by lateral erosion

former positions of point bar

B^3 B^2 B^1

point bar

pool formed adjacent to outside, concave bank where velocity and erosion are greatest

direction of flow

deposition on inside, convex bank where velocity is least: forms a point bar

pool

erosion

bluff line

Figure 3.38

Meanders, point bars and oxbow lakes, showing migration of meanders and changing positions of point bars over time

Figure 3.39

Meanders and oxbow lakes, Alaska, USA

Meanders, point bars and oxbow lakes

A meander has an asymmetrical cross-section (Figure 3.37) formed by erosion on the outside bend, where discharge and velocity are greatest and friction is at a minimum, and deposition on the inside, where discharge and velocity are at a minimum and friction is at its greatest (Figure 3.25). Material deposited on the convex inside of the bend may take the form of a curving **point bar** (Figure 3.38). The particles are usually graded in size, with the largest material being found on the upstream side of the feature (there is rarely any gradation up the slope itself). As erosion continues on the outer bend, the whole meander tends to migrate slowly downstream. Material forming the point bar becomes a contributory factor in the formation of the floodplain. Over time, the sinuosity of the meander may become so pronounced that, during a flood, the river cuts through the narrow neck of land in order to shorten its course. Having achieved a temporary straightening of its channel, the main current will then flow in mid-channel. Deposition can now take place next to the banks and so, eventually, the old curve of the river will be abandoned, leaving a crescent-shaped feature known as an **oxbow lake** or **cutoff** (Figures 3.38 and 3.39).

Following a period of heavy rain and a flow of melted snow from the North York Moors, the River Derwent burst its banks over the weekend beginning 6 March 1999 (Figure 3.40). Over 200 mm of rain had fallen in eight days, with 125 mm in one day (the total March average is only 75 mm). The Derwent rose to over 4 m above its normal level, causing some of the worst flooding this century. Worst affected were the small towns of Stamford Bridge (Figure 3.41) and Malton.

On the Monday (8 March), it was announced that schools were to close for two days. Over 100 houses had been flooded, roads were blocked and no trains were expected to operate between Malton and Scarborough for another week. Parents were warned not to let children play in the water which could have sewage floating in it. Although water levels began to show signs of subsiding on the Tuesday (9 March) and some people were returning home, the Environment Agency forecast that it would be several days before water levels returned to normal. The Agency also claimed that the floodwater was now moving slowly downstream where it was not expected to cause as much disruption, and defended, against some local criticism, its system of flood warnings.

Conditions on that day were described in an article in the Wednesday's *Guardian* (Figure 3.42). By the Friday (12 March), the police said that while some blocked roads were now open and water levels were receding quite quickly, people travelling to Malton should check conditions before starting their journey. Meanwhile, insurance experts were estimating the cost of damage to be £20 million.

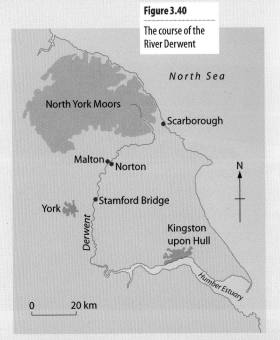

Figure 3.40

The course of the River Derwent

Figure 3.41

Stamford Bridge, North Yorkshire, 13 March 1999

The Waterside café is no longer by the side of the water. The water now flows through the front door and bay window and slops around tables where customers once took their tea. The café is in Stamford Bridge, the village near York where King Harold defeated his half-brother Harald Hardrada in 1066 before being killed at Hastings.

Last night the Environment Agency said that 130 properties had been flooded along the banks of the Derwent. The river has cut Stamford Bridge in two, spreading from its traditional course across the main street, over the village green and onto a picnic site. Police say a plaque outside the Swordsman pub recording the height of the great flood of 1947 now lies half a metre below the surface of the water.

Extract from *The Guardian*, 10 March 1999

Figure 3.42

The effect of the flood at Stamford Bridge on 9 March 1999

Base level and the graded river

Base level

This is the lowest level to which erosion by running water can take place. In the case of rivers, this theoretical limit is sea-level. Exceptions occur when a river flows into an inland sea (e.g. the River Jordan into the Dead Sea) and if there happens to be a temporary **local base level**, such as where a river flows into a lake, where a tributary joins a main river, or where there is a resistant band of rock crossing a valley.

Grade

The concept of **grade** is one of a river forming an open system (Framework 3, page 45) in a state of dynamic equilibrium where there is a balance between the rate of erosion and the rate of deposition. In its simplest interpretation, a graded river has a gently sloping long profile with the gradient decreasing towards its mouth (Figure 3.43a). This balance is always transitory as the slope (profile) has to adjust constantly to changes in discharge and sediment load. These can cause short-term increases in either the rate of erosion or deposition until the state of equilibrium has again been reached. This may be illustrated by two situations:

- The long profile of a river happens to contain a waterfall and a lake (Figure 3.43b). Erosion is likely to be greatest at the waterfall, while deposition occurs in the lake. In time, both features will be eliminated.
- There is a lengthy period of heavy rainfall within a river basin. As the volume of water rises and consequently the velocity and load of the river increase, so too will the rate of erosion. Ultimately, the extra load carried by the river leads to extra deposition further down the valley or out at sea.

In a wider interpretation, grade is a balance not only in the long profile, but also in the river's cross-profile and in the roughness of its channel. In this sense, balance or grade is when all aspects of the river's channel (width, depth and gradient) are adjusted to the discharge and load of the river at a given point in time. If the volume and load change, then the river's channel morphology must adjust accordingly. Such changes, where and when they do occur, are likely to take lengthy periods of geological time.

Changes in base level

There are two groups of factors which influence changes in base level:

- **Climatic:** the effects of glaciation and changes in rainfall.
- **Tectonic:** crustal uplift, following plate movement, and local volcanic activity.

As will be seen in Chapter 6, changes in base level affect coasts as well as rivers. There are two types of base level movement: positive and negative.

- **Positive change** occurs when sea-level rises in relation to the land (or the land sinks in relation to the sea). This results in a decrease in the gradient of the river with a corresponding increase in deposition and potential flooding of coastal areas.
- **Negative change** occurs when sea-level falls in relation to the land (or the land rises in relation to the sea). This movement causes land to emerge from the sea, steepening the gradient of the river and therefore increasing the rate of fluvial erosion. This process is called **rejuvenation**.

Figure 3.43

River profiles

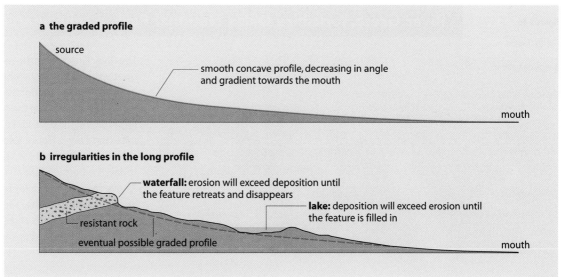

a the graded profile

source

smooth concave profile, decreasing in angle and gradient towards the mouth

mouth

b irregularities in the long profile

waterfall: erosion will exceed deposition until the feature retreats and disappears

lake: deposition will exceed erosion until the feature is filled in

resistant rock

eventual possible graded profile

mouth

Drainage basins and rivers

Figure 3.44

The effect of rejuve-nation on the long profile

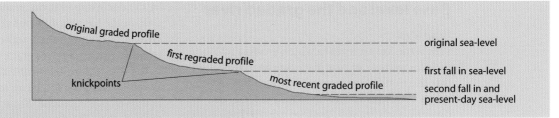

original graded profile

original sea-level

first regraded profile

first fall in sea-level

knickpoints

most recent graded profile

second fall in and present-day sea-level

Figure 3.45

A rejuvenated river, Antalya, Turkey: the land has only recently experienced tectonic uplift and the river has had insufficient time to re-adjust to the new sea-level

Rejuvenation

A negative change in base level increases the potential energy of a river, enabling it to revive its erosive activity; in doing so, it upsets any possible graded long profile. Beginning in its lowest reaches, next to the sea, the river will try to regrade itself.

During the Pleistocene glacial period, Britain was depressed by the weight of ice. Following deglaciation, the land slowly and intermittently rose again (**isostatic uplift**, page 123). Thus rejuvenation took place on more than one occasion, with the result that many rivers today show several

partly graded profiles (Figure 3.44). Where the rise in the land (or drop in sea-level) is too rapid to allow a river sufficient time to erode vertically to the new sea-level, it may have to descend as a waterfall over recently emerged sea cliffs (Figure 3.45). In time, the river will cut downwards and backwards and the waterfall will retreat upstream. The **knickpoint**, usually indicated by the presence of a waterfall, marks the maximum extent of the newly graded profile (Places 13). Should a river become completely regraded, which is unlikely because of the timescale involved, the knickpoint and all of the original graded profile will disappear.

River terraces and incised meanders
River terraces are remnants of former floodplains which, following vertical erosion caused by rejuvenation, have been left high and dry above the maximum level of present-day flooding. They offer excellent sites for the location of towns (e.g. London, Figures 3.47 and 14.9). Above the present floodplain of the Thames at London are two earlier ones forming the Taplow and Boyn Hill terraces. If a river cuts rapidly into its floodplain, a pair of terraces of equal height may be seen flanking the river and creating a **valley-in-valley** feature. However, more often than not, the river cuts down relatively slowly, enabling it to meander at the same time. The result is that the terrace to one side of the river may be removed as

Places 13
River Greta, Yorkshire Dales National Park: a rejuvenated river

Figure 3.46

The River Greta

a before rejuvenation

graded River Greta meandering over a wide floodplain

bluff line

tributary

floodplain

b after rejuvenation

side of Ingleborough Hill

bluff line

floodplain

river terrace

rapids

original graded section of River Greta

Beezley Falls knickpoint

rejuvenated River Greta flowing in a valley-in-valley due to an increase in erosion

After D.S. Walker

The River Greta, in north-west Yorkshire, is a good example of a rejuvenated river. Figure 3.46a is a reconstruction to show what its valley (upstream from the village of Ingleton) might have looked like before the fall in base level. Figure 3.46b is a simplified sketch showing how the same area appears today. The Beezley Falls are a knickpoint. Above the falls, the valley has a wide, open appearance. Below the falls, the river flows over a series of rapids and smaller falls in a deep, steep-sided 'valley-in-valley'.

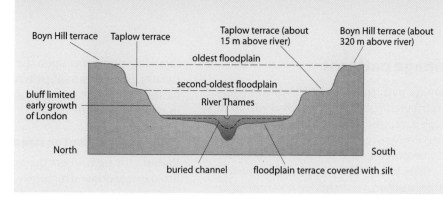

the meanders migrate downstream. Figure 3.49 shows terraces, not paired, on a small stream crossing a beach on southern Arran. In this case, rejuvenation takes place twice daily as the tide ebbs and sea-level falls.

If the uplift of land (or fall in sea-level) continues for a lengthy period, the river may cut downwards to form incised meanders. There are two types of incised meander. **Entrenched meanders** have a symmetrical cross-section and result from either a very rapid incision by the river, or

the valley sides being resistant to erosion (the River Wear at Durham, Figures 3.48a and 14.6). **Ingrown meanders** occur when the uplift of the land, or incision by the river, is less rapid, allowing the river time to shift laterally and to produce an asymmetrical cross-valley shape (the River Wye at Tintern Abbey, Figure 3.48b). As with meanders in the lower course of a normal river, incised meanders can also change their channels to leave an abandoned meander with a central **meander core** (Figure 3.48b).

Figure 3.48

Incised meanders and associated cross-valley profiles

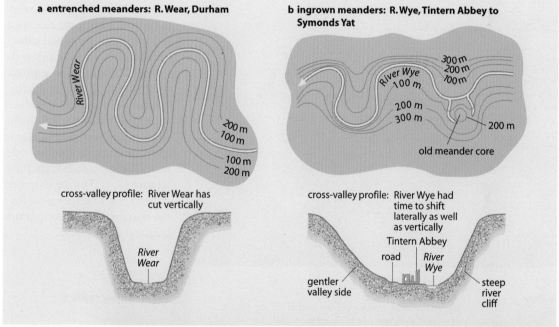

Figure 3.49

Rejuvenation on a micro scale: a small stream crossing a beach at Kildonan, Arran, has cut downwards to the level of the falling tide – note the ingrown meander, river terraces and valley-in-valley features

Drainage patterns

A **drainage pattern** is the way in which a river and its tributaries arrange themselves within their drainage basin (see Horton's Laws, page 65). Most patterns evolve over a lengthy period of time and usually become adjusted to the structure of the basin. There is no widely accepted classification, partly because most patterns are descriptive.

Patterns independent of structure
Parallel   This, the simplest pattern, occurs on newly uplifted land or other uniformly sloping surfaces which allow rivers and tributaries to flow downhill more or less parallel with each other, e.g. rivers flowing south-eastwards from the Aberdare Mountains in Kenya (Figure 3.50a).
Dendritic   Deriving its name from the Greek word *dendron*, meaning a tree, this is a tree-like pattern in which the many tributaries (branches) converge upon the main river (trunk). It is a common pattern and develops in basins having one rock type with no variations in structure (Figure 3.50b).

Patterns dependent on structure

Radial   In areas where the rocks have been lifted into a dome structure (e.g. the batholiths of Dartmoor and Arran) or where a conical volcanic cone has formed (e.g. Mount Etna), rivers radiate outwards from a central point like the spokes of a wheel (Figure 3.50c).
Trellised or rectangular   In areas of alternating resistant and less resistant rock, tributaries will form and join the main river at right-angles (Figure 3.50d). Sometimes each individual segment is of approximately equal length. The main river, called a **consequent river** because it is a consequence of the initial uplift or slope (compare parallel drainage), flows in the same direction as the dip of the rocks (Figure 3.51a). The tributaries which develop, mainly by headward erosion along areas of weaker rocks, are called **subsequent streams** because they form at a later date than the consequents. In time, these subsequents create wide valleys or vales (Figure 3.51b). **Obsequent streams** flow in the opposite direction from the consequent streams, i.e. down the steep scarp slope of the escarpment (Figure 3.51b). It is these obsequents that often provide the sources of water for scarp-foot springline settlements (Figure 14.4). The development of this drainage pattern is also responsible for the formation of the **scarp and vale topography** of south-east England (Figure 8.9).

Figure 3.50

Drainage patterns

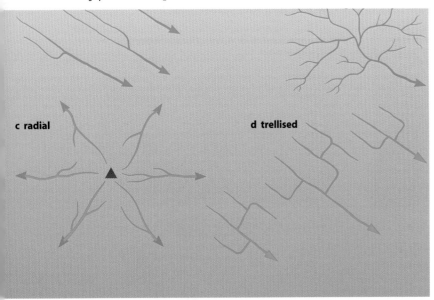

c radial

d trellised

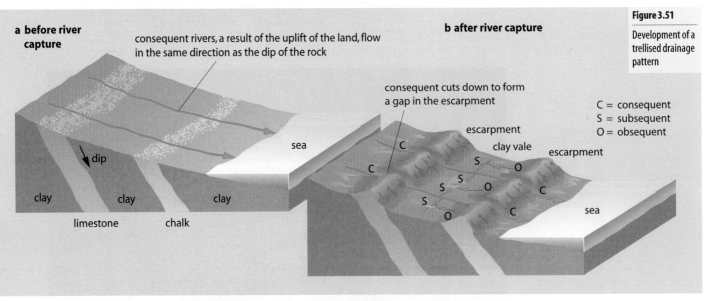

a before river capture

consequent rivers, a result of the uplift of the land, flow in the same direction as the dip of the rock

dip

clay   clay   clay

limestone   chalk

sea

b after river capture

consequent cuts down to form a gap in the escarpment

escarpment

clay vale   escarpment

C
C
S   O
S   O
S   C
O   C
O   sea

Figure 3.51

Development of a trellised drainage pattern

C = consequent
S = subsequent
O = obsequent

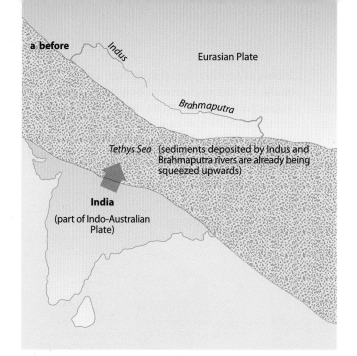

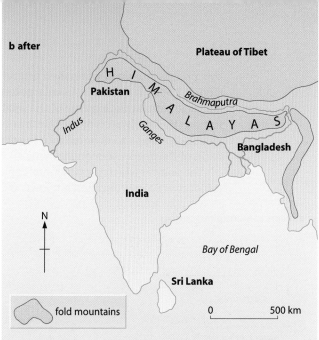

Figure 3.52

Antecedent drainage, Himalayas

Patterns apparently unrelated to structure

Antecedent   Antecedence is when the drainage pattern developed before such structural movements as the uplift or folding of the land, and where vertical erosion by the river was able to keep pace with the later uplift. The Brahmaputra River rises in Tibet, but turns southwards to flow through a series of deep gorges in the Himalayas before reaching the Bay of Bengal (Figure 3.52). It must at one stage have flowed southwards into the Tethys Sea (Figure 1.4) which had existed before the Indo-Australian Plate moved northwards and collided with the Eurasian Plate forming the Himalayas (pages 19 and 20). The Brahmaputra, with an increasing gradient and load, was able to cut downwards through the rising Himalayas to maintain its original course.

Superimposed   In several parts of the world, including the English Lake District, the drainage pattern seems to have no relationship to the present-day surface rocks. When the Lake District was uplifted into a dome, the newly-formed volcanic rocks were covered by sedimen-

tary limestones and sandstones. The radial drainage pattern which developed, together with later glacial processes, cut through and ultimately removed the surface layers of sedimentary rock to superimpose itself upon the underlying volcanic rocks.

River capture

Rivers, in attempting to adjust to structure, may capture the headwaters of their neighbours. For example, most eastward-flowing English rivers between the Humber and central Northumberland have had their courses altered by **river capture** or **piracy** (Figure 3.53).

Figure 3.54a shows a case where there are two consequent rivers with one having a greater discharge and higher erosional activity than the other. Each has a tributary (subsequents X and Y) flowing along a valley of weaker rock, but subsequent X (the tributary of the master, or larger, consequent) is likely to be the more vigorous. Subsequent X will, therefore, cut backwards by headward erosion until it reaches subsequent Y (the tributary of the weaker consequent); then,

Figure 3.53

River capture, Northumberland

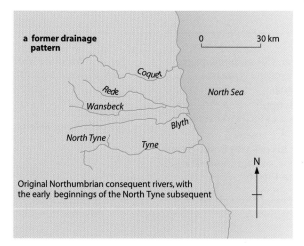

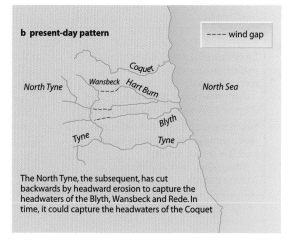

by a process known as **watershed migration** (Figure 3.54b), it will begin to enlarge its own drainage basin at the expense of the smaller river. In time, the headwaters of the minor consequent will be captured and diverted into the drainage basin of the major consequent (Figure 3.54c).

The point at which the headwaters of the minor river change direction is known as the **elbow of capture**. Below this point, a **wind gap** marks the former course of the now **beheaded consequent** (a wind gap is a dry valley which was cut through the hills by a former river). The beheaded river is also known as a **misfit stream**, as its discharge is far too low to account for the size of the valley through which it flows (Figure 3.54c).

a before capture (piracy) occurs

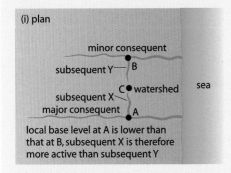

(i) plan

local base level at A is lower than that at B, subsequent X is therefore more active than subsequent Y

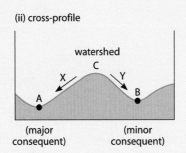

(ii) cross-profile

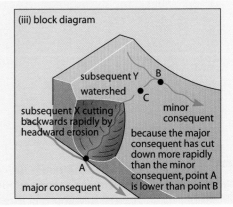

(iii) block diagram

b watershed migration (recession)

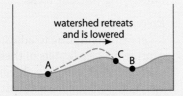

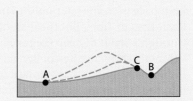

c after capture has taken place

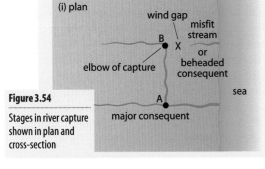

(i) plan

Figure 3.54

Stages in river capture shown in plan and cross-section

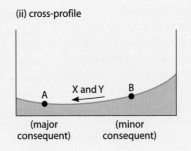

(ii) cross-profile

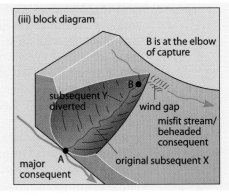

(iii) block diagram

The need for river management

A River flooding: The Mississippi, 1993

Flooding by rivers is a natural event which, because people often choose to live in flood-risk areas, becomes a hazard (page 31). To people living in the Mississippi valley, 'that their river should flood is as natural as sunshine in Florida or snowfall in the Rockies'. Without human intervention, the Mississippi would flood virtually every year. Indeed, it has been this frequency of flooding which has, over many centuries, allowed today's river to flow for much of its course over a wide, fertile, flat, alluvial floodplain (Figures 3.55 and 3.56).

Figure 3.55

The flood hazard and the Mississippi River

"Usually, of course, the great floods occur in the lower river, in the last 1 600 km below Cairo, Illinois. This is where the plain flattens out (the river drops less than 120 m from here to its mouth) and where the Ohio and Tennessee flow into the Mississippi.

Of the water that flows past Memphis, only about 38 per cent comes from the Missouri–Mississippi network. The bulk comes from the Ohio and Tennessee, from the lush Appalachians, rather than the dry mid-West. 'We don't mind too much about the Missouri', says Donna Willett, speaking for the US Army Corps of Engineers (who have the responsibility of flood prevention). "It can rain there for weeks, and we wouldn't mind. We can handle three times the water coming down in those floods. But the Ohio, well, that's another story. When that starts rising, we start watching. . . "

Enquiry route on river flooding	Application to a specific event: the Mississippi, 1993
1 Where is the river/drainage basin located?	The Mississippi – together with its main tributaries, the Missouri and the Ohio – drains one-third of the USA and a small part of Canada (Figure 3.56).
2 What is the frequency of flooding?	Left to its own devices, flooding would be an almost annual event with late spring being the peak period.
3 What is the magnitude of flooding?	Until recently, major floods occurred every 5–10 years (there were six in the 1880s) and a serious/extreme flood occurred approximately once every 40 years.
4 What are the natural causes of flooding?	Usually it results from heavy rainfall (January–May) in the Appalachian Mountains, especially if this coincides with snowmelt (Figure 3.56).
5 What are the consequences of flooding?	Initially, it was to develop the wide, alluvial floodplain. The 1927 flood caused 217 deaths; 700 000 people were evacuated; the river became up to 150 km wide (usual width 1 km); livestock and crops were lost; services were destroyed.
6 What attempts can be made to reduce the flood hazard?	Until the 1927 flood, the main policy was 'hold by levées' – by 1993, some levées were 15 m high (Figure 3.57). After 1927, new schemes included building dams and storage reservoirs (six huge dams and 105 on Missouri); afforestation to reduce/delay runoff; creating diversion spillways (e.g. Bonnet Carré floodway diverts floodwater into Lake Pontchartrain and the sea); cutting through meanders to straighten and shorten the course (Figure 3.57).
7 How successful have the attempts to reduce flooding been?	In 1883, Mark Twain claimed that 'You cannot tame that lawless stream'. By 1973, it appeared that the river had been tamed: there was no further flooding … until 1993. Has human intervention made the danger worse? (page 90)

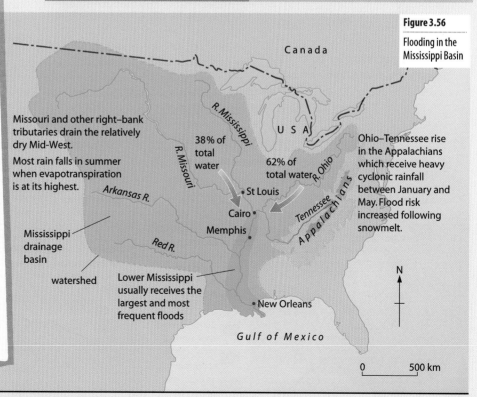

Figure 3.56

Flooding in the Mississippi Basin

Missouri and other right–bank tributaries drain the relatively dry Mid-West.

Most rain falls in summer when evapotranspiration is at its highest.

Ohio–Tennessee rise in the Appalachians which receive heavy cyclonic rainfall between January and May. Flood risk increased following snowmelt.

Mississippi drainage basin

watershed

Lower Mississippi usually receives the largest and most frequent floods

Canada

USA

R. Mississippi

R. Missouri

Arkansas R.

Red R.

38% of total water

62% of total water

R. Ohio

Tennessee R.

Appalachians

St Louis

Cairo

Memphis

New Orleans

Gulf of Mexico

N

0   500 km

1a Height (metres) of levées at Memphis

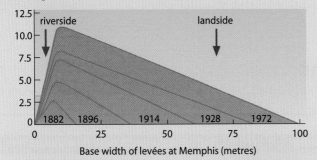

Base width of levées at Memphis (metres)

b The 1993 flood at St Louis

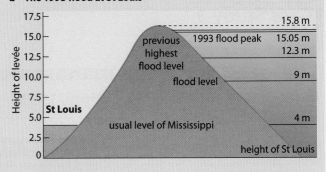

15.8 m
1993 flood peak   15.05 m
previous highest flood level   12.3 m
flood level   9 m
usual level of Mississippi
height of St Louis   4 m

2

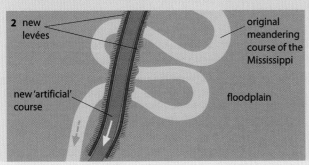

new levées

new 'artificial' course

original meandering course of the Mississippi

floodplain

By making the course straighter and shorter, floodwater could be removed from the river basin as quickly as possible. It was achieved by cutting through the narrow necks of large meanders. Between 1934 and 1945 one stretch of the river alone was reduced from 530 km to almost 230 km. By shortening the distance, the gradient and therefore the velocity of the river increases. (But rivers try to create meanders rather than flow naturally in straight courses.)

Figure 3.57

Two engineering schemes to try to control flooding

Engineering/planning schemes in the Mississippi basin

Prior to the 1993 flood, it was perceived that the flow of the Mississippi had been controlled. This had been achieved through a variety of flood prevention schemes (Figure 3.57).

- Levées had been heightened, in places to over 15 m, and strengthened. There were almost 3000 km of levées along the main river and its tributaries.
- By cutting through meanders, the Mississippi had been straightened and shortened: for 1750 km, it flows in artificial channels.
- Large spillways had been built to take excess water during times of flood.

- The flow of the major tributaries (Missouri, Ohio and Tennessee) had been controlled by a series of dams.

Why did the Mississippi flood in 1993?

The Mid-West was already having a wet year when record-setting spring and summer rains hit. The rain ran off the soggy ground and into rapidly rising rivers. Several parts of the central USA had over 200 per cent more rain than was usual for the time of year (Figure 3.58). It was the ferocity, location and timing of the flood that took everyone by surprise. Normally, river levels are falling in midsummer, the upper Mississippi was not perceived to be the major flood-risk area, and people believed that flooding in the basin had been controlled. Floodwater at St Louis reached an all-time high (Figure 3.58).

Satellite photographs showed the extent of the flooding (Figure 3.59). Figure 3.60 describes some of its effects.

After the flood: should rivers run freer?

Since the first levée was built on the Mississippi in 1718, engineers have been channelling the river to protect farmland and towns from floodwaters. But have the levées, dams and diversion channels actually aggravated the flooding? There are two schools of thought. One advocates accepting that rivers are part of a complex ecological balance and that flooding should be allowed as a natural event (Figure 3.61). The other argues for better defences and a more effective control of rivers (Figure 3.62).

Figure 3.58

Extract from *US Today*, a daily newspaper

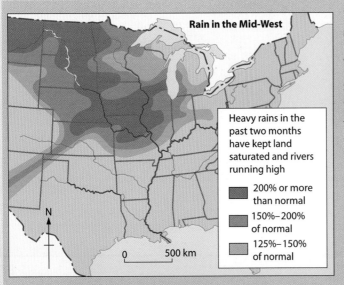

Rain in the Mid-West

Heavy rains in the past two months have kept land saturated and rivers running high

- 200% or more than normal
- 150%–200% of normal
- 125%–150% of normal

N

0   500 km

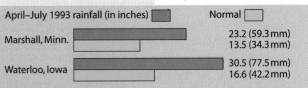

April–July 1993 rainfall (in inches)   Normal

Marshall, Minn.   23.2 (59.3 mm)
13.5 (34.3 mm)

Waterloo, Iowa   30.5 (77.5 mm)
16.6 (42.2 mm)

St Louis

Although there were some nervous moments, the city's massive 11–mile long, 52–foot floodwall protected the downtown from flooding. The river crested here August 1 at a record 49.4 feet, and the amount of water flowing past the Gateway Arch surpassed a record 1 million cubic feet per second.

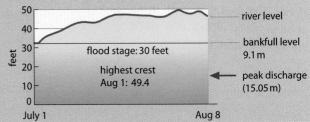

river level

bankfull level 9.1 m

flood stage: 30 feet

highest crest Aug 1: 49.4

peak discharge (15.05 m)

July 1   Aug 8

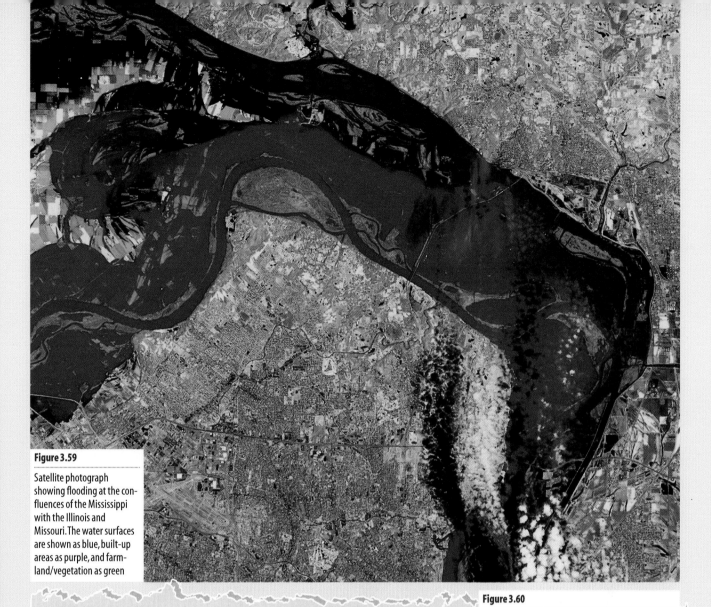

Figure 3.59

Satellite photograph showing flooding at the confluences of the Mississippi with the Illinois and Missouri. The water surfaces are shown as blue, built-up areas as purple, and farmland/vegetation as green

Figure 3.60

The consequences of flooding in the St Louis area

US Today, 9 August 1993

Flood of '93

Damage $10.5 billion	Deaths 45	Evacuated 74 000	Houses 45 000	Crops $6.5 billion

Nearly half of the counties in nine states bordering the upper reaches of the Mississippi and Missouri rivers have been declared federal disaster areas. This is the first step in becoming eligible for federal aid, including direct grants from Congress, Federal Emergency Management Agency and many other groups:

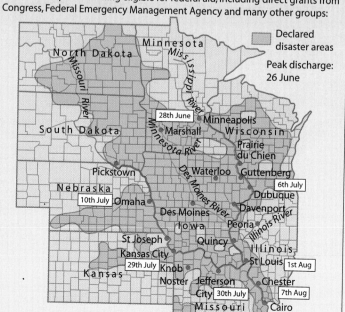

Declared disaster areas

Peak discharge: 26 June

Illinois: In the fight against flooding rivers, 17 levées were breached, including one that flooded the town of Valmeyer and 70 000 acres of surrounding farmland. One flood-related death was reported.

In Alton, the treatment plant was flooded Aug 1, cutting off water to the town's 33 000 residents. "Our levee did not breach, but the water came in through the street, the drains, anywhere there was a hole, at such a rate that pumps couldn't keep up," says Mayor Bob Towse.

Statewide property losses may top $365 million, including damage to 140 miles of roads and eight bridges. Agricultural damage is estimated at more than $610 million. An estimated 4% of the state's cropland—900 000 acres—was flooded. In addition, 15 727 people were displaced, 860 businesses closed and nearly 9000 jobs lost.

Missouri: The highest death toll—25—and the greatest property damage—$1.3 billion—of all flooded states were reported here. Statewide, 13 airports have been closed, and 25 000 residents evacuated. Flooding on 1.8 million acres of farmland has caused about $1.7 billion in crop losses.

Heroic efforts apparently saved historic Ste Genevieve, which has been battling rising waters since the start of July.

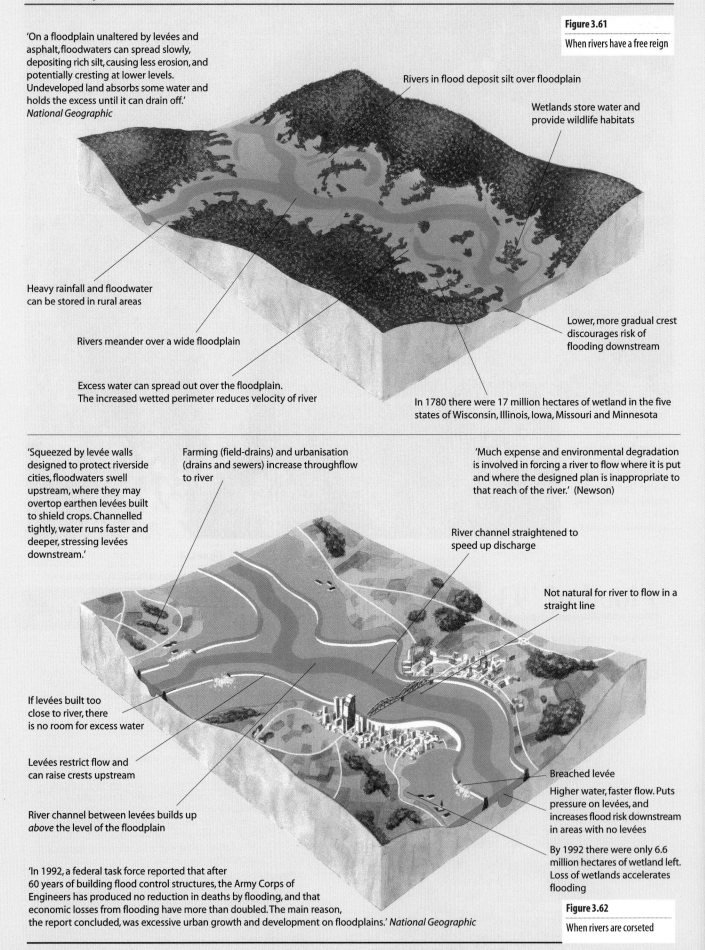

Figure 3.61

When rivers have a free reign

'On a floodplain unaltered by levées and asphalt, floodwaters can spread slowly, depositing rich silt, causing less erosion, and potentially cresting at lower levels. Undeveloped land absorbs some water and holds the excess until it can drain off.' *National Geographic*

Rivers in flood deposit silt over floodplain

Wetlands store water and provide wildlife habitats

Heavy rainfall and floodwater can be stored in rural areas

Rivers meander over a wide floodplain

Excess water can spread out over the floodplain. The increased wetted perimeter reduces velocity of river

Lower, more gradual crest discourages risk of flooding downstream

In 1780 there were 17 million hectares of wetland in the five states of Wisconsin, Illinois, Iowa, Missouri and Minnesota

'Squeezed by levée walls designed to protect riverside cities, floodwaters swell upstream, where they may overtop earthen levées built to shield crops. Channelled tightly, water runs faster and deeper, stressing levées downstream.'

Farming (field-drains) and urbanisation (drains and sewers) increase throughflow to river

'Much expense and environmental degradation is involved in forcing a river to flow where it is put and where the designed plan is inappropriate to that reach of the river.' (Newson)

River channel straightened to speed up discharge

Not natural for river to flow in a straight line

If levées built too close to river, there is no room for excess water

Levées restrict flow and can raise crests upstream

River channel between levées builds up *above* the level of the floodplain

Breached levée

Higher water, faster flow. Puts pressure on levées, and increases flood risk downstream in areas with no levées

By 1992 there were only 6.6 million hectares of wetland left. Loss of wetlands accelerates flooding

'In 1992, a federal task force reported that after 60 years of building flood control structures, the Army Corps of Engineers has produced no reduction in deaths by flooding, and that economic losses from flooding have more than doubled. The main reason, the report concluded, was excessive urban growth and development on floodplains.' *National Geographic*

Figure 3.62

When rivers are corseted

B Flooding and flood control: the Yangtze

The flood hazard

The Yangtze is the third longest river in the world. It rises on the Tibetan Plateau and flows 6380 km to the East China Sea, just north of Shanghai (Figure 3.63). Although always prone to flooding, the frequency and intensity have increased in recent times and the effects have become greater as population and human activity in the basin have grown.

1931 140 000 deaths and an estimated 28 to 40 million people affected.

1954 33 000 deaths.

1981 and **1991** Both floods submerge large areas.

1995 1300 deaths and up to 100 million affected.

1998 Worst flood for over half a century with the river, in places, reaching its highest ever recorded level – a level that was maintained for over two months (Figure 3.66). There were an estimated 3000 deaths, 5.6 million homes washed away, 30 million made homeless, and huge areas of crops destroyed (Figure 3.64).

The Yangtze rose steadily during June and by early July 100 000 soldiers and police were working in teams 24 hours day, seven days a week, strengthening the Jingjiang and other levées (Figure 3.65). On 21 July, an unprecedented 280 mm of rain fell on Wuhan in 24 hours and, the next day, the city experienced the first of several flood peaks which were to pass over two months. By 2 August, Dongting Lake had reached an all-time high (meaning there was nowhere left for surplus water to go) and some cities, like Yueyang, had already been submerged for over 30 days (Figure 3.65). By now, the Chinese authorities admitted, for the first time, that they were allowing the Yangtze to flood many smaller towns and villages in order to protect larger settlements.

1999 By mid-July, 240 deaths, 1.84 million evacuated and 660 000 ha of cropland flooded.

Causes of flooding

- Large areas of the Yangtze basin receive high annual amounts of rain.
- Most of this rain is seasonal and it is usually intensive (summer monsoon – page 239).
- The rains arrive as temperatures rise and there is rapid snow and glacier melt in China's western mountains.
- Large areas of the upper basin were deforested after 1950 (during Chairman Mao's 'Great Leap Forward'). Estimates suggest that one-third of the forest was cleared between then and 1988. Deforestation reduced interception and increased both runoff and soil erosion which, in turn, caused the bed of the river to rise even higher above its floodplain.
- Wetlands, which had previously absorbed extra water, had been drained.

Figure 3.64

Flooding near Wuhan in 1998

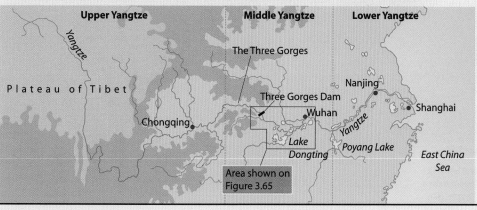

Figure 3.63

The course of the Yangtze River

Flood control

The following account has been adapted from a book published by the China Three Gorges Project Development Corporation in 1999.

'The plains of the middle and lower Yangtze, a region where Chinese industry and agriculture are of major importance, are constantly menaced by the flood hazard. This is mainly because much of the land lies below the level of the river and relies upon 30 000 km of levées for its protection. Even in a normal year, the level of the Yangtze can vary by 30 m (Figure 3.66). Between 185 BC and 1911, more than 200 floods were recorded – about one in every ten years. Since then, the frequency and severity of flooding has increased. In the last 50 years, attempts have been made to control the river against a flood height with:

- a 10- to 20-year frequency by repairing and strengthening existing levées, and

- a 20- to 50-year frequency by **a** diverting water into large nearby lakes such as Lake Dongting and **b** creating flood diversion areas (Figure 3.65). Unfortunately the land to which excess floodwater is directed is also fairly highly populated – e.g. over 400 000 people live in the Jingjiang flood diversion area. (The Chinese have only recently admitted that poor rural areas are sacrificed in order to protect larger, richer urban areas such as Wuhan.)

The region at greatest risk lies between Yichang and Wuhan (Figure 3.65). On the Yangtze's north bank is the Jingjiang levée, a floodbank 180 km long and with a height of 12–16 m. The levée protects 50 000 km² of land, of which two-thirds is high-quality farmland, and where over 7 million people live 7 to 10 m below the flood level of the river. Should the Jingjiang levée be overtopped, there would be heavy economic losses as well as of human life.'

The Three Gorges Dam is, at present, being built upriver from Yichang (Places 82, page

544). One, if not the main, benefit of the dam will be its ability to store water from a flood with a frequency of 1 in 100 years (the last flood of this magnitude was in 1881). Should a 1 in 1000-year-frequency flood occur, any water in excess of the dam's capability could be stored in the existing flood diversion areas. Those in favour of the dam argue that as the river's discharge can be controlled and the floodwater released gradually, then places down river will be safe.

Opponents of the project point out that, once the scheme is completed in 2007, a lake 600 km long and with a water level 70 m higher than that of the present river, will have formed behind the dam (Figure 3.66). The increased water level will drown numerous settlements, forcing 1.3 million people to move home. So to reduce the flood threat in one part of the Yangtze basin, other places will be permanently flooded. (A longer list of pros and cons of the Three Gorges Project is given in Places 82, page 544.)

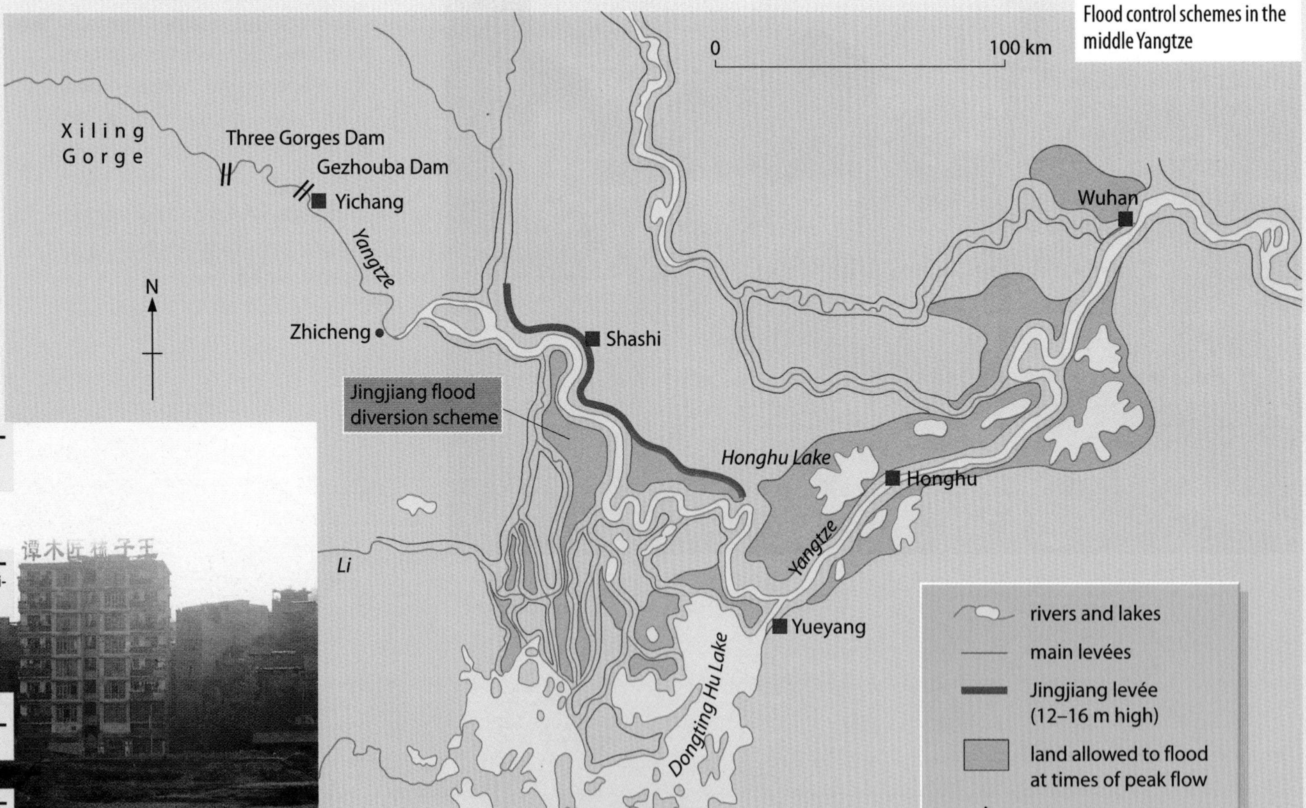

Figure 3.65

Flood control schemes in the middle Yangtze

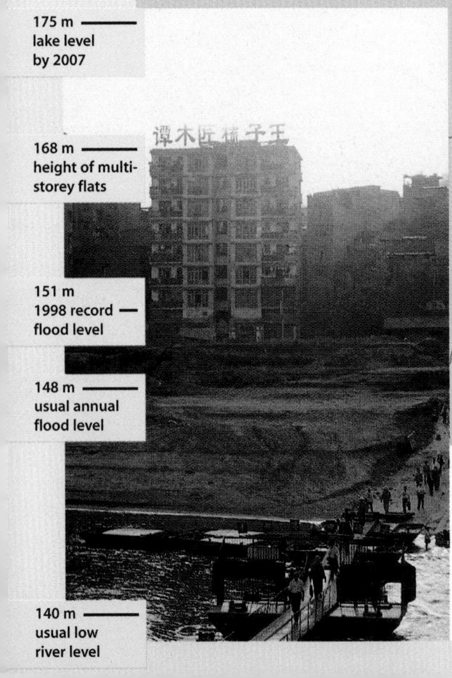

0 —————— 100 km

Xiling Gorge
Three Gorges Dam
Gezhouba Dam
■ Yichang
Yangtze
N
Zhicheng ●
■ Shashi
Jingjiang flood diversion scheme
Honghu Lake
■ Honghu
Li
Yangtze
■ Yueyang
Dongting Hu Lake
Wuhan

rivers and lakes
main levées
Jingjiang levée (12–16 m high)
land allowed to flood at times of peak flow
dams
large cities

175 m lake level by 2007

168 m height of multi-storey flats

151 m 1998 record flood level

148 m usual annual flood level

140 m usual low river level

Figure 3.66

Present low-water and flood levels on the Yangtze, with projected finished lake level

C Managing a river: the Colorado

The River Colorado (not the longest river in the US) is the most dramatic, most used and closer than most basins in the US to using the last drop of available water for man's needs.

National Academy of Sciences, 1968

The Colorado is regarded by modern Americans as a symbol of the development and conflicts in the American West. The Colorado has been used as a source of water for over 2000 years by Native American peoples such as the Hohokam, Anasazi and Mogollon, who have long vanished from the region. During the last 100 years, it has changed from an uncontrolled river passing through a semi-desert, into one supplying the demands of 40 million people. Colorado water is now controlled by 11 major dams in 7 US states and Mexico, produces 120 million kw of electricity and irrigates 800 000 hectares of farmland. Except for the occasional local flood flow, no water from the Colorado has reached the Gulf of California in the last 30 years.

Colorado water users point to major concerns for the next 20 years. These include:

- who has the right to the river water – the Native Americans now demand greater access to water in their tribal lands and irrigated districts
- who should control the river flow
- whether greater concern should be shown for the environment and for endangered species – this may include restricting recreational use still further
- how the water should be allocated
- whether there should be more effective control over the use of the water, especially in urban areas
- who should pay for schemes proposed – should it be state or federal funding?

What is the background to these issues?

1 The characteristics of the river

The river rises in the snowfields of the Rockies, where the annual precipitation may be as much as 2540 mm of snow during the winter – although the mountain states rarely receive more than 500 mm of rainfall a year. Temperatures throughout the

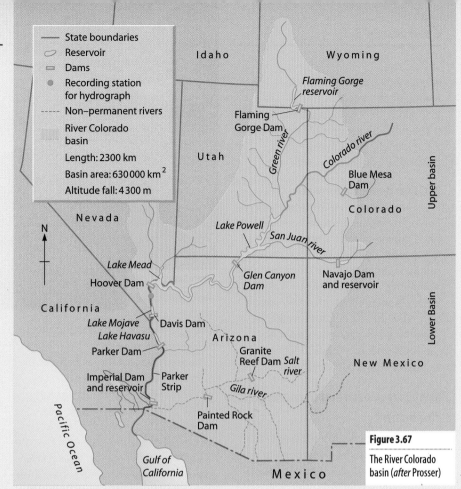

Figure 3.67

The River Colorado basin (*after* Prosser)

Colorado basin are high during the summer, which means that there is a net loss of water through evaporation. Evapotranspiration can be as high as 1830 mm per annum. The snow meltwater is essential to the maintenance of the flow of the river, which is at its highest in April and May. Annual flows vary between a recorded all-time low of 4.9 km^3 (in 1917) and a maximum of 25.3 km^3 (in 1984).

The lower basin had been liable to flood – with disastrous consequences in the period 1905–09, when the river burst its banks, flooding the Imperial Valley agricultural lands and re-creating the Salton Sea.

Some of the highest recorded flows occurred in the El Niño floods of 1983 which affected the Parker region. In 1993 the Gila/Salt river basin flooded following exceptionally high rainfall, again associated with the El Niño Southern Oscillation (ENSO).

The river rarely appears clear except in the mountain reaches. It contains a high sediment load – Colorado means 'red colour'. This is the result of erosion of plateau sandstones during periods of high flow, including snowmelt. The sediment is deposited in the

channel in low flow periods, and also in the dams. The sediment formed rich alluvial soils in the Lower Colorado and on the Mexican border. Before the dams were constructed the sediment helped to form the delta at the head of the Gulf of California.

2 Population growth in the south-west

In the past 50 years, there has been a marked change in population distribution in the USA. The west and south-west have attracted people from the declining industrial areas of the north-east. The south-west is becoming more urban. California's population has increased rapidly, creating a huge demand for water. As much as 70 per cent of the water for San Diego is piped daily from the Colorado. Irrigation districts in California use 75 per cent of the water allocated to the state, and 186 500 ha are irrigated in the

Figure 3.68

Population change in the south-west, 1991–98

	1991	1998 estimate	% change
Arizona	3 811 000	4 668 631	+22
Colorado	3 302 000	3 970 971	+22.25
California	30 316 000	32 666 550	+7.7
Nevada	1 266 000	1 746 898	+37
New Mexico	1 529 000	1 736 931	+13.5
Utah	1 749 000	2 099 758	+20
Wyoming	439 000	480 907	+9.5

US Bureau of Census 1999

Imperial Valley with water from the All American Canal. Towns such as Las Vegas (population 376 906) in Nevada, and Phoenix (1 159 014) and Tucson (449 002) in Arizona, in the Gila/Salt river basin, have grown rapidly; their water needs must be met.

3 Other demands for water

In the upper-basin states of Wyoming, Colorado and Utah, the water is needed for cattle farming and for the irrigation of fodder crops, as well as for human consumption. East of the Great Divide, 283 500 ha of farmland are supplied with water via the Big Thompson and other diversion projects, which also help to meet the demands of the urban populations in Denver, Boulder and Colorado Springs, along the Rocky Mountain Front.

Why is there a need to manage the basin?

The river flows through or between seven states and Mexico. The demands of the states are different, although the greatest perceived need is for hydro-electric power to supply urban areas and farms and to provide irrigation water for agriculture.

River water is used for irrigation throughout the basin. California was the first state to construct canals to carry irrigation water to the farms being developed in the Imperial Valley, in the late 19th century. At the same time, Mormon farmers moved into the remote areas along the river in Utah and into Arizona. In the upper basin, the cattle-ranchers of the interior of Wyoming took water from the river and its tributaries, the Green and the Grand Rivers. As farmers arrived from the east, there was conflict over who had use of the water. The question of water rights is complex, with prior appropriation (first come, first served) being normal,

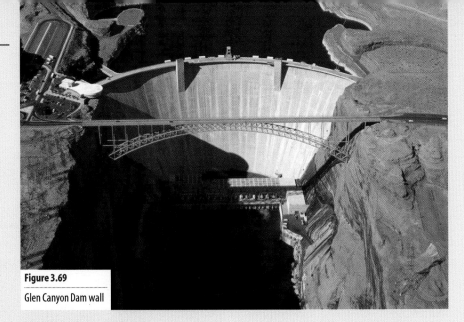

Figure 3.69

Glen Canyon Dam wall

complicated by Native American demands for water to their tribal lands.

The need for an agreement between the states to control the river was evident by the 1920s. Two years of discussion took place and the Colorado Compact was signed in 1922. States that first built a dam and provided irrigation water from this dam may not have their right to the water taken away by another state building upstream at a later date. This system favours California. In 1944, Mexico was granted rights to 2 million acre feet (2465 million m^3) under an extension to the Colorado Compact. This supplies farms in Mexico close to the US border along the Canal Central.

The Great Depression of the early 1930s followed the signing of the Colorado Compact. Many farmers from the Mid-West moved towards the south-west, especially to California. An urgent need for irrigation water developed; but who should pay for it?

In an effort to provide help for the unemployed during the Great Depression, the government set up the Bureau of Reclamation which, together with the Corps of Army Engineers, was to plan and

construct major engineering projects requiring federal (national) funds, throughout the west.

How is the river being managed?

There is no overall control of the river from its source to the Gulf of California. Management is through a number of federal- and state-funded projects, some of which are multi-purpose.

The first of the large schemes was the Hoover Dam. Its primary purpose was to control the flow of the river, with the provision of hydro-electric power (HEP) being seen as a means of paying for the huge costs of construction. This later changed as state and federal governments saw the benefits to be gained from the provision of cheap power in and to a relatively remote area. In the period after 1945 there was a rapid movement to develop the south-west. More great dams were planned on the river and their primary aim was to provide HEP, then flow control and irrigation.

1 Flood control	2 Hydro-electricity	3 Irrigation	4 Urban water supply	5 Recreation
Dams: Fontanelle, Flaming Gorge, Glen Canyon, Hoover, Davis, Imperial	Major stations at Fontanelle, Flaming Gorge, Glen Canyon, Hoover, Davis, Parker, Imperial	Water piped to farmers in Wyoming and Utah from Green River Hoover, Davis, Headgate, Paolo Verde, Imperial, Morelos and Yuma supply farmers in Arizona, Nevada, California and Mexico	Denver, Colorado Springs and towns on Rocky Mountain Front supplied across the Great Divide through Big Thompson Project California – 33% southern California supplied via Colorado River Aqueduct Central Arizona Project (CAP) – canal across the desert to Phoenix and Tucson districts	Glen Canyon, Lake Powell – boating and water sports on the largest water recreation area in the south west Hoover Dam, Lake Mead and Parker Dam, Lake Havasu: water sports Rafting, boating and camping along stretches of the river including Grand Canyon. Numbers are strictly controlled

Figure 3.70

Main projects on the Colorado River

Figure 3.71

The Colorado Compact allocation

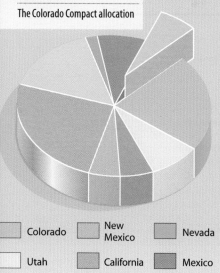

- ☐ Colorado
- ☐ New Mexico
- ☐ Nevada
- ☐ Utah
- ☐ California
- ☐ Mexico
- ☐ Wyoming
- ☐ Arizona
- ☐ Extra

From its confluence with the Green River in southern Utah, the Colorado flows across deeply dissected, level, sandstone plateaus, having cut a series of great gorges, or canyons, in their surfaces. The most famous of these is the Grand Canyon (Figure 7.19). These canyons are valuable sites for storage dams. Glen Canyon Dam (Figure 3.67) is 178 m high and 475 m long, and stores water in Lake Powell which extends for 290 km to the north-east. Hoover Dam, 600 km to the west, was built in Black Canyon. It is 3.5 m higher than Glen Canyon Dam and holds more water, because of the depth of the canyon.

Environmental and conservationist movements believe in the preservation of the wilderness in its original state. There has been successful opposition to the further construction of dams in the Grand Canyon National Park and at Echo Park in the Dinosaur National Monument. The Navajo Indians have the right to the water within their tribal territories alongside the river. Many traditional Native American sites close to the river, including Rainbow Arch in Glen Canyon (Figure 3.72), are in danger from changing levels in Lake Powell. This remote area was one which the conservationists unsuccessfully tried to prevent from being flooded to form Lake Powell in 1962.

South of Hoover Dam the river flows across a lower, wider valley where smaller dams supply irrigation water to farms and small towns close to the river, such as Parker Strip. It was on this reach that the Colorado burst its banks in 1907 and flooded the Salton Sink, part of which is now the valuable agricultural region of Imperial Valley.

Figure 3.72

Rainbow Arch, Utah: the level of Lake Powell is not allowed to reach this traditional Navajo site

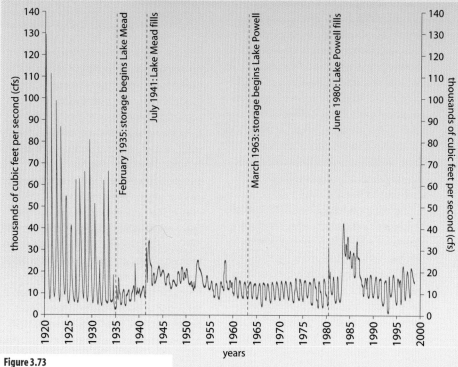

Figure 3.73

Colorado hydrograph: flow below Hoover Dam site 1920–2000, showing changes as a result of control of flows by dams

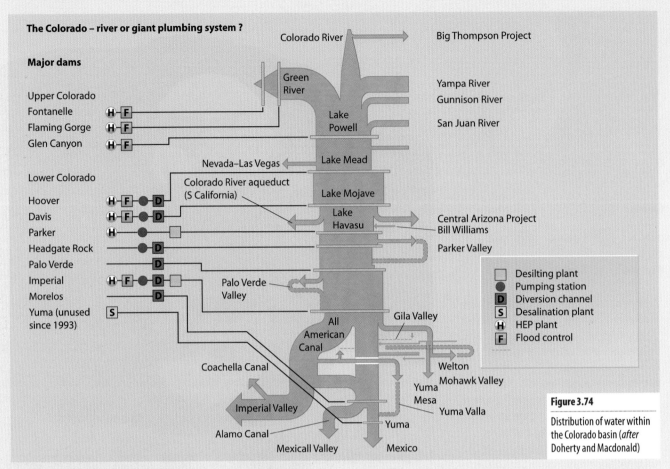

The Colorado – river or giant plumbing system ?

Major dams

Colorado River — Big Thompson Project

Upper Colorado
Fontanelle [H]-[F]
Flaming Gorge [H]-[F]
Glen Canyon [H]-[F]

Green River
Lake Powell
Yampa River
Gunnison River
San Juan River

Lake Mead

Nevada–Las Vegas

Lower Colorado
Colorado River aqueduct (S California)
Lake Mojave

Hoover [H]-[F]-●-[D]
Davis [H]-[F]-●-[D]
Parker [H]-●-[ ]
Headgate Rock ●-[D]
Palo Verde [D]
Imperial [H]-[F]-●-[D]-[ ]
Morelos [D]
Yuma (unused since 1993) [S]

Lake Havasu

Central Arizona Project
Bill Williams
Parker Valley

Palo Verde Valley

Gila Valley

Legend:
[ ] Desilting plant
● Pumping station
[D] Diversion channel
[S] Desalination plant
[H] HEP plant
[F] Flood control

All American Canal

Coachella Canal

Imperial Valley

Alamo Canal

Mexicall Valley

Palo Verde Valley

Welton
Mohawk Valley
Yuma
Mesa
Yuma Valla
Yuma
Mexico

Figure 3.74

Distribution of water within the Colorado basin (*after* Doherty and Macdonald)

Great variations in the seasonal flows along the Colorado system (see Figure 3.73) have been affected by the dams. In winter, the levels of both river and dams are reduced to prepare for the expected high spring runoff from snowmelt and seasonal precipitation. Floods occurred in 1983 because of high and rapid snowmelt and rainfall during the strong El Niño season. This was again expected in the spring of 1998 following the El Niño season, but snowfall was less than expected and the river did not flood. The hydrograph shows that the annual variations in flow are now relatively even. There are daily variations below the power plants related to the demand for electricity from urban areas.

Scientists now recognise that fluctuations of flow are having severe effects on the environment and ecology below the dams. Sediment is being trapped in the lakes and extensive scouring occurs in the channels below the dams. Tributary rivers such as the Paria deposit large amounts of sediment during flood conditions at the confluence with the Colorado. This has not been distributed to sandbanks further downstream.

In June 1992 the *Arizona Republic* reported:

'Wild fluctuations in the Colorado will end as the US Bureau of Reclamation places interim restrictions on the Glen Canyon Dam's hydro-electric plant. The ebb and flow from the dam has been blamed for taking away beaches in the Grand Canyon and for damaging fish populations and other wildlife habitats in and around the Colorado River.'

This was pleasing for environmentalists but power companies opposed it saying that it would mean a rise in prices for consumers, as additional power plants would have to be constructed elsewhere to meet demand.

Following an environmental impact study the Bureau of Reclamation agreed to an artificial flood in March 1996 (see *GAIA Statistical Supplement 1998*). The aim was to redistribute the sediment from the confluence of the Paria and Colorado along the beaches of Marble Canyon in Grand Canyon National Park. Storms had brought down an estimated 1 million tonnes of sediment into the river. The artificial flow redeposited much of this along beaches and sandbars. A second smaller artificial flood took place in November 1997. The 16-day event in 1996 achieved most of its objectives in improving the beaches and providing a better habitat for native fish. The 1997 flood allowed some

reworking of the sediments. The results from these Beach Habitat Building Flows (BHBF) are carefully monitored, and are expected to continue when needed. It is unlikely that there will be any during the 1998/99 water year (Colorado Water Management Plan, US Bureau of Reclamation 1999).

Additional programmes to improve the environment for endangered species such as fish and bird life include restoring and improving wetlands along the river floodlands of the Upper Colorado and Green Rivers.

The Central Arizona Project (CAP)

A 540 km canal has been completed across the Arizona Desert from Parker Dam to the cities of Phoenix and Tucson. It was intended as a domestic water supply for these rapidly growing cities, and to provide irrigation water for the agricultural districts

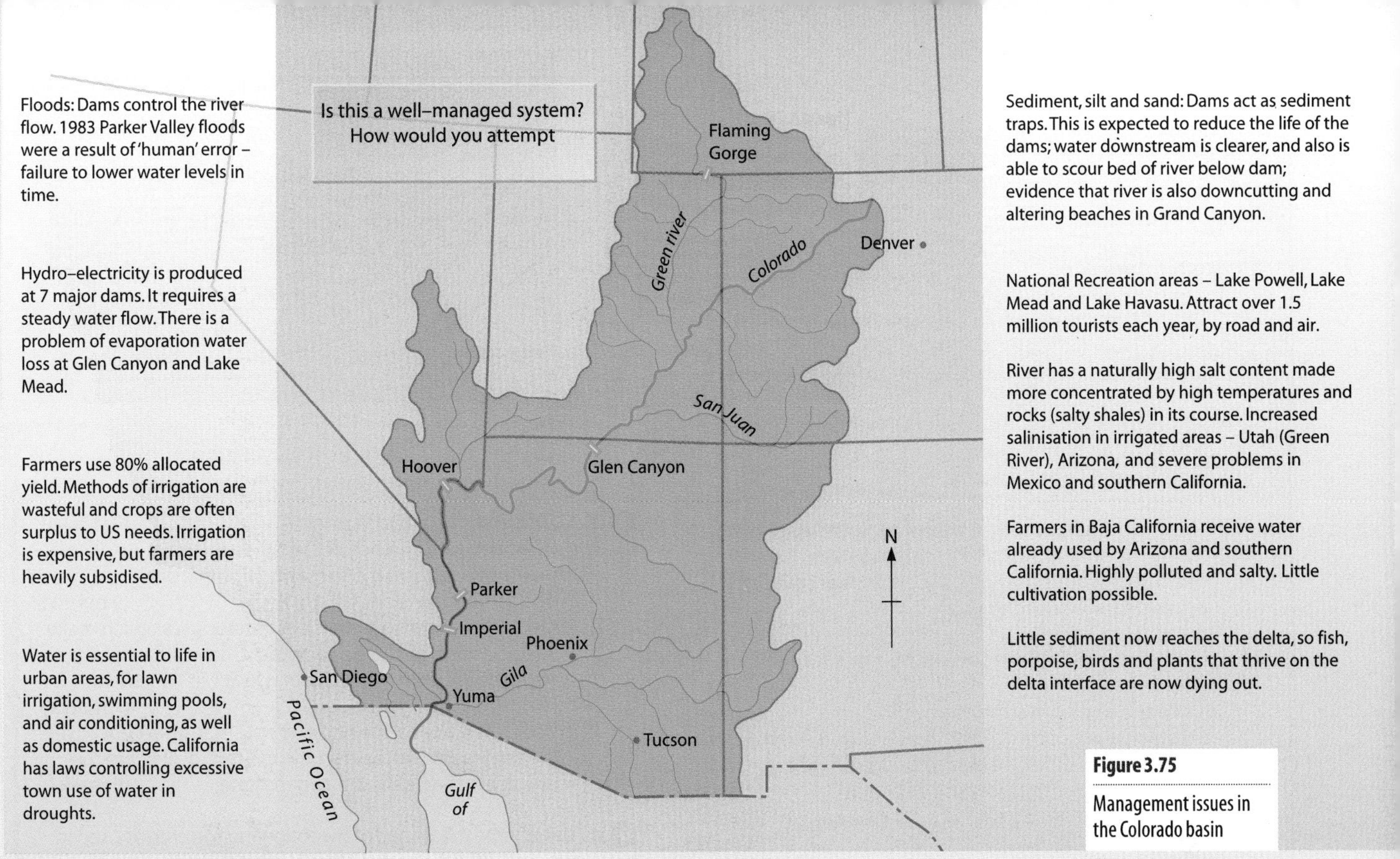
Floods: Dams control the river flow. 1983 Parker Valley floods were a result of 'human' error – failure to lower water levels in time.
Hydro–electricity is produced at 7 major dams. It requires a steady water flow. There is a problem of evaporation water loss at Glen Canyon and Lake Mead.
Farmers use 80% allocated yield. Methods of irrigation are wasteful and crops are often surplus to US needs. Irrigation is expensive, but farmers are heavily subsidised.
Water is essential to life in urban areas, for lawn irrigation, swimming pools, and air conditioning, as well as domestic usage. California has laws controlling excessive town use of water in droughts.
Is this a well–managed system? How would you attempt
Flaming Gorge
Green river
Colorado
Denver
San Juan
Hoover
Glen Canyon
N
Parker
Imperial
Phoenix
Gila
San Diego
Yuma
Tucson
Pacific Ocean
Gulf of
Sediment, silt and sand: Dams act as sediment traps. This is expected to reduce the life of the dams; water downstream is clearer, and also is able to scour bed of river below dam; evidence that river is also downcutting and altering beaches in Grand Canyon.
National Recreation areas – Lake Powell, Lake Mead and Lake Havasu. Attract over 1.5 million tourists each year, by road and air.
River has a naturally high salt content made more concentrated by high temperatures and rocks (salty shales) in its course. Increased salinisation in irrigated areas – Utah (Green River), Arizona, and severe problems in Mexico and southern California.
Farmers in Baja California receive water already used by Arizona and southern California. Highly polluted and salty. Little cultivation possible.
Little sediment now reaches the delta, so fish, porpoise, birds and plants that thrive on the delta interface are now dying out.

Figure 3.75 Management issues in the Colorado basin

along its route. The water is lifted 900 metres to reach the Tucson Plateau. Groundwater from deep aquifers had been the main source of domestic water until the CAP was completed. Problems of water quality including poor colour, taste and smell have forced the Tucson authorities to suspend its use until 2001, in spite of the rapid depletion of the groundwater supplies. There have been fewer problems in Phoenix District, where 25 per cent of drinking water comes from the Central Arizona Project.

How effective is the management of the Colorado?

Baja California

The issue of water supply to the farmers of Baja California was partly resolved after the construction of the Yuma and Morelos Dams. A total of 1848 million m^3 should be supplied through the Compact. Irrigation water from districts close to the Mexican border is returned to the river, but with a high saline content so that farmers in Mexico are disadvantaged. It is calculated that over 9 million tonnes of dissolved salts are washed down the Colorado each year Half the salt comes from natural sources, the remainder from agriculture, industry and mining in the river basin. Downstream users lose $1 billion each year in corroded pipes, damaged crops and increased desalinisation costs.

The Wellton–Mohawk Canal was constructed to take irrigation water to southern Arizona and return surplus saline water to the Colorado below the last US usage point. This meant that the only Colorado water available to Mexican farmers was highly saline. A temporary canal was constructed to take this water to the Gulf of California, and Mexico was promised additional water from US storages as replenishment. An expensive new desalinisation plant was completed at Yuma, on the Mexican border, in 1992. This was operational for less than a year. Following the 1993 Gila basin floods when 6 billion m^3 of water was deposited in the Colorado, the US Bureau of Reclamation took the plant off line 'until the salinity rises again'. It has not been used up to the present (June 1999) because of the costs.

The delta

The Colorado delta is now a wasteland. Until the river flow was controlled, it was home to birds such as the egret; to wild cats, especially the jaguar; to marine species, such as the totoaba, a large fish, and porpoises; and to marine grasses which used to thrive on the interface between land and sea. There are now few communities attempting to make a life in this area.

References

Gregory, K. J. and Walling, D. E. (1973) *Drainage Basin Form and Process*, E. J. Arnold.

Newson, M. (1975) *Flooding and Flood Hazard in the UK*, Oxford University Press.

Newson, M. (1994) *Hydrology and the River Environment*, Oxford University Press.

Prosser, R. (1992) *Natural Systems and Human Responses*, Thomas Nelson.

Small, J. (1999) *A Modern Dictionary of Geography* (4th edition), Hodder and Stoughton.

Weyman, D. (1973) *Runoff Processes and Streamflow Modelling*, Oxford University Press.

Winchester, S. (1998) *The River at the Centre of the World*, Penguin Books.

See page 86 for Websites

Q

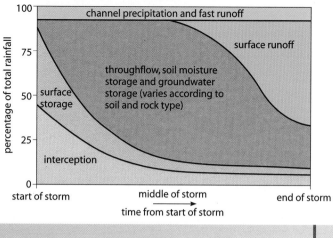

Figure 3.76

The relationship between rainfall and runoff in the course of a typical storm

1 Study Figure 3.76.
  a i What is 'surface storage'? **(2 marks)**
    ii Why does 'interception' decrease during a storm? **(3 marks)**
    iii What happens to 'surface runoff' during the storm? **(4 marks)**
  b What would happen to a river at the following stages:
    i at the start of this storm
    ii at the middle of the storm
    iii at the end of the storm? **(8 marks)**
  c The figure shows the reaction of a vegetated area to a heavy rainstorm. Describe and explain which parts of the model would change if the area were covered in concrete paving and drains. **(8 marks)**

2 a Study Figure 3.3 (page 60) and answer the following questions:
    i What is a 'soil moisture budget'? **(2 marks)**
    ii Explain each of the following terms used in the description of a soil moisture (water) budget: field capacity; water balance; soil moisture utilisation. **(7 marks)**
    iii Why is there no soil moisture deficit shown in Figure 3.3? **(4 marks)**
  b Why would a farmer need to understand the water balance of farmland? **(6 marks)**
  c Why do water companies in Britain depend on *winter* rainfall to maintain reservoirs? **(6 marks)**

3 a i What is a 'storm hydrograph'? **(3 marks)**
    ii What is meant by each of the following terms used in relation to a storm hydrograph: lag time; peak discharge; recession (falling) limb? **(9 marks)**
  b i Identify two drainage basin characteristics which make a river react quickly to a rainstorm (have a 'flashy' regime). For **each** one explain why it has this effect. **(7 marks)**
    ii With reference to specific example/s, suggest how river management strategies may be used to alleviate the problems caused by a 'flashy' regime. **(6 marks)**

4 a Identify the meanings of the following **four** terms used in the study of river basins: inputs; outputs; stores; transfers. **(4 marks)**
  b Explain how and why water moves through a drainage basin. **(12 marks)**
  c Why do many rivers continue to flow for a long time after the end of a rainstorm? **(9 marks)**

AS

5 a Study Figure 3.4a and b (page 60).
    i Describe the **temperature** and the **precipitation** regimes for each of the two stations. **(6 marks)**
    ii Using Figure 3.4b, explain the pattern of soil moisture balance through this year at Dalhart. **(6 marks)**
  b Describe the probable regime of a river in the area near Salisbury (Figure 3.4a). Explain your answer. **(6 marks)**
  c Suggest how a farmer near Salisbury will use a different moisture management system to a farmer near Dalhart. **(7 marks)**

6 a Using annotated diagram/s to help your answer, illustrate the components of a **storm hydrograph**. **(5 marks)**
  b Explain how it is possible to measure the discharge of a stream in the field and how the results collected will be processed. **(10 marks)**
  c Why do lag times differ on the same stream at different times? **(10 marks)**

7 When a housing estate is built on the rural/urban fringe, pre-existing drainage patterns are changed and river systems respond in a different way to storm events.
  a Study of such changes must start before building to establish a 'baseline' for change. Briefly describe **one** technique you could use to measure the discharge of a stream in a rural catchment. **(5 marks)**
  b Describe and account for **two** changes to discharge which may occur once the housing estate is built. **(10 marks)**
  c Describe **two** problems which could occur in the area due to the altered discharge pattern. **(10 marks)**

8 Study the following data associated with Figure 3.13 (page 66).
  a Copy out the table below and complete it by adding data to the information for drainage basin B. **(13 marks)**
  b Compare the two drainage basins in terms of:
    i bifurcation ratios
    ii drainage density.
    In your answer you are comparing drainage basins, not just mathematical answers. **(7 marks)**
  c Using annotated diagram/s only, show how the storm hydrographs at X and Y will differ from the same storm. **(5 marks)**

Basin	A	B	Basin	A	B
Number of streams			**Total length of streams (km)**		
N1 (first order)	26	10	L1 (first order)	12	11.25
N2 (second order)	6	4	L2 (second order)	7	7.5
N3 (third order)	2	☐	L3 (third order)	2.4	☐
N4 (fourth order)	1	☐	L4 (fourth order)	1.25	☐
Total number of streams (ΣN)	35	☐	**Total stream length (ΣL)**	22.65	☐
Bifurcation ratios			**Mean stream length**		
$\frac{N1}{N2}$	4.33	2.5	L1 (first order)	0.46	☐
$\frac{N2}{N3}$	3	☐	L2 (second order)	1.17	☐
$\frac{N3}{N4}$	2	☐	L3 (third order)	1.2	2
			L4 (fourth order)	1.25	2
			Total area of drainage basin	12.5	
Average bifurcation ratio	3.11	☐	**Drainage density**	1.81	☐

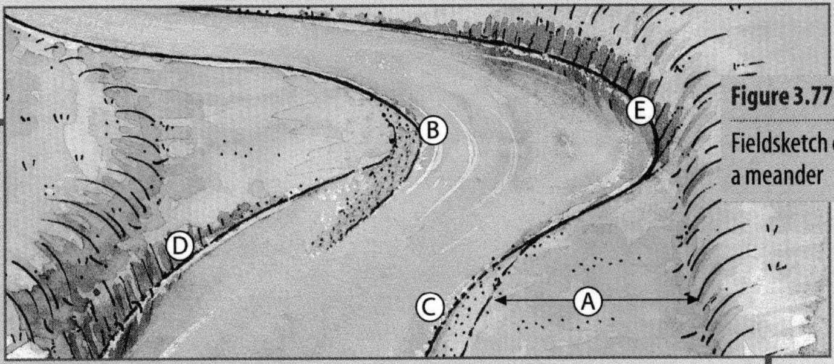

Figure 3.77

Fieldsketch of a meander

9 a i What is the 'wetted perimeter' of a river? **(2 marks)**
   ii How can the cross-sectional area of a river be measured? **(6 marks)**
   iii How is the 'velocity' of a river calculated? **(4 marks)**
   b What is the 'load of a river'? **(5 marks)**
   c Explain the factors which affect the size of particles that are **eroded, transported,** and **deposited** by a river. **(8 marks)**

10 a i Study Figure 3.27 (page 74). Describe the river bed shown in the photograph. **(3 marks)**
   ii Where have the loose boulders shown beside the river come from? **(4 marks)**
   iii How does a river erode a river bed such as the one in the photograph? **(6 marks)**
   b Explain **two** ways in which you would know that loose rocks found on a field trip had been worn away by a river. **(6 marks)**
   c With the aid of diagrams of a waterfall, show how it is being changed over time by river processes. **(6 marks)**

11 Use Places 9 (page 64) to answer the following questions.
   a Study Figure 3.9.
   i What was the monthly rainfall in January? **(1 mark)**
   ii What was the monthly rainfall in September? **(1 mark)**
   iii What was the difference in **each** case between monthly rainfall and total surface runoff? **(4 marks)**
   iv Explain the difference you have identified in iii. **(4 marks)**

   b Study Figure 3.10.
   i Using specific examples from the graph, describe and explain the relationship between rainfall and peak discharge as shown in the data. **(8 marks)**
   ii Choose **two** groups of people who would be interested in using these data. For each group explain how the data could affect their actions. **(7 marks)**

12 a i Study the diagram of a meander (Figure 3.77) and identify the location of the following landforms: inside of the bend; outside of the bend; flood plain; slip-off slope; river cliff. **(5 marks)**
   ii Describe the features of the **cross-section** shape of a meander. **(6 marks)**
   b Choose **one** of the following features of a river: waterfall; cascade; rapids. Making good use of sketches / diagrams, describe the features of your chosen landform and explain how it is eroded by a river. **(8 marks)**
   c i How does a meandering river form an oxbow lake? **(4 marks)**
   ii How could the formation of an oxbow lake lead to management problems on the floodplain of a river? **(4 marks)**

AS

13 a Using annotated diagram/s **only,** illustrate the variation in velocity in the cross-section of a river. **(5 marks)**
   b i Describe the processes by which the load of a river is transported. **(8 marks)**
   ii What factors affect the size of the particles eroded, transported and deposited by a river? **(12 marks)**

14 a i Describe the characteristic features of a **dendritic drainage pattern.** **(3 marks)**
   ii Making good use of **annotated** diagrams, explain the development of a **trellis drainage pattern.** **(8 marks)**
   b i Study Figure 3.53 (page 85). Describe the valley shape you would see if you were walking from the River Wansbeck to the Hart Burn. **(2 marks)**
   ii Explain how the present drainage pattern evolved from the former drainage pattern. **(6 marks)**
   c Choose and name an example of a drainage pattern other than a trellis pattern. Describe it and explain how it has been formed. **(6 marks)**

15 With reference to a complete field investigation which you have carried out concerning the interaction between people and the environment in relation to natural processes in a river basin:
   a Identify how you planned the data you would need to collect. **(5 marks)**
   b Identify the methods and equipment you used to collect and record the data, and explain how you tried to ensure that your data was valid. **(10 marks)**
   c State the results of your investigation and suggest how they contributed to your understanding of the possible management issues identified. **(10 marks)**

16 a Describe and suggest reasons for the **cross-section** shape of a river:
   i near the source of the river
   ii close to the mouth of the river. **(12 marks)**
   b Identify and suggest reasons for **two** variations in the **long profile** of a river. **(13 marks)**

17 a What is the difference between **general base level** and **local base level?** **(8 marks)**
   b i Explain what happens to base level in a river system if sea-level falls. **(6 marks)**
   ii Choose **two** landforms formed in a river valley by a change in base level. Identify the direction of change involved and describe and explain the formation of **each** landform. **(11 marks)**

18 Use Case Study 3A (pages 87–90) to complete the following.
   a Use Figures 3.55 and 3.56 to suggest reasons why most floodwater on the lower Mississippi is from the Ohio system rather than from the Missouri/Mississippi system. **(6 marks)**
   b Using Figures 3.59 and 3.60, describe the 1993 flood at St Louis. In your answer you should refer to:
   ■ the weather contributing to the flood
   ■ the comparative height of the floodwater
   ■ the damage caused in the St Louis area by flooding. **(12 marks)**
   c Explain the arguments for and against human interference in river flow to try to control flooding. **(7 marks)**

A2

19 a Study Figure 3.78a showing the Nile regime before the building of the Aswan High Dam and the management of the area around Musha at the time.
   i Describe the river regime. **(3 marks)**
   ii What effect did this river regime have on the floodplain of the river? **(2 marks)**
   iii Describe **two** ways in which it affected the lives of the people. **(4 marks)**

b Study the regime of the Nile after the building of the Aswan Dam.
   i Describe the changes in the river regime. **(4 marks)**
   ii Identify and explain **two** ways in which the farming of Musha had changed by 1990. **(6 marks)**
   iii Describe and explain two other ways in which the lives of the people changed in Musha. **(6 marks)**

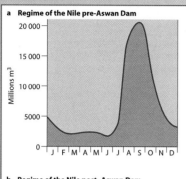

a **Regime of the Nile pre-Aswan Dam**
Millions m³
20 000 / 15 000 / 10 000 / 5000 / 0
J F M A M J J A S O N D

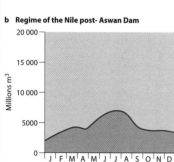

b **Regime of the Nile post-Aswan Dam**
Millions m³
20 000 / 15 000 / 10 000 / 5000 / 0
J F M A M J J A S O N D

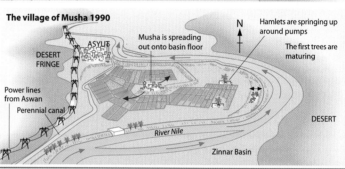

The village of Musha 1960

N
Wide basin levée protected village from flooding since 1902 (called sahel)
ASYUT
DESERT FRINGE
DESERT
Basin flooded annually until 1964
MUSHA
Causeway built in 1948
The farmers all lived together in the village of Musha
Previously boats were only means of communication during 3 months of flood
Traditional site of monasteries, quarries, cemeteries and villages
Flat, treeless, basin
Sluice gates to allow annual inundation
River Nile

The village of Musha 1990

N
Hamlets are springing up around pumps
ASYUT
Musha is spreading out onto basin floor
DESERT FRINGE
The first trees are maturing
Power lines from Aswan
Perennial canal
DESERT
River Nile
Zinnar Basin

Figure 3.78

Changes in river management on the River Nile

20 a Study the photograph and map in Figures 3.63 and 3.64 (page 91). Using material from the photograph, suggest **two** ways in which river flooding can affect the people in a flooded area. **(6 marks)**

b i What is meant by the term 'river management'? **(3 marks)**
   ii Explain, using examples of at least **two** different managed river basins, why it is important for people to 'manage' rivers. **(6 marks)**
   iii With reference to **one named** large drainage basin that you have studied, evaluate the success of schemes devised to manage the problems faced in that environment. **(10 marks)**

21 a Study Figure 3.79.
   i Which region experienced the greatest demand for water in 1980?
   ii Which region is expected to experience the greatest *growth* in competition for water between 1980 and 2100?
   iii Describe the pattern of competition for water as shown in Figure 3.79. **(5 marks)**

b i Identify **two** groups of people who may oppose a water management scheme such as the one in **a** above (Figure 3.78). Suggest why they may oppose it. **(10 marks)**
   ii Using examples of water development schemes which you have studied, suggest ways in which it is possible for objections to be overcome. **(10 marks)**

22 a Why does a river deposit material? **(5 marks)**
b i What is a levée? Why do they occur on natural rivers?
   ii What effect do levées have on the river and its tributaries? **(10 marks)**
c Suggest and explain **two** uses of levées by people. **(10 marks)**

23 a Discuss the role of people in the flow of water in a river in the British Isles. **(13 marks)**
b Suggest how this role may affect the landforms of the river valley. **(12 marks)**

Figure 3.79

Projected change in water competition, 1980–2100

The freshwater available to the areas shown is limited and therefore as populations increase there is an increasing competition for water. The worst problems will be seen in Asia and Africa where the numbers of individuals to be fed on each cubic kilometre of water will increase from the present indices of 200 and 120 to 380 and 620.

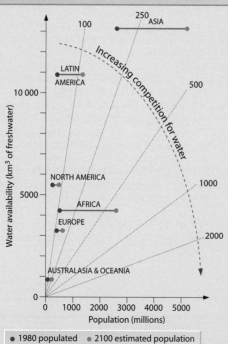

Increasing competition for water
250 / 100 / ASIA
LATIN AMERICA
10 000
500
NORTH AMERICA
1000
AFRICA
5000
EUROPE
2000
AUSTRALASIA & OCEANIA
0
0 1000 2000 3000 4000 5000
Population (millions)
Water availability (km³ of freshwater)
● 1980 populated   ● 2100 estimated population

AS

A2

All of the work done for any subject provides opportunities to learn a whole series of skills and techniques that will be of value to you in years to come. Employers are interested in the ability of their workforces to **communicate** in writing and orally, to be able to use **number skills**, and to be able to use **information technology**. They also like their employees to be able to get on with others, to work in teams, and to be able to learn from their experiences. All of these are included in the Key Skills which are available for accreditation as modules to show exactly what skills an applicant has – as well as the abilities shown through achievement in a subject.

The **Key Skills** are:

1 Application of Number
2 Communication
3 Information Technology
4 Improving Own Learning and Performance
5 Working with Others
6 Problem Solving.

The Key Skills of Application of Number (1), Communication (2), and Information Technology (3) together form the **Key Skills Certificate** which identifies the Level achieved in each Key Skill. Examples of your work are identified and tracked as a 'portfolio' to show the evidence needed for the Key Skills Certificate. The other three Key Skills (4–6) can also be accredited, but as individual units. One piece of work can provide evidence for more than one Key Skill.

At Level 3 these skills are described as needing to be part of a 'substantial activity'. The activities on this page can be used to provide evidence for the Key Skill of **Communication** to Level 3. On page 259 you will find activities to cover the Key Skill of Information Technology, and on page 387 activities to cover Application of Number, also to Level 3.

These activities are a particular way of carrying out tasks that would form a part of a Geography course and its assignments. As you carry out such activities to gather the evidence, your teachers and tutors are expected to monitor the work (as an examination board would require of any coursework to ensure that the work is done by the candidate). Their testimony will provide one source of evidence that *you* did the work. The monitoring also has a teaching role, helping you to strengthen plans and activities to create a final result that achieves all of the outcomes planned for the activities.

The activities suggested below can be used as individual tasks to support Key Skill development; as a completed piece of evidence towards a Key Skills portfolio; or as a learning vehicle to consider the Geography content of the material.

Communication – Level 3

The main areas of communication competence include the understanding of oral and written communications both to gather and to provide information. The specification requires you to take an active part in meetings, to be able to carry out a formal presentation, and to be able to read and write a range of document types, including those that use images.

The tasks suggested below allow you to work in a way that will, in total, present a 'package' of required competences.

Task One

Background

The active processes in accumulation of water from a drainage basin and within flowing water in a river are complex. You can improve your understanding of this through a **group discussion** of the hydrology of a drainage basin, including the processes of flow in river channels.

You will initially need to research the required information – either individually or as a part of an agreed group activity (if documented this could provide evidence for 'Improving Own Learning and Performance' or 'Working with Others'). You would develop evidence of this from **written notes**, selected from a **range of sources** (which could be from more than one type of information source, e.g. books, CD-ROMs, the Internet, etc.) and **synthesise** the resulting material to build up an understanding to inform the discussion. The evidence for the discussion would come either from a video or from an observer checklist.

This activity can either result from a teacher-directed series of tasks, or the group could initially direct research by individuals in particular directions (e.g. surface type, slope type or precipitation conditions) to ensure a range of information being researched.

Each student should find out about, and make notes on, the ways in which water entering a drainage basin fills stores and eventually enters the channels as running water in areas with a range of surface characteristics. Use this information as the basis for a **group discussion** on the drainage basin hydrological cycle.

Analysis of task

A systems approach to **drainage basin hydrology** identifying:
- the **inputs** of water into the system
- **transfers** and **stores** within the system, and
- **outputs** from the system

and …
the patterns of flow of water in a river channel including the variation in **flow rates**:
- over time (storm hydrograph) and
- in space (flow variations within the channel).

Task Two

Background

Landforms within a river basin resulting from the action of running water are well known and well documented. As rivers are natural systems that respond to a wide range of factors, such landforms in ordinary rivers seldom mirror their theoretical or 'textbook' examples exactly.

Choose an accessible river valley feature (channel, long profile or cross profile), and **collect information** about the theory from **information sources** and from detailed **fieldwork**. Plan and develop an **oral presentation** (of perhaps 10–15 minutes) to allow you to communicate your findings to others, and complete a **written document**, in report style (of four or more A4 sides, about 1500–2000 words), as if it were to be included in a geographical publication. It should include an executive summary, footnotes or terminal references, written text, tabulated data and images, as relevant.

Analysis of task

Research into the morphology of a river feature (any physical feature of a river valley – channel feature or slope feature) identifying:
- from secondary sources the theoretical or idealised 'textbook example' feature and its causes
- from fieldwork a real example of a feature (its size, scale and – where relevant – its composition).

Finally, compare the two, giving suggestions of the causes of the features noted in your comparison.

Glaciation

'Great God! this is an awful place.'

The South Pole, Robert Falcon Scott, *Journal*, 1912

Ice ages

It appears that roughly every 200–250 million years in the Earth's history there have been major periods of ice activity (Figure 4.1). Of these, the most recent and significant occurred during the Pleistocene epoch of the Quaternary period (Figure 1.1). In the 2 million years since the onset of the Quaternary, there have been fluctuations in global temperature of between 5° and 6°C which have led to cold phases (**glacials**) and warm phases (**interglacials**). Until recently it was believed that there had been four main glacials, but in the last decade evidence obtained from cores taken from ocean floor deposits and polar icecaps has led glaciologists to claim that there have been more than 20 (Figure 4.2).

When the ice reached its maximum extent, it is estimated that it covered 30 per cent of the Earth's land surface (compared with some 10 per cent today). However, its effect was not only felt in polar latitudes and mountainous areas, for each time the ice advanced there was a change in the global climatic belts (Figure 4.3). Only 18 000 years ago, at the time of the maximum advance within the last glacial, ice covered Britain as far south as the Bristol Channel, the Midlands and Norfolk. The southern part of Britain experienced **tundra conditions** (page 333), as did most of France.

Climatic change

Although it is accepted that climatic fluctuations occur on a variety of timescales, as yet there is no single explanation for the onset of major ice ages or for fluctuations within each ice age. The most feasible of theories to date is that of Milutin Milankovitch, mathematician/astronomer. Between 1912 and 1941, he performed exhaustive calculations which show that the Earth's position in space, its tilt and its orbit around the Sun all change. These changes, he claimed, affect incoming radiation from the Sun and produce three main cycles of 95, 42 and 21 thousand years (Figure 4.6). His theory, and the timescale of each cycle, has been given considerable support by evidence gained, since the mid-1970s, from ocean floor cores. As yet, although the relationship appears to have been established, it is not known precisely how these celestial cycles relate to climatic change.

Figure 4.1

A chronology of ice ages (in **bold**)

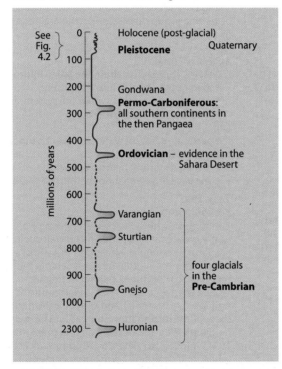

Figure 4.2

Generalised trends in mean global temperatures during the past 1 million years

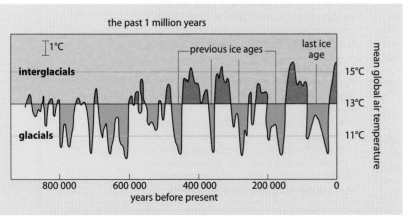

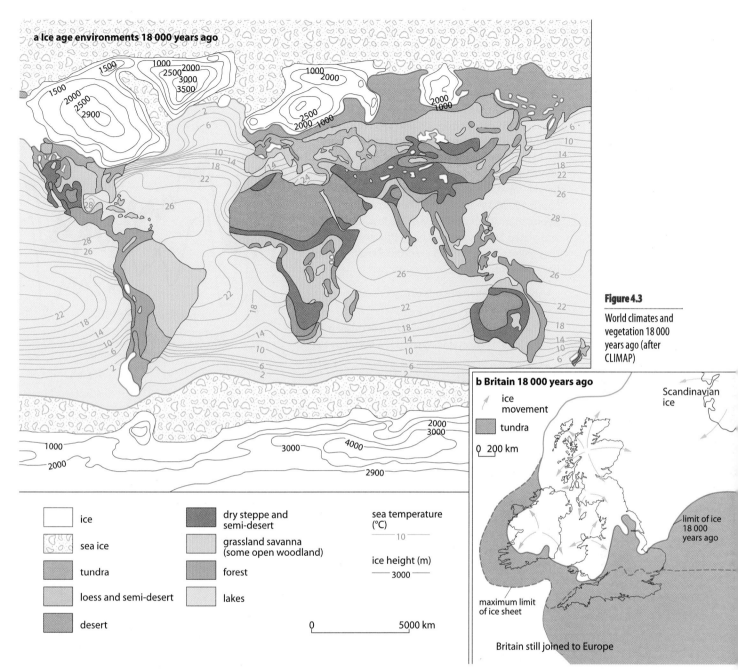

a Ice age environments 18 000 years ago

ice

sea ice

tundra

loess and semi-desert

desert

dry steppe and semi-desert

grassland savanna (some open woodland)

forest

lakes

sea temperature (°C)
— 10 —

ice height (m)
— 3000 —

0   5000 km

Figure 4.3

World climates and vegetation 18 000 years ago (after CLIMAP)

b Britain 18 000 years ago

ice movement

tundra

0   200 km

Scandinavian ice

limit of ice 18 000 years ago

maximum limit of ice sheet

Britain still joined to Europe

Other suggestions have been made as to the causes of ice ages, and it is possible that one or more of these may, at some future date, be shown to be significant (Places 14):

- Variations in sunspot activity may increase or decrease the amount of radiation received by the Earth.
- Injections of volcanic dust into the atmosphere can reflect and absorb radiation from the Sun (page 207 and Figure 1.48).
- Changes in atmospheric carbon dioxide gas could accentuate the greenhouse effect (Case Study 9B). Initially extra CO_2 traps heat in the atmosphere, possibly raising world temperatures by an estimated 3°C. In time, some of this CO_2 will be absorbed by the seas, reducing the amount remaining in the atmosphere and causing a drop in world temperatures and the onset of another ice age (Figure 4.5).
- The movement of plates – either into colder latitudes or at constructive margins, where there is an increase in altitude – could lead to an overall drop in world land temperatures.
- Changes in ocean currents (page 211) or jet streams (page 227).

Antarctica, 1988

The first results of a five-year drilling experiment by the Russians in Antarctica were announced in January 1988. It took from 1980 to 1985 to drill and carefully extract an ice core extending downwards for 2 km. By dating the rings, which show the annual accumulation of snow (Figure 4.4), it has been estimated that ice from the bottom of the core is 160 000 years old. This should provide information about the whole of the last interglacial and glacial. The researchers hope that eventually they will reach ice 500 000 years old. Results to date have indicated that:

- there have been periods of considerable volcanic activity

- the Earth may well wobble on its axis causing the 21 000 year cycle suggested by Milankovitch

- there have been earlier periods when CO_2 built up in the atmosphere, with no plausible suggestions as to why this should occur. It does appear though that the last glacial began at a time when the content of CO_2 was very low.

- There is a very close link between temperature change and the content of CO_2 in the atmosphere (Figure 4.5).

Evidence seems to support the view that the Earth is nearing the end of a brief interglacial and that, following a period of colder, drier climatic conditions (which would seriously affect food crop production), the next glacial is likely to start within the next few thousands years. This might be temporarily delayed by the increase in temperatures resulting from the extra CO_2 content of the atmosphere.

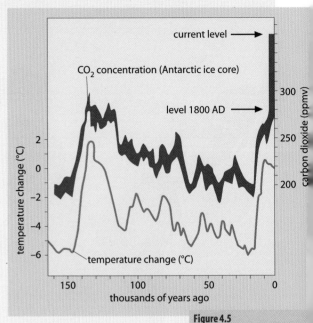

Figure 4.5

Atmospheric CO_2 concentration and temperature change

Figure 4.4

Dirt bands (englacial debris) in an Icelandic glacier: the amount of ice between each dirt band represents one year's accumulation of snow

Greenland, 1998

Two projects conducted from 1989 to 1993 collected parallel cores of ice from two places 30 km apart in the central part of the Greenland ice sheet. Each core is over 3 km deep and has been shown to extend back 110 000 years. During that period snowfall averaged 15–20 cm a year. At the same time as the snow was being compressed into ice (page 105), volcanic dust, wind-blown dust, sea salt, gases and chemicals which were present in the atmosphere, were trapped within the ice. The gases included two types of oxygen isotope, 016 and 018 (page 248). The ratio between these two isotopes changes as the proportion of global water bound up in the ice changes (the amount of 018 in the atmosphere increases as air temperature falls, and decreases as air temperature rises). The changing ratios from the Greenland cores show (even more clearly than the one from Vostok in Antarctica) both short-term and long-term changes in temperature. Indications are that at one time temperatures were 20°C lower than they are at present and that large, rapid global change is more the norm for the Earth's climate than is stability. Indeed research in other scientific fields and in different locations is tending to support the idea that temperatures can change significantly within a decade, and that climate can change dramatically within just a few years rather than over a period of hundreds of years, as thought previously (Case Study 9B).

a the 95 000 year stretch

The Earth's orbit stretches from being nearly circular to an elliptical shape and back again in a cycle of about 95 000 years. During the Quaternary, the major glacial–interglacial cycle was almost 100 000 years. Glacials occur when the orbit is almost circular and interglacials when it is a more elliptical shape.

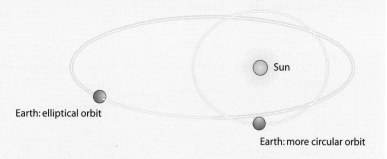

b the 42 000 year tilt

Although the tropics are set at 23.5°N and 23.5°S to equate with the angle of the Earth's tilt, in reality the Earth's axis varies from its plane of orbit by between 21.5° and 24.5°. When the tilt increases, summers will become hotter and winters colder, leading to conditions favouring interglacials.

a = 21.5°

b = 24.5°

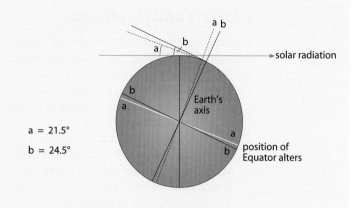

c the 21 000 year wobble

As the Earth slowly wobbles in space, its axis describes a circle once in every 21 000 years.
1 At present, the orbit places the Earth closest to the Sun in the northern hemisphere's winter and furthest away in summer. This tends to make winters mild and summers cool. These are ideal conditions for glacials to develop.
2 The position was in reverse 12 000 years ago, and this has contributed to our present warm 'interglacial'.

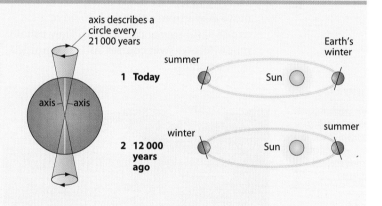

Figure 4.6

The orbital forcing mechanisms of Milankovitch's climatic change theory

Snow accumulation and ice formation

As the climate gets colder, more precipitation is likely to be in the form of snow in winter and there is less time for that snow to melt in the shorter summer. If the climate continues to deteriorate, snow will lie throughout the year forming a permanent **snow line** – the level above which snow will lie all year. In the northern hemisphere, the snow line is at a lower altitude on north-facing slopes, as these receive less insolation than south-facing slopes. The snow line is also lower nearer to the poles and higher nearer to the Equator: it is at sea-level in northern Greenland; at about 1500 m in southern Norway; at 3000 m in the Alps; and at 6000 m at the Equator. It is estimated that the Cairngorms in Scotland would be snow-covered all year had they been 200 m higher.

When snowflakes fall they have an open, feathery appearance, trap air and have a low density. Where snow collects in hollows, it becomes compressed by the weight of subsequent falls and gradually develops into a more compact, dense form called **firn** or **névé** . Firn is compacted snow which has experienced one winter's freezing and survived a summer's melting. It is composed of randomly oriented ice crystals separated by air passages. In temperate latitudes, such as in the Alps, summer meltwater percolates into the firn only to freeze either at night or during the following winter, thus forming an increasingly dense mass. Air is progressively squeezed out and after 20–40 years the firn will have turned into solid ice. This same process may take several hundred years in Antarctica and Greenland where there is no summer melting. Once ice has formed, it may begin to flow downhill, under the force of gravity, as a **glacier**.

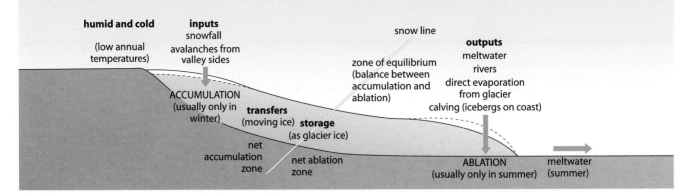

Glaciers and ice masses

Glaciers may be classified (Framework 7, page 167) according to size and shape – characteristics that are relatively easy to identify by field observation.

1 **Niche glaciers** are very small and occupy hollows and gulleys on north-facing slopes in the northern hemisphere.
2 **Corrie** or **cirque glaciers**, although larger than niche glaciers, are small masses of ice occupying armchair-shaped hollows in mountains (Figure 4.14). They often overspill from their hollows to feed valley glaciers.
3 **Valley glaciers** are larger masses of ice which move down from either an icefield or a cirque basin source (Figure 4.8). They usually follow former river courses and are bounded by steep sides.

4 **Piedmont glaciers** are formed when valley glaciers extend onto lowland areas, spread out and merge.
5 **Icecaps** and **ice sheets** are huge areas of ice which spread outwards from central domes. Apart from exposed summits of high mountains, called **nunataks**, the whole landscape is buried. Ice sheets, which once covered much of northern Europe and North America (Figure 4.3) are now confined to Antarctica (86 per cent of present-day world ice) and Greenland (11 per cent).

Glacial systems and budgets

A glacier behaves as a system (Framework 3, page 45), with inputs, stores, transfers and outputs (Figure 4.7). Inputs are derived from snow falling directly onto the glacier or from **avalanches** along valley sides (Case Study 4). The glacier itself is water in storage. Outputs from the glacier system include evaporation, **calving** (the formation of icebergs), and meltwater streams which flow either on top of or under the ice during the summer months.

The upper part of the glacier, where inputs exceed outputs, is known as the **zone of accumulation**; the lower part, where outputs exceed inputs, is called the **zone of ablation**. The **zone of equilibrium** is where the rates of accumulation and ablation are equal, and it corresponds with the snow line (Figures 4.7 and 4.8).

The **glacier budget**, or **net balance**, is the difference between the total accumulation and the total ablation for one year. In temperate glaciers (page 108), there is likely to be a negative balance in summer when ablation exceeds accumulation, and a positive balance in winter when the reverse occurs (Figure 4.9). If the summer and winter budgets cancel each other out, the glacier appears to be stationary. It appears stationary because the **snout** – i.e. the end of the glacier – is neither advancing nor retreating, although ice from the accumulation zone is still moving down-valley into the ablation zone. Glaciers are sensitive indicators of climatic change, both short-term and long-term.

Figure 4.9

The glacial budget or net balance

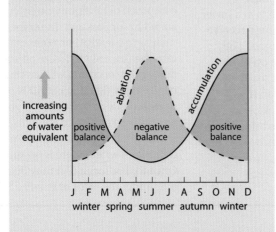

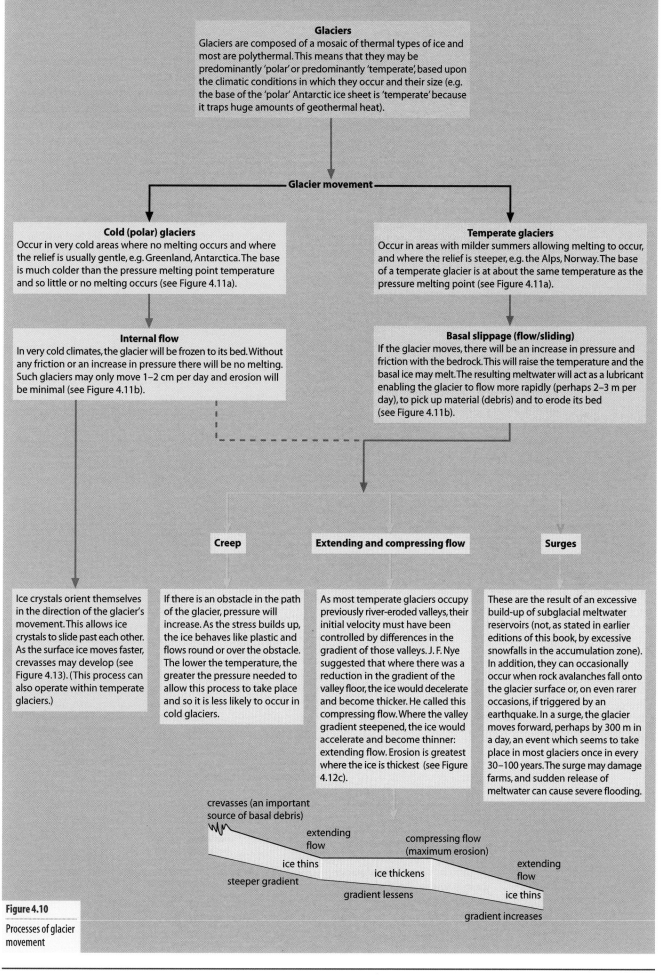

Glaciers

Glaciers are composed of a mosaic of thermal types of ice and most are polythermal. This means that they may be predominantly 'polar' or predominantly 'temperate', based upon the climatic conditions in which they occur and their size (e.g. the base of the 'polar' Antarctic ice sheet is 'temperate' because it traps huge amounts of geothermal heat).

Glacier movement

Cold (polar) glaciers

Occur in very cold areas where no melting occurs and where the relief is usually gentle, e.g. Greenland, Antarctica. The base is much colder than the pressure melting point temperature and so little or no melting occurs (see Figure 4.11a).

Temperate glaciers

Occur in areas with milder summers allowing melting to occur, and where the relief is steeper, e.g. the Alps, Norway. The base of a temperate glacier is at about the same temperature as the pressure melting point (see Figure 4.11a).

Internal flow

In very cold climates, the glacier will be frozen to its bed. Without any friction or an increase in pressure there will be no melting. Such glaciers may only move 1–2 cm per day and erosion will be minimal (see Figure 4.11b).

Basal slippage (flow/sliding)

If the glacier moves, there will be an increase in pressure and friction with the bedrock. This will raise the temperature and the basal ice may melt. The resulting meltwater will act as a lubricant enabling the glacier to flow more rapidly (perhaps 2–3 m per day), to pick up material (debris) and to erode its bed (see Figure 4.11b).

Creep

Extending and compressing flow

Surges

Ice crystals orient themselves in the direction of the glacier's movement. This allows ice crystals to slide past each other. As the surface ice moves faster, crevasses may develop (see Figure 4.13). (This process can also operate within temperate glaciers.)

If there is an obstacle in the path of the glacier, pressure will increase. As the stress builds up, the ice behaves like plastic and flows round or over the obstacle. The lower the temperature, the greater the pressure needed to allow this process to take place and so it is less likely to occur in cold glaciers.

As most temperate glaciers occupy previously river-eroded valleys, their initial velocity must have been controlled by differences in the gradient of those valleys. J. F. Nye suggested that where there was a reduction in the gradient of the valley floor, the ice would decelerate and become thicker. He called this compressing flow. Where the valley gradient steepened, the ice would accelerate and become thinner: extending flow. Erosion is greatest where the ice is thickest (see Figure 4.12c).

These are the result of an excessive build-up of subglacial meltwater reservoirs (not, as stated in earlier editions of this book, by excessive snowfalls in the accumulation zone). In addition, they can occasionally occur when rock avalanches fall onto the glacier surface or, on even rarer occasions, if triggered by an earthquake. In a surge, the glacier moves forward, perhaps by 300 m in a day, an event which seems to take place in most glaciers once in every 30–100 years. The surge may damage farms, and sudden release of meltwater can cause severe flooding.

crevasses (an important source of basal debris)

extending flow

compressing flow (maximum erosion)

ice thins

ice thickens

extending flow

steeper gradient

ice thins

gradient lessens

gradient increases

Figure 4.10

Processes of glacier movement

Glacier movement and temperature

The character and movement of ice depend upon whether it is warm or cold, which in turn depends upon the **pressure melting point (PMP)**. The pressure melting point is the temperature at which ice is on the verge of melting. A small increase in pressure can therefore cause melting. PMP is normally 0°C on the surface of a glacier, but it can be lower within a glacier (due to an increase in pressure caused by either the weight or the movement of ice). In other words, as pressure increases, then the freezing point for water falls below 0°C.

Warm and cold ice

Warm ice has a temperature of around 0°C (PMP) throughout its depth (Figure 4.11a) and consequently is able, especially in summer, to release large amounts of meltwater. Temperatures in cold ice are permanently below 0°C (PMP) and so there is virtually no meltwater (Figure 4.11a). It is the presence of meltwater that facilitates the movement of a glacier. Temperature is therefore an alternative criterion to size or shape for use when categorising glaciers – they may be either **temperate** (mainly warm ice) or **polar** (mainly cold ice) – Figure 4.10. Movement is much faster in temperate glaciers where the presence of meltwater acts as a lubricant and reduces friction

(Figure 4.11b). It can take place by one of four processes of **basal flow** (or **slipping**): **creep**; **extending–compressing flow**; and **surges** (Figure 4.10). Polar glaciers move less quickly as, without the presence of meltwater, they tend to be frozen to their beds. The main process here is **internal flow**, although creep and extending–compressing flow may also occur.

Both types of glacier move more rapidly on the surface and away from their valley sides (Figure 4.12a and b), but it is the temperate one that is the more likely to erode its bed and to carry and deposit most material as **moraine** (page 117). Recent research suggests that any single glacier may exhibit, at different points along its profile, the characteristics of both polar and temperate glaciers.

Movement is greatest:
- at the point of equilibrium – as this is where the greatest volume of ice passes and consequently where there is most energy available
- in areas with high precipitation and ablation
- in small glaciers, which respond more readily to short-term climatic fluctuations
- in temperate glaciers, where there is more meltwater available, and
- in areas with steep gradients.

Figure 4.11

Comparison of temperature and velocity profiles in polar and temperate glaciers

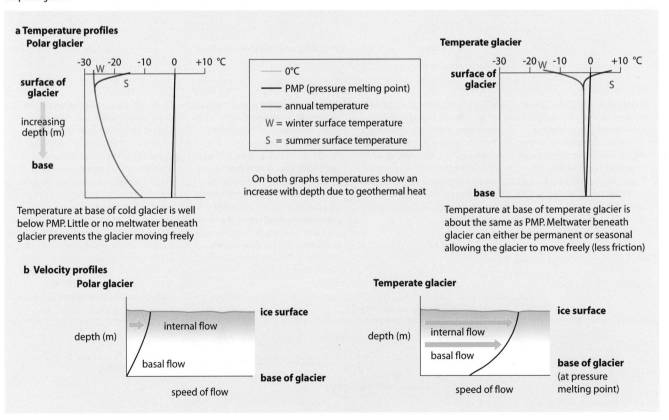

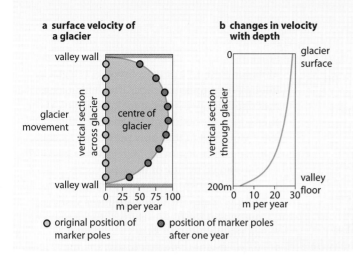

a surface velocity of a glacier

valley wall

glacier movement

vertical section across glacier

centre of glacier

valley wall

0 25 50 75 100
m per year

○ original position of marker poles ● position of marker poles after one year

b changes in velocity with depth

0 — glacier surface

vertical section through glacier

200m — valley floor

0 10 20 30
m per year

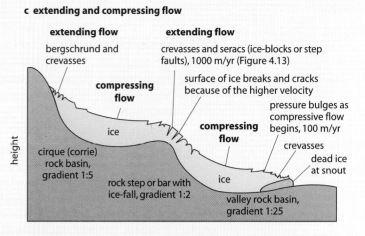

c extending and compressing flow

extending flow
bergschrund and crevasses

extending flow
crevasses and seracs (ice-blocks or step faults), 1000 m/yr (Figure 4.13)

compressing flow

surface of ice breaks and cracks because of the higher velocity

ice

compressing flow

pressure bulges as compressive flow begins, 100 m/yr

crevasses

dead ice at snout

cirque (corrie) rock basin, gradient 1:5

rock step or bar with ice-fall, gradient 1:2

ice

valley rock basin, gradient 1:25

height

Figure 4.12

Velocity and flow in a glacier

Figure 4.13

Crevasses on an icefall, Skafta glacier, Iceland

Transportation by ice

Glaciers are capable of moving large quantities of debris. This rock debris may be transported in one of three ways:

1 **Supraglacial debris** is carried on the surface of the glacier as lateral and medial moraine (page 117). It consists of material that has fallen onto the glacier from the surrounding valley sides. In summer, the relatively small load carried by surface meltwater streams often disappears down crevasses.

2 **Englacial debris** is material carried within the body of the glacier. It may once have been on the surface, only to be buried by later snowfalls or to fall into crevasses (Figure 4.4).

3 **Subglacial debris** is moved along the floor of the valley either by the ice or by meltwater streams formed by pressure melting (page 108).

Glacial erosion

Ice that is stationary or contains little debris has limited erosive power, whereas moving ice carrying with it much debris can drastically alter the landscape. Although ice lacks the turbulence and velocity of water in a river, it has the 'advantage' of being able to melt and refreeze in order to overcome obstacles in its path (Figure 4.10) and consequently has the ability to lower (i.e. erode) the landscape perhaps ten times more quickly than can running water. Virtually all the glacial processes of erosion are physical, as the climate tends to be too cold for chemical reactions to operate (Figure 2.10).

Processes of glacial erosion

The processes associated with glacial erosion are: frost shattering, abrasion, plucking, rotational movement, and extending and compressing flow.

Frost shattering

This process (page 40) produces much loose material which may fall from the valley sides onto the edges of the glacier to form **lateral moraine**, be covered by later snowfall, or plunge down crevasses to be transported as **englacial debris**. Some of this material may be added to rock loosened by frost action as the climate deteriorated (but before glaciers formed) to form **basal debris** (page 117).

Abrasion

This is the sandpapering effect of angular material embedded in the glacier as it rubs against the valley sides and floor. It usually produces smoothened, gently sloping landforms.

Plucking

At its simplest, this process involves the glacier freezing onto rock outcrops, after which ice movement pulls away masses of rock. In reality, as the strength of the bedrock is greater than that of the ice, it would seem that only previously loosened material can be removed. Material may be continually loosened by one of three processes:

1 The relationship between local pressure and temperature (the PMP) produces sufficient meltwater for freeze–thaw activity to break up the ice-contact rock.
2 Water flowing down a **bergschrund** (a large, crevasse-like feature found near the head of some glaciers – Figure 4.14b) or smaller crevasses will later freeze onto rock surfaces.
3 Removal of layers of bedrock by the glacier causes a release in pressure and an enlarging of joints in the underlying rocks (pressure release, page 41).

Plucking generally creates a jagged-featured landscape.

Figure 4.14

Processes in the formation of a cirque

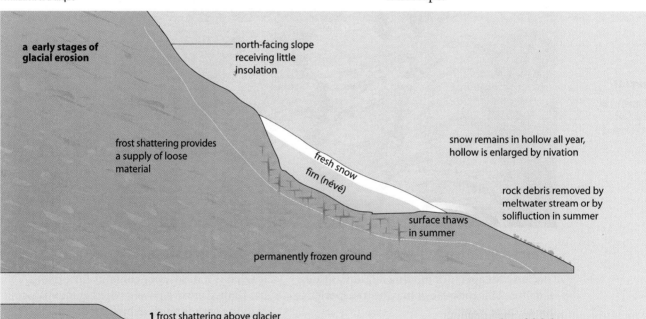

a early stages of glacial erosion

north-facing slope receiving little insolation

frost shattering provides a supply of loose material

fresh snow
firn (névé)

snow remains in hollow all year, hollow is enlarged by nivation

rock debris removed by meltwater stream or by solifluction in summer

surface thaws in summer

permanently frozen ground

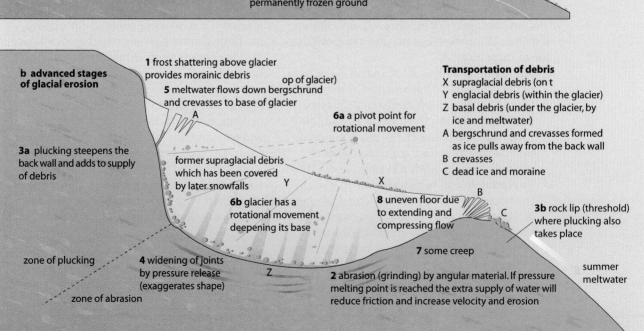

b advanced stages of glacial erosion

1 frost shattering above glacier provides morainic debris
op of glacier)
5 meltwater flows down bergschrund and crevasses to base of glacier

6a a pivot point for rotational movement

Transportation of debris
X supraglacial debris (on t
Y englacial debris (within the glacier)
Z basal debris (under the glacier, by ice and meltwater)
A bergschrund and crevasses formed as ice pulls away from the back wall
B crevasses
C dead ice and moraine

3a plucking steepens the back wall and adds to supply of debris

former supraglacial debris which has been covered by later snowfalls

6b glacier has a rotational movement deepening its base

8 uneven floor due to extending and compressing flow

7 some creep

3b rock lip (threshold) where plucking also takes place

zone of plucking

4 widening of joints by pressure release (exaggerates shape)

zone of abrasion

2 abrasion (grinding) by angular material. If pressure melting point is reached the extra supply of water will reduce friction and increase velocity and erosion

summer meltwater

Rotational movement

This is a downhill movement of ice which, like a landslide (Figure 2.17), pivots about a point. The increase in pressure is responsible for the over-deepening of a cirque floor (Figure 4.14b).

Extending and compressing flow

Figures 4.10 and 4.12c show how this process causes differences in the rate of erosion at the base of a glacier.

Maximum erosion occurs:

- where temperatures fluctuate around 0°C, allowing frequent freeze–thaw to operate
- in areas of jointed rocks which can be more easily frost shattered
- where two tributary glaciers join, or the valley narrows, giving an increased depth of ice, and
- in steep mountainous regions in temperate latitudes, where the velocity of the glacier is greatest.

Landforms produced by glacial erosion

Cirques

These are amphitheatre or armchair-shaped hollows with a steep back wall and a rock basin (Figure 4.15). They are also known as **corries** (Scotland) and **cwms** (Wales – Figures 4.25 and 4.26).

During periglacial times (Chapter 5), before the last glacial, snow collected in hollows, especially on north-facing slopes. A series of processes, collectively known as **nivation** and which included freeze–thaw, solifluction and possibly chemical weathering, operated under and around the snow patch (Figure 4.14a). These processes caused the underlying rocks to disintegrate. The resultant debris was then removed by summer meltwater streams to leave, in the enlarged hollow, an embryo cirque. It has been suggested that the overdeepening process might need several periglacials or interglacials and glacials in which to form. As the snow patch grew, its layers became increasingly compressed to form firn and, eventually, ice (page 105).

It is accepted that several processes interact to form a fully developed cirque (Figure 4.14b). Plucking is one process responsible for steepening the back wall, but this partly relies upon a supply of water for freeze–thaw and partly upon pressure release in well-jointed rocks. A rotational movement, aided by water from pressure point melting and angular subglacial debris from frost shattering, enables abrasion to over-deepen the floor of the cirque. A **rock lip** develops where erosion decreases. This may be increased in height by the deposition of morainic debris at the glacier's snout. When the climate begins to get warmer, the ice remaining in the hollow melts to leave a deep, rounded lake or tarn (Figures 4.15 and 4.26).

In Britain, as elsewhere in the northern hemisphere, cirques are nearly always oriented between the north-west (315°), through the north-east (where the frequency peaks) to the south-east (135°). This is largely attributable to two facts:

- Northern slopes receive least **insolation** and so glaciers remained there much longer than those facing in more southerly directions (less melting on north-facing slopes).
- Western slopes face the sea and, although still cold, the relatively warmer winds which blew from that direction were more likely to melt the snow and ice (more snow accumulated on east-facing slopes).

Lip orientation is the direction of an imaginary line from the centre of the back wall of the cirque to its lip. Of 56 cirques identified in the Snowdon area, 51 have a lip orientation of between 310° and 120°, and of 15 on Arran, 14 have an orientation between 5° and 115°.

These are all types of average (measures of dispersion, Framework 8, page 246).

1 The **mean** (or arithmetic average) is obtained by totalling the values in a set of data and dividing by the number of values in that set. It is expressed by the formula:

$$\bar{x} = \frac{\sum x}{n}$$

where:

$\bar{x}$ = mean, $\sum$ = the sum of, x = the value of the variable, n = the number of values in the set

The mean is reliable when the number of values in the sample is high and their range, i.e. the difference between the highest and lowest values, is low, but it becomes less reliable as the number in the sample decreases, as it is then influenced by extreme values.

2 The **median** is the mid-point value of a set of data. For example, if the sample is of 15 cirques, values must be ranked in descending order. The mid-point is the orientation of the eighth cirque because there are seven values above and seven below. Had there been an even number of values then the median would have been the mean of the two middle values. The median is a less accurate measure of dispersion than the mean because widely differing sets of data can return the same median, but it is less distorted by extreme values.

3 The **mode** is the value or class that occurs most frequently in the data. In the set of values 4, 6, 4, 2, 4 the mode would be 4. Although it is the easiest of the three 'averages' to obtain it has limited value. Some data may not have two values in the same class (e.g. 1, 2, 3, 4, 5), while others may have more than one modal value (e.g. 1, 1, 2, 4, 4).

Relationships between mean, median and mode

When data is plotted on a graph we can often make useful observations about the shape of the curve. For example, we would expect A-level results nationally to show a few top grades, a smaller number of 'unclassifieds' and a large number of average passes. Graphically this would show a normal distribution, with all three averages at the peak. If the distribution is skewed, then by definition only the mode will lie at the peak (Figure 4.16).

Figure 4.16

Normal and skewed distributions

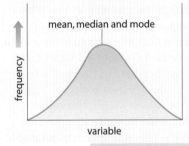

a the normal distribution

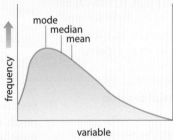

b a positively skewed distribution

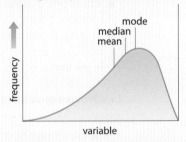

c a negatively skewed distribution

Arêtes and pyramidal peaks

When two adjacent cirques erode backwards or sideways towards each other, the previously rounded landscape is transformed into a narrow, rocky, steep-sided ridge called an **arête**, as at Striding Edge in the Lake District (Figure 4.17) and Crib Goch on Snowdon (Figure 4.25). If three or more cirques develop on all sides of a mountain, a **pyramidal peak**, or horn, may be formed. This feature has steep sides and several arêtes radiating from the central peak (Figures 4.18 and 4.19), e.g. the Matterhorn.

Figure 4.17

An arête: Striding Edge on Helvellyn in the Lake District

Figure 4.18

Arêtes in the Karakoram Mountains, northern Pakistan

Figure 4.19

A pyramidal peak: Machhappuchare, Nepal

Glacial troughs, rock steps, truncated spurs and hanging valleys

These features are interrelated in their formation. Valley glaciers straighten, widen and deepen preglacial valleys, turning the original V-shaped, river-formed feature into the characteristic U shape typical of glacial erosion, e.g. Wast Water in the Lake District (Figure 4.20). These steep-sided, flat-floored valleys are known as **glacial troughs**. The overdeepening of the valleys is credited to the movement of ice which, aided by large volumes of meltwater and subglacial debris, has a greater erosive power than that of rivers. Extending and compressing flow may overdeepen parts of the trough floor, which later may be occupied by long, narrow **ribbon lakes**, such as Wast Water, or may leave less eroded, more resistant **rock steps**.

Theories to explain pronounced overdeepening of valley floors are debated amongst glaciologists and geomorphologists. Suggested causes include: extra erosion following the confluence of two glaciers; the presence of weaker rocks; an area of rock deeply weathered in preglacial times; or a zone of well-jointed rock. Should the deepening of the trough continue below the former sea-level, then during deglaciation and subsequent rises in sea-level the valley may become submerged to form a **fiord** (Figures 4.21 and 6.48).

Abrasion by englacial and subglacial debris and plucking along the valley sides remove the tips of preglacial interlocking spurs leaving cliff-like **truncated spurs** (Figure 4.20, and to the left of Figure 4.27).

Figure 4.20

Glacial trough with ribbon lake: Wast Water in the Lake District

Figure 4.21

A fiord: Milford Sound, New Zealand

Figure 4.22

Hanging valley: Lake Bigden, Norway

Figure 4.23

A roche moutonnée: Yosemite National Park, California

Hanging valleys result from differential erosion between a main glacier and its tributary glaciers. The floor of any tributary glacier is deepened at a slower rate so that when the glaciers melt it is left hanging high above the main valley and its river has to descend by a single waterfall or a series of waterfalls, e.g. Garbh Allt in Glen Rosa, Arran (Figure 4.37) and Cwm Dyli, Snowdonia (Figure 4.27).

Striations, roches moutonnées, rock drumlins and crag and tail

These are all smaller erosion features which help to indicate the direction of ice movement. As a glacier moves across areas of exposed rock, larger fragments of angular debris embedded in the ice tend to leave a series of parallel scratches and grooves called **striations** (e.g. Central Park in New York).

A **roche moutonnée** is a mass of more resistant rock. It has a smooth, rounded upvalley or stoss slope facing the direction of ice flow, formed by abrasion, and a steep, jagged, down-valley or lee slope resulting from plucking (Figures 4.23 and 4.24).

Rock drumlins are more streamlined bedrock which lack the quarried lee face of the roche moutonnée. They are sometimes referred to as **whalebacks** as they resemble the backs of whales breaking the ocean surface.

A **crag and tail** consists of a larger mass of resistant rock or crag (e.g. the basaltic crag upon which Edinburgh Castle has been built) which protected the lee-side rocks from erosion, thus forming a gently sloping tail of deposited material (e.g. the tail down which the Royal Mile extends).

It should be remembered that while many of these erosional landforms may be found together in most glaciated uplands, their arrangement, frequency and presence is likely to change from one area to another. Places 15 describes some of these glacial features as found in one part of Snowdonia.

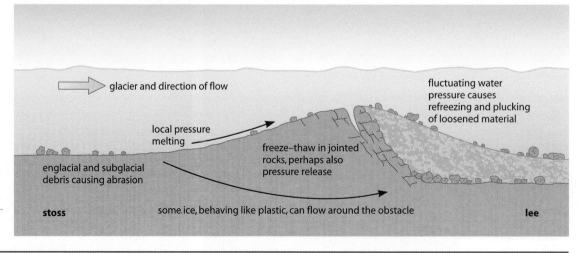

glacier and direction of flow

local pressure melting

englacial and subglacial debris causing abrasion

freeze–thaw in jointed rocks, perhaps also pressure release

fluctuating water pressure causes refreezing and plucking of loosened material

stoss

some ice, behaving like plastic, can flow around the obstacle

lee

Figure 4.24

The formation of a roche moutonnée

Labels on Figure 4.25:
- Llyn Peris - ribbon lake
- Nant Llanberis - glacial trough
- Snowdon - pyramidal peak
- Crib y Ddysgl - arête
- Cwmbrwynog - arête
- Y Lliwedd - arête
- Bwlch Main - arête
- Glaslyn - corrie
- Crib Goch - arête
- hanging valley
- Llyn Llidaw - corrie
- Cwm Dyli - hanging valley
- truncated spur
- waterfall
- Llyn Gwnynant - ribbon lake
- Nant Gwynant glacial trough

Figure 4.25

Landsketch of glacial features in Snowdonia (looking west)

Snowdonia is an example of a glaciated upland area. Although Snowdon itself has the characteristics of a pyramidal peak, the ice age was too short (by several thousand years) for the completed development of the classic pyramidal shape which makes the appearance of the Matterhorn so spectacular (compare Figure 4.19). What *are* well developed are the arêtes, such as Crib Goch and Bwlch Main, which radiate from the central peak. Between these arêtes are up to half a dozen cirques (cwms, as this is Wales), including the eastward-facing Glaslyn and the north-eastward-oriented (page 111) Llyn (lake) Llydaw. Glaslyn, which is trapped by a rock lip, is 170 m higher than Llyn Llydaw (Figure 4.26). Striations and roches moutonnées can be found in several places where the rocks are exposed on the surface. To the north and south-east of Snowdon are the glacial troughs of Nant (valley) Llanberis and Nant Gwynant. Both valleys have the characteristic U shape, with steep valley sides, truncated spurs and a flat valley floor. Located on the valley floors are ribbon lakes, including Llyn Peris and Llyn Gwynant (Figure 3.24). Numerous small rivers, with their sources in hanging valleys, descend by waterfalls, as at Cwm Dyli (Figure 4.27), into the two main valleys. Although the ice has long since gone, the actions of frost and snow, together with that of rain and more recently people, continue to modify the landscape – remember that rarely does a landscape exhibit stereotyped 'textbook' features (see Figure 4.25)!

Figure 4.26

Llyn Glaslyn: a cirque lake formed behind a rock lip, and a hanging valley into Llyn Llydaw

Figure 4.27

Cwm Dyli: a hanging valley with truncated spurs to the left, Snowdon and Crib Goch behind, and Nant Gwynant in the foreground

Figure 4.32

Medial and lateral moraines, Meade Glacier, Alaska

Figure 4.33

Morainic mounds above Haweswater, Cumbria

highest point of the feature is near to the stoss end (Figure 4.34). The shape of drumlins can be described by using the **elongation ratio**:

$$E = \frac{I}{W}$$

where I is the maximum bedform length, and W is the maximum bedform width. Drumlins are always longer than they are wide, and they are usually found in **swarms** or *en echelon*.

There is much disagreement as to how drumlins are formed. Theories suggest they may be an erosion feature, or formed by deposition around a central rock, or even by meltwater. However, none of these accounts for the fact that the majority of drumlins are composed of till which, lacking a central core of rock and consisting of unsorted material, would be totally eroded by moving ice. The most widely accepted view is that they were formed when the ice became overloaded with material, thus reducing the capacity of the glacier. The reduced competence may have been due to the melting of the glacier or to changes in velocity related to the pattern of extending–compressing flow. Once the material had been deposited, it may then have been moulded and streamlined by later ice movement. The most recent theory (1987) is based on evidence that drumlins can be composed of both till and glacifluvial sediments. The most widely accepted view now is that 'they are subglacially deformed masses of pre-existing sediment to which more sediment may be added by the melting out of debris from the glacier base' (D. Evans, 1999).

Drumlins

These are smooth, elongated mounds of till with their long axis parallel to the direction of ice movement. Drumlins may be over 50 m in height, over 1 km in length and nearly 0.5 km in width. The steep stoss end faces the direction from which the ice came, while the lee side has a more gentle, streamlined appearance. The

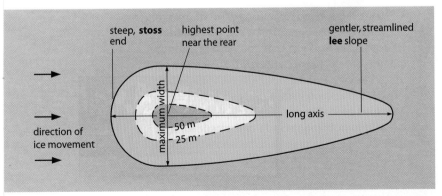

Figure 4.34

Plan of a drumlin showing typical dimensions

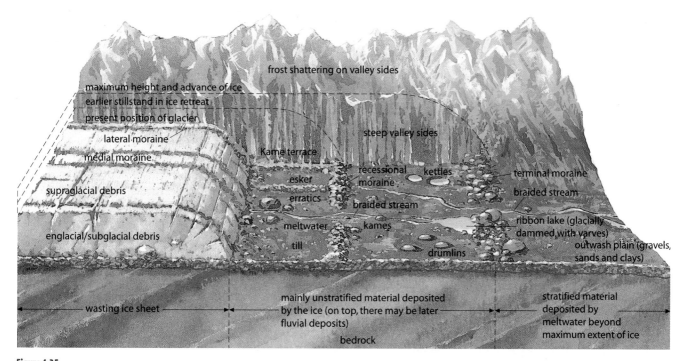

Figure 4.35

Features of lowland glaciation

Labels on Figure 4.35: frost shattering on valley sides; maximum height and advance of ice; earlier stillstand in ice retreat; present position of glacier; lateral moraine; medial moraine; supraglacial debris; englacial/subglacial debris; steep valley sides; Kame terrace; recessional moraine; esker; erratics; kettles; braided stream; meltwater; kames; till; drumlins; terminal moraine; braided stream; ribbon lake (glacially dammed, with varves); outwash plain (gravels, sands and clays); wasting ice sheet; mainly unstratified material deposited by the ice (on top, there may be later fluvial deposits); bedrock; stratified material deposited by meltwater beyond maximum extent of ice

Glacifluvial landforms

Glacifluvial landforms are those moulded by glacial meltwater and have, in the past, been considered to be mainly depositional. More recently it has been realised that meltwater plays a far more important role in the glacial system than was previously thought, especially in temperate glaciers and in creating erosion features as well as depositional landforms. Most meltwater is derived from ablation. The discharge of glacial streams, both supraglacial and subglacial, is high during the warmer, if not warm, summer months. As the water often flows under considerable pressure, it has a high velocity and is very turbulent. It is therefore able to pick up and transport a larger amount of material than a normal river of similar size. This material can erode vertically, mainly through abrasion but partly by solution, to create subglacial valleys and large potholes, some of the latter being up to 20 m in depth. Deposition occurs whenever there is a decrease in discharge, and it is responsible for a group of landforms (Figures 4.35 and 4.37).

Outwash plains (sandur)
These are composed of gravels, sands and, uppermost and furthest from the snout, clays. They are deposited by meltwater streams issuing from the ice either during summer or when the glacier melts. The material may originally have been deposited by the glacier and later picked up, sorted and dropped by running water beyond the maximum extent of the ice sheets. In parts of the North German Plain, deposits are up to 75 m deep. Outwash material may also be deposited on top of till following the retreat of the ice (Figure 4.35).

Glacilacustrine sediments (varves)
A varve is a distinct layer of silt lying on top of a layer of sand, deposited annually in lakes found near to glacial margins. The coarser, lighter-coloured sand is deposited during late spring when meltwater streams have their peak discharge and are carrying their maximum load. As discharge decreases towards autumn when temperatures begin to drop, the finer, darker-coloured silt settles. Each band of light and dark materials represents one year's accumulation (Figure 4.36). By counting the number of varves, it is possible to date the origin of the lake; variations in the thickness of each varve indicate warmer and colder periods (e.g. greater melting causing increased deposition).

Figure 4.36

The formation of varves in a postglacial lake

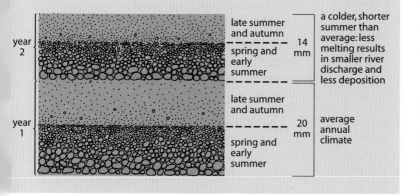

Labels on Figure 4.36: year 2; year 1; late summer and autumn; spring and early summer; 14 mm; 20 mm; a colder, shorter summer than average: less melting results in smaller river discharge and less deposition; average annual climate

Kames and kame terraces

Kames are undulating mounds of sand and gravel deposited unevenly by meltwater, similar to a series of deltas, along the front of a stationary or slowly melting ice sheet (Figure 4.35). As the ice retreats, the unsupported kame often collapses. Kame terraces, also of sand and gravel, are flat areas found along the sides of valleys. They are deposited by meltwater streams flowing in the trough between the glacier and the valley wall. Troughs occur here because, in summer, the valley side heats up faster than the glacier ice and so the ice in contact with it melts. Kame terraces are distinguishable from lateral moraines by their sorted deposits.

Eskers

These are very long, narrow, sinuous ridges composed of sorted coarse sands and gravel. It is thought that eskers are the fossilised courses of subglacial meltwater streams. As the channel is restricted by ice walls, the hydrostatic pressure and the transported load are both considerable. As the bed of the channel builds up (there is no floodplain), material is left above the surrounding land following the retreat of the ice. Like kames, eskers usually form during times of deglaciation (Figure 4.35).

Kettles

These form from detached blocks of ice, left by the glacier as it retreats, and then partially buried by the glacifluvial deposits left by meltwater streams. When the ice blocks melt, they leave enclosed depressions which often fill with water to form kettle-hole lakes and 'kame and kettle' topography (Figure 4.35).

Braided streams

Channels of meltwater rivers often become choked with coarse material as a result of the marked seasonal variations in discharge (compare Figure 3.32 and 5.16).

Places 16   Arran: glacial landforms

Using fieldwork to answer an Advanced GCE question: 'Describe the landforms found near the snout of a former glacier.'

Figure 4.28 lists the types of feature formed by glacial deposition, subdividing them into those composed of unsorted material, left by the glacier, and sorted material deposited by glacifluvial activity. If the snout of a glacier had remained stationary for some time, indicating a balance between accumulation and ablation, and had then slowly retreated, several of these landforms might be visible following deglaciation. One such site studied by a sixth form was the lower Glen Rosa valley on the Isle of Arran (Figure 4.37).

The dominant feature was a mound **A**, 14 m high, into which the Rosa Water had cut, giving a fine exposed section of the deposited material. As the mound was a long, narrow, ridge-like feature extending across the valley, it was suggested that it might be either a terminal or a recessional moraine. It was concluded that the feature was ice-deposited because the material was unsorted: many of the largest boulders were high up in the exposure; also, most of the stones were sub-angular (not more rounded as might be expected in glacifluvial deposits).

However, an observation downstream at point **B** revealed that material there was also unsorted and this, together with some large granite erratics seen earlier nearer the coast, seemed to indicate that the mound could not be a terminal moraine as it did not mark the maximum advance of the ice. When a till fabric analysis was carried out, it was noted that the average dip of the stones was about 25°, suggesting that the feature might instead have been a push moraine resulting from a minor re-advance during deglaciation. The orientation of 50 sample stones (Figure 4.29) showed that the ice must have come either from the north-north-west (probable, as this was the highland) or the south-south-east (unlikely, as the lower ground would not be the source of a glacier). An examination of the geology of the stones showed that 80 per cent were granite, and therefore were erratics carried from the upper Rosa valley; 15 per cent were schists (the local rock); and 5 per cent were other igneous rocks not found on the island. It was inferred from the presence of these other rocks that some of the ice must have originated on the Scottish mainland. Also at point **B**, an investigation of river banks showed a mass of sand and gravel with some level of sorting – as might be expected in an outwash area.

Upstream from **A** was a second mound, **C**, filling much of the valley floor (Figure 4.38). Student suggestions as to the nature of the feature included its being a drumlin, a lateral, a medial, a recessional or even another push moraine. When measured, it was found that its length was slightly greater than its width (an elongation ratio of 1.25:1) and the highest point was nearest the up-valley end; it had neither the streamlined shape nor a sufficiently high elongation ratio to be a drumlin (and there

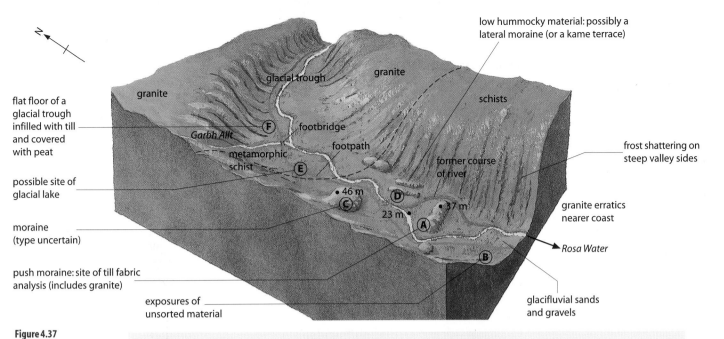

low hummocky material: possibly a lateral moraine (or a kame terrace)

glacial trough

granite

schists

granite

flat floor of a glacial trough infilled with till and covered with peat

Garbh Allt

metamorphic schist

footbridge

footpath

former course of river

frost shattering on steep valley sides

possible site of glacial lake

(F)

(E)

• 46 m

(C)

(D)

23 m •

(A)

• 37 m

granite erratics nearer coast

Rosa Water

moraine (type uncertain)

push moraine: site of till fabric analysis (includes granite)

(B)

exposures of unsorted material

glacifluvial sands and gravels

Figure 4.37

Sketch to show features of glacial deposition in the lower Glen Rosa valley, Arran

were no signs of a swarm!). It appeared to be too far from the valley side to be a lateral moraine; and as two glaciers could not have met here, neither could it have been a medial moraine. It was concluded that it *was* a moraine – perhaps formed during a stillstand in the glacier's retreat, or if the glacier lost momentum after having negotiated a bend in the glacial trough.

Across the river (**D**), was an area of low hummocky material winding along the foot of the valley side to as far as **A**. It was speculated that the feature may have been formed in one of three ways: meltwater depositing sands and gravel between the valley side and the former glacier as a kame terrace; a lateral moraine from frost shattering on the valley sides; or solifluction deposits (page 47) formed as the climate grew milder and the glacier retreated (the feature was not flat enough for a river terrace to be seriously considered).

Upstream, the valley floor was extremely flat (**E**). This could be the remains of a former glacial lake, formed when meltwater from the retreating glacier had become trapped behind the moraine at **C** and before it had had time to cut through the deposits. It was impossible to gain a profile to prove or disprove the existence of a lake. Had there been a lake, the profile might have shown a chronological sequence (since the last glacial) as depicted in Figure 4.39.

After crossing the Garbh Allt (a hanging valley), the steep-sided, flat-floored U-shape of the glacial trough through which the Rosa Water flows was visible. The flatness of the floor was probably due to the deposition of subglacial debris – although the till has since been covered by peat, a symptom of the cold, wet conditions.

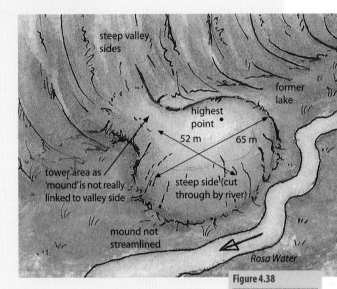

steep valley sides

former lake

highest point •

52 m

65 m

tower area as 'mound' is not really linked to valley side

steep side (cut through by river)

mound not streamlined

Rosa Water

Figure 4.38

Fieldsketch of landform at **C** in Figure 4.37

Although not every feature of glacial deposition was present – there was no evidence of eskers or kettles – this small area did contain several of the landforms and deposits that might be expected at, or near to, the snout of a former glacier.

Figure 4.39

Possible profile of ground at point **E** in Figure 4.37

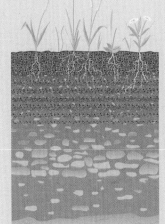

Juncus (rushes), cotton grass, heather

thick layer of peat formed under cold, wet conditions

varves and silt deposited on bed of post-glacial lake

stratified sands and gravels deposited by meltwater stream from the retreating glacier

unsorted till deposited by glacier

bedrock (schists)

Glaciation

Other effects of glaciation

Drainage diversion and proglacial lakes
Where ice sheets expand, may divert the courses of rivers. For example, the preglacial River Thames flowed in a north-easterly direction. It was progressively diverted southwards by advancing ice (Figure 4.40).

Where ice sheets expand and dam rivers, proglacial lakes are created, e.g. Lakes Lapworth and Harrison (Figure 4.40). Before the ice age, the River Severn flowed northwards into the River Dee, but this route became blocked during the Pleistocene by Irish Sea ice. A large lake, Lapworth, was impounded against the edge of the ice until the waters rose high enough to breach the lowest point in the southern water-shed. As the water overflowed through an **overspill channel**, there was rapid vertical erosion which formed what is now the Ironbridge Gorge. When the ice had completely melted, the level of this new route was lower than the original course (which was also blocked by drift), forcing the present-day River Severn to flow southwards.

Other rivers, e.g. the Warwickshire Avon (Figure 4.40) and the Yorkshire Derwent (Places 17), have also been diverted as a consequence of glacial activity. Sometimes the glacial overspill channels have been abandoned, e.g. at Fenny Compton, where the Warwickshire Avon temporarily flowed south-east into the Thames (O[1] in Figure 4.40). Proglacial lakes are also found behind eskers and recessional moraines.

Figure 4.40

Glacial diversion of drainage and proglacial lakes in England and Wales

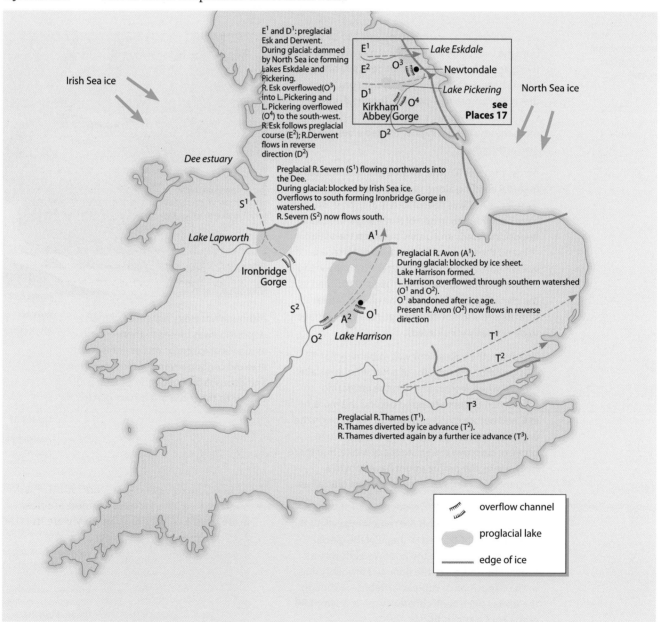

Figure 4.41

Proglacial lakes and overflow channels in North Yorkshire

Lake Eskdale, a proglacial lake, formed when the North Sea ice sheet blocked the mouth of the River Esk. The level of the lake rose until its water found a new route over a low point in its southern watershed on the North Yorkshire Moors. The overflow river flowed through Lake Glaisdale before cutting the deep, narrow, steep-sided, flat-floored Newtondale valley. At the end of this valley, the river formed a delta where it flowed into another proglacial lake – Lake Pickering. Lake Pickering, also dammed by North Sea ice, found an outlet to the south-west where it formed an overflow channel – the present-day Kirkham Gorge. After the ice melted, the Esk reverted to its original course, entering the sea near Whitby; Newtondale became virtually a dry valley; and the River Derwent, its eastward exit from Lake Pickering blocked by glacial deposits, continued to follow its new south-westerly course. Today, the site of Lake Pickering forms the fertile, flat-floored Vale of Pickering.

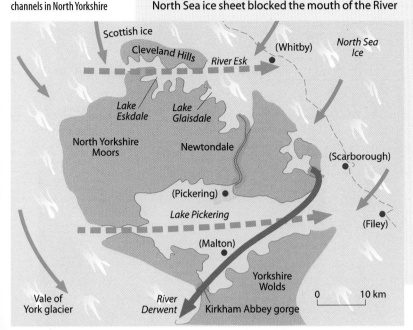

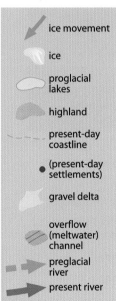

ice movement

ice

proglacial lakes

highland

present-day coastline

(present-day settlements)

gravel delta

overflow (meltwater) channel

preglacial river

present river

Changes in sea-level

The expansion and contraction of ice sheets affected sea-level in two different ways. **Eustatic** (also now called **glacio-eustatic**) refers to a worldwide fall (or rise) in sea-level due to changes in the hydrological cycle caused by water being held in storage on land in ice sheets (or released following the melting of ice sheets). **Isostatic** (or **glacio-isostatic**) adjustment is a more local change in sea-level resulting from the depression (or uplift) of the Earth's crust by the increased (or decreased) weight imposed upon it by a growing (or a declining) ice sheet. Evans (1991) claims that 'Because of their great weight, ice sheets depress the Earth's crust below them by approximately 0.3 times their thickness. So, at the centre of an ice sheet 700 m thick, there will be a maximum of 210 m of depression'. The sequence of events resulting from eustatic and isostatic changes during and after the last glacial can be summarised as follows:

1 At the beginning of the glacial, water in the hydrological cycle was stored as ice on the land instead of returning to the sea. There was a universal (eustatic) fall in sea-level, giving a negative change in base level (page 81).

2 As the glacial continued towards its peak, the weight of ice increased and depressed the Earth's crust beneath it. This led to a local (isostatic) rise in sea-level relative to the land and a positive change in base level.

3 As the ice sheets began to melt, large quantities of water, previously held in storage, were returned to the sea causing a worldwide (eustatic) rise in sea-level (a positive change in base level). This formed fiords, rias and drowned estuaries (page 163 and Places 22, page 164).

4 Finally, and still continuing in several places today, there was a local (isostatic) uplift of the land as the weight of the ice sheets decreased (a negative change in base level). This change created raised beaches (Places 23, page 166) and caused rejuvenation of rivers (page 82).

Looking into the future:

■ If the ice sheets continue to melt at their present rate, caused by global warming (Case Study 9B) or a milder climate, sea-levels could rise by up to 5 m by the end of the 21st century.

■ If isostatic uplift continues in Britain, it will increase the tilt that has already resulted in north-west Scotland rising by an estimated 10 m in the last 9000 years, and south-east England sinking. Tides in London are now more than 4 m higher than they were in Roman times – hence the need for the Thames Barrier.

Avalanches

An avalanche is a sudden downhill movement of snow, ice and/or rock (Case Study 2A). It occurs, like a landslide, when the weight (mass) of material is sufficient to overcome friction (Figure 4.42). This allows the debris to descend at a considerable speed under the force of gravity (mass movement). The average speed of descent is 40–60 km/hr, but video-recordings have shown extreme speeds in excess of 200 km/hr.

There are several different types of avalanche, which makes a simple classification difficult. Figure 4.43 gives a mainly descriptive classification put forward in the 19th century, while Figure 4.44 gives a modern classification based more upon genetic and morphological characteristics.

Figure 4.43

A late 19th-century classification of avalanches

a **Staublawinen (airborne powder snow)**	Pure (completely airborne)
	Common (some contact with the ground)
b **Grundlawinen (ground-hugging)**	Rolling
	Sliding

a **Avalanche break-away point**	single point – loose snow avalanche	easier (not easy) to predict and manage; originates from a single point, usually soon after the snow falls
	large area, or 'slab'	often localised, hardest to predict, greatest threat to off-piste skiers; originates from a wider area and after the snow has had time to develop cohesion
b **Depth**	total snow depth	total mass of snow moves
	top layers of snow move over lower layers	alpine inhabitants regard this as the most dangerous
c **Channel (track) width**	unconfined – no channel	wide area, hard to manage
	gulley – confined to narrow track	dangerous, as it can reach higher speeds, but easier to manage
d **Nature of snow (water content)**	dry snow – mainly rolling	above ground-level so friction is reduced; can reach speeds of 200 km/hr – very destructive
	wet snow – mainly sliding	follows ground topography, occurs under föhn conditions (page 241), limited protection, much damage

Figure 4.44

A more recent classification of avalanches (1979)

Causes

- Heavy snowfall compressing and adding weight to earlier falls, especially on windward slopes.
- Steep slopes of over 25° where stability is reduced and friction is more easily overcome.
- A sudden increase in temperature, especially on south-facing slopes and, in the Alps, under föhn wind conditions (page 241).
- Heavy rain falling upon snow (more likely in Scotland than the Alps).
- Deforestation, partly for new ski-runs, which reduces slope stability.
- Vibrations triggered by off-piste skiers (Figure 4.47), any nearby traffic and, more dangerously, earth movements (Case Study 2A).
- Very long, cold, dry winters followed by heavy snowfalls in spring. Under these conditions, earlier falls of snow will turn into ice over which later falls will slide (some local people perceive this to pose the greatest avalanche risk).

Consequences

Avalanches can block roads and railways, cut off power supplies and telecommunications and, under extreme conditions, destroy buildings and cause loss of life. Between 1980 and 1991 there were, in Alpine Europe alone, 1210 recorded avalanche deaths, of whom nearly half were skiers (virtually all in off-piste areas – Figure 4.45). This death rate is increasing as the popularity of skiing grows and alpine weather becomes less predictable (a record total of 145 deaths in 1998–99).

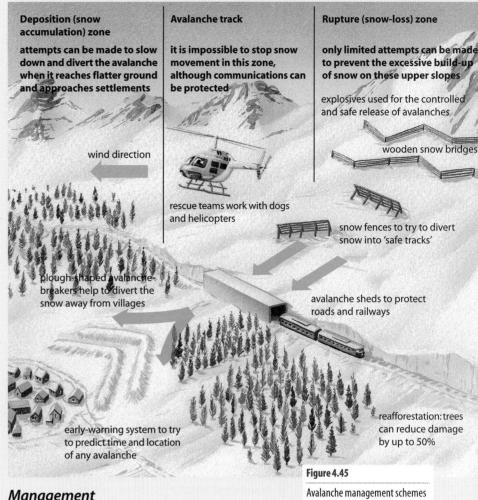

Deposition (snow accumulation) zone

attempts can be made to slow down and divert the avalanche when it reaches flatter ground and approaches settlements

Avalanche track

it is impossible to stop snow movement in this zone, although communications can be protected

Rupture (snow-loss) zone

only limited attempts can be made to prevent the excessive build-up of snow on these upper slopes

explosives used for the controlled and safe release of avalanches

wooden snow bridges

wind direction

rescue teams work with dogs and helicopters

snow fences to try to divert snow into 'safe tracks'

plough-shaped avalanche-breakers help to divert the snow away from villages

avalanche sheds to protect roads and railways

reafforestation: trees can reduce damage by up to 50%

early-warning system to try to predict time and location of any avalanche

Figure 4.45
Avalanche management schemes

Management

There is a close link between avalanches and:
- time of year – almost 80 per cent of avalanches in the French Alps occur between January and March, the 'avalanche season'
- altitude – over 90 per cent occur between 1500 and 3000 m.

Although it is possible to predict **when** and in which regions avalanches are most likely

to occur, it is less easy to predict exactly **where** an event is likely to happen. It is this unpredictability that makes avalanches a major environmental hazard in alpine areas (Figure 4.48). However, despite this uncertainty, many avalanches do tend to follow certain 'tracks'. Consequently, as well as setting up early warning systems and training rescue teams (Figure 4.46), it is possible to take some measures to try to protect life and property (Figure 4.45).

Figure 4.46
Avalanche protection and rescue schemes

The Alps – February 1999

Experts claimed that the avalanche risk in February 1999 was unprecedented in living memory. Villages, hamlets and Alpine resorts are built in places that have, in the past, proved to be safe. Some of the winter's first avalanches were triggered by skiers and snowboarders straying off-piste into danger zones (Figure 4.47). However, by late February the victims were tourists and residents in their chalets and houses – places normally considered among the safest in times of avalanche danger (Figure 4.48). Alpine experts agree that the reason for the chaos was excess snow, with more falling in many villages than at any time this century. The position was made worse during mid-February when the weather turned milder. Rain fell on many lower slopes which, as temperatures fell again, turned to snow and

froze. Further snowfalls, falling onto the icy base, were whipped up by winds of up to 120 km/hr. The result was that some parts of the Alps, especially western Austria where

the snow had been heaviest, experienced avalanches on an unprecedented scale, and in places where they had never been seen before (Figure 4.49).

THE British woman who died in the latest Alpine avalanche was swept for half a mile down a forested hillside after she and her friends ignored safety warnings and held a snowboard race off-piste.

Safety experts issued a warning yesterday that winter holidaymakers are taking too many risks. There have been 17 deaths in the past week in the Alps.

Snowboarding is the biggest danger. One in five Britons had swapped skis for boards, and it was a snowboard race that triggered Friday's avalanche above the French resort of Val d'Isère. Police estimate the 150 ft-high [50 m] wall of snow that engulfed the snowboarders could have been moving at 80 mph [130 km/hr]. Trees snapped like matchsticks before it.

In a separate avalanche yesterday, three Frenchmen were also killed when they were engulfed along with their guide while skiing off-piste at Les Arcs, near Val d'Isère.

The accidents come just three days after 12 people, including four children, were killed when a huge avalanche struck two mountain villages near Chamonix.

Rescue workers yesterday continued to search the village of Le Tour, near Chamonix, for survivors of Tuesday's avalanche, the worst disaster in this popular resort in 91 years.

Yesterday they recovered the body of a youth, 17, the 12th victim.

From the *Newcastle Journal*, 12 February 1999

Figure 4.47

Avalanche events

Figure 4.48

Avalanches in the news, February 1999

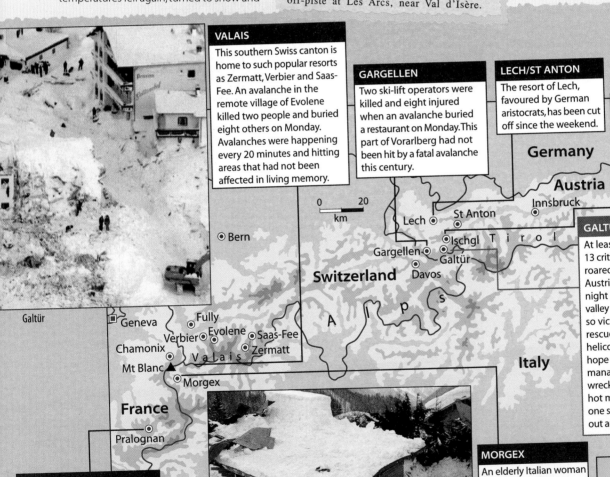

VALAIS
This southern Swiss canton is home to such popular resorts as Zermatt, Verbier and Saas-Fee. An avalanche in the remote village of Evolene killed two people and buried eight others on Monday. Avalanches were happening every 20 minutes and hitting areas that had not been affected in living memory.

GARGELLEN
Two ski-lift operators were killed and eight injured when an avalanche buried a restaurant on Monday. This part of Vorarlberg had not been hit by a fatal avalanche this century.

LECH/ST ANTON
The resort of Lech, favoured by German aristocrats, has been cut off since the weekend.

ISCHGL
One of the most popular resorts with British skiers, Ischgl has been cut off since last week. Tour operators have evacuated thousands of their clients by helicopter.

GALTUR
At least seven people were killed and 13 critically injured when an avalanche roared into this tiny town in the Austrian Tirol at 4 pm yesterday. Last night 35 people were still missing. The valley was isolated by other avalanches so victims were reliant on local rescuers. The weather was so bad that helicopters could not fly in. The best hope for the missing was that they had managed to find air pockets in the wreckage of houses. 'We were drinking hot mulled wine when it started,' said one survivor. 'Suddenly the lights went out and there was only dust and snow.

MORGEX
An elderly Italian woman was killed on Monday when an avalanche hit this hamlet near the resort of Courmayeur at the foot of Mont Blanc.

Le Tour, near Chamonix

PRALOGNAN
Blizzards and the threat of avalanches yesterday stranded rescuers trying to reach three French skiers trapped for a week in a makeshift igloo 3000 metres (10 000 ft) up in the Alps. The rescuers set out early from a refuge but had to return after wading just 150 metres (yards) through snow 5 metres (16 ft) deep above the ski resort.

Adapted from *The Independent*, 24 February

Galtür, Western Austria

The small town of Galtür is located in western Austria close to the border with Switzerland (Figure 4.49). It, together with its 700 inhabitants and estimated 3000 tourists, had already been cut off by heavy snowfalls and avalanches for several days prior to Tuesday 23 February. Figure 4.50 summarises several reports, taken from the Internet, which describe the events of that afternoon and of several subsequent days.

Galtür, situated 200 m from the base of the mountains, was considered safe from the avalanche threat. Indeed, computer – simulated tests showed that even a 1 in 150 year event would not reach the village. So why was this avalanche so much worse than anything previously recorded? Six months of investigations have shown there was an exceptional series of events before the disaster, and new processes affecting avalanches have been discovered.

1 From 20 January onwards, a sequence of three severe storms brought warm, moist air from the Atlantic which, on meeting with very cold Arctic air (Figure 9.41), gave record snowfalls of up to 4 m. Very strong winds, up to 120 km/hr and from an unusual north-west direction, increased the depth of snow on the mountains above Galtür.

2 Stability depends on the type of snow (there are some 20 main types dependent on crystal shape and spacings). As snow falls, it often forms two layers separated, as in a sandwich, by a weaker layer. Normally the weaker layer soon collapses to cause, on steep slopes, avalanches. At Galtür, warmer weather at the end of January caused melting and re-freezing until there was a much greater accumulation of snow.

3 Avalanche warnings are given on a 5-point scale. Level 5, the highest, had previously only been recorded three times in the area. That February, 16 such warnings were given, but with thousands of slopes it was impossible to predict the exact locations of avalanches.

4 Initial estimates suggest the 'slab' of ice and snow that broke away from the mountainside (Figure 4.44) was

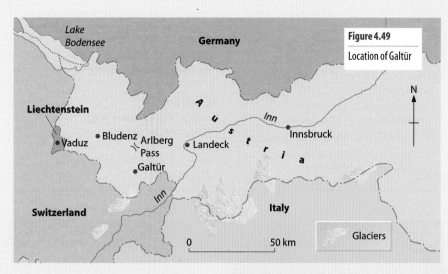

Figure 4.49
Location of Galtür

N

500 m wide and had a mass (weight) of 170 000 tonnes (yet even with this data, the computer showed that the avalanche could not have reached the village).

5 Scientists already knew that wet snow is slow-moving ('slab' avalanche) and that dry, light snow, with denser snow at its base, is fast-moving ('powder' avalanche). What they only discovered in 1999 (too late for Galtür) was that powder avalanches can also have a saltation layer (compare Figure 3.21), consisting of different sized particles, at their base, and that it is this layer that enables an avalanche to travel further.

6 Two weeks before the Galtür event, scientists made a controlled experiment by dynamiting an area of snow and waiting in a reinforced building to take recordings as the resultant avalanche passed over them. They were horrified at the initial impact and the subsequent findings, as the latter showed that the avalanche increased in volume considerably as it moved downhill.

7 Using the new information, it was shown that the weight of snow that hit Galtür was up to 400 000 tonnes, that the avalanche was 100 m in height and travelled at 300 km/hr, and that it was the saltation layer that destroyed the buildings and the powder snow that suffocated people.

Based on 'Horizon' TV programme,
25 November 1999

Figure 4.50
Internet news: the Galtür avalanche

Tuesday 23 February 1999
A wall of snow, 5 m high, smashed through the centre of Galtür just after 4 pm. It crushed cars, hurling them across roads. Many houses were completely buried, and the snow and ice cleanly sliced off the top of one building. Snow was still falling heavily at night, with another 50 cm expected by the morning.

Wednesday 24 February
The avalanche which hit Galtür last night is reported to have killed at least 8 people, with up to 30 others missing. Residents managed to dig out alive about 20 people, although several of the survivors were said to be in a critical condition. Outside help could not at first reach the town as the main road had been blocked by an earlier avalanche, and bad weather prevented helicopters from flying in.

Thursday 25 February
The death toll from Tuesday's avalanche was put at 16, with 29 people still missing. Austrian television showed scores of rescuers using either long metal probes or specially trained sniffer dogs in an attempt to detect survivors buried under the masses of snow. A steady stream of helicopters took pallets of fresh fruit, vegetables and other foodstuffs into Galtür, returning with survivors and tourists. Estimates suggest that 2500 tourists had been helicoptered out of the town.

Saturday 27 February
Rescuers today recovered the body of a German girl believed to be the last of 38 people killed by Tuesday's avalanche. Roads to the town were open for the first time in over a week. Weather forecasters said the threat of further avalanches was diminishing as higher temperatures had reduced the amount of snow on mountainsides. Officials in Galtür claimed $20 million had been spent on avalanche protection structures. Local records showed that the town had been destroyed by an avalanche in 1689 when 250 people were killed.

References

Dawson, A. G. (1992) *Ice Age Earth*, Routledge.

Evans, D. and Benn, D. (1998) *Glaciers and Glaciation*, E. J. Arnold.

Glacial Deposits Geography Today: A Search for Order, BBC Television.

Glaciers, BBC Television/Open University.

Goudie, A. (1993) *The Nature of the Environment*, Basil Blackwell.

McCullagh, P. (1978) *Modern Concepts in Geomorphology*, Oxford University Press.

Valley Glaciers Geography Today: A Search for Order, BBC Television.

Websites

Glacier Project home page:
http://glacier.rice.edu

Glaciers:
http://www.educ.wsu.edu/esl/glacier.html

Cyberspace Snow and Avalanche Center – CSAC
http://www.csac.org/

Alaska Science Forum – Water, Snow and Ice Index:
http://dogbert.gi.alaska.edu/Science Forum/water.html

See also for more links:
http://www.nelsonthornes.com/gaia

Questions

1 a Define the terms 'interglacial' and 'interstadial'. **(4 marks)**
 b Describe the extent of ice across the British Isles at the height of the last ice advance (18 000 years ago). **(4 marks)**
 c Suggest and explain **one** theory for the cause of ice ages. **(4 marks)**
 d How is glacier ice formed? **(6 marks)**
 e Explain the **difference** in movement processes between **temperate** and **polar** glaciers. **(7 marks)**

2 Study Figure 4.25 (page 115).
 a For **one** of the features named in the figure:
  i Name it and identify the kind of glacial landform it is.
  ii Making good use of a labelled diagram, describe its shape and size. **(5 marks)**
  iii Explain how a glacier created the feature you have chosen. **(5 marks)**
  iv Describe and explain **one** change which has probably happened to the feature since the ice age. **(4 marks)**
 b Many hollows in a glaciated upland are filled with water. Where does the water come from? **(2 marks)**
 c Suggest **two** pieces of evidence you would look for to suggest the direction of movement of a glacier if you were to carry out a study of a glaciated valley. **(4 marks)**
 d Describe and explain **one** difference between a glaciated upland area and one that has not been glaciated. **(5 marks)**

3 A glacier erodes, transports and deposits material using a range of methods.
 a i Name *two* types of glacial erosion. **(2 marks)**
  ii For **one** of the types of erosion in a i, explain how the glacier erodes. **(4 marks)**
 b Some loose material is carried on top of the glacier. Making good use of diagrams, show where, on the surface, this material is carried. **(4 marks)**
 c Where else is material carried by a glacier? **(2 marks)**
 d Choose **one** of the following landforms created by glacial deposition: drumlin; end moraine; kame terrace.
  i Describe its shape, size and composition. **(6 marks)**
  ii Explain how it was created by the glacier. **(7 marks)**

4 a Study Figure 4.9 (page 106). Define the terms 'accumulation' and 'ablation' when used in relation to the budget of a glacier. **(4 marks)**
 b What reasons would explain the **advance** of the snout of a glacier? **(4 marks)**
 c What does the term 'glacial retreat' mean? **(4 marks)**
 d i What is 'pressure melting'?
  ii How is 'pressure melting' involved in the movement of glacier ice? **(6 marks)**
 e i Suggest how the pattern of movement of a glacier may be measured by fieldwork.
  ii What movement pattern would you expect to find from such field measurements, and why? **(7 marks)**

AS

5 a Identify **two** pieces of evidence to suggest that climatic change in an area has included at least **one** glacial period. For one of these pieces of evidence, show how it suggests a past glacial period. **(5 marks)**
 b i Describe how a glacier operates as an 'open system'. **(8 marks)**
  ii How and why does a glacier budget vary between *winter* and *summer* seasons? **(12 marks)**

6 a Why would geographers classify glaciers into different types? Suggest **one** system of classification. **(5 marks)**
 b i Describe the ways in which ice moves in a glacier.
  ii Why does movement of glacier ice vary across and within the glacier? **(12 marks)**
 c Explain the difference in movement between glaciers in polar and temperate latitudes. **(8 marks)**

7 For any one drainage diversion system you have studied, discuss the role of glacial ice and other factors in its formation. **(25 marks)**

8 Study the diagram of Snowdon in Figure 4.25 (page 115) and the 1:25 000 OS map of the same area (*Outdoor Leisure Map. 17 1:25 000* between map references [2]580[3]470; [2]580[3]600; [2]650[3]470; [2]650[3]600).
 a Use the OS map of the Snowdon area to locate by 6-figure grid references the following six locations marked on the sketch: **(6 marks)**
  i Cwm Dyli ii Llyn Llydaw iii Snowdon summit iv Crib Goch v Glaslyn vi Cwm Llan waterfalls.
 b Identify the **advantages** *and* **disadvantages** of using an OS map such as this to study landforms. **(6 marks)**
 c Identify **one** landform created by glacial erosion which is visible on the map and/or sketch.
  i Name and locate the landform. **(2 marks)**
  ii Making good use of a diagram, describe its shape. **(5 marks)**
  iii Explain how glacial action helped to create the landform. **(6 marks)**

A2

9 a Suggest **two** ways in which geographers could use a study of **glacial deposits**. (4 marks)

b How can glacial deposits be recognised in the field? (4 marks)

c i Describe how a **till fabric analysis** can be carried out. (4 marks)

c ii What technique could you use to best present data from a till fabric analysis? (5 marks)

d Describe what **erratics** look like. (4 marks)

e How were erratics formed? (4 marks)

10 The area in front of a glacier is a **glacifluvial** landform often called a **sandur** or an **outwash plain**.

a i Describe the characteristic deposits (shape and composition) of this area. (4 marks)

ii Explain how glacifluvial processes helped to create the characteristics you have identified. (4 marks)

b Choose **one** of the following features of a sandur: lakebed deposits; esker; kame; braided stream. Describe the shape and characteristics of the feature. (4 marks)

c i What is a **kettle lake**? (2 marks)

ii How is a kettle lake formed? (5 marks)

iii Suggest how a kettle lake may disappear after the glacial period. (6 marks)

11 a What is a 'valley glacier'? (2 marks)

b Describe and explain the origins of **two** *surface* features of a moving glacier. (6 marks)

c Explain how you could measure the movement of a valley glacier. (4 marks)

d Why does the snout of a glacier sometimes retreat even though the ice always moves forward? (6 marks)

e What feature may mark where the snout of a retreating glacier was in the past? Describe the shape and composition of the feature. (7 marks)

12 Ice movement during the last ice age had **indirect** as well as **direct** effects on the landscape. Indirect effects occur where the ice *itself* was not involved in the effect.

a i Explain what is meant by the term 'drainage diversion'. (2 marks)

ii Choose **one** example of drainage diversion. Draw a sketch map to show the diversion and explain the role of glacier ice in the cause of the diversion. (6 marks)

b Why did the land experience an **isostatic change** of sea-level during the ice age? (4 marks)

c Why are 'raised beaches' found in coastal areas where glacial ice caused an isostatic change in sea-level? (6 marks)

d Choose **one** landform (other than a raised beach) which has been affected by sea-level change. Describe it and explain how it was formed. (7 marks)

13 Study Case Study 4 (pages 124–127) and complete the following exercise.

a What is an **avalanche**? (2 marks)

b Choose **one** cause of avalanche movement and explain how it can lead to an avalanche. (4 marks)

c List **four** consequences of avalanches. (4 marks)

d Choose **one** study of an avalanche and name it.

i Describe the events and their consequences. (4 marks)

ii Suggest **one** method which could be used to reduce the risks at that place in future. (4 marks)

e 'Avalanches are more common than they used to be.' Discuss this statement in the light of your study of avalanche hazards. (7 marks)

14 a Identify **two** pieces of evidence to suggest that loose material could be of glacial origin. For one of those pieces of evidence explain how it could show a glacial origin. (5 marks)

b i Explain the difference between **lateral moraine** and **medial moraine**.

ii How is moraine spread across the surface of a glacier? (10 marks)

c Choose **one** landform created mainly by glacial deposition.

i Describe its shape, size and composition.

ii Explain the role of glacial deposition in its formation

iii Suggest how the shape may be modified by postglacial processes. (10 marks)

15 a i Define the terms 'eustatic' and 'isostatic' in relation to sea-level change. (4 marks)

ii How is glacial ice involved in sea-level change? (9 marks)

b i Describe the shape and scale of a fjord.

ii Explain the roles of glacial processes and sea level change in the formation of a fjord. (12 marks)

16 Using your own case studies, discuss the effects of the increasing development of ski resorts in glacial areas on the local environment and economy. (25 marks)

Degrees	No. of clasts	Degrees	No. of clasts	Degrees	No. of clasts
0	0	120	2	240	8
15	0	135	3	255	3
30	10	150	1	270	1
45	12	165	1	285	1
60	8	180	0	300	2
75	3	195	0	315	3
90	1	210	10	330	1
105	1	225	12	345	1

17 In a field survey (till fabric analysis) the orientation of clasts (stones) showed the following data. Orientation shows **two** possible directions (e.g. **NW/SE**).

a i Draw a graph to illustrate the data. (6 marks)

ii Using the data, suggest an interpretation of the ice movement in this area. (7 marks)

b Why do glacial deposits have a particular orientation? (7 marks)

c Suggest **two** other sources of data to indicate the direction of ice movement in an area. For **one** of these sources, explain how it shows the direction of ice movement. (5 marks)

AS

A2

Chapter 5

Periglaciation

'Perennially frozen material lurks beneath at least one-fifth, and perhaps as much as one-fourth, of the Earth's land surface.'

Frederick Nelson, 1999

The term **periglacial**, strictly speaking, means 'near to or at the fringe of an ice sheet', where frost and snow have a major impact upon the landscape. However, the term is often more widely used to include any area that has a cold climate – e.g. mountains in temperate latitudes – or which has experienced severe frost action in the past – e.g. southern England during the Quaternary ice age (Figure 4.3b). Today, the most extensive periglacial areas lie in the Arctic regions of Canada, the USA and Russia. These areas, which have a tundra climate, soils and vegetation (pages 333–334), exhibit their own characteristic landforms.

Permafrost

Permafrost is permanently frozen ground. It occurs where soil temperatures remain below 0°C for at least two consecutive years. Permafrost covers almost 25 per cent of the Earth's land surface (Figure 5.1) although its extent changes over periods of time. Its depth and continuity also vary (Figure 5.2).

Figure 5.1

Permafrost zones of the Arctic

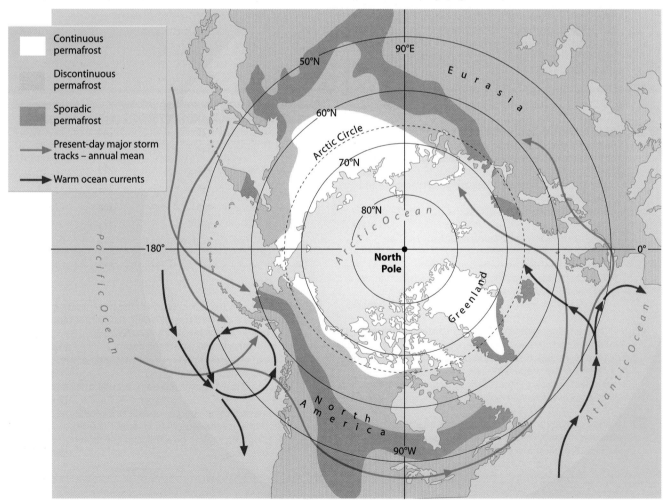

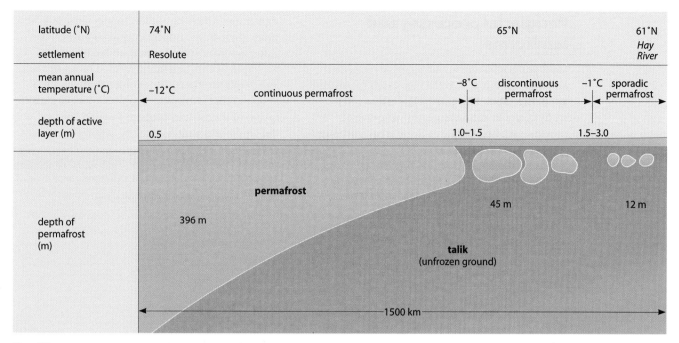

latitude (°N)	74°N			65°N		61°N
settlement	Resolute					Hay River
mean annual temperature (°C)	–12°C	continuous permafrost		–8°C discontinuous permafrost	–1°C sporadic permafrost	
depth of active layer (m)	0.5			1.0–1.5	1.5–3.0	
depth of permafrost (m)	396 m	permafrost		45 m	12 m	
			talik (unfrozen ground)			
		1500 km				

Figure 5.2

Transect through part of the permafrost zone in northern Canada

Figure 5.3

Soil temperatures in permafrost at Yakutsk, Siberia

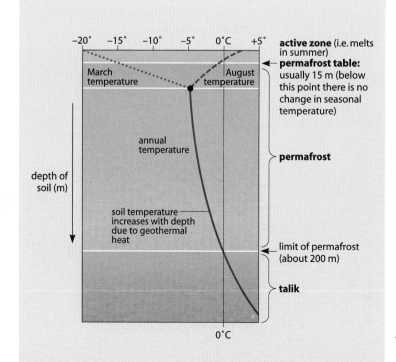

Continuous permafrost is found mainly within the Arctic Circle where the mean annual air temperature is below –5°C. Here winter temperatures may fall to –50°C and summers are too cold and too short to allow anything but a superficial melting of the ground. The permafrost has been estimated to reach a depth of 700 m in northern Canada and 1500 m in Siberia. As Figure 5.1 shows, continuous permafrost extends further south in continental interiors than in coastal areas which are subject to the warming influence of the sea, e.g. the North Atlantic Drift in north-west Europe.

Discontinuous permafrost lies further south in the northern hemisphere, reaching 50°N in central Russia, and corresponds to those areas with a mean annual temperature of between –1°C and –5°C. As is shown in Figure 5.2, discontinuous permafrost consists of islands of permanently frozen ground, separated by less cold areas which lie near to rivers, lakes and the sea.

Sporadic permafrost is found where mean annual temperatures are just below freezing point and summers are several degrees above 0°C. This results in isolated areas of frozen ground (Figure 5.2).

In areas where summer temperatures rise above freezing point, the surface layer thaws to form the **active layer**. This zone, which under some local conditions can become very mobile for a few months before freezing again, can vary in depth from a few centimetres (where peat or vegetation cover protects the ground from insolation) to 5 m. The active layer is often saturated because meltwater cannot infiltrate downwards through the impermeable permafrost. Meltwater is unlikely to evaporate in the low summer temperatures or to drain downhill since most of the slopes are very gentle. The result is that permafrost regions contain many of the world's few remaining wetland environments.

The unfrozen layer beneath, or indeed any unfrozen material within, the permafrost is known as *talik*. The lower limit of the permafrost is determined by geothermal heat which causes temperatures to rise above 0°c (Figure 5.3).

Temperatures taken over a period of years in the discontinuous and continuous permafrost suggest that, in both Alaska and Russia, there is a general thawing of the frozen ground, an event accredited to global warming (Case Study 5).

Periglacial processes and landforms

Most periglacial regions are sparsely populated and underdeveloped. Until the search for oil and gas in the 1960s, there had been little need to study or understand the geomorphological processes which operate in these areas. Although significant strides have been made in the last 30 years, there is still uncertainty as to how certain features have developed and, indeed, whether such features are still being formed today or are a legacy of a previous, even colder climate – i.e. a fossil or relict landscape. Figure 5.4 gives a classification of the various processes which operate, and the landforms which develop, in periglacial areas.

	Processes	Landforms
Ground ice	Ice crystals and lenses (frost-heave)	Sorted stone polygons (stone circles and stripes: patterned ground)
	Ground contraction	Ice wedges with unsorted polygons: patterned ground
	Freezing of groundwater	Pingos
Frost weathering	Frost shattering/Freeze–thaw	Blockfields, talus (scree), tors (Chapter 8)
Snow	Nivation	Nivation hollows
Meltwater	Solifluction	Solifluction sheets, rock streams
	Streams	Braiding, dry valleys in chalk (Chapter 8)
Wind	Windblown	Loess (limon)

Ground ice

Frost-heave: ice crystals and lenses

Frost-heave includes several processes which cause either fine-grained soils such as silts and clays to expand to form small domes, or individual stones within the soil to be moved to the surface (Figure 5.5). It results from the direct formation of ice – either as crystals or as lenses. The **thermal conductivity** of stones is greater than that of soil. As a result, the area under a stone becomes colder than the surrounding soil, and ice crystals form. Further expansion by the ice widens the capillaries in the soil, allowing more moisture to rise and to freeze. The crystals, or the larger ice lenses which form at a greater depth, force the stones above them to rise until eventually they reach the surface. (Ask a gardener in northern Britain to explain why a plot that was left stoneless in the autumn has become stone-covered by the spring, following a cold winter.)

During periods of thaw, meltwater leaves fine material under the uplifted stones, preventing them from falling back into their original positions. In areas of repeated freezing (ideally where temperatures fall to between –4°C and –6°C) and thawing, frost-heave both lifts and sorts material to form **patterned ground** on the surface (Figure 5.6). The larger stones, with their extra weight, move outwards to form, on almost flat areas, stone circles or, more accurately, **stone polygons**. Where this process occurs on slopes with a gradient in excess of 2°, the stones will slowly move downhill under gravity to form elongated **stone stripes**.

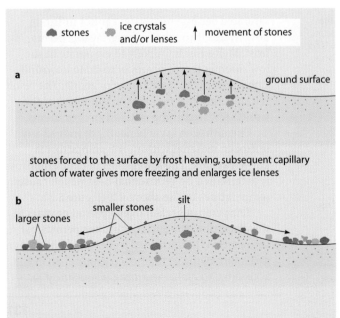

a stones forced to the surface by frost heaving, subsequent capillary action of water gives more freezing and enlarges ice lenses

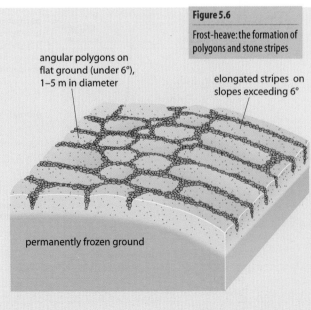

angular polygons on flat ground (under 6°), 1–5 m in diameter

elongated stripes on slopes exceeding 6°

permanently frozen ground

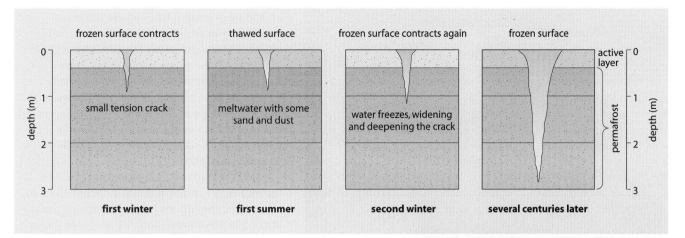

frozen surface contracts | thawed surface | frozen surface contracts again | frozen surface

depth (m) 0 1 2 3

small tension crack

meltwater with some sand and dust

water freezes, widening and deepening the crack

active layer

permafrost

depth (m) 0 1 2 3

first winter | **first summer** | **second winter** | **several centuries later**

Figure 5.7

The formation of ice wedges

Figure 5.9

Fossil ice wedge

Ground contraction

The refreezing of the active layer during the severe winter cold causes the soil to contract. Cracks open up which are similar in appearance to the irregularly shaped polygons found on the bed of a dried-up lake. During the following summer, these cracks open close or fill with meltwater and, sometimes, also with water and wind-blown deposits. When the water refreezes, during the following winter the cracks widen and deepen to form **ice wedges** (Figure 5.7). This process is repeated annually until the wedges, which underlie the perimeters of the polygons, grow to as much as 1 m in width and 3 m in depth. **Fossil ice wedges**, i.e. cracks filled with sands and silt left by meltwater, are a sign of earlier periglacial conditions (Figure 5.9). Patterned ground (Figure 5.8) can, therefore, be produced by two processes: frost-heaving (Figure 5.6) and ground contraction (Figure 5.7). Frost-heaving results in small dome-shaped polygons with larger stones found to the outside

Figure 5.8

Ice wedge polygons near Barrow, Alaska. The patterned ground is formed by polygons up to 30 m in diameter. The polygon boundaries mark the position of the ice wedges.

of the circles, whereas ice contraction produces larger polygons with the centre of the circles depressed in height and containing the bigger stones. The diameter of an individual polygon can reach over 30 m.

Freezing of groundwater

Pingos are dome-shaped, isolated hills which interrupt the flat tundra plains (Figure 5.10). They can have a diameter of up to 500 m and may rise 50 m in height to a summit that is sometimes ruptured to expose an icy core. As they occur mainly in sand, they are not susceptible to frost-heaving. American geographers recognise two types of pingo (Figure 5.11a and b), although recent investigations have led to the suggestion of a third type: **polygenetic** (or mixed) pingos.

Figure 5.10

A pingo, Mackenzie Delta, Canada

Open-system (hydraulic) pingos occur in valley bottoms and in areas of thin or discontinuous permafrost. Surface water is able to infiltrate into the upper layers of the ground where it can circulate in the unfrozen sediments before freezing. As the water freezes, it expands and forms localised masses of ice. The ice forces any overlying sediment upwards into a dome-shaped feature, in the same way that frozen milk lifts the cap off its bottle. This type of pingo, referred to as the **East Greenland type**, grows from below (Figure 5.11a).

Figure 5.11

Formation of pingos

Closed-system (hydrostatic) pingos are more characteristic of flat, low-lying areas where the permafrost is continuous. They often form on the sites of small lakes where water is trapped (**enclosed**) by freezing from above and by the advance of the permafrost inwards from the lake margins. As the water freezes it will expand, forcing the ground above it to rise upwards into a dome shape. This type of pingo is known as the **Mackenzie type** as over 1400 have been recorded in the delta region of the River Mackenzie. It results from the downward growth of the permafrost (Figure 5.11b).

As the surface of a pingo is stretched, the summit may rupture and crack. Where the ice-core melts, the hill may collapse leaving a meltwater-filled hollow (Figure 5.11c). Later, a new pingo may form on the same site, and there may be a repeated cycle of formation and collapse.

Frost weathering

Mechanical weathering is far more significant in periglacial areas than is chemical weathering, with freeze–thaw being the dominant process (Figure 2.10). On relatively flat upland surfaces, e.g. the Scafell range in the Lake District and the Glyders in Snowdonia, the extensive spreads of large, angular boulders, formed *in situ* by frost action, are known as **blockfields** or **felsenmeer** (literally, a 'rock sea').

Scree, or **talus**, develops at the foot of steep slopes, especially those composed of well-jointed rocks prone to frost action. Freeze–thaw

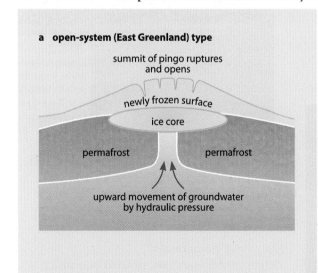

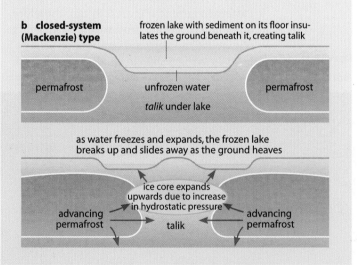

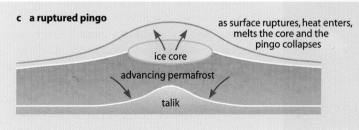

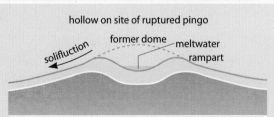

may also turn well-jointed rocks, such as granite, into **tors** (page 202). One school of thought on tor formation suggests that these landforms result from frost shattering, with the weathered debris later having been removed by solifluction. If this is the case, tors are therefore a relict (fossil) of periglacial times.

Snow

Snow is the agent of several processes which collectively are known as **nivation** (page 111). These nivation processes, sometimes referred to as 'snowpatch erosion', are believed to be responsible for enlarging hollows on hillsides. Nivation hollows are still actively forming in places like Iceland, but are relict features in southern England (as on the scarp slope of the South Downs behind Eastbourne).

Figure 5.12

Solifluction sheet in the Ogilvie Mountains, Yukon, Canada

Meltwater

During periods of thaw, the upper zone (active layer) melts, becomes saturated and, if on a slope, begins to move downhill under gravity by the process of solifluction (page 47). Solifluction leads to the infilling of valleys and hollows by sands and clays to form **solifluction sheets** (Figures 5.12 and 5.13a) or, if the source of the flow was a nivation hollow, a rock stream (Figure 5.20). Solifluction deposits, whether they have in-filled valleys or have flowed over cliffs, as in southern England, are also known as **head** or, in chalky areas, **coombe** (Figure 5.13b).

The chalklands of southern England are characterised by numerous dry valleys (Figure 8.11). The most favoured of several hypotheses put forward to explain their origin suggests that the valleys were carved out under periglacial conditions. Any water in the porous chalk at this time would have frozen, to produce permafrost, leaving the surface impermeable. Later, meltwater rivers would have flowed over this frozen ground to form V-shaped valleys (page 200).

Rivers in periglacial areas have a different regime from those flowing in warmer climates. Many may stop flowing altogether during the long and very cold winter (Figure 5.14) and have a peak discharge in late spring or early summer when melting is at its maximum (Places 18). With their high velocity, these rivers are capable of transporting large amounts of material when at their peak flow. Later in the year, when river levels fall rapidly, much of this material will be deposited, leaving a braided channel (Figures 3.32 and 5.16).

Figure 5.13

Formation of solifluction sheet and head

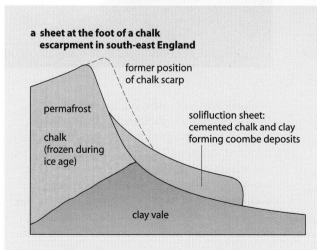

a sheet at the foot of a chalk escarpment in south-east England

former position of chalk scarp

permafrost

chalk (frozen during ice age)

solifluction sheet: cemented chalk and clay forming coombe deposits

clay vale

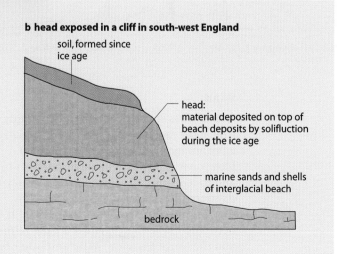

b head exposed in a cliff in south-west England

soil, formed since ice age

head: material deposited on top of beach deposits by solifluction during the ice age

marine sands and shells of interglacial beach

bedrock

Figure 5.14

Model of a river regime in a periglacial area

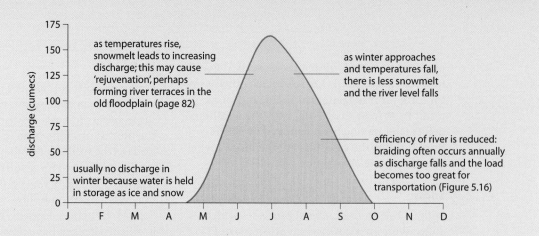

as temperatures rise, snowmelt leads to increasing discharge; this may cause 'rejuvenation', perhaps forming river terraces in the old floodplain (page 82)

as winter approaches and temperatures fall, there is less snowmelt and the river level falls

efficiency of river is reduced: braiding often occurs annually as discharge falls and the load becomes too great for transportation (Figure 5.16)

usually no discharge in winter because water is held in storage as ice and snow

Places 18   Alaska: periglacial river regimes

Permafrost also affects the hydrological regimes of subarctic rivers. Figure 5.15 shows the regime of two Alaskan rivers, both of which flow in first order drainage basins (page 65). One river, however, is located in northern Alaska where over 50 per cent of the basin is underlain with continuous permafrost. The other river, in contrast, is located further south where most of the basin consists of discontinuous permafrost and only 3 per cent is continuous permafrost. The northern river, flowing over more impermeable ground (more permafrost giving increased surface runoff and reduced

throughflow) responds much more readily to changes in both temperature (increased snowmelt or freezing) and rainfall (amounts and seasonal distribution). It has a more extreme regime showing that it is more likely to flood in summer and to have a higher peak discharge and then to dry up sooner, and for a longer period, in winter or during dry spells. Figure 5.16 was taken on 7 August 1996 in the Dynali National Park. The river level had already fallen (as had the first snow of winter!), and the large load carried by the early summer meltwaters had already been deposited.

Figure 5.15

Contrasting regimes of rivers flowing over continuous and discontinuous permafrost

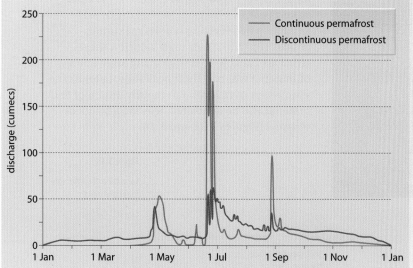

Figure 5.16

A river in the Dynali National Park

Wind

A lack of vegetation and a plentiful supply of fine, loose material (i.e. silt) found in glacial environments enabled strong, cold, out-blowing winds to pick up large amounts of dust and to redeposit it as **loess** in areas far beyond its source. Loess covers large areas in the Mississippi–Missouri valley in the USA. It also occurs across France (where it is called **limon**) and the North European Plain and into north-west China (where in places it exceeds 300 m

in depth and forms the yellow soils of the Huang He valley – see Case Study 10). In all areas, it gives an agriculturally productive, fine-textured, deep, well-drained and easily worked soil which is, however, susceptible to further erosion by water and wind if not carefully managed (Figure 10.35). Large tracts of central Europe, other than those consisting of loess, are covered in dunes (coversands) which were formed by wind deposition during peri-glacial times.

Thermokarst

Thermokarst is a landscape where the ground surface is very uneven. It develops when masses of ground ice melt. This arises from an increase in the depth of the active layer which in turn causes parts of the land surface to subside. Thermokarst is, therefore, the general name given to irregular, hummocky terrain with marshy or lake-filled hollows created by the disruption of the thermal equilibrium of the permafrost (Figure 5.17). Russian geographers claim that thermokarst only develops where:

1 there are either large amounts of ground ice (ice wedges, ice lenses and ice sheets) or unconsolidated sediments with a high ice content (where ice forms over 70 per cent of the volume)
2 the depth of seasonal or perennial thawing is greater **a** than the depth of ice wedges and **b** where it reaches ice-rich rocks.

Although thermokarst has, in the past – and presumably again in the future – developed due to climatic change, increasingly in the last few decades the major agent has been the human occupancy and development of periglacial areas. The thermal equilibrium, which is very delicate, may be upset by a range of human activities (Figure 5.19) which may increase or reduce the thermokarst.

a Increase in the thermokarst – i.e. the level of permafrost table is lowered

- Removal of mosses and other tundra vegetation (especially in the continuous permafrost) and forest (more likely in the discontinuous permafrost) for construction purposes means that in summer more heat penetrates the soil and so the depth of thaw increases, as does the likelihood of flooding.
- Construction of centrally heated buildings has warmed the ground underneath them, causing the buildings to subside.
- Siting of oil, sewerage and water pipes in the active zone has increased the rate of thaw, sometimes causing fracturing of the pipes as the ground moves. Similar earth movements have caused roads and railways to lose alignment and dams and bridges to crack.
- Drilling for oil and gas poses problems because the heat from the drilling fluid melts the permafrost. Having enlarged the drilling hole, the machinery vibrates and is less effective.
- Road construction upsets the delicate equilibrium of many slopes produced by solifluction during different climatic conditions.

Figure 5.18

Inuvik, Mackenzie Delta, Canada

Figure 5.17

Thermokarst scenery on the North Slope, Alaska, with the Brooks Range behind

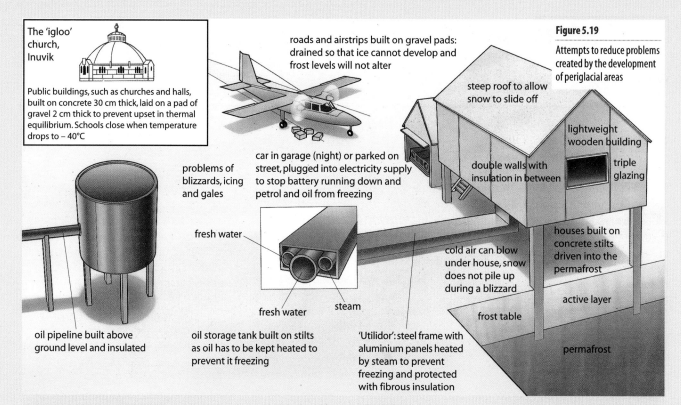

The 'igloo' church, Inuvik

Public buildings, such as churches and halls, built on concrete 30 cm thick, laid on a pad of gravel 2 cm thick to prevent upset in thermal equilibrium. Schools close when temperature drops to – 40°C

roads and airstrips built on gravel pads: drained so that ice cannot develop and frost levels will not alter

steep roof to allow snow to slide off

lightweight wooden building

double walls with insulation in between

triple glazing

problems of blizzards, icing and gales

car in garage (night) or parked on street, plugged into electricity supply to stop battery running down and petrol and oil from freezing

fresh water

cold air can blow under house, snow does not pile up during a blizzard

houses built on concrete stilts driven into the permafrost

active layer

frost table

fresh water

steam

permafrost

oil pipeline built above ground level and insulated

oil storage tank built on stilts as oil has to be kept heated to prevent it freezing

'Utilidor': steel frame with aluminium panels heated by steam to prevent freezing and protected with fibrous insulation

Figure 5.19

Attempts to reduce problems created by the development of periglacial areas

b Reduction in the thermokarst – i.e. the level of permafrost table is raised

Development may also upset the thermal balance in the opposite direction. The construction of unheated buildings, for example, reduces the already low amounts of heat received during the short summer. This causes the upper surface of permafrost to rise and buildings to tilt. Similarly, early road construction increased the permafrost, and several Arctic highways now run nearly 2 m above the surrounding land. To offset this, all large, modern unheated buildings and airstrips are constructed on thick gravel pads, as gravel is relatively immune to frost-heaving and has less tendency to thaw the underlying permafrost. Heat extractors are installed to try to prevent the extension of the thermokarst.

Human attempts to exploit periglacial regions commercially have only been made relatively recently. They have, so far, often been made without sufficient attention being paid to the seemingly universal problem of trying to balance short-term economic gain with the often longer-term environmental loss. The result has been an upset in the thermal equilibrium of many places, and destruction of the fragile environment.

c The potential effects of global warming

Temperature measurements have been taken since the late 1970s in drill holes in the undisturbed, discontinuous permafrost near Fairbanks in central Alaska. The measurements showed that temperatures at the permafrost table (the height of the permafrost) had remained steady at about –3.5°C from 1983 to 1990 but then increased by +1.5°C, to –2°C, between 1990 and 1993. Scientists suggest that this increase is a direct result of global warming (Case Study 9B). Other measurements, taken along a north–south transect adjacent to the Alaskan pipeline, suggest that the depth of the active layer is increasing and the depth of the permafrost table is decreasing. This could mean an increase in:

- the thermokarst
- the release of organic carbon (the precursor for carbon dioxide and methane emissions) which at present is stored in the permafrost
- the depth of the active layer by 20 to 30 per cent by the year 2050.

Scientists admit, however, that they are unsure as to how global warming and higher temperatures might affect the permafrost, or how thawing ground ice could affect climate. Warmer, wetter weather might, for example, increase vegetation growth which could insulate the permafrost and keep the near-surface ground layers frozen.

References

French, H. M. (1996) *The Periglacial Environment*, Addison Wesley Longman Higher Education.

Goudie, A. (1993) *The Nature of the Environment*, Basil Blackwell.

McCullagh, P. (1978) *Modern Concepts in Geomorphology*, Oxford University Press.

Websites

Permafrost in Canada:
http://ellesmere.ccm.emr.ca/wwwnais/select/pfrost/english/html/epfrost.html

See also for more links:
http://www.nelsonthornes.com/gaia

1 a What is **permafrost**? **(4 marks)**

b Why is permafrost found at 50°N in North America but only at 66°N in Western Europe? **(6 marks)**

c Explain the **difference between** 'continuous' permafrost and 'discontinuous' permafrost. **(4 marks)**

d i Why is loose earth formed into mounds in areas of permafrost? **(4 marks)**

ii Making good use of diagrams, explain how individual stones are moved to the surface of areas with permafrost. **(7 marks)**

2 Study Figure 5.1 (page 130), which shows where there is permafrost in the northern hemisphere, and Figure 5.2 (page 131).

a i Where is the place *closest* to the North Pole where there is no permafrost?

ii How close to the North Pole is this place? **(2 marks)**

b i From the diagram suggest **two** reasons why there is no permafrost in some places while there is in other places. Give examples from the diagram to support your answer. **(6 marks)**

iii Identify the cause/s of the 'pocket' of permafrost in north-west Scandinavia. **(2 marks)**

c What is the 'active layer' in permafrost like? **(3 marks)**

d i What is meant by the term 'mean annual temperature'? **(3 marks)**

ii How deep is **a** the active layer and **b** the permafrost at Resolute Bay? **(2 marks)**

iii Use data from Figure 5.2 to suggest the relationship between depth of permafrost and latitude. **(2 marks)**

e Why does the permafrost not occur throughout the crustal rocks? **(5 marks)**

3 a Describe the shape and scale of **two** of the following periglacial landforms: ice wedge polygons; scree; nivation hollow; solifluction terracettes. **(6 marks)**

b For **one** of the landforms you have described in a, explain how periglacial processes have led to its formation. **(6 marks)**

c Figure 5.10 (page 134) shows a pingo in northern Canada. Write a description of the pingo from the photograph, including the area around it and its scale. **(6 marks)**

d How is a pingo formed? **(7 marks)**

4 Study Figure 5.14 (page 136) which shows the flow of a river (its regime) in a periglacial area.

a i When does water *not* flow in this river? **(2 marks)**

ii Why does water not flow during this time? **(3 marks)**

iii How would you recognise 'river terraces cut into the old floodplain' by such a river? **(5 marks)**

b Using diagrams in your answer, explain the meaning of the term 'braiding' as used in the diagram. **(5 marks)**

c Give **two** reasons why the wind has a greater erosional effect in periglacial environments than in most other areas. **(5 marks)**

d How could you recognise that the wind had:

i removed material from one area and

ii deposited the material elsewhere? **(5 marks)**

6 'Changes to soil stability due to frost are a major problem for development in regions where there is a periglacial climate.' Using examples you have studied, explain why this could be the case, and describe methods people use to overcome the problems of living in such areas. **(25 marks)**

7 Study Figure 5.20 which shows a range of periglacial landforms and their locations.

a Explain the processes which are operating in the *snow patch* (A). **(5 marks)**

b Choose **one** of the landforms labelled B to H. Describe its size and location in the field and suggest how it has been formed. **(8 marks)**

c Explain the role of i wind and ii meltwater in the formation of landforms in areas of periglacial landscape. **(12 marks)**

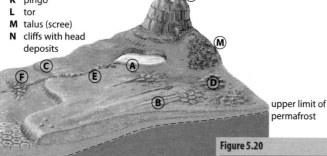

y

nivation hollow with snow patch
stone polygons, garlands and stripes
solifluction sheets/benches
blockfield
rock stream
debris fan

G braided stream
H ice-wedge polygons
K pingo
L tor
M talus (scree)
N cliffs with head deposits

0   500 m
horizontal scale

upper limit of permafrost

sea

upper limit of bedrock

Figure 5.20

Landsketch showing typical landforms found in a periglacial area

5 a Describe and explain the formation of stone polygons (patterned ground) in areas which experience permafrost. **(12 marks)**

b Why does loose material move down slopes rapidly in periglacial areas? **(8 marks)**

c Identify **two** pieces of evidence of periglacial action which may be found in areas which today have no permafrost. Explain how they show periglacial action. **(5 marks)**

8 a Identify **two** pieces of evidence that periglacial conditions once existed in the British Isles. For **each** explain how the processes operating in these areas in past periglacial conditions helped to cause the evidence. **(10 marks)**

b i Using Figure 5.10 (page 134) and other information you possess, describe the vegetation of modern permafrost areas. **(7 marks)**

ii Explain how the vegetation is adapted to live in this environment. **(8 marks)**

Chapter 6

Coasts

'A recent estimate of the coastline of England and Wales is 2750 miles and it is very rare to find the same kind of coastal scenery for more than 10 to 15 miles together.'

J. A. Steers, *The Coastline of England and Wales*, 1960

'I do not know what I may appear to the world; but to myself I seem to have been only a boy playing on the sea-shore, and diverting myself in now and then finding a smoother pebble or a prettier shell than ordinary, while the great ocean of truth lay all undiscovered before me.'

Isaac Newton, *Philosophiae Naturalis Principia Mathematica*, 1687

The coast is a narrow zone where the land and the sea overlap and directly interact. Its development is affected by terrestrial, atmospheric, marine and human processes (Figure 6.1) and their interrelationships. The coast is the most varied and rapidly changing of all landforms and ecosystems.

Waves

Waves are created by the transfer of energy from the wind blowing over the surface of the sea. (An exception to this definition is those waves – **tsunamis** – that result from submarine shock waves generated by earthquake or volcanic activity.) As the strength of the wind increases, so too does **frictional drag** and the size of the waves. Waves that result from local winds and travel only short distances are known as **sea**, whereas those waves formed by distant storms and travelling large distances are referred to as **swell**.

The energy acquired by waves depends upon three factors: the wind velocity, the period of time during which the wind has blown, and the length of the fetch. The **fetch** is the maximum distance of open water over which the wind can blow, and so places with the greatest fetch potentially receive the highest-energy waves. Parts of south-west England are exposed to the Atlantic Ocean and when the south-westerly winds blow it is possible that some waves may have originated several thousand kilometres away. The Thames estuary, by comparison, has less open water between it and the Continent and consequently receives lower-energy waves.

Figure 6.1

Factors affecting coasts

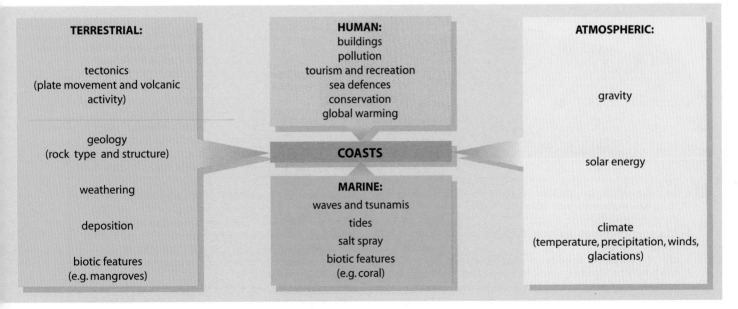

TERRESTRIAL:

tectonics
(plate movement and volcanic activity)

geology
(rock type and structure)

weathering

deposition

biotic features
(e.g. mangroves)

HUMAN:
buildings
pollution
tourism and recreation
sea defences
conservation
global warming

COASTS

MARINE:
waves and tsunamis
tides
salt spray
biotic features
(e.g. coral)

ATMOSPHERIC:

gravity

solar energy

climate
(temperature, precipitation, winds, glaciations)

Wave terminology

The **crest** and the **trough** are respectively the highest and lowest points of a wave (Figure 6.2).

Wave height (H) is the distance between the crest and the trough. The height has to be estimated when in deep water. Wave height rarely exceeds 6 m although freak, unexpected waves can be a hazard to life, property and shipping.

Wave period (T) is the time taken for a wave to travel through one wave length. This can be timed either by counting the number of crests per minute or by timing 11 waves and dividing by 10 – i.e. the number of intervals.

Wave length (L) is the distance between two successive crests. It can be determined by the formula:

$$L = 1.56\ T^2$$

Wave velocity (C) is the speed of movement of a crest in a given period of time.

Wave steepness ($H \div L$) is the ratio of the wave height to the wave length. This ratio cannot exceed 1:7 (0.14) because at that point the wave will break. Steepness determines whether waves will build up or degrade beaches. Most waves have a steepness of between 0.005 and 0.05.

The **energy** (E) of a wave in deep water is expressed by the formula:

$$E \propto \text{(is proportional to)}\ LH^2$$

This means that even a slight increase in wave height can generate large increases in energy. It is estimated that the average pressure of a wave in winter is 11 tonnes per m^2, but this may be three times greater during a storm – it is little wonder that under such conditions sea defences may be destroyed and that wave power is a potential source of renewable energy (page 541).

Swell is characterised by waves of low height, gentle steepness, long wave length and a long period. **Sea**, with opposite characteristics, usually has higher-energy waves.

Waves in deep water

Deep water is when the depth of water is greater than one-quarter of the wave length:

$$(D\ =\ > \tfrac{L}{4})$$

The drag of the wind over the sea surface causes water and floating objects to move in an **orbital motion** (Figure 6.3). Waves are surface features (submerged submarines are unaffected by storms) and therefore the sizes of the orbits decrease rapidly with depth. Any floating object in the sea has a small net horizontal movement but a much larger vertical motion.

Waves in shallow water

As waves approach shallow water, i.e. when their depth is less than one-quarter of the wave length,

$$(D\ =\ < \tfrac{L}{4})$$

friction with the seabed increases. As the base of the wave begins to slow down, the circular oscillation becomes more elliptical (Figure 6.4). As the water depth continues to decrease, so does the wave length.

Meanwhile the height and steepness of the wave increase until the upper part spills or plunges over. The point at which the wave breaks is known as the **plunge line**. The body of foaming water which then rushes up the beach is called the **swash**, while any water returning down to the sea is the **backwash**.

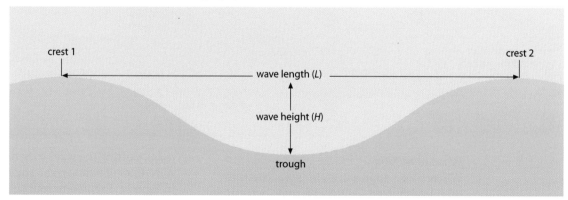

Figure 6.2

Wave terminology

Figure 6.3

Movement of an object in deep water: the diagrams show the circular movement of a ball or piece of driftwood through five stages in the passage of one wave length (crest 1 to crest 2); although the ball moves vertically up and down and the wave moves forward horizontally, there is very little horizontal movement of the ball until the wave breaks; the movement is orbital and the size of the orbit decreases with depth

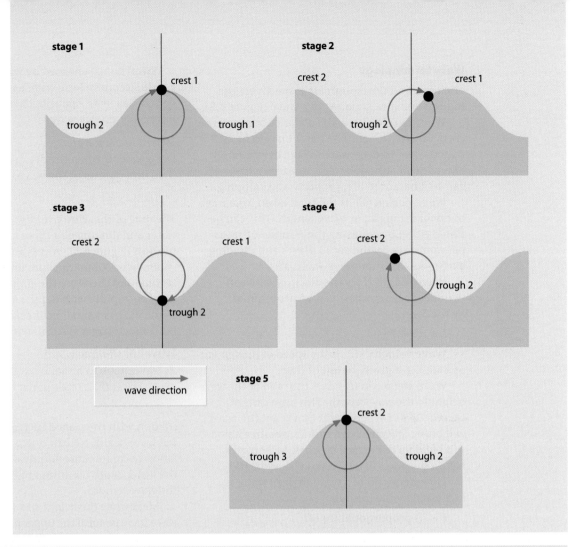

stage 1

crest 1

trough 2   trough 1

stage 2

crest 2   crest 1

trough 2

stage 3

crest 2   crest 1

trough 2

stage 4

crest 2

trough 2

wave direction

stage 5

crest 2

trough 3   trough 2

Figure 6.4

Why a wave breaks

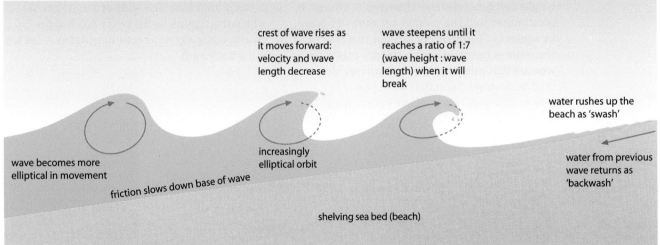

crest of wave rises as it moves forward: velocity and wave length decrease

wave steepens until it reaches a ratio of 1:7 (wave height : wave length) when it will break

water rushes up the beach as 'swash'

wave becomes more elliptical in movement

increasingly elliptical orbit

water from previous wave returns as 'backwash'

friction slows down base of wave

shelving sea bed (beach)

Wave refraction

Where waves approach an irregular coastline, they are refracted, i.e. they become increasingly parallel to the coastline. This is best illustrated where a headland separates two bays (Figure 6.5). As each wave crest nears the coast, it tends to drag in the shallow water near to a headland, or indeed any shallow water, so that the portion of the crest in deeper water moves forward while that in shallow water is retarded (by frictional drag), causing the wave to bend. The **orthogonals** (lines drawn at right-angles to wave crests) in Figure 6.5 represent four stages in the advance of a particular wave crest. It is apparent from the convergence of lines S^1, S^2, S^3 and S^4 that wave energy becomes concentrated upon, and so accentuates erosion at, the headland. The diagram also shows the formation of **longshore** (littoral) **currents**, which carry sediment away from the headland.

Figure 6.5

Wave refraction at a
headland

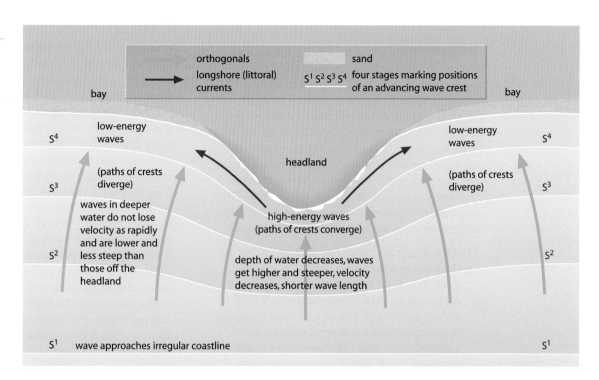

orthogonals

longshore (littoral)
currents

sand

S^1 S^2 S^3 S^4 four stages marking positions
of an advancing wave crest

bay

bay

low-energy
waves

S^4

S^3

(paths of crests
diverge)

waves in deeper
water do not lose
velocity as rapidly
and are lower and
less steep than
those off the
headland

S^2

headland

low-energy
waves

S^4

(paths of crests
diverge)

S^3

S^2

high-energy waves
(paths of crests converge)

depth of water decreases, waves
get higher and steeper, velocity
decreases, shorter wave length

S^1

wave approaches irregular coastline

S^1

Beaches

Beaches may be divided into three sections –
backshore (upper), **foreshore** (lower) and
nearshore – based upon the influence of waves
(Figure 6.6). A beach forms a buffer zone
between the waves and the coast. If the beach
proves to be an effective buffer, it will dissipate
wave energy without experiencing any net
change itself. Because it is composed of loose
material, a beach can rapidly adapt its shape to
changes in wave energy. It is, therefore, in
dynamic equilibrium with its environment
(Framework 3, page 45).

Beach profiles fall between two extremes:
those that are wide and relatively flat; and those
that are narrow and steep. The gradient of natural
beaches is dependent upon the interrelationship
between two main variables:

■ **Wave energy** Field studies have shown a
close relationship between the profile of a
beach and the action of two types of wave:
constructive and destructive (page 144).
However, the effect of wave steepness on beach
profiles is complicated by the second variable.

■ **Particle size** There is also, due to differences
in the relative dissipation of wave energy, a
distinct relationship between beach slope and
particle size. This relationship is partly due to
grain size and partly to percolation rates, both
of which are greater on shingle beaches than
on sand (pages 145–46). Consequently,
shingle beaches are steeper than sand
beaches (Figure 6.6).

Figure 6.6

Wave zones and beach morphology
(*after* King, 1980)

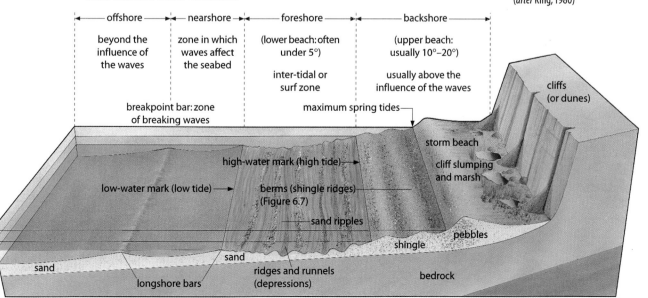

offshore

nearshore

foreshore

backshore

beyond the
influence of
the waves

zone in which
waves affect
the seabed

(lower beach: often
under 5°)

inter-tidal or
surf zone

(upper beach:
usually 10°–20°)

usually above the
influence of the waves

cliffs
(or dunes)

breakpoint bar: zone
of breaking waves

maximum spring tides

storm beach

cliff slumping
and marsh

high-water mark (high tide)

low-water mark (low tide)

berms (shingle ridges)
(Figure 6.7)

sand ripples

pebbles

shingle

sand

sand

ridges and runnels
(depressions)

bedrock

longshore bars

Types of wave

It is widely accepted that there are two extreme wave types that affect the shape of a beach. However, whereas the extreme types have, in the past, been labelled **constructive** and **destructive** (Figure 6.7, and Andrew Goudie *The Nature of the Environment*), it is now becoming more usual to use the terms **high energy** and **low energy** (Figure 6.8, and John Pethick *An Introduction to Coastal Geomorphology*). Note that 'high-energy waves' and 'low-energy waves' are *not* synonymous terms for 'constructive waves' and 'destructive waves'.

Constructive and destructive waves

■ **Constructive waves** often form where the fetch distance is long. They are usually small (or low) waves, flat in form and with a long wave length (up to 100 m) and a low frequency (a wave period of 6 to 8 per minute). On approaching a beach, the wave front steepens relatively slowly until the wave gently 'spills' over (Figure 6.7a). As the resultant swash moves up the beach, it rapidly loses volume and energy due to water percolating through the beach material. The result is that the back-wash, despite the addition of gravity, is weak and has insufficient energy either to transport sediment back down the beach or to impede the swash from the following wave. Consequently sand and shingle is slowly, but constantly, moved up the beach. This will gradually increase the gradient of the beach and leads to the formation of **berms** at its crest (Figures 6.9 and 6.10) and, especially on sandy beaches, **ridges** and **runnels** (Figure 6.6).

■ **Destructive waves** are more common where the fetch distance is shorter. They are often large (or high) waves, steep in form and with a short wave length (perhaps only 20 m) and a high frequency (10 to 14 per minute). These waves, on approaching a beach, steepen rapidly until they 'plunge' over (Figure 6.7b). The near-vertical breaking of the wave creates a powerful backwash which can move considerable amounts of sediment down the beach and, at the same time, reduce the effect of the swash from the following wave. Although some shingle may be thrown up above the high-water mark by very large waves, forming a **storm beach**, most material is moved downwards to form a **longshore (breakpoint) bar** (Figures 6.6 and 6.7b).

High-energy waves and low-energy waves

Recent opinion appears to support the view that beach shape is more dependent upon, and linked to, wave energy. The correlation between the two types of wave energy and beach profile is given in Figure 6.8 (after J. Pethick).

Figure 6.7

Constructive and destructive waves

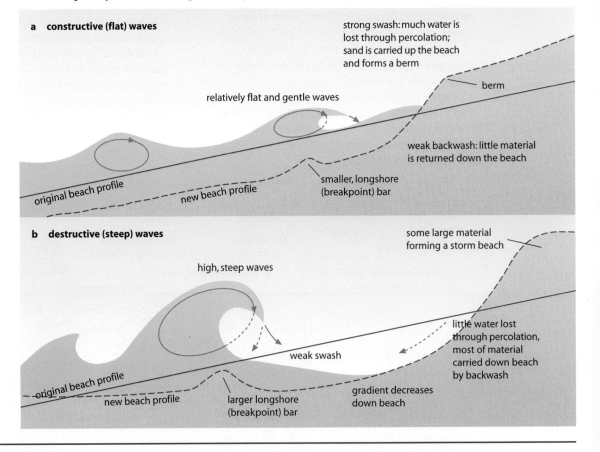

a constructive (flat) waves

strong swash: much water is lost through percolation; sand is carried up the beach and forms a berm

berm

relatively flat and gentle waves

weak backwash: little material is returned down the beach

original beach profile

new beach profile

smaller, longshore (breakpoint) bar

b destructive (steep) waves

some large material forming a storm beach

high, steep waves

little water lost through percolation, most of material carried down beach by backwash

weak swash

original beach profile

new beach profile

larger longshore (breakpoint) bar

gradient decreases down beach

Figure 6.8

High-energy and low-energy waves (*after* J. Pethick)

High-energy waves		Low-energy waves
Produced by distant storms	**Source**	Formed more locally
Large	**Fetch distance**	Short
Long (up to 100 m)	**Wave length**	Short (perhaps only 20 m)
High and short	**Wave height**	Low and flat
Move quickly and so lose little energy	**Speed of wave movement**	Move less quickly and so lose more energy
Spilling	**Type of breaker**	Surging
Long	**Dissipation distance**	Shorter
Flat and wide	**Beach shape**	Steeper and narrower

Particle size

This factor complicates the influence of wave steepness on the morphology of a beach. The fact that shingle beaches have a steeper gradient than sandy beaches is due mainly to differences in percolation rates resulting from differences in particle size – i.e. water will pass through coarse-grained shingle more rapidly than through fine-grained sand (Figure 8.2).

Figure 6.9

Storm beaches and berms: berms mark the limits of successively lower high tides

S + 1 = high tide after the spring high tide
S + 2 = second high tide after the spring high tide
S + 3 = third high tide after spring high tide

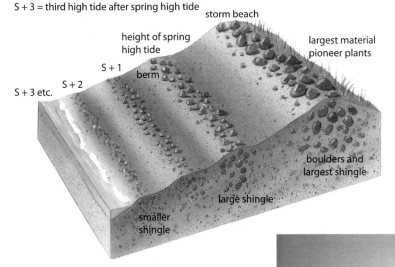

Shingle beaches

Shingle may make up the whole, or just the upper part, of the beach and, like sand, it will have been sorted by wave action. Usually, the larger the size of the shingle, the steeper the gradient of the beach, i.e. the gradient is in direct proportion to shingle size. This is an interesting hypothesis to test by experiment in the field (Framework 10, page 299).

Regardless of whether waves on shingle beaches are constructive or destructive, most of the swash rapidly percolates downwards leaving limited surface backwash. This, together with the loss of energy resulting from friction caused by the uneven surface of the shingle (compare this with the effects of bed roughness of a stream, page 70), means that under normal conditions, very little shingle is moved back down the beach. Indeed, the strong swash will probably transport material up the beach forming a berm at the spring high-tide level. Above the berm there is often a storm beach, composed of even bigger boulders thrown there by the largest of waves, while below may be several smaller ridges, each marking the height of the successively lower high tides which follow the maximum spring tide (Figures 6.9 and 6.10).

Figure 6.10

Berms and storm beaches in north-east Anglesey, Wales

Sand beaches

Sand usually produces beaches with a gentle gradient. This is because the small particle size allows the sand to become compact when wet, severely restricting the rate of percolation. Percolation is also hindered by the storage of water in pore spaces in sand which enables most of the swash from both constructive and destructive waves to return as backwash. Relatively little energy is lost by friction (sand presents a smoother surface than shingle) so material will be carried down the beach. The material will build up to form a longshore bar at the low-tide mark (Figure 6.6). This will cause waves to break further from the shore, giving them a wider beach over which to dissipate their energy. The lower parts of sand beaches are sometimes crossed by shore-parallel ridges and runnels (Figure 6.6). The ridges may be broken by channels which drain the runnels at low tide.

The interrelationship between wave energy, beach material and beach profiles may be summarised by the following generalisations which refer to net movements:

■ Destructive waves carry material down the beach.
■ Constructive waves carry material up the beach.
■ Material is carried upwards on shingle beaches.
■ Material is carried downwards on sandy beaches.

Tides

The position at which waves break over the beach, and their range, are determined by the state of the tide. It has already been seen that the levels of high tides vary (berms are formed at progressively lower levels following spring high tides; Figure 6.9). Tides are controlled by gravitational effects, mainly of the moon but partly of the sun, together with the rotation of the Earth and, more locally, the geomorphology of sea basins.

The moon has the greatest influence. Although its mass is much smaller than that of the sun, this is more than compensated for by its closer proximity to the Earth. The moon attracts, or pulls, water to the side of the Earth nearest to it. This creates a bulge or **high tide** (Figure 6.11a), with a complementary bulge on the opposite side of the Earth. This bulge is compensated for by the intervening areas where water is repelled and which experience a **low tide**. As the moon orbits the Earth, the high tides follow it.

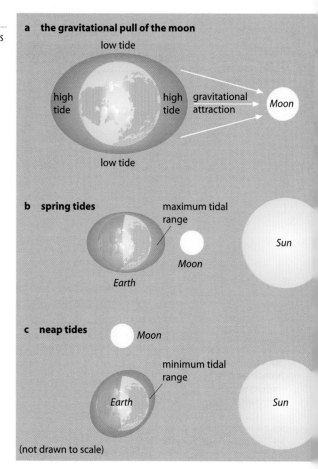

Figure 6.11

Causes of tides

a the gravitational pull of the moon

b spring tides

c neap tides

(not drawn to scale)

A lunar month (the time it takes the moon to orbit the Earth) is 29 days and the tidal cycle (the time between two successive high tides) is 12 hours and 25 minutes, giving two high tides, near enough, per day. The sun, with its smaller gravitational attraction, is the cause of the difference in tidal range rather than of the tides themselves. Once every 14/15 days (i.e. twice in a lunar month), the moon and sun are in alignment on the same side of the Earth (Figure 6.11b). The increase in gravitational attraction generates the **spring tide** which produces the highest high tide, the lowest low tide and the maximum tidal range.

Midway between the spring tides are the **neap tides**, which occur when the sun, Earth and moon form a right-angle, with the Earth at the apex (Figure 6.11c). As the sun's attraction partly counterbalances that of the moon, the tidal range is at a minimum with the lowest of high tides and the highest of low tides (Figure 6.12). Spring and neap tides vary by approximately 20 per cent above and below the mean high-tide and low-tide levels.

So far, we have seen how tides might change on a uniform or totally sea-covered Earth. In practice, the tides may differ considerably from the above scenario due to such factors as: the Earth's rotation (and the effect of the Coriolis force, page 224); the distribution of land masses; and the size, depth and configuration of ocean and sea basins.

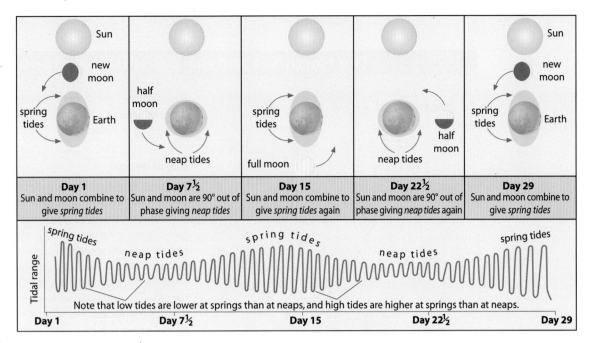

Figure 6.12

Tidal cycles during
the lunar month

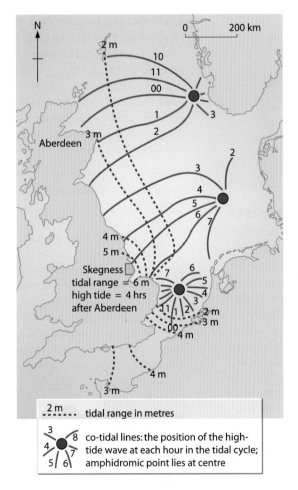

Figure 6.13

Tidal range and
difference in times
of high tide in the
North Sea

The morphology of the seabed and coastline affects tidal range. In the example of the North Sea, as the tidal wave travels south it moves into an area where both the width and the depth of the sea decrease. This results in a rapid accumulation, or funnelling, of water to give an increasingly higher tidal range – the range at Dover is several metres greater than in northern Scotland (Figure 6.13). Estuaries where incoming tides are forced into rapidly narrowing valleys also have considerable tidal ranges, e.g. the Severn estuary with 13 m, the Rance (Brittany) with 11.6 m and the Bay of Fundy (Canada) with 15 m. It is due to these extreme tidal ranges that the Rance has the world's first tidal power station, while the Bay of Fundy and the Severn have, respectively, experimental and proposed schemes for electricity generation (page 542). Extreme narrowing of estuaries can concentrate the tidal rise so rapidly that an advancing wall of water, or **tidal bore**, may travel upriver, e.g. the Rivers Severn and Amazon. In contrast, small enclosed seas have only minimal tidal ranges, e.g. the Mediterranean with 0.01 m.

Storm surges

Storm surges are rapid rises in sea-level caused by intense areas of low pressure, i.e. depressions (page 230) and tropical cyclones (page 235). For every drop in air pressure of 10 mb (page 224), sea-level can rise 10 cm. In tropical cyclones, pressure can fall by 100 mb causing the sea-level to rise by 1 m. Areas at greatest risk are those where sea basins become narrower and more shallow (e.g. southern North Sea and the Bay of Bengal) and where tropical cyclones move from the sea and cross low-lying areas (e.g. Bangladesh and Florida). When these storms coincide with hurricane-force winds and high tides, the surge can be topped by waves reaching 8 m in height. Where such events occur in densely populated areas, they pose a major natural hazard as they can cause considerable loss of life and damage to property (Places 19 and 31, page 238).

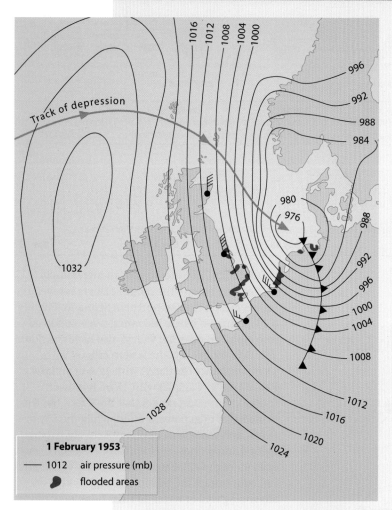

1 February 1953

— 1012   air pressure (mb)

🌑   flooded areas

Figure 6.14

The North Sea
storm surge of
1 February 1953

Delta Scheme have since been constructed. Both
schemes needed considerable capital and
technology to implement.

Bay of Bengal

The south of Bangladesh includes many flat islands
formed by deposition from the Rivers Ganges and
Brahmaputra. This delta region is ideal for rice
growing and is home to an estimated 40 million
people. However, during the autumn, tropical
cyclones (tropical low pressure storms) funnel
water northwards up the Bay of Bengal which
becomes increasingly narrower and shallower
towards Bangladesh. The water sometimes builds
up into a surge which may exceed 4 m in height and
which may be capped by waves reaching a further
4 m. The result can be a wall of water which sweeps
over the defenceless islands. Three days after one
such surge in 1985, the Red Cross suggested that
over 40 000 people had probably been drowned,
many having been washed out to sea as they slept
(Places 31, page 238). The only survivors were those
who had climbed to the tops of palm trees and
managed to cling on despite the 180 km/hr winds.
The Red Cross feared outbreaks of typhoid and
cholera in the area because fresh water had been
contaminated. Famine was a serious threat as the
rice harvest had been lost under the salty waters.

There is increasing international concern about the
possible effect of global warming on Bangladesh.
Estimates suggest that a 1 m rise in sea-level could
submerge 25 per cent of the country, affecting over
one-half of the present population (page 169).
Because Bangladesh lacks the necessary capital
and technology, for the last two decades the World
Bank has been helping in the construction of
cyclone early warning systems, providing flood
shelters and improving coastal defences. However,
recent thinking (Case Studies 3a and 6) suggests
that flooding, being a natural event, should be
allowed to take its course and that only smaller,
local and sustainable sea-defence projects should
be undertaken.

North Sea, 31 January – 1 February 1953

A deep depression to the north of Scotland, instead
of following the usual track which would have taken
it over Scandinavia, turned southwards into the
North Sea (Figure 6.14). As air is forced to rise in a
depression (page 230), the reduced pressure tends
to raise the surface of the sea area underneath it. If
pressure falls by 56 mb, as it did on this occasion, the
level of the sea may rise by up to 0.5 m. The gale-
force winds, travelling over the maximum fetch,
produced storm waves over 6 m high. This caused
water to pile up in the southern part of the North
Sea. This event coincided with spring tides and with
rivers discharging into the sea at flood levels. The
result was a high tide, excluding the extra height of
the waves, of over 2 m in Lincolnshire, over 2.5 m in
the Thames estuary and over 3 m in the Netherlands.
The immediate result was the drowning of 264
people in south-east England and 1835 people in
the Netherlands. To prevent such devastation by
future surges, the Thames Barrier and the Dutch

Year	Height of storm surge	Death toll (estimated)
1963	5.2	22 000
1966	6.1	80 000
1985	5.7	40 000
1988	4.8	25 000
1990	6.3	140 000
1991	6.1	150 000
1994	5.8	25 000

Figure 6.15

Waves breaking on Filey Brigg, Yorkshire: wave energy is absorbed by a band of residual rock and so the cliff behind is protected

Processes of coastal erosion

Subaerial According to J. Pethick, 'Cliff recession is primarily the result of mass failure'. Mass failure may be caused by such non-marine processes as: rain falling directly onto the cliff face; by throughflow or, under extreme conditions, surface runoff of water from the land; and the effects of weathering by the wind and frost. These processes, individually or in combination, can cause mass movement either as soil creep on gentle slopes or as slumping and landslides on steeper cliffs (Figures 2.17 and 2.18).

Wave pounding Steep waves have considerable energy. When they break as they hit the foot of cliffs or sea walls, they may generate shock-waves of up to 30 tonnes per m^2. Some sea walls in parts of eastern England need replacing within 25 years of being built, due to wave pounding (Case Study 6).

Hydraulic pressure When a parcel of air is trapped and compressed, either in a joint in a cliff or between a breaking wave and a cliff, then the resultant increase in pressure may, over a period of time, weaken and break off pieces of rock or damage sea defences.

Abrasion/corrasion This is the wearing away of the cliffs by sand, shingle and boulders hurled against them by the waves. It is the most effective method of erosion and is most rapid on coasts exposed to storm waves.

Attrition Rocks and boulders already eroded from the cliffs are broken down into smaller and more rounded particles.

Corrosion/solution This includes the dissolving of limestones by carbonic acid in sea water (compare Figure 2.8), and the evaporation of salts to produce crystals which expand as they form and cause the rock to disintegrate (Figure 2.2). Salt from sea water or spray is capable of corroding several rock types.

Factors affecting the rate of erosion

Breaking point of the wave A wave that breaks as it hits the foot of a cliff releases most energy and causes maximum erosion. If the wave hits the cliff before it breaks, then much less energy is transmitted, whereas a wave breaking further offshore will have had its energy dissipated as it travelled across the beach (Figure 6.15).

Wave steepness Highest-energy waves, associated with longer fetch distances, have a high, steep appearance. They have greater erosive power than low-energy waves, which are generated where the fetch is shorter and have a lower and flatter form (Figure 6.8).

Depth of sea, length and direction of fetch, configuration of coastline A steeply shelving beach creates higher and steeper waves than one with a more gentle gradient. The longer the fetch, the greater the time available for waves to collect energy from the wind. The existence of headlands with vertical cliffs tends to concentrate energy by wave refraction (page 142).

Supply of beach material Beaches, by absorbing wave energy, provide a major protection against coastal erosion.

Beach morphology Beaches, by dissipating wave energy, act as a buffer between waves and the land. As they receive high-energy inputs at a rapid rate from steep waves, and low-energy inputs at a slower rate from flat waves, they must adopt a morphology (shape) to counteract the different energy inputs. High, rapid energy inputs are best dissipated by wide, flat beaches which spread out the oncoming wave energy. In contrast, the lower-energy inputs of flatter waves can easily be dissipated by narrow, steep beaches which act rather like a wall against which the waves flounder. An exception is when steep waves break onto a shingle beach. As energy is rapidly dissipated through friction and percolation, then a wide, narrow beach profile is unnecessary (page 145).

Rock resistance, structure and dip The strength of coastal rocks influences the rate of erosion (Figure 6.16). In Britain, it is coastal areas where glacial till was deposited that are being worn back most rapidly (Places 20). When Surtsey first arose out of the sea off the southwest coast of Iceland in 1963 (Places 3, page 16), it consisted of unconsolidated volcanic ash. It was only when the ash was covered and protected by a lava flow the following year that the island's survival was seemingly guaranteed.

Rocks that are well-jointed (Figure 8.1) or have been subject to faulting have an increased vulnerability to erosion. The steepest cliffs are usually where the rock's structure is horizontal or vertical and the gentlest where the rock dips upwards away from the sea. In the latter case, blocks may break off and slide downwards (Figure 2.17). Erosion is also rapid where rocks of different resistance overlie one another, e.g. chalk and Gault clay in Kent.

Figure 6.16

Rock type and average rates of cliff recession

Rock type	Location	Rate of erosion (m/yr)
Volcanic ash	Krakatoa	40
Glacial till	Holderness	2
Glacial till	Norfolk	1
Chalk	South-east England	0.3
Shale	North Yorkshire	0.09
Granite	South-west England	0.001

Human activity The increase in pressure resulting from building on cliff tops and the removal of beach material which may otherwise have protected the base of the cliff both contribute to more rapid coastal erosion.

Although rates of erosion may be reduced locally by the construction of sea defences, such defences often lead to increased rates of erosion in adjacent areas. Human activity therefore has the effect of disturbing the equilibrium of the coast system (Case Study 6).

Places 20 Holderness: coastal processes

The coastline at Holderness is retreating by an average of 1.8 m a year. Since Roman times, the sea has encroached by nearly 3 km, and some 50 villages mentioned in the Domesday Book of 1086 have disappeared.

The following extract was taken from a management report, 'Humber Estuary & Coast' (1994) prepared by Professor J. S. Pethick (then of the University of Hull and now at the University of Newcastle) for Humberside County Council.

'The soft glacial till cliffs of Holderness are eroding at a rapid rate. The reasons for such erosion are, however, less to do with the soft sediment of the cliff than with the lack of beach material and the poorly developed nearshore zone [Figure 6.6]. Retreat of the cliff line here is matched by progressive lowering of the seabed to give a wide shallow platform stretching several kilometres seaward. Eventually this platform will be so extensive that most of the incident wave energy will be expended here rather than at the cliff so that erosion rates will decrease or even halt. Since this may take several thousand years, it cannot form part of any management plan for this coast – yet it is important to recognise that the natural erosional

processes here are neither random nor pernicious.

The process of cliff retreat along the Holderness coast is more complex than appears at first sight. Mass failures of the cliff are triggered by wave action at the cliff toe. Such failures may be 50 to 100 m wide and up to 30 m deep giving a scalloped edge to the cliff. The retreat rate varies temporarily; a large failure may produce a 10 m retreat in one year but no further retreat will then occur for 3 or 4 years – giving a periodicity of 4 or 5 years in total. This means that attempts to measure erosion rates over periods of less than 10 years, that is over 2 cycles, can be extremely misleading, resulting in massive over- or under-estimates of the long-term retreat rate which is remarkably constant at 1.8 m per year [Figure 6.17]. Three issues may be highlighted here.

- The beaches of Holderness are thin veneers covering the underlying glacial tills. The beaches do not increase in volume since, south of Hornsea, a balance exists between the input of sand by erosion and the removal of the sand by wave action, principally from the north-east, which drives sands south.

- The sediment balance on the Holderness coast is maintained by the action of storm waves from the north-east. These waves approach the coast obliquely, the angle between wave crest and shore being critical for the sediment transport rate. A clockwise movement would increase the transport and erosion rate while an anti-clockwise swing would decrease both of these. Random changes in the orientation of the shore are quickly eradicated by changes in the sediment balance, but any permanent change in the orientation of the coastline, such as that caused by the introduction of hard sea defences as at Hornsea, Mappleton and Withernsea, means that the sediment balance is disturbed.

Figure 6.17

Houses collapsing into the sea, Holderness

- Hard defences can have two long-term effects: first, although erosion is halted at the defence itself, several kilometres to the north erosion continues as before. This causes an anti-clockwise re-orientation of the coast, sand transport is reduced and sand accumulates immediately north of the defences – as can be seen north of Hornsea. Second, the accumulation of sand north of the defences starves the beaches to the south causing an increase in erosion there. The fine-grained sediments from the Holderness cliff and seabed erosion are not transported along the beaches as are the sands and shingle but are moved in suspension. Research is presently under way which is intended to chart the precise movement of this material but it is clear that its dominant movement is south towards the Humber. A large proportion may enter the estuary and become deposited there. The remainder is moved south and east into the North Sea where the transport pathway is towards the Dutch and German coast.'

Figure 6.18

Wave-cut notch at Coromandel Peninsula, New Zealand

Figure 6.19

Abrasion or wave-cut platform at Flamborough Head, Yorkshire

Erosion landforms

Headlands and bays

These are most likely to be found in areas of alternating resistant and less resistant rock. Initially, the less resistant rock experiences most erosion and develops into **bays**, leaving the more resistant outcrops as **headlands**. Later, the headlands receive the highest-energy waves and so become more vulnerable to erosion than the sheltered bays (Figure 6.5). The latter now experience low-energy breakers which allow sand to accumulate and so help to protect that part of the coastline.

Abrasion or wave-cut platforms

Wave energy is at its maximum when a high, steep wave breaks at the foot of a cliff. This results in undercutting of the cliff to form a **wave-cut notch** (Figure 6.18). The continual undercutting causes increased stress and tension in the cliff until eventually it collapses. As these processes are repeated, the cliff retreats leaving, at its base, a gently sloping, **abrasion** or **wave-cut platform** which has a slope angle of less than 4° (Figure 6.19). The platform, which appears relatively even when viewed from a distance, cuts across rocks regardless of their type and structure. A closer inspection of this inter-tidal feature usually reveals that it is deeply dissected by abrasion, resulting from material carried across it by tidal movements, and corrosion. As the cliff continues to retreat, the widening of the platform means that incoming waves break further out to sea and have to travel over a wider area of beach. This dissipates their energy, reduces the rate of erosion of the headland, and limits the further extension of the platform. It has been hypothesised that wave-cut platforms cannot exceed 0.5 km in width.

Where there has been negative change in sea-level (page 81), former wave-cut platforms remain as **raised beaches** above the present influence of the sea (Figure 6.51).

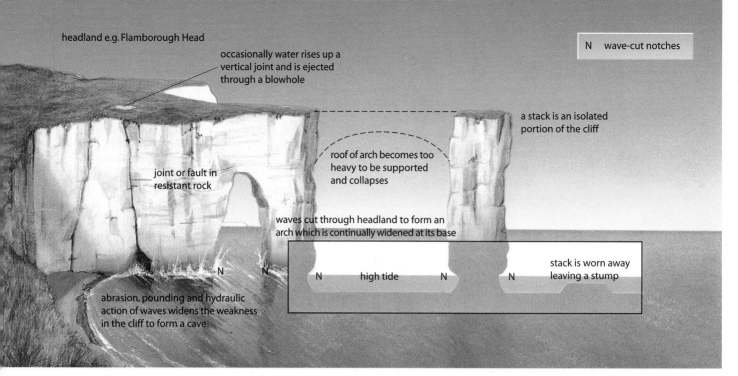

headland e.g. Flamborough Head

occasionally water rises up a vertical joint and is ejected through a blowhole

a stack is an isolated portion of the cliff

roof of arch becomes too heavy to be supported and collapses

joint or fault in resistant rock

waves cut through headland to form an arch which is continually widened at its base

stack is worn away leaving a stump

N   N   N   high tide   N   N

abrasion, pounding and hydraulic action of waves widens the weakness in the cliff to form a cave

Figure 6.20

The formation of caves, blowholes, arches and stacks

Caves, blowholes, arches and stacks

Where cliffs are of resistant rock, wave action attacks any line of weakness such as a joint or a fault. Sometimes the sea cuts inland, along a joint, to form a narrow, steep-sided inlet called a **geo**, or at other times, it can undercut part of the cliff to form a **cave**. As shown in Figure 6.20, caves are often enlarged by several combined processes of marine erosion. Erosion may be vertical, to form **blowholes**, but is more typically backwards

through a headland to form **arches** and **stacks** (Figures 6.20 and 6.21).

These landforms, which often prove to be attractions to sight-seers and mountaineers, can be found at The Needles (Isle of Wight), Old Harry (near Swanage) and Flamborough Head (Yorkshire, Figure 6.19), which are all cut into chalk, and at The Old Man of Hoy (Orkneys) which is Old Red Sandstone (Figure 8.12).

Figure 6.21

Icelandic coastline

Transportation of beach material

Up and down the beach

As we have already seen, flat, constructive waves tend to move sand and shingle up the beach, whereas the net effect of steep, destructive waves is to comb the material downwards.

Longshore (littoral) drift

Usually wave crests are not parallel to the shore, but arrive at a slight angle. Only rarely do waves approach a beach at right-angles. The wave angle is determined by wind direction, the local configuration of the coastline, and refraction at headlands and in shallow water. The oblique wave angle creates a nearshore current known as **longshore** (or **littoral**) **drift** which is capable of moving large quantities of material in a down-drift direction (Figure 6.22). On many coasts, longshore drift is predominantly in one direction; for example, on the south coast of England, where the maximum fetch and prevailing wind are both from the south-west, there is a predominantly eastward movement of beach material. However, brief changes in wind – and therefore wave – direction can cause the movement of material to be reversed.

Of lesser importance, but more interesting and easier to observe, is the movement of material along the shore in a zigzag pattern. This is because when a wave breaks, the swash carries material up the beach at the same angle as that at which the wave approached the shore. As the swash dies away, the backwash and any material carried by it returns straight down the beach, at right-angles to the waterline, under the influence of gravity. If beach material is carried a considerable distance, it becomes smaller, more rounded and better sorted.

Where beach material is being lost through longshore drift, the coastline in that locality is likely to be worn back more quickly because the buffering effect of the beach is lessened. To counteract this process, wooden breakwaters or **groynes** may be built (Figure 6.23). Groynes encourage the local accumulation of sand (important in tourist resorts) but can result in a depletion of material, and therefore an increase in erosion, further along the coast (Case Study 6).

Figure 6.22

The effects of longshore drift

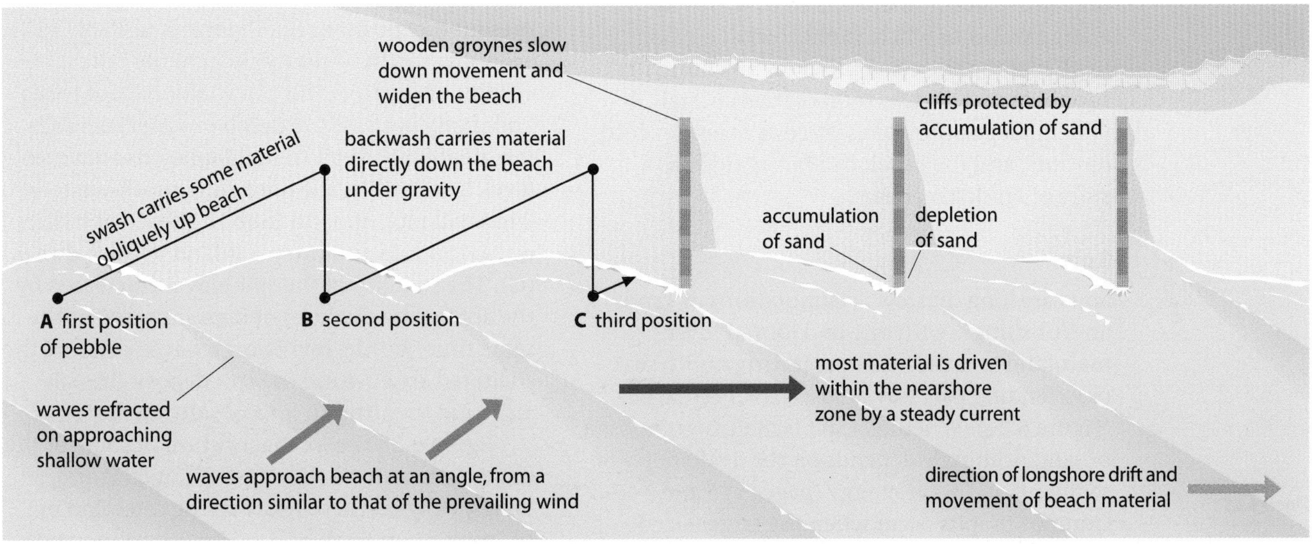

wooden groynes slow down movement and widen the beach

cliffs protected by accumulation of sand

swash carries some material obliquely up beach

backwash carries material directly down the beach under gravity

accumulation of sand

depletion of sand

A first position of pebble

B second position

C third position

waves refracted on approaching shallow water

waves approach beach at an angle, from a direction similar to that of the prevailing wind

most material is driven within the nearshore zone by a steady current

direction of longshore drift and movement of beach material

Figure 6.23

The effect of groynes on longshore drift, Southwold, Suffolk: this type of coastal management is usually undertaken at holiday resorts where sandy beaches are a major tourist attraction

Figure 6.24

A spit: Dawlish Warren at the mouth of the River Exe, Devon

Coastal deposition

Deposition occurs where the accumulation of sand and shingle exceeds its depletion. This may take place in sheltered areas with low-energy waves or where rapid coastal erosion further along the coast provides an abundant supply of material. In terms of the coastal system, deposition takes place as inputs exceed outputs, and the beach can be regarded as a store of eroded material.

Spits

Spits are long, narrow accumulations of sand and/or shingle with one end joined to the mainland and the other projecting out to sea or extending part way across a river estuary (Figure 6.24). Whether a spit is mainly composed of sand or shingle depends on the availability of sediment and wave energy (pages 145 and 146). Composite spits occur when the larger-sized shingle is deposited before the finer sands.

Figure 6.25

Stages in the formation of a spit

In Figure 6.25, the line X–Y marks the position of the original coastline. At point **A**, because the prevailing winds and maximum fetch are from the south-west, material is carried eastwards by longshore drift. When the orientation of the old coastline began to change at **B**, some of the larger shingle and pebbles were deposited in the slacker water in the lee of the headland. As the spit continued to grow, storm waves threw some larger material above the high-water mark (**C**), making the feature more permanent; while, under normal conditions, the finer sand was carried towards the end of the spit at **D**. Many spits develop a hooked or curved end. This may be for two reasons: a change in the prevailing wind to coincide with the second-most-dominant wave direction and second-longest fetch, or wave refraction at the end of the spit carrying some material into more sheltered water.

Eventually the seaward side of the spit will retreat, while longshore drift continues to extend the feature eastwards. A series of recurved ends may form (**E**) each time there is a series of storms from the south-east giving a lengthy period of altered wind direction. Having reached its present-day position (**F**), the spit is unlikely to grow any further – partly because the faster current of the river will carry material out to sea and partly because the depth of water becomes too great for the spit to build upwards above sea-level. Meanwhile, the prevailing south-westerly wind will pick up sand from the beach as it dries out at low tide and carry it inland to form **dunes** (**G**). The stability of the spit may be increased by the anchoring qualities of marram grass. At the same time, gentle, low-energy waves entering the sheltered area behind the spit deposit fine silt and mud, creating an area of **saltmarsh** (**H**).

Figure 6.28 shows the location of some of the larger spits around the coast of England and Wales. How do these relate to the direction of the maximum fetch and of the prevailing and dominant winds?

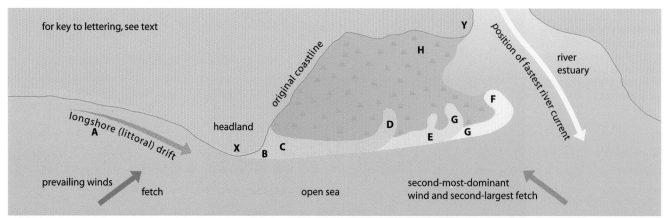

for key to lettering, see text

original coastline

position of fastest river current

river estuary

longshore (littoral) drift

headland

prevailing winds

fetch

open sea

second-most-dominant wind and second-largest fetch

Figure 6.27

A bar: Slapton Ley, Devon

Figure 6.26

A tombolo: Loch Eriboll, Highland, Scotland

Tombolos, bars and barrier islands

A **tombolo** is a beach that extends outwards to join with an offshore island (Figure 6.26). Chesil Beach, in Dorset, links the Isle of Portland to the mainland. Some 30 km long and up to 14 m high, it presents a gently smoothed face to the prevailing winds in the English Channel.

If a spit develops in a bay into which no major river flows, it may be able to build across that bay, linking two headlands, to form a **bar**. Bars straighten coastlines and trap water in lagoons on the landward side. Bars, such as that at Slapton Ley, in Devon (Figure 6.27), may also result in places where constructive waves lead to the landward migration of offshore, seabed material.

Barrier islands are a series of sandy islands totally detached from, but running almost parallel to, the mainland. Between the islands, which may extend for several hundred kilometres, and the mainland is a tidal lagoon (Figures 6.29 and 6.31). Although relatively uncommon in Britain, they are widespread globally, accounting for 13 per cent of the world's coastlines. They are easily recognisable on maps of the eastern USA (Places 21), the Gulf of Mexico, the northern Netherlands, West Africa and southern and western Australia. While their origin is uncertain, they do tend to develop on coasts with relatively high-energy waves and a low tidal range. One theory suggests that they formed, below the low-tide mark, as offshore bars of sand and have moved progressively landwards. An alternative theory suggests that rises in post-glacial sea-level may have partly submerged older beach ridges. In either case, the breaches between the islands seem likely to have been caused by storm waves.

Figure 6.28

Location of some major spits, tombolos and bars in England and Wales

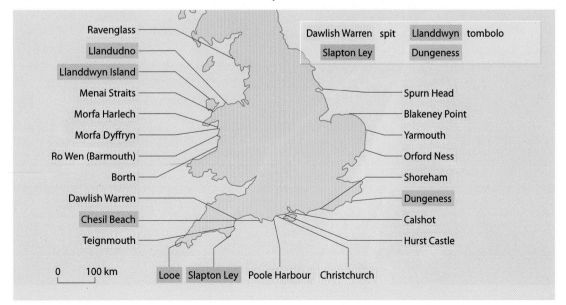

Ravenglass
Llandudno
Llanddwyn Island
Menai Straits
Morfa Harlech
Morfa Dyffryn
Ro Wen (Barmouth)
Borth
Dawlish Warren
Chesil Beach
Teignmouth

Dawlish Warren spit   Llanddwyn tombolo
Slapton Ley   Dungeness

Spurn Head
Blakeney Point
Yarmouth
Orford Ness
Shoreham
Dungeness
Calshot
Hurst Castle

0   100 km

Looe   Slapton Ley   Poole Harbour   Christchurch

Barrier islands have a unique morphology, flora and fauna. The smooth, straight, ocean edge is characterised by wide, sandy beaches which slope gently upwards to sand dunes which are anchored by high grasses (Figure 6.30). Behind the dunes, the 'island' interior may contain shrubs and woods, deer and snakes, insects and birds. The landward side is punctuated by sheltered bays, quiet tidal lagoons, saltmarshes and, towards the tropics, mangrove swamps. These wetlands provide a natural habitat for oysters, fish and birds. Although barrier islands form the interface between the land and the ocean, they seem fragile in comparison with the power that the wind and sea brings to them. It is virtually impossible for a tropical storm or hurricane to move ashore without first crossing either of the two longest stretches of barrier islands in the world: either that which extends for 2500 km from New Jersey to the southern tip of Florida (Figure 6.29); or the one stretching for 2100 km along the Gulf Coast states to Mexico.

Barrier islands are subject to a process called 'wash over'. This process, which might occur up to 40 times in some years, is when storm waves carry large quantities of sand over the island from the seaward face to the landward side. This results in the seaward side being eroded and pushed backwards. The landward marshes and mangrove swamps become suffocated, and the tidal lagoons are narrowed. From a human viewpoint, barrier islands form an essential natural defence against hurricanes and their storm-force waves.

Sand dunes

Sand dunes are a dynamic landform whose equilibrium depends upon the interrelationship between mineral content (sand) and vegetation.

Longshore drift may deposit sand in the intertidal zone. As the tide ebbs, the sand will dry out allowing winds from the sea to move material up the beach by saltation (page 183). This process is most likely to occur when the prevailing winds come from the sea and where there is a large tidal range which exposes large expanses of sand at low tide. Sand may become trapped by seaweed and driftwood on berms or at the point of the highest spring tides. Plants begin to colonise the area (Figure 11.10), stabilising the sand and encouraging further accumulation. The regolith has a high pH value due to calcium carbonate from seashells.

Embryo dunes are the first to develop (Figure 6.31). They become stabilised by the growth of lyme and marram grasses. As these grasses trap more sand, the dunes build up and, due to the high rate of percolation, become increasingly arid. Plants need either succulent leaves to store water (sand couch), or thorn-like leaves to reduce transpiration in the strong winds (prickly saltwort), or long tap-roots to reach the water table (marram grass). As more sand accumulates, the embryo dunes join to form **foredunes** which can attain a height of 5 m (Figures 6.31 and 6.32). Due to a lack of humus, their colour gives them the name **yellow dunes**. The dunes become increasingly grey as humus and bacteria from plants and animals are added, and they gradually become more vegetation-covered and acidic. These **grey (mature) dunes** may reach a height of 10–30 m before the supply of fresh sand is cut off by their increasing distance from the beach (Figure 11.11). There may be several parallel ridges of old dunes (as at Morfa Harlech, Figure 6.33), separated by low-lying, damp slacks. Heath plants begin to dominate the area as acidity, humus and moisture content all increase (Figure 11.9). Paths cut by humans and animals expose areas of sand. As the wind funnels along these tracks, **blowouts** may form in the now **wasting dunes**. To combat further erosion at Morfa Harlech, parts of the dunes have been fenced off and marram grass has been planted to try to re-stabilise the area and to prevent any inland migration of the dunes.

The above idealised scheme can be interrupted at any stage by storms or human use. If the supply of sand is cut off, then new embryo dunes cannot form and yellow dunes may be degraded so that it is the older, grey dunes that line the beach.

Figure 6.31

A transect across sand dunes, based on fieldwork at Morfa Harlech, North Wales

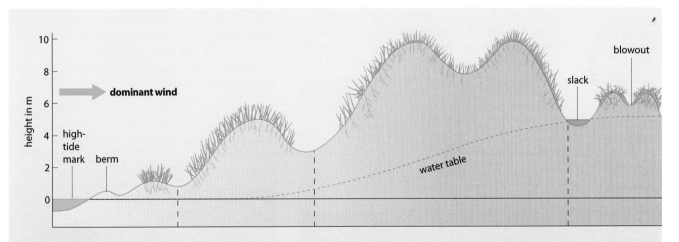

Type of dune	Embryo	Fore or yellow dunes	Grey dunes and dune ridges	Wasting dunes with blowouts
Dune height (m)	1	5	8–10	6–8
Percentage of exposed sand	80	20	less than 10	over 40 on dunes
Humus and moisture content	very little mixed salt and fresh water	some humus, very little moisture fresh water	humus increases inland, water content still low fresh water	high humus, brackish water in slacks
pH	over 8	slightly alkaline	increasingly acid inland: pH 6.5–7	acid: pH 5–6
Plant types	sand couch, lyme grass	marram, xerophytic species	creeping fescue, sea spurge, some marram, cotton grass and heather	heather, gorse on dunes, *Juncus* in slacks

Figure 6.32

Embryo and foredunes at Morfa Harlech, North Wales (refer also to Figures 11.10 and 11.11)

Figure 6.34

Llanrhidian saltmarsh, Gower peninsula, South Wales (refer also to Figures 11.13 and 11.14)

Figure 6.33

Morfa Harlech from Harlech Castle showing foredunes, grey or wasting dunes, old cliff-line and, in the distance, saltmarsh

Saltmarshes

Where there is sheltered water in river estuaries or behind spits, silt and mud will be deposited either by the gently rising and falling tide or by the river, thus forming a zone of **inter-tidal mudflats**. Initially, the area may only be uncovered by the sea for less than 1 hour in every 12-hour tidal cycle. Plants such as algae and *Salicornia* can tolerate this lengthy submergence and the high levels of salinity. They are able to trap more mud around them, creating a surface that remains exposed for increasingly longer periods between tides (Figure 6.34). *Spartina* grows throughout the year and since its introduction into Britain has colonised, and become dominant in, many estuaries. The landward side of the inter-tidal mudflats is marked by a small cliff (Figure 11.13), above which is the flat **sward zone**. This zone may only be covered by the sea for less than 1 hour in each tidal cycle (Figure 6.12). Seawater collects in hollows which become increasingly saline as the water evaporates. The hollows often enlarge into **saltpans** (Figure 11.13) which are devoid of vegetation except for certain algae and the occasional halophyte (page 291). As each tide retreats, water drains into **creeks** which are then eroded rapidly both laterally and vertically (Figure 6.35). The upper sward zone may only be inundated by the highest of spring tides.

Figure 6.35

Llanrhidian saltmarsh showing the sward zone, creeks and saltpan

Figure 6.36

A sample population in relation to the total population

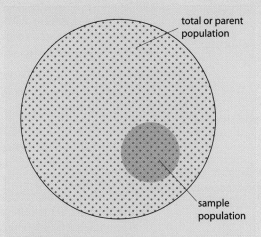

total or parent population

sample population

Why sample?

Geographers are part of a growing number of people who find it increasingly useful and/or necessary to use data to quantify the results of their research. The problem with this trend is that the amount of data may be very expensive, too time-consuming, or just impracticable to collect – as it would be, for example, to investigate everybody's shopping patterns in a large city, to find the number of stones on a spit, or to map the land use of all the farms in Britain.

Sampling is the method used to make statistically valid inferences when it is impossible to measure the **total population** (Figure 6.36). It is essential, therefore, to find the most accurate and practical method of obtaining a **representative sample**. If that sample can be made with the minimum of bias, then statistically significant conclusions may be drawn. However, even if every effort is made to achieve precision, it must be remembered that any sample can only be a close estimate.

Sampling basics

Most sampling procedures assume that the total population has a **normal distribution** (Figure 4.16a) which, when plotted on a graph, produces a symmetrical curve on either side of the mean value. This shows that a large proportion of the values are close to the average, with few extremes. Figure 6.37 shows a normal distribution curve and the **standard deviation** (page 247) – the measure of dispersion from the mean. Where most of the values are clustered near to the mean, the standard deviation is low.

The larger the sample, the more accurate it is likely to be, and the more likely it is to resemble the parent population; it is also more likely to conform to the normal distribution curve. While the generally accepted minimum size for a sample is 30, there is no upper limit – although there is a point beyond which the extra time and cost involved in increasing the sample size do not give a significant improvement in accuracy (an example of the law of diminishing returns, page 462).

Figure 6.37 shows that, in a normal distribution, 68.27 per cent of the values in the sample occur within a range of ±1 standard deviations (SDs) from the mean; 95 per cent of the values fall within ±2 SDs; and 99 per cent within ±3 SDs. These percentages are known as **confidence limits**, or **probability levels**. Geographers usually accept the 95 per cent probability level when sampling. This means that they accept the chance that, in 5 cases out of every 100, the true mean will lie outside 2 SDs to either side of their sample mean.

Figure 6.37

A normal distribution showing standard deviations from the mean

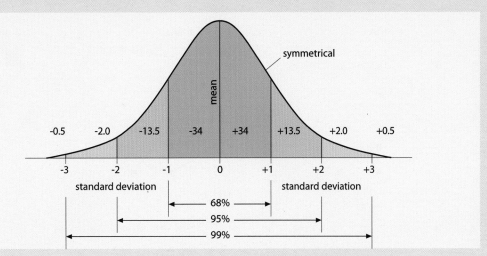

symmetrical

mean

| -0.5 | -2.0 | -13.5 | -34 | +34 | +13.5 | +2.0 | +0.5 |

-3   -2   -1   0   +1   +2   +3

standard deviation   standard deviation

68%
95%
99%

Sampling techniques

Several different methods may be used according to the demands of the required sample and the nature of the parent population. There are two major types with one refinement:

- **Random sampling** This is the most accurate method as it has no bias.
- **Systematic sampling** This method is often quicker and easier to use, although some bias or selection is involved.
- **Stratified sampling** This method is often a very useful refinement for geographers; it can be used with either a random or a systematic sample.

Random sampling

Under normal circumstances, this is the ideal type of sample because it shows no bias. Every member of the total population has an equal chance of being selected, and the selection of one member does not affect the probability of selection of another member. The ideal random sample may be obtained using **random numbers**. These are often generated by computer and are available in the form of printed tables of random numbers, but if necessary they can be obtained by drawing numbers out of a hat. Random number tables usually consist of columns of pairs of digits. Numbers can be chosen by reading either along the rows or down the columns, provided only one method is used. Similarly, any number of figures may be selected – six for a grid reference, four for a grid square, three for house numbers in a long street, etc. Using the grid shown in Figure 6.38, the random number table given above yields eight 6-figure grid references: 927114; (986691 has to be excluded because the grid does not contain these numbers); 906126; etc.

One feature of a genuine random sample is that the same number can be selected more than once – so remember that if you are pulling numbers from a hat, they should be replaced immediately after they have been read and recorded.

There are three alternative ways of using random numbers to sample areal distributions (patterns over space) (Figure 6.38).

1 **Random point** A grid is superimposed over the area of the map to be sampled. Points, or map references, are then identified using random number tables, and plotted on the map. The eight points identified earlier (in the random number table) have been plotted on Figure 6.38a. A large number of points may be needed to ensure coverage of the whole area – see Figure 6.40.

2 **Random line** Random numbers are used to obtain two end points which are then joined by a line, as in Figure 6.38b which uses the same eight random points, in the order in which they occurred in the table. Several random lines are needed to get a representative sample (e.g. lines across a city to show transects of variation in land use).

3 **Random area** Areas of constant size, e.g. grid squares or quadrats, are obtained using random numbers. By convention, the number always identifies the south-west corner of a grid square. If sample squares one-quarter the size of a grid square are used, together with the same sample points, their locations are as shown on Figure 6.38c – note that the point in the north-east cannot be used because part of the sample square lies outside the study area. This method can be used to sample land-use areas or the distribution of plant communities over space.

Part of a random number table			
9271	0143	2141	9381
1498	3796	4413	1405
6691	4294	6077	9091
9061	1148	9493	1940
2660	7126	7126	4591
3459	7585	4897	8138
6090	7962	5766	7228
2191	9271	9042	5884

Figure 6.38

Random sampling using point, line and area techniques

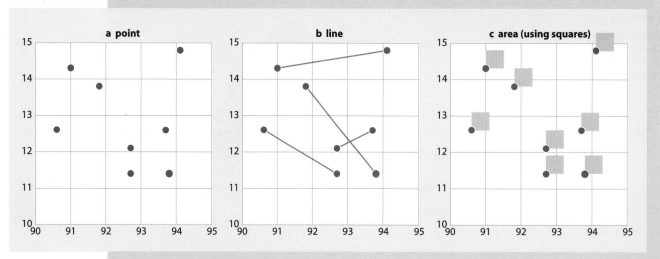

The advantages of random sampling include its ability to be used with large populations and its avoidance of bias. Careful **sample design** is needed, however, to avoid the possibility of achieving misleading results, for example when sampling small populations, and when sampling over a large area. Also, when used in the field, it may involve considerable time and energy in visiting every point.

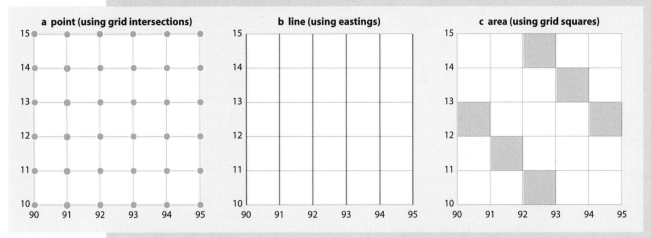

Figure 6.39

Systematic sampling using point, line and area techniques

Systematic sampling

A systematic sample is one in which values are selected in a regular way, e.g. choosing every 10th person on a list, or every 20th house in a street. This can be an easier method in terms of time and effort than random sampling. Like random sampling, it can be operated using individual points, lines or areas (Figure 6.39).

1 **Systematic point** This can show changes over distance, e.g. by sampling the land use every 100 m. It can also show change through time, e.g. by sampling from the population censuses (taken every 10 years).

2 **Systematic line** This may be used to choose a series of equally spaced transects across an area of land, e.g. a shingle spit.

3 **Systematic area** This is often used for land-use sampling, to show change with distance or through time (if old maps or air photographs are available). Quadrats, positioned at equal intervals, are used for assessing plant distributions.

The main advantage of systematic sampling lies in its ease of use. However, its main disadvantage is that all points do not have an equal chance of selection – it may either overstress or miss an underlying pattern (Figure 6.40).

Stratified sampling

When there are significant groups of known size within the parent population, in order to ensure adequate coverage of all the sub-groups it may be advisable to stratify the sample, i.e. to divide the population into categories and sample within each. Although categorising into groups (layers or strata) may be a subjective decision, the practical application of this technique has considerable advantages for the geographer. Once the groups have been decided, they can be sampled either systematically or randomly, using point, line or area techniques.

1 **Stratified systematic sampling** This method can be useful in many situations – when interviewing people, sampling from maps, and during fieldwork. For example, in political opinion polls, the total population to be sampled can be divided (stratified) into equal age and/or socio-economic groups, e.g. 10–19, 20–29, etc. The number interviewed in each category should be in proportion to its known size in the parent population. This is most easily achieved by sampling at a regular interval (systematically) throughout the entire population, so that the required total sample size is obtained. For example, if a sample size of 800 is required from a total population of 8000 (i.e. a 10 per cent sample), every 10th person would be interviewed.

Figure 6.40

Poor sample design and selection can lead to inaccurate results: an area of woodlands is completely missed in this example

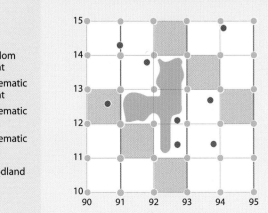

KEY

● random point

○ systematic point

| systematic line

▨ systematic area

woodland

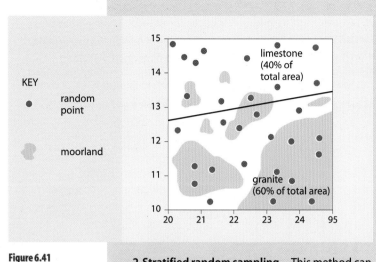

Figure 6.41

A random point
sample, stratified
by area

rock types: granite occupies 60% of the total area and limestone 40%. To discover whether the proportion of moorland cover varies with rock type, the sampling must be in proportion to their relative extents. Thus, if a sample size of 30 points is derived using random numbers, 18 are needed within the granite area (18 is 60 per cent of 30) and 12 within the limestone area (12 is 40 per cent of 30). If it was decided to area sample, 18 quadrats would have to fall within the granite area, and 12 in the limestone.

2 Stratified random sampling This method can be used to cover a wide range of data, both in interviewing and in geographical fieldwork and map work. For example, Figure 6.41 shows the distribution of moorland on two contrasting

The advantages of stratified sampling include its potential to be used either randomly or systematically, and in conjunction with point, line or area techniques. This makes it very flexible and useful, as many populations have geographical sub-groups. Care must be taken, however, to select appropriate strata

Changes in sea-level

Although the daily movement of the tide alters the level at which waves break onto the foreshore, the average position of sea-level in relation to the land has remained relatively constant for nearly 6000 years (Figure 6.42). Before that time there had been several major changes in this mean level, the most dramatic being a result of the Quaternary ice age and of plate movements.

During times of maximum glaciation, large volumes of water were stored on the land as ice – probably three times more than today. This modification of the **hydrological cycle** meant that there was a worldwide, or **eustatic** (glacio-eustatic, page 123), fall in sea-level of an estimated 100–150 m.

As ice accumulated, its weight began to depress those parts of the crust lying beneath it. This caused a local, or **isostatic** (glacio-isostatic, page 123), change in sea-level.

Figure 6.42

Eustatic changes in sea-level since 18 000 BC

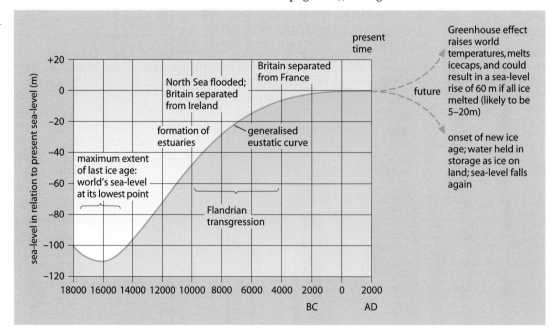

The world's sea-level was at its minimum 18 000 years ago when the ice was at its maximum (Figure 6.42). Later, as temperatures began to rise and icecaps melted, there was first a eustatic rise in sea-level followed by a slower isostatic uplift which is still operative in parts of the world today. This sequence of sea-level changes may be summarised as follows:

1 Formation of glaciers and ice sheets. Eustatic fall in sea-level gives rise to a negative change in base level (page 81).
2 Continued growth of ice sheets. Isostatic depression of the land under the ice produces a positive change in base level.
3 Ice sheets begin to melt. Eustatic rise in sea-level with a positive change in base level.
4 Continued decline of ice sheets and glaciers. Isostatic uplift of the land under former ice sheets results in a negative change in base level.

During this deglaciation, there may have been a continuing, albeit small, eustatic rise in sea-level but this has been less rapid than the isostatic uplift so that base level appears to be falling. Measurements suggest that parts of north-west Scotland are still rising by 4 mm a year and some northern areas of the Gulf of Bothnia (Scandinavia) by 20 mm a year (Places 23, page 166). The uplift in northern Britain is causing the British Isles to tilt and the land in south-east England to be depressed. This process is of utmost importance to the future natural development and human management of British coasts.

Tectonic changes have resulted in:
■ the uplift (**orogeny**) of new mountain ranges, especially at destructive and collision plate margins (pages 17 and 19)
■ local tilting (**epeirogeny**) of the land, as in south-east England, which has increased the flood risk, and in parts of the Mediterranean, leading to the submergence of several ancient ports and leaving others stranded above the present-day sea-level
■ local volcanic and earthquake activity, as in Iceland.

Landforms created by sea-level changes

Changes in sea-level have affected:
■ the shape of coastlines and the formation of new features by increased erosion or deposition
■ the balance between erosion and deposition by rivers (page 81) resulting in the drowning of lower sections of valleys or in the rejuvenation of rivers, and
■ the migration of plants, animals and people.

Landforms resulting from submergence

Eustatic rises in sea-level following the decay of the ice sheets led to the drowning of many low-lying coastal areas.

Estuaries are the tidal mouths of rivers, most of which have inherited the shape of the former river valley (Figure 6.29). In many cases, estuaries have resulted from the lower parts of the valleys being drowned by the post-glacial rise of sea-level. Being tidal, estuaries are subject to the ebbs and flows of the tide, and usually large expanses of mud are revealed at low tide (Figure 6.43). Many estuaries widen towards the sea and narrow to a meandering section inland (Figure 6.44).

Estuaries are affected by processes that are very different from those at work along rivers and coasts, because of particular features.

■ **Residual currents** are created by the mixing of fresh water (from rivers) and saline water (sea water brought in by the tides). Mixing tends to take place only when discharge and velocities are high; otherwise the fresh river water, being less dense, tends to rise and flow over the saline water.
■ **Tidal currents** have a two-way flow associated with the incoming (flood) and outgoing (ebb) tide.
■ Continuous variations in both **discharge** and **velocity** resulting from the tidal cycle. Tidal velocities are highest at mid-tide and reduce to zero around high and low water. Times of zero velocity result in the deposition of fine-grained sediments, especially in upper estuary channels, which form mudflats and saltmarsh.

Figure 6.43

The Humber estuary

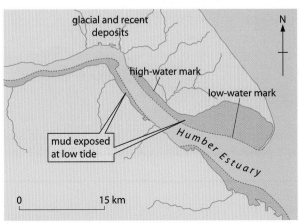

Figure 6.44

Estuary morphology (*after* Pethick, 1984)

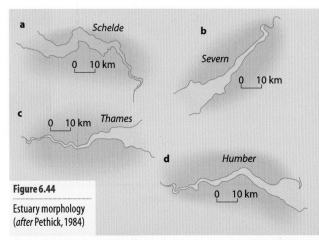

Classification of estuaries

a **According to origin** This traditional method divides estuaries into different shapes but on the basis of their river valley origins.

- **Drowned river valleys**, resulting from post-glacial rises in sea-level, includes most estuaries.
- **Rias**, formed when valleys in a dissected upland are submerged, are one type of drowned river valley (Places 22).
- **Dalmatian coasts** are similar to rias except that their rivers flow almost parallel to the coast, in contrast to rias where they flow more at right-angles, e.g. Croatia.
- **Fiords**, formed by the drowning of glacial troughs (page 113), are extremely deep and steep-sided estuaries (Places 22).
- **Fiards** are drowned, glaciated lowland areas, e.g. Strangford Lough, Northern Ireland.

b **According to tidal process and estuary shape** This modern approach, supported by Pethick, acknowledges that it is tidal range that determines the tidal current, the residual current velocities and, therefore, the amount and source of sediment.

- **Micro-tidal** estuaries, which have a tidal range of less than 2 m, are dominated by freshwater river discharge and wind-driven waves from the sea. They tend to be long, wide and shallow, often with a fluvial delta or coastal spits and bars.
- **Meso-tidal** estuaries have a tidal range of between 2 m and 4 m. This fairly limited range means that, although fresh water has less influence, the tidal flow does not extend far upstream and the resultant shape is said to be stubby, with the presence of tidal meanders in the landward section.
- **Macro-tidal** estuaries have a tidal range in excess of 4 m and a tidal influence that extends far inland. They have a characteristic trumpet shape (Figure 6.44) and long, linear sand bars formed parallel to the tidal flow.

Places 22 Devon and Norway: a ria and a fiord

Kingsbridge estuary

During the last ice age, rivers in south-west England were often able to flow during the warmer summer months (compare Figure 5.14), cutting their valleys downwards to the then lower sea-level (page 163). When, following the ice age, sea-levels rose (Figure 6.42), the lower parts of many main rivers and their tributaries were drowned to form sheltered, winding inlets called **rias**. The Kingsbridge estuary (Figures 6.45 and 6.46) is a natural harbour produced by the drowning of a dendritic drainage system (Figure 3.50b). The deepest water is at the estuary mouth, a characteristic of a ria, with depth decreasing inland. The result is a fine natural harbour with an irregular shoreline and, at low tide, 800 hectares of tidal creeks and mudflats.

Apart from south-west England, rias are also found in south-west Wales, south-west Ireland, western Brittany and north-west Spain.

Figure 6.45

Kingsbridge estuary

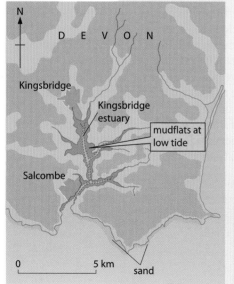

Figure 6.46

Kingsbridge estuary, looking north

Sognefjorden

Fiords (fjords) such as the Sogne Fiord (Sognefjorden) were formed by glaciers eroding their valleys to form deep glacial troughs (page 113). When the ice melted, the glacial troughs were flooded by a eustatic rise in sea-level (page 163) to form long, deep, narrow inlets with precipitous sides, a U-shaped cross-section, and hanging valleys (Figure 4.21). Glaciers seem to have followed lines of weakness, such as a pre-glacial river valley or, as suggested by their rectangular pattern, a major fault line (Figure 6.47). Unlike rias, fiords are deeper inland and have a pronounced shallowing towards their seaward end. The shallow entrance, comprising a rock bar, is known as a **threshold**.

The Sogne Fiord extends 195 km inland and, at its deepest, has a depth of 1308 m (Figure 6.48). One description of the Sogne Fiord is given in Figure 6.49.

Apart from Norway, fiords are also found on the west coasts of the South Island of New Zealand, British Columbia, Alaska, Greenland and southern Chile.

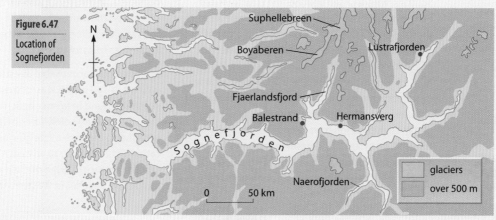

Figure 6.47

Location of Sognefjorden

Suphellebreen
Boyaberen
Lustrafjorden
Fjaerlandsfjord
Balestrand
Hermansverg
Sognefjorden
Naerofjorden

0   50 km

glaciers
over 500 m

Figure 6.48

Sognefjorden

Figure 6.49

Extract from *Blue Ice*, a novel by Hammond Innes

As we sailed up the fjord, the wind died away leaving the water as flat as glass. The view was breathtakingly beautiful. Mountains rose to snow-covered, jagged peaks. The dark green of the pines covered the lower slopes, but higher up the vegetation vanished leaving sheer cliffs of bare rock which seems to rise to the blue sky. In the distance, on a piece of flat land, was Balestrand, with a steamer moving to the quay. Beyond was the hotel on a delta of green and fertile land.

'Isn't it lovely?' Dahler said. 'It is the sunniest place in all the Sogne Fjord. The big hotel you see is built completely of wood. Here the fjord is friendly, but when you reach Fjaerlandsfjord you will find the water like ice, the mountains dark and terrible, rising to 1300 metres in precipitous cliffs. High above you will see the Boya and Suphelle glaciers, and from these rivers from the melting snow plunge as giant waterfalls into the calm, cold, green coloured fjord.'

Figure 6.50

Erosion surfaces (marine peneplanation) at St David's, Dyfed, South Wales

Landforms resulting from emergence
Following the global rise in sea-level, and still occurring in several parts of the world today, came the isostatic uplift of land as the weight of the ice sheets decreased. Landforms created as a result of land rising relative to the sea include erosion surfaces and raised beaches.

Erosion surfaces   In Dyfed, the Gower peninsula (South Wales) and Cornwall, flat planation surfaces dominate the scenery. Where their general level is between 45 m and 200 m, the surfaces are thought to have been cut during the Pleistocene period when sea-levels were higher – hence the alternative name of **marine platforms** (Figure 6.50).

Raised beaches As the land rose, former wave-cut platforms and their beaches were raised above the reach of the waves. Raised beaches are characteristic of the west coast of Scotland (Figure 6.51). They are recognised by a line of degraded cliffs fronted by what was originally a wave-cut platform. Within the old cliff-line may be relict landforms such as wave-cut notches, caves, arches and stacks (Figure 6.52). The presence of such features indicates that isostatic uplift could not have been constant. It has been estimated that it would have taken an unchanging sea-level up to 2000 years to cut each wave-cut platform. (This evidence has been used to show that the climate did not ameliorate steadily following the ice age.)

Places 23   Arran: raised beaches

The Isle of Arran is one of many places in western Scotland where raised beaches are clearly visible. Early workers in the field claimed that there were three levels of raised beach on the west coast of Scotland, found at 25, 50 and 100 feet above the present sea-level. These are now referred to as the 8-m, 15-m and 30-m raised beaches. However, this description is now considered too simplistic, since it has been accepted that places nearest to the centre of the ice depression have risen the most and that the amount of uplift decreases with distance from that point. Thus, for example, the much-quoted '8-m raised beach' on Arran in fact lies at heights of 4-6 m.

Where the raised beach is extensive, there is a considerable difference in height between the old cliff on its landward side and the more recent cliff to the seaward side, e.g. the 30-m beach in south-east Arran rises from 24 to 38 m.

It is now more acceptable to estimate the time at which a raised beach was formed by carbon-dating seashells found in former beach deposits, rather than by referring solely to its height above sea-level (i.e. to indicate a 'late glacial raised beach' rather than a '100-ft/30-m beach'). Figure 6.53 is a labelled transect, based on fieldwork, showing the two raised beaches in western Arran.

Figure 6.51

Raised beaches on the Isle of Arran: the lower one relates to the younger '8-m beach'; the upper one to the older '30-m beach'

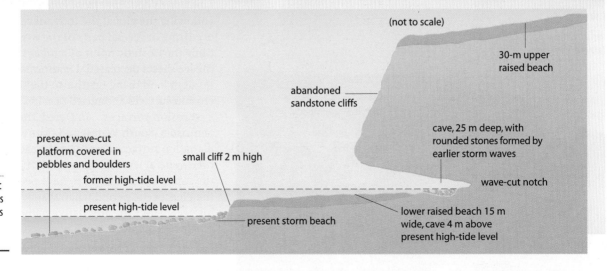

Figure 6.52

The abandoned cliff-line at King's Cave, Arran, with its '8-m raised beach' (see Figure 6.53)

Figure 6.53

Diagrammatic transect across raised beaches of Arran

(not to scale)

30-m upper raised beach

abandoned sandstone cliffs

cave, 25 m deep, with rounded stones formed by earlier storm waves

present wave-cut platform covered in pebbles and boulders

small cliff 2 m high

former high-tide level

wave-cut notch

present high-tide level

present storm beach

lower raised beach 15 m wide, cave 4 m above present high-tide level

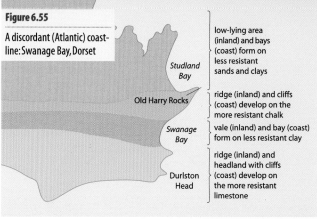

Figure 6.55

A discordant (Atlantic) coast-line: Swanage Bay, Dorset

Studland Bay

low-lying area (inland) and bays (coast) form on less resistant sands and clays

Old Harry Rocks

ridge (inland) and cliffs (coast) develop on the more resistant chalk

Swanage Bay

vale (inland) and bay (coast) form on less resistant clay

Durlston Head

ridge (inland) and headland with cliffs (coast) develop on the more resistant limestone

Figure 6.54

A concordant (Pacific) coastline: Lulworth Cove, Dorset

Rock structure

Concordant coasts and **discordant coasts** are located where the natural relief is determined by rock structure (geology). They form where the geology consists of alternate bands of resistant and less resistant rock which form hill ridges and valleys (page 199). Concordant coasts occur where the rock structure is parallel to the coast, as at Lulworth Cove, Dorset (Figure 6.54). Should there be local tectonic movements, a eustatic rise in sea-level, or a breaching of the coastal ridge, then summits of the ridge may be left as islands and separated from the mainland by drowned valleys. These can be seen on atlas maps showing Croatia/the former Yugoslavia (Dalmatian coast) or San Francisco and southern Chile (Pacific coasts). Discordant coasts occur where the coast 'cuts across' the rock structure, as in Swanage Bay, Dorset (Figure 6.55). Here the ridges end as cliffs at headlands, while the valleys form bays.

Framework 7   Classification

Why classify?

Geographers frequently utilise classifications, e.g. types of climate, soil and vegetation, forms and hierarchy of settlement, and types of landform. This is done to try to create a sense of order by grouping together into classes features that have similar, if not identical, characteristics into identifiable categories. For example, no two stretches of coastline will be exactly the same, yet by describing Kingsbridge estuary as a ria, and Sognefjorden as a fiord (Places 22), it may be assumed that their appearance and the processes leading to their formation are similar to those of other rias and fiords, even if there are local differences in detail.

How to classify

When determining the basis for any classification, care must be taken to ensure that:

- only meaningful data and measures are used
- within each group or category, there is the maximum number of similarities
- between each group, there is the maximum number of differences
- there are no exceptions, i.e. all the features should fit into one group or another, and
- there is no duplication, i.e. each feature should fit into one category only.

As classifications are used for convenience and to assist understanding, they should be easy to use. They should not be oversimplified (too generalised), or too complex (unwieldy); but they should be appropriate to the purpose for which they are to be used.

No classification is likely to be perfect, and several approaches may be possible.

An example

The following landforms have already been referred to in this book:

> arch; braided river; corrie; delta; esker; hanging valley; knickpoint; moraine; raised beach; rapids; spit; wave-cut platform.

Can you think of at least three different ways in which they may be categorised? The following are some possibilities:

a Perhaps the simplest classification is a two-fold division based upon whether they result from erosion, or from deposition.

b They could be reclassified into two different categories: those formed under a previous climate (i.e. relict features), and those still being formed today.

c The most obvious may be a three-fold division into coastal, glacial, and fluvial landforms.

d A more complex classification would result from combining either **a** and **b**, or **a** and **c**, to give six groups.

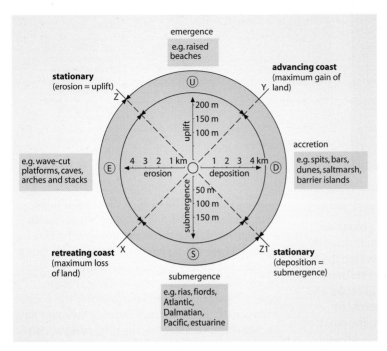

emergence
e.g. raised beaches

stationary
(erosion = uplift)

advancing coast
(maximum gain of land)

accretion
e.g. spits, bars, dunes, saltmarsh, barrier islands

e.g. wave-cut platforms, caves, arches and stacks

uplift

200 m
150 m
100 m

4 3 2 1 km   1 2 3 4 km
erosion   deposition

50 m
100 m
150 m

submergence

retreating coast
(maximum loss of land)

stationary
(deposition = submergence)

submergence
e.g. rias, fiords, Atlantic, Dalmatian, Pacific, estuarine

Figure 6.56

Valentin's classification of coasts, showing the conditions under which the maximum advance or retreat of the coast will occur, and how it is possible for a coast to be in equilibrium (stationary)

Classification of coasts

Considering the wide variety of coastlines, it should not be surprising that several different attempts have been made to classify coasts, each based upon a range of criteria.

Changes in sea-level (Johnson, 1919)

This, the earliest method, divided coasts into four categories:

1 **Emergent:** those resulting from a fall in sea-level and/or uplift of the land. Coastlines formed in this way are basically straight in appearance (e.g. barrier islands – Places 21).
2 **Submergent:** those showing a recent rise in sea-level and/or a fall in the land surface. This type is characterised by an indented coastline (e.g. rias and fiords – Places 22).
3 **Stable or neutral: a** those showing no signs of changes in sea-level or in the land; **b** those produced by non-marine processes (e.g. a river delta, page 77; or a volcanic island, Places 3 page 16).
4 **Compound:** those with a mixture of at least two of the three previously listed groups.

This classification came to be regarded as too simplistic when new evidence accumulated to show that there had been several changes in sea-level (local and global) and that many coasts exhibited signs both of submergence and emergence.

Relationship between coastal and other processes of erosion and deposition (Shepard, 1963)

This has been shown to be overgeneralised, having only two categories. These are:

1 **Primary:** where the influence of the sea has been minimal, e.g. fiords (ice), rias and deltas (rivers), and islands (volcanic or coral).
2 **Secondary:** where marine processes have been dominant, e.g. stacks and spits.

Advancing and retreating coasts (Valentin, 1952)

This method also assumes that all coasts can be fitted into a two-fold classification (Figure 6.56).

1 **Advancing:** where marine deposition or the uplift of the land is dominant.
2 **Retreating:** where marine erosion or the submergence of the land is more significant.

Tectonic coasts (Inman and Nordstrom, 1971)

This classification was based upon plate tectonic theory. It recognised four types of coastal plate boundaries (compare Figure 1.9):

1 **Diverging plates** (the Red Sea).
2 **Converging plates** (the island arcs of Japan and the Philippines).
3 **Major transform faults** (California).
4 **Stable plate boundaries** (India and Australia).

Energy-produced coastlines (Davis, 1980)

This, the most recent attempt to categorise coasts, is gaining credibility because it relates directly to the amount of energy expended by different types of wave upon a particular stretch of coastline (Figure 6.8).

1 **High-energy environments:** where destructive storm waves breaking on shingle beaches are most typical.
2 **Low-energy environments:** where constructive swell waves breaking upon sandy beaches are more frequent.
3 **Protected environments:** where wave action is limited in small, sheltered, sea areas.

Future sea-level rise and coastal stability

We have already seen (page 162) that over long periods of geological time (tens of millions of years) sea-level has been controlled by the movement of ocean plates and the uplift of continents, and that over shorter periods (the last million years) it has been controlled by the volume of ice on the land. Since the 'Little Ice Age' in the 17th century, when glaciers in the Alps and Arctic regions advanced, the world has been slowly warming (Figure 11.18). This seems to explain why sea-levels are now between 15 and 25 cm higher than a century ago, why they are at present rising by 2 mm a year and – accepting that global warming also results from the release of greenhouse gases (Case Study 9B) – why it is predicted to rise at an increasing rate in the next 100 years (Figure 6.57). Sea-levels are expected to rise for two specific reasons.

- The most significant factor is the thermal expansion of the oceans (warmer water is less dense and so occupies a greater volume). The Arctic Ocean warmed by 1°C or more between 1987 and 1999.

- A less significant, but increasing, contribution is from melting ice – mainly alpine glaciers (including the 1500 or so in the Himalayas) and, to a lesser extent, melting sea-ice (5 per cent decrease in the Arctic between 1984 and 1999). (The polar icecaps are not forecast to melt significantly in the next century or two partly because higher global temperatures are expected to increase evaporation from the oceans, which, could increase snowfall in Greenland and Antarctica.)

The present rate of sea-level rise of 2 mm per year (0.2 m/100 years) is expected to increase to 5 mm per year (0.5 m/100 years) in the next two or three decades. The extreme scenario, based upon the continued increase in the rate of global warming, is 10 mm per year (1 m/100 years).

Until recently it was believed that coastal ecosystems had the ability to adjust to sea-level changes. Saltmarshes and mangrove swamps could trap mud in sufficient quantities to enable them to grow upwards by up to 10 mm per year, while coral and sand dunes also seemed capable of making any necessary response. However, in early 1999, the Worldwide Fund for Nature (WWF) stated that '80 per cent of the world's beaches are eroding', while English Nature estimated that 'at least 13 000 hectares of English shoreline, much of it vital wildlife habitat, will disappear in the next 20 years'.

At present, some 10 million people are at constant risk of coastal flooding, while another 46 million could be affected by storm surges. A rise of 1 m over the next 100 years could inundate 25 per cent of Bangladesh, affecting 60 per cent of the population, and it would flood 30 per cent of Egypt's arable land (Figure 6.58). Cities at increasing risk from sea-level rises include Tokyo, Shanghai, Lagos, London, Bangkok, Calcutta, Alexandria, Hong Kong and Miami – despite some of the more wealthy of these trying to protect themselves, but at an astronomical cost, behind ever higher coastal defences. Several low-lying ocean states, such as the Maldives in the Indian Ocean and Tuvalu and the Marshall Islands in the Pacific Ocean, could be totally submerged.

Figure 6.57

Projections of future sea-level rise resulting from global warming: the extreme values cover the 95 per cent probability range (*after* Clayton, 1992)

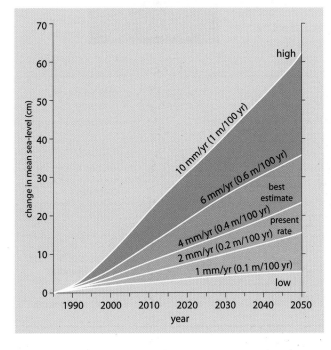

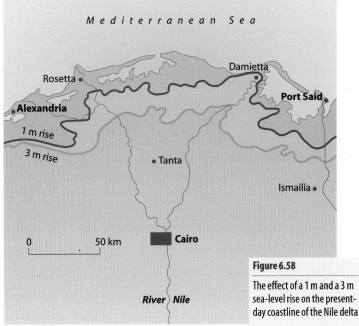

Figure 6.58

The effect of a 1 m and a 3 m sea-level rise on the present-day coastline of the Nile delta

The management of coastal erosion and flooding in the UK

The need for coastal management

- 23% total land area lies within 10 km of the coast
- 16.9 million people live within the coastal zone (7% within 10 km of the Thames estuary)

Land use
33% pasture land
25% arable land
 7% woodland
 2% heathland
 3% cliffs, beaches and mudflats
30% covered by buildings, roads and recreation facilities

40% UK manufacturing industry lies close to the coast

Source: Environment Agency 1999

Figure 6.59

Coastal zone facts – England and Wales

The facts in Figure 6.59 help to illustrate why there is a continuing demand for protection against marine erosion or coastal flooding by rivers and the sea. In recent years there have been spectacular cliff falls at Beachy Head, the collapse of Holbeck Hall Hotel onto the beach at Scarborough, storm damage and erosion at Hurst Castle Spit, as well as new cliff falls at Barton on Sea. Arable land in East Anglia and on Holderness continues to be lost to the sea at a rapid rate (Places 20, page 150). At Covehithe near Southwold in Suffolk, there has been a loss of 8.5 m each year in just five years (Figure 6.60). One estimate says that on the east coast an area the size of Jersey could be submerged in the next 20 years (*The Times*, 18 January 1999 – see Figure 6.73). Sea defences around the UK coast cost £300 million to maintain annually. Clearly there is a need to assess the risks of erosion and flooding and the strategy for protection of the coast zone where required.

Who is responsible for coastal protection?

At present in England and Wales, no single government department has overall responsibility for coastal protection. It is divided amongst the Ministry of Agriculture, Fisheries and Food (MAFF), the Environment Agency (responsible for protection against coastal flooding), maritime local authorities (both urban and rural district councils and county councils are responsible for protection against erosion), English Nature, the Countryside Commission which is responsible for Heritage Coasts, the National Trust, National Parks, and National Nature Reserves. There is a total of 1259 km of defences along the English and Welsh coasts: 805 km are provided by the Environment Agency, 242 km by local authorities and county councils, and 212 km by other organisations. Consequently when a coastal protection issue arises, the complexity of organisations involved leads to a division of authority, with no overall approach. The effects of the construction of sea defences at Aldeburgh in Suffolk and along the coast between Hastings and Pett Level in Sussex (see below) show the need for a more integrated approach (Figure 6.61).

Until relatively recently little weight was given to social impacts, so some proposed schemes failed to gain grant aid: as a result, some people lost their homes, with little or no compensation.

Figure 6.60

Erosion at Covehithe, Suffolk

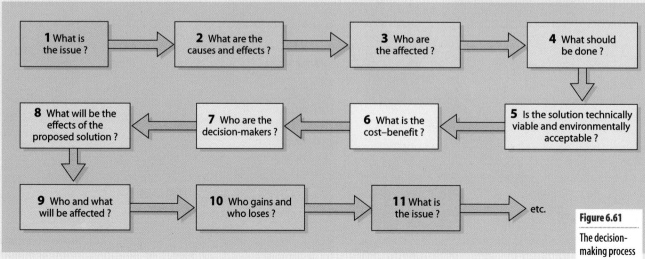

1 What is the issue ?

2 What are the causes and effects ?

3 Who are the affected ?

4 What should be done ?

5 Is the solution technically viable and environmentally acceptable ?

6 What is the cost–benefit ?

7 Who are the decision-makers ?

8 What will be the effects of the proposed solution ?

9 Who and what will be affected ?

10 Who gains and who loses ?

11 What is the issue ?

etc.

Figure 6.61

The decision-making process

How is the coast protected?

Hard defences involve the construction of major engineering works such as sea walls, revetments, groynes and breakwaters.

Some of the earliest protection was provided in the late 19th and early 20th centuries, as seaside towns realised the importance of promenades as part of the visitor and tourist attractions. Brighton, Eastbourne and Hastings built sea walls which provided additional space along the seafront as well as protecting the hotels and boarding houses there from storm waves. Walls repel the waves but do little to reduce their power. The base of the wall may be undermined as the beach is lowered by the waves. These hard defences are expensive to build and to maintain, and many are now coming to the end of their design lives, so further protection may be needed (page 150).

Wooden groynes are a familiar sight on beaches, where they are constructed at right-angles to the coastline to reduce the effects of longshore drift. For example, they are important on beaches in Christchurch Bay where they help to reduce movement of sediments away from popular tourist areas. This has resulted in loss of protection to the base of cliffs further east at Barton. Concrete breakwaters are sometimes used to protect small harbours from strong wave action. Their extension out into the main wave zone, as at Hastings, can prevent the natural movement of materials along the coast.

Soft defences include the use of beach replenishment at the base of structures, as at Pett Level and Aldeburgh (see below), as well as attempts to provide structures such as artificial headlands and offshore banks and reefs. Beach replenish-ment is expensive and needs to be maintained for long periods. Most artificial structures are constructed from rocks brought from Norway or, in some cases, from hardwoods from sustainable forests. Hurst Castle Spit (Figure 6.28) has been repaired using rocks from Norway as well as shingle from Shingles Bank at the northern edge of the Solent. Some engineers regard the use of fishtail groynes, which are longer than the normal timber groynes and made of large

boulders, as a soft solution. Fewer of them are required and they appear more natural. Most soft solutions are an attempt to keep protection as natural as possible.

In practice, a combination of hard and soft engineering solutions is common.

Aldeburgh, Suffolk

Aldeburgh, in Suffolk, at the northern end of Orford Ness (Figure 6.28), was protected by a sea wall and timber groynes to reduce the loss of beach material. Six streets to the east of the town have been lost to the sea since the 16th century, and the only visible remains of the former village of Slaughden, 1 km to the south, are a Martello tower and what is now a marina.

Following the partial failure of the sea wall in 1988, Anglian Water and the National Rivers Authority (now the Environment Agency) devised a £4.9 million plan to provide sea defences that would also protect the tidal freshwater areas of the River Alde immediately to the west of the town. The existing sea wall was extended at its base in the section considered most threatened. Several 10-tonne rock blocks were placed in front of the sea wall to absorb the wave energy; 200 m of wall originally protecting the northern end of Orford Ness was demolished, and a rock armour bank put in its place. A total of 24 new groynes were built, stretching south beyond the Martello tower, and 75 000 m³ of shingle were deposited as beach replenishment. More rocks were brought in to make a 400 m bank between the existing sea wall to the south and the shingle bank (Figures 6.62 and 6.63). The scheme was completed in 1992. It took into account the risk that storm damage could cause to an important natural area.

The coast between Hastings and Pett Level

Hastings is a large coastal resort town in East Sussex. The cliff coastline extends approximately 10 km eastwards from Hastings to the cliff-top coastal village of Fairlight and as far as Cliff End. Beyond Cliff End, artificial sea defences protect the low-lying Pett Level marsh, behind which the cliffs of the former coastline now lie several

kilometres inland (Figure 6.64). The geology along this section of the coast consists of unconsolidated Tertiary deposits of sands, gravels and sandstones. Natural rates of erosion are 0.5–1.0 m per year, with high-energy waves from the Atlantic having a fetch of 5000 km (page 140).

Figure 6.62

Sea defences at Aldeburgh

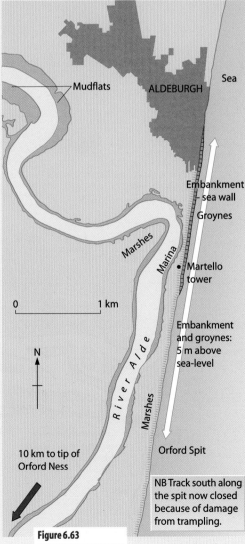

Figure 6.63

Aldeburgh sea defences

Coastal works at Hastings

Coastal land use changes over the last 100 years at Hastings have had a profound effect on this whole stretch of coastline.

- The cliff face has been protected to prevent rock falls and cliff retreat, with the result that the natural input of sediment into the coastal system has been slowed.
- Groynes were built to the west in Victorian times to encourage the beach to accrete (build up), thus providing protection for the town.

- The harbour, built in 1896, accelerated this beach-building by providing an effective sediment trap for the longshore drift, which flows in an easterly direction. This has had the benefit of creating a well built-up beach up to 300 m wide on the updrift (western) side of the harbour arm. (Any surplus material is carried eastwards by the longshore drift, well beyond Hastings.) An effective barrier to the largest storm waves, the beach makes sea walls and other hard engineered defences unnecessary.

- A £1 million improvement and extension to the harbour was built in the 1970s to provide a larger area for fishing boats. This has increased the sediment trap effect.
- The town has built outwards over 150 years and the buildings, the promenade and the wide beach ensure that waves no longer break behind the beach at the former cliff base, as the whole seafront is now artificial.

However, these changes at Hastings have been to the detriment of areas downdrift, i.e. towards the east at Fairlight and Pett Level.

Figure 6.64

The Sussex coast east of Hastings

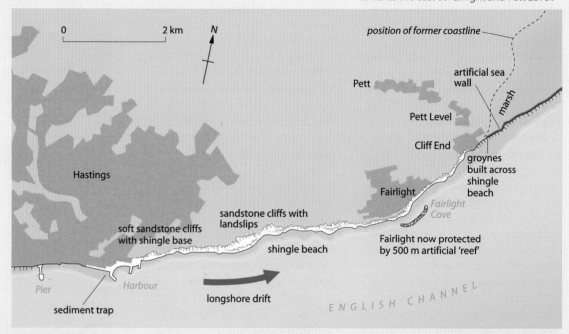

Figure 6.65

Sketch map of the sea front at East Hastings

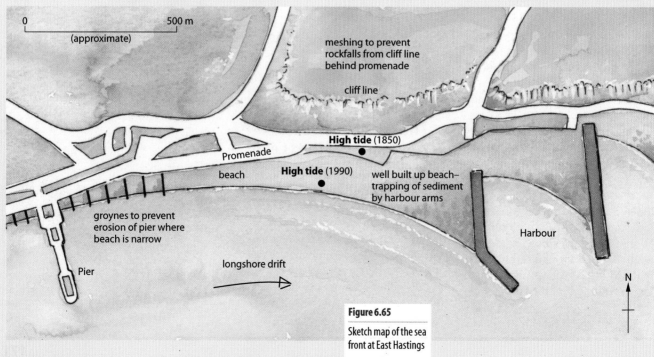

Figure 6.66

The cliffs at Fairlight village looking west, and showing evidence of mass movement

Figure 6.67

The artificial reef at Fairlight

The effects on Fairlight

The coastal village of Fairlight has experienced erosion of its 30 m high, near-vertical cliffs and the loss of land and homes. The main cause has been basal undercutting followed by mass movements resulting from sub-aerial processes (page 149) in the form of slides and slumping (Figures 2.17 and 6.66); this has led to rapid cliff retreat (compare Holderness, Places 20, page 150). High-energy waves at the cliff foot ensured that eroded material was quickly dispersed, thus encouraging the continual movement of slides above the tidal zone. Throughout the 1970s, rates of retreat were 0.5–1.0 m a year. However, there were extreme events, such as the loss of 66 m in one night at Fairlight Cove in December 1979 when heavy rainfall produced saturated groundwater conditions, mobilising landslides that had been initiated by winter storms. In 1989, houses at the cliff top were abandoned and the adjacent 'Sea View' road was closed to traffic.

It was believed that the 1970s extension of the harbour arm at Hastings had reduced the sediment input into the coastal system west of Fairlight and had exacerbated cliff erosion by reducing the width of the beach. With little sand in the environment, the base of the cliffs was thus exposed to higher wave energies with the result that the rate of cliff retreat had increased to more than 1 m per year by the late 1980s. More and more properties were coming under the threat of cliff fall.

In 1987 a scheme to protect the base of Fairlight cliff costing £2.5 million was proposed. A grant of 70 per cent would be paid by MAFF, and the rest paid by the local authorities. Rother District Council had decided on a 'do nothing' policy. The local residents who were affected decided to fight this, and appointed their own consultants who re-assessed costs and benefits, taking inflation into account when valuing property. The accelerated rate of erosion and the probable loss of 57 homes within 75 years, plus the costs of damage to roads, power supplies and sewerage systems, eventually persuaded Rother District Council to change their view and support the application for grant aid.

The scheme involved the construction of an offshore reef 500 m long, with its crest approximately 20 m from the cliff base (Figure 6.67). Waves now break offshore, thus reducing their effect on the base of the cliff, which is protected naturally by weathered material from the cliff face. This now appears to be eroding more slowly.

The problems created at Pett Level

This low-lying marshy area east of the small village of Cliff End has been drained since medieval times. It provides rough grazing for sheep and cattle, and there is some summer tourism. Most of Pett Level now lies below sea-level because of the contraction of sediments over 600 years. With an increase in the height of high tides and the incidence of storminess around this coast, there is danger of flooding, which will be increased as sea-levels rise with global warming.

The marsh is protected by an embankment maintained by the Environment Agency (see Figure 6.68). Hard defences, including groynes and wooden revetments and a sea wall, protect the embankment (see Figure 6.69). Wave energy is reduced by concrete stepping on the western end. Beach nourishment is provided near Cliff End to protect the embankment from further erosion and to prevent flooding of the low-lying area (see Figure 6.70). Costs of protection of a mainly agricultural area are increasingly being questioned.

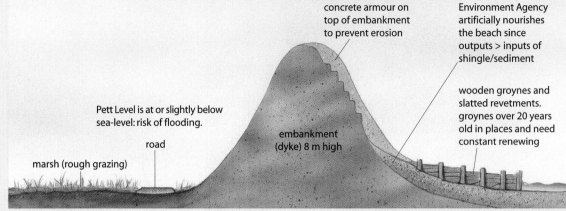

Figure 6.68

Coastal defences at Pett Level

concrete armour on top of embankment to prevent erosion

Environment Agency artificially nourishes the beach since outputs > inputs of shingle/sediment

wooden groynes and slatted revetments. groynes over 20 years old in places and need constant renewing

Pett Level is at or slightly below sea-level: risk of flooding.

road

embankment (dyke) 8 m high

marsh (rough grazing)

Figure 6.69

The flood wall at Pett Level caravan park

Figure 6.70

Beach nourishment by the Environment Agency at Pett Level

Coastal management strategies

It is now recognised that **shoreline management plans** provide a sound basis for the provision of coastal protection.

In 1991 the Anglian Sea Defence Management Study assessed defences from the Thames to the Humber. This showed clearly the need to link provision in one area with the effects in adjoining areas. As a result the MAFF published guidelines to be used in preparation of plans. Research had shown the existence of 11 sediment cells and 46 subcells around the English and Welsh coast (Figure 6.71). These are now used as limits for study and planning.

There are four strategic coast defence options:
1 Do nothing.
2 Hold the present defences.
3 Advance existing defences.
4 Retreat existing defences.

As far as possible, defences must be sustainable, which may allow the coast to reach its natural position, leading to the loss of farmland, buildings, roads and other infrastructure. Decisions have to be made between defending the coast at all costs or allowing nature to take its course.

Shoreline management plans are developed by groups from adjoining areas. These groups, including planners, engineers and geomorphologists, have specialist local knowledge. This has led to a high level of public consultation in affected areas and a greater knowledge of the physical processes involved. Alternative solutions are recognised as being necessary in areas similar to Fairlight. The issue of compensation has to be faced. At present in the UK there is no provision under law for this. Should the law be altered, then local authorities face a heavy cost for relatively small areas of sea defences.

Some important nature conservation sites may have been protected in the past but now decisions need to be made on the economic effectiveness of maintaining this protection, or letting nature take its course (Figures 6.72 and 6.73).

Urban areas too should consider whether to continue maintenance of expensive hard defences, or whether planners need to look to future long-term developments that allow the coast to realign itself naturally.

It has to be accepted that the coast is a dynamic system. Management may involve short-term losses for a relatively small number of people, but in the long term the gain both to people and to the environment will be greater.

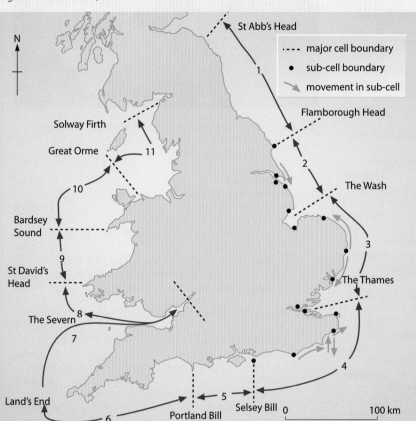

Figure 6.71

Major sediment cells around the coasts of England and Wales

'Millions are wasted' on flood barriers

By Charles Clover, Environment Editor

MILLIONS of pounds of public money are mis-spent defending farmland from the sea, according to a report yesterday. It said building concrete coastal defences often destroyed beaches and saltmarshes, Britain's richest habitat for wildlife.

The study, commissioned from Leeds University by the Wildlife Trusts and the World Wide Fund for Nature, says some of the £315 million a year spent by the Environment Agency on flood and coastal defences is unjustifiable.

It says it would be far cheaper and better for wildlife to abandon agricultural land around the Humber, Wash, Essex and Kent estuaries to the sea, allowing saltmarshes to retreat inland.

It finds that unless sea defences are realigned, as Britain continues to tilt down in the east and as sea-levels rise because of global warming, 124 square miles of wildlife habitat protected under European law will be lost in the next 50 years.

The report's author, John Bowers, criticises a number of proposed schemes including a £2.52 million scheme for renewing 11 miles of sea defences around Wallasea Island in Essex.

He calculates that the public benefit from the 2,000 acres of agricultural land over the 25-year lifetime of the defences is £43,669.

Figure 6.72

Article in the *Daily Telegraph*, July 1999

NO SHORE BET

The cost of holding back the rising tides

For years, politicians have been arguing about the erosion of Britain's sovereignty by the European Union. Yet the gradual disappearance of Britain itself has never received the same attention. Much like the gradual seepage of power from Westminster to Brussels, Britain has been slipping beneath the waves for decades. That rate of disappearance is now set to accelerate. On the east coast, an area the size of Jersey could be submerged in the next 20 years. Meeting this threat with a Maginot line of sea walls would cost taxpayers millions of pounds. A debate is needed on how Britain might best fight on its beaches.

The conventional wisdom used to be that sea walls were the best. This has been usurped by the concept of 'managed retreat'. Sea defences are to be abandoned at certain points, so natural habitats - such as saltmarshes - can slow down the sea's advance as well as create new habitats for wildlife. The members of the Agriculture Select Committee endorse this idea, considering it 'time to declare an end to the centuries-old war with the sea and seek a peaceful accommodation'. So too does the government, delighted by the savings it might make: sea walls cost up to £5 million per kilometre to build.

Figure 6.73

From *The Times*, 18 January 1999

References

Clayton, K. (1992) *Coastal Geomorphology*, Thomas Nelson.

Coastal Dunes: Geography Today, ITV.

Defence Against the Sea: Place and People, ITV.

Goudie, A. (1993) *The Nature of the Environment*, Blackwell.

Pethick, J. (1984) *An Introduction to Coastal Geomorphology*, Edward Arnold.

Waves and Beaches: Geography Today, ITV.

Wilson, J. (1995) *Statistics in Geography: For A Level Students*, Schofield & Sims.

Pethick, J. (ed.) (1994) *Humber Estuary and Coast, Management Issues*, University of Hull.

Websites

The LOIS (Land Ocean Interaction Study, of the Holderness coastline) home page:

http://wwwl.npm.ac.uk/lois/Education/case.htm

See also for more links:

http://www.nelsonthornes.com/gaia

Figure 6.74
Marsden Roc

1 a i Making good use of diagram/s, illustrate the meaning of the following terms used in relation to waves in the sea: crest; trough; wave length; wave height; wave period. **(5 marks)**
   ii What is meant by the term 'fetch' as used in discussions of wave energy? **(4 marks)**
 b Explain what happens as a wave approaches a beach. **(5 marks)**
 c What is the difference between a *constructive* wave and a *destructive* wave? **(6 marks)**
 d Why are waves *refracted* when they approach a coastline of headlands and bays? **(5 marks)**

2 a Study the photograph in Figure 6.74 and answer the following questions.
   i Describe the material found between the two stacks. **(3 marks)**
   ii Describe the beach material found in the foreground of the photograph. **(3 marks)**
   iii Describe the main stack. **(4 marks)**
 b How is a feature like this stack formed? **(6 marks)**
 c Some cliff coastlines, such as Old Harry Rock near Swanage (Figure 6.21, page 152), have no beach while others, such as Marsden Rock (Figure 6.74), have. Suggest a reason for this difference. **(4 marks)**
 d Marine erosion is concentrated at the base of a cliff. Suggest **two** ways in which the rest of the cliff is eroded. **(5 marks)**

3 a Making good use of diagrams, describe **two** landforms which may be found on a beach. **(6 marks)**
 b Why are large stones and boulders found at the **back** of a beach? **(4 marks)**
 c Making good use of diagrams, explain how sand and other material is moved along a beach by the action of waves. **(5 marks)**
 d Why are shingle beaches steeper, on average, than sandy beaches? **(5 marks)**
 e How and why may human activity change this marine transport process? **(5 marks)**

4 a What is meant by each of the following terms used in relation to the effects of waves on a coastline:
   i abrasion (sometimes called corrasion)
   ii attrition
   iii hydraulic action? **(6 marks)**
 b Explain how the processes identified in **a** cause a cliff to change its shape. **(6 marks)**
 c Study Figure 6.17 (page 150).
   i Describe and suggest reasons for the shape of the cliff shown in the photograph. **(6 marks)**
   ii Although there are houses on top of this cliff it has been decided not to attempt to protect this coastline. Suggest **two** reasons for this decision. **(7 marks)**

5 a Why is erosion concentrated at the foot of a marine cliff? **(4 marks)**
 b Identify and explain **two** processes by which waves erode a cliff. **(4 marks)**
 c Identify and explain **two** processes other than waves which can erode a cliff. **(4 marks)**
 d i Identify and describe **two** ways in which people can manage the erosion of a cliff foot. **(6 marks)**
   ii Using specific example/s, evaluate the success of **one** of these management strategies. **(7 marks)**

AS

6 a On a coastline with cliffs, **deposition** can cause the shape of the coastline to change. Suggest where there will be *deposition* on such a coastline and the reasons for deposition there. **(10 marks)**
 b i Study Figure 6.74. Draw an annotated diagram to identify the main features of the landform in the photograph. **(5 marks)**
   ii With reference to evidence from the photograph, explain how marine processes may have created this landform. **(10 marks)**

7 a Explain the difference in water movement in a wave between deep water and on a shelving beach. In your response you should indicate the difference between waves of differing period. **(12 marks)**
 b Using examples of coastal areas you have studied, explain the difference to wave energy caused by the *fetch* of the wave. **(8 marks)**
 c Using diagram/s **only**, explain the process of *wave refraction*. **(5 marks)**

8 'The interface between the sea and the land is an area of conflict in nature and for people.' Using examples from your studies, explain this statement. **(25 marks)**

9 a Choose **two** of the following micro-morphological features of a beach: berm; beach cusps; ridge and runnels; longshore bar.
   i For **each** feature you have chosen, making use of annotated diagrams describe its shape and location on a beach. **(6 marks)**
   ii Explain how it is formed. **(8 marks)**
 b What effect do storm waves have on a beach profile? **(6 marks)**
 c Describe **one** method you could use to survey the profile of a beach. **(5 marks)**

10 a Why do 'tides' occur on many beaches? **(8 marks)**
 b Study Figure 6.13 (page 147). Describe and explain the difference in tidal range and time of high tide between Aberdeen and Skegness. **(8 marks)**
 c Using examples you have studied, explain how tides influence the way of life of people living in areas with different tidal ranges. **(9 marks)**

A2

11 a Explain the terms 'eustatic' and 'isostatic' used when studying sea-level change. **(4 marks)**
b Explain how **i** an ice age and **ii** one other mechanism could cause sea-level change. **(7 marks)**
c Choose **one** landform which has been created by or significantly changed by a **fall** in sea-level. Describe the landform and explain the role of sea-level change in its formation. **(7 marks)**
d Choose **one** landform which has been created or changed significantly by a rise in sea-level. Describe the landform and explain the role of sea-level change in its formation. **(7 marks)**

12 a Describe the material which may occur naturally on a beach. **(4 marks)**
b i What is 'wave refraction'? **(4 marks)**
ii How does wave refraction help to explain the formation of beaches in coastal bays? **(4 marks)**
c i In many places people try to manage the beaches. Explain, using examples, one reason why they may do this. **(4 marks)**
ii One method of controlling the movement of beach material is the use of groynes. Making good use of diagrams, explain how groynes work. **(4 marks)**
iii Why do some people think that groynes do not provide the answer? **(5 marks)**

13 a Making good use of annotated diagrams, explain the process of long shore drift. **(5 marks)**
b i Study Figure 6.23 (page 153). Suggest, with reasons, the direction of long shore drift on this coastline. **(3 marks)**
ii Why were the sea defences put along this shoreline? **(6 marks)**
iii What effect would you expect there to be further down the coast as a result of the building of these sea defences? Explain your answer. **(6 marks)**
c Choose **one** landform created by marine deposition. Describe the size and shape of the landform and suggest how marine deposition has helped to create it. **(5 marks)**

14 Use Figure 6.31 (page 157).
a Identify the meaning of the following terms used: dominant wind; dune slack. **(4 marks)**
b Describe the location of the water table. **(4 marks)**
c i Draw a line graph to show (i) the percentage of exposed sand and (ii) the dune height. **(8 marks)**
ii Describe the relationship shown by your graph. **(3 marks)**
iii Suggest reasons for the relationship you have identified. **(6 marks)**

15 a Using an annotated diagram **only**, explain the process by which beach material is moved along the coastline. **(5 marks)**
b Choose **one** landform which is created when beach material is deposited. Name and describe the landform. Explain the processes by which the landform is created. **(10 marks)**
c Why do people try to reduce the movement of beach material on some coastlines? Suggest and explain **two** methods for reducing such movement. **(10 marks)**

16 a Using your own case studies, choose **two** examples of hazards which occur on marine coasts. For **each** hazard:
i Identify the hazard and its location. **(2 marks)**
ii Explain how the action of the sea leads to danger on the coast. **(12 marks)**
b Describe **one** way in which the people prepare to face marine hazards and evaluate their success when the danger occurs. **(11 marks)**

17 a Making good use of diagrams, identify the morphological zones of a beach. For each zone describe one piece of evidence you could see in the field to show that it is a distinct zone. **(8 marks)**
b Explain the effect of wave type and particle size on the profile of a beach. **(10 marks)**
c Explain **one** way in which human use of a beach can alter its profile. **(7 marks)**

18 Discuss the causes and results of possible future sea-level changes. In your answer you should refer to the long-term and immediate changes and identify groups and areas which will be most directly affected. **(25 marks)**

19 a Using an example from your studies, explain why a particular coastal management scheme was felt to be necessary. **(6 marks)**
b Describe the planning and decision-making process involved in the creation of the management plan for the area. **(6 marks)**
c Outline the plan and suggest why the changes outlined should overcome the identified problem/s. **(6 marks)**
d Evaluate the success of the project. **(7 marks)**

20 Choose **one** system of coastal classification. Describe and explain the principles on which it is based and, making use of examples, describe some of the problems of applying your classification system to cover all coastal areas. **(25 marks)**

21 Study the sand dune area in Figure 6.75.
a i Identify and locate **one** feature of the photograph which indicates that this area is popular with people. Explain how it shows the presence of people. **(4 marks)**
ii Explain **one** piece of evidence from the photograph which show that this popularity is causing damage to the environment. **(4 marks)**
b i Suggest **one** possible effect of the environmental damage caused in this area. **(7 marks)**
ii Explain how conservation work could overcome the damage done to this sand dune belt. **(10 marks)**

Figure 6.75

Damaged sand dunes in Gower, Wales

Deserts

'Now the wind grew strong and hard and it worked at the rain crust in the cornfields. Little by little the sky was darkened by the mixing dust, and the wind felt over the earth, loosened the dust and carried it away.'

J. Steinbeck, *The Grapes of Wrath*, 1939

What is a desert?

'The deserts of the world, which occur in every continent including Antarctica, are areas where there is a great deficit of moisture, predominantly because rainfall levels are low. In some deserts this situation is in part the result of high temperatures, which mean that evaporation rates are high. It is the shortage of moisture which determines many of the characteristics of the soils, the vegetation, the landforms, the animals, and the activities of humans' (Goudie and Watson, 1990).

A desert environment has conventionally been described in terms of its deficiencies – water, soils, vegetation and population. Deserts include those parts of the world that produce the smallest amount of organic matter and have the lowest net primary production (NPP, page 306). In reality, many desert areas have potentially fertile soils, evidenced by successful irrigation schemes; all have some plant and animal life, even if special adaptations are necessary for their survival; and some are populated by humans, occasionally only seasonally by nomads but elsewhere permanently, e.g. in large cities like Cairo and Karachi.

The traditional definition of a desert is an area receiving less than 250 mm of rain per year. While very few areas receive no rain at all (Places 24, page 180), amounts of precipitation are usually small and occurrences are both infrequent and unreliable. Climatologists have sometimes tried to differentiate between cold deserts where for at least one month a year the mean temperature is below 6°C, and hot deserts. Several geomorphologists have used this to distinguish the landforms found in the hot sub-tropical deserts – our usual mental image of a desert – from those found in colder latitudes, e.g. the Gobi Desert and the tundra.

Modern attempts to define deserts are more scientific and are specifically linked to the water balance (page 60). This approach is based on the relationship between the input of water as precipitation (P), the output of moisture resulting from evapotranspiration (E), and changes in water held in storage in the ground. In parts of the world where there is little precipitation annually or where there is a seasonal drought, the **actual evapotranspiration** (AE) is compared with **potential evapotranspiration** (PE) – the amount of water loss that would occur if sufficient moisture was always available to the vegetation cover. C. W. Thornthwaite (1931) was the first to define an **aridity index** using this relationship (Figure 7.1).

Figure 7.1

The index of aridity

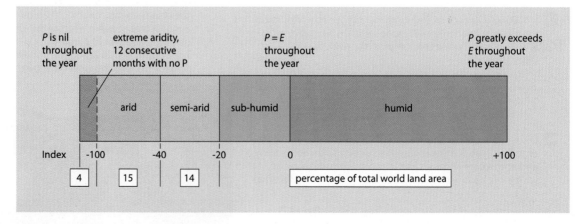

Location and causes of deserts

On the basis of climatic characteristics, including Thornthwaite's aridity index, one-third of the world's land surface can be classified as desert, i.e. arid and semi-arid. Alarmingly, this figure, and therefore the extent of deserts, may be increasing (Case Study 7).

As shown in Figure 7.2, the majority of deserts lie in the centre or on the west coast of continents between 15° and 30° north and south of the Equator. This is the zone of sub-tropical high pressure where air is subsiding (the descending limb of the Hadley cell, Figure 9.34). On page 226 there is an explanation of how warm, tropical air is forced to rise at the Equator, producing convectional rain, and how later that air, once cooled and stripped of its moisture, descends at approximately 30° north and south of the Equator. As this air descends it is compressed, warmed and produces an area of permanent high pressure. As the air warms, it can hold an increasing amount of water vapour which causes the lower atmosphere to become very dry. The low relative humidity, combined with the fact that there is little surface water for evaporation, gives clear skies.

A second cause of deserts is the rainshadow effect produced by high mountain ranges. As the prevailing winds in the sub-tropics are the trade winds, blowing from the north-east in the northern hemisphere and the south-east in the southern hemisphere, then any barrier, such as the Andes, prevents moisture from reaching the western slopes. Where plate movements have pushed up mountain ranges in the east of a con-

tinent, the **rainshadow effect** creates a much larger extent of desert (e.g. 82 per cent of the land area of Australia) than when the mountains are to the west, as in South America.

Aridity is increased as the trade winds blow towards the Equator, becoming warmer and therefore drier. Where the trade winds blow from the sea, any moisture which they might have held will be precipitated on eastern coasts leaving little moisture for mid-continental areas. The three major deserts in the northern hemisphere which lie beyond the sub-tropical high pressure zone (the Gobi and Turkestan in Asia and the Great Basins of the USA) are mid-continental regions far removed from any rain-bearing winds, and surrounded by protective mountains.

A third combination of circumstances giving rise to deserts is also shown in Figure 7.2. Several deserts lie along western coasts where the ocean water is cold. In each case, the prevailing winds blow parallel to the coastline and, due to the Earth's rotation, they tend to push surface water seaward at right-angles to the wind direction. The Coriolis force (page 224) pushes air and water coming from the south towards the left in the southern hemisphere and water from the north to the right in the northern hemisphere. Consequently, very cold water is drawn upwards to the ocean surface, a process called **upwelling**, to replace that driven out to sea. Any air which then crosses this cold water is cooled and its capacity to hold moisture is diminished. Where these cooled winds from the sea blow on to a warm land surface, advection fogs form (page 222 and Places 24).

⬛		extreme aridity
⬛		arid
⬜		semi-arid

H high pressure
R rainshadow
M mid-continent
U upwelling of cold water

1 Australia (e.g. Simpson Great Sandy) **H R M**
2 Gobi **M**
3 Thar **H**
4 Iran **H M**
5 Turkestan **M**
6 Arabia **H**
7 Somalia **H**
8 Kalahari **H M**
9 Namib **H U**
10 Sahara **H** (**U** in west, **M** in centre)
11 Patagonia **R**
12 Monte **H**
13 Atacama **H R U**
14 Sonora **H**
15 Mojave **H R** (**U** on coasts)
16 Great Basins **R**

Figure 7.2

Arid lands of the world

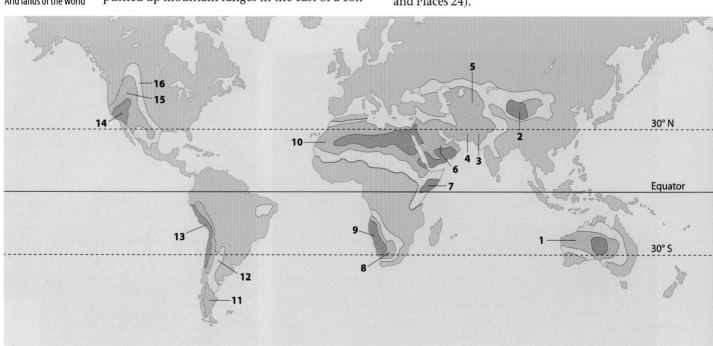

Figure 7.3

The Atacama Desert

The prevailing winds in the Atacama, which lies in the sub-tropical high pressure belt, blow northwards along the South American coast. These winds, and the northward-flowing Humboldt (Peruvian) current over which they blow, are pushed westwards (to the left) and out to sea by the Coriolis force as they approach the Equator. This allows the upwelling of cold water from the deep Peru–Chile sea trench (Figure 1.12) that provides the rich nutrients to nourish the plankton which form the basis of Peru's fishing industry. The upwelling also cools the air above which then drifts inland and over the warmer desert. The meeting of warm and cold air produces advection fogs (page 222) which provide sufficient moisture for a limited vegetation cover. Inland, parts of the Atacama are alleged to be the only truly rainless desert in the world, but even here the occasional rainfall event does occur.

Desert landscapes: what does a desert look like?

Deserts provide a classic example of how easy it is to portray or to accept an inaccurate mental picture of different places (or people) in the world. What is your image of a desert? Is it a landscape of sand dunes similar to those shown in Figures 7.15–7.18, perhaps with a camel or palm tree somewhere in the background? Large areas of dunes, known as **erg**, do exist – but they cover only about one-quarter of the world's deserts. Most deserts consist either of bare rock, known as **hammada** (Figure 7.4), or stone-covered plains, called **reg** (Figure 7.5). Deserts contain a great diversity of landscapes. This diversity is due to geological factors (tectonics and rock type) as well as to climate (temperature, rainfall and wind) and resultant weathering processes.

Figure 7.4

A rocky (hammada) desert, Wadi Rum, Jordan

Figure 7.5

A stony (reg) des[ert] Sahara

Arid processes and landforms

In their attempts to understand the development of arid landforms, geographers have come up against three main difficulties:

- How should the nature of the weathering processes be assessed? Desert weathering was initially assumed to be largely mechanical and to result from extreme diurnal ranges in temperature. More recently, the realisation that water is present in all deserts in some form or other has led to the view that chemical weathering is far more significant than had previously been thought. Latest opinions seem to suggest that the major processes, e.g. exfoliation and salt weathering, may involve a combination of both mechanical and chemical weathering.
- What is the relative importance of wind and water as agents of erosion, transportation and deposition in deserts?
- How important have been the effects of climatic change on desert landforms? During some phases of the Quaternary, and previously when continental plates were in different latitudes, the climate of present arid areas was much wetter than it is today. How many of the landforms that we see now are, therefore, relict and how many are still in the process of being formed?

Mechanical weathering

Traditionally, weathering in deserts was attributed to mechanical processes resulting from extremes of temperature. Deserts, especially those away from the coast, are usually cloudless and are characterised by daily extremes of temperature. The lack of cloud cover can allow day temperatures to exceed 40°C for much of the year; while at night, rapid radiation often causes temperatures to fall to zero. Although in some colder, more mountainous deserts, frost shattering is a common process, it was believed that the major process in most deserts was **insolation weathering**. Insolation weathering occurs when, during the day, the direct rays of the sun heat up the surface layers of the rock. These surface layers, lacking any protective vegetation cover, may reach 80°C. The different types and colours of minerals in most rocks, especially igneous rocks, heat up and cool down at different rates, causing internal stresses and fracturing. This process was thought to cause the surface layers of exposed rock to peel off – **exfoliation** – or individual grains to break away – **granular disintegration** (page 41). Where surface layers do peel away, newly exposed surfaces experience pressure release (page 41). This is believed to be a contributory process in the formation of rounded exfoliation domes such as Ayers Rock (Figure 7.6) and Sugarloaf Mountain (Figure 2.3).

Doubts about insolation weathering began when it was noted that the 4500-year-old ancient monuments in Egypt showed little evidence of exfoliation, and that monuments in Upper Egypt, where the climate is extremely arid, showed markedly fewer signs of decay than those located in Lower Egypt, where there is a limited rainfall. D. T. Griggs (1936) conducted a series of laboratory experiments in which he subjected granite blocks to extremes of temperature in excess of 100°C. After the equivalent of almost 250 years of diurnal temperature change, he found no discernible difference in the rock. Later, he subjected the granite to the same temperature extremes while at the same time spraying it with water. Within the equivalent of two and a half years of diurnal temperature change, he found the rock beginning to crack. His conclusions, and those of later geomorphologists, suggest that some of the weathering previously attributed to insolation can now be ascribed to chemical changes caused by moisture. Although rainfall in deserts may be limited, the rapid loss of temperature at night frequently produces dew (175 nights a year in Israel's Negev) and the mingling of warm and cold air on coasts (e.g. of the Atacama) causes advection fog (page 222). There is sufficient moisture, therefore, to combine with certain minerals to cause the rock to swell (hydration) and the outer layers to peel off (exfoliation). At present, it would appear that the case for insolation weathering is neither proven nor disproven and that it may be a consequence of either mechanical weathering, or chemical weathering, or both.

Figure 7.6

An exfoliation dome: Ayers Rock, Australia (compare with Figure 2.3)

The second mechanical process in desert environments, **salt weathering**, is more readily accepted although the action of salt can cause chemical, as well as physical, changes in the rock. Salts in rainwater, or salts brought to the surface by capillary action, form crystals as the moisture is readily evaporated in the high temperatures and low relative humidities. Further evaporation causes the salt crystals to expand and mechanically to break off pieces of the rock upon which they have formed (page 40). Subsequent rainfall, dew or fog may be absorbed by salt minerals causing them to swell (hydration) or chemically to change their crystal structure (page 42). Where salts accumulate near or on the surface, particles may become cemented together to form **duricrusts**. These hard crusts are classified according to the nature of their chemical composition. (Students with a special interest in geology or chemistry may wish to research the meaning of the terms **calcretes**, **silcretes** and **gypcretes**.) Another form of crust, **desert varnish**, is a hard, dark glazed surface found on exposed rocks which have been coated by a film composed largely of oxides of iron and manganese (Figure 7.7) and, possibly, bacterial action. It is hoped that the dating of desert varnish may help to establish a chronology of climatic changes in arid and semiarid environments.

Figure 7.7

Carvings in desert varnish, Wadi Rum, Jordan

The importance of wind and water

Geomorphologists working in Africa at the end of the last century believed the wind to be responsible for most desert landforms. Later fieldwork, carried out mainly in the higher and wetter semi-arid regions of North America, recognised and emphasised the importance of running water and, in doing so, de-emphasised the role of wind. Today, it is more widely accepted that both wind and water play a significant, but locally varying, part in the development of the different types of desert landscape.

Aeolian (wind) processes

Transport

The movement of particles is determined by several factors. Aeolian movement is greatest where winds are strong (usually over 20 km/hr), turbulent, come from a constant direction and blow steadily for a lengthy period of time. Of considerable importance, too, is the nature of the regolith. It is more likely to be moved if there is no vegetation to bind it together or to absorb some of the wind's energy; if it is dry and unconsolidated; if particles are small enough to be transported; and if material has been loosened by farming practices. While such conditions do occur locally in temperate latitudes, e.g. coastal dunes, summits of mountains and during dry summers in arable areas, the optimum conditions for transport by wind are in arid and semi-arid environments.

Wind can move material by three processes: suspension, saltation, and surface creep. The effectiveness of each method is related to particle size (Figure 7.8).

Suspension Where material is very fine, i.e. less than 0.15 mm in diameter, it can be picked up by the wind, raised to considerable heights and carried great distances. There have been occasions, though perhaps recorded only once a decade, when red dust from the Sahara has been carried northwards and deposited as 'red rain' over parts of Britain. Visibility in deserts is sometimes reduced to less than 1000 m and this is called a **dust storm** (Figure 7.9). The number of recorded dust storms on the margins of the Sahara has increased rapidly in the last 25 years as the drought of that region has intensified. In Mauritania, there was an average of only 5 days/yr with dust storms during the early part of the 1960s compared with an average of 80 days/yr over a similar period in the early 1990s.

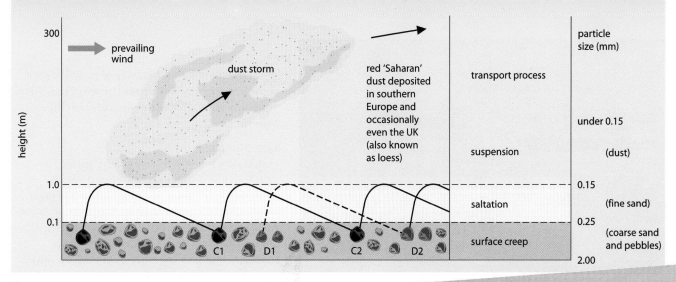

particle size (mm)	transport process
under 0.15 (dust)	suspension
0.15 (fine sand)	saltation
0.25 (coarse sand and pebbles)	surface creep
2.00	

red 'Saharan' dust deposited in southern Europe and occasionally even the UK (also known as loess)

Figure 7.8

Processes of wind transportation

Saltation When wind speeds exceed the threshold velocity (the speed required to initiate grain movement), fine and coarse-grained sand particles are lifted. They may rise almost vertically for several centimetres before returning to the ground in a relatively flat trajectory of less than 12° (Figure 7.8). As the wind continues to blow, the sand particles bounce along, leap-frogging over one another. Even in the worst storms, sand grains are rarely lifted higher than 2 m above the ground.

Surface creep Every time a sand particle, transported by saltation, lands, it may dislodge and push forward larger particles (more than 0.25 mm in diameter) which are too heavy to be uplifted. This constant bombardment gradually moves small stones and pebbles over the desert surface.

Figure 7.10

A desert pavement with ventifacts in Jordan, created by deflation

Figure 7.9

A dust storm in Egypt

Erosion

There are two main processes of wind erosion: deflation and abrasion.

Deflation This is the progressive removal of fine material by the wind leaving pebble-strewn desert pavements or reg (Figures 7.10 and 7.11). Over much of the Sahara, and especially in Sinai in Egypt, vast areas of monotonous, flat and colourless pavement are the product of an earlier, wetter climate. Pebbles were transported by water from the surrounding highlands and deposited with sand, clay and silt on the lowland plains. Later, the lighter particles were removed by the wind, causing the remaining pebbles to settle and to interlock like cobblestones.

Elsewhere in the desert, dew may collect in hollows and material may be loosened by chemical weathering and then removed by wind to leave **closed depressions** or **deflation hollows**. Closed depressions are numerous and vary in size from a few metres across to the extensive Qattara Depression in Egypt which reaches a depth of 134 m below sea-level. Closed depressions may also have a tectonic origin (the south-west of the

Figure 7.11

The process of deflation

deflation: silt and sand removed by wind leaving stones

land surface is lowered

leaving desert pavement: a coarse mosaic of stones resembling a cobbled street which protects against further erosion

183

Figure 7.12

Landshore yardangs, Western Desert, Egypt

USA) or a solution origin (limestone areas in Morocco). The Dust Bowl, formed in the American Mid-West in the 1930s, was a consequence of deflation following a severe drought in a region where inappropriate farming techniques had been introduced. Vast quantities of valuable topsoil were blown away, some of which was deposited as far away as Washington, DC.

Abrasion is a sandblasting action effected by materials as they are moved by saltation. This process smooths, pits, polishes and wears away rock close to the ground. Since sand particles cannot be lifted very high, the zone of maximum erosion tends to be within 1 m of the Earth's surface. Abrasion produces a number of distinctive landforms which include ventifacts, yardangs and zeugen.

Ventifacts are individual rocks with sharp edges and, due to abrasion, smooth sides. The white rock in the foreground of Figure 7.10 has a long axis of 25 cm.

Yardangs are extensive ridges of rock, separated by grooves (troughs), with an alignment similar to that of the prevailing winds (Figure 7.12). In parts of the Sahara, Arabian and Atacama Deserts, they are large enough to be visible on air photographs and satellite imagery.

Figure 7.13

The movement of a crescent-shaped barchan

Zeugen are tabular masses of resistant rock separated by trenches where the wind has cut vertically through the cap into underlying softer rock.

Deposition

Dunes develop when sand grains, moved by saltation and surface creep, are deposited. Although large areas of dunes, known as ergs, cover about 25 per cent of arid regions, they are mainly confined to the Sahara and Arabian Deserts, and are virtually absent in North America. Much of the early fieldwork on dunes was carried out by R. A. Bagnold in North Africa in the 1920s. He noted that some, but by no means all, dunes formed around an obstacle – a rock, a bush, a small hill or even a dead camel; and most dunes were located on surfaces that were even and sandy and not on those which were irregular and rocky. He concentrated on two types of dune: the barchan and the seif. The **barchan** is a small, crescent-shaped dune, about 30 m high, which is moved by the wind (Figures 7.13 and 7.15). The **seif**, named after an Arab curved sword, is much larger (100 km in length and 200 m in height) and more common (Figure 7.17), although the process of its formation is more complex than initially thought by Bagnold. Textbooks often over-emphasise these two dunes, especially the barchan which is a relatively uncommon feature.

While Bagnold had to travel the desert in specially converted cars, modern geographers derive their picture of desert landforms from aerial photographs and Landsat images. These new techniques have helped to identify several types of dune, and the modern classification, still based on morphology, contains several additional types (Figure 7.14). Dune morphology depends upon the supply of sand, wind direction, availability of vegetation and the nature of the ground surface.

a in plan

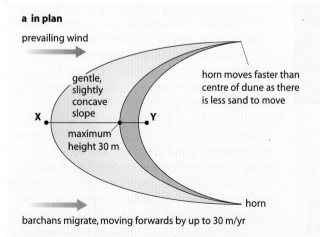

prevailing wind

gentle, slightly concave slope

horn moves faster than centre of dune as there is less sand to move

X — Y

maximum height 30 m

horn

barchans migrate, moving forwards by up to 30 m/yr

b in profile

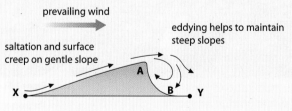

prevailing wind

saltation and surface creep on gentle slope

eddying helps to maintain steep slopes

A

X — B — Y

A steep, upper slip slope of coarse grains and with continual sand avalanches due to unconsolidated material (unlike a river, coarse grains are at the top)

B gentle, basal apron with sand ripples: the finer grains, as on a beach, give a gentler gradient than coarser grains

Type of dune		Description	Supply of sand	Wind direction and speed	Vegetation cover	Speed of dune movement
barchan 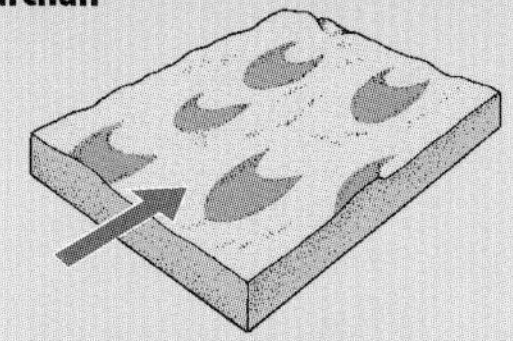		individual dunes, crescent shape with horns pointing downwind (Figures 7.13 and 7.15)	limited	constant direction, at right-angles to dune	none	highly mobile
barchanoid ridges 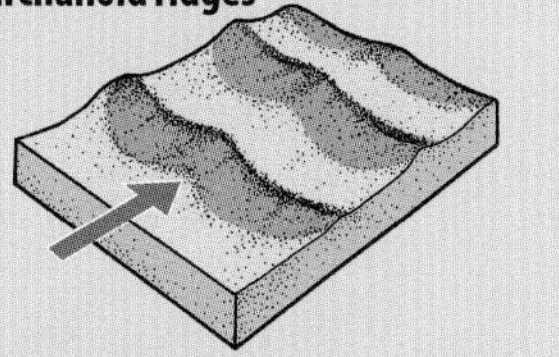		asymmetrical, oriented at right-angles to wind, rows of barchans forming parallel ridges	limited	constant direction, at right-angles to dune	none	mobile
transverse 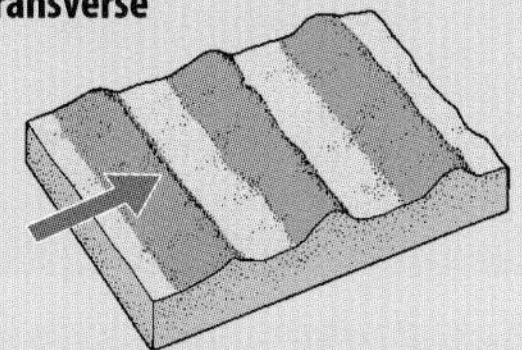		oriented at right-angles to wind but lacking barchanoid structure, resemble ocean waves (Figure 7.16)	abundant (thick) sand cover	steady winds (trades), constant direction but with reducing speeds, at right-angles to dune	vegetation stabilises sand	sand checked by barriers, limited mobility
dome 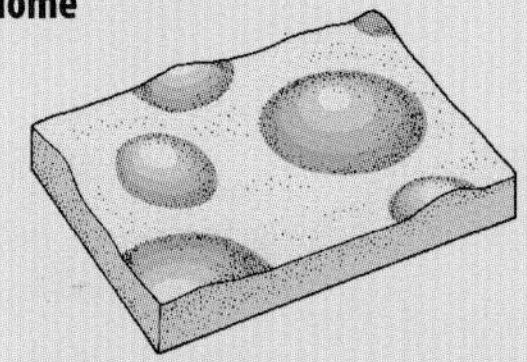		dome-shaped (height restricted by wind)	appreciable amounts of coarse sand	strong winds limit height of dune	none	virtually no movement
seif (linear) 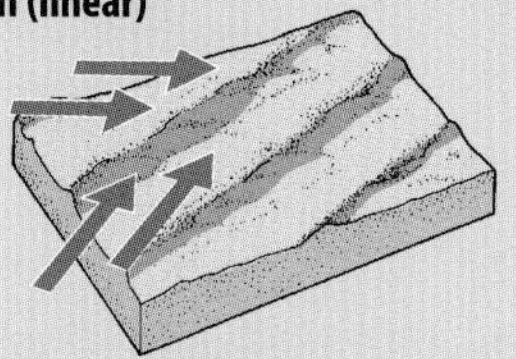		longitudinal, parallel dunes with slip faces on either side, can extend for many km (Figure 7.17)	large	persistent, steady winds (trades), with slight seasonal or diurnal changes in direction	none	regular (even) surface, virtually no movement
parabolic 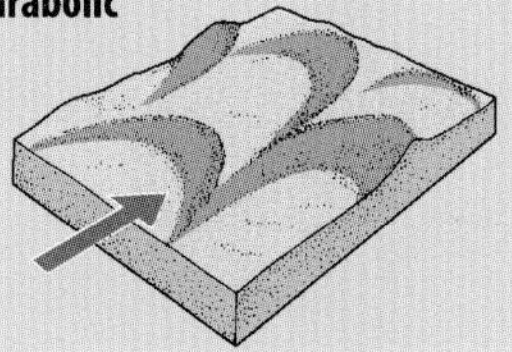		hairpin-shaped with noses pointing downwind, a type of blowout (eroded) dune where middle section has moved forward, may occur in clusters	limited	constant direction	where present, can anchor sand	highly mobile (by blowouts in nose of dune)
star 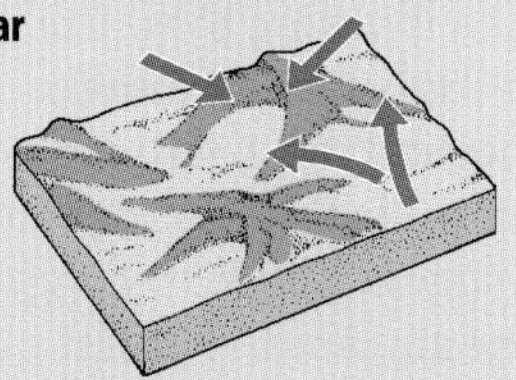		complex dune with a star (starfish) shape (compare arêtes radiating from central peak) (Figure 7.18)	limited	effective winds blow from several directions	none	virtually no movement
reversing 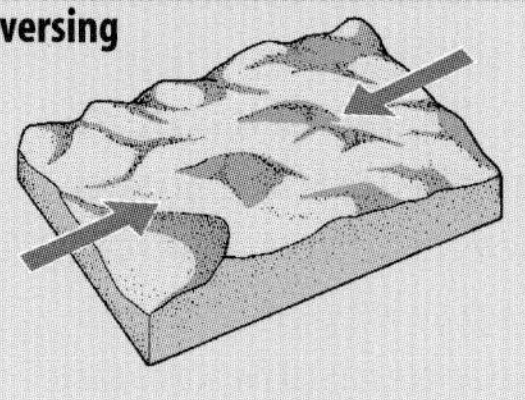		undulating, haphazard shape	limited	winds of equal strength and duration from opposite directions	none	virtually no movement

Figure 7.14

Classification of sand dunes (*after* Goudie)

Figure 7.15

Barchan dunes near Lüderitz, Namibia

Figure 7.16

Transverse dunes near Djanet, Algeria

Figure 7.17

Seif (linear) dunes, Sossusvlei, Namibia

Figure 7.18

Star dunes, Sossusvlei, Namibia

The effects of water

It has already been noted that, in arid areas, moisture must be present for processes of chemical weathering to operate. We have also seen that often rainfall is low, irregular and infrequent, with long-term fluctuations. Although most desert rainfall occurs in low-intensity storms, the occasional sudden, more isolated, heavy downpour, does occur. There are records of several extreme desert rainfall events, each equivalent to the 3-monthly mean rainfall of London. The impact of water is, therefore, very significant in shaping desert landscapes.

Rivers in arid environments fall into three main categories.

Exogenous Exogenous rivers are those like the Colorado, Nile, Indus, Tigris and Euphrates, which rise in mountains beyond the desert margins. These rivers continue to flow throughout the year even if their discharge is reduced by evaporation when they cross the arid land. (The last four rivers mentioned provided the location for some of the earliest urban settlements – page 388). The Colorado has, for over 300 km of its course, cut down vertically to form the Grand Canyon. The canyon, which in places is almost 2000 m (over 1 mile) deep, has steep sides partly due to rock structure and partly due to insufficient rainfall to degrade them (Figure 7.19).

Figure 7.19

The Grand Canyon,
Arizona, USA

Endoreic Endoreic drainage occurs where rivers terminate in inland lakes. Examples are the River Jordan into the Dead Sea and the Bear into the Great Salt Lake.

Ephemeral Ephemeral streams, which are more typical of desert areas, flow intermittently, or seasonally, after rainstorms. Although often shortlived, these streams can generate high levels of discharge due to several local characteristics. First, the torrential nature of the rain exceeds the infiltration capacity of the ground and so most of the water drains away as surface runoff (overland flow, page 59). Secondly, the high temperatures and the frequent presence of duricrust combine to give a hard, impermeable surface which inhibits infiltration. Thirdly, the lack of vegetation means that no moisture is lost or delayed through interception and the rain is able to hit the ground with maximum force. Fourthly, fine particles are displaced by rainsplash action and, by infilling surface pore spaces, further reduce the infiltration capacity of the soil. It is as a result of these minimal infiltration rates that slopes of less than 2° can, even under quite modest storm conditions, experience extensive overland flow.

Studies in Kenya, Israel and Arizona suggest that surface runoff is likely to occur within 10 minutes of the start of a downpour (Figure 7.20). This may initially be in the form of a **sheet flood** where the water flows evenly over the land and is not confined to channels. Much of the sand, gravel and pebbles covering the desert floor is thought to have been deposited by this process; yet, as the event has rarely been witnessed, it is assumed that deposition by sheet floods occurred mainly during earlier wetter periods called **pluvials**.

Very soon, the collective runoff becomes concentrated into deep, steep-sided ravines known as **wadis** (Figure 7.21) or **arroyos**. Normally dry, wadis may be subjected to irregular flash floods (Figure 7.20 and Places 25). The average occurrence of these floods is once a year in the semi-arid margins of the Sahara, and once a decade in the extremely arid interior. This infrequency of floods compared with the great number and size of wadis, suggests that they were created when storms were more frequent and severe – i.e. they are a relict feature.

Figure 7.20

A flash flood, Ethiopia

Pediments and playas

Stretching from the foot of the highlands, there is often a gently sloping area either of bare rock or of rock covered in a thin veil of debris (Figures 7.21 and 7.24). This is known as a **pediment**. There is often an abrupt break of slope at the junction of the highland area and the pediment. Two main theories suggest the origin of the pediment, one involving water. This theory proposes that weathered material from cliff faces, or debris from alluvial fans, was carried during pluvials by sheet floods. The sediment planed the lowlands before being deposited, leaving a gently concave slope of less than 7° (Figure 7.24). The alternative theory involves the parallel retreat of slopes resulting from weathering (King's hypothesis, Figure 2.24c).

Figure 7.21

Pediment at foot of highlands, Wadi Rum, Jordan

Playas are often found at the lowest point of the pediment. They are shallow, ephemeral, saline lakes formed after rainstorms. As the rain water rapidly evaporates, flat layers of either clay, silt or salt are left. Where the dried-out surface consists of clay, large **desiccation cracks**, up to 5 m deep, are formed. When the surface is salt-covered, it produces the 'flattest landform on land'. Rogers Lake, in the Mojave Desert, California, has been used for spacecraft landings, while the Bonneville saltflats in Utah have been the location for land-speed record attempts.

Places 25   Wadis: flash floods

Figure 7.22

A wadi near Qumran, Israel

Camping in a wadi is something that experienced desert travellers avoid. It is possible to be swept away by a flash flood which occurs virtually without warning – there may have been no rain at your location, and perhaps nothing more ominous than a distant rumble of thunder. Indeed, the first warning may be the roar of an approaching wall of water. One minute the bed of the wadi is dry, baked hard under the sun and littered with weathered debris from the previous flood or from the steep valley sides (Figure 7.22), and the next minute it is a raging torrent.

The energy of the flood enables large boulders to be moved by traction, and enormous amounts of coarse material to be taken into suspension – some witnesses have claimed it is more like a mudflow. Friction from the roughness of the bed, the large amounts of sediment and the high rates of evaporation soon cause a reduction in the stream's velocity. Deposition then occurs, choking the channel, followed by braiding as the water seeks new outlets. Within hours, the floor of the wadi is dry again (Figure 7.23).

The rapid runoff does not replenish groundwater supplies, and without the groundwater contribution to base flow, characteristic of humid climates, rivers cease to flow. At the mouth of the wadi, where the water can spread out and energy is dissipated, material is deposited to form an **alluvial fan** or **cone** (Figure 7.24). If several wadis cut through a highland close to each other, their semi-circular fans may merge to form a **bahada** (**bajada**), which is an almost continuous deposit of sand and gravel.

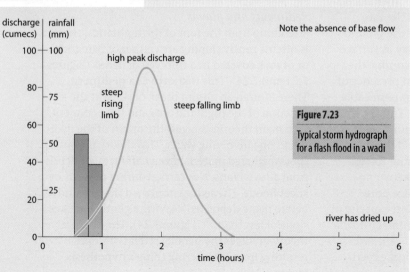

Note the absence of base flow

high peak discharge

steep rising limb

steep falling limb

Figure 7.23

Typical storm hydrograph for a flash flood in a wadi

river has dried up

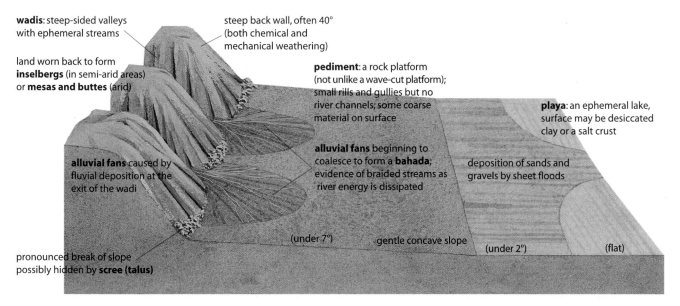

wadis: steep-sided valleys with ephemeral streams

steep back wall, often 40° (both chemical and mechanical weathering)

land worn back to form **inselbergs** (in semi-arid areas) or **mesas and buttes** (arid)

pediment: a rock platform (not unlike a wave-cut platform); small rills and gullies but no river channels; some coarse material on surface

playa: an ephemeral lake, surface may be desiccated clay or a salt crust

alluvial fans caused by fluvial deposition at the exit of the wadi

alluvial fans beginning to coalesce to form a **bahada**; evidence of braided streams as river energy is dissipated

deposition of sands and gravels by sheet floods

(under 7°)   gentle concave slope   (under 2°)   (flat)

pronounced break of slope possibly hidden by **scree (talus)**

Figure 7.24

Pediments and playas

Occasionally, isolated, flat-topped remnants of former highlands, known as **mesas**, rise sheer from the pediment. Some mesas, in Arizona, have summits large enough to have been used as village sites by the Hopi Indians. **Buttes** are smaller versions of mesas. The most spectacular mesas and buttes lie in Monument Valley National Park in Arizona (Figure 7.25).

Relationship between wind and water

Some desert areas are dominated by wind, others by water. Areas where wind appears to be the dominant geomorphological agent are known as **aeolian domains**. The effectiveness of the wind increases where, and when, amounts of rainfall decrease. As rainfall decreases, so too does any vegetation cover. This allows the wind to transport material unhindered, and rates of erosion (abrasion and deflation) and deposition (dunes)

to increase. **Fluvial domains** are those where water processes are dominant or, as evidence increasingly suggests, have been dominant in the past. Vegetation, which stabilises material, increases as rainfall increases or where coastal fog and dew are a regular occurrence.

Evidence also suggests that wind and water can interact in arid environments and that landforms produced by each do co-exist within the same locality. However, the balance between their relative importance has often altered, mainly due to climatic change either over lengthy periods of time (e.g. the 18 000 years since the time of maximum glaciation) or during shorter fluctuations (e.g. since the mid-1960s in the Sahel). At present, and especially in Africa, the decrease in rainfall in the semi-arid desert fringes means that the role of water is probably declining, while that of the wind is increasing.

Figure 7.25

Mesas and buttes, Monument National Park, Arizona, USA

Climatic change

There have already been references to pluvials within the Sahara Desert (page181). Prior to the Quaternary era, these may have occurred when the African Plate lay further to the south and the Sahara was in a latitude equivalent to that of the present-day savannas. In the Quaternary era, the advance of the ice sheets resulted in a shift in windbelts which caused changes in precipitation patterns, temperatures and evaporation rates. At the time of maximum glaciation (18 000 BP), desert conditions appear to have been more extensive than they are today (Figure 7.26). Since then, as suggested by radio-carbon dating (page 248), there have been frequent, relatively short-lived pluvials, the last occurring about 9000 BP. Evidence for a once-wetter Sahara is given in Figure 7.27.

Herodotus, a historian living in Ancient Greece, described the Garamantes civilisation which flourished in the Ahaggar Mountains 3000 years ago. This people, who recorded their exploits in cave paintings at Tassili des Ajjers, hunted elephants, giraffes, rhinos and antelope. Twenty centuries ago, North Africa was the 'granary of the Roman Empire'. Wadis are too large and deep and alluvial cones too widespread to have been formed by today's occasional storms, while sheet floods are too infrequent to have moved so much material over pediments. Radiating from the Ahaggar and Tibesti Mountains, aerial photographs and satellite imagery have revealed many dry valleys which once must have held permanent rivers (Figure 16.43). Lakes were also once much larger and deeper. Around Lake Chad, shorelines 50 m above the present level are visible, and research suggests that lake levels might once have been over 100 m higher. (Lake Bonneville in the USA is only one-tenth of its former maximum size and, like Lake Chad, is drying up rapidly.) Small crocodiles found in the Tibesti must have been trapped in the slightly wetter uplands as the desert advanced. Also, pollen analysis has shown that oak and cedar forests abounded in the same region 10 000 years ago. Groundwater in the Nubian sandstone has been dated, by radio-isotope methods, to be over 25 000 years old, and may have accumulated at about the same time as fossil laterite soils (page 321).

Figure 7.26

Extent of sand dunes in Africa

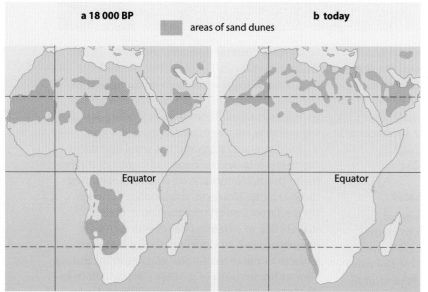

a 18 000 BP   areas of sand dunes   b today

Equator

Equator

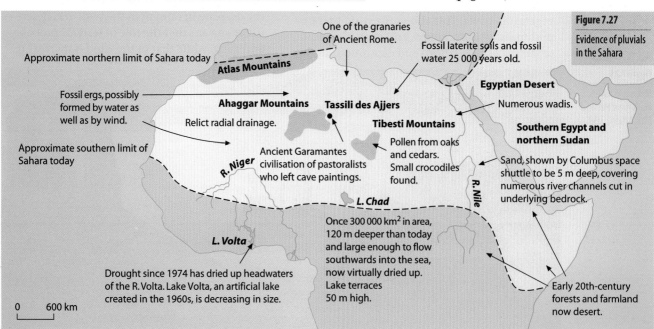

Figure 7.27

Evidence of pluvials in the Sahara

Approximate northern limit of Sahara today

One of the granaries of Ancient Rome.

Fossil laterite soils and fossil water 25 000 years old.

Atlas Mountains

Fossil ergs, possibly formed by water as well as by wind.

Ahaggar Mountains

Tassili des Ajjers

Egyptian Desert

Numerous wadis.

Relict radial drainage.

Tibesti Mountains

Southern Egypt and northern Sudan

Approximate southern limit of Sahara today

R. Niger

Ancient Garamantes civilisation of pastoralists who left cave paintings.

Pollen from oaks and cedars. Small crocodiles found.

Sand, shown by Columbus space shuttle to be 5 m deep, covering numerous river channels cut in underlying bedrock.

R. Nile

L. Chad

L. Volta

Once 300 000 km² in area, 120 m deeper than today and large enough to flow southwards into the sea, now virtually dried up. Lake terraces 50 m high.

Drought since 1974 has dried up headwaters of the R. Volta. Lake Volta, an artificial lake created in the 1960s, is decreasing in size.

Early 20th-century forests and farmland now desert.

0   600 km

Desertification: fact or fiction?

'Literally, desertification means the making of a desert, but although the word has been in use for more than 40 years, few can agree on exactly what it means: there are more than 100 different definitions of the term' (Middleton, 1993). Later (1996), Zhu, a Chinese scientist, suggested that 'desertification is an environmental degradation process created as a result of the influence of excessive human activities that, owing to the emergence of desert-like landscapes, leads to the decline of productive land'. The divergence of definitions is due largely to the uncertainty as to the causes of desertifi-

cation. Goudie says that 'the question has been asked whether this process is caused by temporary drought periods of high magnitude, is due to longer-term climatic change towards aridity, is caused by man-induced climatic change, or is the result of human action through man's degradation of the biological environments in arid zones. Most people now believe that it is produced by a combination of increasing human and animal populations, which cause the effects of drought years to become progressively more severe so that the vegetation is placed under increasing stress.'

Those places perceived to be at **greatest risk** from desertification are shown in Figure 7.28. (This map and the four levels of risk were devised at the 1977 Nairobi Conference on Desertification.) It is generally agreed that the desert is encroaching into semi-arid, desert margins, especially in the Sahel – a broad belt of land on the southern side of the Sahara (**2–4** in Figure 7.28).

Some of the main interrelationships between the believed causes of desertification are shown in Figure 7.29.

Worst areas	Percentage of population at risk
1 Ethiopia	18
2 Sudan	23
3 Chad	30
4 Niger	42
5 Somalia	26

Level of risk

- very severe
- severe
- moderate
- slight

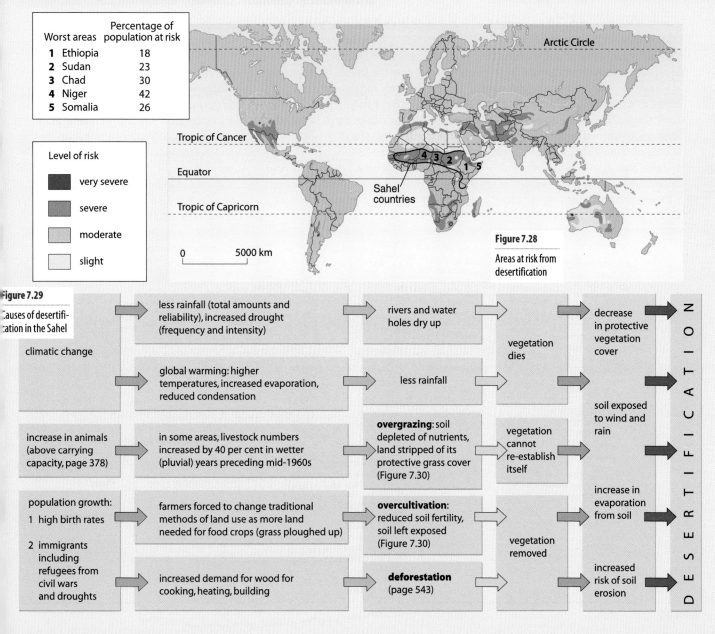

Figure 7.28

Areas at risk from desertification

Figure 7.29

Causes of desertification in the Sahel

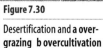

Figure 7.30

Desertification and **a** overgrazing **b** overcultivation

In 1975, Hugh Lamprey, a bush pilot and environmentalist, claimed that, since his previous study 17 years earlier, the desert in the Sudan had advanced southwards by 90–100 km. In 1982 and at the height of one of Africa's worst-ever recorded droughts, UNEP (United Nations Environmental Programme) claimed that the Sahara was advancing southwards by 6–10 km a year and that, globally, 21 million hectares of once-productive soil were being reduced each year to zero productivity, that 850 million people were being affected, and

that 35 per cent of the world's land surface was at risk. This last figure was again quoted by UNEP at the 1992 Rio Earth Summit.

Recent scientific studies, made by using satellite imagery and through more detailed fieldwork, have thrown considerable doubt upon both the causes and the effects of desertification. Three examples are given in Figure 7.31.

However, such doubts do not mean that the risk of desertification has disappeared. They do suggest that we should be wary of

overgeneralisation, and that our assessment of risk should be based on detailed evidence.

The semi-arid lands are fragile environments. Their boundaries are subject to change resulting from variations in rainfall, from variations in human use, and perhaps also from global warming. It is often difficult to separate natural causes from human ones, and short-term fluctuations from long-term trends. Regardless of whether desertification is already a major environmental hazard or whether it poses a future environmental risk, one thing is certain: any increase in desertification is likely to be the result of people's mismanagement of their natural resources.

Researchers at the University of Lund in Sweden have carried out field surveys and examined satellite pictures of the Sudan in an effort to confirm Lamprey's findings. Earlier this year (1993) in the journal *Ambio*, Ulf Hellden, head of the university's remote-sensing laboratory and working for UNEP, reported "no major shifts in the northern cultivation limit, no major sand dune transformation, no major changes in vegetation cover" beyond the dramatic but short-term effects of variable rainfall.

A belt of sand dunes that Lamprey said formed the advancing front of the Sahara in the Northern Kordofan province of Sudan showed no sign of movement between 1962 and 1984. Nor was there any evidence of patches of desert growing round boreholes or villages – a phenomenon frequently claimed to be the result of overgrazing by herds of cattle [Places 65, page 479]. Hellden claims that "As long as the effects of desertification in Africa and elsewhere are not documented according to scientific standards, there is a risk that the desertification issue will become a political and development fiction rather than a scientific fact".

Ridley Nelson, a researcher for the World Bank (which has spent heavily on planting trees to halt the deserts), says that "advancing sand in not a major global problem. Claims that the Sahara is expanding at some horrendous rate are unacceptable today". Andrew Warren and Clive Agnew of the ecology and conservation unit at University College, London, agree that "active sand dunes seldom threaten valuable land".

Nelson also questions UNEP's figures concerning the rate of, and the areas affected by, desertification. He says that "they are meaningless. They come largely from answers to a questionnaire sent out to governments in 1982, at the height of a major African drought. Nobody knows how these governments arrived at their figures."

Amid the shimmering sands of the desert margins, dogma often stands in for reality. One such dogma holds that desert margins have a fixed 'carrying capacity' of humans and animals. According to a recent executive director of UNEP, "when the number is exceeded, the whole piece of land will quickly degenerate. Population pressure is definitely one of the major causes of desertification."

True? New research suggests not. Long-term studies show that in the desert margins of two of Africa's largest and fastest-growing countries – Nigeria and Kenya – the opposite has happened. Rapidly-increasing populations emerge from these studies as saviours of the landscape, rescuing it from rampant soil erosion, and protecting trees and conserving water.

Figure 7.31

Alternative views on desertification

References

Desert Processes, Open University, BBC Television.

Goudie, A. (1993) *The Nature of the Environment*, Blackwell.

Goudie, A. and Watson, A. (1990) *Desert Morphology*, Thomas Nelson.

Planet Earth: Arid Lands (1984) Time-Life Books.

Thomas, D. S. G. (1997) *Arid Zone Geomorphology*, John Wiley & Sons.

Websites

The Desert Environment:
http://www.desertusa.com/desert.html

The United Nations Convention to Combat Desertification:
http://www.un.org/ecosocdev/geninfo/sustdev/desert.htm

The Unitarian Service Committee of Canada:
http://www.usc-canada.org/

See also for more links:
http://www.nelsonthornes.com/gaia

Questions

1
a Describe the characteristics which define a hot desert climate. **(4 marks)**
b Study Figure 7.2 (page 179) and describe the location of the world's deserts. **(4 marks)**
c Explain **two** causes of a desert climate. **(4 marks)**
d Write a paragraph to explain to someone why the typical view of a desert as a 'sea of sand' is often not true. **(4 marks)**
e What is 'exfoliation' weathering? **(4 marks)**
f Explain one other denudation process which operates in hot desert areas. **(5 marks)**

2
a Describe how wind transports material in a desert environment. **(6 marks)**
b Why is wind transportation a more important method of movement in deserts than in wet environments? **(3 marks)**
c Choose **one** type of sand dune.
  i Draw an annotated diagram to show its main features.
  ii Explain how the dune has been formed. **(6 marks)**
d Choose **one** desert landform created by wind erosion.
  i Describe its shape and size.
  ii Explain the processes which have formed it. **(6 marks)**
e What is a 'playa'? Describe how a playa is formed. **(4 marks)**

3
a i Draw a sketch of the mesas and buttes in Figure 7.25 (page 189). **(2 marks)**
  ii Annotate it by adding the following labels to describe the different slopes shown: caprock; free face; bare rock; rectilinear slope; loose scree; gently sloping plain. **(6 marks)**
b Explain why the loose material you can see in the photograph has not been moved away. **(5 marks)**
c i In the Sahara Desert in North Africa there is evidence that the climate has not always been like this. Choose **one** piece of evidence to show that the climate has changed, state it and explain how it shows climate change. **(6 marks)**
  ii Choose **one** piece of evidence to suggest that the climate of North Africa is changing now. State it and explain how it shows climate is changing. **(6 marks)**

4
a Describe and name an example of a wadi. **(4 marks)**
b i Sometimes a 'flash flood' rushes through a wadi. Explain what a flash flood is. **(3 marks)**
  ii Why is there little or no warning that a flash flood is about to happen? **(3 marks)**
  iii Why do rivers stop flowing very soon after a flood in a desert area? **(3 marks)**
c In the area where a wadi opens onto lowland there is often an alluvial fan. Describe an alluvial fan and explain how it is formed. **(6 marks)**
d In deserts there are often areas where the soil is very salty (sometimes pure salt). Describe these areas and explain where this salt has come from. **(6 marks)**

AS

5
a Why do arid conditions occur in continental areas in the tropics? **(10 marks)**
b Making good use of examples, describe **two** ways in which plants adapt to drought conditions in desert areas. **(8 marks)**
c Explain the term 'water balance' used to identify the extent of tropical desert climates. **(7 marks)**

6
a i Define the term 'potential evapotranspiration'. **(5 marks)**
  ii How is vegetation adapted to survive the climatic conditions of hot desert areas? **(10 marks)**
b Explain how human activities on the fringes of hot deserts are leading to the movement of the margins of the desert. **(10 marks)**

7 'Semi-arid lands are fragile environments.' Discuss this statement in the light of Case Study 7 (pages 191–192). In your answer you should refer to:
■ the location of semi-arid lands
■ the risk of desertification and
■ the differing attitudes of people to risks in semi-desert environments. **(25 marks)**

8 Using Figures 7.3, 7.4 and 7.5 (page 180), describe and account for the range of surface conditions found in desert areas. **(25 marks)**

A2

Chapter 8

Rock types and landforms

'At first sight it may appear that rock type is the dominant influence on most landscapes …
As geomorphologists, we are more concerned with the ways in which the characteristics of rocks respond to the processes or erosion and weathering than with the detailed study of rocks themselves.'

Roy Collard, *The Physical Geography of Landscape*, 1988

Previous chapters have demonstrated how landscapes at both local and global scales have developed from a combination of processes. Plate tectonics, weathering and the action of moving water, ice and wind both create and destroy landforms. Yet these processes, however important they are at present or have been in the past, are insufficient to explain the many different and dramatic changes of scenery which can occur within short distances, especially in the British Isles.

Lithology refers to the physical characteristics of a rock. As each individual rock type has different characteristics, so it is capable of producing its own characteristic scenery. Land-forms are greatly influenced by a rock type's vulnerability to weathering, its permeability and its structure.

To show how these three factors affect different rocks and to explain their resultant landforms and potential economic use, five rock types have been selected as exemplars. Carboniferous limestone, chalk and sandstone (sedimentary rocks), and granite and basalt (both igneous) have been chosen because, arguably, these produce some of the most distinctive types of landform and scenery.

Lithology and geomorphology

Vulnerability to weathering

Mechanical weathering in Britain occurs more readily in rocks that are jointed. Water can penetrate either down the **joints** or along the **bedding planes** (Figure 8.1) of Carboniferous limestone, or into cracks resulting from pressure release or contraction on cooling within granite and basalt (page 41 and Figure 1.31). Subsequent freezing and thawing along these lines of weakness causes frost shattering (page 40).

Chemical weathering is a major influence in limestone and granite landforms. Limestone, composed mostly of calcium carbonate, is slowly dissolved by the carbonic acid in rainwater, i.e. the process of carbonation (pages 42–43). Granite consists of quartz, feldspar and mica. It is susceptible to hydration, where water is incorporated into the rock structure causing it to swell and crumble (page 42), and to hydrolysis, when the feldspar is chemically changed into clay (pages 42–43). Quartz, in comparison with other minerals, is one of the least prone to chemical weathering.

Mottershead has emphasised that 'the mechanical resistance of rocks depends on the strength of the individual component minerals and the bonds between them, and that chemical resistance depends on the individual chemical resistances of the component minerals. Mechanical strength decreases if just one of these component minerals becomes chemically altered.'

Figure 8.1

Bedding planes with joints and angle of dip

a massively bedded Carboniferous limestone

vertical joints at right-angles to the bedding planes

horizontal bedding planes separating different layers/ strata in a sedimentary rock and indicating different phases of deposition

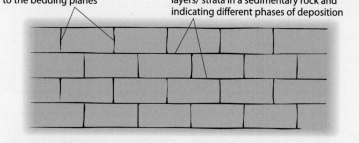

b thinly bedded chalk

the angle of dip is the difference between the actual inclination of the rock and the horizontal

horizontal

10° dip

joints still at right-angles to bedding planes

gently dipping bedding planes

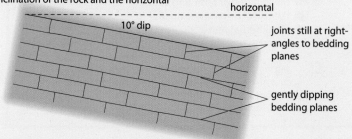

Permeability

Permeability is the rate at which water may be stored within a rock or is able to pass through it. Permeability can be divided into two types.

1 **Primary permeability or porosity** This depends upon the texture of the rock and the size, shape and arrangement of its mineral particles. The areas between the particles are called **pore spaces** and their size and alignment determine how much water can be absorbed by the rock. Porosity is usually greatest in rocks that are coarse-grained, such as gravels, sands, sandstone and oolitic lime-stone, and usually lowest in those that are fine-grained, such as clay and granite. (It is possible to have fine-grained sandstone and coarse-grained granite.)

 Infiltration capacity is the maximum rate at which water percolates into the ground. The infiltration capacity of sands is estimated to average 200 mm/hr, whereas in clay it is only 5 mm/hr. Pore spaces are larger where the grains are rounded rather than angular and compacted (Figure 8.2). Porosity can be given as an index value based upon the percentage of the total volume of the rock which is taken up by pore space, e.g. clay 20 per cent, gravel 50 per cent. When all the pore spaces are filled with water, the rock is said to be **saturated**. The water table marks the upper limit of saturation (Figure 8.9). Perm-eable rocks which store water are called **aquifers**.

2 **Secondary permeability** or **perviousness** This occurs in rocks that have joints and fissures along which water can flow. The most pervious rocks are those where the joints have been widened by solution, e.g. Carboniferous limestone, or by cooling, e.g. basalt. A rock may be pervious because of its structure, though water may not be able to pass through the rock mass itself.

 Where rocks are porous or pervious, water rapidly passes downwards to become groundwater, leaving the surface dry and without evident drainage – chalk and limestone regions have few surface streams. **Impermeable rocks**, e.g. granite, neither absorb water nor allow it to pass through them. These rocks therefore have a higher drainage density (page 67).

Figure 8.2

Pore spaces and infiltration capacity

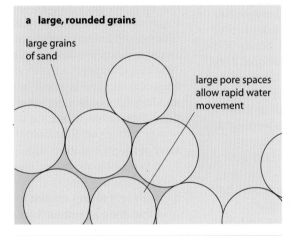

a large, rounded grains

large grains of sand

large pore spaces allow rapid water movement

b small, rounded grains

although there are more pore spaces, they are much smaller: water clings to grains (surface tension) preventing the passage of moisture (Figure 10.12)

c crystals in granite:

these fit together more closely than rounded grains, limiting the amount of water held and inhibiting the movement of moisture

Structure

Resistance to erosion depends upon whether the rock is massive and stratified, folded or faulted. Usually the more massive the rock and the fewer its joints and bedding planes, the more resistant it is to weathering and erosion. Conversely, the softer, more jointed and less compact the rock, the more vulnerable it is to denudation processes. Usually, more resistant rocks remain as upland areas (granite), while those that are less resistant form lowlands (clay).

However, there are exceptions. Chalk, which is relatively soft and may be well-jointed, forms rolling hills because it allows water to pass through it and so fluvial activity is limited. Carboniferous or Mountain limestone, having joints and bedding planes, produces jagged karst scenery because although it is pervious it has a very low porosity.

Limestone

Limestone is a rock consisting of at least 80 per cent calcium carbonate. In Britain, most limestone was formed during four geological periods, each of which experienced different conditions. The following list begins with the oldest rocks. Use an atlas to find their location.

Carboniferous limestone This is hard, grey, crystalline and well-jointed. It contains many fossils, including corals, crinoids and brachiopods. These indicate that the rock was formed on the bed of a warm, clear sea and adds to the evidence that the British Isles once lay in warmer latitudes. Carboniferous limestone has developed its own unique landscape, known as **karst**, which in Britain is seen most clearly in the Peak District and Yorkshire Dales National Parks.

Magnesian limestone This is distinctive because it contains a higher proportion of magnesium carbonate. In Britain, it extends in a belt from the mouth of the River Tyne to Nottingham. In the Alps, it is known as **dolomite**.

Jurassic (oolitic) limestone This forms a narrow band extending southwards from the North Yorkshire Moors to the Dorset coast. Its scenery is similar to that typical of chalk.

Cretaceous chalk This is a pure, soft, well-jointed limestone. Stretching from Flamborough Head in Yorkshire (Figure 6.19), it forms the escarpment of the Lincoln Wolds, the East Anglian Heights and the North and South Downs, before ending up as the 'White Cliffs' at Dover and at Beachy Head, the Needles and Swanage. Cretaceous chalk is assumed to be the remains of small marine organisms which lived in clear, shallow seas.

The most distinctive of the limestone landforms are found in Carboniferous limestone and chalk.

Carboniferous limestone

This rock develops its own particular type of scenery primarily because of three characteristics. First, it is found in thick beds separated by almost horizontal bedding planes and with joints at right-angles (Figure 8.1). Secondly, it is pervious but not porous, meaning that water can pass along the bedding planes and down joints but not through the rock itself. Thirdly, calcium carbonate is soluble. Carbonic acid in rainwater together with humic acid from moorland plants, dissolve the limestone and widen any weaknesses in the rock, i.e. the bedding planes and joints. Acid rain also speeds up carbonation and solution (page 43). As there is minimum surface drainage and little breakdown of bedrock to form soil, the vegetation cover tends to be thin or absent. In winter, this allows frost shattering to produce scree at the foot of steep cliffs.

It is possible to classify Carboniferous limestone landforms into four types:

1 **Surface features caused by solution**
   Limestone pavements are flat areas of exposed rock. They are flat because they represent the base of a dissolved bedding plane, and exposed because the surface soil may have been removed by glacial activity and never replaced. Where joints reach the surface, they may be widened by the acid rainwater (carbonation, page 43) to leave deep gashes called **grikes**. Some grikes at Malham in north-west Yorkshire are 0.5 m wide and up to 2 m deep. Between the grikes are flat-topped yet dissected blocks referred to as **clints** (Figure 2.8). In time, the grikes widen and the clints are weathered down until a lower bedding plane is exposed and the process of solution–carbonation is repeated.

2 **Drainage features** Rivers which have their source on surrounding impermeable rocks, such as the shales and grits of northern England, may disappear down **swallow holes** or **sinks** as soon as they reach the limestone (Figure 8.3). The streams flow underground finding a pathway down enlarged joints, forming **potholes**, and along bedding planes. Where solution is more active, underground **caves** may form. While most caves develop above the water table (**vadose caves**, Figure 8.8), some may form beneath it (**phreatic caves**).

Figure 8.3

A stream disappearing down a swallow hole near Hunt Pot, Penyghent Yorkshire Dales National Park

Figure 8.4

The resurgence at the foot of Malham Cove, Yorkshire Dales

Figure 8.6

Stalactites, stalagmites and pillars, Carlsbad Caverns, New Mexico, USA

Corrosion often widens the caverns until parts of the roof collapse, providing the river with angular material ideal for corrasion. Heavy rainfall very quickly infiltrates downwards, so caverns and linking passages may become water-filled within minutes. The resultant turbulent flow can transport large stones and the floodwater may prove fatal to cavers and potholers. Rivers make their way downwards, often leaving caverns abandoned as the water finds a lower level, until they reach underlying impermeable rock. A **resurgence** occurs where the river reappears on the surface, often at the junction of permeable and impermeable rocks (Figure 8.4).

3 **Surface features resulting from underground drainage** Steep-sided valleys are likely to have been formed as rivers flowed over the surface of the limestone, probably

Figure 8.5

The Watlowes dry valley above Malham Cove

during periglacial times when permafrost acted as an impermeable layer. When the rivers were able to revert to their subterranean passages, the surface valleys were left dry (Figure 8.5). Many dry valley sides are steep and gorge-like, e.g. Cheddar Gorge. If the area above an individual cave collapses, a small surface depression called a **doline** is formed. **Shakeholes** are smaller doline-like features found in the northern Pennines where glacial drift has subsided into underground cavities (Figure 8.8). In the former Yugoslavia, where the term 'karst' originated, huge depressions called **poljes** may have formed in a similar way. Poljes may be up to 400 km^2 in area. In the tropics, the landscape may be composed of either cone-shaped hills and polygonal depressions known as 'cockpit country' (e.g. Jamaica) or tall isolated 'towers' rising from wide plains (e.g. near Guilin, China – Places 26).

4 **Underground depositional features** Groundwater may become saturated with calcium bicarbonate, which is formed by the chemical reaction between carbonic acid in rainwater and calcium carbonate in the rock. However, when this 'hard' water reaches a cave, much of the carbon dioxide bubbles out of solution back into the air – i.e. the process of carbonation in reverse. Aided by the loss of some moisture by evaporation, calcium carbonate (calcite) crystals are subsequently precipitated. Water dripping from the ceiling of the cave initially forms pendant soda straws which, over a very long period of time, may grow into icicle-shaped **stalactites** (Figure 8.6). Experiments in Yorkshire caves suggest that stalactites grow at about 7.5 mm per year. As water drips on to the floor, further deposits of calcium carbonate form the more rounded, cone-shaped **stalagmites** which may, in time, join the stalactites to give **pillars**.

Figure 8.7

The karst towers of Guilin, south China

The limestones that outcrop near Guilin have formed a unique karst landscape. The massively bedded, crystalline rock, which in places is 300 m thick, has been slowly pushed upwards from its seabed origin by the same tectonic movements that formed the Himalayas and the Tibetan Plateau far to the west. The heavy summer monsoon rain, sometimes exceeding 2000 mm, has led to rapid fluvial erosion by such rivers as the Li Jiang (Li River). The availability of water together with the high sub-tropical temperatures (Guilin is at 25°N) encourage highly active chemical weathering (solution–carbonation, page 43).

Limestone covers some 300 000 km² of China – an area larger than that of the UK. Its scenery is seen at its most spectacular in the Three Gorges section of the Yangtze River (Case Study 3B) and where it forms the karst towers in the Guilin region of Guangxi Province.

The result has been the formation of a landscape which for centuries has inspired Chinese artists and, recently, has attracted growing numbers of tourists. To either side of the river are natural domes and towers, some of which rise almost vertically 150 m from surrounding paddy fields (Figure 8.7), giving the valley its gorge-like profile. Caves, visible on the sides of the towers, were formed by underground tributaries to the Li Jiang when the main river was flowing at levels considerably higher than those of today.

Figure 8.8

Characteristic features of Carboniferous limestone (karst) scenery

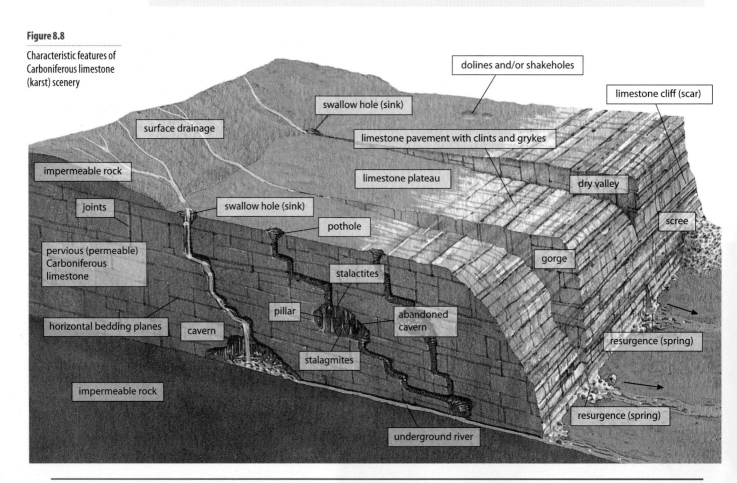

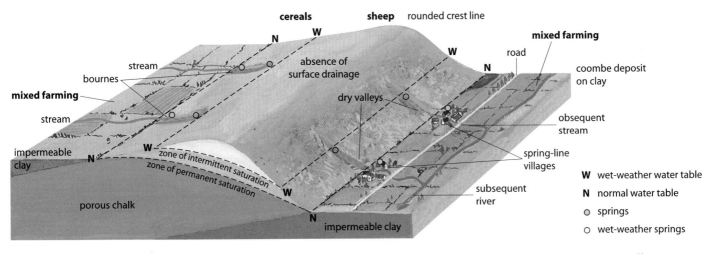

escarpment or cuesta

dip slope — scarp slope — clay vale

cereals   **sheep**   rounded crest line

mixed farming

road

coombe deposit on clay

stream

absence of surface drainage

bournes

dry valleys

mixed farming

stream

impermeable clay

zone of intermittent saturation

zone of permanent saturation

obsequent stream

spring-line villages

porous chalk

subsequent river

impermeable clay

W wet-weather water table

N normal water table

⊙ springs

○ wet-weather springs

Figure 8.9

Scarp and vale scenery: an idealised section through a chalk escarpment in south-east England

Economic value of Carboniferous limestone
Human settlement on this type of rock is usually limited and dispersed (page 397) due to limited natural resources, especially the lack of water and good soil. Villages such as Castleton (Derbyshire) and Malham (Yorkshire) have grown up near to a resurgence.

Limestone is often quarried as a raw material for the cement and steel industries or as ornamental stone, but the resultant scars have led to considerable controversy (Case Study 8). The conflict is between the economic advantages of extracting a valuable raw material and providing local jobs, versus the visual eyesore, noise, dust and extra traffic resulting from the operations, e.g. the Hope valley, Derbyshire.

Farming is hindered by the dry, thin, poorly developed soils for, although most upland limestone areas of Britain receive high rainfall totals, water soon flows underground. The rock does not readily weather into soil-forming particles, such as clay or sand, but is dissolved and the residue is then leached (page 261). On hard limestones, rendzina soils may develop (page 274). These soils are unsuitable for ploughing and their covering of short, coarse, springy grasses favours only sheep grazing. In the absence of hedges and trees, drystone walls were commonly built as

field boundaries. The scenery attracts walkers and school parties, while underground features lure cavers, potholers and **speleologists** (scientists who study caves).

Chalk

Chalk, in contrast to Carboniferous limestone scenery, consists of gently rolling hills with rounded crest lines. Typically, chalk has steep, rather than gorge-like, dry valleys and is rarely exposed on the surface (Figure 8.9).

The most distinctive feature of chalk is probably the **escarpment**, or **cuesta**, e.g. the North Downs and South Downs (Figures 8.10 and 14.4). Here the chalk, a pure form of limestone, was gently tilted by the earth movements associated with the collision of the African and Eurasian Plates. Subsequent erosion has left a steep scarp slope and a gentle dip slope. In south-east England, clay vales are found at the foot of the escarpment (Figure 3.51b).

Although chalk – like Carboniferous limestone – has very little surface drainage, its surface is covered in numerous dry valleys (Figure 8.11). Given that chalk can absorb and allow rainwater to percolate through it, how could these valleys have formed?

Figure 8.10

South Downs chalk escarpment, Poynings, Sussex

Figure 8.11

A dry valley in chalk: Devil's Dyke, South Downs, Sussex

Goudie lists 16 different hypotheses that have been put forward regarding the origins of dry valleys. These he has grouped into three categories:

1 **Uniformitarian** These hypotheses assume that there have been no major changes in climate or sea-level and that 'normal' – i.e. fluvial – processes of erosion have operated without interruption. A typical scenario would be that the drainage system developed on impermeable rock overlying the chalk, and subsequently became superimposed upon it (page 85).

2 **Marine** These hypotheses are related to relative changes in sea-level or base level (page 81). One, which has a measure of support, suggests that when sea-levels rose eustatically at the end of the last ice age (page 123), water tables and springs would also have risen. Later, when the base level fell, so too did the water table and spring line, causing valleys to become dry.

3 **Palaeoclimatic** This group of hypotheses, based on climatic changes during and since the ice age, is the most widely accepted. One hypothesis claims that under periglacial conditions any water in the pore spaces would have been frozen, causing the chalk to behave as an impermeable rock (page 135). As temperatures were low, most precipitation would fall as snow. Any meltwater would have to flow over the surface, forming valleys that are now relict landforms (Figure 8.11).

An alternative hypothesis stems from occasions when places receive excessive amounts of rainfall and streams temporarily reappear in dry valleys. Climatologists have shown that there have been times since the ice age when rainfall was considerably greater than it is today. Figure 8.9 shows the normal water table with its associated spring line. If there is a wetter than average winter, or longer period, when moisture loss through evaporation is at its minimum, then the level of permanent saturation will rise. Notice that the wet-weather water table causes a rise in the spring line and so seasonal rivers, or **bournes**, will flow in the normally dry valleys. Remember also that there will be a considerable lag time (Figure 3.5 and page 61) between the peak rainfall and the time when the bournes will begin to flow (throughflow rather than surface run-off on chalk). The springs are the source of obsequent streams (page 84).

The presence of coombe deposits, resulting from solifluction (pages 47 and 135), also links chalk landforms with periglacial conditions.

Economic value of chalk

The main commercial use of chalk is in the production of cement, but there are objections on environmental grounds to both quarries and the processing works. Settlement tends to be in the form of nucleated villages strung out in lines along the foot of an escarpment, originally to take advantage of the assured water supply from the springs (Figures 8.9, 8.10 and 14.4). Water-storing chalk aquifers have long been used as a natural, underground reservoir by inhabitants of London, although recent increases in demand have exceeded the rate at which this artesian water has been replaced. The result has been the lowering of the water table.

Chalk weathers into a thin, dry, calcareous soil with a high pH. Until this century, the springy turf of the Downs was mainly used to graze sheep and to train race horses. Horse racing is still important locally, as at Epsom and Newmarket, but much of the land has been ploughed and converted to the growing of wheat and barley. In places, the chalk is covered by a residual deposit of **clay-with-flints** which may have been an insoluble component of the chalk or may have been left from a former overlying rock. This soil is less porous and more acidic than the calcareous soil and several such areas are covered by beech trees – or were, before the violent storm of October 1987 (Places 29, page 232). Flint has been used as a building material and was the major source for Stone Age tools and weapons.

Figure 8.12

Bedding planes in Old Red Sandstone, Old Man of Hoy, Orkney

Sandstone

Sandstone is the most common rock in Britain. It is a sedimentary rock composed mainly of grains of quartz, and occasionally feldspar and even mica, which have been compacted by pressure and cemented by minerals such as calcite and silica. This makes it a more coherent and resistant, but less porous, rock than sands. The sands, before compaction, may have been deposited in either **a** shallow seas, **b** estuaries and deltas, or **c** hot deserts. The presence of bedding planes (Figure 8.12) indicates the laying down of successive layers of sediment. Sandstone can vary in colour from dark brown or red through to yellow, grey and white (Figure 6.52), depending on the degree of oxidation or hydration (page 42). Like limestone (page 196), sandstone has formed in several geological periods (Figure 8.13), of which perhaps the most significant have been the following:

- The **Devonian**, or **Old Red Sandstone** (Figure 1.1), when sand was deposited in a shallow sea which covered present-day south-west England, South Wales and Herefordshire. These deposits, which were often massively bedded, were contorted and uplifted by subsequent earth movements. Landforms, indicative of an often resistant rock, vary from spectacular coastal cliffs to the plateau-like Exmoor, the north-facing scarp slope of the Brecon Beacons and the flatter lowlands of Herefordshire.
- The **Carboniferous** period, during part of which **Millstone Grit** was formed under river delta conditions. This is a darker, coarser and more resistant rock interbedded with shales. In the southern Pennines it can form either a plateau (Kinder Scout) or steep escarpments (Stanage Edge).
- The **Permian**, or **New Red Sandstone**, when sand was deposited under hot desert conditions, often in shallow water (i.e. when Britain lay in the latitude of the present-day Sahara). The rock is red, due to oxidation, and, being less resistant than the Old Red Sandstone, tends to form valleys (Exe and Eden) or low-lying hills (English Midlands).

Economic value of sandstone

Sandstone is the most common building material in Britain. In the past it was often used as stone for castles and cathedrals and, later, converted into brick for housing. Much of the New Red Sandstone has weathered into a warm, red light and easily worked soil of high agricultural value, in contrast to the Old Red Sandstone which,

Figure 8.13

Geological periods of various British sandstones

Geological period	Type of sandstone	Examples: location in the UK
Post-Eocene	See Figure 1.1	
Eocene		London and Hampshire basins
Cretaceous	Greensand	The Weald (south-east England)
Jurassic		
Triassic	Bunter and Keuper sandstone	English Midlands, Cheshire
Permian	New Red Sandstone	Exe and Eden valleys, south Arran
Carboniferous	Millstone Grit	Southern Pennines
Devonian	Old Red Sandstone	South-west England, South Wales, Herefordshire, central and north-east Scotland
Silurian, Ordovician and Cambrian		
Pre-Cambrian	Torridon	Wester Ross, Scotland

being more resistant, weathers to form uplands that have largely been left as moorland. Millstone Grit areas provided grindstones for Sheffield's cutlery industry in the past, and today these areas are popular for walking, rock-climbing, grouse moors and reservoirs.

Granite

Granite was formed when magma was intruded into the Earth's crust. Initially, as on Dartmoor and in northern Arran, the magma created deep-seated, dome-shaped batholiths (page 29). Since then the rock has been exposed by various processes of weathering and erosion. Having been formed at a depth and under pressure, the rate of cooling was slow and this enabled large crystals of quartz, mica and feldspar to form. As the granite continued to cool, it contracted and a series of cracks were created vertically and horizontally, at irregular intervals. These cracks may have been further enlarged, millions of years later, by pressure release as overlying rocks were removed (Figure 8.14).

The coarse-grained crystals render the rock non-porous but, although many texts quote granite as an example of an impermeable rock, water can find its way along the many cracks making some areas permeable. Despite this, most granite areas usually have a high drainage density and, as they occur in upland parts of Britain which have a high rainfall, they are often covered by marshy terrain.

Although a hard rock, granite is susceptible to both physical and chemical weathering. The joints, which can hold water, are widened by frost shattering (page 40), while the different rates of expansion and cooling of the various minerals within the rock cause granular disintegration (page 41). The feldspar and, to a lesser extent, mica can be changed chemically by hydrolysis (page 42). This means that calcium, potassium, sodium, magnesium and, if the pH is less than 5.0, iron and aluminium, are released from the chemical structure. Where the feldspar is changed near to the surface it forms a whitish clay called **kaolinite**. Where the change occurs at a greater depth (perhaps due to hydrothermal action), it produces **kaolin**. Quartz, which is not affected by chemical weathering, remains as loose crystals (Figure 2.7).

The most distinctive granite landform in temperate countries is the **tor** (Figure 8.14) and, in tropical regions, the **inselberg** (Figures 2.3 and 7.6). There are two major theories concerning their formation, based on physical and chemical weathering respectively. Both, however, suggest the removal of material by solifluction and hence lead to the opinion that tors and inselbergs are relict features.

The first hypothesis suggests that blocks of exposed granite were broken up, sub-aerially, by frost shattering during periglacial times. The weathered material was then moved downhill by solifluction to leave the more resistant rock upstanding on hill summits and valley sides.

The second, proposed by D. L. Linton, suggests that joints in the granite were widened by sub-surface chemical weathering (Figure 8.15). He suggested that deep weathering occurred during the warm Pliocene period (Figure 1.1) when rainwater penetrated the still-unexposed granite. As the joints widened, roughly rectangular blocks or core-stones were formed. The weathered rock is believed to have been removed by solifluction during periglacial times to leave outcrops of granite tors, separated by shallow depressions. The spacing of the joints is believed to be critical in tor formation: large, resistant core-stones have been left where joints were spaced far apart; where they were closely packed and weathering was more active, clay-filled depressions have developed. The rounded nature of the core-stones (Figure 8.15), especially in tropical regions, is caused by **spheroidal weathering**, a form of exfoliation (page 41).

Figure 8.14

Hound Tor, Dartmoor

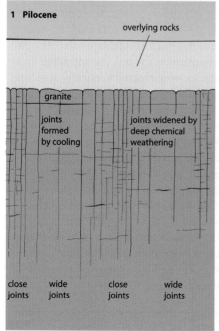

1 Pliocene

overlying rocks

granite

joints formed by cooling

joints widened by deep chemical weathering

close joints   wide joints   close joints   wide joints

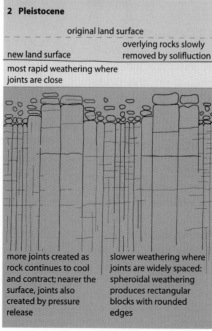

2 Pleistocene

original land surface

overlying rocks slowly removed by solifluction

new land surface

most rapid weathering where joints are close

more joints created as rock continues to cool and contract; nearer the surface, joints also created by pressure release

slower weathering where joints are widely spaced: spheroidal weathering produces rectangular blocks with rounded edges

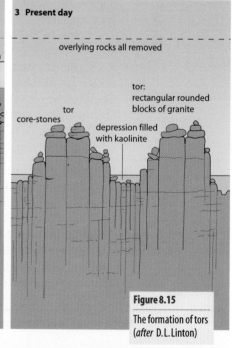

3 Present day

overlying rocks all removed

tor: rectangular rounded blocks of granite

tor

core-stones

depression filled with kaolinite

Figure 8.15

The formation of tors (*after* D. L. Linton)

Economic value of granite

As a raw material, granite can be used for building purposes; Aberdeen, for example, is known as 'the granite city'. Kaolin, or china clay, is used in the manufacture of pottery. Peat, which overlies large areas of granite bedrock, is an acidic soil which is often severely gleyed (page 275) and saturated with water, forming blanket bogs. The resultant heather-covered moorland is often unsuitable for farming but provides ideal terrain for grouse, and for army training. With so much surface water and heavy rainfall, granite areas provide ideal sites for reservoirs. Tors, such as Hound Tor on Dartmoor (Figure 8.14), may become tourist attractions, but granite environments tend to be inhospitable for settlement.

Basalt

Unlike granite, basalt formed on the Earth's surface, usually at constructive plate margins. The basic lava, on exposure to the air, cooled and solidified very rapidly. The rapid cooling produced small, fine-grained crystals and large cooling cracks which, at places like the Giant's Causeway in Northern Ireland (Figure 1.27) and Fingal's Cave on the Isle of Staffa, are charac-terised by perfectly shaped hexagonal, columnar jointing. Basalt can be extruded from either fissures or a central vent (page 25). When extruded from fissures, the lava often covers large areas of land – hence the term flood basalts – to produce flat plateaus such as the Deccan Plateau in India and the Drakensbergs in South Africa. Successive eruptions often build upwards to give, sometimes aided by later erosion, stepped hillsides beneath flat, tabular summits (e.g. the Drakensbergs, Lanzarote and Antrim). When extruded from a central vent, the viscous lava produces gently sloping shield volcanoes (Figure 1.22b). Shield volcanoes can reach considerable heights – Mauna Loa (Hawaii) rises over 9000 m from the Pacific seabed making it, from base to summit, the highest mountain on Earth.

Economic value of basalt

Basaltic landforms can sometimes be monotonous, such as places covered in flood basalts, and sometimes scenic and spectacular, as the Giant's Causeway, the Hawaiian volcanoes and the Iguaçu Falls in Brazil (Places 11, page 76). Basaltic lava can weather relatively quickly into a deep, fertile soil as on the Deccan in India and in the coffee-growing region of south-east Brazil. It can also be used for road foundations.

Case Study 8

Superquarries in Scotland

Quarrying can be an important source of local employment and, in poorer areas and countries, of regional and national wealth. However, any social and economic gain can easily be offset by environmental loss. In Britain, it has been the quarrying of limestone, especially in National Parks such as the Peak District, the Lake District and Snowdonia, that has caused most public concern. Yet these quarries are often small compared with the superquarries proposed in Scotland in the 1990s.

An article in the *New Scientist* of 8 January 1994 stated that 'Any week now, the Scottish Office will rule on a controversial proposal, already approved by the Western Isles Council, for a superquarry on the Hebridean island of Harris that will gouge the heart out of Roineabhal mountain, leaving behind a new loch'. At that time, it was thought that, due to an increasing demand for aggregates, between 15 and 20 superquarries would be needed in western Europe. Aggregate is hard rock that is crushed and used, along with sand and gravel, for road building and cement making. Demand, by 1994, had trebled in 30 years and was expected to double by 2010. The ideal sites were perceived to be in Scotland (Figure 8.16), Norway and Spain.

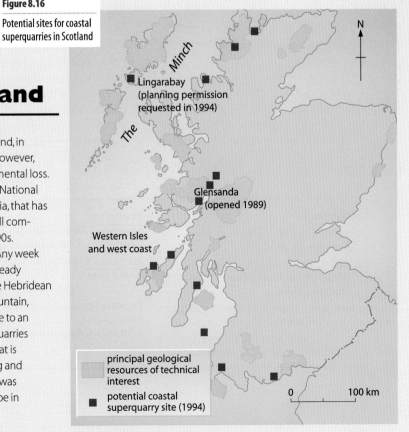

Figure 8.16

Potential sites for coastal superquarries in Scotland

principal geological resources of technical interest

potential coastal superquarry site (1994)

In 1994, Scotland's only superquarry was at Glensanda, near Fort William. It was producing 5 million tonnes of granite a year, blasted in a series of giant steps above the loch. That year, Redland Aggregates Ltd applied for permission to start quarrying at Roineabhal mountain near Lingarabay (Figure 8.17). The £50 million development would cover 4 km², create a hole up to 500 m deep and produce up to 10 million tonnes a year. In 1999, after a debate lasting five years, that decision was still awaited (see Charles Warren, *Geography Review*, March 1999).

Arguments in favour (mainly local people and Redland Aggregates Ltd)

- Lingarabay has deep water (a depth of 24 m) close to the shore, is protected from the open sea, and has room to manoeuvre the large (up to 60 000 tonnes) bulk-carrying vessels (Figure 8.17).
- The area is sparsely populated and so disturbance to local people should be limited.
- Job opportunities in the area are limited (often to fishing, crofting and tourism) and a superquarry would provide employment and generate income.
- Employment opportunities and an associated improvement in services (e.g. health, education, retailing and transport) could reverse the decline in depopulation.
- The rock is ideal, and the estimated reserve would last for over 100 years.
- The development of Western style-economies depends upon the extraction of finite resources that include oil and hard rock.

Figure 8.17

Lingarabay/Rodel and Roineabhal Mountain, Harris, Western Isles

- If Scotland does not develop superquarries, other countries will (Norway has since developed three sites).

Arguments against (mainly environmentalists and Scottish Natural Heritage)

- The quarry would ruin a quiet, scenic area by creating a visual eyesore and generating dust and noise (from blasting) and air and water pollution.
- The increase in shipping in the Minch – the waterway between Harris and the mainland – could lead to collisions with oil tankers which already use that route (en route to and from Sullom Voe in the Shetland Islands).
- Either an oil spillage or aggregate-laden ships discharging their ballast, could destroy the local marine and coastal ecosystems.
- Many of the new jobs would probably go to skilled people from outside the Hebrides.

- The hard rock being quarried is a finite resource and its extraction is non-sustainable (Figure 18.1).
- The aggregate is mainly for use in England and on the Continent, so why spoil the Scottish environment?
- With an EU recession and the British Government curtailing road building, then the growth in demand for aggregates may be less than previously thought.

As Charles Warren concludes: 'Should Scotland capitalise on its natural resources of coastal bedrock before other nations grab the market, or should these remote shores be preserved in all their pristine wildness? As with many environmental controversies, it comes down to a tug of war between values – the value of economic development for struggling communities, the value of unspoilt nature, and the contrasting perceptions of different groups in society.'

References

Goudie, A. (1993) *The Nature of the Environment*, Blackwell.

Ollier, C. (1997) *Aspects of Geography: Weathering and Landforms*, Thomas Nelson.

Planet Earth: Underground World (1982) Time-Life Books.

Small, J. (1999) *A Modern Dictionary of Geography*, Hodder & Stoughton.

Websites

Maritmine Stoneworks Inc home page:
http://www.dorchestersandstone.com/

Michigan Karst Conservancy group home page:
http://www.cyberspace.org/~mkc/

Pretoria Portland Cement Co. Ltd home page:
http://www.ppc.co.za/index.html

See also for more links:
http://www.nelsonthornes.com/gaia

1 a Describe the characteristics of each of the following rock types in terms of chemical composition, rock structure and origin: Carboniferous limestone; chalk; granite; basalt. **(12 marks)**

b Choose **one** of the rock types in **a** and draw an annotated diagram to identify the characteristic landscape features associated with it. **(9 marks)**

c For **each** of the rock types identified in **a**, suggest **one** reason why it may be of value as a resource for human use. **(4 marks)**

2 Study the OS map extract of the area around Malham in Figure 8.18.

a i How high above sea-level is the minor road at GR 907649? **(1 mark)**

ii What is the feature at GR 906655? **(1 mark)**

b Identify and locate **two** pieces of evidence that this area consists of limestone rock. **(6 marks)**

c i Much of the area is managed by the National Trust and it all lies inside the North York Moors National Park. What is a 'National Park'? **(3 marks)**

ii Why are some areas, even inside a National Park, managed by the National Trust? **(5 marks)**

d Describe and explain, making good use of diagrams, the cause of the areas of bare flat rocks shown on the map extract. **(9 marks)**

3 a Making good use of annotated diagrams, describe the surface features of a chalk cuesta. **(6 marks)**

b Describe and explain the location of the water table within an area of chalk hills. **(6 marks)**

c Describe and suggest reasons for the location of settlements close to the foot of a chalk cuesta. **(4 marks)**

d Suggest **two** reasons why some chalk downs have prehistoric carved figures on them. **(4 marks)**

e Chalk escarpments may have 'hangers' (areas of beech woodland on the brow of the scarp). Suggest why these woodlands are found here. **(5 marks)**

4 a i Study Figure 8.3 (page 196) of a swallow hole near Alum Pot (Ingleborough). Describe the physical features shown in the photograph. **(4 marks)**

ii Explain how the swallow hole was formed. **(6 marks)**

b Describe a 'limestone pavement' and explain how it was formed. **(5 marks)**

c i Describe the features you would expect to find in a limestone cavern. **(4 marks)**

ii Choose **one** of the features you have described and explain how it was formed. **(6 marks)**

Figure 8.18

OS map of Malham area, 1:25 000

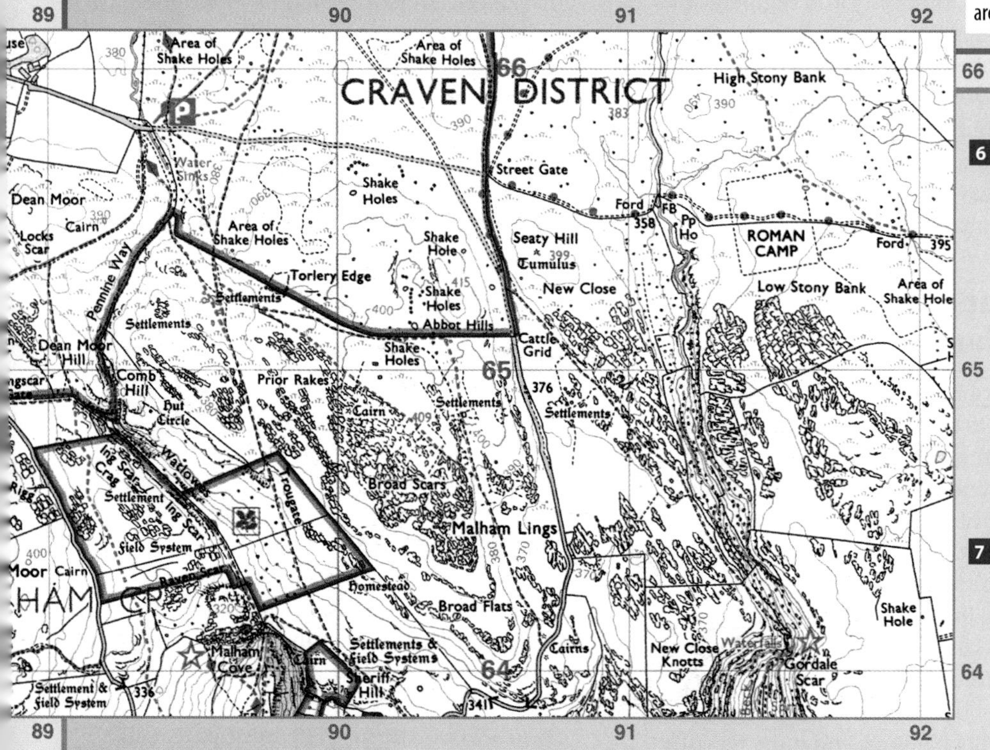

6 a What are 'primary permeability' and 'secondary permeability' in rocks? How does the difference affect natural processes? **(5 marks)**

b Choose **two** of the following rock types: Carboniferous limestone; chalk; granite; basalt. For **each** one discuss the relationship between the rock type and climate. Your answer should refer to characteristic features of the landscape which evolve on each rock type. **(20 marks)**

7 Study the OS map extract of the area around Malham in Figure 8.18.

a Identify and locate **two** pieces of evidence to suggest that this area is limestone rock. For **each** explain how the evidence shows it to be limestone. **(10 marks)**

5 a Explain what happens to a river flowing from an area of impervious rock onto an area of limestone. **(8 marks)**

b Describe and explain the characteristics of **two** surface features of an area of Carboniferous limestone landscape. **(10 marks)**

c Identify **two** dangers to limestone areas which occur as a result of human activity. **(7 marks)**

b Why is there so much settlement and other ancient remains visible in an area such as this? **(7 marks)**

c This area is both a tourist area and a working farming area. Identify **one** way these two land uses are in conflict and explain the reasons for this conflict. **(8 marks)**

AS

A2

Chapter 9

Weather and climate

'There is really no such thing as bad weather, only different types of good weather.'

John Ruskin, Quote from Lord Avebury

'When two Englishmen meet, their first talk is of the weather.' **Samuel Johnson, *The Idler***

The science of meteorology is the study of atmospheric phenomena; it includes the study of both weather and climate. The distinction between climate and weather is one of scale. **Weather** refers to the state of the atmosphere at a local level, usually on a short timescale of minutes to months. It emphasises aspects of the atmosphere that affect human activity, such as sunshine, cloud, wind, rainfall, humidity and temperature. **Climate** is concerned with the long-term behaviour of the atmosphere in a specific area. Climatic characteristics are represented by data on temperature, pressure, wind, precipitation, humidity, etc. which are used to calculate daily, monthly and yearly averages (Framework 8, page 246) and to build up global patterns (Chapter 12).

Structure and composition of the atmosphere

The atmosphere is an envelope of transparent, odourless gases held to the Earth by gravitational attraction. While the furthest limit of the atmosphere is said by international convention to be at 1000 km, most of the atmosphere, and therefore our climate and weather, is concentrated within 16 km of the Earth's surface at the Equator and 8 km at the poles. Fifty per cent of atmospheric mass is within 5.6 km of sea-level and 99 per cent is within 40 km. Atmospheric pressure decreases rapidly with height but, as recordings made by radiosondes, weather balloons and more recently weather satellites have shown, temperature changes are more complex. Changes in temperature mean that the atmosphere can be conveniently divided into four distinctive layers (Figure 9.1); moving outwards from the Earth's surface, these are:

1   **Troposphere**   Temperatures in the troposphere decrease by 6.4°C with every 1000 m increase in altitude (environmental lapse rate, page 216). This is because the Earth's surface is warmed by incoming solar radiation which in turn heats the air next to it by conduction, convection and radiation. Pressure falls as the effect of gravity decreases, although wind speeds usually increase with height. The layer is unstable and contains most of the atmosphere's water vapour, cloud, dust and pollution. The tropopause, which forms the upper limit to the Earth's climate and weather, is marked by an isothermal layer where temperatures remain constant despite any increase in height.

2   **Stratosphere**   The stratosphere is characterised by a steady increase in temperature (temperature inversion, page 217) caused by a concentration of **ozone** (O_3) (Places 27, page 209). This gas absorbs incoming **ultra-violet (UV) radiation** from the sun. Winds are light in the lower parts, but increase with height; pressure continues to fall and the air is dry. The stratosphere, like the two layers above it, acts as a protective shield against meteorites which usually burn out as they enter the Earth's gravitational field. The **stratopause** is another isothermal layer where temperatures do not change with increasing height.

3   **Mesosphere**   Temperatures fall rapidly as there is no water vapour, cloud, dust or ozone to absorb incoming radiation. This layer experiences the atmosphere's lowest temperatures (–90°C) and strongest winds (nearly 3000 km/hr). The **mesopause**, like the tropopause and stratopause, shows no change in temperature.

4   **Thermosphere**   Temperatures rise rapidly with height, perhaps to reach 1500°C. This is due to an increasing proportion of atomic oxygen in the atmosphere which, like ozone, absorbs incoming ultra-violet radiation.

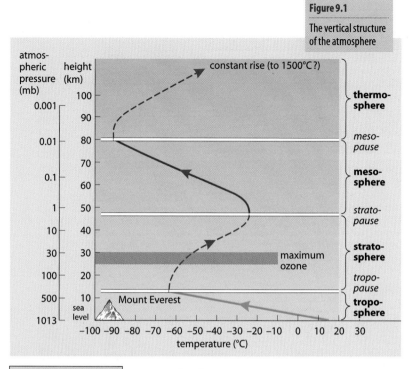

Figure 9.1

The vertical structure of the atmosphere

changes in temperature with height

← — fall
— constant
→ -- rise

Atmospheric gases

The various gases which combine to form the atmosphere are listed in Figure 9.2. Of these, nitrogen and oxygen together make up 99 per cent by volume. Of the others, water vapour (lower atmosphere), ozone (O_3) (upper atmosphere) and carbon dioxide (CO_2) have an importance far beyond their seemingly small amounts. It is the depletion of O_3 (Places 27) and the increase in CO_2 (Case Study 9) which are causing concern to scientists.

Energy in the atmosphere

The sun is the Earth's prime source of energy. The Earth receives energy as incoming **short-wave** solar radiation (also referred to as **insolation**). It is this energy that controls our planet's climate and weather and which, when converted by photosynthesis in green plants, supports all forms of life. The amount of incoming radiation received by the Earth is determined by four astronomical factors (Figure 9.3): the solar constant, the distance from the sun, the altitude of the sun in the sky, and the length of night and day. Figure 9.3 is theoretical in that it assumes there is no atmosphere around the Earth. In reality, much insolation is absorbed, reflected and scattered as it passes through the atmosphere (Figure 9.4).

Absorption of incoming radiation is mainly by ozone, water vapour, carbon dioxide and particles of ice and dust. Clouds and, to a lesser extent, the Earth's surface **reflect** considerable amounts of radiation back into space. The ratio between incoming radiation and the amount reflected, expressed as a percentage, is known as the **albedo**. The albedo varies with cloud type from 30–40 per cent in thin clouds, to 50–70 per cent in thicker stratus and 90 per cent in cumulo-nimbus (when only 10 per cent reaches the atmosphere below cloud level). Albedos also vary over different land surfaces, from less than 10 per cent over oceans

Figure 9.2

The composition of the atmosphere

Gas		Percentage by volume	Importance for weather and climate	Other functions/source
Permanent gases:	nitrogen	78.09	Mainly passive	Needed for plant growth.
	oxygen	20.95		Produced by photosynthesis; reduced by deforestation.
Variable gases:	water vapour	0.20–4.0	Source of cloud formation and precipitation, reflects/absorbs incoming radiation. Keeps global temperatures constant. Provides majority of natural 'greenhouse effect'.	Essential for life on Earth. Can be stored as ice/snow.
	carbon dioxide	0.03	Absorbs long-wave radiation from Earth and so contributes to 'greenhouse effect'. Its increase due to human activity is a major cause of global warming.	Used by plants for photosynthesis; increased by burning fossil fuels and by deforestation.
	ozone	0.00006	Absorbs incoming ultra-violet radiation.	Reduced/destroyed by chlorofluorocarbons (CFCs).
Inert gases:	argon	0.93		
	helium, neon, krypton	trace		
Non-gaseous:	dust	trace	Absorbs/reflects incoming radiation. Forms condensation nuclei necessary for cloud formation.	Volcanic dust, meteoritic dust, soil erosion by wind.
Pollutants:	sulphur dioxide, nitrogen oxide, methane	trace	Affects radiation. Causes acid rain.	From industry, power stations and car exhausts.

Note: the figures refer to dry air and so the variable amount of water vapour is not usually taken into consideration.

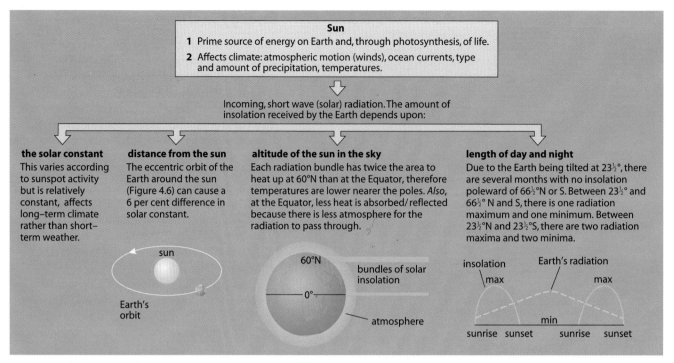

Figure 9.3

Incoming radiation received by the Earth (assuming that there is no atmosphere)

and dark soil, to 15 per cent over coniferous forest and urban areas, 25 per cent over grasslands and deciduous forest, 40 per cent over light-coloured deserts and 85 per cent over reflecting fresh snow. Where deforestation and overgrazing occur, the albedo increases. This reduces the possibility of cloud formation and precipitation and increases the risk of **desertification** (Case Study 7). **Scattering** occurs when incoming radiation is diverted by particles of dust, as from volcanoes and deserts, or by molecules of gas. It takes place in all directions and some of the radiation will reach the Earth's surface as **diffuse** radiation.

As a result of absorption, reflection and scattering, only about 24 per cent of incoming radia-tion reaches the Earth's surface directly, with a further 21 per cent arriving at ground-level as diffuse radiation (Figure 9.4). Incoming radiation is converted into heat energy when it reaches the Earth's surface. As the ground warms, it radiates energy back into the atmosphere where 94 per cent is absorbed (only 6 per cent is lost to space), mainly by water vapour and carbon dioxide – the greenhouse effect (Case Study 9). Without the natural greenhouse effect, which traps so much of the outgoing radiation, world temperatures would be 33°C lower than they are at present and life on Earth would be impossible. (During the ice age, it was only 4°C cooler.) This outgoing (terrestrial) radiation is **long-wave** or **infra-red** radiation.

Figure 9.4

The solar energy cascade

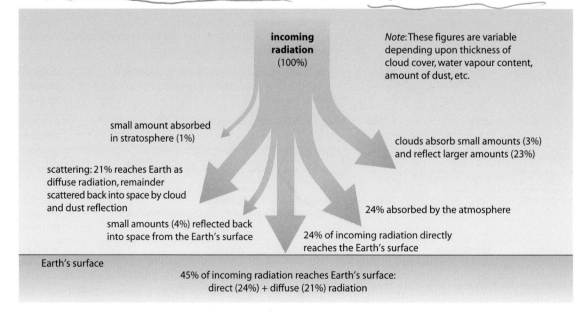

The major concentration of ozone is in the stratosphere, 25–30 km above sea-level (Figure 9.1). Ozone acts as a shield protecting the Earth from the damaging effects of ultra-violet (UV) radiation from the sun. However, there is growing concern that this protective shield appears to be breaking down. An increase in ultra-violet radiation means increases in sunburn and skin-cancer (fair skin is at greater risk than dark skin), snow blindness, cataracts and eye damage, ageing and the wrinkling of skin (in 1997, some doctors suggested a link with blood cancer).

A depletion in ozone above the Antarctic was first observed by the British Antarctic Survey in 1977, and the first 'hole' noted in 1985. The term 'hole' is misleading, as it really means a depletion of over 50 per cent (not 100 per cent!). Each spring over the Antarctic (September–November), the very low temperatures cause ozone to be destroyed in a chemical reaction with chlorine. The main sources of chlorine are:

- the release of chlorofluorocarbons, especially from aerosols such as hairsprays, deodorants and fly-killers, refrigerator coolant, and manufacturing processes that produce foam packagings – all of which have been used increasingly in recent decades (a long-term effect)
- from major volcanic eruptions, e.g. Mount Pinatubo, (Case Study 1 – a short-term effect).

By 1993, ozone over Antarctica had been reduced over an area the size of the USA to between one-half and two-thirds of its 1970 amount, most of it since 1990. The first observed 'hole' over the Arctic followed the coldest-ever January (1989). In the decade since then, levels over the Arctic have fallen by 10 per cent, with the March 1997 levels being 40 per cent lower than those of March 1996. In Britain, the increase in ultra-violet rays is said to have been 6.8 per cent per decade since 1979 – significant when scientists claim that a 1 per cent depletion causes a 5 per cent increase in skin cancer.

In contrast, vehicle exhaust systems generate dangerous quantities of ozone close to the Earth's surface, causing damage to plant tissues and affecting the health of people and animals. For example, the high levels of asthma in children in London during the hot summer of 1994 and in Paris in 1995 were linked with traffic pollution. The problem was so acute that health warnings were broadcast in both countries. Since then traffic has continued to increase and summers have become warmer.

Ground-level ozone increases during warm, sunny, anticyclonic conditions (page 234) and can, under extreme conditions, form petrochemical smog. This is because nitrogen oxides, from vehicle exhausts, react with VOCs (volatile organic compounds) in sunlight.

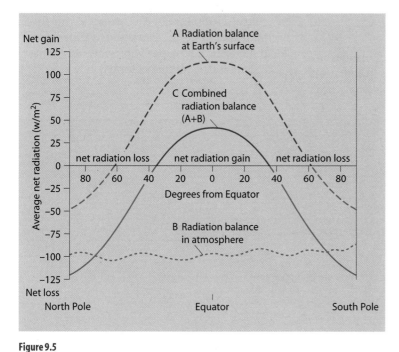

Figure 9.5

The heat budget

The heat budget

Since the Earth is neither warming up nor cooling down, there must be a balance between incoming insolation and outgoing terrestrial radiation. Figure 9.5 shows that:

- there is a net gain in radiation everywhere on the Earth's surface (curve A) except in polar latitudes which have high albedo surfaces
- there is a net loss in radiation throughout the atmosphere (curve B)
- after balancing the incoming and outgoing radiation, there is a net surplus between 35°S and 40°N (the difference in latitude is due to the larger land masses of the northern hemisphere) and a net deficit to the poleward sides of those latitudes (curve C).

This means that there is a **positive heat balance** within the tropics and a **negative heat balance** both at high latitudes (polar regions) and high altitudes. Two major **transfers** of heat, therefore, take place to prevent tropical areas from overheating (Figure 9.6).

1 **Horizontal heat transfers** Heat is transferred away from the tropics, thus preventing the Equator from becoming increasingly hotter and the poles increasingly colder. Winds (air movements including jet streams, page 227; hurricanes, page 235; and depressions, page 230) are responsible for 80 per cent of this heat transfer, and ocean currents for 20 per cent (page 211).

2 **Vertical heat transfers** Heat is also transferred vertically, thus preventing the Earth's surface from getting hotter and the atmosphere colder. This is achieved through **radiation, conduction, convection** and the transfer of **latent heat**. Latent heat is the amount of heat energy needed to change the state of a substance without affecting its temperature. When ice changes into water or water into vapour, heat is taken up to help with the processes of melting and evaporation. This absorption of heat results in the cooling of the atmosphere. When the process is reversed – i.e. vapour condenses into water or water freezes into ice – heat energy is released and the atmosphere is warmed.

Variations in the radiation balance occur at a number of spatial and temporal scales. Regional differences may be due to the uneven distribution of land and sea, altitude, and the direction of prevailing winds. Local variations may result from **aspect** and amounts of cloud cover. Seasonal and diurnal variations are related to the altitude of the sun and the length of night and day.

Global factors affecting insolation

Factors that influence the amount of insolation received at any point, and therefore its radiation balance and heat budget, vary considerably over time and space.

Long-term factors
These are relatively constant at a given point.

■ **Height above sea-level** The atmosphere is not warmed directly by the sun, but by heat radiated from the Earth's surface and distributed by conduction and convection. As the height of mountains increases, they present a decreasing area of land surface from which to heat the surrounding air. In addition, as the density or pressure of the air decreases, so too does its ability to hold heat (Figure 9.1). This is because the molecules in the air which receive and retain heat become fewer and more widely spaced as height increases.

■ **Altitude of the sun** As the angle of the sun in the sky decreases, the land area heated by a given ray and the depth of atmosphere through which that ray has to pass both increase. Consequently, the amount of insolation lost through absorption, scattering and reflection also increases. Places in lower latitudes therefore have higher temperatures than those in higher latitudes.

■ **Land and sea** Land and sea differ in their ability to absorb, transfer and radiate heat energy. The sea is more transparent than the land, and is capable of absorbing heat down to a depth of 10 metres. It can then transfer this heat to greater depths through the movements of waves and currents. The sea also has a greater **specific heat capacity** than that of land. Specific heat capacity is the amount of energy required to raise the temperature of 1 kg of a substance by 1°C, expressed in kilojoules per kg per °C. Expressed in kilocalories, the specific heat capacity of water is 1.0, that of land is 0.5 and that of sand 0.2.

Figure 9.6

Heat transfers in the atmosphere

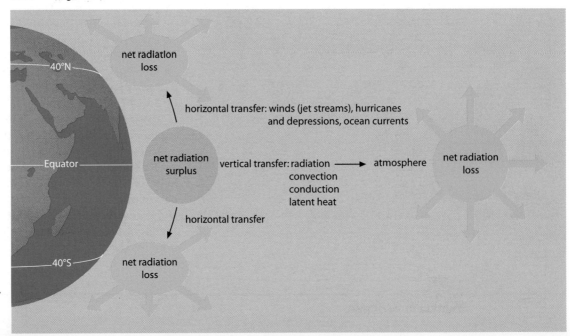

Figure 9.7

Mean annual ranges in global temperature (°C)

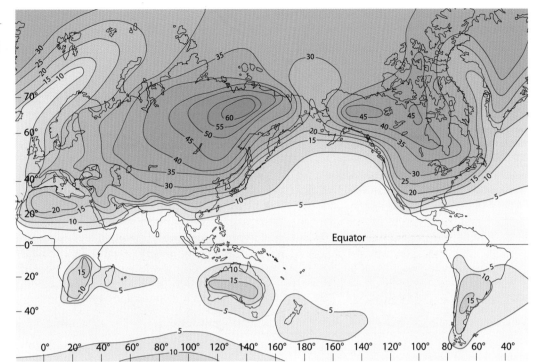

This means that water requires twice as much energy as soil and five times more than sand to raise an equivalent mass to the same temperature. During summer, therefore, the sea heats up more slowly than the land. In winter, the reverse is the case and land surfaces lose heat energy more rapidly than water. The oceans act as efficient 'thermal reservoirs'. This explains why coastal environments have a smaller annual range of temperature than locations at the centres of continents (Figure 9.7).

- **Prevailing winds**   The temperature of the wind is determined by its area of origin and by the characteristics of the surface over which it subsequently blows (Figure 9.8). A wind blowing from the sea tends to be warmer in winter and cooler in summer than a corresponding wind coming from the land.

- **Ocean currents**   These are a major component in the process of horizontal transfer of heat energy. Warm currents carry water polewards and raise the air temperature of the maritime environments where they flow. Cold currents carry water towards the Equator and so lower the temperatures of coastal areas (Figure 9.9). The main ocean currents follow circular routes – clockwise in the northern hemisphere, anticlockwise in the southern hemisphere.

Figure 9.10 shows the difference between the mean January temperature of a place and the mean January temperatures of other places with the same latitude; this difference is known as a **temperature anomaly**. (The term 'temperature anomaly' is used specifically to describe temperature differences from a mean. It should not be confused with the more general definition of 'anomaly' which refers to something that does not fit into a general pattern.) For example, Stornoway (Figure 9.10) has a mean January temperature of 4°C, which is 20°C higher than the average for other locations lying at 58°N. Such anomalies result primarily from the uneven heating and cooling rates of land and sea and are intensified by the horizontal transfer of energy by ocean currents and prevailing winds. Remember that the sun appears overhead in the southern hemisphere at this time of year (January) and isotherms have been reduced to sea-level – i.e. temperatures are adjusted to eliminate some of the effects of relief, thus emphasising the influence of prevailing winds, ocean currents and continentality.

Figure 9.8

Simplified diagram showing the effect of prevailing winds on land and sea temperatures

Season	SEA	West coast	LAND	East coast	SEA	Season
Winter	*Warm*	warm wind →	**COLD**	cool wind →	*Warm*	Winter
Summer	*Cool*	cool wind →	**WARM**	warm wind →	*Cool*	Summer

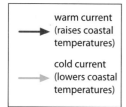

Figure 9.9

Major ocean currents

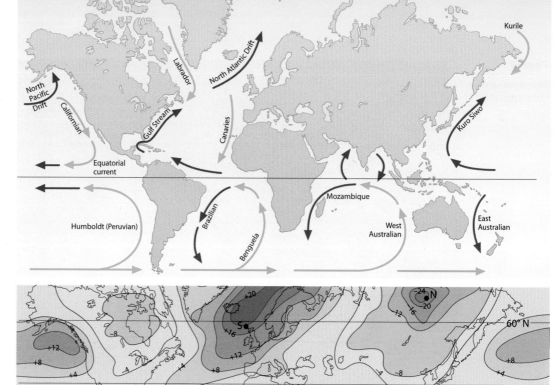

warm current (raises coastal temperatures)

cold current (lowers coastal temperatures)

Figure 9.10

Temperature anomalies for January (*after* D.C. Money)

S = Stornoway
N = North-east Siberia

Short-term factors

■ **Seasonal changes** At the spring and autumn equinoxes (21 March and 22 September) when the sun is directly over the Equator, insolation is distributed equally between both hemispheres. At the summer and winter solstices (21 June and 22 December) when, due to the Earth's tilt, the sun is overhead at the tropics, the hemisphere experiencing 'summer' will receive maximum insolation.

■ **Length of day and night** Insolation is only received during daylight hours and reaches its peak at noon. There are no seasonal variations at the Equator, where day and night are of equal length throughout the year. In extreme contrast, polar areas receive no insolation during part of the winter when there is continuous darkness, but may receive up to 24 hours of insolation during part of the summer when the sun never sinks below the horizon ('the lands of the midnight sun').

Local influences on insolation

■ **Aspect** Hillsides alter the angle at which the sun's rays hit the ground (Places 28). In the northern hemisphere, north-facing slopes, being in shadow for most or all of the year, are cooler than those facing south. The steeper the south-facing slope, the higher the angle of the sun's rays to it and therefore the higher will be the temperature. North- and south-facing slopes are referred to, respectively, as the **adret** and **ubac**.

■ **Cloud cover** The presence of cloud reduces both incoming and outgoing radiation. The thicker the cloud, the greater the amount of absorption, reflection and scattering of insolation, and of terrestrial radiation. Clouds may reduce daytime temperatures, but they also act as an insulating blanket to retain heat at night. This means that tropical deserts, where skies are clear, are warmer during the day and cooler at night than humid equatorial regions with a greater cloud cover. The world's greatest diurnal ranges of temperature are therefore found in tropical deserts.

■ **Urbanisation** This alters the albedo (page 207) and creates urban 'heat islands' (page 242).

Atmospheric moisture

Water is a liquid compound which is converted by heat into vapour (gas) and by cold into a solid (ice). The presence of water serves three essential purposes:

1 It maintains life on Earth: flora, in the form of natural vegetation (biomes) and crops; and fauna, i.e. all living creatures, including humans.

2 Water in the atmosphere, mainly as a gas, absorbs, reflects and scatters insolation to keep our planet at a habitable temperature (Figure 9.4).

3 Atmospheric moisture is of vital significance as a means of transferring surplus energy from tropical areas either horizontally to polar latitudes or vertically into the atmosphere to balance the heat budget (Figure 9.5). Despite this need for water, its existence in a form readily available to plants, animals and humans is limited. It has been estimated that 97.2 per cent of the world's water is in the oceans and seas; in this form, it is only useful to plants tolerant of saline conditions (**halophytes**, page 291) and to the populations of a few wealthy countries that can afford desalinisation plants (the Gulf oil states).

Approximately 2.1 per cent of water in the hydrosphere is held in storage as polar ice and snow. Only 0.7 per cent is fresh water found either in lakes and rivers (0.1 per cent), as soil moisture and groundwater (0.6 per cent), or in the atmosphere (0.001 per cent).

Places 28 An alpine valley: aspect

Many alpine valleys in Switzerland and Austria have an east–west orientation which means that their valley sides face either north or south. South-facing adret slopes are much warmer and drier than those facing north (Figure 9.11). The south-facing slopes have more plant species, a higher tree-line, and a greater land use with alpine pastures at higher altitudes and fruit and hay lower down; also, they usually provide the best sites for settlement. In contrast, north-facing ubac slopes are snow-covered for a much longer period, they are less suited to farming, the tree-line is lower, and they tend to be left forested. However, on the valley floors, as severe frosts are likely to occur during times of temperature inversion (page 217), sensitive plants and crops do not flourish.

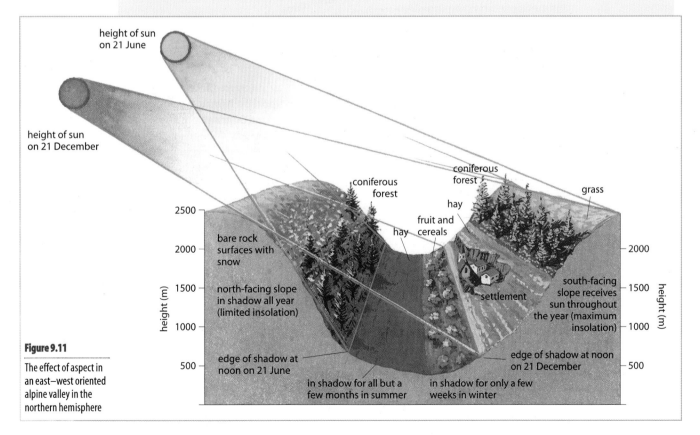

Figure 9.11

The effect of aspect in an east–west oriented alpine valley in the northern hemisphere

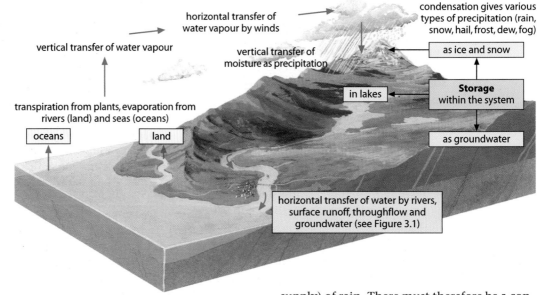

Figure 9.12

The hydrological cycle (compare with Figure 3.1)

horizontal transfer of water vapour by winds

vertical transfer of water vapour

condensation gives various types of precipitation (rain, snow, hail, frost, dew, fog)

vertical transfer of moisture as precipitation

as ice and snow

Storage within the system

in lakes

as groundwater

transpiration from plants, evaporation from rivers (land) and seas (oceans)

oceans

land

horizontal transfer of water by rivers, surface runoff, throughflow and groundwater (see Figure 3.1)

At any given time, the atmosphere only holds, on average, sufficient moisture to give every place on the Earth 2.5 cm (about 10 days' supply) of rain. There must therefore be a constant recycling of water between the oceans, atmosphere and land. This recycling is achieved through the **hydrological cycle** (Figure 9.12).

Figure 9.13

The world's water balance

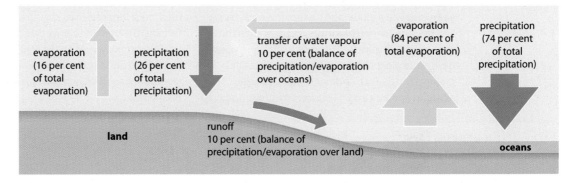

evaporation (16 per cent of total evaporation)

precipitation (26 per cent of total precipitation)

transfer of water vapour 10 per cent (balance of precipitation/evaporation over oceans)

evaporation (84 per cent of total evaporation)

precipitation (74 per cent of total precipitation)

land

runoff 10 per cent (balance of precipitation/evaporation over land)

oceans

Humidity

Humidity is a measure of the water vapour content in the atmosphere. **Absolute humidity** is the mass of water vapour in a given volume of air measured in grams per cubic metre (g/m^3). **Specific humidity** is similar but is expressed in grams of water per kilogram of air (g/kg). Humidity depends upon the temperature of the air. At any given temperature, there is a limit to the amount of moisture that the air can hold. When this limit is reached, the air is said to be **saturated**. Cold air can hold only relatively small quantities of vapour before becoming saturated but this amount increases rapidly as temperatures rise (Figure 9.14). This means that the amount of precipitation obtained from warm air is generally greater than that from cold air. **Relative humidity** (RH) is the amount of water vapour in the air at a given temperature expressed as a percentage of the maximum amount of vapour that the air could hold at that temperature. If the RH is 100 per cent, the air is saturated. If it lies between 80 and 99 per cent, the air is said to be 'moist' and the weather is humid or clammy. When the RH drops to 50 per cent, the air is 'dry'– figures as low as 10 per cent have been recorded over hot deserts.

Figure 9.14

Air temperatures and absolute humidity for saturated air

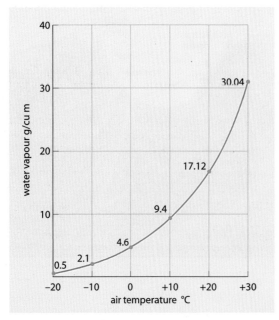

If unsaturated air is cooled and atmospheric pressure remains constant, a critical temperature will be reached when the air becomes saturated (i.e. RH = 100 per cent). This is known as the **dew point**. Any further cooling will result in the condensation of excess vapour, either into water droplets where condensation nuclei are present, or into ice crystals if the air temperature is below 0°C. This is shown in the following worked example.

1   The early morning air temperature was 10°C. Although the air could have held 100 units of water at that temperature, at the time of the reading it held only 90. This meant that the RH was 90 per cent.

2   During the day, the air temperature rose to 12°C. As the air warmed it became capable of holding more water vapour, up to 120 units. Owing to evaporation, the reading reached a maximum of 108 units which meant that the RH remained at 90 per cent – i.e. $(108 \div 120) \times 100$.

3   In the early evening, the temperature fell to 10°C at which point, as stated above, it could hold only 100 units. However, the air at that time contained 108 units so, as the temperature fell, dew point was reached and the 8 excess units of water were lost through condensation.

Condensation

This is the process by which water vapour in the atmosphere is changed into a liquid or, if the temperature is below 0°C, a solid. It usually results from air being cooled until it is saturated. Cooling may be achieved by:

1   **Radiation** (contact) **cooling**   This typically occurs on calm, clear evenings. The ground loses heat rapidly through terrestrial radiation and the air in contact with it is then cooled by conduction. If the air is moist, some vapour will condense to form radiation fog, dew, or – if the temperature is below freezing point – hoar frost (page 221).

2   **Advection cooling**   This results from warm, moist air moving over a cooler land or sea surface. Advection fogs (Places 24 page 180, and page 222) in California and the Atacama Desert (Places 24, page 180) are formed when warm air from the land drifts over cold off-shore ocean currents (Figure 9.9).

As both radiation and advection involve horizontal rather than vertical movements of air, the amount of condensation created is limited.

3   **Orographic** and **frontal uplift**   Warm, moist air is forced to rise either as it crosses a mountain barrier (orographic ascent, page 220) or when it meets a colder, denser mass of air at a front (page 229).

4   **Convective** or **adiabatic cooling**   This is when air is warmed during the daytime and rises in pockets as **thermals** (Figure 9.15). As the air expands, it uses energy and so loses heat and the temperature drops. Because air is cooled by the reduction of pressure with height rather than by a loss of heat to the surrounding air, it is said to be adiabatically cooled (see lapse rates, page 216).

As both orographic and adiabatic cooling involve vertical movements of air, they are more effective mechanisms of condensation.

Condensation does not occur readily in clean air. Indeed, if air is absolutely pure, it can be cooled below its dew point to become **supersaturated** with an RH in excess of 100 per cent. Laboratory tests have shown that clean, saturated air can be cooled to –40°C before condensation or, in this case, **sublimation**. Sublimation is when vapour condenses directly into ice crystals without passing through the liquid state. However, air is rarely pure and usually contains large numbers of condensation nuclei. These microscopic particles, referred to as **hygroscopic nuclei** because they attract water, include volcanic dust (heavy rain always accompanies volcanic eruptions); dust from windblown soil; smoke and sulphuric acid originating from urban and industrial areas; and salt from sea spray. Hygroscopic nuclei are most numerous over cities, where there may be up to 1 million per cm^3, and least common over oceans (only 10 per cm^3). Where large concentrations are found, condensation can occur with an RH as low as 75 per cent – as in the smogs of Los Angeles (Figure 9.25 and Case Study 15A).

Figure 9.15

Convective cooling

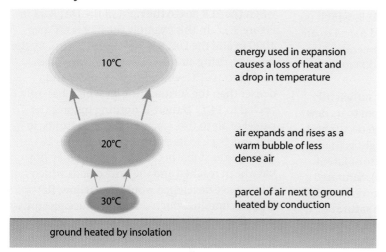

energy used in expansion causes a loss of heat and a drop in temperature

air expands and rises as a warm bubble of less dense air

parcel of air next to ground heated by conduction

10°C

20°C

30°C

ground heated by insolation

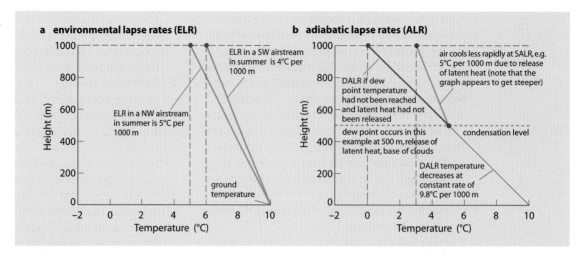

a environmental lapse rates (ELR)

ELR in a SW airstream in summer is 4°C per 1000 m

ELR in a NW airstream in summer is 5°C per 1000 m

ground temperature

b adiabatic lapse rates (ALR)

air cools less rapidly at SALR, e.g. 5°C per 1000 m due to release of latent heat (note that the graph appears to get steeper)

DALR if dew point temperature had not been reached and latent heat had not been released

dew point occurs in this example at 500 m, release of latent heat, base of clouds

condensation level

DALR temperature decreases at constant rate of 9.8°C per 1000 m

Lapse rates

The **environmental lapse rate** (ELR) is the decrease in temperature usually expected with an increase in height through the troposphere (Figure 9.1). The ELR is approximately 6.5°C per 1000 m, but varies according to local air conditions. It may vary due to several factors: **height** – ELR is lower nearer ground-level; **time** – it is lower in winter or during a rainy season; over different **surfaces** – it is lower over continental areas; and between different **air masses** (Figure 9.16a).

The **adiabatic lapse rate** (ALR) describes what happens when a parcel of air rises and the decrease in pressure is accompanied by an associated increase in volume and a decrease in temperature (Figure 9.15). Conversely, descending air will be subject to an increase in pressure causing a rise in temperature. In either case, there is negligible mixing with the surrounding air. There are two adiabatic lapse rates.

1   If the upward movement of air does not lead to condensation, the energy used by expansion will cause the temperature of the parcel of air to fall at the **dry adiabatic lapse rate** (DALR on Figure 9.16b). The DALR, which is the rate at which an unsaturated parcel of air cools as it rises or warms as it descends, remains constant at 9.8°C per 1000 m (i.e. approximately 1°C per 100 m).

2   When the upward movement is sufficiently prolonged to enable the air to cool to its dew point temperature, condensation occurs and the loss in temperature with height is then partly compensated by the release of latent heat (Figure 9.16b and page 210). Saturated air, which therefore cools at a slower rate than unsaturated air, loses heat at the **saturated adiabatic lapse rate** (SALR). The SALR can vary because the warmer the air the more moisture it can hold, and so the greater the amount of latent heat released following con-

densation. The SALR may be as low as 4°C per 1000 m and as high as 9°C per 1000 m. It averages about 5.4°C per 1000 m (i.e. approximately 0.5°C per 100 m). Should temperatures fall below 0°C, then the air will cool at the **freezing adiabatic lapse rate** (FALR). This is the same as the DALR as very little moisture is present at low temperatures.

Air stability and instability

Parcels of warm air which rise through the lower atmosphere cool adiabatically. The rate and maintenance of any vertical uplift depend upon the temperature–density balance between the rising parcel and the surrounding air. In a simplified form, this balance is the relationship between the environmental lapse rate and the dry and saturated adiabatic lapse rates.

Stability

The state of **stability** is when a rising parcel of unsaturated air cools more rapidly than the air surrounding it. This is shown diagrammatically when the ELR lies to the right of the DALR, as in Figure 9.17. In this example the ELR is 6°C per 1000 m and the DALR is 9.8°C per 1000 m. By the time the rising air has reached 1000 m, it has cooled to 10.2°C which leaves it colder and denser than the surrounding air which has only cooled to 14°C. If there is nothing to force the parcel of air to rise, e.g. mountains or fronts, it will sink back to its starting point. The air is described as stable because dew point may not have been reached and the only clouds which might have developed would be shallow, flat-topped cumulus which do not produce precipitation (Figure 9.20). Stability is often linked with anticyclones (page 234), when any convection currents are suppressed by sinking air to give dry, sunny conditions.

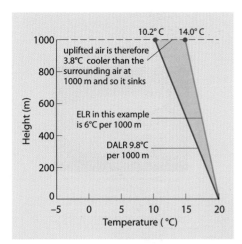

Figure 9.17

Stability: changes in lapse rates and air temperature with height

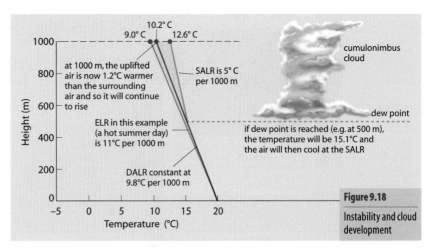

Figure 9.18

Instability and cloud development

Instability

Conditions of **instability** arise in Britain on hot days. Localised heating of the ground warms the adjacent air by conduction, creating a higher lapse rate. The resultant parcel of rising unsaturated air cools less rapidly than the surrounding air. In this case, as shown in Figure 9.18, the ELR lies to the left of the DALR. The rising air remains warmer and lighter than the surrounding air. Should it be sufficiently moist and if dew point is reached, then the upward movement may be accelerated to produce towering cumulus or cumulo-nimbus type cloud (Figure 9.20). Thunderstorms are likely (Figure 9.21) and the saturated air, following the release of latent heat, will cool at the SALR.

Conditional instability

This type of instability occurs when the ELR is lower than the DALR but higher than the SALR. In Britain, it is the most common of the three conditions. The rising air is stable in its lower layers and, being cooler than the surrounding air, would normally sink back again. However, if the mechanism which initially triggered the uplift remains, then the air will be cooled to its dew point. Beyond this point, cooling takes place at the slower SALR and the parcel may become warmer than the surrounding air (Figure 9.19). It

will now continue to rise freely, even if the uplifting mechanism is removed, as it is now in an unstable state. Instability is conditional upon the air being forced to rise in the first place, and later becoming saturated so that condensation occurs. The associated weather is usually fine in areas at altitudes below condensation level, but cloudy and showery in those above.

Temperature inversions

As the lapse rate exercises have shown, the temperature of the air usually decreases with altitude, but there are certain conditions when the reverse occurs. **Temperature inversions**, where warmer air overlies colder air, may occur at three levels in the atmosphere. Figure 9.1 showed that temperatures increase with altitude in both the stratosphere and the thermosphere. Inversions can also occur near ground-level and high in the troposphere. High-level inversions are found in depressions where warm air overrides cold air at the warm front or is undercut by colder air at the cold front (page 229). Low-level, or ground, inversions usually occur under anticyclonic conditions (page 234 and Figure 9.24) when there is a rapid loss of heat from the ground due to radiation at night, or when warm air is advected over a cold surface. Under these conditions, fog and frost (page 221 and Figure 9.23) may form in valleys and hollows.

Figure 9.19

Conditional instability

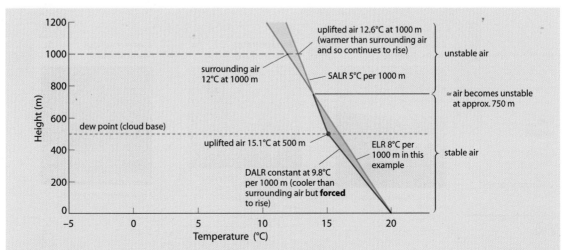

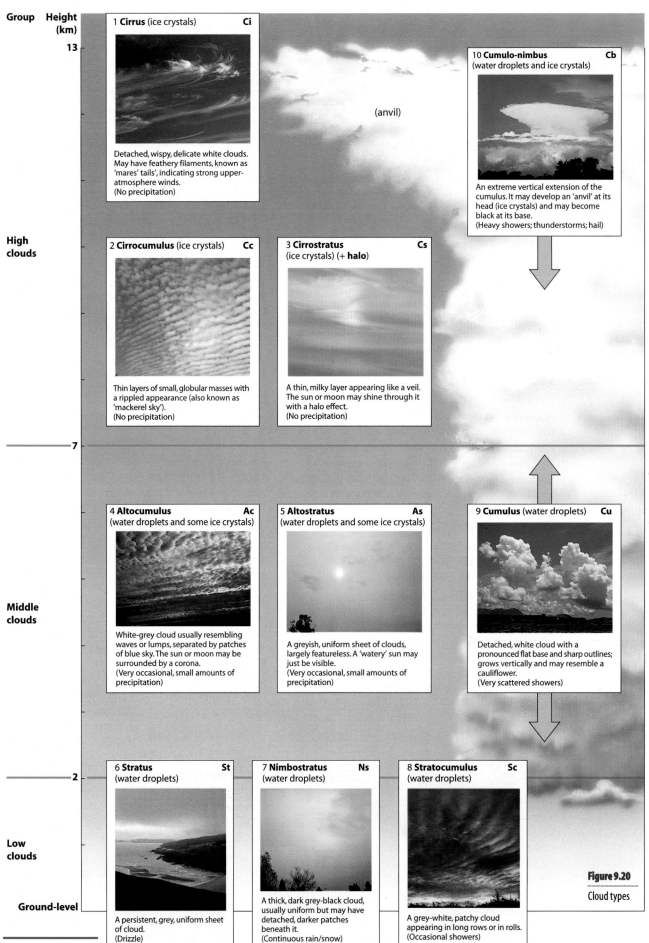

Group	Height (km)

13

High clouds

7

Middle clouds

2

Low clouds

Ground-level

1 Cirrus (ice crystals) **Ci**

Detached, wispy, delicate white clouds. May have feathery filaments, known as 'mares' tails', indicating strong upper-atmosphere winds.
(No precipitation)

(anvil)

10 Cumulo-nimbus **Cb**
(water droplets and ice crystals)

An extreme vertical extension of the cumulus. It may develop an 'anvil' at its head (ice crystals) and may become black at its base.
(Heavy showers; thunderstorms; hail)

2 Cirrocumulus (ice crystals) **Cc**

Thin layers of small, globular masses with a rippled appearance (also known as 'mackerel sky').
(No precipitation)

3 Cirrostratus **Cs**
(ice crystals) (+ **halo**)

A thin, milky layer appearing like a veil. The sun or moon may shine through it with a halo effect.
(No precipitation)

4 Altocumulus **Ac**
(water droplets and some ice crystals)

White-grey cloud usually resembling waves or lumps, separated by patches of blue sky. The sun or moon may be surrounded by a corona.
(Very occasional, small amounts of precipitation)

5 Altostratus **As**
(water droplets and some ice crystals)

A greyish, uniform sheet of clouds, largely featureless. A 'watery' sun may just be visible.
(Very occasional, small amounts of precipitation)

9 Cumulus (water droplets) **Cu**

Detached, white cloud with a pronounced flat base and sharp outlines; grows vertically and may resemble a cauliflower.
(Very scattered showers)

6 Stratus **St**
(water droplets)

A persistent, grey, uniform sheet of cloud.
(Drizzle)

7 Nimbostratus **Ns**
(water droplets)

A thick, dark grey-black cloud, usually uniform but may have detached, darker patches beneath it.
(Continuous rain/snow)

8 Stratocumulus **Sc**
(water droplets)

A grey-white, patchy cloud appearing in long rows or in rolls.
(Occasional showers)

Clouds with vertical height development

Figure 9.20

Cloud types

Clouds

Clouds form when air cools to dew point and vapour condenses into water droplets and/or ice crystals. There are many different types of cloud, but they are often difficult to distinguish as their form constantly changes. The general classification of clouds was proposed by Luke Howard in 1803. His was a descriptive classification, based upon cloud shape and height (Figure 9.20). He used four Latin words: **cirrus** (a lock of curly hair); **cumulus** (a heap or pile); **stratus** (a layer); and **nimbus** (rain-bearing). He also compiled composite names using these four terms, such as cumulo-nimbus, cirrostratus; and added the prefix 'alto-' for middle-level clouds.

Precipitation

Condensation produces minute water droplets, less than 0.05 mm in diameter, or, if the dew point temperature is below freezing, ice crystals. The droplets are so tiny and weigh so little that they are kept buoyant by the rising air currents which created them. So although condensation forms clouds, clouds do not necessarily produce precipitation. As rising air currents are often strong, there has to be a process within the clouds which enables the small water droplets and/or ice crystals to become sufficiently large to overcome the uplifting mechanism and fall to the ground.

There are currently two main theories which attempt to explain the rapid growth of water droplets.

The **ice crystal mechanism** is often referred to as the Bergeron–Findeisen mechanism. It appears that when the temperature of air is between –5°C and –25°C, supercooled water droplets and ice crystals exist together. Super-cooling takes place when water remains in the atmosphere after temperatures have fallen below 0°C – usually due to a lack of condensation nuclei. Ice crystals are in a minority because the **freezing nuclei** necessary for their formation are less abundant than condensation nuclei. The relative humidity of air is ten times greater above an ice surface than over water. This means that the water droplets evaporate and the resultant vapour condenses (sublimates) back onto the ice crystals which then grow into hexagonal-shaped snowflakes. The flakes grow in size – either as a result of further condensation or by fusion as their numerous edges interlock on collision with other flakes. They also increase in number as ice splinters break off and form new nuclei. If the air temperature rises above freezing point as the snow falls to the ground, flakes melt into raindrops. Experiments to produce rainfall artificially by cloud-seeding are based upon this process.

The Bergeron–Findeisen theory is supported by evidence from temperate latitudes where rainclouds usually extend vertically above the freezing level. Radar and high-flying aircraft have reported snow at high altitudes when it is raining at sea-level. However, as clouds rarely reach freezing point in the tropics, the formation of ice crystals is unlikely in those latitudes.

The **collision and coalescence** process was suggested by Longmuir. 'Warm' clouds (i.e. those containing no ice crystals), as found in the tropics, contain numerous water droplets of differing sizes. Different-sized droplets are swept upwards at different velocities and, in doing so, collide with other droplets. It is thought that the larger the droplet, the greater the chance of collision and subsequent coalescence with smaller droplets. When coalescing droplets reach a radius of 3 mm, their motion causes them to disintegrate to form a fresh supply of droplets. The thicker the cloud (cumulo-nimbus), the greater the time the droplets have in which to grow and the faster they will fall, usually as thundery showers.

Latest opinions suggest that these two theories may complement each other, but that a major process of raindrop enlargement has yet to be understood.

Types of precipitation

Although the definition of precipitation includes sleet, hail, dew, hoar frost, fog and rime, only rain and snow provide significant totals in the hydrological cycle.

Rainfall

There are three main types of rainfall, distinguished by the mechanisms which cause the initial uplift of the air. Each mechanism rarely operates in isolation.

1 **Convergent** and **cyclonic** (**frontal**) rainfall results from the meeting of two air streams in areas of low pressure. Within the tropics, the trade winds, blowing towards the Equator, meet at the inter-tropical convergence zone or ITCZ (page 226). The air is forced to rise and, in conjunction with convection currents, produces the heavy afternoon thunderstorms associated with the equatorial climate (page 316). In temperate latitudes, depressions form at the boundary of two air masses. At the associated fronts, warm, moist, less dense air is forced to rise over colder, denser air, giving periods of prolonged and sometimes intense rainfall. This is often augmented by orographic precipitation.

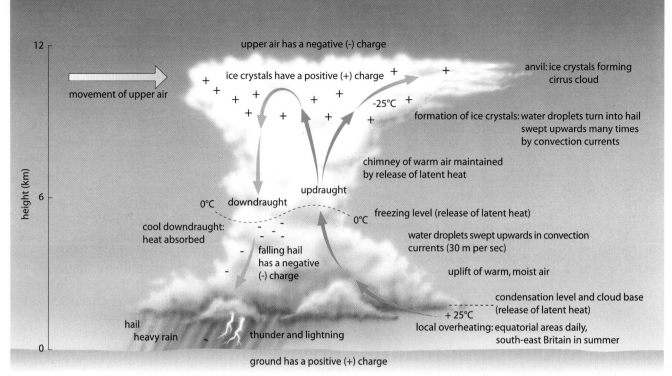

height (km)

12

6

0

movement of upper air

upper air has a negative (-) charge

ice crystals have a positive (+) charge

anvil: ice crystals forming cirrus cloud

-25°C

formation of ice crystals: water droplets turn into hail swept upwards many times by convection currents

chimney of warm air maintained by release of latent heat

updraught

downdraught

0°C

freezing level (release of latent heat)

0°C

cool downdraught: heat absorbed

water droplets swept upwards in convection currents (30 m per sec)

falling hail has a negative (-) charge

uplift of warm, moist air

condensation level and cloud base (release of latent heat)

+ 25°C

local overheating: equatorial areas daily, south-east Britain in summer

hail
heavy rain

thunder and lightning

ground has a positive (+) charge

Figure 9.21

Convectional rainfall: the development of a thunderstorm

2 **Orographic** or **relief** rainfall results when near-saturated, warm maritime air is forced to rise where confronted by a coastal mountain barrier. Mountains reduce the water-holding capacity of rising air by enforced cooling and can increase the amounts of cyclonic rainfall by retarding the speed of depression movement. Mountains also tend to cause air streams to converge and funnel through valleys. Rainfall totals increase where mountains are parallel to the coast, as is the Canadian Coast Range, and where winds have crossed warm offshore ocean currents, as they do before reaching the British Isles. As air descends on the leeward side of a mountain range, it becomes compressed and warmed and condensation ceases, creating a **rainshadow** effect where little rain falls.

3 **Convectional** rainfall occurs when the ground surface is locally overheated and the adjacent air, heated by conduction, expands and rises. During its ascent, the air mass remains warmer than the surrounding environmental air and it is likely to become unstable (page 217) with towering cumulonimbus clouds forming. These unstable conditions, possibly augmented by frontal or orographic uplift, force the air to rise in a 'chimney' (Figure 9.21). The updraught is maintained by energy released as latent heat at both condensation and freezing levels. The cloud summit is characterised by ice crystals in an anvil shape, the top of the cloud being flattened by upper-air movements. When the ice crystals and frozen water droplets, i.e. hail,

become large enough, they fall in a downdraught. The air through which they fall remains cool as heat is absorbed by evaporation. The downdraught reduces the warm air supply to the 'chimney' and therefore limits the lifespan of the storm. Such storms are usually accompanied by thunder and lightning. How storms develop immense amounts of electric charge is still not fully understood. One theory suggests that as raindrops are carried upwards into colder regions, they freeze on the outside. This ice-shell compresses the water inside it until the shell bursts and the water freezes into positively-charged ice crystals while the heavier shell fragments, which are negatively charged, fall towards the cloud base inducing a positive charge on the Earth's surface (Figure 9.21). **Lightning** is the visible discharge of electricity between clouds or between clouds and the ground. **Thunder** is the sound of the pressure wave created by the heating of air along a lightning flash. Convection is one process by which surplus heat and energy from the Earth's surface are transferred vertically to the atmosphere in order to maintain the heat balance (Figure 9.6).

Thunderstorms associated with the so-called **Spanish plume** can affect southern England several times during a hot, sultry summer. They occur when very hot air over the Sierra Nevada mountains (southern Spain) moves northwards over the Bay of Biscay where it draws in cooler, moist air. Should the resultant storm reach Britain, it can cause flash flooding, landslips and electricity blackouts.

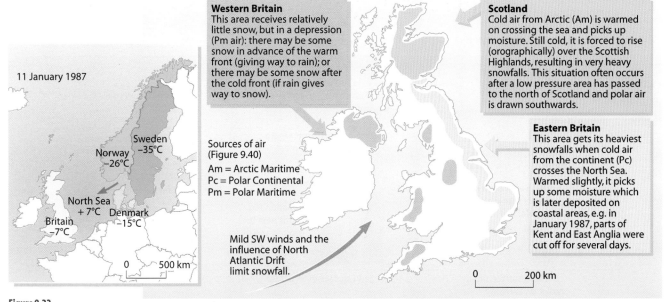

Western Britain
This area receives relatively little snow, but in a depression (Pm air): there may be some snow in advance of the warm front (giving way to rain); or there may be some snow after the cold front (if rain gives way to snow).

Scotland
Cold air from Arctic (Am) is warmed on crossing the sea and picks up moisture. Still cold, it is forced to rise (orographically) over the Scottish Highlands, resulting in very heavy snowfalls. This situation often occurs after a low pressure area has passed to the north of Scotland and polar air is drawn southwards.

11 January 1987

Sweden −35°C
Norway −26°C
North Sea + 7°C   Denmark −15°C
Britain −7°C

0   500 km

Sources of air
(Figure 9.40)
Am = Arctic Maritime
Pc = Polar Continental
Pm = Polar Maritime

Mild SW winds and the influence of North Atlantic Drift limit snowfall.

Eastern Britain
This area gets its heaviest snowfalls when cold air from the continent (Pc) crosses the North Sea. Warmed slightly, it picks up some moisture which is later deposited on coastal areas, e.g. in January 1987, parts of Kent and East Anglia were cut off for several days.

0   200 km

Figure 9.22

Causes of uneven snowfall patterns across Britain

Snow, sleet, glazed frost and hail

Snow forms under similar conditions to rain (Bergeron–Findeisen process) except that as dew point temperatures are under 0°C, then the vapour condenses directly into a solid (sublimation, page 215). Ice crystals will form if hygroscopic or freezing nuclei are present and these may aggregate to give snowflakes. As warm air holds more moisture than cold, snowfalls are heaviest when the air temperature is just below freezing. As temperatures drop, it becomes 'too cold for snow'. Figure 9.22 shows the typical conditions under which snow might fall in Britain.

Sleet is a mixture of ice and snow formed when the upper air temperature is below freezing, allowing snowflakes to form, and the lower air temperature is around 2 to 4°C, which allows their partial melting.

Glazed frost is the reverse of sleet and occurs when water droplets form in the upper air but turn to ice on contact with a freezing surface. When glazed frost forms on roads, it is known as 'black ice'.

Hail is made up of frozen raindrops which exceed 5 mm in diameter. It usually forms in cumulo-nimbus clouds, resulting from the uplift of air by convection currents, or at a cold front. It is more common in areas with warm summers where there is sufficient heat to trigger the uplift of air, and less common in colder climates. Hail frequently proves a serious climatic hazard in cereal-growing areas such as the American Prairies.

Dew, hoar frost, fog and rime

Dew, hoar frost and radiation fog all form under calm, clear, anticyclonic conditions when there is rapid terrestrial radiation at night. Dew point is reached as the air cools by conduction and moisture in the air, or transpired from plants, condenses. If dew point is above freezing, **dew** will form; if it is below freezing, **hoar frost** develops. Frost may also be frozen dew. Dew and hoar frost usually occur within 1 m of ground-level.

If the lower air is relatively warm, moist and contains hygroscopic nuclei, and if the ground cools rapidly, **radiation fog** may form. Where visibility is more than 1 km it is mist, if less than 1 km, fog. In order for radiation fog to develop, a gentle wind is needed to stir the cold air adjacent to the ground so that cooling affects a greater

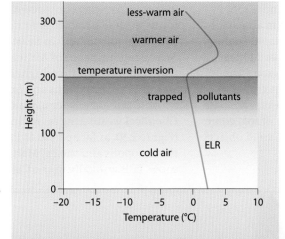

Figure 9.23

Temperature inversion: radiation fog in a valley, Iceland

Figure 9.24

A low-level temperature inversion

less-warm air
warmer air
temperature inversion
trapped   pollutants
cold air
ELR

Height (m)
300
200
100
0
−20   −15   −10   −5   0   5   10
Temperature (°C)

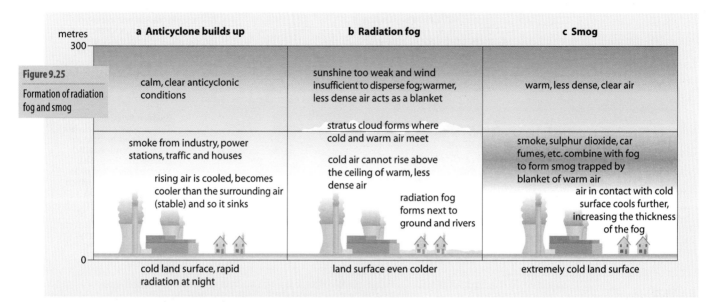

Figure 9.25

Formation of radiation fog and smog

metres 300	**a Anticyclone builds up**	**b Radiation fog**	**c Smog**
	calm, clear anticyclonic conditions	sunshine too weak and wind insufficient to disperse fog; warmer, less dense air acts as a blanket	warm, less dense, clear air
		stratus cloud forms where cold and warm air meet	
	smoke from industry, power stations, traffic and houses	cold air cannot rise above the ceiling of warm, less dense air	smoke, sulphur dioxide, car fumes, etc. combine with fog to form smog trapped by blanket of warm air
	rising air is cooled, becomes cooler than the surrounding air (stable) and so it sinks	radiation fog forms next to ground and rivers	air in contact with cold surface cools further, increasing the thickness of the fog
0	cold land surface, rapid radiation at night	land surface even colder	extremely cold land surface

Figure 9.26

Rime frost, Woodlands St Mary, Berkshire

thickness of air. Radiation fogs usually occur in valleys, are densest around sunrise, and consist of droplets which are sufficiently small to remain buoyant in the air. Fog is likely to thicken if temperature inversion takes place (Figures 9.23 and 9.24), i.e. when cold surface air is trapped by overlying warmer, less dense air. It is under such conditions, in urban and industrial areas, that smoke and other pollutants released into the air are retained as smog (Figures 9.25 and 15.58).

Advection fog forms when warm air passes over or meets with cold air to give rapid cooling. In the coastal Atacama Desert (Places 24, page 180), sufficient droplets fall to the ground as 'fog-drip' to enable some vegetation growth.

Rime (Figure 9.26) occurs when supercooled droplets of water, often in the form of fog, come into contact with, and freeze upon, solid objects such as telegraph poles and trees.

Acid rain

This is an umbrella term for the presence in rainfall of a series of pollutants which are produced mainly by the burning of fossil fuels. Coal-fired power stations, heavy industry and vehicle exhausts emit sulphur dioxide and nitrogen oxides. These are carried by prevailing winds across seas and national frontiers to be deposited either directly onto the Earth's surface as dry deposition or to be converted into acids (sulphuric and nitric acid) which fall to the ground in rain as wet deposition. Clean rainwater has a pH value of between 5 and 6 (Figure 10.15). Today, rainfall over most of north-west Europe has a pH value of between 4 and 5; the lowest ever recorded was 2.4.

The effects of acid rain include an increase in levels of water acidity. This has caused the death of fish and plant life in many Scandinavian rivers and lakes, and the pollution of fresh water supplies. Forests are being destroyed as important soil nutrients (calcium and potassium) are washed away, to be replaced by manganese and aluminium which are harmful to root growth. In time, trees shed their needles and leaves as they become less resistant to drought, frost and disease. Forecasts suggest that 90 per cent of Germany's trees may have died by early in the 21st century, with consequential effects on forest wildlife habitats. Acid rain has been linked with a decline in human health as seen by the increasing incidence of Alzheimer's disease (which may result from higher concentrations of aluminium), bronchitis and lung cancer. As soils become more acidic, crop yields are likely to fall. Chemical weathering is eroding buildings (Figure 2.9) as far apart as St Paul's Cathedral, the Acropolis and the Taj Mahal, while 'black snow' falls annually in the Cairngorms.

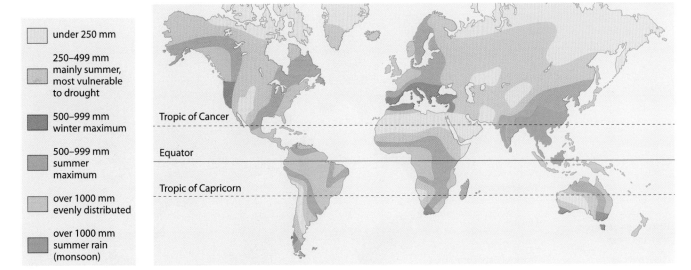

under 250 mm	
250–499 mm mainly summer, most vulnerable to drought	
500–999 mm winter maximum	
500–999 mm summer maximum	
over 1000 mm evenly distributed	
over 1000 mm summer rain (monsoon)	

Tropic of Cancer

Equator

Tropic of Capricorn

Figure 9.27

World precipitation: mean annual totals and seasonal distribution

World precipitation: distribution and reliability

Geographers are interested in describing distributions and in identifying and accounting for any resultant patterns. Where precipitation is concerned, geographers have, in the past, concentrated upon long-term distributions which show either mean annual amounts or seasonal variations. Long-term fluctuations vary considerably across the globe but, nevertheless, a map showing world precipitation does show identifiable patterns (Figure 9.27).

Equatorial areas have high annual rainfall totals due to the continuous uplift of air resulting from the convergence of the trade winds and strong convectional currents (page 226). The presence of the ITCZ ensures that rain falls throughout the year. Further away from the Equator, rainfall totals decrease and the length of the dry season increases. These tropical areas, especially those inland, experience convectional rainfall in summer, when the sun is overhead, followed by a dry winter. Latitudes adjacent to the tropics receive minimal amounts as they correspond to areas of high pressure caused by subsiding, and therefore warming, air (Figure 7.2).

To the poleward side of this arid zone, rainfall quantities increase again and the length of the dry season decreases. These temperate latitudes receive large amounts of rainfall, spread evenly throughout the year, due to cyclonic conditions and local orographic effects. Towards the polar areas, where cold air descends to give stable conditions, precipitation totals decrease and rain gives way to snow. Between 30° and 40° north and south (in the west of continents) the Mediterranean climate is characterised by winter rain and summer drought. This general latitudinal zoning of rainfall is interrupted locally by the apparent movement of the overhead sun,

the presence of mountain ranges or ocean currents, the monsoon, and continentality (distance from the sea).

More recently, geographers have become increasingly concerned with shorter-term variations. In many parts of the world, economic development and lifestyles are more closely linked to the duration, intensity and reliability of rainfall than to annual amounts. Precipitation is more valuable when it falls during the growing season (Canadian Prairies) and less effective if it occurs when evapotranspiration rates are at their highest (Sahel countries). In the same way, lengthy episodes of steady rainfall as experienced in Britain provide a more beneficial water supply than storms of a short and intensive duration which occur in tropical semi-arid climates. This is because moisture penetrates the soil more gradually and the risks of soil erosion, flooding and water shortages are reduced.

Of utmost importance is the reliability of rainfall. There appears to be a strong positive correlation (Framework 19, page 635) between rainfall totals and rainfall reliability – i.e. as rainfall totals increase, so too does rainfall reliability. In Britain and the Amazon Basin, rainfall is reliable with relatively little variation in annual totals from year to year (Figure 9.28).

Elsewhere, especially in monsoon or tropical continental climates, there is a pronounced wet and dry season. Consequently, if the rains fail one year, the result can be disastrous for crops, and possibly also for animals and people. The most vulnerable areas, such as north-east Brazil and the Sahel countries, lie near to desert margins (Figure 9.28). Here, where even a small variation of 10 per cent below the mean can be critical, many places often experience a variation in excess of 30 per cent.

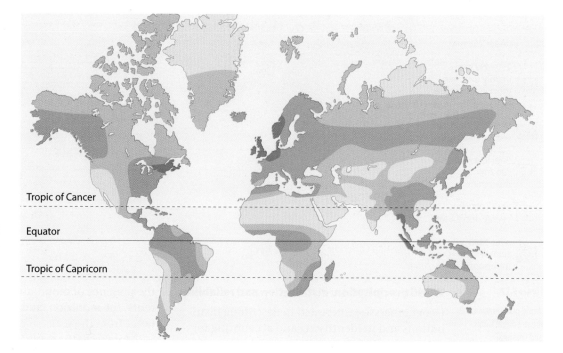

Figure 9.28

World rainfall
reliability

Tropic of Cancer

Equator

Tropic of Capricorn

Percentage departure
from the mean

over 30

21–30

11–20

10 and under

Atmospheric motion

The movement of air in the atmospheric system
may be vertical (i.e. rising or subsiding) or horizontal; in the latter case it is commonly known as
wind. Winds result from differences in air pressure which in turn may be caused by differences in
temperature and the force exerted by gravity, as
pressure decreases rapidly with height (Figure 9.1).
An increase in temperature causes air to heat,
expand, become less dense and rise, creating an
area of low pressure below. Conversely, a drop in
temperature produces an area of high pressure.
Differences in pressure are shown on maps by
isobars, which are lines joining places of equal
pressure. To draw isobars, pressure readings are
normally reduced to represent pressure at sea-
level. Pressure is measured in **millibars** (mb) and
it is usual for isobars to be drawn at 4-mb inter-

vals. Average pressure at sea-level is 1013 mb.
However, the isobar pattern is usually more
important in terms of explaining the weather
than the actual figures. The closer together the
isobars, the greater the difference in pressure – **the
pressure gradient** – and the stronger the wind.
Wind is nature's way of balancing out differences
in pressure as well as temperature and humidity.

Figure 9.29 shows the two basic pressure
systems which affect the British Isles. In addition
to the differences in pressure, wind speed and
wind direction, the diagrams also show that winds
blow neither directly at right-angles to the isobars
along the pressure gradient, nor parallel to them.
This is due to the effects of the Coriolis force and
of friction, which are additional to the forces
exerted by the pressure gradient and gravity.

The Coriolis force

If the Earth did not rotate and was composed
entirely of either land or water, there would be
one large convection cell in each hemisphere
(Figure 9.30). Surface winds would be parallel to
pressure gradients and would blow directly from
high to low pressure areas. In reality, the Earth
does rotate and the distribution of land and sea is
uneven. Consequently, more than one cell is
created (Figures 9.34 and 9.35) as rising air,
warmed at the Equator, loses heat to space – there
is less cloud cover to retain it – and as it travels
further from its source of heat. A further consequence is that moving air appears to be deflected
to the right in the northern hemisphere and to
the left in the southern hemisphere. This is a
result of the Coriolis force.

Figure 9.29

The two basic pressure
systems affecting
Britain

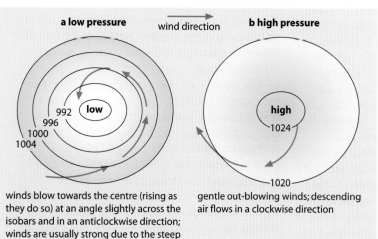

a low pressure   wind direction   **b high pressure**

992
996
1000
1004
low

high
1024

1020

winds blow towards the centre (rising as
they do so) at an angle slightly across the
isobars and in an anticlockwise direction;
winds are usually strong due to the steep
pressure gradient

gentle out-blowing winds; descending
air flows in a clockwise direction

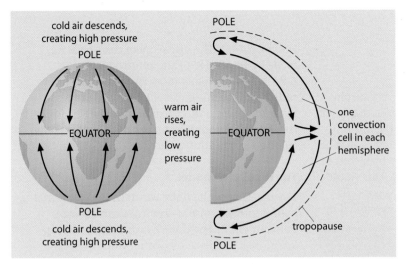

Figure 9.30

Air movement on a rotation-free Earth

Imagine that Person A stands in the centre of a large rotating disc and throws a ball to Person B, standing on the edge of that moving disc. As Person A watches, the ball appears to take a curved path away from Person B – due to the fact that, while the ball is in transit, Person B has been moved to a new position by the rotation of the disc (Figure 9.31). Similarly, the Earth's rotation through 360° every 24 hours means that a wind blowing in a northerly direction in the northern hemisphere appears to have been diverted to the right on a curved trajectory by 15° of longitude for every hour (though to an astronaut in a space

shuttle, the path would look straight). This helps to explain why the prevailing winds blowing from the tropical high pressure zone approach Britain from the south-west rather than from the south. In theory, if the Coriolis force acted alone, the resultant wind would blow in a circle.

Winds in the upper troposphere, unaffected by friction with the Earth's surface, show that there is a balance between the forces exerted by the pressure gradient and the Coriolis deflection. The result is the **geostrophic wind** which blows parallel to isobars (Figure 9.32). The existence of the geostrophic wind was recognised in 1857 by a Dutchman, Buys Ballot, whose law states that 'if you stand, in the northern hemisphere, with your back to the wind, low pressure is always to your left and high pressure to your right'.

Friction, caused by the Earth's surface, upsets the balance between the pressure gradient and the Coriolis force by reducing the effect of the latter. As the pressure gradient becomes relatively more important when friction is reduced with altitude, the wind blows across isobars towards the low pressure (Figure 9.29). Deviation from the geostrophic wind is less pronounced over water because its surface is smoother than that of land.

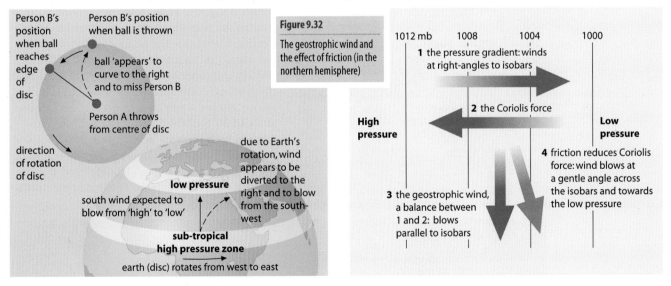

Figure 9.32

The geostrophic wind and the effect of friction (in the northern hemisphere)

Figure 9.31

The Coriolis force in the northern hemisphere

A hierarchy of atmospheric motion

An appreciation of the movement of air is fundamental to an understanding of the workings of the atmosphere and its effects on our weather and climate. The extent to which atmospheric motion influences local weather and climate depends on winds at a variety of scales and their interaction in a hierarchy of patterns. One such hierarchy, which is useful in studying the influence of atmospheric motion, was suggested by B. W. Atkinson in 1988.

Although defining four levels, he stressed that there were important interrelationships between each (Figure 9.33).

Scale	Characteristic horizontal size (km)	Systems
1 Planetary	5000–10 000	Rossby waves, ITCZ
2 Synoptic (macro)	1000–5000	Monsoons, hurricanes, depressions, anticyclones
3 Meso-scale	10–1000	Land and sea breezes, mountain and valley winds, föhn, thunderstorms
4 Small (micro)	0.1–10	Smoke plumes, urban turbulence

Weather and climate

Figure 9.33

A hierarchy of atmosphere motion systems (*after* Atkinson, 1988)

Planetary scale: atmospheric circulation

It has already been shown that there is a surplus of energy at the Equator and a deficit in the outer atmosphere and nearer to the poles (Figure 9.6). Therefore, theoretically, surplus energy should be transferred to areas with a deficiency by means of a single convective cell (Figure 9.30). This would be the case for a non-rotating Earth, a concept first advanced by Halley (1686) and expanded by Hadley (1735). The discovery of three cells was made by Ferrel (1856) and refined by Rossby (1941). Despite many modern advances using **radiosonde** readings, satellite imagery and computer modelling, this tricellular model still forms the basis of our understanding of the general circulation of the atmosphere.

The tricellular model

The meeting of the trade winds in the equatorial region forms the Inter-tropical Convergence Zone, or **ITCZ**. The trade winds, which pick up latent heat as they cross warm, tropical oceans, are forced to rise by violent convection currents. The unstable, warm, moist air is rapidly cooled adiabatically to produce the towering cumulonimbus clouds, frequent afternoon thunderstorms and low pressure characteristic of the equatorial climate (page 316). It is these strong upward currents that form the 'powerhouse of the general global circulation' and which turn latent heat first into sensible heat and later into potential energy. At ground-level, the ITCZ experiences only very gentle, variable winds known as the **doldrums**.

As rising air cools to the temperature of the surrounding environmental air, uplift ceases and it begins to move away from the Equator. Further cooling, increasing density, and diversion by the Coriolis force cause the air to slow down and to subside, forming the descending limb of the **Hadley cell** (Figures 9.34 and 9.35). In looking at the northern hemisphere (the southern is its mirror image), it can be seen that the air subsides at about 30°N of the Equator to create the sub-tropical high pressure belt with its clear skies and dry, stable conditions (Figure 9.36). On reaching the Earth's surface, the cell is completed as some of the air is returned to the Equator as the northeast trade winds.

Figure 9.34

Tricellular model showing atmospheric circulation in the northern hemisphere

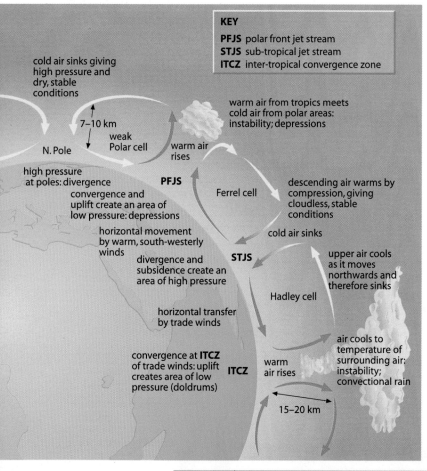

KEY
PFJS polar front jet stream
STJS sub-tropical jet stream
ITCZ inter-tropical convergence zone

cold air sinks giving high pressure and dry, stable conditions

7–10 km

N. Pole   weak Polar cell

warm air rises

warm air from tropics meets cold air from polar areas: instability; depressions

high pressure at poles: divergence

PFJS

convergence and uplift create an area of low pressure: depressions

Ferrel cell

descending air warms by compression, giving cloudless, stable conditions

horizontal movement by warm, south-westerly winds

cold air sinks

divergence and subsidence create an area of high pressure

STJS

upper air cools as it moves northwards and therefore sinks

Hadley cell

horizontal transfer by trade winds

convergence at **ITCZ** of trade winds: uplift creates area of low pressure (doldrums)

ITCZ

warm air rises

air cools to temperature of surrounding air: instability; convectional rain

15–20 km

Figure 9.35

Tricellular model to show atmospheric circulation in the northern hemisphere and within the tropopause

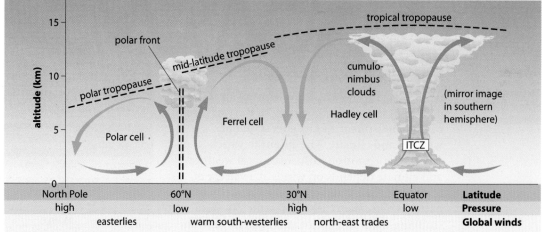

tropical tropopause

polar front

mid-latitude tropopause

cumulo-nimbus clouds

polar tropopause

altitude (km)

15

10

5

0

Polar cell

Ferrel cell

Hadley cell

(mirror image in southern hemisphere)

ITCZ

North Pole	60°N	30°N	Equator	**Latitude**
high	low	high	low	**Pressure**
easterlies	warm south-westerlies	north-east trades		**Global winds**

The remaining air is diverted polewards, forming the warm south-westerlies which collect moisture when they cross sea areas. These warm winds meet cold Arctic air at the polar front (about 60°N) and are uplifted to form an area of low pressure and the rising limb of the **Ferrel** and **Polar cells** (Figures 9.34 and 9.35). The resultant unstable conditions produce the heavy cyclonic rainfall associated with mid-latitude **depressions**. Depressions are another mechanism by which surplus heat is transferred. While some of this rising air eventually returns to the tropics, some travels towards the poles where, having lost its heat, it descends to form another stable area of high pressure. Air returning to the polar front does so as the cold easterlies.

This overall pattern is affected by the apparent movement of the overhead sun to the north and south of the Equator. This movement causes the seasonal shift of the heat Equator, the ITCZ, the equatorial low pressure zone and global wind and rainfall belts. Any variation in the characteristics of the ITCZ – i.e. its location or width – can have drastic consequences for the surrounding climates, as seen in the Sahel droughts of the early 1970s and most of the 1980s (Case Study 7).

Figure 9.36

Image taken by the Meteosat geosynchronous satellite. Notice the clouds resulting from uplift at the ITCZ (not a continuous belt), the clear skies over the Sahara, the polar front over the north Atlantic, and a depression over Britain

Rossby waves and jet streams

Evidence of strong winds in the upper troposphere first came when First World War Zeppelins were blown off-course, and several inter-war balloons were observed travelling at speeds in excess of 200 km/hr. Pilots in the Second World War, flying at heights above 8 km, found eastward flights much faster and their return westward journeys much slower than expected, while north–south flights tended to be blown off-course. The explanation was found to be a belt of upper-air westerlies, the **Rossby waves**, which often follow a meandering path (Figure 9.37a). The number of meanders, or waves, varies season-ally, with usually four to six in summer and three in winter. These waves form a complete pattern around the globe (Figure 9.37b and c).

Further investigation has shown that the velocity of these upper westerlies is not internally uniform. Within them are narrow bands of extremely fast-moving air known as **jet streams**. Jet streams, which help in the rapid transfer of energy, can exceed speeds of 230 km/hr, which is sufficient to carry a balloon, or ash from a volcano, around the Earth within a week or two (Figure 9.39 and Case Study 1). Of five recognisable jet streams, two are particularly significant, with a third having seasonal importance.

The **polar front jet stream** (PFJS, Figure 9.34) varies between latitudes 40° and 60° in both hemispheres and forms the division between the Ferrel and Polar cells, i.e. the boundary between warm tropical and cold polar air. The PFJS varies in extent, location and intensity and is mainly responsible for giving fine or wet weather on the Earth's surface. Where, in the northern hemisphere, the jet stream moves south (Figure 9.38a), it brings with it cold air which descends in a clockwise direction to give dry, stable conditions associated with areas of high pressure (**anticyclones**, page 234). When the now-warmed jet stream backs northwards, it takes with it warm air which rises in an anticlockwise direction to give the strong winds and heavy rainfall associated with areas of low pressure (**depressions**, page 230). As the usual path of the PFJS over Britain is oblique – i.e. towards the north-east – this accounts for our frequent wet and windy weather. Occasionally, this path may be temporarily altered by a stationary or **blocking anticyclone** (Figures 9.38b and 9.48) which may produce extremes of climate such as the hot, dry summers of 1976 and 1989 or the cold January of 1987.

The **subtropical jet stream**, or STJS, occurs about 25° to 30° from the Equator and forms the boundary between the Hadley and Ferrel cells (Figures 9.34 and 9.35). This meanders less than the PFJS, has lower wind velocities, but follows a similar west–east path.

The **easterly equatorial jet stream** is more seasonal, being associated with the summer monsoon of the Indian subcontinent (page 239).

Figure 9.37

Rossby waves and jet streams (northern hemisphere)

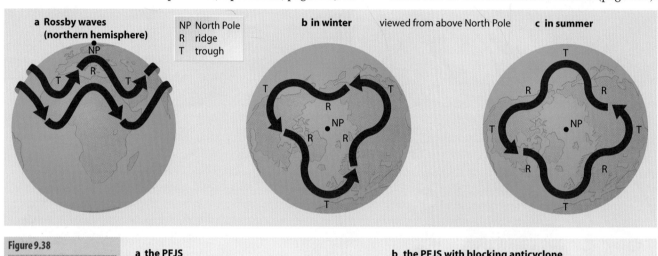

Figure 9.38

The polar front jet stream (PFJS) (northern hemisphere)

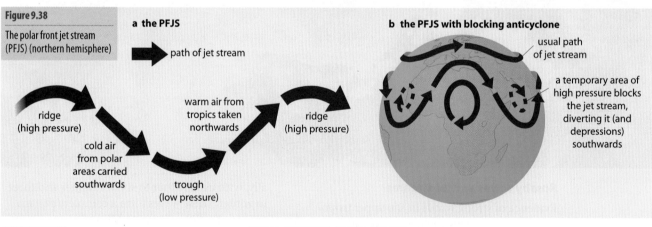

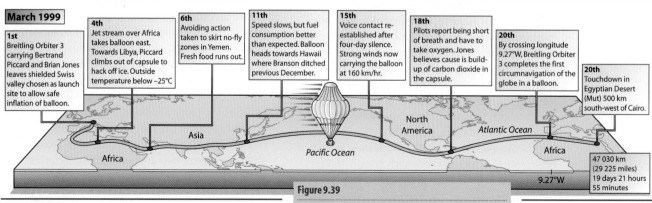

Figure 9.39

Orbiter 3's record breaking (in terms of distance, duration and time) first balloon flight around the world

Macro-scale: synoptic systems

The concept of air masses is important because air masses help to categorise world climate types (Chapter 12). In regions where one air mass is dominant all year, there is little seasonal variation in weather, for example at the tropics and at the poles. Areas such as the British Isles, where air masses constantly interchange, experience much greater seasonal and diurnal variation in their weather.

Air masses and fronts: how they affect the British Isles

If air remains stationary in an area for several days, it tends to assume the temperature and humidity properties of that area. Stationary air is mainly found in the high pressure belts of the subtropics (the Azores and the Sahara) and in high latitudes (Siberia and northern Canada). The areas in which homogeneous air masses develop are called **source regions**. Air masses can be classified according to:

- the **latitude** in which they originate, which determines their temperature – Arctic (*A*), Polar (*P*), or Tropical (*T*)
- the nature of the **surface** over which they develop, which affects their moisture content – maritime (*m*), or continental (*c*).

The five major air masses which affect the British Isles at various times of the year (*Am*, *Pm*, *Pc*, *Tm* and *Tc*) are derived by combining these characteristics of latitude and humidity (Figure 9.40). When air masses move from their source region they are modified by the surface over which they pass and this alters their tem-perature, humidity and stability. For example, tropical air moving northwards is cooled and becomes more stable, while polar air moving south becomes warmer and increasingly unstable. Each air mass therefore brings its own characteristic weather conditions to the British Isles. The general conditions expected with each air mass are given in Figure 9.41. However, it should be remembered that each air mass is unique and dependent on: the climatic conditions in the source region at the time of its development; the path which it subsequently follows; the season in which it occurs; and, since it has a three-dimensional form, the vertical characteristics of the atmosphere at the time.

When two air masses meet, they do not mix readily, due to differences in temperature and density. The point at which they meet is called a **front**. A **warm front** is found where warm air is advancing and being forced to override cold air. A **cold front** occurs when advancing cold air undercuts a body of warm air. In both cases, the rising air cools and usually produces clouds, easily seen on satellite weather photographs (Figures 9.67 and 9.68); these clouds often generate precipitation. Fronts may be several hundred kilometres wide and they extend at relatively gentle gradients up into the atmosphere. The most notable type of front, the **polar front**, occurs when warm, moist, *Tm* air meets colder, drier, *Pm* air. It is at the polar front that **depressions** form. Depressions are areas of low pressure. They form most readily over the oceans in mid-latitudes, and track eastwards bringing cloud and rain to western margins of continents.

Figure 9.40

Air masses that affect the British Isles

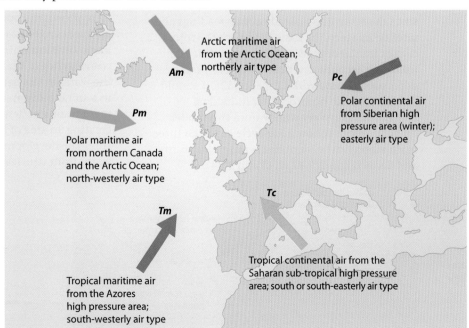

Arctic maritime air from the Arctic Ocean; northerly air type

Am

Pc

Polar continental air from Siberian high pressure area (winter); easterly air type

Pm

Polar maritime air from northern Canada and the Arctic Ocean; north-westerly air type

Tc

Tm

Tropical continental air from the Saharan sub-tropical high pressure area; south or south-easterly air type

Tropical maritime air from the Azores high pressure area; south-westerly air type

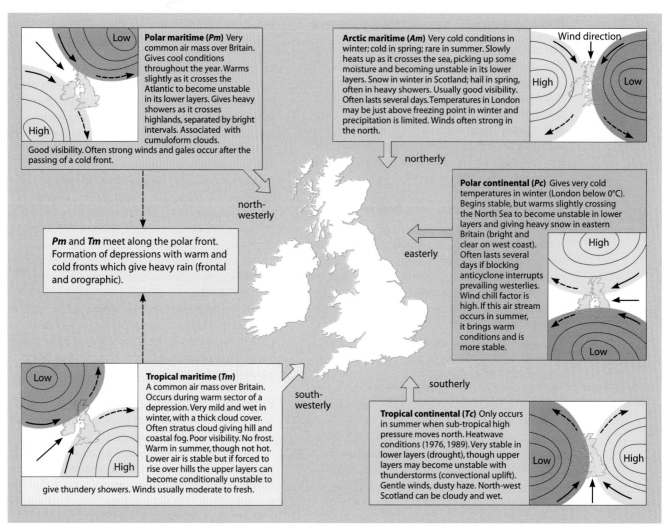

Polar maritime (Pm) Very common air mass over Britain. Gives cool conditions throughout the year. Warms slightly as it crosses the Atlantic to become unstable in its lower layers. Gives heavy showers as it crosses highlands, separated by bright intervals. Associated with cumuloform clouds. Good visibility. Often strong winds and gales occur after the passing of a cold front.

Arctic maritime (Am) Very cold conditions in winter; cold in spring; rare in summer. Slowly heats up as it crosses the sea, picking up some moisture and becoming unstable in its lower layers. Snow in winter in Scotland; hail in spring, often in heavy showers. Usually good visibility. Often lasts several days. Temperatures in London may be just above freezing point in winter and precipitation is limited. Winds often strong in the north.

Wind direction

north-westerly

northerly

Pm and **Tm** meet along the polar front. Formation of depressions with warm and cold fronts which give heavy rain (frontal and orographic).

easterly

Polar continental (Pc) Gives very cold temperatures in winter (London below 0°C). Begins stable, but warms slightly crossing the North Sea to become unstable in lower layers and giving heavy snow in eastern Britain (bright and clear on west coast). Often lasts several days if blocking anticyclone interrupts prevailing westerlies. Wind chill factor is high. If this air stream occurs in summer, it brings warm conditions and is more stable.

Tropical maritime (Tm) A common air mass over Britain. Occurs during warm sector of a depression. Very mild and wet in winter, with a thick cloud cover. Often stratus cloud giving hill and coastal fog. Poor visibility. No frost. Warm in summer, though not hot. Lower air is stable but if forced to rise over hills the upper layers can become conditionally unstable to give thundery showers. Winds usually moderate to fresh.

south-westerly

southerly

Tropical continental (Tc) Only occurs in summer when sub-tropical high pressure moves north. Heatwave conditions (1976, 1989). Very stable in lower layers (drought), though upper layers may become unstable with thunderstorms (convectional uplift). Gentle winds, dusty haze. North-west Scotland can be cloudy and wet.

Depressions

The polar front theory was put forward by a group of Norwegian meteorologists in the early 1920s. Although some aspects have been refined since the innovation of radiosonde readings and satellite imagery, the basic model for the formation of frontal depressions remains valid. The following account describes a 'typical' or 'model' depression (Framework 12, page 352). It should be remembered, however, that individual depressions may vary widely from this model.

Depressions follow a life-cycle in which three main stages can be identified: embryo, maturity and decay (Figures 9.42, 9.43 and 9.44).

1 The **embryo depression** begins as a small wave on the polar front. It is here that warm, moist, tropical (*Tm*) air meets colder, drier, polar (*Pm*) air (Figure 9.42). Recent studies have shown the boundary between the two air masses to be a zone rather than the simple linear division claimed in early models. The convergence of the two air masses results in the warmer, less dense air being forced to rise in a spiral movement. This upward movement results in 'less' air at the Earth's surface, creating an area of below-average or low pressure. The developing depression, with its warm front (the leading edge of the tropical

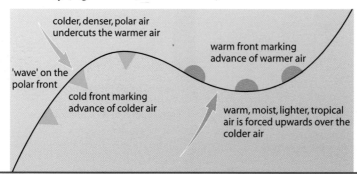

colder, denser, polar air undercuts the warmer air

warm front marking advance of warmer air

'wave' on the polar front

cold front marking advance of colder air

warm, moist, lighter, tropical air is forced upwards over the colder air

air) and cold front (the leading edge of the polar air), usually moves in a north-easterly direction under the influence of the upper westerlies, i.e. the polar front jet stream.

2 A **mature depression** is recognised by the increasing amplitude of the initial wave (Figure 9.43). Pressure continues to fall as more warm air, in the warm sector, is forced to rise. As pressure falls and the pressure gradient steepens, the inward-blowing winds increase in strength. Due to the Coriolis force (page 224), these anticlockwise-blowing winds come from the south-west. As the relatively warm air of the warm sector continues to rise along the warm front, it eventually cools to dew point. Some of its vapour will condense to release large amounts of latent heat, and clouds will develop. Continued uplift and cooling will cause precipitation as the clouds become both thicker and lower.

Satellite photographs have shown that there is likely to be a band of 'meso-scale precipitation' extending several hundred kilometres in length and up to 150 km in width along, and just in front of, a warm front. As temperatures rise and the uplift of air decreases within the warm sector, there is less chance of precipitation and the low cloud may break to give some sunshine. The cold front moves faster and has a steeper gradient than the warm front (Figure 9.45).

Progressive undercutting by cold air at the rear of the warm sector gives a second episode of precipitation – although with a greater intensity and a shorter duration than at the warm front. This band of meso-scale precipitation may be only 10–50 km in width. Although the air behind the cold front is colder than that in advance of the warm front (having originated in and travelled through more northerly latitudes), it becomes unstable, forming cumulo-nimbus clouds and heavy showers. Winds often reach their maximum strength at the cold front and change to a more north-westerly direction after its passage (Figure 9.45).

3 The depression begins to **decay** when the cold front catches up the warm front to form an **occlusion** or **occluded front** (Figure 9.44). By this stage, the *Tm* air will have been squeezed upwards leaving no warm sector at ground-level. As the uplift of air is reduced, so too are (or will be) the amount of condensation, the release of latent heat and the amount and pattern of precipitation – there may be only one episode of rain. Cloud cover begins to decrease, pressure rises and wind speeds decrease as the colder air replaces the uplifted air and 'infills' the depression.

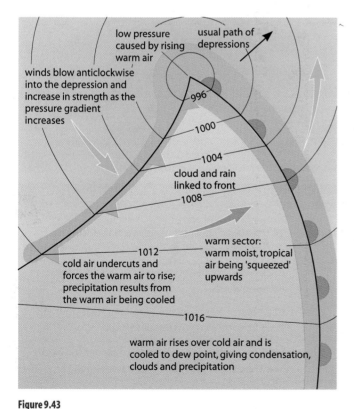

Figure 9.43

Life-cycle of a depression: Stage 2 – maturity

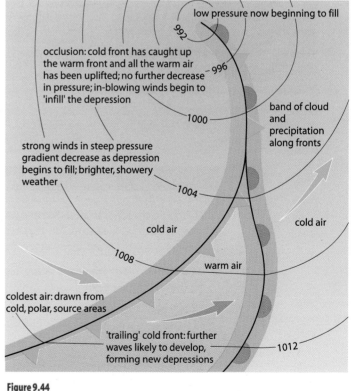

Figure 9.44

Life-cycle of a depression: Stage 3 – decay

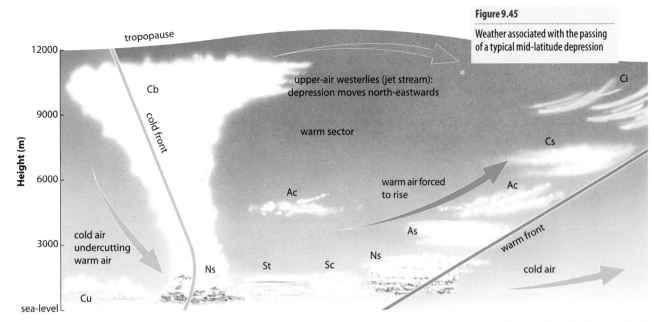

read from right to left (i.e. from 1 to 5)

	5. Behind the cold front	4. Passing of the cold front	3. Warm sector	2. Passing of the warm front	1. Approach of depression
Pressure	rise continues more slowly	sudden rise	steady	fall ceases	steady fall
Wind direction	NW	veers from SW to NW	SW	veers from SSE to SW	SSE
Wind speed	squally; speed slowly decreases (e.g. force 3–6)	very strong to gale force (e.g. force 6–8)	decreases (e.g. force 2–4)	strong (e.g. force 5–6)	slowly increases (e.g. force 1–3)
Temperature (e.g. winter)	cold (e.g. 3°C)	sudden decrease	warm/mild (e.g. 10°C)	sudden rise	cool (e.g. 6°C)
Relative humidity	rapid fall	high during precipitation	steady and high	high during precipitation	slow rise
Cloud (Figure 9.20)	decreasing; in succession, Cb and Cu	very thick and towering Cb	low or may clear; St, Sc, Ac	low and thick Ns	high and thin; in succession, Ci, Cs, Ac, As
Precipitation	heavy showers	short period of heavy rain or hail	drizzle or stops raining	continuous rainfall, steady and quite heavy	none
Visibility	very good; poor in showers	poor	often poor	decreases rapidly	good but beginning to decrease

Places 29   South-east England: 'The Great Storm', 16 October 1987

This storm, the worst to affect south-east England since 1703, developed so rapidly that its severity was not predicted in advance weather forecasts.

11 October: High winds and heavy rain forecast for the end of the week.

15 October 1200 hrs: Depression expected to move along the English Channel with fresh to strong winds.

2130 hrs: TV weather forecast: strong winds gusting to 50 km/hr.

16 October 0030 hrs: Radio weather forecast: warning of severe gales.

0130 hrs: Police and fire services alerted about extreme winds.

0500 hrs: Winds reach 94 km/hr at Heathrow and 100 km/hr on parts of the south coast.

0800 hrs: Centre of depression reaches the North Sea. Winds over southern England dropped to 50–70 km/hr.

1200 hrs: 'The Great Storm' was over.

Figure 9.46

'The Great Storm',
0000 hrs,
16 October 1987

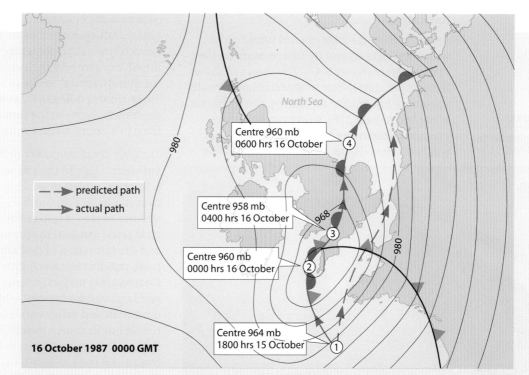

The storm began on 15 October as a small wave on a cold front in the Bay of Biscay, where the few weather ships give only limited information. It was caused by contact between very warm air from Africa and cold air from the North Atlantic. It appeared to be a 'typical' depression until, at about 1800 hrs on 15 October, it unexpectedly deepened giving a central pressure reading of 958 mb and creating an except-ionally steep pressure gradient. The exact cause of this is unknown but it was believed to result from a combination of an exceptionally strong jet stream (initiated on 13 October by air spiralling upwards along the east coast of North America in Hurricane Floyd) and extreme warming over the Bay of Biscay (see hurricanes, page 235). Together, these could have caused an excessive release of latent heat energy which North American meteorologists compare with the effect of detonating a bomb. It was this unpredicted deepening, combined with the change of direction from the English Channel towards the Midlands, which caught experts by surprise.

The depression moved rapidly across southern England, clearing the country in six hours (Figure 9.46). Winds remained light in and around the centre (Birmingham 13 km/hr), but the strong pressure gradient on its southern flank resulted in severe winds from Portland Bill (102 km/hr, gusting to 141 km/hr) to Dover (115 km/hr, gusting to 167 km/hr).

Although the storm passed within a few hours, and luckily during the night when most people were asleep, it left a trail of death and destruction. There were 16 deaths; several houses collapsed and many others lost walls, windows and roofs; an estimated 15 million trees were blown over, blocking railways and roads; one-third of the trees in Kew Gardens were destroyed; power lines were cut and, in some remote areas, not restored for several days; few commuters managed to reach London the next day; a ferry was blown ashore at Folkestone; and insurance claims set an all-time record.

Once every 50 years, winds exceeding 100 km/hr with gusts of over 165 km/hr can be expected north of a line from Cornwall to Durham, and even stronger winds, gusting to 185 km/hr, once in 20 years in western and northern Scotland. The winds associated with 'the Great Storm' were remarkable not so much for their strength as for their occurrence over south-east England. Here, the predicted return period can be measured in centuries rather than decades.

Anticyclones

An anticyclone is a large mass of subsiding air which produces an area of high pressure on the Earth's surface (Figure 9.47). The source of the air is the upper atmosphere, where amounts of water vapour are limited. On its descent, the air warms at the DALR (page 216), so dry conditions result. Pressure gradients are gentle, resulting in weak winds or calms (Figure 9.29b). The winds blow outwards and clockwise in the northern hemisphere. Anticyclones may be 3000 km in diameter – much larger than depressions – and, once established, can give several days or, under extreme conditions, several weeks, of settled weather. There are also differences, again unlike in a depression, between the expected weather conditions in a summer and a winter anticyclone.

Weather conditions over Britain

Summer   Due to the absence of cloud, there is intense insolation which gives hot, sunny days (up to 30°C in southern England) and an absence of rain. Rapid radiation at night, under clear skies, can lead to temperature inversions and the formation of dew and mist, although these rapidly clear the following morning. Coastal areas may experience advection fogs and land and sea breezes, while highlands have mountain and valley winds (pages 240–241). If the air has its source over North Africa – that is, if it is a *Tc* air mass (Figure 9.40) – then heatwave conditions tend to result. Often, after several days of increasing thermals, there is an increased risk of thunderstorms and the so-called Spanish plume (page 220).

Winter   Although the sinking air again gives cloudless skies, there is little incoming radiation during the day due to the low angle of the sun. At night, the absence of clouds means low temperatures and the development of fog and frost. These may take a long time to disperse the next day in the weak sunshine. Polar continental (*Pc*) air (Figure 9.40), with its source in central Asia and a slow movement over the cold European land mass, is cold, dry and stable until it reaches the North Sea where its lower layers acquire some warmth and moisture. This can cause heavy snowfalls on the east coast (Figure 9.22).

Blocking anticyclones

These occur when cells of high pressure detach themselves from the major high pressure areas of the subtropics or poles (Figure 9.38b). Once created, they last for several days and 'block' eastward-moving depressions (Figure 9.48) to create anomalous conditions such as extremes of temperature, rainfall and sunshine – as in Britain in the summer of 1995 and the winter of 1987.

Figure 9.47

Anticyclone over the British Isles

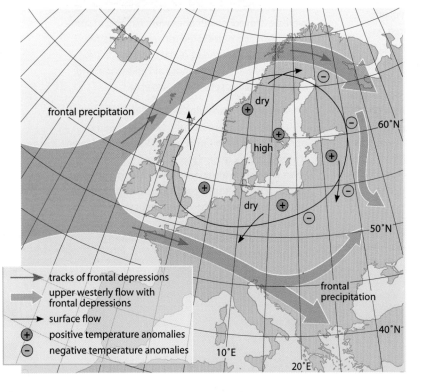

tracks of frontal depressions
upper westerly flow with frontal depressions
surface flow
⊕ positive temperature anomalies
⊖ negative temperature anomalies

Figure 9.48

A blocking anticyclone over Scandinavia: the upper westerlies divide upwind of the block and flow around it with their associated rainfall; there are positive temperature anomalies within the southerly flow to the west of the block and negative anomalies to the east

Tropical cyclones

Tropical cyclones are systems of intense low pressure known locally as **hurricanes**, **typhoons** and **cyclones** (Figure 9.49). They are characterised by winds of extreme velocity and are accompanied by torrential rainfall – two factors that can cause widespread damage and loss of life (Places 31, page 238). As yet, there is still insufficient conclusive evidence as to the process of their formation, although knowledge has been considerably improved recently due to airflights through and over individual systems, and the use of weather satellites. Tropical cyclones tend to develop:

- over warm tropical oceans, where sea temperatures exceed 26°C and where there is a considerable depth of warm water
- in autumn, when sea temperatures are at their highest
- in the trade wind belt, where the surface winds warm as they blow towards the Equator
- between latitudes 5° and 20° north or south of the Equator (nearer to the Equator the Coriolis force is insufficient to enable the feature to 'spin' – page 225).

Once formed, they move westwards –often on erratic, unpredictable courses – swinging poleward on reaching land, where their energy is rapidly dissipated (Figure 9.49). They are another mechanism by which surplus energy is transferred away from the tropics (Figure 9.6).

Hurricanes

Hurricanes are the tropical cyclones of the Atlantic. They form after the ITCZ has moved to its most northerly extent enabling air to converge at low levels, and can have a diameter of up to 650 km. Unlike depressions, hurricanes occur when temperatures, pressure and humidity are uniform over a wide area in the lower troposphere for a lengthy period, and anticyclonic conditions exist in the upper troposphere. These conditions are essential for the development, near the Earth's surface, of intense low pressure and strong winds. To enable the hurricane to move, there must be a continuous source of heat to maintain the rising air currents. There must also be a large supply of moisture to provide the latent heat, released by condensation, to drive the storm and to provide the heavy rainfall. It is estimated that in a single day a hurricane can release an amount of energy equivalent to that released by 500 000 atomic bombs the size of the one dropped on Hiroshima in the Second World War. Only when the storm has reached maturity does the central **eye** develop. This is an area of subsiding air, some 30–50 km in diameter, with light winds, clear skies and anomalous high temperatures (Figure 9.50). The descending air increases instability by warming and exaggerates the storm's intensity.

The hurricane rapidly declines once the source of heat is removed, i.e. when it moves over colder water or a land surface; these increase friction and cannot supply sufficient moisture. The average lifespan of a tropical cyclone is 7–14 days. The characteristic weather conditions associated with the passage of a typical hurricane are shown diagrammatically in Figure 9.50, and from space in Figure 9.51.

Figure 9.49

Global location and mean frequency of tropical cyclones

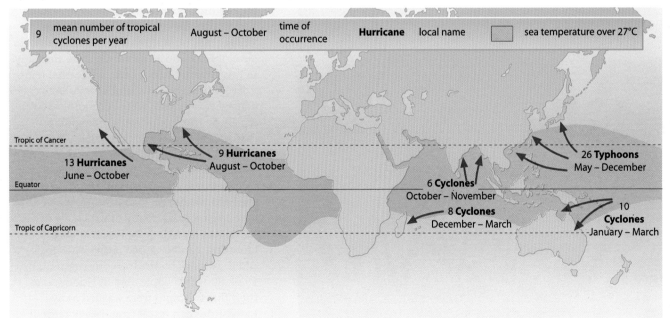

9 mean number of tropical cyclones per year | August – October time of occurrence | **Hurricane** local name | sea temperature over 27°C

Tropic of Cancer

13 Hurricanes June – October

9 Hurricanes August – October

26 Typhoons May – December

Equator

6 Cyclones October – November

8 Cyclones December – March

10 Cyclones January – March

Tropic of Capricorn

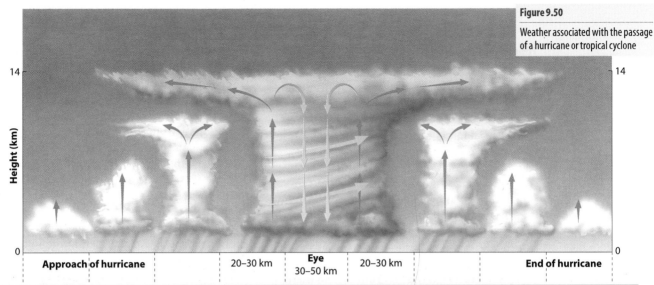

Figure 9.50

Weather associated with the passage of a hurricane or tropical cyclone

Height (km) — 14 ... 0

| Approach of hurricane | | | | 20–30 km | Eye 30–50 km | 20–30 km | | | End of hurricane |

Vertical movement	updraughts increasing →		spiral uplift		subsiding air	spiral uplift		updraughts decreasing →	
Clouds (Figure 9.20)	few Cu	Cu	Cu and some Cb	giant CB and Ci	none	giant Cb and Ci	Cu and some Cb	Cu	small Cu
Precipitation	none	showers	heavy showers	torrential rain 250 mm/day	none	torrential rain 250 mm/day	heavy showers	showers	none
Wind speed	gentle	fresh, gusty	locally very strong	hurricane force 160 km/hr	calm	hurricane force 160 km/hr	locally very strong	fresh gusty	gentle
Wind direction	NNW	NW	WNW	WNW	calm	SSE	SSE	SE	ESE
Temperatures (plus examples)	high (30°C)	still high (30°C)	falling (26°C)	low (24°C)	high (32°C)	low (24°C)	rising (26°C)	high (28°C)	high (30°C)
Pressure	average, 1012 mb	steady, 1010 mb	slowly falling, 1006 mb	rapid fall	low, 960 mb	rapid rise	slowly rising, 1004 mb	steady, 1010 mb	average, 1012 mb

Figure 9.51

Satellite image of Hurricane Mitch, October 1998. The 'eye' is very noticable

Tropical cyclones are a major natural hazard which often cause considerable loss of life and damage to property and crops (Places 30). There are four main causes of damage.

1 **High winds**, which often exceed 160 km/hr and, in extreme cases, 300 km/hr. Whole villages may be destroyed in economically less developed countries (of which there are many in the tropical cyclone belt), while even reinforced buildings in the south-east USA may be damaged. Countries whose economies rely largely on the production of a single crop (bananas in Nicaragua) may suffer serious economic problems. Electricity and communications can also be severed.

2 **Ocean storm (tidal) surges**, resulting from the high winds and low pressure, may inundate coastal areas, many of which are densely populated (Bangladesh, Places 19 page 148).

3 **Flooding** can be caused either by a storm (tidal) surge or by the torrential rainfall. In 1974, 800 000 people died in Honduras as their flimsy homes were washed away.

4 **Landslides** can result from heavy rainfall where buildings have been erected on steep, unstable slopes (Hong Kong, Figure 2.33).

'The Number 8 signal may be raised today as Typhoon Leo moves closer to Hong Kong. Its approach forced the Hong Kong observatory to hoist the strong wind signal Number 3 yesterday afternoon (Figure 9.52) – the first time it had ever been raised in April (Figure 9.55). Leo intensified into a typhoon yesterday, with central wind speeds of up to 130 km/hr. At midnight, it was 310 km south-south-east of Hong Kong, and was moving at about 8 km/hr (Figure 9.53). The typhoon is expected to be closest to Hong Kong early tomorrow morning, by

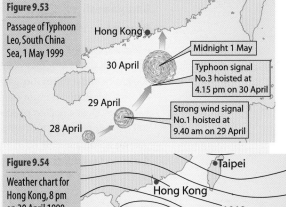

Figure 9.53

Passage of Typhoon Leo, South China Sea, 1 May 1999

Hong Kong

30 April

Midnight 1 May

29 April

Typhoon signal No.3 hoisted at 4.15 pm on 30 April

28 April

Strong wind signal No.1 hoisted at 9.40 am on 29 April

Figure 9.52

Typhoon warning system, Hong Kong

Typhoon No. 1 is hoisted.
一號風球現正懸掛

which time weather will deteriorate further and average rainfall could exceed 500 mm (Figure 9.54).

The Education Department has ordered kindergartens, schools for the mentally and physically handicapped, and nursing schools to remain closed. The Home Affairs Department's temporary shelters will open if Signal 8 goes up. People in need of shelter can make enquiries by calling the hotline.'

Source: *South China Morning Post*

Figure 9.55

Typhoon warning signals, Hong Kong

Figure 9.54

Weather chart for Hong Kong, 8 pm on 30 April 1999

Taipei

Hong Kong

1012

1008

T 1000

1008

Manila

Bangkok

1004

1008

Ho Chi Minh City

°C	**Saturday:** Overcast with rain, heavy at times.
21 18	Wind: east to northeast force 6 to 7, becoming force 7 to 8 later.
	Sunday: Overcast with frequent heavy rain.
22 20	Wind: east to southeast force 7 to 8.

Coastal waters

Hong Kong adjacent waters: East to northeast force 7 to 8, up to force 10 later today in the south. Occasional heavy rain. Rough to very rough seas becoming high.

Signal (four categories of tropical cyclone based on wind speed)		Meaning of the signal	What you should do Specific advice is contained in weather broadcasts, but the following general precautions can be taken
Stand by	**1**	A tropical cyclone is centred within about 800 km of Hong Kong. Hong Kong is placed on a state of alert because the tropical cyclone is a potential threat and may cause destructive winds later.	**Listen to weather broadcasts.** Some preliminary precautions are desirable and you should take the existence of the tropical cyclone into account in planning your activities.
Strong wind – A Tropical depression	**3**	Strong wind expected or blowing, with a sustained speed of 41–62 km/hr and gusts that may exceed 110 km/hr. The timing of the hoisting of the signal is aimed to give about 12 hours' advance warning of a strong wind in Victoria Harbour but the warning period may be shorter for more exposed waters.	**Take all necessary precautions.** Secure all loose objects, particularly on balconies and rooftops. Secure hoardings, scaffolding and temporary structures. Clear gutters and drains. Take full precautions for the safety of boats. Ships in port normally leave for typhoon anchorages or buoys. Ferry services may soon be affected by wind or waves. Even at this stage heavy rain accompanied by violent squalls may occur.
Gale or storm – B Tropical storm	**8**	Gale or storm expected or blowing, with a sustained wind speed of 63–117 km/hr from the quarter indicated and gusts that may exceed 180 km/hr. The timing of the replacement of the Strong Wind Signal No.3 by the appropriate one of these four signals, is aimed to give about 12 hours' advance warning of a gale in Victoria Harbour, but the sustained wind speed may reach 63 km/hr within a shorter period over more exposed waters. Expected changes in the direction of the wind will be indicated by corresponding changes of these signals.	**Complete all precautions** as soon as possible. It is extremely dangerous to delay precautions until the hoisting of No.9 or No.10 signals as these are signals of great urgency. Windows and doors should be bolted and shuttered. **Stay indoors** when the winds increase to avoid flying debris, but if you must go out, keep well clear of overhead wires and hoardings. All schools and law courts close and ferries will probably stop running at short notice. The sea-level will probably be higher than normal, particularly in narrow inlets. If this happens near the time of normal high tide then low-lying areas may have to be evacuated very quickly. Heavy rain may cause flooding, rockfalls and mudslips.
Increasing gale or storm – C Severe tropical storm	**9**	Gale or storm expected to increase significantly in strength. This signal will be hoisted when the sustained wind speed is expected to increase and come within the range 88–117 km/hr during the next few hours.	**Stay where you are** if reasonably protected and **away from exposed windows and doors.** These signals imply that the centre of a severe tropical storm or a typhoon will come close to Hong Kong. If the eye passes over there will be a lull lasting from a few minutes to some hours, but be prepared for a sudden resumption of destructive winds from a different direction.
Hurricane – D Typhoon	**10**	Hurricane-force winds expected or blowing, with a sustained wind speed reaching upwards from 118 km/hr and with gusts that may exceed 220 km/hr.	

Bangladesh

November 1970

A tropical cyclone moved northwards up the narrowing, shallowing Bay of Bengal. Winds of over 200 km/hr and a storm surge 8 m high hit the densely populated Ganges delta (Places 19, page 148). Over 4 million people were affected; 300 000 people died and 1 million were left homeless; half a million cows and oxen were drowned; two-thirds of the fishing fleet was lost; and 80 per cent of the rice crop was ruined.

May 1985

Three days after a tropical cyclone hit the coastal islands of Bangladesh, countless bodies could still be seen floating in the sea while hundreds of survivors, on bamboo rafts and floating rooftops, were awaiting rescue from the floodwaters. The Red Cross estimated that the tidal surge, 9 m in height and penetrating 150 km inland across the flat delta region, may have claimed the lives of 40 000 people. An official source feared that on the island of Hatia alone, 6000 people, many in their sleep, were washed out to sea and that the only survivors were the few who managed to cling to the tops of palm trees in the 180 km/hr winds. Links had still to be established with several of the more remote islands. The Red Cross were fearing, in the short term, an outbreak of typhoid and cholera, as fresh water supplies had been contaminated; and, in the long term, a food shortage, as the rice crop had been lost and it would take next year's monsoon rains to wash the salt out of the soil. Thousands of animals and most of the coastal fishing fleet were also believed to have been lost.

April 1991

The latest tropical cyclone to hit Bangladesh brought with it winds of 225 km/hr and waves 7 m in height. It swept over the unprotected offshore islands and the flat delta of the Ganges–Brahmaputra rivers where the land is never more than 2 m above sea-level. Over 150 000 people and half a million cattle were drowned, entire villages swept away, thousands of hectares of crops lost, electricity supplies cut off and roads and fishing boats destroyed. When the floodwater subsided, people were faced with food shortages and disease caused by water supplies being contaminated with seawater, sewage and dead bodies.

Central America – Hurricane Mitch, 1998

Hurricane Mitch developed in the southern Caribbean Sea on 22 October (Figure 9.51). In its early days it showed the characteristic unpredictable, but generally westward, course of most tropical storms in that region. Forecasters did not, however, predict first that it would suddenly turn southwards to affect the north Honduras coast, and secondly that it would then become almost stationary. It was the resultant torrential rain, rather than the severe winds, that led to the hurricane becoming the most destructive storm in 200 years.

Most of the damage occurred on or before 1 November, yet the world barely took notice. Three days later, it heard that hundreds, perhaps thousands, had died in Nicaragua, mostly when one side of the Casitas volcano collapsed, burying villages (Case Study 2). But it was not until a Honduran official spoke of 7000 dead in his country, many in the capital city of Tegucigalpa, that the world awoke to the tragedy. Reports issued on 11 and 12 November (Figure 9.56) claimed that between 11 000 and 12 000 had died, another 16 000 were missing, up to 150 000 had been made homeless, and more than 3 million were affected either by the loss of their crops or serious damage to the infrastructure (roads, bridges, electricity, water and sewerage systems, schools and hospitals). It took several days before international aid organisations began to make urgent appeals for help, and even longer before national governments began a more organised, co-ordinated relief programme. Years of economic growth had been wiped out. It would take more than a decade to recover fully.

Figure 9.56

Passage of Hurricane Mitch, October–November 1998

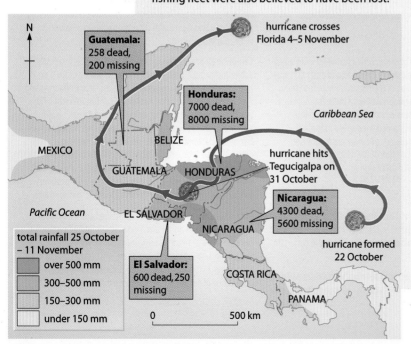

N

Guatemala: 258 dead, 200 missing

hurricane crosses Florida 4–5 November

Honduras: 7000 dead, 8000 missing

Caribbean Sea

MEXICO

BELIZE

GUATEMALA   HONDURAS

hurricane hits Tegucigalpa on 31 October

Pacific Ocean

EL SALVADOR

Nicaragua: 4300 dead, 5600 missing

NICARAGUA

hurricane formed 22 October

El Salvador: 600 dead, 250 missing

COSTA RICA

PANAMA

total rainfall 25 October – 11 November

over 500 mm

300–500 mm

150–300 mm

under 150 mm

0   500 km

Figure 9.57

The monsoon in the Indian subcontinent

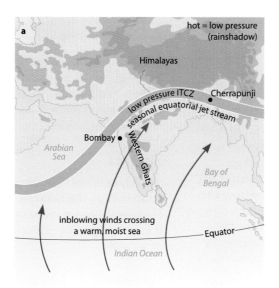

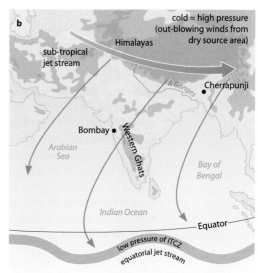

The monsoon

The word **monsoon** is derived from the Arabic word for 'a season', but the term is more commonly used in meteorology to denote a seasonal reversal of wind direction.

The major monsoon occurs in south-east Asia and results from three factors.

1   The extreme heating and cooling of large land masses in relation to the smaller heat changes over adjacent sea areas (page 210). This in turn affects pressure and winds.

2   The northward movement of the ITCZ (page 226) during the northern hemisphere summer.

3   The uplift of the Himalayas which, some 6 million years ago, became sufficiently high to interfere with the general circulation of the atmosphere (Places 5, page 20).

The south-west or summer monsoon

As the overhead sun appears to move northwards to the Tropic of Cancer in June, it draws with it the convergence zone associated with the ITCZ (Figure 9.57a). The increase in insolation over northern India, Pakistan and central Asia means that heated air rises, creating a large area of low pressure. Consequently, warm moist *Em* (equatorial maritime) and *Tm* air, from over the Indian Ocean, is drawn first northwards and then, because of the Coriolis force, is diverted north-eastwards (page 224). The air is humid, unstable and conducive to rainfall. Amounts of precipitation are most substantial on India's west coast, where the air rises over the Western Ghats, and on the windward slope of the Himalayas: Bombay has 2000 mm and Cherrapunji 13 000 mm in four summer months. The advent of monsoon storms allows the planting of rice (Places 67, page 481). Rainfall totals are accentuated as the air rises by both orographic and convectional uplift and the 'wet' monsoon is maintained by the release of substantial amounts of latent heat. The average arrival date is 10 May in Sri Lanka and 5 July at the Pakistan border – a time-lapse of seven weeks (Places 32).

The north-east or winter monsoon

During the northern winter, the overhead sun, the ITCZ and the subtropical jet stream all move southwards (Figure 9.57b). At the same time, central Asia experiences intense cooling which allows a large high pressure system to develop. Airstreams that move outwards from this high pressure area are dry because their source area is semi-desert. They become even drier as they cross the Himalayas and adiabatically warmer as they descend to the Indo-Gangetic plain. Bombay receives less than 100 mm of rain during these eight months. The south-west monsoon usually begins its retreat from the extreme north-west of India on 1 September and takes until 15 November, i.e. 11 weeks, to clear the southern tip.

The monsoon, which in reality is much more complex than the model described above, affects the lives of one-quarter of the world's population. Unfortunately, monsoon rainfall, especially in the Indian subcontinent, is unreliable (Figure 9.28). If the rains fail, then drought and famine ensue: 1987 was the ninth year in a decade when the monsoon failed in north-west India. If, conversely, there is excessive rainfall then large areas of land experience extreme flooding (Bangladesh in 1987, 1988 and 1998).

June 1994

'Rain brought welcome relief to the Indian capital yesterday, a day after 18 people collapsed and died on the streets in the blistering heat, pushing the summer death toll in northern India to nearly 350. Heavy showers cooled the furnace-like city, reeling under a three-week heatwave that has kept daytime temperatures at an almost constant 45°C and which had, the previous day, experienced its hottest day in 50 years when the mercury soared to 42.6°C. It was the first pre-monsoon rain of the season to lash Delhi, and children celebrated by soaking themselves in the rain, with many elderly citizens joining them in the belief that monsoon rains help cure blisters and skin diseases caused by extreme heat. More thunderstorms are expected by the weekend, which should mark the onset of the summer monsoon.'

July 1994

'The July death toll from relentless monsoon rains across India and Pakistan rose to more than 590 as a several waves of severe storms passed across the subcontinent. Many streets in Delhi are still under water.'

Meso-scale: local winds

Of the three meso-scale circulations described here, two – **land and sea breezes** and **mountain and valley winds** – are caused by local temperature differences; the third – the **föhn** – results from pressure differences on either side of a mountain range.

The land and sea breeze

This is an example, on a diurnal timescale, of a circulation system resulting from differential heating and cooling between land surfaces and adjacent sea areas. The resultant pressure differences, although small and localised, produce gentle breezes which affect coastal areas during calm, clear anticyclonic conditions. When the land heats up rapidly each morning, lower pressure forms and a gentle breeze begins to blow from the sea to the land (Figure 9.58a). By early afternoon, this breeze has strengthened sufficiently to bring a freshness which, in the tropics particularly, is much appreciated by tourists at the beach resorts. Yet by sunset, the air and sea are both calm again.

Although the circulation cell rarely rises above 500 m in height or reaches more than 20 km inland in Britain, the sea breeze is capable of lowering coastal temperatures by 15°C and can produce advection fogs such as the 'sea-fret' or 'haar' of eastern Britain.

At night, when the sea retains heat longer than the land, there is a reversal of the pressure gradient and therefore of wind direction (Figure 9.58b). The land breeze, the gentler of the two, begins just after sunset and dies away by sunrise.

The mountain and valley wind

This wind is likely to blow in mountainous areas during times of calm, clear, settled weather. During the morning, valley sides are heated by the sun, especially if they are steep, south-facing (in the northern hemisphere) and lacking in vegetation cover. The air in contact with these slopes will heat, expand and rise (Figure 9.59a), creating a pressure gradient. By 1400 hours, the time of maximum heating, a strong uphill or **anabatic wind** blows up the valley and the valley sides – ideal conditions for hang-gliding! The air becomes conditionally unstable (Figure 9.19), often producing cumulus cloud and, under very warm conditions, cumulo-nimbus with the possibility of thunderstorms on the mountain ridges. A compensatory sinking of air leaves the centre of the valley cloud-free.

Figure 9.58

Land and sea breezes in Britain

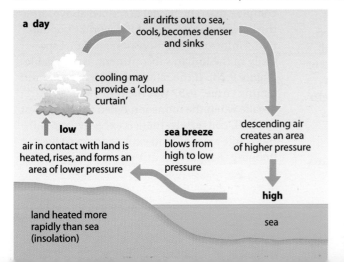

a day

air drifts out to sea, cools, becomes denser and sinks

cooling may provide a 'cloud curtain'

low

air in contact with land is heated, rises, and forms an area of lower pressure

sea breeze blows from high to low pressure

descending air creates an area of higher pressure

high

land heated more rapidly than sea (insolation)

sea

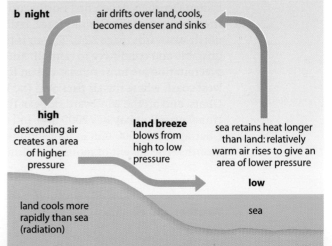

b night

air drifts over land, cools, becomes denser and sinks

high

descending air creates an area of higher pressure

land breeze blows from high to low pressure

sea retains heat longer than land: relatively warm air rises to give an area of lower pressure

low

land cools more rapidly than sea (radiation)

sea

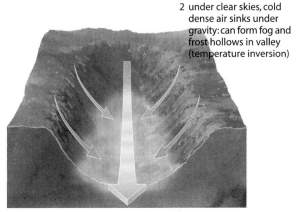

a day (anabatic flow)

updraughts may produce cloud on hills

descending air gives clear skies

3 winds less strong if valley sides face north (less heating)

2 wind blows up valley sides

1 wind blows up-valley

b night (katabatic flow)

2 under clear skies, cold dense air sinks under gravity: can form fog and frost hollows in valley (temperature inversion)

1 wind blows down-valley

Figure 9.59

Mountain and valley winds

During the clear evening, the valley loses heat through radiation. The surrounding air now cools and becomes denser. It begins to drain, under gravity, down the valley sides and along the valley floor as a mountain wind or **katabatic wind** (Figure 9.59b). This gives rise to a temperature inversion (Figure 9.24) and, if the air is moist enough, in winter may create fog (Figure 9.23) or a **frost hollow**. Maximum wind speeds are generated just before dawn, normally the coldest time of the day. Katabatic winds are usually gentle in Britain, but are much stronger if they blow over glaciers or permanently snow-covered slopes. In Antarctica, they may reach hurricane force.

The föhn

The föhn is a strong, warm and dry wind which blows periodically to the lee of a mountain range. It occurs in the Alps when a depression passes to the north of the mountains and draws in warm, moist air from the Mediterranean. As the air rises (Figure 9.60), it cools at the DALR of

1°C per 100 m (page 216). If, as in Figure 9.60, condensation occurs at 1000 m, there will be a release of latent heat and the rising air will cool more slowly at the SALR of 0.5°C per 100 m. This means that when the air reaches 3000 m it will have a temperature of 0°C instead of the –10°C had latent heat not been released. Having crossed the Alps, the descending air is compressed and warmed at the DALR so that, if the land drops sufficiently, the air will reach sea-level at 30°C. This is 10°C warmer than when it left the Mediterranean. Temperatures may rise by 20°C within an hour and relative humidity can fall to 10 per cent.

This wind, also known as the **chinook** on the American Prairies, has considerable effects on human activity. In spring, when it is most likely to blow, it lives up to its Native American name of 'snow-eater' by melting snow and enabling wheat to be sown; and in Switzerland it clears the alpine pastures of snow. Conversely, its warmth can cause avalanches, forest fires and the premature budding of trees.

Figure 9.60

The föhn

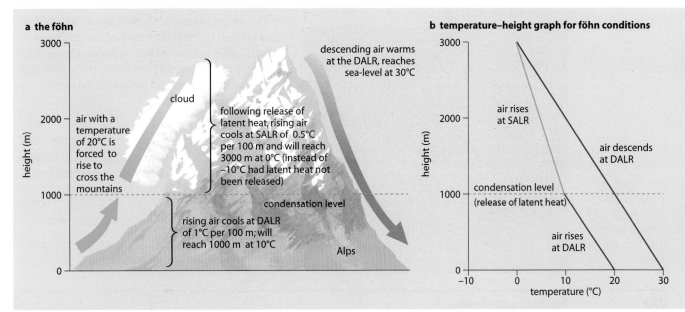

a the föhn

air with a temperature of 20°C is forced to rise to cross the mountains

cloud

following release of latent heat, rising air cools at SALR of 0.5°C per 100 m and will reach 3000 m at 0°C (instead of –10°C had latent heat not been released)

descending air warms at the DALR, reaches sea-level at 30°C

condensation level

rising air cools at DALR of 1°C per 100 m; will reach 1000 m at 10°C

Alps

b temperature–height graph for föhn conditions

air rises at SALR

air descends at DALR

condensation level (release of latent heat)

air rises at DALR

Microclimates

Microclimatology is the study of climate over a small area. It includes changes resulting from the construction of large urban centres as well as those existing naturally between different types of land surface, e.g. forests and lakes.

Urban climates

Large cities and conurbations experience climatic conditions that differ from those of the surrounding countryside. They generate more dust and condensation nuclei than natural environments; they create heat; they alter the chemical composition and the moisture content of the air above them; and they affect both the albedo and the flow of air. Urban areas therefore have distinctive climates.

Temperature

Although tower blocks cast more shadow, normal building materials tend to be non-reflective and so absorb heat during the daytime. Dark-coloured roofs, concrete or brick walls and tarmac roads all have a high thermal capacity which means that they are capable of storing heat during the day and releasing it slowly during the night. Further heat is obtained from car fumes, factories, power stations, central heating and people themselves. The term **urban heat island** acknowledges that, under calm conditions, temperatures are highest in the more built-up city centre and decrease towards the suburbs and open countryside (Figure 9.61). In urban areas:

- daytime temperatures are, on average, 0.6°C higher
- night-time temperatures may be 3° or 4°C higher as dust and cloud act like a blanket to reduce radiation and buildings give out heat like storage radiators

- the mean winter temperature is 1° to 2°C higher (rural areas are even colder when snow-covered as this increases their albedo)
- the mean summer temperature may be 5°C higher
- the mean annual temperature is higher by between 0.6°C in Chicago and 1.3°C in London compared with that of the surrounding area.

Note how, in Figure 9.61, temperatures not only decrease towards London's boundary but also beside the Thames and Lea rivers. The urban heat island explains why large cities have less snow, fewer frosts, earlier budding and flowering of plants and a greater need, in summer, for air-conditioning than neighbouring rural areas.

Sunlight

Despite having higher mean temperatures, cities receive less sunshine and more cloud than their rural counterparts. Dust and other particles may absorb and reflect as much as 50 per cent of insolation in winter, when the sun is low in the sky and has to pass through more atmosphere, and 5 per cent in summer. High-rise buildings also block out light (Figure 9.62).

Wind

Wind velocity is reduced by buildings which create friction and act as windbreaks. Urban mean annual velocities may be up to 30 per cent lower than in rural areas and periods of calm may be 10–20 per cent more frequent. In contrast, high-rise buildings, such as the skyscrapers of New York and Hong Kong (Figure 9.62), form 'canyons' through which wind may be channelled. These winds may be strong enough to cause tall buildings to sway and pedestrians to be blown over and troubled by dust and litter. The heat island effect may cause local thermals and reduce the wind chill factor. It also tends to generate considerable small-scale turbulence and eddies. In 19th-century Britain, the most sought-after houses were usually on the western and south-western sides of cities, to be up-wind of industrial smoke and pollution (Mann's model, pages 422–423).

Relative humidity

Relative humidity is up to 6 per cent lower in urban areas where the warmer air can hold more moisture and where the lack of vegetation and water surface limits evapotranspiration.

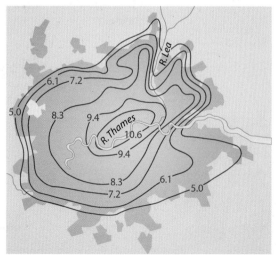

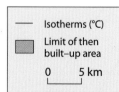

Figure 9.61

An urban heat island: minimum temperatures over London, 14 May 1959 (*after* Chandler)

Cloud

Urban areas appear to receive thicker and up to 10 per cent more frequent cloud cover than rural areas. This may result from convection currents generated by the higher temperatures and the presence of a larger number of condensation nuclei.

Precipitation

The mean annual precipitation total and the number of days with less than 5 mm of rainfall are both between 5 and 15 per cent greater in major urban areas. Reasons for this are the same as for cloud formation. Strong thermals increase the likelihood of thunder by 25 per cent and the occurrence of hail by up to 400 per cent. The higher urban temperatures may turn the snow of rural areas into sleet and limit, by up to 15 per cent, the number of days with snow lying on the ground. On the other hand, the frequency, length and intensity of fog, especially under anticyclonic conditions, is much greater – there may up to 100 per cent more in winter and 25 per cent more in summer, caused by the concentration of condensation nuclei (Figure 9.63).

Figure 9.62

Narrow streets with high-rise buildings are more likely to develop micro-climates than those that are wider and have lower buildings; New York City

Figure 9.63

Fog frequencies in London and south-east England (hours per year)

Location	Visibility less than 40 m (very dense fog)	Visibility less than 1000 m (less dense fog)
Kingsway (central London)	19	940
Kew (middle suburbs)	79	633
London Airport (outer suburbs)	46	562
South-east England (mean of 7 stations)	20	494

Atmospheric composition

There may be three to seven times more dust particles over a city than in rural areas. Large quantities of gaseous and solid impurities are emitted into urban skies by the burning of fossil fuels, by industrial processes and from car exhausts. Urban areas may have up to 200 times more sulphur dioxide and 10 times more nitrogen oxide (the major components of acid rain) than rural areas, as well as 10 times more hydrocarbons and twice as much carbon dioxide. These pollutants tend to increase cloud cover and precipitation, cause smog (Figure 9.25), give higher temperatures and reduce sunlight.

Forest and lake microclimates

Different land surfaces produce distinctive local climates. Figure 9.64 summarises and compares some of the characteristics of microclimates found in forests and around lakes. As with urban climates, research and further information are still needed to confirm some of the statements.

Figure 9.64

Microclimates of forests and water surfaces

Microclimate feature	Forest (coniferous and deciduous)	Water surface (lake, river)
Incoming radiation and albedo	Much incoming radiation is absorbed and trapped. Albedo for coniferous forest is 15%; deciduous 25% in summer and 35% in winter; and desert scrub 40%.	Less insolation absorbed and trapped. Albedo may be over 60%, i.e. higher than over seas/oceans (page 207). Higher on calm days.
Temperature	Small diurnal range due to blanket effect of canopy. Forest floor is protected from direct sunlight. Some heat lost by evapotranspiration.	Small diurnal range because water has a higher specific heat capacity. Cooler summers and milder winters. Lakesides have a longer growing season.
Relative humidity	Higher during daytime and in summer, especially in deciduous forest. Amount of evapotranspiration depends on length of day, leaf surface area, wind speed, etc.	Very high, especially in summer when evaporation rates are also high.
Precipitation	Heavy rain can be caused by high evapotranspiration rates, e.g. in tropical rainforests. On average, 30–35% of rain is intercepted: more in deciduous woodland in summer than in winter.	Air is humid. If forced to rise, air can be unstable and produce cloud and rain. Amounts may not be great due to fewer condensation nuclei. Fogs form in calm weather.
Wind speed and direction	Trees reduce wind speeds, especially at ground level. (They are often planted as windbreaks.) Trees can produce eddies.	Wind may be strong due to reduced friction. Large lakes (e.g. L. Victoria) can create land and sea breezes (page 240).

Weather maps and forecasting in Britain

A weather map or **synoptic chart** shows the weather for a particular area at one specific time (Figures 9.67 and 9.68). It is the result of the collection and collation of a considerable amount of data at numerous weather stations, i.e. from a number of sample points (Framework 6, page 159). These data are then refined, usually as quickly as possible and now using computers, and are plotted using internationally accepted weather symbols. A selection of these symbols is shown in Figure 9.65. Weather maps are produced for different purposes and at various scales.

1 The daily weather map, as seen on television or in a national newspaper, aims to give a clear, but highly simplified, impression of the weather.

2 At a higher level, a synoptic map shows selected meteorological characteristics for specific **weather stations**. The **station model** in Figure 9.66 shows six elements: temperature, pressure, cloud cover, present weather (e.g. type of precipitation), wind direction and wind speed.

3 At the highest level, the Meteorological Office produces maps showing finite detail, e.g. amounts of various types of cloud at low, medium and high levels, dew point temperatures, barometric tendency (i.e. trends of pressure change), etc.

The role of the weather forecaster is to try to determine the speed and direction of movement of various air masses and any associated fronts, and to try to predict the type of weather these movements will bring. Forecasters now make considerable use of **satellite images** (Figures 9.67 and 9.68). Satellite images are photos taken by weather satellites as they continually orbit the Earth. These photos, which are relayed back to Earth, are invaluable in the prediction of short-term weather trends. Although forecasting is increasingly assisted by information from satellites, radar and computers, which show upper air as well as surface air conditions in a three-dimensional model, the complexity and unpredictability of the atmosphere can still catch the forecaster by surprise (Places 29, page 232). Part of this problem is related to the fact that meteorological information is a sample (Framework 6, page 159) rather than a total picture of the atmosphere, and so there is always a risk of the anomaly becoming the reality.

Figure 9.65

Weather symbols for cloud, precipitation, wind speed, temperature, pressure and wind direction

Cloud		Weather (present)		Wind speed			Temperature
Symbol	Cloud amount (oktas)	Symbol	Weather	Symbol	Speed (knots)	Force	3 — 3°Celsius
○	0	═	mist	○	calm	0	
	1 or less	═	fog		1 - 2	1	**Pressure**
	2	,	drizzle		3 - 7	2	Pressure is shown by isobars and is measured in millibars
	3	;	rain and drizzle		8 - 12	3	
	4	•	rain		13 - 17	4	
	5	✳	rain and snow				——1012——
	6	✳	snow	For each additional half-feather add 5 knots or add an extra force up to			mean sea level pressure
	7	▽	rain shower				L = centre of an area of low pressure
	8	✳▽	snow shower	◄	48 - 52	11	
⊗	sky obscured	◊	hail shower		**Wind direction**		H = centre of an area of high pressure
⊗	missing or doubtful data	⬞	thunderstorm		Indicates a north-westerly wind direction		

Figure 9.66

A weather station model and an example

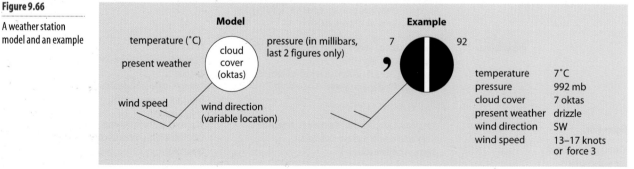

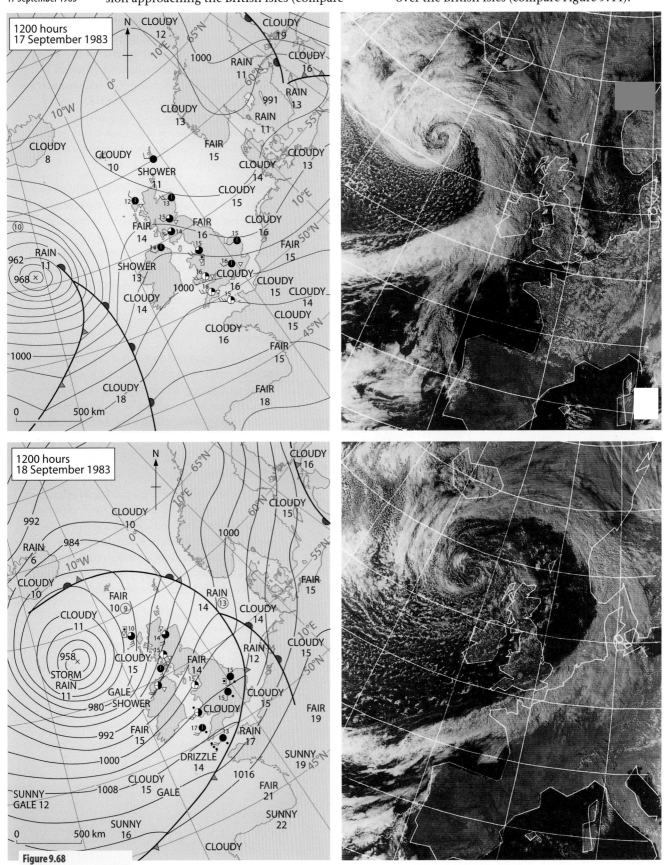

Figure 9.67

Synoptic chart and
satellite image,
17 September 1983

Figure 9.67 shows the synoptic chart (weather map) and satellite (infrared) image of a depression approaching the British Isles (compare Figure 9.43). Figure 9.68 shows the same depression 24 hours later, by which time it had passed over the British Isles (compare Figure 9.44).

Figure 9.68

Synoptic chart and satellite
image, 18 September 1983

Throughout this chapter on weather and climate, mean climatic figures have been quoted. To build up these pictures of global, regional and local climate patterns, statistics have been obtained by averaging readings, usually for temperature and precipitation, over a 30-year timescale. However, these averages themselves are often not as significant as the **range** or the degree to which they vary from, or are dispersed about, the mean.

For example, two tropical weather stations may have equal annual rainfall totals when measured over 30 years. Station A may lie on the Equator and experience reliable rainfall with little variation from one year to the next. Station B may experience a monsoon climate where in some years the rains may fail entirely while in others they cause flooding.

The measure of dispersion from the mean can be obtained by using any one of three statistical techniques:

- the range
- the interquartile range, or
- the standard deviation.

These techniques are included here because meteorological data both require and benefit from their use, but they may be applied to most branches of geography where there is a danger that the mean, taken alone, may be misleading (the problems of overgeneralisation are discussed in Framework 11, page 347). Again, it must be stressed that use of a quantitative technique does not guarantee objective interpretation of data: great care must be taken to ensure that an appropriate method of manipulating the data is chosen.

It has already been seen how it is possible, given a data set, to calculate the mean and the median (Framework 5, page 112). However, neither statistic gives any idea of the spread, or range, of that data. As the example above of two tropical weather stations shows, mean values on their own give only part of the full picture. The spread of the data around the mean should also be considered.

Range

This very simple method involves calculating the difference between the highest and lowest values of the sample population, e.g. the annual range in temperature for London is 14°C (July 18°C, January 4°C). The range emphasises the extreme values and ignores the distribution of the remainder.

Interquartile range

The interquartile range consists of the middle 50 per cent of the values in a distribution, 25 per cent each side of the median (middle value). This calculation is useful because it shows how closely the values are grouped around the median (Figure 9.69). It is easy to calculate; it is unaffected by extreme values; and it is a useful way of comparing sets of similar data.

The example in Figure 9.69 gives temperatures for 19 weather stations in the British Isles at 0600 on 14 January 1979. These temperatures have been ranked in the table.

Figure 9.69

The interquartile range

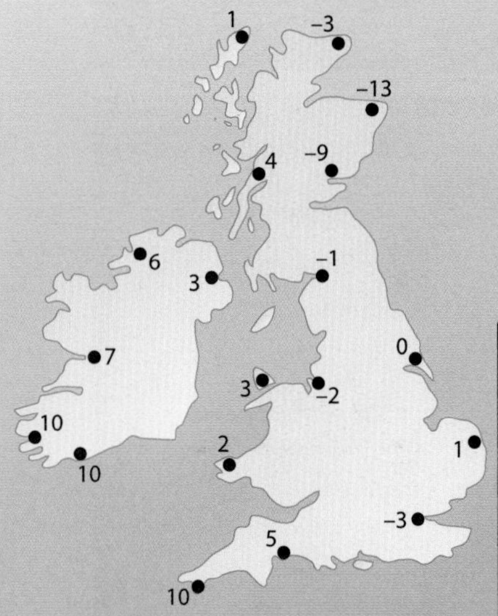

Rank	Temperatures 0°C (ranked)	
	10	
	10	
	10	
	7	
5	6	upper quartile
	5	
	4	
	3	
	3	
10	2	median (middle quartile)
	1	
	1	
	0	
	−1	
15	−2	lower quartile
	−3	
	−3	
	−9	
	−13	

50 per cent of values fall into the interquartile range

The **upper quartile** (UQ) is obtained by using the formula:

$$UQ = \left(\frac{n+1}{4}\right) \qquad \text{i.e.} \quad \left(\frac{19+1}{4}\right) = 5$$

This means that the UQ is the fifth figure from the top of the ranking order, i.e. 6°C. The **lower quartile** (LQ) is found by using a slightly different formula:

$$LQ = \left(\frac{n+1}{4}\right) \times 3 \qquad \text{i.e.} \quad \left(\frac{19+1}{4}\right) \times 3 = 5$$

This shows the LQ to be the 15th figure in the ranking order, i.e. –2°C. You will notice that the middle quartile is the same as the median. The **interquartile range** is the difference between the upper and lower quartiles, i.e. 6°C – –2°C = 8°C.

Another measure of dispersion, the **quartile deviation**, is obtained by dividing the interquartile range by two, i.e. 8°C ÷ 2 = 4°C

The smaller the interquartile range, or quartile deviation, the greater the grouping around the median and the smaller the dispersion or spread.

Standard deviation

This is the most commonly used method of measuring dispersion and although it may involve lengthy calculations it can be used with the arithmetic mean and it removes extreme values. The formula for the standard deviation is:

$$\sigma = \sqrt{\frac{\Sigma (x - \bar{x})^2}{n}}$$

where: σ = standard deviation

x = each value in the data set

$\bar{x}$ = mean of all values in the data set, and

n = number of values in the data set.

Let us suppose that the minimum temperatures for 10 weather stations in Britain on a winter's day were, in °C, 5, 8, 3, 2, 7, 9, 8, 2, 2 and 4. The standard deviation of this data set is worked out in Figure 9.70, proceeding as follows:

1 Find the mean ($\bar{x}$).

2 Subtract the mean from each value in the set: $x - \bar{x}$.

3 Calculate the square of each value in **2**, to remove any minus signs: $(\bar{x} - \bar{x})^2$.

4 Add together all the values obtained in **3**: $\Sigma (\bar{x} - \bar{x})^2$.

5 Divide the sum of the values in **4** by n:

$$\frac{\Sigma (x - \bar{x})^2}{n}$$

6 Take the square root of **5** to obtain the standard deviation:

$$\sqrt{\frac{\Sigma (x - \bar{x})^2}{n}}$$

The resulting standard deviation of σ = 2.65 is a low value, indicating that the data are closely grouped around the mean.

Figure 9.70

Finding the standard deviation

Minimum temperatures for 10 weather stations in Britain on a winter's day

The mean of 5, 8, 3, 2, 7, 9, 8, 2, 2, 4:

$$\bar{x} = \frac{50}{10} = 5$$

Weather station	Temperature at each station (x)	x − x̄	(x × x̄)²
1	5	5 − 5 = 0	0
2	8	8 − 5 = 3	9
3	3	3 − 5 = −2	4
4	2	2 − 5 = −3	9
5	7	7 − 5 = 2	4
6	9	9 − 5 = 4	16
7	8	8 − 5 = −3	9
8	2	2 − 5 = −3	9
9	2	2 − 5 = −3	9
10	4	4 − 5 = −1	1
			$\Sigma (x - \bar{x})^2 = 70$

$$\sigma = \sqrt{\frac{50}{10}}$$

$$\sigma = \sqrt{7} \qquad \therefore \text{ standard deviation} = 2.65$$

Climatic change

Climates have changed and still are constantly changing at all scales, from local to global, and over varying timespans, both long-term and short-term (Case Study 9). However, there have been surges of change over time which meteorologists and earth scientists are continually trying to clarify and explain.

Evidence of past climatic changes

- **Rocks** are found today which were formed under climatic conditions and in environments that no longer exist (Figure 1.1). In Britain, for example, coal was formed under hot, wet tropical conditions; sandstones were laid down during arid times; various limestones accumulated on the floors of warm seas; and glacial deposits were left behind by retreating ice sheets.
- **Fossil landscapes** exist, produced by certain geomorphological processes which no longer operate. Examples include glacially eroded highlands in north and west Britain (Chapter 4), granite tors on Dartmoor (page 202) and wadis formed during wetter periods (pluvials) in deserts (Places 25 page 188).
- Evidence exists of **changes in sea-level** (both isostatic as on Arran – Places 23, page 166) and eustatic (as at present in the Maldives – page 169) and changes in **lake levels** (Sahara, Figure 7.27).
- **Vegetation belts** have shifted through some 10° of latitude, e.g. changes in the Sahara Desert (Figure 7.27).
- **Pollen analysis** shows which plants were dominant at a given time. Each plant species has a distinctively-shaped pollen grain. If these grains land in an oxygen-free environment, such as a peat bog, they resist decay. Although pollen can be transported considerable distances by the wind and by wildlife, it is assumed that grains trapped in peat form a representative sample of the vegetation that was growing in the surrounding area at a given time; also, that this vegetation was a response to the climatic conditions prevailing at that time. Vertical sections made through peat show changes in pollen (i.e. vegetation), and these changes can be used as evidence of climatic change (the vegetation–climatic timescale in Figures 11.18 and 11.19).
- **Dendrochronology**, or tree-ring dating, is the technique of obtaining a core from a tree-trunk and using it to determine the age of the tree. Tree growth is rapid in spring, slower by the autumn and, in temperate latitudes, stops in winter. Each year's growth is shown by a single ring. However, when the year is warm and wet, the ring will be larger because the tree grows more quickly than when the year is cold and dry. Tree-rings therefore reflect climatic changes. Recent work in Europe has shown that tree growth is greatest under intense cyclonic activity and is more a response to moisture than to temperature. Tree-ring timescales are being established by using the remains of oak trees, some nearly 10 000 years old, found in river terraces in south-central Europe. Bristlecone pines, still alive after 5000 years, give a very accurate measure in California (page 294).
- **Chemical methods** include the study of oxygen and carbon isotopes. An isotope is one of two or more forms of an element which differ from each other in atomic weight (i.e. they have the same number of protons in the nucleus, but a different number of neutrons). For example, two isotopes in oxygen are O-16 and O-18. The O-16 isotope, which is slightly lighter, vaporises more readily; whereas O-18, being heavier, condenses more easily. During warm, dry periods, the evaporation of O-16 will leave water enriched with O-18 which, if it freezes into polar ice, will be preserved as a later record (Places 14, page 104). Colder, wetter periods will be indicated by ice with a higher level of O-16. The most accurate form of dating is based on C-14, a radioactive isotope of carbon. Carbon is taken in by plants during the carbon cycle (Figure 11.25). Carbon-14 decays radioactively at a known rate and can be compared with C-12, which does not decay. Using C-12 and C-14 from a dead plant, scientists can determine the date of death to a standard error of ± 5 per cent. This method can accurately date organic matter up to 50 000 years old.
- **Historical records** of climatic change include:
  - cave paintings of elephants in central Sahara (Figure 7.27)
  - vines growing successfully in southern England between AD 1000 and 1300
  - graves for human burial in Greenland which were dug to a depth of 2 m in the 13th century, but only 1 m in the 14th century, and could not be dug at all in the 15th century due to the extension of permafrost – in contrast to its retreat in the 1990s (Case Study 5)
  - fairs held on the frozen River Thames in Tudor times

– the measurement of recent advances and retreats of alpine glaciers and polar sea-ice.

Causes of climatic change

Several theories, covering varying timescales, have been advanced to try to explain climatic change. At present there is no unanimous consensus of opinion as to its causes: climatic change may be explained by one of these theories, several in combination, or by a theory yet to be propounded. The suggested theories include the following.

1 **Variations in solar energy** Although it was initially believed that solar energy output did not vary over time (hence the term 'solar constant' in Figure 9.3), increasing evidence suggests that sunspot activity, which occurs in cycles, may significantly affect our climate – times of high annual temperatures on Earth appear to correspond to periods of maximum sunspot activity.

2 **Astronomical relationships between the sun and the Earth** There is increasing evidence supporting Milankovitch's cycles of change in the Earth's orbit, tilt and wobble (Figure 4.6), which would account for changes in the amounts of solar radiation reaching the Earth's surface. This evidence is mainly from cores that have been drilled through undisturbed ocean-floor sediment which has accumulated over thousands of years (compare Places 14, page 104).

3 **Changes in oceanic circulation** Changes in oceanic circulation affect the exchange of heat between the oceans and the atmosphere. This can have both long-term effects on world climate (where currents at the onset of the Quaternary ice age flowed in opposite directions to those at the end of the ice age) and short-term effects (El Niño, Case Study 9). The latest theory compares the North Atlantic Drift with a conveyor belt that brings water to north-west Europe. Should this conveyor belt be closed down, possibly by a huge influx of fresh water into the sea, then the climate will become dramatically colder.

4 **Meteorites** A major extinction event, which included the dinosaurs, took place about 65 million years ago. This event was believed to have been caused by one or more meteors colliding with the Earth. This seems to have caused a reduction in incoming radiation, a depletion of the ozone layer and a lowering of global temperatures.

5 **Volcanic activity** It has been accepted for some time that volcanic activity has influenced climate in the past, and continues to do so. World temperatures are lowered after any large single eruption, e.g. Mount Pinatubo (Case Study 1) and Krakatoa (Figure 1.29 and Places 35 page 289) or after a series of volcanic eruptions. This is due to the increase in dust particles in the lower atmosphere which will absorb and scatter more of the incoming radiation (Figure 9.4). Evidence suggests that these major eruptions may temporarily offset the greenhouse effect. Precipitation also increases due to the greater number of hygroscopic nuclei (dust particles) in the atmosphere (page 215).

6 **Plate tectonics** Plate movements have led to redistributions of land masses and to long-term effects on climate. These effects may result from a land mass 'drifting' into different latitudes (British Isles, page 21); or from the seabed being pushed upwards to form high fold mountains (page 19). The presence of fold mountains can lead to a colder climate (a suggested cause of the Quaternary ice age, page 103) and can act as a barrier to atmospheric circulation – the Asian monsoon was established by the creation of the Tibetan Plateau (page 239).

7 **Composition of the atmosphere** Gases in the atmosphere can be increased and altered following volcanic eruptions. At present there is increasing concern at the build-up of CO_2 and other greenhouse gases in the atmosphere (Case Study 9), together with the use of aerosols and the release of CFCs (Places 27, page 209), which are blamed for the depletion of ozone in the upper atmosphere.

Climatic change in Britain

Britain's climate has undergone changes in the longest term (page 21 and Figure 1.1); during and since the onset of the Quaternary (Figure 4.2); and in the more recent short term. Following the 'little ice age' (which lasted from about AD 1540 to 1700), temperatures generally increased to reach a peak in about 1940. After that time, there was a tendency for summers to become cooler and wetter, springs to be later, autumns milder and winters more unpredictable. However, since the onset of the 1980s there appears to have been a considerable warming of our climate, with seven of the ten warmest years this century occurring during the 1990s. This, together with the apparent increase in variations from the norm for Britain's expected autumn, winter and spring weather, tends to add evidence supporting the concept of global warming (Case Study 9).

Short-term and long-term climatic changes

Short-term change: El Niño and La Niña

The oceans, as we have seen, have a considerable heat storage capacity which makes them a major influence on world climates. If ocean temperatures change, this will have a considerable effect upon weather patterns in adjacent land masses. Interactions between the ocean and the atmosphere have become, recently, a major scientific study.

The most important and interesting example of the ocean–atmosphere interrelationship is provided by the El Niño and La Niña events which occur periodically in the Pacific Ocean. Under normal atmospheric conditions, pressure rises over the eastern Pacific Ocean (off the coast of South America) and falls over the western Pacific Ocean (towards Indonesia and the Philippines). The descending air over the eastern Pacific gives the clear, dry conditions that create the Atacama Desert in Peru (Figure 7.2 and Places 24 page 180), while the warm, moist ascending air over the western Pacific gives that region its heavy convectional rainfall (page 226). This movement of air creates a circulation cell, named after Walker who first described it, in which the upper air moves from west to east, and the surface air from east to west as the trade winds (Figure 9.71). The trade winds:

- push surface water westwards so that sea-level in the Philippines is normally 60 cm higher than in Panama and Colombia

- allow water, flowing westward as the equatorial current, to remain near to the ocean surface where it can gradually heat. This gives the western Pacific the world's highest ocean temperature, usually above 28°C. In contrast, as warm water is pushed away from South America, it is replaced by an upwelling of colder, nutrient-rich water. This colder water lowers temperatures, sometimes to below 20°C, but does provide a plentiful supply of plankton which forms the basis of Peru's fishing industry.

Figure 9.71

The Walker circulation cell

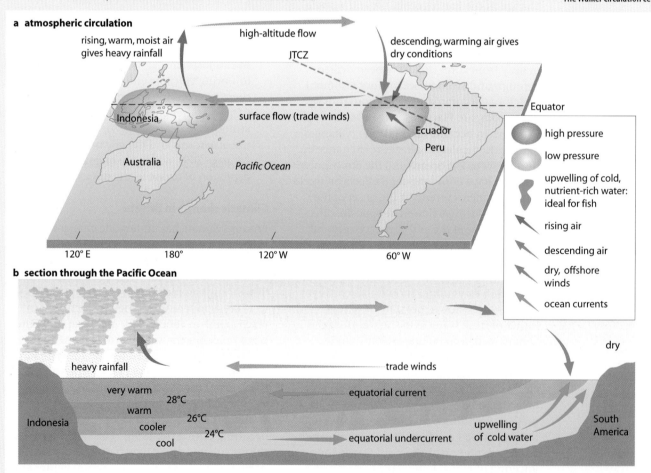

a atmospheric circulation

rising, warm, moist air gives heavy rainfall

high-altitude flow

JTCZ

descending, warming air gives dry conditions

Indonesia

surface flow (trade winds)

Ecuador
Peru

Equator

Australia

Pacific Ocean

120° E   180°   120° W   60° W

high pressure

low pressure

upwelling of cold, nutrient-rich water: ideal for fish

rising air

descending air

dry, offshore winds

ocean currents

b section through the Pacific Ocean

heavy rainfall

trade winds

dry

very warm   28°C
warm
cooler   26°C
Indonesia   24°C
cool

equatorial current

equatorial undercurrent

upwelling of cold water

South America

El Niño

An El Niño event, scientifically referred to as an El Niño Southern Oscillation (ENSO), occurs periodically – on average every three to four years. It is called 'El Niño', which means 'little child' in Spanish, because, in those years that it does occur, it appears just after Christmas. An El Niño event usually lasts for 12–18 months.

In contrast to normal conditions (Figure 9.71) there is a reversal, in the equatorial Pacific region, in pressure, precipitation and, often, winds and ocean currents (Figure 9.72). Pressure rises over the western Pacific and falls over the eastern Pacific. This allows the ITCZ (Figure 9.34) to migrate

southwards and causes the trade winds to weaken in strength, or, sometimes, even to be reversed in their direction. The descending air, now over South-east Asia, gives that region much drier conditions than it usually experiences and, on extreme occasions, even causing drought. In contrast the air over the eastern Pacific is now rising, giving much wetter conditions in places, like Peru, that normally experience desert conditions. The change in the direction of the trade winds means that:

• surface water tends to be pushed eastwards so that sea-level in South-east Asia falls, while it rises in tropical South America

• surface water temperatures in excess of 28°C extend much further eastwards and the upwelling of cold water off South America is reduced, allowing sea temperatures to rise by up to 6°C. The warmer water in the eastern Pacific lacks oxygen, nutrients and, therefore, plankton and so has an adverse effect on Peru's fishing industry.

NASA-Mir astronauts were able, during the record-breaking 1997–98 El Niño Southern Oscillation, to observe, photograph and document the global impacts of the event. These, together with ground observations and recordings, are summarised in Figure 9.73.

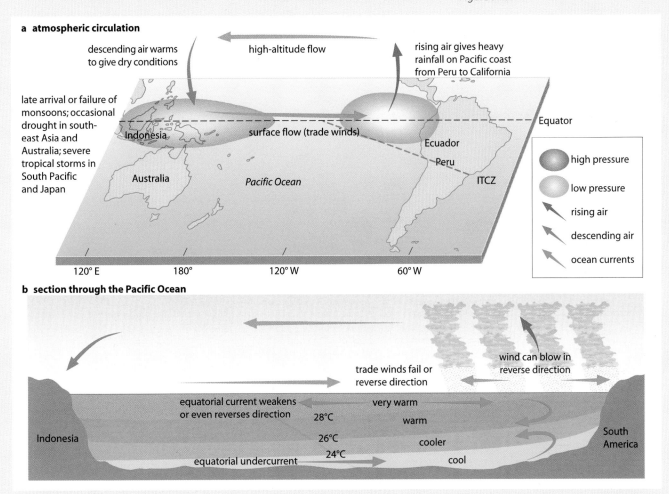

Figure 9.72

An El Niño event

Evidence collected during the El Niño events of 1982–83 (at the time the biggest ever recorded), 1986 and 1992–93, increasingly suggested that the ENSO had a major effect on places far beyond the Pacific margins as well as on those bordering the ocean itself in its low latitudes. Apart from the drier conditions in South-east Asia and the wetter conditions in South America:

- severe droughts were experienced in the Sahel (Case Study 7) and southern Africa as well as across the Indian subcontinent

- there were extremely cold winters in central North America, and stormy conditions with floods in California
- exceptionally wet, mild and windy winters were experienced in Britain and north-west Europe.

The 1997–98 event – the biggest yet experienced

Early 1997	Evidence of a rapid rise in sea temperatures in the eastern Pacific
July	El Niño conditions intense
September	Over 24 million km² of warm water (size of North and Central America) extended from the International Dateline to South America
1998 April	Evidence of El Niño weakening
June	NASA satellite surveillance showed a significant drop in sea temperatures in the eastern Pacific
Autumn	Signs of a La Niña event (page 253)

Figure 9.73

The effects of the 1997–98 El Niño event

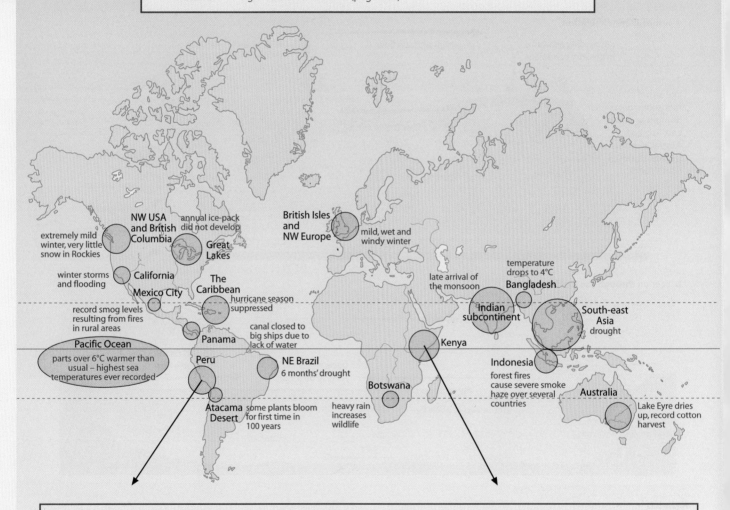

PERU For *each* of 12 days in early March, Peru received the equivalent of six months of normal rain. Over several months, flash flooding caused 292 deaths, injured more than 16 000 people, left 400 missing, destroyed 13 200 houses, wrecked 250 000 km of roads, swept away bridges, damaged crops and schools and disrupted the lives of up to half a million Peruvians.

KENYA Parts of Kenya received over 1000 mm of rainfall during six months (up to 50 times more than the average) at a time normally considered to be the 'dry season'. Roads and the mainline railway were swept away, the latter causing the derailment of the Nairobi–Mombasa train. Later, more than 500 people died of malaria as the receding floodwaters created ideal mosquito-spawning pools.

La Niña

Just as El Niño was ending in June 1998, forecasters were predicting – based on an 8°C fall in sea temperatures in the eastern Pacific in May, the arrival that winter of a La Niña event. La Niña, or 'little girl', has climatic conditions that are the reverse of those of El Niño. However, although when La Niña does appear it is just before or just after El Niño, its occurrence has been less frequent (the last was between June 1988 and February 1989) and, consequently, it is less easy to predict its possible effects because there is less evidence.

Figure 9.74

A La Niña event

In a La Niña event, in contrast to normal conditions in the Pacific Ocean (Figure 9.71), the low pressure over the western Pacific becomes even lower and the high pressure over the eastern Pacific even higher (Figure 9.74). This means that rainfall increases over South-east Asia (was the La Niña event of 1988 responsible for the severe flooding at that time in Bangladesh?), there are drought conditions in South America and, due to the increased difference in pressure between the two places, the trade winds strengthen. The stronger trade winds:

- push large amounts of water westwards, giving a higher than normal sea-level in Indonesia and the Philippines
- increase the equatorial undercurrent and significantly enhance the upwelling of cold water off the Peruvian coast.

Meteorologists suggest that La Niña can be linked with increased hurricane activity in the Caribbean – a prediction made before Hurricanes George and Mitch of autumn 1998 (Case Study 2 and Places 31 page 238), and that it can disrupt the jet streams to bring stormy, colder weather to Britain, e.g. the Boxing Day storm of 1998 (Figure 9.75).

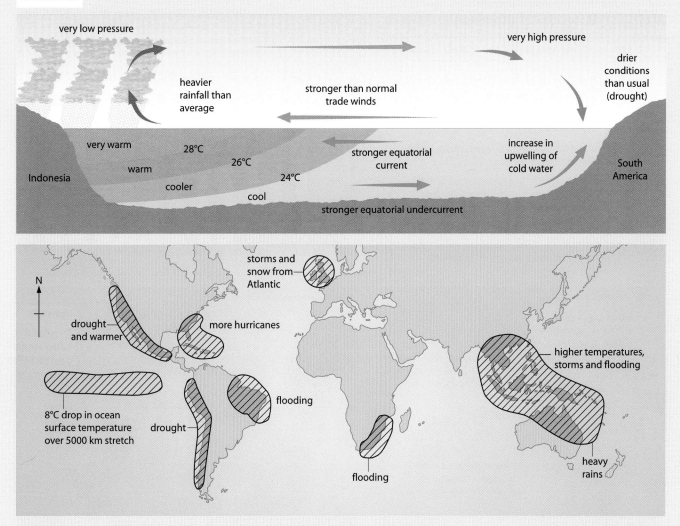

Figure 9.75

Possible effects of a La Niña event on world weather

Long-term change: global warming – an update

1998 – the warmest year on record

Global temperatures in 1998 were the highest since reliable instrumental records began 119 years earlier and, according to palaeoclimatologists, probably the highest in over 1200 years. The previous warmest year had been 1997, and 1998 was the 20th consecutive year when the annual global mean surface temperature exceeded the long-term average (Figure 9.76). What was more staggering about the 1998 figure, collected by the NOAA (National Oceanic and Atmospheric Administration), was not just that the year was 0.64°C above the long-term average of 13.8°C, but that nine of the twelve months were all-time highs (Figure 9.77). Also, it is worth noting that the previous monthly records had all been established in the previous ten years, i.e. between 1988 and 1997. The record high temperatures of early 1998 have been partly attributed to the effects of El Niño, and the lower temperatures towards the end of the year to La Niña (pages 250–253). However, the main reason for the rise in global temperature as shown on Figure 9.76 is the longer-term effect of the continued release of greenhouse gases into the atmosphere.

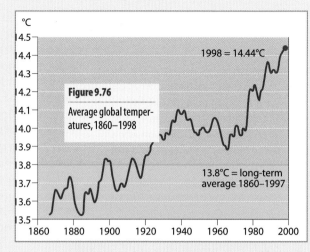

Figure 9.76

Average global temperatures, 1860–1998

1998 = 14.44°C

13.8°C = long-term average 1860–1997

Figure 9.78

The major greenhouse gases

Gas	Sources (natural and man-made)
water vapour	evaporation from the ocean, evapotranspiration from land
carbon dioxide	burning of fossil fuels (power houses, industry, transport), burning rainforests, respiration
methane	decaying vegetation (peat and in swamps), farming (fermenting animal dung and rice-growing), sewage disposal and landfill sites
nitrous oxide	vehicle exhausts, fertiliser, nylon manufacture, power stations
CFCs	refrigerators, aerosol sprays, solvents and foams

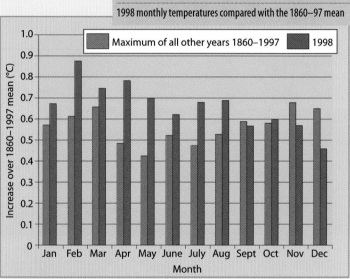

Figure 9.77

1998 monthly temperatures compared with the 1860–97 mean

Maximum of all other years 1860–1997   1998

Increase over 1860–1997 mean (°C)

Month

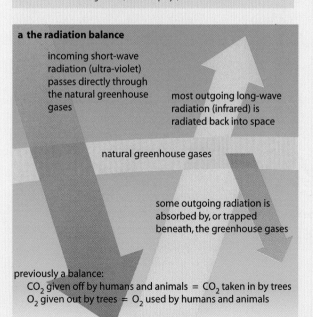

a the radiation balance

incoming short-wave radiation (ultra-violet) passes directly through the natural greenhouse gases

most outgoing long-wave radiation (infrared) is radiated back into space

natural greenhouse gases

some outgoing radiation is absorbed by, or trapped beneath, the greenhouse gases

previously a balance:
CO_2 given off by humans and animals = CO_2 taken in by trees
O_2 given out by trees = O_2 used by humans and animals

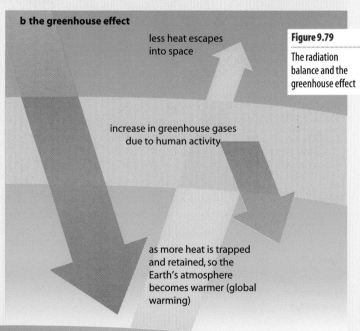

b the greenhouse effect

less heat escapes into space

Figure 9.79

The radiation balance and the greenhouse effect

increase in greenhouse gases due to human activity

as more heat is trapped and retained, so the Earth's atmosphere becomes warmer (global warming)

short-wave radiation is transformed into long-wave radiation (heat) on contact with the Earth's surface

The Earth is warmed during the day by incoming, short-wave radiation (insolation) from the sun and cooled at night by out-going, longer-wave, infrared radiation (page 207). As, over a lengthy period of time, the Earth is neither warming up nor cooling down, there must be a balance between incoming and outgoing radiation (page 209). While incoming radiation is able to pass through the atmosphere (which is 99 per cent nitrogen and oxygen, Figure 9.2), some of the outgoing radiation is trapped by a blanket of trace gases. Because they trap heat as in a greenhouse, these are referred to as **greenhouse gases**. Without these natural greenhouse gases, the Earth's average temperature would be 33°C lower than it is today – far too cold for life in any form. (During the last ice age, temperatures were only 4°C lower.) Water vapour provides the majority of the natural greenhouse effect, with lesser contributions from carbon dioxide, methane, nitrous oxide and ozone.

During the last 150 years there has been, with the exception of water vapour which remains a constant in the system, a rise in greenhouse gas concentrations (Figure 9.78). This has been due largely to the increase in world population and a corresponding growth in human activity, especially agri-cultural and industrial activities.

By adding these gases to the atmos-phere, we are increasing its ability to trap heat (Figure 9.79). Most scientists now accept that the greenhouse effect is causing global warming. World tempera-tures have risen by 0.5°C since the middle of the last century. Latest predictions suggest that temperatures could increase by between 1°C and 3.5°C by the year 2100, with scientists' best guess being about 2°C.

Britain's weather forecast for the next 100 years

'The Met Office's Hadley Centre for Climate Change, and the Climate Research Unit at East Anglia University, predict a grim forecast for the next 100 years. Heavy winter rains will lead to frequent flooding, and destructive gales will be more frequent and severe. With a predicted rise in sea-level of between 2 and 10 cm, storm surges and higher tides will threaten coastal areas. The chance of extreme cold winters will decrease. Days with more than 25 mm rain, at present an extreme event, could occur three or four times a year. Summers will be drier, with fre-quent droughts, particularly in the south and east of England, and summer rainfall of less

than half the present average will be likely every 10 years instead of every 100 years. With a suggested increase in temperature of up to 3°C, heatwaves will be more frequent and there will be many more days when the temperature exceed 25°C.'

February, 1999

Effects of climate change in the UK

A review group, set up by the Department of the Environment, reported on the potential effects of climate change in the UK in November 1997. The group's claims, based on predicted milder, wetter, stormier winters and warmer, drier summers are summarised in Figure 9.80. Its two main concerns were related to:

- the potential effects of changing rainfall patterns on hydrology and ecosystems
- rising sea-levels and more frequent storms in coastal areas where there is a large proportion of Britain's popul-ation, its manufacturing industry, energy production, mineral extrac-tion, valued natural environments and recreational amenities.

Soils	Higher temperatures could reduce water-holding capacities and increase soil moisture deficits, affecting the types of crops and trees. Less organic matter due to drier summers (less produced) and wetter winters (more lost).
Flora/fauna	Higher temperatures and increased water deficit could mean loss of several native species. Warmer climate would allow plants to grow further north and at higher altitudes.
Agriculture	Grasses helped by longer growing season (extra 15 days) but cereals hit by drier summers. Increase in number of pests. Maize and vines in the south.
Forestry	Certain trees able to grow at higher altitudes. New species could be introduced from warmer climates.
Coastal regions	Rise in sea-level plus increase in frequency/number of gales and frequency/height of storm surges would mean more flooding, especially around estuaries. Major impact on housing, industry, farming, energy, transport and wildlife.
Water resources	Water resources would benefit from wetter winters; but hotter, drier summers would increase demands/pressures. Need for irrigation in summer.
Energy	Space heating demand would fall in winter but need for air-conditioning would rise in summer. Probable overall fall in demand. Many power stations are in coastal areas.
Manufacturing/construction	Problem for coastal industries. Fewer days lost in construction due to less snow/frost.
Transport	Many types of transport are sensitive to extreme weather conditions. Benefit of less snow, ice and perhaps fog. Loss due to more storms and flooding.
Recreation/tourism	Tourism would benefit from longer, warmer, drier summers, but perhaps insufficient snow for skiing in Scotland.

Figure 9.80

Specific effects of climate change in the UK

Source: Department of the Environment

References

Barry, R. G. and Chorley, R. J. (1998) *Atmosphere, Weather and Climate*, Routledge.

Chandler, T. J. (1981) *Modern Meteorology and Climatology*, Thomas Nelson.

Goudie, A. (1993) *The Nature of the Environment*, Blackwell.

Money, D. C. (1985) *Climate, Soils and Vegetation*, University Tutorial Press.

Musk, L. (1988) *Weather Systems*, Cambridge University Press.

O'Hare, G. and Sweeney, J. (1986) *The Atmospheric System*, Oliver & Boyd.

Websites

The US Environmental Protection Agency's ozone science information:
http://www.epa.gov/ozone/science/science.html

The Union of Concerned Scientists' ozone depletion site:
http://www.ucsusa.org/resources/ozone.html

The US National Oceanographic and Atmospheric Administration's (NOAA) hurricane main page:
http://hurricanes.noaa.gov/

The Earth Space Research Group's comprehensive Indian monsoon site:
http://www.crseo.ucsb.edu/1OM2/Wiring_Tree.html

The UK Meteorological Office home page:
http://www.meto.gov.uk/home.html

The US NOAA Climate Prediction Center site:
http://www.cpc.noaa.gov/index.html

The US NOAA Climate Prediction Center's El Niño/La Niña (ENSO) main page:
http://www.cpc.noaa.gov/products/analysis_monitoring/ensostuff/

The US Environmental Protection Agency global warming site:
http://www.epa.gov/globalwarming/index.html

The Union of Concerned Scientists' global warming site:
http://ucsusa.org/warming/index.html

The greenhouse effect and global warming:
http://www.geocities.com/Athens/Parthenon/5173/greenhouse_effect.html

See also for more links:
http://www.nelsonthornes.com/gaia

Questions

Q

1
a What is the 'atmosphere' of the Earth? **(3 marks)**
b What is the *difference* between 'weather' and 'climate'? **(4 marks)**
c Describe the 'solar cascade of energy' to the Earth. **(4 marks)**
d What is the importance of **i** carbon dioxide and **ii** clouds in the energy balance of the Earth? **(4 marks)**
e Ozone in the troposphere is a danger to health. Why is there concern that ozone in the *stratosphere* is being depleted? **(5 marks)**
f What measures can be taken to restrict the potential damage due to ozone depletion? **(5 marks)**

2
a Explain how each of the following factors affects the winds that cross them:
  i a large body of water (e.g. a sea) **(4 marks)**
  ii a mountain range. **(6 marks)**
b On a field course in Switzerland a geography student noted: 'On the north-facing side of the valley the forests came close to the valley floor while the settlement huddled at the foot of the south-facing slope and here there were ploughed fields. There were forests but they started higher up the slope.' Suggest the cause of these *differences* in land use. **(6 marks)**
c A January weather forecast for the UK stated: 'Although it will be cool today, temperatures will stay above freezing tonight because of the cloud cover'. Explain the effect of cloud on temperature. **(4 marks)**
d Why is it warmer in summer than in winter? **(5 marks)**

3
a i Using an annotated diagram **only**, illustrate the variation of temperature and pressure with altitude in the atmosphere. **(6 marks)**
  ii Explain what happens to incoming solar radiation as it passes through the Earth's atmosphere. **(6 marks)**
b i Study Figure 9.5 (page 209). Making good use of the data, explain why there is a general trend of movement of heat energy from the Equator to the poles. **(6 marks)**
  ii Describe how heat is transferred from the tropics towards the poles. **(7 marks)**

4
a Describe and explain what happens to *incoming solar radiation* (**insolation**) once it reaches the edge of the Earth's atmosphere. **(10 marks)**
b Explain the importance of **each** of the following in relation to heat energy in the atmosphere:
  i latitude
  ii altitude
  iii land and sea. **(10 marks)**
c The greatest amount of insolation is experienced close to the Equator. Why does this area not become increasingly hot? **(5 marks)**

5
a Suggest **one** way you could test the hypothesis that the temperatures in an urban area are different from those in the surrounding countryside. Describe the method you would use to collect and record the data to carry out the proposed test. **(7 marks)**
b Explain **two** reasons why temperatures in urban areas may be higher than those in surrounding rural areas. **(10 marks)**
c Suggest **two** ways in which planning policies can reduce the problems caused by microclimatic features of urban areas. **(8 marks)**

AS

A2

6 a What is the 'water cycle'? **(4 marks)**
  b Why is there, generally, the same amount of water in the oceans over a period of time? **(6 marks)**
  c Choose **one** mechanism for the formation of rain and describe the processes involved. **(6 marks)**
  d What is 'humidity'? **(4 marks)**
  e Making good use of diagrams to describe the instrument, describe how humidity is measured. **(5 marks)**

7 a How does a meteorologist get information to forecast the weather? **(4 marks)**
  b Use Places 29 (page 232) to answer the following questions:
    i What was the weather forecast on 11–15 October 1987? **(3 marks)**
    ii Describe the meteorological conditions over the Western Approaches and Bay of Biscay at 6.00 pm on 15 October. **(3 marks)**
    iii Describe the track of the storm over the next 12 hours. **(4 marks)**
    iv What happened to the weather over southern England during this 12-hour period? **(4 marks)**
    v Describe **three** effects of the storm on people. **(3 marks)**
  c Explain **two** reasons why meteorologists failed to forecast the very strong winds of 15 October. **(4 marks)**

8 a i What is 'stratus' cloud? **(3 marks)**
    ii What is 'cumulo-nimbus' cloud? **(3 marks)**
  b Making good use of diagrams, explain why rain falls when an onshore wind blows over an upland area. **(7 marks)**
  c Why does fog often form over a coastal area in the autumn? **(6 marks)**
  d Explain the formation of *smog* over an urban area **(6 marks)**

9 a Describe **three** mechanisms which are likely to trigger upward movement of a parcel of air from sea level. **(6 marks)**
  b Study Figure 9.81.
    i What is meant by the term 'ELR'? **(4 marks)**
    ii Identify the height of the base of clouds. **(1 marks)**
    iii Explain why this height is the cloud base. **(4 marks)**
    iv Identify the air stream(s) which would have cloud cover. State why this is so. **(2 marks)**
    v At what height would condensation in a cloud be in the form of ice? **(2 marks)**
  c Choose **either** stability **or** instability. Describe and explain the weather conditions normally associated with that atmospheric condition. **(6 marks)**

10 a Explain the **difference** between absolute humidity and relative humidity. **(8 marks)**
  b Making good use of diagrams, show how condensation occurs as air rises through the atmosphere. **(10 marks)**
  c Explain the cause of low-level clouds (mist) as shown in Figure 9.23 (page 221). **(7 marks)**

11 a i Describe the causes of the ITCZ.
    ii What weather conditions are associated with the ITCZ?
    iii Why does the ITCZ move with the seasons? **(13 marks)**
  b What is a jet stream and how is it associated with ground-level weather conditions? **(12 marks)**

12 The following are meteorological conditions which develop a range of weather conditions over the British Isles:
  a an anticyclone centred over the English Midlands in winter
  b a mature depression with its centre over the Central Valley of Scotland in summer
  c a depression centred over Paris and an anticyclone to the north of Scotland in January.
  In each case describe how and explain why weather conditions vary in two contrasting locations in the British Isles. **(25 marks)**

13 a Study Figure 9.49 (page 235). Describe the distribution of source areas, directions of movement, frequency of occurrence and timing of tropical cyclones. **(10 marks)**
  b Choose any **one** type of tropical storm. Describe and explain the sequence of weather associated with the passage of the storm. **(10 marks)**
  c Suggest **two** different responses of people to the hazard posed by tropical cyclones in the areas where they occur. **(5 marks)**

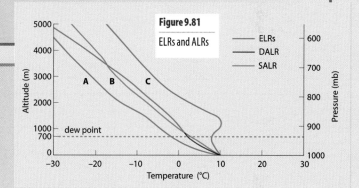

Figure 9.81
ELRs and ALRs
— ELRs
— DALR
— SALR

14 Study the following data and answer the questions below.

Lower atmosphere conditions up to 2000 m

Sea-level temperature, in all cases, is 10°C

Condensation level (dew point), in all cases, is at 500 m

Dry adiabatic lapse rate is 10°C per 1000 m (1°C per 100 m)

Saturated adiabatic lapse rate is 5°C per 1000 m (0.5°C per 100 m)

Airstream 1 – ELR 4°C per 1000 m (0.4°C per 100 m)

Airstream 2 – ELR 12°C per 1000 m (1.2°C per 100 m)

Airstream 3 – ELR 10°C per 1000 m (1.0°C per 100 m)

  a i On **one side of one piece of graph paper** create a graph to compare the conditions shown above. **(7 marks)**
    ii Label each airstream 'stable', 'unstable' or 'conditionally unstable' as relevant. **(3 marks)**
    iii For the airstream you have labelled 'conditionally unstable', identify the height at which it becomes unstable. **(1 mark)**
  b Explain the term 'conditional instability'. **(6 marks)**
  c i Explain what is meant by the term 'air mass'. **(4 marks)**
    ii Explain what is meant by the term 'airstream'. **(4 marks)**

A2

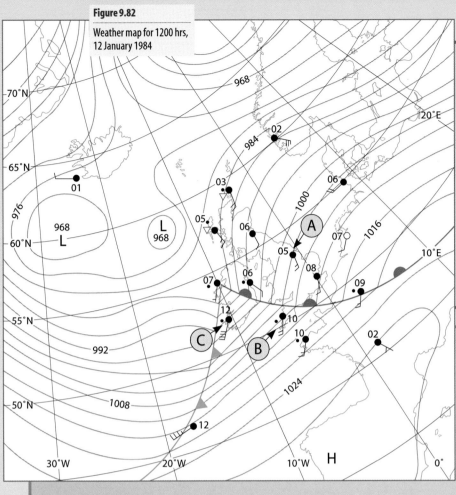

Figure 9.82

Weather map for 1200 hrs, 12 January 1984

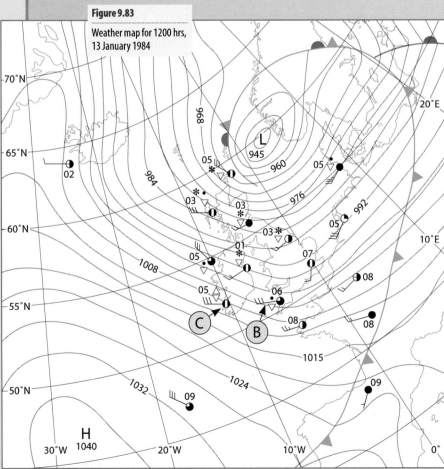

Figure 9.83

Weather map for 1200 hrs, 13 January 1984

15 Study Figure 9.82 and answer the following questions.

   a What is the name of the pressure system shown? **(2 marks)**

   b What is the weather like at place A (Doncaster)? **(4 marks)**

   c What is the red line with half circles on it? **(5 marks)**

   d Locate the warmest and the coolest place in the British Isles. **(2 marks)**

   e i Over the next 12 hours the pressure system moves so that it is in the North Sea. Give a weather forecast for place A (Doncaster) over this period. **(6 marks)**

       ii Why would you expect this to happen? **(6 marks)**

16 a Study Figures 9.82 and 9.83. Describe the changes in the weather being experienced at Limerick (place C) over this 24-hour period. **(8 marks)**

   b Explain what has happened to the frontal system over this period of time. **(8 marks)**

   c Describe, and explain the causes of, the types and distribution of the precipitation shown in Figure 9.83. **(9 marks)**

17 a i What is 'barometric pressure' and how is it measured?

       ii As an example of pressure patterns explain the meaning of the pattern of isobars shown on Figure 9.83. **(10 marks)**

   b What are 'fronts' such as those shown on Figure 9.82? **(8 marks)**

   c Using examples from Figures 9.82 and 9.83, explain the importance of synoptic charts in producing a weather forecast. **(7 marks)**

18 The passage of a depression over the British Isles leads to predictable changes in the weather over a period of time. Describe and explain the sequence of weather experienced in Liverpool over a 12-hour period as a mature depression passes from west to east. In your answer you should refer to:

  ■ changes in wind speed and direction, cloud cover and type, temperature, and precipitation.

  ■ The relationship between these changes and the sectors of the depression. **(25 marks)**

For information on **Key Skills** and the **Key Skills Certificate** see page 101.

At Level 3 the Key Skills are described as needing to be part of a 'substantial activity'. The activities on this page can be used to provide evidence for the Key Skill of **Information Technology** to Level 3. On page 101 you will find activities to cover the Key Skill of Communication and on page 387 activities to cover Application of Number, also to Level 3.

The activities suggested below can be used as individual tasks to support Key Skill development; as a completed piece of evidence towards a Key Skills portfolio; or as a learning vehicle to consider the Geography content of the material.

Information Technology – Level 3

You need to show competence in the use of a range of IT programmes in the creation of a completed activity. The specification requires *two* different activities (Task One and Task Two below) to show the ability to:
- plan a project and obtain information from a range of electronic sources
- develop the information:
  1 by bringing together information from several sources
  2 by sorting information into a useful form
  3 by testing data (e.g. using 'what if' queries)
  4 by comparing data from different sources and
  5 by exchanging information with others electronically.

The tasks suggested below use research into climatic data to allow you to work in a way that will, in total, present a 'package' of required competences.

Task One: Research data about one world climate zone

Choose any *one* **world climate zone** and complete the following activities.

Carry out the required research to identify the main characteristics of one climate zone for a report of 1500 words including tabulated data, graphs and images. The information will be collated and processed using a database and a spreadsheet and will be finally presented, using a word processor program, as a piece of work assessed by a Geography tutor.

Activity One

Design a database structure to store climatic data for weather stations in different parts of the world with the same climate. Your database will need to be able to handle alphabetical and numerical data as identified in Activity Two below.

Activity Two

Collect data for no fewer than **15 places** and enter it in your database. Check that you have correctly entered the data at each stage of your work. The data you collect should include:
- name of weather station
- altitude
- latitude
- longitude
- continent/country
- average temperature for each of the 12 months
- average monthly precipitation for each of the 12 months.

These weather stations can be input in any order – to show that you can sort data you should print out your data and then sort your weather stations into alphabetical order.

Activity Three

Save your work with a recognisable file name and print out a copy of your database including all headings.

Activity Four

Retrieve your saved file and query the database to find the **temperature data** for all the weather stations in *one* continent.

Produce a **report** which shows the temperature in January and July and the altitude of each weather station for the climate zone in your chosen continent. Print the report and save your data with a new file name.

Activity Five

Export all the names, temperature and precipitation data to a spreadsheet. Make sure your spreadsheet can be presented to show all of the data by changing font, size of rows and columns, etc. as appropriate.

Save your spreadsheet. Print out all of the data in your spreadsheet, including the column headings.

Activity Six

Use your spreadsheet data to produce a **line graph** of temperature data and a **bar chart** of precipitation data for *one* weather station. On your graph identify the altitude, temperature range and total annual precipitation for the station.

Print the graph and chart *and also* a spreadsheet showing the formulae used to calculate temperature range and precipitation total. Make sure you save the relevant files.

Activity Seven

Write a word-processed report to describe the main features of the climate you have chosen and any variations in different parts of the world. Your report should contain *one or more* **tables** imported from your database and *one or more* **graphs** imported from your spreadsheet. Save your work and print it out.

Your report should be produced in **draft** form initially and submitted to your Geography tutor, who will make comments on the **geography** and the **presentation** of your work.

Your **final report** should make appropriate use of style, layout, images (clipart or diagrams produced in a drawing package). It should be page-numbered and use appropriate labelling in the header and/or footer boxes.

Task Two: The effect of altitude on climate

Collect information from several sources for a piece of research into the effects of altitude on climate. The research will generate support material (graphical, diagrammatic, statistical and background notes) to inform a class discussion about this aspect of the causes of climate.

Activity Eight

For at least *one* other student, provide data from your database/spreadsheet to give information about the altitude and average annual temperature of every weather station in your sample. This information should be transferred electronically (with evidence that a tutor has seen the transfer system used).

Collect database or spreadsheet data for **30 weather stations** from *two* other climate regions from other students, and use the 45 data items to find out whether there is an impact from altitude on the climate in any of the three climate regions. You should produce **notes** comparing the effect of altitude in the three climate zones, for a student discussion concerning the effects of altitude on climate. The whole student group should prepare the discussion and ensure that the information, its presentation format and style, allow for common use by the whole group.

Chapter 10

Soils

'*To many people who do not live on the land, soil appears to be an inert, uniform, dark-brown coloured, uninteresting material in which plants happen to grow. In fact little could be further from the truth.*'

Brian Knapp, *Soil Processes*, 1979

Soil forms the thin surface layer of the Earth's crust. It provides the foundation for plant and, consequently, animal life on land. The most widely accepted scientific definition is that by J. Joffe (1949) who stated that:

'the soil is a natural body of animal, mineral and organic constituents differentiated into horizons [Figure 10.5] of variable depth which differ from the material below in morphology, physical make-up, chemical properties and composition, and biological characteristics.'

A simpler definition is that soil results from the interrelationships between, and interaction of, several physical, chemical and biological processes, all of which vary according to different natural environments.

The study of soil, its origins and characteristics (**pedology**) is a science in itself.

Soil formation

The first stage in the formation of soil is the accumulation of a layer of loose, broken, unconsolidated parent material known as **regolith**. Regolith may be derived from either the *in situ* weathering of bedrock (i.e. the parent or underlying rock) or from material that has been transported from elsewhere and deposited, e.g. as alluvium, glacial drift, loess or volcanic ash. The second stage, the formation of **true soil** or **topsoil**, results from the addition of water, gases (air), living organisms (biota) and decayed organic matter (humus).

Pedologists have identified five main factors involved in soil formation (Figure 10.1). As all of these are closely interconnected and interdependent, their relationship may be summarised as follows:

soil = f (parent material + climate + topography + organisms + time)
where: f = function of.

Parent material

When a soil develops from an underlying rock, its supply of minerals is largely dependent on that rock. The minerals are susceptible to different rates and processes of weathering – see the example of granite, Figure 10.2. Parent material contributes to control of the depth, texture, drainage (permeability) and quality (nutrient content) of a soil and also influences its colour. In most of Britain, parent material is the major factor in determining the soil type, e.g. limestone, granite or, most commonly, drift.

Figure 10.1

Factors affecting the formation of soil

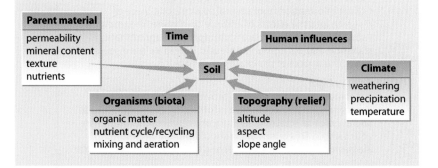

Parent material
- permeability
- mineral content
- texture
- nutrients

Time

Human influences

Soil

Climate
- weathering
- precipitation
- temperature

Organisms (biota)
- organic matter
- nutrient cycle/recycling
- mixing and aeration

Topography (relief)
- altitude
- aspect
- slope angle

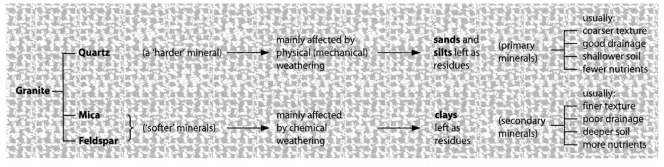

Figure 10.2

The influence of a parent rock – granite – on soil formation

Climate

Climate determines the type of soil at a global scale. The distribution of world soil types corresponds closely to patterns of climate and vegetation. Climate affects the rate of weathering of the parent rock, with the most rapid breakdown being in hot, humid environments. Climate also affects the amount of humus (organic material) in the soil. The amount is a balance between the input and output, the input and output being a function of the effects of temperature and moisture on biological activity. One might expect tropical rainforest soils to have more humus than tundra soils because of the greater mass of vegetation. However, it is possible for some tundra soils to have more humus accumulation due to a lower output, and some tropical rainforest soils to have less because of greater humus breakdown.

Rainfall totals and intensity are also important. Where rainfall is heavy, the downward movement of water through the soil transports mineral salts (i.e. soluble minerals) with it, a process known as **leaching**. Where rainfall is light or where evapotranspiration exceeds precipitation, water and mineral salts may be drawn upwards towards the surface by the process of **capillary action**.

Temperatures determine the length of the growing season and affect the supply of humus. The speed of vegetation decay is fastest in hot, wet climates as temperatures also influence (i) the activity and number of soil organisms and (ii) the rate of evaporation, i.e. whether leaching or capillary action is dominant.

Topography (relief)

As the height of the land increases, so too do amounts of precipitation, cloud cover and wind, while temperatures and the length of the growing season both decrease. Aspect is an important local factor in mid-latitudes (page 212), with south-facing slopes in the northern hemisphere being warmer and drier than those facing north. The angle of slope affects drainage and soil depth. Greater moisture flows and the increased effect of gravity on steeper slopes can accelerate mass movement and the risk of soil erosion. Soils on steep slopes are likely to be thin, poorly developed and relatively dry. The more gentle the slope, the slower the rate of movement of water through the soil and the greater the likelihood of waterlogging and the formation of peat on plateau-like surfaces at the top of the slope (Figure 10.3). There is little risk of soil erosion but the increased rate of weathering, due to the extra water, and the receipt of material moved downslope, tend to produce deep soils at the foot of the slope. A **catena** is where soils are related to the topography of a hillside and is a sequence of soil types down a slope. The catena (Figure 10.3) is described in more detail on page 276.

Organisms (biota)

Plants, micro-organisms such as bacteria and fungi, and animals all interact in the **nutrient cycle** (page 300). Plants take up mineral nutrients from the soil and return them to it after they die. This recycling of plant nutrients (Figure 12.7) is achieved by the activity of micro-organisms, which assist in nitrogen fixation (page 268) and the decomposition and decay of dead vegetation. At the same time, macro-organisms, which include worms and termites, mix and aerate the soil. Human activity is increasingly affecting soil development through the addition of fertiliser, the breaking up of horizons by ploughing, draining or irrigating land, and by unwittingly accelerating or deliberately controlling soil erosion.

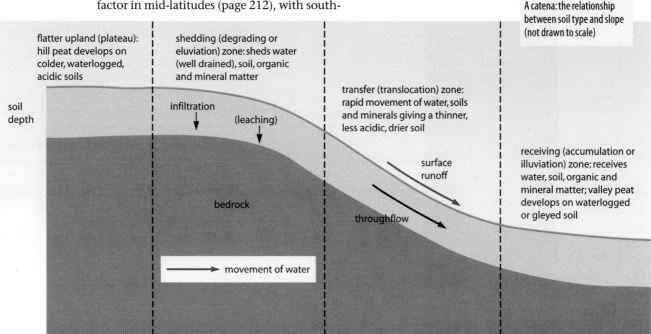

Figure 10.3

A catena: the relationship between soil type and slope (not drawn to scale)

flatter upland (plateau): hill peat develops on colder, waterlogged, acidic soils

shedding (degrading or eluviation) zone: sheds water (well drained), soil, organic and mineral matter

transfer (translocation) zone: rapid movement of water, soils and minerals giving a thinner, less acidic, drier soil

soil depth

infiltration

(leaching)

surface runoff

receiving (accumulation or illuviation) zone: receives water, soil, organic and mineral matter; valley peat develops on waterlogged or gleyed soil

bedrock

throughflow

movement of water

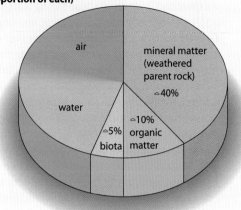

pore space containing **air** and/or water = **45%**
(can be **45%** water, or **45%** air, but is more
usually a proportion of each)

mineral matter + organic matter
+ biota = **55%**

air

water

mineral matter
(weathered
parent rock)
⁓40%

⁓5%
biota

⁓10%
organic
matter

Figure 10.4

Relative proportions,
by volume, of compo-
nents in a 'normal'
soil (*after* Courtney
and Trudgill)

Figure 10.5

An idealised soil
profile in Britain

Time

Soils usually take a long time to form, perhaps
up to 400 years for 10 mm and, under extreme
conditions, 1000 years for 1 mm. It can take
3000–12 000 years to produce a sufficient depth
of mature soil for farming, although agriculture
can be successful on newly deposited alluvium
and volcanic ash. Newly forming soils tend to
retain many characteristics of the parent material from which they are derived. With time,
they acquire new characteristics resulting from
the addition of organic matter, the activity of
organisms, and from leaching. **Horizons**, or
layers (Figure 10.5), reflect the balance between
soil processes and the time that has been available for their development. In northern Britain,
upland soils must be less than 10 000 years old,
as that was the time of the last glaciation, when
any existing soil cover was removed by ice. The
time taken for a mature soil to develop depends
primarily upon parent material and climate.
Soils develop more rapidly where parent mate-

rial derived from *in situ* weathering consists of
sands rather than clays, and in hot, wet climates
rather than in colder and/or drier environments.

A mature, fully-developed soil consists of four
components: mineral matter, organic matter
including biota (page 268), water and air. The relative proportions of these components in a
'normal' soil, by volume, is given in Figure 10.4.

The soil profile

The **soil profile** is a vertical section through the
soil showing its different horizons (Figure 10.5).
It is a product of the balance between soil system
inputs and outputs (Figure 10.6) and the redistribution of, and chemical changes in, the various
soil constituents. Different soil profiles are
described in Chapter 12, but an idealised profile
is given here to aid familiarisation with several
new terms.

The three major soil horizons, which may be
subdivided, are referred to by specific letters to
indicate their genetic origin.

- The upper layer, or **A horizon**, is where
  biological activity and humus content are
  at their maximum. It is also the zone that
  is most affected by the leaching of soluble
  materials and by the downward movement,
  or **eluviation**, of clay particles. Eluviation is
  the washing out of material, i.e. the removal
  of organic and mineral matter from the *A*
  horizon (Figure 10.5).

- Beneath this, the **B horizon** is the zone of
  accumulation, or **illuviation**, where clays
  and other materials removed from the *A*
  horizon are redeposited. Illuviation is the
  process of inwashing, i.e. the redeposition of
  organic and mineral matter in the *B* horizon.
  The *A* and *B* horizons together make up the
  true soil.

- The **C horizon** consists mainly of recently
  weathered parent material (regolith) resting
  on the bedrock.

While this threefold division is useful and
convenient, it is, as will be seen later, over-
simplified. Several examples show this:

- Humus may be mixed throughout the depth
  of the soil, or it may form a distinct layer.
  Where humus is incorporated within the soil
  to give a crumbly, black, nutrient-rich layer it
  is known as **mull** (page 266). Where humus
  is slow to decompose, as in cold, wet upland
  areas, it produces a fibrous, acidic and
  nutrient-deficient surface horizon known
  as **mor** (page 266) (peat moorlands).

- The junctions of horizons may not always
  be clear.

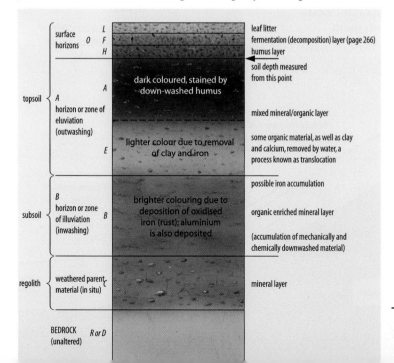

surface
horizons
O

L
F
H

leaf litter
fermentation (decomposition) layer (page 266)
humus layer

soil depth measured
from this point

topsoil

A
horizon or zone of
eluviation
(outwashing)

A

dark coloured, stained by
down-washed humus

mixed mineral/organic layer

E

lighter colour due to removal
of clay and iron

some organic material, as well as clay
and calcium, removed by water, a
process known as translocation

subsoil

B
horizon or zone
of illuviation
(inwashing)

B

brighter colouring due to
deposition of oxidised
iron (rust); aluminium
is also deposited

possible iron accumulation

organic enriched mineral layer

(accumulation of mechanically and
chemically downwashed material)

regolith

weathered parent
material (in situ)

C

mineral layer

BEDROCK
(unaltered)

R or D

- All horizons need not always be present.
- The depth of soil and of each horizon vary at different sites. Local conditions produce soils with characteristic horizons differing from the basic *A*, *B*, *C* pattern: for example, a water-logged soil, having a shortage of oxygen, develops a gleyed (*G*) horizon (page 275).

The soil system

Figure 10.6 is a model showing the soil as an open system where materials and energy are gained and lost at its boundaries. The system comprises inputs, stores, outputs and recycling or feedback loops (Framework 3, page 45). Inputs include:
- water from the atmosphere or throughflow from higher up the slope
- gases from the atmosphere and the respiration of soil animals and plants
- mineral nutrients from weathered parent material, which are needed as plant food
- organic matter and nutrients from decaying plants and animals, and
- solar energy and heat.

Outputs include:
- water lost to the atmosphere through evapo-transpiration
- nutrients lost through leaching and through-flow, and
- loss of soil particles through soil creep and erosion.

Recycling

Plants, in order to live, take up nutrients from the soil (page 268). Some of the nutrients may be stored until:
- either the vegetation sheds its leaves (during the autumn in Britain), or
- the plants die and, over time, decompose due to the activity of micro-organisms (biota, page 268).

These two processes release the stored nutrients, allowing them to be returned to the soil ready for future use – the so-called **nutrient** (or humus) **cycle**.

Soil properties

The four major components of soil – water, air, mineral and organic matter (Figure 10.4) – are all closely interlinked. The resultant interrelation-ships produce a series of 'properties', ten of which are listed and described below.

1. mineral (inorganic) matter
2. texture
3. structure
4. organic matter (including humus)
5. moisture
6. air
7. organisms (biota)
8. nutrients
9. acidity (pH value)
10. temperature.

It is necessary to understand the workings of these properties to appreciate how a particular soil can best be managed.

1 Mineral (inorganic) matter

As shown in Figure 10.2, soil minerals are obtained mainly by the weathering of parent rock. Weathering is the major process by which nutrients, essential for plant growth, are released. **Primary minerals** are minerals that were present in the original parent material and which remain unaltered from their original state. They are present throughout the soil-forming process, mainly because they are insoluble, e.g. quartz. **Secondary minerals** are produced by weathering reactions and are therefore produced within the soil. They include oxides and hydroxides of primary minerals (e.g. iron) which result from the exposure to air and water (page 40).

Figure 10.6

The 'open' soil system

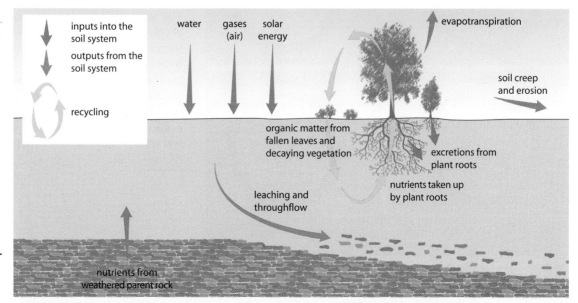

inputs into the soil system

outputs from the soil system

recycling

water   gases (air)   solar energy   evapotranspiration

soil creep and erosion

organic matter from fallen leaves and decaying vegetation

excretions from plant roots

nutrients taken up by plant roots

leaching and throughflow

nutrients from weathered parent rock

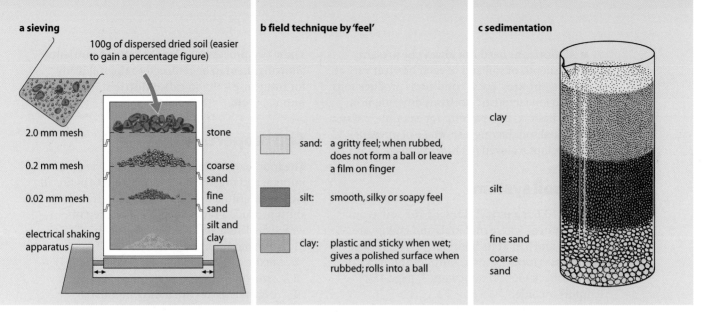

a sieving

100g of dispersed dried soil (easier to gain a percentage figure)

2.0 mm mesh — stone

0.2 mm mesh — coarse sand

0.02 mm mesh — fine sand

silt and clay

electrical shaking apparatus

b field technique by 'feel'

sand: a gritty feel; when rubbed, does not form a ball or leave a film on finger

silt: smooth, silky or soapy feel

clay: plastic and sticky when wet; gives a polished surface when rubbed; rolls into a ball

c sedimentation

clay

silt

fine sand

coarse sand

Figure 10.7

Measuring soil texture (*after* Courtney and Trudgill)

Figure 10.8

The texture of different soil types

2 Soil texture

The term 'texture' refers to the degree of coarseness or fineness of the mineral matter in the soil. It is determined by the proportion of **sand**, **silt** and **clay** particles. Particles larger than sand are grouped together and described as stones. In the field, it is possible to decide whether a soil sample is mainly sand, silt or clay by its 'feel'. As shown in Figure 10.7b, a sandy soil feels gritty and lacks cohesion; a silty soil has a smoother, soaplike feel as well as having some cohesion; and a clay soil is sticky and plastic when wet and, being very cohesive, may be rolled into various shapes.

This method gives a quick guide to the texture, but it lacks the precision needed to determine the proportion of particles in a given soil with any accuracy. This precision may be obtained from either of two laboratory measure-ments, both of which are dependent upon particle size. The Soil Survey of England and Wales uses the British Standards classification, which gives the following diameter sizes:

coarse sand	between 2.0 and 0.6 mm
medium sand	between 0.6 and 0.2 mm
fine sand	between 0.2 and 0.06 mm
silt	between 0.06 and 0.002 mm
clay	less than 0.002 mm.

One method of measuring texture involves the use of sieves with different meshes (Figure 10.7a). The sample must be dry and needs to be well-shaken. A mesh of 0.2 mm, for example, allows fine sand, silt and clay particles to pass through it, while trapping the coarse sand. The weight of particles remaining in each sieve is expressed as a percentage of the total sample.

In the second method, sedimentation (Figure 10.7c), a weighed sample is placed in a beaker of water, thoroughly shaken and then allowed to settle. According to **Stoke's Law**: 'the settling rate of a particle is proportional to the diameter of that particle'. Consequently, the larger, coarser, sand grains settle quickly at the bottom of the beaker and the finer, clay particles settle last, closer to the surface (compare Figure 3.22). The Soil Survey tends to use both methods because sieving is less accurate in measuring the finer material, and sedimentation is less accurate with coarser particles.

The results of sieving and sedimentation are usually plotted either as a pie chart (Figure 10.8) or as a triangular graph (Figure 10.9). As the proportions of sand, silt and clay vary considerably, it is traditional to have 12 texture categories (Figure 10.9).

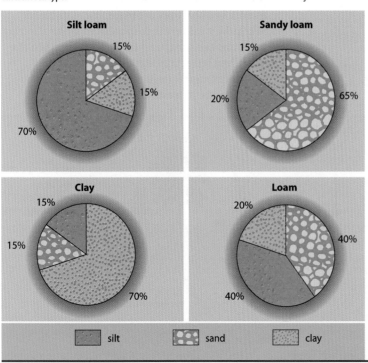

Silt loam
15%
15%
70%

Sandy loam
15%
20%
65%

Clay
15%
15%
70%

Loam
20%
40%
40%

silt   sand   clay

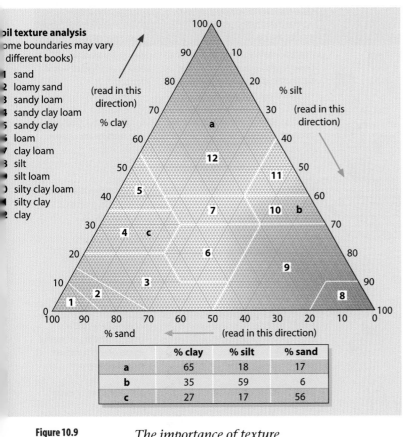

Soil texture analysis
(some boundaries may vary
in different books)

1 sand
2 loamy sand
3 sandy loam
4 sandy clay loam
5 sandy clay
6 loam
7 clay loam
8 silt
9 silt loam
10 silty clay loam
11 silty clay
12 clay

(read in this direction)
% clay

% silt
(read in this direction)

% sand ← (read in this direction)

	% clay	% silt	% sand
a	65	18	17
b	35	59	6
c	27	17	56

Figure 10.9

Soil texture analysis:
the use of a
triangular graph

The importance of texture

As texture controls the size and spacing of soil pores, it directly affects the soil water content, water flow and extent of aeration. Clay soils tend to hold more water and are less well drained and aerated than sandy soils (page 267).

Texture also controls the availability and retention of nutrients within the soil. Nutrients stick to – i.e. are adsorbed onto – clay particles and are less easily leached by infiltration or throughflow than in sandy soils (page 268).

Plant roots can penetrate coarser soils more easily than finer soils, and 'lighter' sandy soils are easier to plough for arable farming than 'heavier' clays.

Texture greatly influences soil structure.

How does texture affect farming?

The following comments are generalised as it must be remembered that soils vary enormously.

Sandy soils, being well drained and aerated, are easy to cultivate and permit crop roots (e.g. carrots) to penetrate. However, they are vulnerable to drought, mainly because, due to their relatively large particle size (Figure 8.2a), they lack the micropores that would retain moisture (page 267) and partly because they usually

contain limited amounts of organic matter. They also need considerable amounts of fertiliser because nutrients and organic matter are often leached out and not replaced.

Silty soils also tend to lack mineral and organic nutrients. The smaller pore size means that more moisture is retained than in sands but heavy rain tends to 'seal' or cement the surface, increasing the risk of sheetwash and erosion.

Clay soils tend to contain high levels of nutrient and organic matter but they are difficult to plough and, after heavy rain and due to their small particle size (Figure 8.2b) which helps to retain water (page 267), are prone to waterlogging and may become gleyed (pages 272 and 275). Plant roots find difficulty in penetration. Clays expand when wet, shrink when dry and take the longest time to warm up.

The ideal soil for agriculture is a **loam** (Figures 10.8 and 10.9). This has sufficient clay (20 per cent) to hold moisture and retain nutrients; sufficient sand (40 per cent) to prevent waterlogging, to be well aerated and to be light enough to work; and sufficient silt (40 per cent) to act as an adhesive, holding the sand and clay together. A loam is likely to be least susceptible to erosion.

3 Soil structure

It is the aggregation of individual particles that gives the soil its structure. In undisturbed soils, these aggregates form different shapes known as **peds**. It is the shape and alignment of the peds which, combined with particle size/texture, determine the size and number of the pore spaces through which water, air, roots and soil organisms can pass. The size, shape, location and suggested agricultural value of each of the six ped types are given in Figure 10.10. It should be noted, however, that some soils may be structureless (e.g. sands), some may have more than one ped structure (Figure 10.11), and most are likely to have a distinctive ped in each horizon. It is accepted that soils with a good crumb structure give the highest agricultural yield, are more resistant to erosion and develop best under grasses – which is why fallow should be included in a farming crop rotation. Sandy soils have the weakest structures as they lack the clays, organic content and secretions of organisms needed to cause the individual particles to aggregate. A crumb structure is ideal as it provides the optimum balance between air, water and nutrients.

Type of structure (ped)	Size of structure (mm)	Description of peds	Shape of peds	Location (horizon: texture) and formation	Agricultural value
crumb	1–5	small individual particles similar to breadcrumbs; porous		A horizon: loam soil; formed by action of soil fauna (e.g. earthworms, mites and termites), high content of fibrous roots (grasses) and excretion of micro-organisms	the most productive; well aerated and drained – good for roots
granular	1–5	small individual particles; usually non-porous	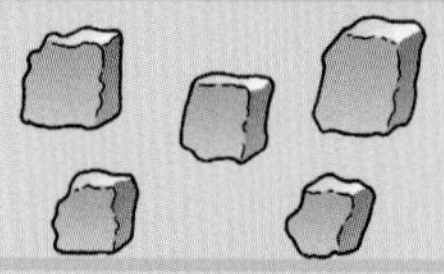	A horizon: clay soil; formation as for crumb structure	fairly productive; problems with drainage and aeration
platy	1–10	vertical axis much shorter than horizontal, like overlapping plates; restrict flow of water	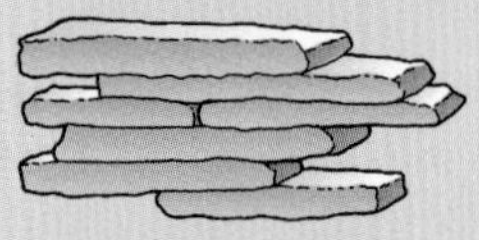	B horizon: silts and clays; formed by contraction by tree roots, especially when trees (e.g. Scots pine) sway in wind. Also due to ice lens, and compaction due to farm machinery	the least productive; hinders water and air movement; restricts roots
blocky	10–75	irregular shape with horizontal and vertical axes about equal; may be rounded or angular but closely fitting	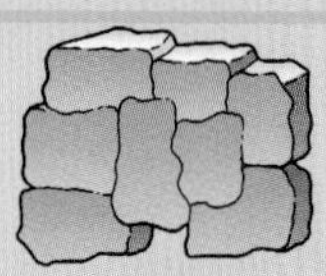	B horizon: clay-loam soils; formation associated with wetting – drying and freeze–thaw processes	productive: usually well drained and aerated
prismatic	20–100	vertical axis much larger than horizontal; angular caps and sides to columns	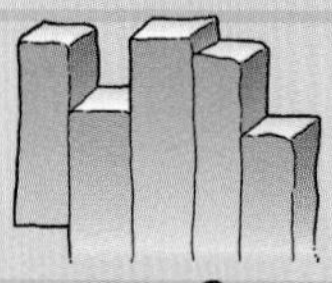	B and C horizons: often limestones or clays; formation associated with wetting–drying and freeze-thaw processes	usually quite productive: formed by wetting and drying; adequate water movement and root development
columnar	20–100	vertical axis much larger than horizontal; rounded caps and sides to columns	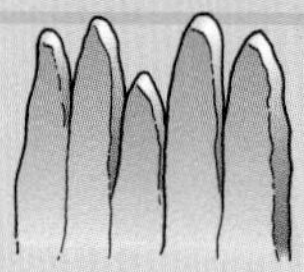	B and C horizons; alkaline soils; formation associated with accumulation of sodium	quite productive (if water available)

Figure 10.10

Different soil structures

Figure 10.11

Differences in peds (*after* Courtney and Trudgill)

4 Organic matter

Organic matter, which includes humus, is derived mainly from decaying plants and animals, or from the secretions of living organisms. Fallen leaves and decaying grasses and roots are the main source of organic matter. Soil organisms, such as bacteria and fungi, break down the organic matter and, depending on the nature of the soil-forming processes (Figure 10.17), help develop up to three distinct organic layers at the surface of the soil profile (Figure 10.5).

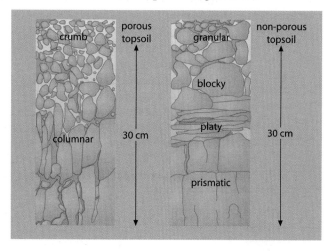

1 *L* or **leaf litter** layer: plant remains are still visible.
2 *F* or **fermentation (decomposition)** layer: decay is most rapid, although some plant remains are still visible.
3 *H* or **humus** layer: primarily organic in nature where, following decomposition, all recognisable plant and animal remains have been broken down into a black, slimy, amorphous organic material.

Wherever soil biological activity is low (due to one or a combination of acidity, low temperatures, wetness or the difficulty in decomposing organic matter), soil organism activity is greatly reduced or absent. As the litter layer cannot be mixed into the soil, then organic horizons build up to give the distinct *L*, *F* and *H* layers of a **mor**.

Where soil organisms are active, they will readily mix the litter into the soil, dispersing it throughout the *A* horizon where it decomposes into an *A* horizon rich in humus – the **mull** layer. Where organic material and mineral matter do mix, mainly due to earthworm activity, the result is the **clay–humus complex** (page 268). The clay–humus complex is essential for a fertile soil as it provides it with a high water- and nutrient-holding capacity and, by binding particles together, helps reduce the risk of erosion.

Humus gives the soil a black or dark-brown colour. The highest amounts are found in the **chernozems**, or black earths (page 327), of the North American Prairies, Russian Steppes and Argentinean Pampas. In tropical rainforests, heavy rainfall and high biological activity cause the rapid decomposition of organic matter which releases nutrients ready for their uptake and storage by plants (Figure 10.6) or, if the forest is cleared, for leaching out of the system. In drier climates there may be insufficient vegetation to give an adequate supply.

5 Soil moisture

Soil moisture is important because it affects the upward and downward movement of water and nutrients. It helps in the development of horizons; it supplies water for living plants and organisms; it provides a solvent for plant nutrients; it influences soil temperature; and it determines the incidence of erosion. The amount of water in a soil at a given time can be expressed as:

$$W \propto R - (E + T + D)$$
(input) − (outputs)

where: W = water in the soil
$\propto$ = proportional to
R = rainfall/precipitation
T = transpiration
E = evaporation
D = drainage.

Drainage depends upon the balance between the **water retention capacity** (water storage in a soil) and the infiltration rate. This is controlled by porosity and permeability which in turn is controlled by the soil's texture and structure. It has already been shown how texture and structure affect the size and distribution of pore spaces. Clays have numerous small pores (**micropores**) which can retain water for long periods, giving it a high water retention capacity, but which also restrict infiltration rates (page 59). Sands have fewer but much larger **macropores** which permit water to pass through

more quickly (a rapid infiltration rate), but have a low water retention capacity. A loam provides a more balanced supply of water, in the micropores, and air, in the macropores.

The presence of moisture in the soil does not necessarily mean that it is available for plant use. Plants growing in clays may still suffer from water stress even though clay has a high water-holding capacity. Soil water can be classified according to the tension at which it is held. Following a heavy storm or a lengthy episode of rain or snowmelt, all the pore spaces may be filled, with the result that the soil becomes saturated. When infiltration ceases, water with a low surface tension drains away rapidly under gravity. This is called **gravitational** or **free** water which is available to plants when the soil is wet, but unavailable when water has drained away. Once this excess water has drained away, the remaining moisture that the soil can hold is said to be its **field capacity** (Figures 3.3 and 10.12).

Moisture at field capacity is held either as **hygroscopic water** or as **capillary water**. Hygroscopic water is always present, unless the soil becomes completely dry, but is unavailable for plant use. It is found as a thin film around the soil particles to which it sticks due to the strength of its surface tension. Capillary water is attracted to, and forms a film around, the hygroscopic water, but has a lower cohesive strength. It is capillary water that is freely available to plant roots. However, this water can be lost to the soil by evapotranspiration. When a plant loses more water through transpiration than it can take up through its roots it is said to suffer **water stress** and it begins to wilt. At **wilting point**, photosynthesis (page 295) is reduced but, provided water can be obtained relatively soon or if the plant is adapted to drought conditions, this need not be fatal. Figure 10.12 shows the different water-holding characteristics of soil.

Figure 10.12

Availability of soil moisture for plant use

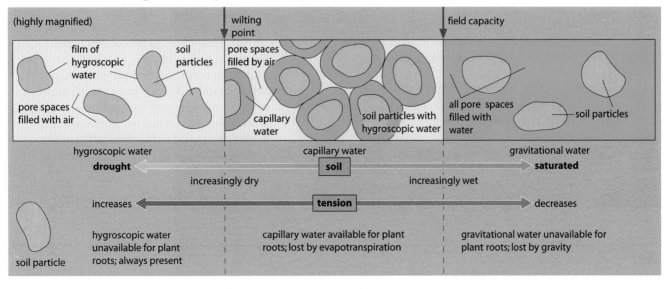

6 Air

Air fills the pore spaces left unoccupied by soil moisture. It is oxygen in the air that is essential for plant growth and living organisms. Compared with atmospheric air, air in the soil contains more carbon dioxide, released by plants and soil biota, and more water vapour; but less oxygen, as this is consumed by bacteria. Biota need oxygen and give off carbon dioxide by respiration and through the oxidation of organic matter. These gases are exchanged through the process of diffusion.

7 Soil organisms (biota)

Soil organisms include bacteria, fungi and earthworms. They are more active and plentiful in warmer, well-drained and aerated soils than they are in colder, more acidic and less well-drained and aerated soils.

Figure 10.13

Nutrients needed by plants

Macro-nutrients — Needed in large quantities	Carbon	C	Needed for basic cell construction. Obtained from air and water.
	Hydrogen	H	
	Oxygen	O	
Macro-nutrients — Needed in smaller quantities	Nitrogen	N	Basis of plant proteins. Promotes rapid growth. Improves quality and quantity of leaf growth.
	Phosphorus	P	Encourages rapid seedling growth and early root formation. Helps in flowering and with seed formation.
	Sulphur	S	Especially important for root crops.
	Potassium	K	Helps with production of proteins and in overcoming disease. Strengthens stems and stalks.
	Calcium	Ca	Reduces acidity. Helps with growth of roots and new shoots.
	Magnesium	Mg	Used in photosynthesis, being a basic constituent of chlorophyll. Important for arable crops.
Micro-nutrients (trace elements) — Needed in very small quantities	Sodium	Na	Helps to increase yields.
	Manganese	Mn	Used in respiration, protein synthesis and enzyme reactions.
	Copper	Cu	Reduces toxicity of other elements in soil. Helps enzyme reactions.
	Zinc	Zn	Helps in fruit production.
	Molybdenum	Mo	Needed in nitrogen fixation by activating enzymes.
	Silicon	Si	Important constituent of grasses.
	Boron	B	Helps growth.
	Chlorine	Cl	Can increase yields of some crops.
	Cobalt	Co	Helps fruit trees and bushes.

Organisms are responsible for three important soil processes:

- **Decomposition:** detritivores, such as earthworms, ants, termites, mites, woodlice and slugs, begin this process by burying leaf litter (detritus), which hastens its decay, and eating some of it. Their faeces (wormcasts, etc.) increase the surface area of detritus upon which fungi and bacteria can act. Fungi and bacteria secrete enzymes which break down the organic compounds in the detritus. This releases nutrient ions essential for plant growth (soil nutrients, Figure 10.13), into the soil while some organic compounds remain as humus.
- **Fixation:** by this process, bacteria can transform nitrogen in the air into nitrate, which is an essential nutrient for plant growth.
- **Development of structure:** fungi help to bind individual soil particles together to give a crumb structure, while burrowing animals create passageways that help the circulation of air and water and facilitate root penetration.

8 Soil nutrients

Nutrient is the term given to chemical elements found in the soil which are essential for plant growth and the maintenance of the fertility of a soil (Figure 10.13). The two main sources of nutrients are:

1 the weathering of minerals in the soil, and
2 the release of nutrients on the decomposition of organic matter and humus by soil organisms.

Nutrients can also be obtained through:

3 rainwater, and
4 the artificial application of fertiliser.

Nutrients occur in the soil solution as positively charged (+) ions called **cations** and negatively charged (−) ions known as **anions**. It is largely in the ionic form that plants can utilise nutrients in the soil. Both clay and humus, which have negative charges, attract the positively charged minerals in the soil solution, notably Ca^{2+}, Mg^{2+}, K^+ and Na^+. This results in the cations being adsorbed (i.e. they become attached) to the clay and humus particles. The process of **cation exchange** allows cations to be moved between:

- soil particles of clay and/or humus and the soil solution
- plant roots and either the surface of the soil particles or from the soil solution (Figure 10.14).

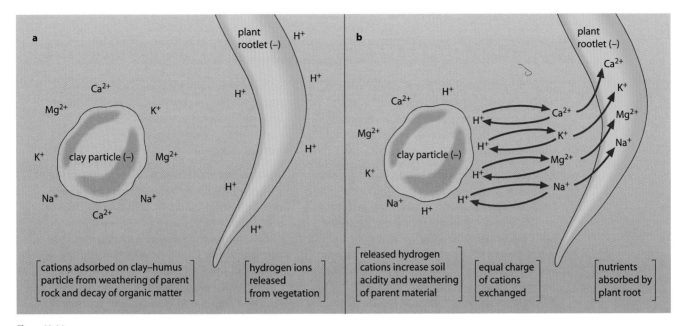

plant rootlet (−)

H⁺

Ca²⁺

Mg²⁺   K⁺

K⁺   clay particle (−)   Mg²⁺

Na⁺   Na⁺

Ca²⁺

H⁺
H⁺
H⁺
H⁺
H⁺
H⁺

b

plant rootlet (−)

Ca²⁺

H⁺

Ca²⁺   Ca²⁺

Mg²⁺   K⁺

H⁺   K⁺

Mg²⁺   Na⁺

K⁺   clay particle (−)   H⁺   Mg²⁺

H⁺   Na⁺

Na⁺   H⁺

[cations adsorbed on clay–humus particle from weathering of parent rock and decay of organic matter]   [hydrogen ions released from vegetation]   [released hydrogen cations increase soil acidity and weathering of parent material]   [equal charge of cations exchanged]   [nutrients absorbed by plant root]

Figure 10.14

The process of cation exchange (*after* Courtney and Trudgill)

As well as providing nutrients for plant roots, the cation exchange releases hydrogen which in turn increases acidity in the soil (see next section). Acidity accelerates weathering of parent rock, releasing more minerals to replace those used by plants or lost through leaching. The **cation exchange capacity** (CEC) is a measure of the ability of a soil to retain cations for plant use. Soils with a low CEC, such as sands, are less able to keep essential plant nutrients than those with a high CEC, like clays and humus; consequently they are less fertile.

9 Acidity (pH)

As mentioned in the previous section, soil contains positively charged hydrogen cations. **Acidity** or **alkalinity** is a measure of the degree of concentration of these cations It is measured on the pH scale (Figure 10.15), which is logarithmic (compare the Richter scale, Figure 1.3). This means that a reading of 6 is 10 times more acidic than a reading of 7 (which is neutral), and 100 times more acidic than one of 8 (which is alkaline). Most British soils are slightly acidic,

Figure 10.15

The pH scale showing soil acidity and alkalinity

although in upland Britain acidity increases as the heavier rainfall leaches out elements such as calcium faster than they can be replaced by weathering. Acid soils therefore tend to need constant liming if they are to be farmed successfully.

A slightly acid soil is the optimum for farming in Britain as this helps to release secondary minerals. However, if a soil becomes too acidic it releases iron and aluminium which, in excess, may become toxic and poisonous to plants and organisms. Increased acidity makes organic matter more soluble and therefore vulnerable to leaching; and it discourages living organisms, thus reducing the rate of breakdown of plant litter and so is a factor in the formation of peat.

In areas where there is a balance between precipitation and evapotranspiration, soils are often neutral, as in the American Prairies (page 327); while in areas with a water deficiency, as in deserts (page 323), soils are more alkaline.

10 Soil temperature

Incoming radiation can be absorbed, reflected or scattered by the Earth's surface (Figure 9.4). The topsoil, especially if vegetation cover is limited, heats up more rapidly than the subsoil during the daytime and loses heat more rapidly at night. A 'warm', moist soil will have greater biota activity, giving a more rapid breakdown of organic matter; it will be more likely to contain nutrients because the chemical weathering of the parent material will be faster; and seeds will germinate more readily in it than in a 'cold', dry soil.

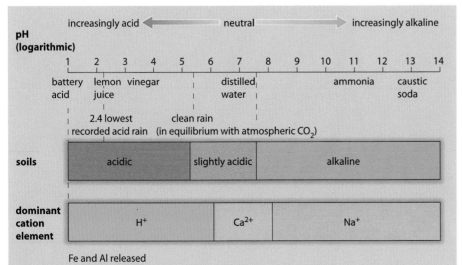

increasingly acid ⟵   neutral   ⟶ increasingly alkaline

pH (logarithmic)

1   2   3   4   5   6   7   8   9   10   11   12   13   14

battery acid   lemon juice   vinegar   distilled water   ammonia   caustic soda

2.4 lowest recorded acid rain   clean rain (in equilibrium with atmospheric CO_2)

soils   acidic   slightly acidic   alkaline

dominant cation element   H⁺   Ca²⁺   Na⁺

Fe and Al released

Begin by reading a book that describes in detail how to dig a soil pit and how to describe and explain the resultant profile (e.g. Courtney and Trudgill, 1984, or O'Hare, 1988; see References at end of chapter).

First, make sure you obtain permission to dig a pit. The site must be carefully chosen. You will need to find an undisturbed soil – so avoid digging near to hedges, trees, footpaths or on recently ploughed land. Ideally, make the surface of the pit approximately 0.7 m², and the depth 1 m (unless you hit bedrock first). Carefully lay the turf and soil on plastic sheets. Clear one face of the pit, preferably one facing south as this will get the maximum light, to get a 'clean' profile so that you can complete your recording sheet. (The one in Figure 10.16 is a very detailed example.)

Sometimes you will not be able to take all the readings due to problems such as lack of clarity between boundaries, time and equipment; sometimes some details will not be relevant to a particular enquiry.

Make a detailed fieldsketch before replacing the soil and turf. You may have to complete several tasks in the laboratory before writing up your description. You can gather information from a soil without needing to know how it formed or what type it is. Remember, it is unlikely that your answer will exactly fit a model profile. It may show the characteristics of a podsol (Figure 12.40) if you live in a cooler, wetter and/or higher part of Britain; or of a brown earth (Figure 12.34) if you live in a warmer, drier and/or lower part of the country – but you must not *force* your profile to fit a model.

a soil site

Figure 10.16

Soil recording sheets

Recorded by	Date	Locality		Six-figure grid reference
Parent rock (geological map)	Altitude (estimated from Ordnance Survey map)	Angle of slope (Abney level)	Aspect (bearing or compass point)	Relief (uniform, concave or convex slope, terrace)
Exposure (exposed, sheltered)	Drainage (shedding or receiving site, floodplain, terrace, boggy)	Natural vegetation or type of farming (tree species, ground vegetation, crops, animals)	Previous few days' weather (warm, cold, wet, dry)	Other local details (remember your labelled fieldsketch)

b soil profile

Horizon	Depth of horizon (cm)	Lower boundary of horizon	Colour	Texture	Stoni-ness	Structure (peds)	Consist-ency	pH	Moisture content	Porosity	Organic matter	Roots	Carbon-ates	Soil biota and/or animals
How to read, estimate and measure	measure from top of soil surface	sharp, abrupt, clear, indistinct, gradual, irregular, smooth, broken	use Munsell colour chart	percent-age clay, silt or sand; 'feel'; sieves; sedimen-tation	size of stones, number of stones, shape of stones	structure-less crumb, etc.	loose, friable, firm, hard, plastic, sticky, soft	pH paper or soil-testing kit	weigh sample, evaporate water, reweigh sample, or use a moisture meter	time taken for a beakerful of water to infiltrate	type, estimate percent-age, measure depth	weigh, burn sample (and roots), reweigh sample, calculate percent-age	add dilute (10%) hydro-chloric acid; if it effer-vesces, sample is over 1% carbonate	number, types
A														
B														
C														

Figure 10.17

Soil-forming processes

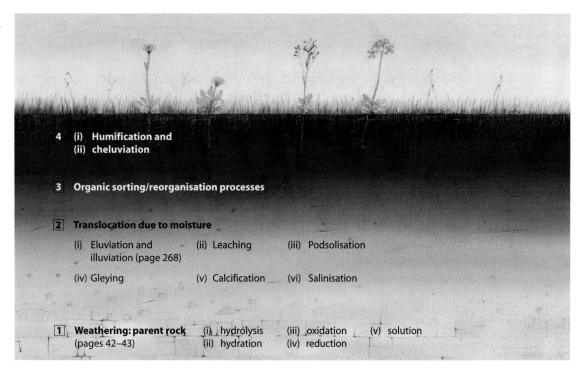

Figure 10.17 Soil-forming processes

4 (i) **Humification and**
(ii) **cheluviation**

3 **Organic sorting/reorganisation processes**

2 **Translocation due to moisture**

(i) Eluviation and   (ii) Leaching   (iii) Podsolisation
illuviation (page 268)

(iv) Gleying   (v) Calcification   (vi) Salinisation

1 **Weathering: parent rock**   (i) hydrolysis   (iii) oxidation   (v) solution
(pages 42–43)   (ii) hydration   (iv) reduction

Processes of soil formation

Numerous processes are involved in the formation of soil and the creation of the profiles, structures and other features described above. Soil-forming processes depend on all the five factors described on pages 260–262. Some of the more important processes are shown in Figure 10.17.

1 Weathering

As described on page 263 and in Figure 10.2, weathering leaves primary minerals as residues and produces secondary minerals as well as determining the rates of release of nutrients and the soil depth, texture and drainage. In systems terms, this means that minerals are released as inputs into the soil system from the bedrock store and transferred into the soil store (Figure 10.6).

2 Humification and cheluviation

Humification is the process by which organic matter is decomposed to form humus (page 266) – a task performed by soil organisms. Humification is most active either in the H horizon of the soil profile (Figure 10.5) where it can result in mull (pH 5.5 to 6.5), or in the upper A horizon where it can produce mor (pH 3.5 to 4.5) (page 266). Moder (pH 4.5 to 5.5) is transitional between the mor and mull (page 262).

As organic matter decomposes, it releases nutrients and organic acids. These acids, known as **chelating agents**, attack clays and other minerals, mainly in the A horizon, releasing iron and aluminium. The chelating agents then combine with the cations of the iron and aluminium to form organic-metal compounds known as **chelates**. Chelates are soluble and are readily transported downwards through the soil profile – the process of **cheluviation**. The iron and aluminium may be deposited in the lower profile as they become less soluble in the slightly higher pH levels found there (Figure 10.5).

3 Organic sorting

Several processes operate within the soil to re-organise mineral and organic matter into horizons, and to contribute to the aggregation of particles and the formation of peds.

4 Translocation of soil materials

Translocation is the movement of soil components in any form (solution, suspension, or by animals) or direction (downward, upward). It usually takes place in association with soil moisture.

In Britain, there is:

- usually a soil moisture budget surplus due to an annual excess of precipitation over evapotranspiration (water balance – Figure 3.3)
- locally, an increase in soil moisture due to poor drainage.

The increase in soil moisture, resulting from these two factors, can lead to:

- either the translocation processes of leaching and podsolisation, or
- gleying associated with areas of poor drainage.

(i) Eluviation and illuviation

See page 262.

(ii) Leaching

Leaching is the removal of soluble material in solution. Where precipitation exceeds evapotranspiration and soil drainage is good, rainwater – containing oxygen, carbonic acid and organic acids, collected as it passes through the surface vegetation – causes chemical weathering, the breakdown of clays and the dissolving of soluble salts (bases). Ca and Mg are eluviated from the *A* horizon, making it increasingly acid as they are replaced by hydrogen ions, and are subsequently illuviated to the underlying *B* horizon, or are leached out of the system (Figure 10.18).

(iii) Podsolisation

Podsolisation is more common in cool climates where precipitation is greatly in excess of evapotranspiration and where soils are well drained or sandy. Podsolisation is also defined as the removal of iron and aluminium oxides, together with humus. As the surface vegetation is often coniferous forest, heathland or moors, rain percolating through it becomes progressively more acidic and may reach a pH of 5.0 or less (Figure 10.15). This in turn dissolves an increasing amount and number of bases (Ca, Mg, Na and K), silica and, ultimately, the sesquioxides of iron and aluminium (Figure 10.19). The resultant **podsol soil** (Figure 12.40) therefore has two distinct horizons: the bleached *A* horizon, drained of coloured minerals by leaching; and the reddish-brown *B* horizon where the sesquioxides have been illuviated. Often the iron deposits form an **iron pan** which is a characteristic of a podsol.

(iv) Gleying

This occurs when the output of water from the soil system is restricted, giving **anaerobic** or **waterlogged** conditions (page 275). This is most likely to occur on gentle slopes, in depressions where the underlying rock is impermeable, where the water table is high enough to enter the soil profile (e.g. along river floodplains) or in areas with very heavy rainfall and poor drainage. Under such conditions the pore spaces fill with stagnant water which becomes de-oxygenised. The reddish-coloured oxidised iron, iron III (Fe^{3+} or ferric iron), is chemically reduced to form iron II (Fe^{2+} or ferrous iron) which is grey-blue in colour. Occasionally, pockets of air re-oxygenise the iron II to give scatterings of red mottles (Figure 10.26). Although many British soils show some evidence of gleying, the conditions develop most extensively on moorland plateaus.

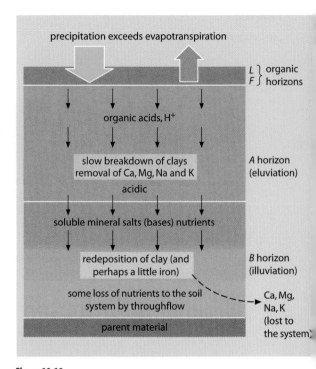

Figure 10.18

The processes of leaching

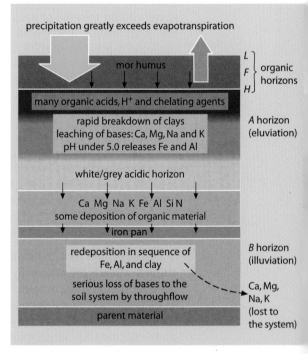

Figure 10.19

The process of podsolisation

Courtney and Trudgill (Figure 10.20) have summarised the relationship between leaching, podsolisation and gleying, and precipitation and drainage.

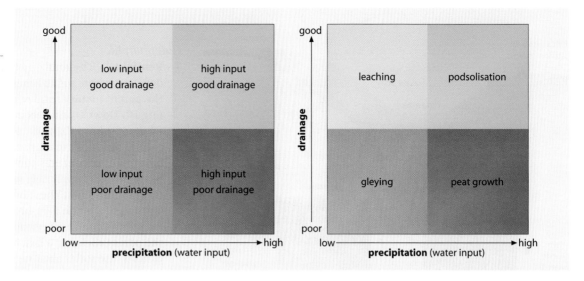

Figure 10.20

Soil-forming processes and the water balance (Figure 3.3) (*after* Courtney and Trudgill)

(v) Calcification

Calcification is a process typical of low-rainfall areas where precipitation is either equal to, or slightly higher than, evapotranspiration. Although there may be some leaching, it is insufficient to remove all the calcium which then accumulates, in relatively small amounts, in the *B* horizon (Figure 10.21; and chernozems, page 327).

(vi) Salinisation

This occurs when potential evapotranspiration is greater than precipitation in places where the water table is near to the surface. It is therefore found locally in dry climates and is not a characteristic of desert soils. As moisture is evaporated from the surface, salts are drawn upwards in solution by capillary action. Further evaporation results in the deposition of salt as a hard crust (Figure 10.22). Salinisation has become a critical problem in many irrigated areas, such as California (Figure 16.53).

Zonal, azonal and intrazonal soils

Zonal soils

Zonal soils are mature soils. They result from the maximum effects of climate and living matter (vegetation) upon parent rock in areas where there are no extremes of weathering, relief or drainage and where the landscape and climate have been stable for a long time. Consequently, zonal soils have had time to develop distinctive profiles and, usually, clear horizons. However, it is misleading to imply that all zonal soils have distinct horizons; brown earths (page 329), chernozems (page 327) and prairie soils (page 327) have indistinct horizons which merge into each other. A description of the major zonal soils, and how their formation can be linked to climate and vegetation, is given in Chapter 12 and Figure 12.2. It should be stressed that this linkage is regarded by soil scientists as greatly outdated and a grossly simplified model – but it is still the one used in all the latest AS, A-level and Scottish Higher syllabuses that examine soils!

Azonal soils

Azonal soils, in contrast to zonal soils, have a more recent origin and occur where soil-forming processes have had insufficient time to operate fully. As a consequence, these soils usually show the characteristics of their origin (i.e. parent material, which may have resulted from *in situ* weathering of parent rock or have been transported from elsewhere and deposited), do not have well defined horizons, and are not associated with specific climatic–vegetational zones. Azonal soils, in Britain, include **scree** (weathering), **alluvium** (fluvial), **till** (glacial), **sands and gravels** (glacifluvial), **sand dunes** (aeolian and marine), **saltmarsh** (marine), and **volcanic** (tectonic) **soils**.

re 10.21

process of calcification

Figure 10.22

The process of salinisation

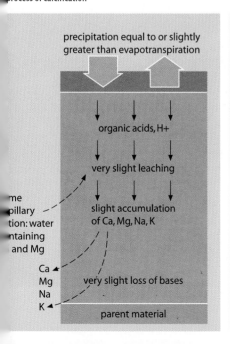

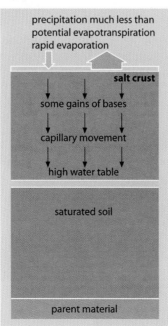

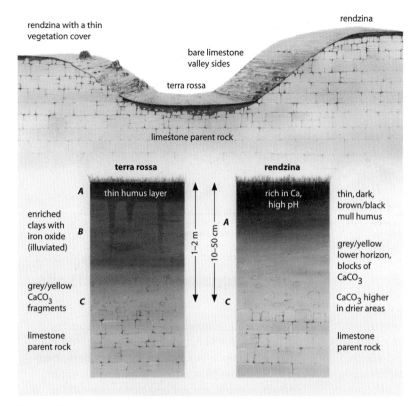

rendzina with a thin
vegetation cover

bare limestone
valley sides

terra rossa

rendzina

limestone parent rock

terra rossa

A

thin humus layer

enriched
clays with
iron oxide
(illuviated)

B

grey/yellow
CaCO₃
fragments

C

limestone
parent rock

1–2 m

10–50 cm

rendzina

rich in Ca,
high pH

A

thin, dark,
brown/black
mull humus

grey/yellow
lower horizon,
blocks of
CaCO₃

C

CaCO₃ higher
in drier areas

limestone
parent rock

Figure 10.23

Calcimorphic soils: terra
rossa and rendzina

Intrazonal soils

Intrazonal soils reflect the dominance of a single local factor, such as parent rock or extremes of drainage. As they are not related to general climatic controls, they are not found in zones. They can be divided into three types:

- **Calcimorphic** or **calcareous** soils develop upon a limestone parent rock (rendzina and terra rossa, Figure 10.23).
- **Hydromorphic** soils are those having a constantly high water content (gleyed soils and peat – Figures 10.26 and 10.27).
- **Halomorphic** soils have high levels of soluble salts which render them saline.

Calcimorphic

1 **Rendzina**   The rendzina (Figure 10.24) develops where softer limestones or chalk are the parent material and where grasses (the English Downs) and beech woodland (the Chilterns) form the surface vegetation. The grasses produce a leaf litter that is rich in bases. This encourages considerable activity by organisms which help with the rapid recycling of nutrients. The *A* horizon therefore consists of a black/dark-brown mull humus. Due to the continual release of calcium from the parent rock and a lack of hydrogen cations, the soil is alkaline with a pH of between 7.0 and 8.0. The calcium-saturated clays, with a crumb or blocky structure, tend to limit the movement of water and so there is relatively little leaching. Consequently there is no *B* horizon. The underlying limestones, affected by chemical weathering, leave very little insoluble residue and this, together with the permeable nature of the bedrock, results in a thin soil with limited moisture reserves.

2 **Terra rossa**   As its name suggests, terra rossa (Figure 10.25) is a red-coloured soil (it has been called a 'red rendzina'). It is found in areas of heavy, even if seasonal, rainfall where the calcium carbonate parent rock is chemically weathered (carbonation) and silicates are leached out of the soil to leave a residual deposit rich in iron hydroxides. It usually occurs in depressions within the limestone and in Mediterranean areas where the vegetation is garrigue (Figure 12.24).

Figure 10.24

A rendzina, Kent

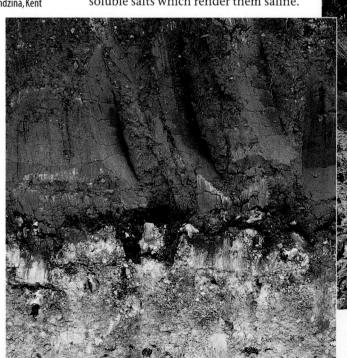

Figure 10.25

Terra rossa, Cuba

Hydromorphic

1 **Gley soils** Gleying occurs in saturated soils when the pore spaces become filled with water to the exclusion of air. The lack of oxygen leads to anaerobic conditions (page 272) and the reduction (chemical weathering) of iron compounds from a ferric (Fe^{3+}) to a ferrous (Fe^{2+}) form. The resultant soil has a grey-blue colour with scatterings of red mottles (Figure 10.26). Because gleying is a result of poor drainage and is almost independent of climate, it can occur in any of the zonal soils. Pedologists often differentiate between **surface gleys**, caused by slow infiltration rates through the topsoil, and **groundwater gleys**, resulting from a seasonal rise in the water table or the presence of an impermeable parent rock.

2 **Peat** Where a soil is waterlogged and the climate is too cold and/or wet for organisms to break down vegetation completely, layers of peat accumulate (Figure 10.27). These conditions mean that litter input (supply) is greater than the rate of decomposition by organisms whose activity rates are slowed down by the low temperatures and the anaerobic conditions. Peat is regarded as a soil in its own right when the layer of poorly decomposed material exceeds 40 cm in depth. Peat can be divided according to its location and acidity. **Blanket peat** is very acidic; it covers large areas of wet upland plateaus in Britain (Kinder Scout in the Peak District); and it is believed to have formed 5000–8000 years ago during the Atlantic climatic phase (Figure 11.18). **Raised bogs**, also composed of acidic peat, occur in lowlands with a heavy rainfall. Here the peat accumulates until it builds up above the surrounding countryside. **Valley**, or **basin**, **peat** may be almost neutral or only slightly acidic if water has drained off surrounding calcareous uplands (the Somerset Levels and the Fens); otherwise, it too will be acid (Rannoch Moor in Scotland). Fen peat is a high-quality agricultural soil.

Halomorphic

Halomorphic soils contain high levels of soluble salts and have developed through the process of salinisation (page 273 and Figure 16.53). They are most likely to occur in hot, dry climates where, in the absence of leaching, mineral salts are brought to the surface by capillary action and where the parent rock or groundwater contains high levels of carbonates, bicarbonates and sulphates, especially as salts of calcium and magnesium and some sodium chloride (common salt). The water, on reaching the surface, evaporates to leave a thick crust (e.g. Bonneville saltflats in Utah, page 188) in which only salt-resistant plants (halophytes, page 291) can grow.

Figure 10.26

Gleying: a hydromorphic soil

Figure 10.27

Peat in the Flow Country, Sutherland, Scotland

The soil catena

A **catena** (Latin for 'chain') is a sequence of soil types down a slope where each **soil type**, or **facet** is different from, but linked to, its adjacent facets (Figure 10.3). Catenas therefore illustrate the way in which soils can change down a slope where there are no marked changes in climate or parent material. Each catena is an example of a small-scale, open system involving inputs, processes and outputs. The slope itself is in a delicate state of dynamic equilibrium (Figure 2.12) with the soils and landforms being in a state of flux and where the ratio of erosion and deposition varies between the different slope facets. Soils on lower slopes tend to be deeper and wetter than those on upper slopes, as well as being more enriched by a range of leached materials. The thinnest and driest soils are likely to be found on central parts of the slope. It takes a considerable period of time for catenary relationships to become established and therefore the best catenas can be found in places with a stable environment, such as in parts of Africa, where there have been relatively few recent changes in either the landscape or the climate.

Places 34 Arran: a soil catena

Figure 10.28 shows a catena based on fieldwork conducted on the Isle of Arran. The transect was taken from a relatively flat, peat-covered upland area above the glaciated Glen Rosa valley, down a steep valley side to the Rosa Water (parallel to, and south of, the Garbh Allt tributary located on Figure 4.37).

Notice, with reference to Figure 10.3, the location on the transect of the shedding (eluviation or input), transfer (translocation) and receiving (illuviation or output) zones, and the relationships between the angles of slope and (i) soil depth, (ii) pH and (iii) soil moisture.

Figure 10.28

Readings taken along a catena in Glen Rosa

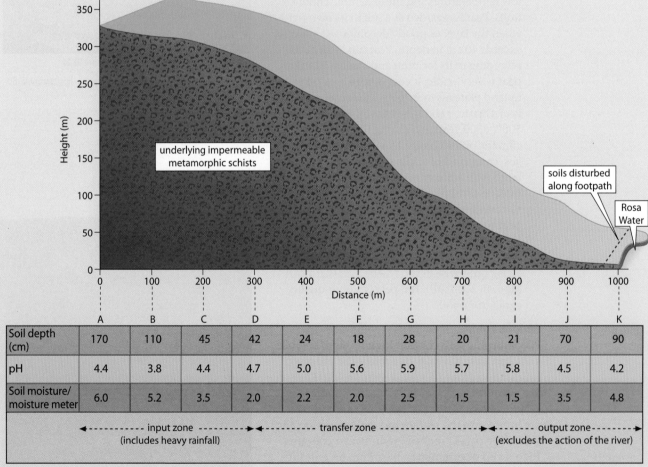

	A	B	C	D	E	F	G	H	I	J	K
Soil depth (cm)	170	110	45	42	24	18	28	20	21	70	90
pH	4.4	3.8	4.4	4.7	5.0	5.6	5.9	5.7	5.8	4.5	4.2
Soil moisture/ moisture meter	6.0	5.2	3.5	2.0	2.2	2.0	2.5	1.5	1.5	3.5	4.8

◀---------- input zone ----------▶◀---------- transfer zone --------------------▶◀------- output zone --------▶
(includes heavy rainfall)   (excludes the action of the river)

Geographic Information Systems (GIS) are concerned with the handling of geographical data, collected from a variety of sources and stored in a digital form, which allows for easy retrieval and display at a later date, in order to make informed decisions. This can be achieved through the development of new and exciting ways of manipulating spatial data (maps) and by performing either simple or complex spatial analysis on the geographical data, quickly and efficiently. GIS need not be limited to IT (a geographical library is a form of GIS) but, due to the amount of data involved, it is ideally suited to computers. Figure 10.29 gives four definitions of GIS.

Figure 10.29

Definitions of Geographic Information Systems (GIS)

- A powerful set of tools for collecting, storing and retrieving at will, transforming and displaying spatial data from the real world. **Burrough, 1986**
- Designed to facilitate sorting, selective retrieval, calculation and spatial analysis and modelling. **Mitchell, 1989**
- A computer-assisted information system to collect, store, manipulate and display spatial data within the context of an organisation, which also functions as a decision support system. **Kraak and Ormeling, 1996**
- A methodology that allows users to ask a set of questions with respect to geographical data, and to visualise the answers to these questions. **Walford, 1999**

Inputs	Stores	Processes	Outputs
Data from a variety of sources: • satellite images, aerial photos, Landsat images • different types of map – OS maps, maps of soils, relief, settlement • digital data and graphs.	**a** Data stored in digital form. **b** Vast quantities of data.	**a** Retrieval, transformation and analysis of data. **b** Data in digital form can be easily updated (unlike atlases or textbooks). **c** Can select and enlarge material for display. **d** Display colours can be changed at the press of a button.	**a** Several maps or sources of data can be displayed on the screen at the same time. **b** A final composite map is achieved by superimposing several maps on each other (overlays – Figure 10.33). **c** Maps, graphs, tables.

As with other systems (Framework 3 page 45), GIS has inputs, processes, stores and outputs (Figure 10.30). A map, drawn on a two-dimensional flat sheet of paper, is the traditional way of showing spatial data. It records and displays data by using (x, y) co-ordinates together with a (z) value that records the features found at each (x, y) location (Figure 10.31). On a two-dimensional map, however, each (x, y) co-ordinate can only store and display, at the most, two (z) features, e.g. height of land and either soils, or settlement, or transport routes or vegetation, etc. A computerised GIS can store hundreds of (z) features for each (x, y) location as well as displaying maps in three dimensions with each (x, y) location recording a specific (z) feature.

Figure 10.30

The GIS

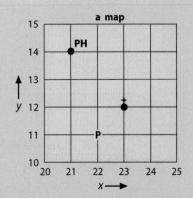

Figure 10.31

Relationship between a map and a database

b database

x	y	z
21	14	public house
23	12	church with spire
22	11	post office

There are two types of data input into GIS: **vector** data and **raster** data.

- In **vector-based** systems, real-world features (Figure 10.32a) are represented by points, lines or polygons/areas (Figure 10.32b). Vector data – which are usually entered into the computer manually although automated systems are increasingly used – include digitised information from existing maps.

- In **raster-based** systems, real-world features are represented by cells referenced in terms of rows and columns (Figure 10.32c). Each feature, e.g. height, soils, settlement or vegetation, has a separate raster representation known as a **layer**. A raster map usually consists of a series of layers which may be viewed individually or collectively

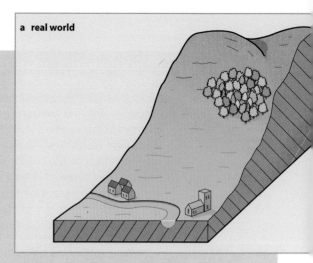

a real world

(Figure 10.33). Raster data is more likely to be received directly from a remote sensing organisation, either 'on-line' or via a computer disk or tape.

Point

A point feature is displayed at one particular (x, y) location on the map, e.g. house. Individual co-ordinate pairs, e.g. 2213 2810

Line

A line consists of a series of (x, y) co-ordinates that join together with each having the same (z) feature, e.g. a railway line or river. Groups of two co-ordinate pairs, e.g. 2013 2112 2612 2711 2710

Polygon/Area

A polygon consists of a series of (x, y) co-ordinates that join together to complete a boundary. Everything inside this boundary is assigned the same (z) feature, e.g. a wood. A series of co-ordinates having the identical starting and ending co-ordinates, e.g. 2916 2816 2717 2718 2918 2916

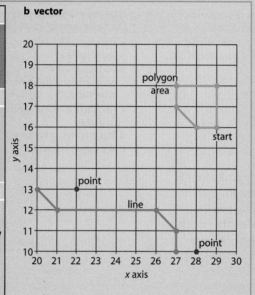

b vector

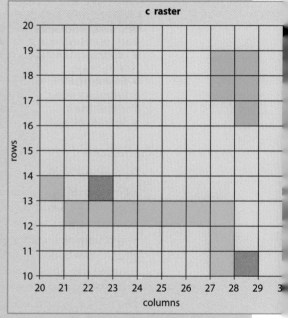

c raster

Figure 10.32

Vector and raster data

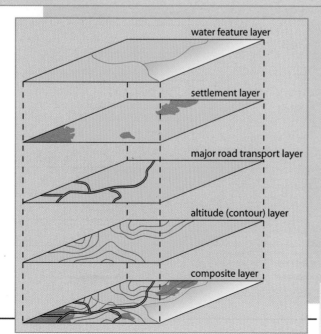

water feature layer

settlement layer

major road transport layer

altitude (contour) layer

composite layer

As to the application of GIS, Walford (*Geography*, 1999) suggested that 'its flexibility derives from the relevance of generic questions to a seemingly infinite range of particular situations in business, commerce, public administration and research. The techniques associated with answering them include:

- spatial data interrogation or querying
- spatial research and buffering
- locational analysis
- network analysis
- socio-economic or customer analysis
- overlay analysis
- spatial statistics and calculation.'

Figure 10.33

The process of overlaying

Soil erosion and soil management

As we have seen (page 262), soil can takes thousands of years to become sufficiently deep and developed for economic use (exceptions include alluvium deposited by rivers and ash ejected from volcanoes). During that time, there is always some natural loss through leaching, mass movement and erosion by either water or wind. Normally there is an equilibrium, however fragile, between the rate at which soil forms and that at which it is eroded or degraded. That natural balance is being disturbed by human mismanagement with increasing frequency and with serious consequences.

Recent estimates suggest that 7 per cent of the world's topsoil is lost each year. The World Resources Institute claims that Burkina Faso loses 35 tonnes of soil per hectare per year. Other comparable figures are Ethiopia 42, Nepal 70, and the loess plateau of North China 251 (Figure 10.35). Soil removed during a single rain storm or dust storm may never be replaced. The Soil Survey of England and Wales claims that 44 per cent of arable soils in the UK, an area once considered not to be under threat, are now at risk (Figure 10.34).

Soil degradation

Degradation is the result of human failures to understand and manage the soil. The major cause of soil erosion is the removal of the natural vegetation cover, leaving the ground exposed to the elements. The most serious of such removals is deforestation. In countries such as Ethiopia (Places 76, page 520), the loss of trees, resulting from population growth and the extra need for farmland and fuelwood, means that the heavy rains, when they do occur, are no longer intercepted by the vegetation. Rainsplash (the direct impact of raindrops, Figure 2.12) loosens the topsoil and prepares it for removal by sheetwash (overland flow). Water flowing over the surface has little time to infiltrate into the soil or recharge the soil moisture store (pages 59–60). More topsoil tends to be carried away where there is little vegetation because there are neither plant roots nor

Soil erosion sweeps shires

"Soil erosion, the scourge of Third World countries, is hitting Britain. Official figures show that nearly half of Britain's topsoil is in danger of being blown and washed away by wind and rain.

This emerging crisis adds weight to Prince Charles's recent attack on conventional agricultural methods and to his support for organic farming. He cited 'large-scale soil erosion' as one of the 'unacceptable' side-effects of modern agriculture.

One-third of the world's arable land is expected to turn to dust by the year 2000. Around 80 per cent of Africa's topsoil is in danger and India loses some 12 billion tons of soil every year, while the American Mid-West seems to be eroding as fast as in the Dust Bowl era of the Thirties.

However, until recently, Britain was thought to be unaffected, thanks to its stable soils and climate, and so the problem has largely been ignored by the government. Several reports now show that this attitude is misguided.

The Government's own Soil Survey of England and Wales reported that 44 per cent of the country's arable soil was at risk, while a recent European Parliament report said that five million acres (2 million hectares) of the UK were threatened by erosion. In addition, a survey of farmers has revealed that one-third believe that some of the fields on their farms are affected.

It now seems certain that vast amounts of soil are being lost. Detailed studies have recorded annual losses of 80 tons from every acre in Norfolk, 73 tons an acre in West Sussex and 60 tons an acre in Shropshire."

The Observer, 29 January 1989

Figure 10.34

Soil erosion in Britain

organic matter to bind it together. Small channels or rills may be formed which, in time, may develop into large gulleys, making the land useless for agriculture (Figure 10.35).

Even where the soil is not actually washed away, heavy rain may accelerate leaching and remove nutrients and organic matter at a rate faster than that at which they can be replaced by the weathering of bedrock and parent material and the decomposition of vegetation (e.g. the Amazon Basin, Figure 12.7 and Places 66, page 480). The loss of trees also reduces the rate of transpiration and therefore the amount of moisture in the air. There are fears that large-scale deforestation will turn areas at present under rainforest into deserts.

Although the North American Prairies and the African savannas were grassland when the European settlers first arrived, it is now believed that these areas too were once forested and were cleared by fire – mainly natural due to lightning, but partly by the local people (Case Study 12). The burning of vegetation initially provides nutrients for the soil, but once these have been leached by the rain or utilised by crops there is little replacement of nutrients. Where the grasslands have been ploughed up for cereal cropping, the breakdown of soil structure (peds) has often led to their drying out and becoming easy prey to wind erosion (Figure 10.34). Large quantities of topsoil were blown away to create the American Dust Bowl in the 1930s, while a similar fate has more recently been experienced by many of the Sahel countries. In Britain, the removal of hedges to create larger fields – easier for modern machinery – has led to accelerated soil erosion by wind (page 495).

Loess plateau of North China

This region experiences the most rapid soil loss in the world. During and following the ice age, Arctic winds transported large amounts of loess and deposited this fine, yellow material to a depth of 200 m in the Huang He basin. Following the removal of the subsequent vegetation cover of trees and grasses to allow cereal farming (especially under the directions of Chairman Mao), the unconsolidated material has been washed away by the heavy summer monsoon rains at the rate of 1 cm per year. It is estimated that 1.6 bn tonnes of soil reach the Huang He River during each annual summer flood. This material, the most carried by any river in the

Figure 10.35

Loess in China

world, has given the Huang He its name – i.e. the 'Yellow River'. A further problem is that 6 cm of silt settles annually on the river's bed so that it now flows 10 m above its floodplain. Should the large flood banks be breached, the river can drown thousands of people (over 1 million in the 1939 flood) and ruin all crops.

Ploughing can have adverse effects on soils. Deep ploughing destroys the soil structure by breaking up peds (page 265) and burying organic material too deep for plant use. It also loosens the topsoil for future wind and water erosion. The weight of farm machinery can compact the soil surface or produce platy peds, both of which reduce infiltration capacity and inhibit aeration of the soil. Ploughing up- and down-hill creates furrows which increase the rate of surface runoff and the process of gullying.

Overgrazing, especially on the African savannas, also accelerates soil erosion. Many African tribes have long measured their wealth in terms of the numbers, rather than the quality, of their animal herds. As the human populations of these areas continue to expand rapidly, so too do the numbers of herbivorous animals needed to support them. This almost inevitably leads to overgrazing and the reduction of grass cover (Case Study 7). When new shoots appear after the rains, they are eaten immediately by cattle, sheep, goats and camels. The arrival of the rains causes erosion; the failure of the rains results in animal deaths.

Where there is a rapid population growth, land that was previously allowed a fallow resting period now has to be cultivated each year (Figure 10.36) – as are other areas that were previously considered to be too marginal for crops. Monoculture – the cultivation of the same crop each year on the same piece of land – repeatedly uses up the same soil nutrients.

Burkina Faso

As the size of cattle and goat herds has grown, the already scant dry scrub savanna vegetation on the southern fringes of the Sahara has been totally removed over increasingly large areas. As the Sahara 'advances', the herders are forced to move southwards into moister environments where they compete for land with sedentary farmers who are already struggling to produce sufficient food for their own increasing numbers. This disruption of equilibrium further reduces the land carrying capacity (page 378) – i.e. the number of people that the soil and climate of an area can permanently support when the land is planted with staple crops. These farmers have long been aware that three years' cropping had to be followed by at least eight fallow years in order for grass and trees to re-establish themselves and organic matter to be replenished. The arrival of the herders has brought a land shortage resulting in crops being grown on the same plots every year, and the nutrient-deficient soil, typical of most of tropical Africa, is rapidly becoming even less productive. This overcropping, a problem in many of the world's subsistence areas, uses up organic matter and other nutrients, weakens soil structures and leaves the surface exposed and thus susceptible to accelerated erosion.

Figure 10.36

Overgrazing: Burkina Faso

Figure 10.37

Eutrophication in a British river

In many parts of the world where livestock are kept and firewood is at a premium, dung has to be used as a fuel instead of being applied to the land. In parts of Ethiopia, the sale of dung – mixed with straw and dried into 'cakes' – is often the only source of income for rural dwellers. If this dung were to be applied to the fields, rather than sold to the towns, harvests could be increased by over 20 per cent.

Water is essential for a productive soil. The early civilisations, which grew up in river valleys (Figure 14.1), relied upon irrigation, as do many areas of the modern world. Unfortunately, irrigation in a hot, dry climate tends to lead to salinisation, with dissolved salts being brought, by capillary action, into the root zone of agricultural trees and crops (Figure 16.53). Wells, sunk in dry climates, use up reserves of groundwater which may have taken many centuries to accumulate and which cannot be replaced quickly (fossil water stores, page 190). The resultant lowering of the water table makes it harder for plant roots to obtain moisture. The sinking of wells in sub-Saharan Africa, following the drought of the early 1980s, has unintentionally created difficulties. The presence of an assured water supply has attracted numerous migrants and their animals and this has accelerated the destruction of the remaining trees and exacerbated the problems of overgrazing (Places 65, page 479). Even well-intentioned aid projects may therefore be environmentally damaging.

Fertiliser and pesticides are not always beneficial if applied repeatedly over long periods. Chemical fertiliser does not add organic material and so fails to improve or maintain soil structure. There is considerable concern over the leaching of nitrate fertiliser into streams and underground water supplies. Where nitrates reach rivers they enrich the water and encourage the rapid growth of algae and other aquatic plants which use up oxygen, through the process of eutrophication, to leave insufficient for plant life (Figures 10.37 and 16.50). The use of pesticides (including insecticides and fungicides) can increase yields by up to 100 per cent by killing off insect pests. However, their excessive and random use also kills vital soil organisms, which means organic matter decomposes more slowly and the release of nutrients is retarded. Chemical pesticides are blamed for the decline in Britain's bee population.

Soil management

Fertility refers to the ability of a soil to provide for the unconstrained or optimum growth of plants. The capacity to produce high or low yields depends upon the nutrient content, structure, texture, drainage, acidity and organic content of a particular soil as well as the relief, climate and farming techniques. For ideal growth, plants must have access to nine macro-nutrients and nine micro-nutrients (Figure 10.13). Under normal recycling (Figure 10.6), these nutri-

ents will be returned to the soil as the vegetation dies and decomposes. When a crop is harvested there is less organic material left to be recycled. As nutrients are taken out of the soil system and not replaced, there will be an increasing shortage of macro-nutrients, particularly nitrogen, calcium, phosphorus and potassium. Where this occurs, and when other nutrients are dissolved and leached from the soil, fertiliser is essential if yields are to be maintained.

Soils need to be managed carefully if they are to produce maximum agricultural yields and cause least environmental damage.

If the most serious cause of erosion is the removal of vegetation cover, the best way to protect the soil is likely to be by the addition of vegetation. Afforestation provides a long-term solution because, once the trees have grown, their leaves intercept rainfall while their roots help to bind the soil together and reduce surface runoff. The growing of ground-cover crops reduces rainsplash and surface runoff, and can protect newly ploughed land from exposure to climatic extremes. Marram grass anchors sand, while gulleys can be seeded and planted with brushwood. Certain crops and plants, especially leguminous species such as peas, beans, clover and gorse, are capable of fixing atmospheric nitrogen in the soil, thus improving its quality. Trees can also be planted to act as windbreaks and shelterbelts. This reduces the risk of wind erosion as well as providing habitats for wildlife.

Soil can also be managed by improving farming methods. Most arable areas benefit from a rotation of crops, including grasses, which improve soil structures and reduce the likelihood of soil-borne diseases which may develop under monoculture. Many tropical soils need a recovery period of 5–15 years under shrub or forest for each 3–6 years under crops. In areas where slopes reach up to 12°, ploughing should follow the contours to prevent excessive erosion. On even steeper slopes (Figures 10.41 and 16.29), terracing helps to slow down runoff, giving water more time to infiltrate and thus reducing its erosive ability.

Strip cropping can involve either the planting of crops in strips along the contours or the intercropping of different crops in the same field. Both methods are illustrated in Figure 10.38. The crops may differ in height, time of harvest and use of nutrients.

Where evapotranspiration exceeds precipitation, dry farming can be adopted. This entails covering the soil with a mulch of straw and/or weeds to reduce moisture loss and limit erosion. In the Sahel countries, the drastic depopulation of cattle following the droughts of the 1980s has given herders a chance to restock with smaller (reducing overgrazing), better-quality (giving more meat and milk) herds so that incomes do not fall and the soils are given time to recover.

The addition of organic material helps to bind loose soil and so reduces its vulnerability to erosion (Figures 10.38 and 10.39). Soil structure and texture may be improved, theoretically, by adding lime to acid soils, which reduces their acidity and helps to make them warmer; by adding humus, clay or peat to sands, to give body and to

Organic farming in Washington State, USA

Two adjacent farms in the American state of Washington share similar soil properties, relief and climate. Since 1948 one of the farms has been managed organically, the other conventionally. By 1990, the organic farm had, in comparison with the conventional farm:

- A much greater mass of microbes and enzyme activity. Microbes help in the breakdown of organic matter into humus and in the release of nutrients into a form usable by plants. They also stabilise soil structures, fix nitrogen and break down some pesticides.
- Nearly two-thirds more organic matter on the surface of the soil. Organic matter improves soil structure and increases the moisture retention capacity of the soil, the cation exchange capacity, and the amounts of available nitrogen and potassium.
- A topsoil 16 cm thicker, mainly due to the conventional farm losing 32.4 tonnes of soil per hectare to water erosion compared to only 8.3 tonnes on the organic farm.
- A soil that was much easier to work (plough), in which plants germinated and grew more readily, and which sustained higher crop yields.

Figure 10.39

Organic farming in Washington State, USA

Figure 10.38

Strip farming along contours in the southern USA

improve their water-holding capacity; and by adding sand to heavy clays, so improving drainage and aeration and making them lighter to work. In practice, such methods are rarely used due to the expense involved.

Chemical (inorganic) fertilisers help to replenish deficient nutrients, especially nitrogen, potassium and phosphorus. However, their use is expensive, especially to farmers in economically less developed countries, and can cause environmental damage. Many farmers in poorer countries cannot afford such fertilisers and have to rely upon organic fertiliser. Animal dung and straw left after the cereal harvest are mixed together and spread over the ground. This improves soil structure and, as it decays, returns nutrients to the soil. Where crop rotations are practised, grasses add organic matter, and legumes provide

Stone lines in Burkina Faso

Figure 10.40

Stone lines in Burkina Faso

This project, begun by Oxfam in 1979, aimed to introduce water-harvesting techniques for tree planting. It met with resistance from local people who were reluctant to divert land and labour from food production, or to risk wasting dry-season water needed for drinking.

Attention was therefore diverted to improving food production by using the traditional local technique of placing lines of stones across slopes to reduce runoff (Figures 10.40 and 16.64). When aligned with the contours, these lines dammed rainfall giving it time to infiltrate. Unfortunately, most slopes were so gentle, under 2°, that local farmers could not determine the contours. A device costing less than £3 solved the problem. A calibrated transparent hose, 15 m long, is fixed at each end to the tops of stakes of equal length and filled with water. When the water level is equal at both ends of the hose, the bottom of the stakes must be on the same contour. The lines can be made during the dry season when labour is not needed for farming. Although they take up only 1 or 2 per cent of cropland, they can increase yields by over 50 per cent. They also help to replenish falling water tables and can regenerate the barren, crusted earth because soil, organic matter and seeds collect on the upslope side of the stone lines and plants begin to grow again.

nitrogen. In Britain and North America, a growing number of farmers are turning to organic farming for environmental reasons (Figure 10.39 and page 498).

Many soils suffer from either a shortage or a surfeit of water. In irrigated areas, water must be continually flushed through the system to prevent salinisation. In areas of heavy and/or seasonal rainfall, dams may be built to control flooding and to store surplus water. The drainage of waterlogged soils can be improved by adding field drains.

In several Sahelian countries, people use stones to build small dams which trap water for long enough for some to infiltrate into the ground; they also collect the soil carried away by surface runoff (Figures 10.40 and 16.64).

Soil conservation in northern Shaanxi (China)

According to historic records, the northern province of Shaanxi was once a region with plenty of water, fertile loess soil, lush grass and livestock. Since then, overcultivation and internal wars have led to severe soil erosion (Figure 10.35). This has in turn caused serious desertification (Case Study 7), creating drifting sand dunes which have buried farmland and villages, while frequent droughts, floods and windstorms have hindered the development of the local farming economy. Agriculture fell into a vicious circle: people, because of their poverty, reclaimed land but the more land they reclaimed, the poorer they became because this land was also subject to erosion.

Since the early 1980s, however, the Central Government has encouraged and supported a comprehensive programme for erosion control on the loess plateau. The two main aims have been to control and stabilise drifting sand in northern Shaanxi and to transform the soil throughout the province. This has involved the development of irrigation projects, the terracing of hillsides (Figure 10.41), the planting of trees as a shelter-forest network against the shifting sand (Figure 10.42) and the construction of check-dams (Figure 10.43).

Figure 10.41 Soil – terraced hillsides in Shaanxi

The dams trap silt carried by streams and small rivers, the surplus water being drained away through pipes. The countryside is again looking green, yields have increased considerably, there is a surplus of grain, farming is more diversified with the production of tobacco, fruit and vegetables as cash crops, rainfall has increased and droughts are less serious, the average income of farmers has trebled, and many people who had previously left due to poverty have returned.

Figure 10.42 The shelter-forest network

By the late 1990s, over 50 per cent of the eroded land had been converted back into farmland. The shelter-forest network was stabilising the sand, reducing sandstorms and protecting crops, restricting the amount of soil washed into the Huang He River and encouraging the return of birds and other wildlife. Check-dams, built with pulverised loess to a height of 6 m, have been constructed in deep gullies and narrow valleys.

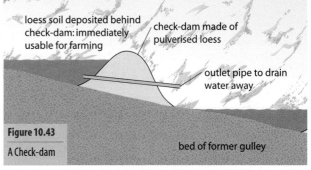

loess soil deposited behind check-dam: immediately usable for farming
check-dam made of pulverised loess
outlet pipe to drain water away
bed of former gulley

Figure 10.43 A Check-dam

Adapted from *China Pictorial Publications*, 1999

References

Bradshaw, M. (1977) *Earth, the Living Planet*, Hodder & Stoughton.

Bridges, D. (1997) *World Soils*, Cambridge University Press.

Briggs, D. (1977) *Soils*, Butterworth.

Courtney, F. M. and Trudgill, S. T. (1984) *The Soil*, Edward Arnold.

Goudie, A. (1993) *The Nature of the Environment*, Blackwell.

Knapp, B. (1979) *Soil Processes*, Allen & Unwin.

Money, D. C. (1978) *Climate, Soils and Vegetation*, University Tutorial Press.

O'Hare, G. (1988) *Soils, Vegetation and the Ecosystem*, Oliver & Boyd.

Prosser, R. (1992) *Natural Systems and Human Responses*, Thomas Nelson.

Timberlake, L. (1987) *Only One Earth*, BBC/ Earthscan.

Websites

A general soils site:
http://www.agri.upm.edu.my/jst/soilinfo/html

Soil salinity and erosion control in Alberta, Canada:
http://www.agric.gov.ab.ca/sustain/soil/salinity/

Department of Environment (Malaysia) site on controlling soil erosion:
http://www.jas.sains.my/doe/new/index.html

China Environment site, Land/soil and Forestry main page:
http://www.chinaenvironment.com/soil/index.html

See also for more links:
http://www.nelsonthornes.com/gaia

1 a i What are the two main components of a soil? **(2 marks)**
      ii Study Figure 10.1 (page 260) and describe how **two** of these factors affect the formation of a soil. **(4 marks)**
      iii Why does the water content of a soil vary from the top of a slope to the bottom? **(4 marks)**
   b What is a 'soil horizon'? **(4 marks)**
   c Choose **one** soil which you have studied.
      i Name the soil.
      ii Draw an annotated soil profile to show the main characteristics of the soil. **(6 marks)**
   d Why do farmers plough their arable land? **(5 marks)**

2 a What can happen to water when it lands on the surface of a soil? **(4 marks)**
   b i What does it mean when 'precipitation exceeds evapotranspiration'? **(4 marks)**
      ii What happens to the soil when *leaching* occurs? **(5 marks)**
   c Name and describe a soil which results from the process of leaching. **(4 marks)**
   d i Why would a farmer want to change soil acidity? **(2 marks)**
      ii What can a farmer do to change the pH of a soil? **(2 marks)**
      iii How does the activity you have described in **ii** change the pH? **(4 marks)**

3 a i What is 'soil acidity'?
      ii Explain how soil acidity can be measured. **(5 marks)**
   b Making good use of a diagram, describe the characteristic features of a *gley* soil. **(5 marks)**
   c How would a farmer try to improve a gleyed soil? **(5 marks)**
   d i Describe the process of *leaching*. **(5 marks)**
      ii Why is a sandy soil more likely to become leached than a clay soil? **(5 marks)**

4 a What is a 'soil horizon'? **(3 marks)**
   b Draw an **annotated** diagram to show the main features of a *brown earth* soil. **(5 marks)**
   c What natural vegetation type and climatic type is associated with formation of a brown earth soil? **(3 marks)**
   d Explain the processes by which a brown earth is formed. **(6 marks)**
   e In what type of area would you expect to find a brown earth within the British Isles? **(3 marks)**
   f What effect is a farmer trying to achieve when ploughing a brown earth? **(5 marks)**

5 a Identify and explain the **five** main factors affecting the formation of a soil. **(10 marks)**
   b What is:
      i soil texture
      ii soil structure? **(8 marks)**
   c For *either* soil structure *or* soil texture, describe how you would identify it in a soil. In your answer you should identify equipment used and explain how to interpret the results. **(7 marks)**

6 a Why are soils classified? **(5 marks)**
   b i Choose **one** soil classification system; describe and explain the basis of the classification. **(10 marks)**
      ii Explain how the classification applies to **two** soils from different environments. **(10 marks)**

7 a Study Figure 10.9 on page 265.
      i Identify the constituents of soils a, b and c, and suggest a name for each soil. **(6 marks)**
      ii Plot the soil textures from Figure 10.44 onto a triangular graph. **(5 marks)**
   b Explain the role of the clay–humus micelle in soil fertility. **(6 marks)**
   c Identify **two** ways in which a farmer can improve the fertility of the soil. In your answer you should explain the effect of the activity on the farmer's output. **(8 marks)**

8 a Why does soil move downhill? **(5 marks)**
   b Describe **two** unintended effects of human activity on soils. **(10 marks)**
   c Explain **two** ways in which farmers can combat accelerated soil erosion. **(10 marks)**

9 a i What is an *azonal soil*?
      ii Choose **one** azonal soil you have studied and draw an annotated diagram to show the characteristics of the soil. Explain why it is classified as azonal. **(10 marks)**
   b Why do geographers and others classify soils? **(5 marks)**
   c Identify **one** scientific soil classification system you have studied. Making use of example soils, explain the basis on which the classification is made. **(10 marks)**

10 Study Places 34 on the Isle of Arran (page 276).
   a On a copy of Figure 10.28 identify the 'shedding', 'transfer' and 'receiving' sites on this slope. **(3 marks)**
   b Describe and account for the differences in **i** soil depth and **ii** acidity shown in Figure 10.28. **(12 marks)**
   c Suggest and explain **two** located uses of parts of this slope for farming. **(10 marks)**

11 a i What is a 'soil catena'? **(5 marks)**
      ii For a soil catena you have studied, identify the area, describe the soil pattern and explain its causes. **(10 marks)**
   b How does an understanding of the properties of soils help farmers to plan their management of the land? **(10 marks)**

12 Describe and explain the variations in soil at a world scale. **(25 marks)**

Figure 10.44	Sample	Clay (%)	Silt (%)	Sand (%)
Five soil samples	d	61	–26	13
	e	33	7	60
	f	–8	79	13
	g	5	5	90
	h	34	36	30

13 Study Figures 10.45 and 10.46 which show four soils and their locations.

  a **i** For *either* profile A *or* profile C, describe the kinds of natural vegetation you would expect to find growing there and explain why these plants would be there. **(5 marks)**

    ii Explain how the *black peat* in profile D was formed. **(4 marks)**

  b **i** Study soil B. How deep is this soil? **(1 mark)**

    ii What is 'humus'? **(4 marks)**

    iii What is the 'texture' of this soil? **(2 marks)**

    iv How does this texture affect farming? **(3 marks)**

  c Explain why a farm on the Charnwood Forest would be different from one on the Lincoln Edge. **(6 marks)**

14 Choose one example of soil you have studied in the field.

  a **i** Identify the aims and objectives of the study. **(3 marks)**

    ii Describe the main features of the area where the fieldwork was carried out. **(3 marks)**

    iii Explain how the fieldwork was planned before the trip took place. **(3 marks)**

  b Describe the methods used to collect the data (your response should include 'what', 'why', 'where', 'how' and 'how it was recorded'). **(8 marks)**

  c **i** For **one** piece of analysis you have carried out, explain how the data were sorted to prepare them for analysis. **(4 marks)**

    ii How were results prepared for presentation after the fieldwork trip? **(4 marks)**

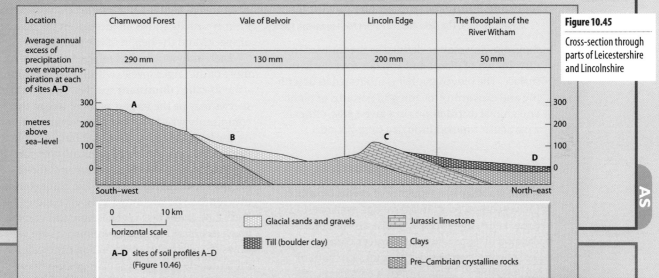

Figure 10.45

Cross-section through parts of Leicestershire and Lincolnshire

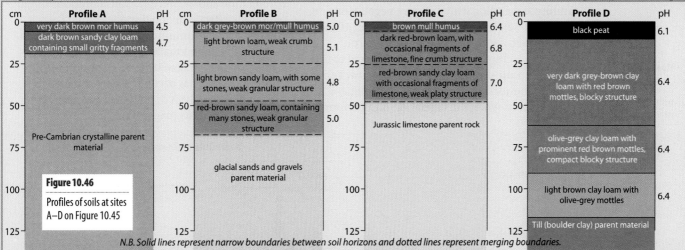

Figure 10.46

Profiles of soils at sites A–D on Figure 10.45

N.B. Solid lines represent narrow boundaries between soil horizons and dotted lines represent merging boundaries.

15 **a** Using Figure 10.45 and 10.46, identify which of the soils are zonal, azonal and interzonal. **(4 marks)**

  b **i** Study soil C. Describe this soil.

    ii Suggest what vegetation would grow as a climax on this soil type. Explain your answer. **(10 marks)**

  c Soil D is a peaty gley. Explain **two** aspects of this soil that make it difficult for a farmer to cultivate. **(6 marks)**

  d Suggest **two** reasons why soil A is a very shallow soil. **(5 marks)**

16 Use the information on Burkina Faso in Figures 10.36 (pages 280) and 10.40 (page 282).

  a Identify the climatic and natural vegetation characteristics of Burkina Faso. **(7 marks)**

  b Explain how human activity is causing soil degradation. **(10 marks)**

  c Describe **two** techniques used to reduce soil degradation in Burkina Faso. **(8 marks)**

Biogeography

> 'The Earth's green cover is a prerequisite for the rest of life. Plants alone, through the alchemy of photosynthesis, can use sunlight energy, and convert it to the chemical energy animals need for survival.'
>
> James Lovelock, *The Gaia Atlas of Planet Management*, 1985

Biogeography may be defined as the study of the distribution of plants and animals over the Earth's surface. The biogeographer is interested in describing and explaining meaningful patterns of plant and animal distributions in a given area, either at a particular time or through a time-period.

Seres and climax vegetation

A **sere** is a stage in a sequence of events by which the vegetation of an area develops over a period of time. The first plants to colonise an area and develop in it are called the **pioneer community** (or **species**). A **prisere** is the complete chain of successive seres beginning with a pioneer community and ending with a **climax vegetation** (Figure 11.1a). F. E. Clements suggested, in 1916, that for each climatic zone only one type of climax vegetation could evolve. He referred to this as the **climatic climax vegetation**; we now know it as the **monoclimax concept**. The climatic climax occurs when the vegetation is in harmony

or equilibrium with the local environment, i.e. when the natural vegetation has reached a delicate but stable balance with the climate and soils of an area (Chapter 12). Each successive seral community usually shows an increase in the number of species and the height of the plants.

Each individual sere is referred to by one or more of the larger species within that community – the so-called **dominant species**. The dominant species may be the **largest** plant or tree in the community which exerts the maximum influence on the local environment or habitat, or the most **numerous** species in the community. In parts of the world where the climatic climax is forest – i.e. areas with higher rainfall – the plant community tends to be structured in layers (Figures 11.2 and 12.4). It can take several thousand years to reach a climatic climax. Communities are, however, relatively ephemeral on timescales of millennia. When climatic change does occur, temperature and/or precipitation alterations often only affect individual species rather than changing the community as a whole. This concept, the 'individualistic concept of plant association', was originated by H. A. Gleason in 1928. In recent years it has become widely accepted as a result of the analysis of pollen taken from lake sediments and peats (page 294).

Figure 11.1

A seral progression, with possible interruptions

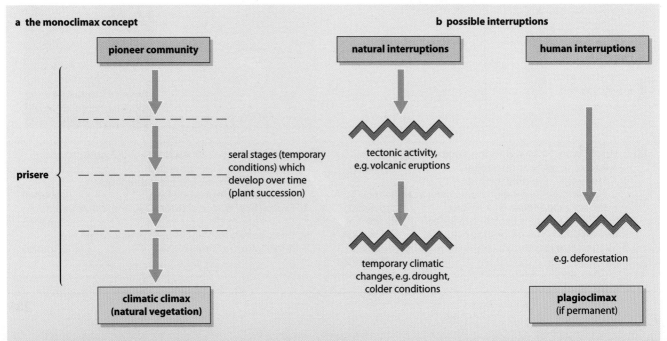

a the monoclimax concept

pioneer community

seral stages (temporary conditions) which develop over time (plant succession)

prisere

climatic climax
(natural vegetation)

b possible interruptions

natural interruptions

tectonic activity, e.g. volcanic eruptions

temporary climatic changes, e.g. drought, colder conditions

human interruptions

e.g. deforestation

plagioclimax
(if permanent)

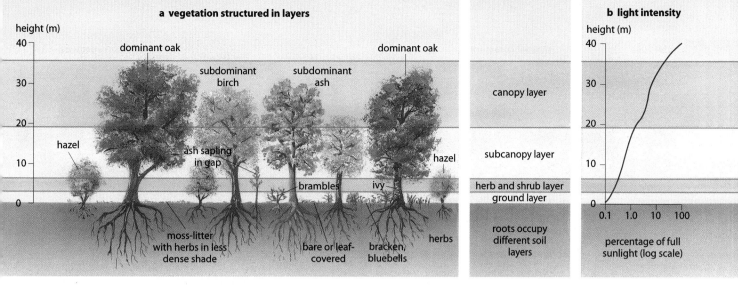

a vegetation structured in layers

height (m)

- dominant oak
- subdominant birch
- subdominant ash
- dominant oak
- hazel
- ash sapling in gap
- brambles
- ivy
- hazel
- moss-litter with herbs in less dense shade
- bare or leaf-covered
- bracken, bluebells
- herbs

canopy layer

subcanopy layer

herb and shrub layer
ground layer

roots occupy different soil layers

b light intensity

height (m)

percentage of full sunlight (log scale)

Figure 11.2

Vegetation structure and light intensity typical of a temperate deciduous woodland (*after* O'Hare)

There are, however, very few parts of today's world with a climatic climax. This is partly because few physical environments remain stable sufficiently long for the climax to be reached: most are affected by tectonic or temporary climatic changes (an area becomes warmer, colder, wetter or drier). More recently, however, instability has resulted from such human activities as deforestation, the ploughing of grassland, and acid rain. Where human activity has permanently arrested and altered the natural succession and the ensuing vegetation is maintained through management, the resultant community is said to be a **plagioclimax** (Figure 11.1b) – examples of which include heather moorlands in Britain, and the temperate grasslands (page 326).

While it is still accepted that climate exerts a major influence upon vegetation, the linear monoclimax concept has been replaced by the **polyclimax theory**. This theory acknowledges the importance not only of climate, but of several (poly) local factors including drainage,

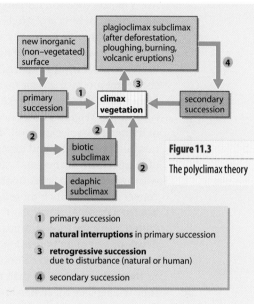

Figure 11.3

The polyclimax theory

1. primary succession
2. **natural interruptions** in primary succession
3. **retrogressive succession** due to disturbance (natural or human)
4. secondary succession

parent rock, relief and microclimate. The polyclimax theory, therefore, relates the climax vegetation to a variety of factors. Figure 11.3 shows how the climax vegetation may result from a **primary** or a **secondary succession**. A primary succession occurs on a new or previously sterile land surface, or in water. Figure 11.4 shows how the four more commonly accepted non-vegetated environments in Britain develop until they all reach the same climax vegetation: the oak woodland. A secondary succession is more likely to occur on land on which the previous management has been discontinued, e.g. abandoned farmland due to shifting cultivation in the tropical rainforest (Places 66, page 480). A **subclimax** occurs when the vegetation is prevented from reaching its climax due to interruptions by local factors such as soils and human interference.

Figure 11.4

Primary successions

non-vegetated surfaces (i.e. initially unsuitable for vegetation)

land

water

1. **lithosere** (rock)
2. **psammosere** (sand)
3. **halosere** (salt water)
4. **hydrosere** (fresh water)

e.g. new volcanic island, emerging raised beach, retreat of a glacier

e.g. sand dunes

e.g. salt marsh

e.g. infilling lake, pond

increasing soil depth and nutrients

increasing stability and humus

decreasing influence of the sea

decreasing influence of open water

several stages **(seres)**

mesophytic (transitional: adapted to neither very dry conditions [xerophytic] nor very wet conditions [hydrophytic]) e.g. oak climax

Biogeography

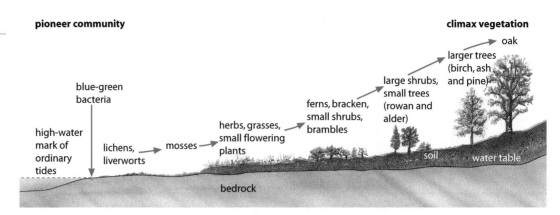

Four basic seres forming a primary succession

1 Lithoseres

Areas of bare rock will initially be colonised by blue-green bacteria and single-celled photosynthesisers that have no root system and can survive where there are few mineral nutrients. Blue-green bacteria are autotrophs (page 296), photosynthesising and producing their own food source. Lichens and mosses also make up the pioneer community (Figure 11.5). These plants are capable of living in areas lacking soil, devoid of a permanent supply of water and experiencing extremes of temperature. Lichen and various forms of weathering help to break up the rock to form a veneer of soil in which more advanced plant life can then grow. As these plants die, they are converted by bacteria into humus which helps in the development of an increasingly richer soil. Seeds, mainly of grasses, then colonise the area. As these plants are taller than the pioneer species, they will replace the lichen and mosses as the dominants, although the lichens and mosses will still continue to grow in the community. As the plant succession evolves over a period of time, the grasses will give way as dominants to fast-growing shrubs, which in turn will be replaced by relatively fast-growing trees (rowan). These will eventually face competition from slower-growing trees (ash) and, finally, the oak which forms the climax vegetation. It should be noted that although each stage of the succession is marked by a new dominant, many of the earlier species continue to grow there, although some are shaded out.

Figure 11.5 shows an idealised primary succession across a newly emerging rocky coastline. It excludes the increasing number of species found at each stage of the seral succession. The species are determined by local differences in rainfall, temperature and sunlight, bedrock and soil type, aspect and relief. Lithoseres can develop on bare rock exposed by a retreating glacier (page 294), on ash or lava following a volcanic eruption on land (Krakatoa, Places 35) or forming a new island (Surtsey, Places 3, page 16), or, as in Figure 11.5, on land emerging from the sea as a result of isostatic uplift following the melting of an icecap (page 163).

Over time, the area shown to have the pioneer community passes through several stages until the climatic climax is reached – assuming that the land continues to rise, that there is no significant change in the local climate, and that there is no human interference. Figures 11.6 and 11.7 are photos showing two stages in the succession,

Figure 11.6

Primary succession on a lithosere on the Isle of Arran: lichens, mosses and grasses on a rocky coastline

Figure 11.7

Primary succession on the same lithosere in Arran: bracken and deciduous woodland behind the rocky coastline

taken on a raised beach on the east coast of Arran. Figure 11.6 shows lichen, favouring a south-facing aspect on gently dipping rocks, and mosses, growing in darker north-facing hollows. Beyond, where soil has begun to form and where the water table is high, grasses and bog myrtle have entered the succession. Figure 11.7 was taken where the soil depth and amount of humus have increased and the water table is lower, as indicated by the presence of bracken. To the right, but not clearly visible on the photo, reeds are growing in a hollow where the water table is nearer to the surface. In the middle distance are small deciduous trees with, behind them, taller oaks indicating a climax vegetation.

Places 35 Krakatoa: a lithosere

In August 1883, a series of volcanic eruptions reduced the island of Krakatoa to one-third of its previous size and left a layer of ash over 50 m deep. No vegetation or animal life was left on the island or in the surrounding sea. Yet within three years (Figure 11.8), 26 species had reappeared and, in 1933, 271 plant and 720 insect species, together with several reptiles, were recorded. The first recolonisers arrived in three ways. Most were seeds blown from surrounding islands by the wind, while others drifted in from the sea or were carried by birds. However, in this example, the concept of plant succession, put forward later by F. E. Clements (page 286), includes a variable, as many of the plants which recolonised Krakatoa arrived there by chance – for example, a piece of driftwood with a particular seed type just happened to be washed ashore, whereas it could just as easily have missed the island altogether.

Figure 11.8

Primary succession, Krakatoa: vegetation distribution according to height above sea-level, 1983

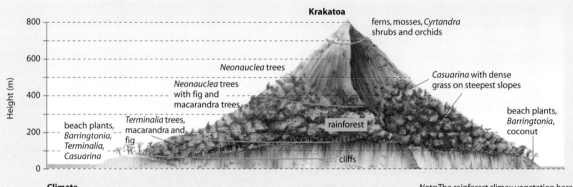

Climate
Temperatures are high and constant. Most months average 28°C, giving a very low annual range. Rain is heavy, falling in convectional storms most afternoons throughout the year.

Note: The rainforest climax vegetation here does not contain as many species as the rainforests on surrounding islands.

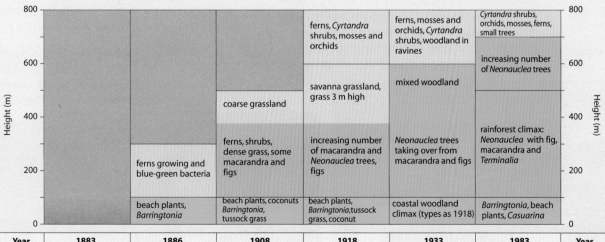

Year	1883	1886	1908	1918	1933	1983	Year
Number of plant species	0	26	115	132	271	?	Number of plant species

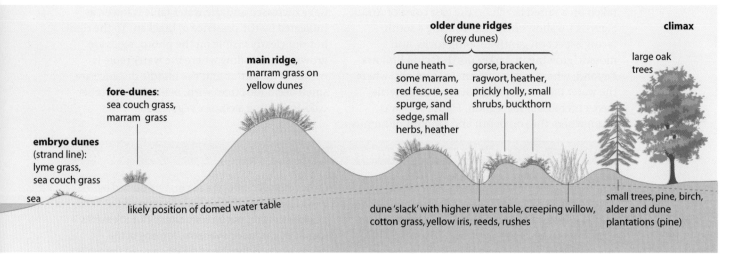

older dune ridges
(grey dunes)

climax

large oak
trees

main ridge,
marram grass on
yellow dunes

dune heath –
some marram,
red fescue, sea
spurge, sand
sedge, small
herbs, heather

gorse, bracken,
ragwort, heather,
prickly holly, small
shrubs, buckthorn

fore-dunes:
sea couch grass,
marram grass

embryo dunes
(strand line):
lyme grass,
sea couch grass

sea

likely position of domed water table

dune 'slack' with higher water table, creeping willow,
cotton grass, yellow iris, reeds, rushes

small trees, pine, birch,
alder and dune
plantations (pine)

2 Psammoseres

A psammosere succession develops on sand and is best illustrated by taking a transect across coastal dunes (Figure 11.9). The first plants to colonise, indeed to initiate dune formation, are usually lyme grass, sea couch grass and marram grass. Sea couch grass grows on berms around the tidal high-water mark and is often responsible for the formation of embryo dunes (Figure 6.31). On the yellow fore-dunes, which are arid, being above the highest of tides and experiencing rapid percolation by rainwater, marram grass becomes equally important.

The main dune ridge, which is extremely arid and exposed to wind, is likely to be vegetated exclusively by marram grass. Marram has adapted to these harsh conditions by having leaves that can fold to reduce surface area, which are shiny and which can be aligned to the wind direction: three factors capable of limiting evapotranspiration. Marram also has long roots to tap underground water supplies and is able to grow upwards as fast as sand deposition can cover it. Grey dunes, behind the main ridge, have lost their supply of sand and are sheltered from the prevailing wind. Their greater humus content, from the decomposition of earlier marram grass, enables the soil to hold more moisture. Although marram is still present, it faces increasing competition from small flowering plants and herbs such as sea spurge (with succulent leaves to store water) and heather.

The older ridges, further from the water, have both more and taller species. Dune slacks may form in hollows between the ridges if the water table reaches the surface. Plants such as creeping willow, yellow iris, reeds and rushes and shrubs are indicators of a deeper and wetter soil. On the landward side of the dunes, perhaps 400 m from the beach, are small deciduous trees including ash and hawthorn and, as the soil is sandy, pine plantations. Furthest inland comes the oak climax. Figure 11.9 shows a psammosere based on sand dunes at Morfa Harlech, north Wales. Figures 11.10 and 6.32 show marram and lyme grass forming the yellow fore-dunes, with gorse and heather on the greyer dunes behind. Figures 11.11 and 6.33 show vegetation on the inland ridges.

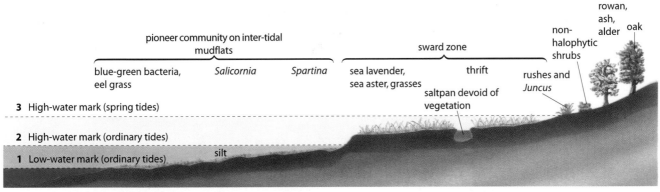

pioneer community on inter-tidal mudflats

sward zone

rowan, ash, alder oak

non-halophytic shrubs

blue-green bacteria, eel grass   *Salicornia*   *Spartina*

sea lavender, sea aster, grasses   thrift   rushes and *Juncus*

saltpan devoid of vegetation

3 High-water mark (spring tides)

2 High-water mark (ordinary tides)

1 Low-water mark (ordinary tides)   silt

Figure 11.12

Transect showing a primary succession in a halosere, Llanrhidian Marsh, Gower Peninsula, south Wales

3 Haloseres

In river estuaries, large amounts of silt are deposited by the ebbing tide and inflowing rivers. The earliest plant colonisers are green algae and eel grass which can tolerate submergence by the tide for most of the 12-hour cycle and which trap mud, causing it to accumulate. Two other colonisers are *Salicornia* and *Spartina* which are **halophytes** – i.e. plants that can tolerate saline conditions. They grow on the inter-tidal mudflats (Figure 6.34), with a maximum of 4 hours' exposure to the air in every 12 hours. *Spartina* has long roots enabling it to trap more mud than the initial colonisers of algae and *Salicornia*, and so, in most places, it has become the dominant vegetation. The inter-tidal flats receive new sediment daily, are waterlogged to the exclusion of oxygen, and have a high pH value.

The sward zone, in contrast, is inhabited by plants that can only tolerate a maximum of 4 hours' submergence in every 12 hours. Here the dominant species are sea lavender, sea aster and grasses, including the 'bowling green turf' of the Solway Firth. However, although the vegetation here tends to form a thick mat, it is not continuous. Hollows may remain where the seawater becomes trapped leaving, after evaporation, saltpans in which the salinity is too great for plants (Figure 11.13). As the tide ebbs, water draining off the land may be concentrated into creeks (Figure 6.35). The upper sward zone is only covered by spring tides and here *Juncus* and other rushes grow. Further inland, non-halophytic grasses and shrubs enter the succession, to be followed by small trees and ultimately by the climax oak vegetation. Figure 11.12 is a transect based on the saltmarshes on the north coast of the Gower Peninsula in south Wales. Figure 11.14 shows several stages in the halosere succession.

Figure 11.13

Primary succession in a halosere: a saltpan on the Suffolk coast, covered only by the highest of tides

Figure 11.14

Primary succession in a halosere: Blackney Point, Norfolk

Figure 11.15

Idealised primary succession in a hydrosere

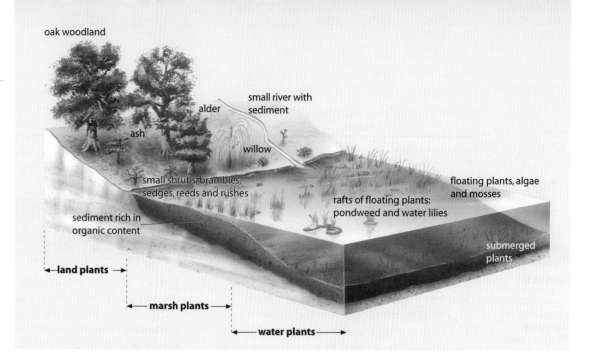

Figure 11.15

Idealised primary succession in a hydrosere

oak woodland

small river with sediment

alder

ash

willow

small shrubs, brambles, sedges, reeds and rushes

sediment rich in organic content

rafts of floating plants: pondweed and water lilies

floating plants, algae and mosses

submerged plants

←land plants→

← marsh plants →

← water plants →

Figure 11.16

Primary succession in a hydrosere at the head of a reservoir in Cumbria

4 Hydroseres

Lakes and ponds originate as clear water which contains few plant nutrients. Any sediment carried into the lake will enrich its water with nutrients and begin to infill it. The earliest colonisers will probably be algae and mosses whose spores have been blown onto the water surface by the wind. These grow to form vegetation rafts which provide a habitat for bacteria and insects. Next will be water-loving plants which may either grow on the surface, e.g. water lilies and pondweed, or be totally submerged (Figure 11.15). Bacteria recycle the nutrients from the pioneer community, and marsh plants such as bulrushes, sedges and reeds begin to encroach into the lake. As these marsh plants grow outwards into the lake and further sediment builds upwards at the expense of the water, small trees will take root forming a marshy thicket. In time, the lake is likely to contract in size, to become deoxygenised by the decaying vegetation and eventually to disappear and be replaced by the oak climax vegetation. This primary succession is shown in Figure 11.15. Figure 11.16 shows land plants encroaching at the head of a reservoir, while Figure 11.17 illustrates the water, marsh and land plant succession in and around a small lake.

Incidentally, it is not necessary to be an expert botanist to recognise the plants named in these primary successions; you just need access to a good plant recognition book!

Figure 11.17

Primary succession in a small lake, Sussex

Secondary succession

A climatic climax occurs when there is stability in transfers of material and energy in the ecosystem (page 295) between the plant cover and the physical environment. However, there are several factors that can arrest the plant succession before it has achieved this dynamic equilibrium, or which may alter the climax after it has been reached. These include:

- a mudflow or landslide (Places 36)
- deforestation or afforestation
- overgrazing by animals or the ploughing-up of grasslands
- burning grasslands, moorlands, forests and heaths
- draining wetlands
- disease (e.g. Dutch elm), and
- changes in climate (page 286).

Places 36   Arran: secondary plant succession

The mudflow shown in Figure 2.16 occurred in October 1981 and completely covered all the existing vegetation. Twelve months later it was estimated that 20 per cent of the flow had been recolonised, a figure that had grown to 40 per cent in 1984 and 70 per cent in 1988. Had this been a primary succession, lichens and mosses would have formed the pioneer community and they would probably have covered only a small area. The pioneer plants would probably also have been randomly distributed and, even after seven years, the species would have been few in number and small in height.

Instead, by 1988, much of the flow had already been recolonised. It could be seen that most of the plants were found near the edges of the flow and were not randomly distributed, and there were already several species including grasses, heather, bog myrtle and mosses, some of which exceeded 50 cm in height.

These observations suggest a secondary succession with plants from the surrounding climax area having invaded the flow, mainly due to the dispersal of their seeds by the wind.

The effect of fire

The severity of a fire and its effect on the ecosystem depend largely upon the climatic conditions at the time. The fire is likely to be hottest in dry weather and, in the northern hemisphere, on sunny south-facing slopes where the vegetation is driest. The spread of a fire is fastest when the wind is strong and blowing uphill and where there is a build-up of combustible material. The extent of disruption also depends upon the type and the state of the vegetation. The following is a list of examples, in rank order of severity.

1 Areas with a Mediterranean climate, where the chaparral of California and the maquis/garrigue of southern Europe are densest and tinder-dry in late summer after the seasonal drought. Recent examples include bush fires in south-east Australia (in 1994, 200 houses were destroyed and four people were killed on the outskirts of Sydney), in the south of France (1999), and around Los Angeles in 1993 (Case Study 15A).
2 Coniferous forests where the leaf litter burns readily.
3 Ungrazed grasslands and, especially, the savannas, which have a low biomass but a thick litter layer (Figure 11.28). **Biomass** is the total mass of living organisms present in a community at any given time, expressed in terms of oven-dry weight (mass) per unit area.

4 Intensively grazed grasslands which have a lower biomass and a limited litter layer.
5 Deciduous woodlands which, despite the presence of a thick litter layer, are often slow to burn.

Following a fire, the blackened soil has a lower albedo and absorbs heat more readily and, without its protective vegetation cover, the soil is more vulnerable to erosion. Ash initially increases considerably the quantity of inorganic nutrients in the soil and bacterial activity is accelerated. Any seedlings left in the soil will grow rapidly as there is now plenty of light, no smothering layer of leaf litter, plenty of nutrients, a warmer soil and, at first, less competition from other species. Heaths and moors that have been fired are conspicuous by their greener, more vigorous growth. A fire climax community, known as **pyrophic** vegetation, contains plants with seeds which have a thick protective coat and which may germinate because of the heat of the fire. The community may have a high proportion of species that can sprout quickly after the fire – plants that are protected by thick, insulating bark (cork oak in the chaparral (page 324) and baobab in the savannas (Figure 12.14)), or which have underground tubers or rhizomes insulated by the soil. It has been suggested that the grasslands of the American Prairies and the African savannas are not climatic climax vegetation, but are the result of firing by indigenous Indian and African tribes (Case Study 12).

Vegetation changes in the Holocene

The Holocene is the most recent of the geological periods (Figures 1.1 and 11.18). The last glacial advance in Britain ended about 18 000 BP. Although the extreme south of England remained covered with hardy tundra plants, most of northern Britain was left as bare rock or glacial till. Had the climate gradually and constantly ameliorated, a primary succession would have taken place, from south to north, as previously described for a lithosere. However, we know that there have been several major fluctuations in climate during those 18 000 years which have resulted in significant changes in the climax vegetation (Figure 11.18).

There are several techniques that help to determine vegetation change: pollen analysis, dendrochronology, radio-carbon dating, and historical evidence (page 248). Families of plants have the same pollen grain in terms of its shape and pattern. Where pollen is blown by the wind onto peat bogs, such as at Tregaron in west Wales, the grains are trapped by the peat. As more peat accumulates over the years, the pollen of successively later times indicates which were the dominant and subdominant plants of the period (Figures 11.18 and 11.19). As each plant grows best within certain defined temperature and precipitation limits, it is possible to determine when the climate either improved (ameliorated) or deteriorated. Dendrochronology – dating by means of the annual growth-rings of trees – has shown that the bristlecone pine of California can be dated back some 5000 years, while British dendrochronology, based on bog oaks in Ireland, extends back some 10 000 years. Radio-carbon dating is based on changing amounts of radioactivity in the atmosphere and in plants. Notice in Figure 11.18, which links climatic and vegetation changes, how forests increase as the climate ameliorates, and how heathland and peat moors take over when the climate deteriorates.

Figure 11.19

Changes in the surface of lowland England, Wales and Scotland over the last 12 000 years (*after* Wilkinson)

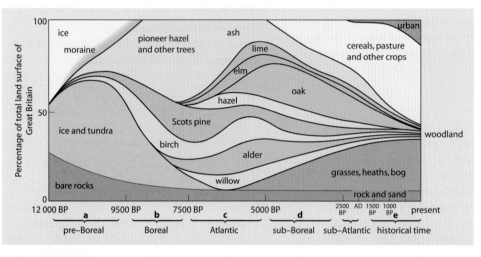

Figure 11.18

Climatic and vegetation change in Britain since the Holocene (BP = Before Present)

Date BP	Phase/period	Climate	Vegetation	Cultures
pre-17 000	final glaciation	glacial	none in northern Britain; tundra in southern England	none
17 000–14 000	periglacial	cold, 6°C in summer	tundra	Palaeolithic
14 000–12 000	Allerød	warming slowly to 12°C in summer	tundra with hardy trees, e.g. willow and birch	Palaeolithic
12 000–10 000	pre-Boreal	glacial advance: colder, 4°C in summer	Arctic/Alpine plants, tundra	Mesolithic
10 000–8000	Boreal	continental: winters colder and drier, summers warmer than today	forests: juniper first then pine and birch and finally oak, elm and lime	Mesolithic
8000–5000	Atlantic	maritime: warm summers, 20°C; mild winters, 5°C; wet	our 'optimum' climate and vegetation: oak, alder, hazel, elm and lime (too cold for lime today); peat on moors	beginning of Neolithic; first deforestation about 3500 BC
5 000–2500	sub-Boreal	continental: warmer and drier	elm and lime declined; birch flourished; peat bogs dried out	Neolithic period, settled agriculture; beginning of Bronze Age
2500–2000	sub-Atlantic	maritime: cooler, stormy and wet	peat bogs re-formed; decline in forests due to climate and farming	settled agriculture
2000–1000	historical times	improvement: warmer and drier	clearances for farming	Roman occupation during early part
1000–450		decline: much cooler and wetter	further clearances: little climax vegetation left; medieval farming	
450–300		'little ice age': colder than today		
post-300– present		gradual improvement	recently some afforestation: coniferous trees	Agrarian and Industrial Revolutions

Ecology and ecosystems

The term **ecology**, which comes from the Greek word *oikos* meaning 'home', refers to the study of the interrelationships between organisms and their habitats. An organism's home or **habitat** lies in the biosphere, i.e. the surface zone of the Earth and its adjacent atmosphere in which all organic life exists. The scale of each home varies from small **micro-habitats**, such as under a stone or a leaf, to **biomes**, which include tropical rainforests and deserts (Figure 11.20). Fundamental to the four ecological units listed in Figure 11.20 is the concept of the **environment**. The environment is a collective term to include all the conditions in which an organism lives. It can be divided into:

a the physical, non-living or **abiotic environment**, which includes temperature, water, light, humidity, wind, carbon dioxide, oxygen, pH, rocks and nutrients in the soil, and

b the living or **biotic environment**, which comprises all organisms: plants, animals, humans, bacteria and fungi.

The ecosystem

An ecosystem is a natural unit in which the life-cycles of plants, animals and other organisms are linked to each other and to the non-living constituents of the environment to form a natural system (Framework 3, page 45). The **community** consists of all the different species within a habitat or ecosystem. The **population** is all the individuals of a particular species in a habitat. An ecosystem depends on two basic processes: **energy flows** and **material cycling**. As the flow of energy is only in one direction and because it crosses the system boundaries, this aspect of the ecosystem behaves as an **open** system. Nutrients, which are constantly recycled for future use, are circulated in a series of **closed** systems.

1 Energy flows

The sun is the primary source of energy for all living things on Earth. As energy is retained only briefly in the biosphere before being returned to space, ecosystems have to rely upon a continual supply. The sun provides *heat energy* which cannot be captured by plants or animals but which warms up the communities and their non-living surroundings. The sun is also a source of *light energy* which can be captured by green plants and transformed into chemical energy through the process of **photosynthesis**. Without photosynthesis, there would be no life on Earth. Light, chlorophyll, warmth, water and carbon dioxide are required for this process to operate. Carbon dioxide, which is absorbed through stomata in the leaves of higher plants, reacts indirectly with water taken up by the roots when temperatures are suitably high, to form carbohydrate. The energy needed for this comes from sunlight which is 'trapped' by chlorophyll. Oxygen is a by-product of the process. The carbohydrate is then available as food for the plant.

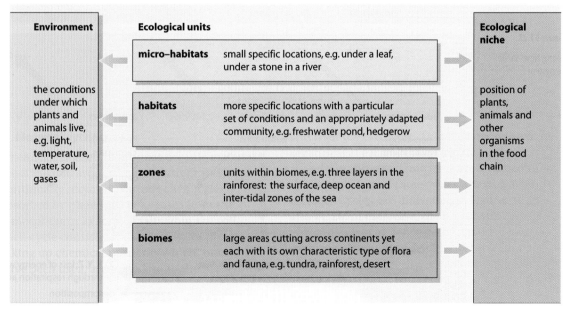

Figure 11.20

A hierarchical structure of ecological units

Figure 11.25

The carbon cycle (*after* M. B. V. Roberts)

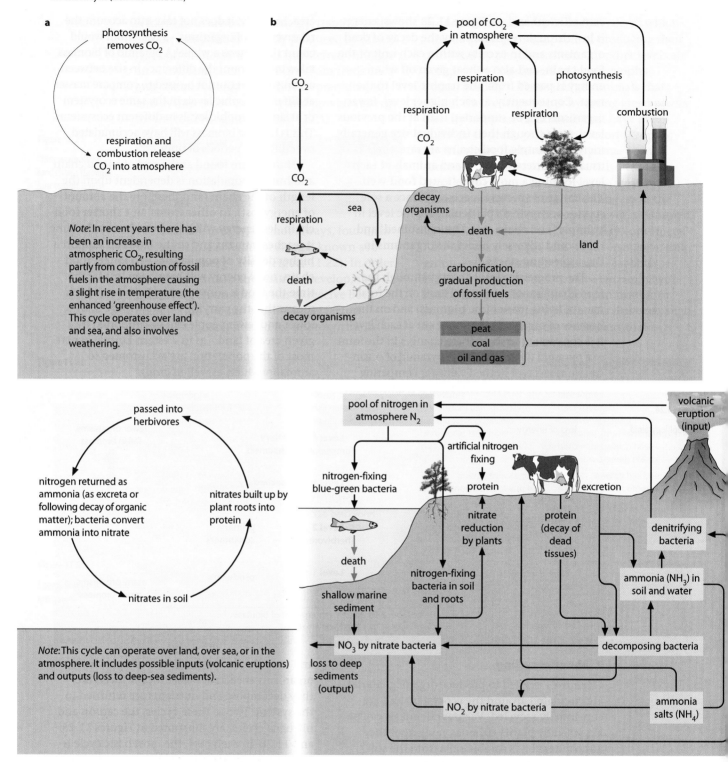

a

photosynthesis removes CO_2

respiration and combustion release CO_2 into atmosphere

Note: In recent years there has been an increase in atmospheric CO_2, resulting partly from combustion of fossil fuels in the atmosphere causing a slight rise in temperature (the enhanced 'greenhouse effect'). This cycle operates over land and sea, and also involves weathering.

b

pool of CO_2 in atmosphere

CO_2

CO_2

respiration

respiration

respiration

photosynthesis

combustion

sea

respiration

decay organisms

death

land

death

decay organisms

carbonification, gradual production of fossil fuels

peat
coal
oil and gas

passed into herbivores

nitrogen returned as ammonia (as excreta or following decay of organic matter); bacteria convert ammonia into nitrate

nitrates built up by plant roots into protein

nitrates in soil

Note: This cycle can operate over land, over sea, or in the atmosphere. It includes possible inputs (volcanic eruptions) and outputs (loss to deep-sea sediments).

pool of nitrogen in atmosphere N_2

volcanic eruption (input)

artificial nitrogen fixing

nitrogen-fixing blue-green bacteria

protein

excretion

denitrifying bacteria

nitrate reduction by plants

protein (decay of dead tissues)

ammonia (NH_3) in soil and water

death

nitrogen-fixing bacteria in soil and roots

shallow marine sediment

NO_3 by nitrate bacteria

decomposing bacteria

loss to deep sediments (output)

NO_2 by nitrate bacteria

ammonia salts (NH_4)

Figure 11.26

The nitrogen cycle (*after* M. B. V. Roberts)

Recent investigations, mainly in New Zealand and the Andes, have shown that nitrogen from seawater, or released by plants and animals as they die on the seabed, can be channelled upwards, together with magma, at subduction (destructive) plate margins. The nitrogen can later be released back into the atmosphere, either as water or as a gas, through volcanic eruptions. Once in the atmosphere, the nitrogen can return to Earth and the sea in rainwater – so completing another nitrogen cycle.

Since the 1960s, geographers have felt an increasing need to adopt a more scientific approach to their studies. This stemmed from a number of changes that were taking place in attitudes to the study of geography and to science in a broader sense:

- The increasing scale and complexity of the subject's material and the data available.

- The rapid development of theory, often using computer modelling, from which predictions could be made.

- The realisation that, despite great care, all human observers have their own, subjective, opinions which influence an assessment or conclusion (i.e. scientific objectivity could not be guaranteed).

The scientific approach to geography involved a series of logical steps, already practised in the physical sciences, which enabled conclusions to be drawn from precise and unbiased data (Framework 8, page 246). This approach is summarised in the flow diagram (Figure 11.27).

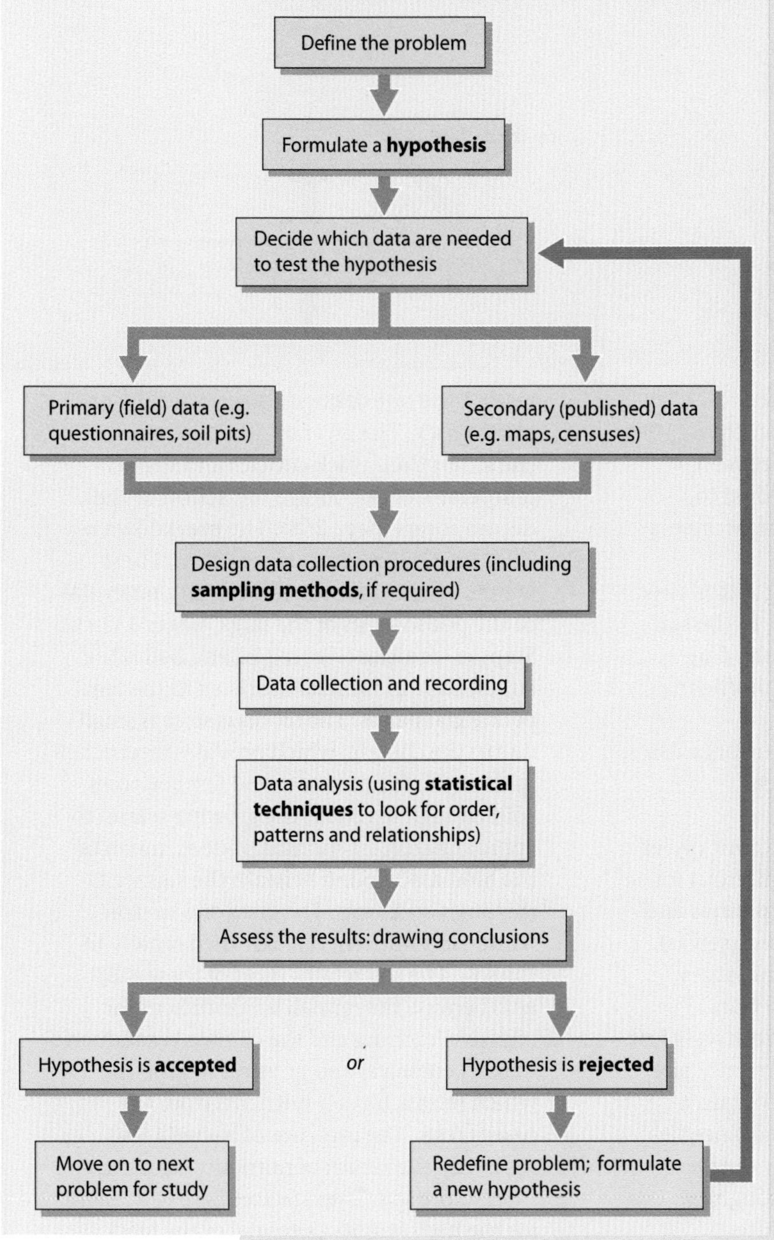

During a sixth-form field week on the Isle of Arran, one day was set aside for hypothesis testing. This involved seeking possible relationships between several variables on Goatfell (Figure 11.37). The hypotheses included:

- Vegetation density decreases as altitude increases.

- Soil acidity increases as altitude increases.

- Soil acidity increases as the angle of slope increases.

- Soil moisture increases as the angle of slope increases.

- Depth of soil increases as altitude decreases.

- Height of vegetation increases as altitude decreases.

- Number of species increases as altitude increases.

- Soil temperature increases as altitude decreases.

Data collection required the taking of readings at a minimum of 15 sites from sea-level to the top of Goatfell. It is important that the selection of sites is made without introducing bias (see Framework 6, page 159).

Data analysis may include drawing a scattergraph to investigate the possibility of any correlation between the two variables; calculating the strength of the relationship between the variables by using the Spearman's rank correlation coefficient (Framework 19, page 635); and then testing the result to see how likely it is that the correlation occurred by chance (page 637).

It should then be possible to determine whether the original hypothesis is acceptable as an explanation of the data, or not. If it is rejected, then a new hypothesis should be formulated.

Figure 11.27

Hypothesis testing

Define the problem

Formulate a **hypothesis**

Decide which data are needed to test the hypothesis

Primary (field) data (e.g. questionnaires, soil pits)

Secondary (published) data (e.g. maps, censuses)

Design data collection procedures (including **sampling methods**, if required)

Data collection and recording

Data analysis (using **statistical techniques** to look for order, patterns and relationships)

Assess the results: drawing conclusions

Hypothesis is **accepted**   or   Hypothesis is **rejected**

Move on to next problem for study

Redefine problem; formulate a new hypothesis

Figure 11.28

A model of the mineral nutrient cycle (*after* Gersmehl)

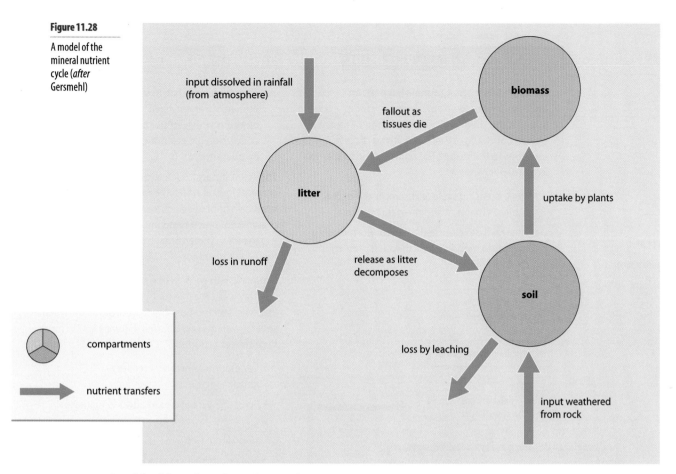

Model of the mineral nutrient cycle

This model, developed by P. F. Gersmehl in 1976, attempts to show the differences between ecosystems in terms of nutrients stored in, and transferred between, three compartments (Figure 11.28).

- **Litter** – the total amount of organic matter, including humus and leaf litter, in the soil (it is, therefore, more than just the *L* or litter layer as shown in the soil profile in Figure 10.5).
- **Biomass** – the total mass of living organisms, mainly plant tissue, per unit area.
- **Soil**.

Figure 11.29 shows the mineral nutrient cycles for three selected biomes: **a** the coniferous forest (taiga), **b** the temperate grassland (prairies and steppes), and **c** the tropical rainforest (selvas).

a   **Taiga** (Figure 11.29a) Litter is the largest store of mineral nutrients in the taiga. Although forest, the biomass is relatively low because the coniferous trees form only one layer, have little undergrowth, contain a limited variety of species, and have needle-like leaves. The soil contains few nutrients because, following their loss through leaching and as surface runoff (after snowmelt when the ground is still frozen), replacement is slow: the low temperatures restrict the rate of chemical weathering of parent rock. The layer of needles is often thick, but their thick cuticles and the low temperatures discourage the action of the decomposers (page 268). The breakdown of litter into humus is thus very slow. These factors account for the low fertility potential of the podsol soils of the taiga (pages 331–32).

b   **Steppes/prairies** (Figure 11.29b) Soil is the largest store of mineral nutrients in the temperate grasslands. The biomass store is small due to the climate, which provides insufficient moisture to support trees and temperatures cold enough to reduce the growing season to approximately six months. Indeed, much of the biomass is found beneath the surface as rhizomes and roots. The grass dies back in winter and nutrients are returned rapidly to the soil. The soil retains most of these nutrients because the rainfall is insufficient for effective leaching and the climate is conducive to both chemical and physical weathering which release further nutrients from the parent rock. The presence of bacteria also speeds up the return of nutrients from the litter to the soil. These factors help to account for the high fertility potential of the black chernozem soils associated with temperate grasslands (page 327).

Figure 11.29

Mineral nutrient cycles in three different environments (*after* Gersmehl)

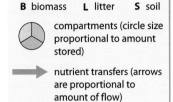

B biomass   L litter   S soil

compartments (circle size proportional to amount stored)

nutrient transfers (arrows are proportional to amount of flow)

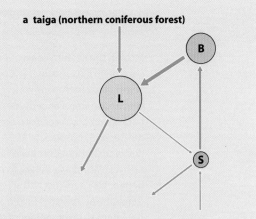

a taiga (northern coniferous forest)

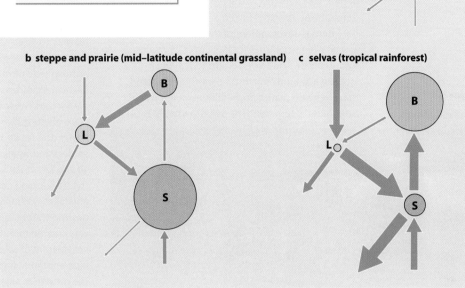

b steppe and prairie (mid–latitude continental grassland)   **c selvas (tropical rainforest)**

c **Selvas** (Figure 11.29c) The tropical rainforests have, of all the major environments, the highest rates of transfer – an annual rate ten times greater than that of the taiga. The biomass is the largest store of mineral nutrients in the tropical rainforests. High annual temperatures, the heavy, evenly distributed rainfall and the year-long growing season all contribute to the tall, dense and rapid growth of vegetation. The biomass is composed of several layers of plants and countless different species. The many plant roots take up vast amounts of nutrients. In comparison, the litter store is limited, despite the continuous fall of leaves, because the hot, wet climate provides the ideal environment for bacterial action (both in numbers and type) and the decomposition of dead vegetation. In areas where the forest is cleared, the heavy rain soon removes the nutrients from the soil by leaching or surface runoff. The litter content is rapidly reduced so that the soil has to rely upon the replacement of nutrients from the bedrock, a process that is usually rapid as climatic conditions are optimal for chemical weathering (Figure 2.9b). Initially nutrients such as phosphorus may increase if the forest is burnt, but deforestation usually leads to a rapid decline in soil fertility (pages 317–18). Figure 11.30 compares the storage and transfer of nutrients in four major biomes (i.e. ecosystems on a large scale). Remember that these figures refer to natural cycles which, in reality, have often been interrupted or modified by human activity.

Figure 11.30

Storage and transfer of nutrients within selected biomes

			Nutrient storage			Annual nutrient transfer		
		Ecosystem type	Stored in biomass	Stored in litter	Stored in soil	Soil to biomass	Biomass to litter	Litter to soil
Forest cycles	**A**	Equatorial rainforest	11 081	178	352	2 028	1 540	4 480
		Coniferous forest	3 350	2 100	142	178	145	86
Grassland cycles	**B**	Tropical savanna	978	300	502	319	312	266
		Temperate prairie/steppe	540	370	5 000	422	426	290
All measurements in kg/ha								

Most of the eastern coast of Africa is protected by coral reefs (Case Study 17). Coral, which live in clear, warm, shallow tropical waters, are small organisms that exude lime. For centuries, coast-dwellers have hacked out blocks of dead coral to build their houses and mosques. In 1954, the Bamburi Portland Cement Company built a factory 10 km north of Mombasa, Kenya, to produce cement, and began the open-cast extraction of coral (page 522). Cement was essential to Kenya, partly to help in the internal development of the country and partly as a vital export earner. By 1971, over 25 million tonnes of coral had been quarried, leaving a sterile wasteland covering 3.5 km². On that land there were no plants, no wildlife, no soil: it was a degraded ecosystem. The Swiss-owned trans-national cement company then appointed Dr Rene Haller to restore the environment from what he himself described as 'a lunar landscape filled with saline pools' (Figure 11.31).

After trying 26 different types of tree, Dr Haller found the key to be the *Casuarina* tree (Figure 11.32). This pioneer tree grew by 3 m a year, flourished in the sterile coral rubble, and was able to withstand both the high salinity and the high ground temperatures (up to 40°C). The constant fall of the needle-type leaves provided a habitat for red-legged millipedes which, together with the *Casuarina*'s ability to 'fix' atmospheric nitrogen, helped with the formation of the first soil and provided the base for a new ecosystem. The soil was collected and more trees were planted. Over the next few years, indigenous herbs, grasses and tree species, as well as beetles, spiders and small animals, were introduced into the young forest, each with its own function (niche) in the developing ecosystem. The depth of the ponds and lakes was increased until they reached the ground water table so that a freshwater habitat was created for fish (initially the local tilapia which are tolerant of saline water), crocodiles and hippopotami. Hippopotami excrement stimulated the growth of algae which oxygenated the water, preventing eutrophication. After only 20 years, the soil depth had reached 20 cm and the rainforest, with over 220 tree species, had become sufficiently restored to be home for over 180 recorded species of bird. The ecosystem was completed with the introduction of grazing animals (herbivores) such as the buffalo, oryx, antelope and giraffe. The re-creation of the rainforest (Figure 11.33) had been completed without the use of artificial fertiliser and insecticides, as Dr Haller considered these to be incompatible with his concept of a complex, balanced ecosystem.

The project has not only been an environmental success, it has also become a sustainable commercial venture with income derived from, for example, the sale of timber, bananas, vegetables, crocodiles and honey. The main source of the economy is the integrated aquaculture system (Figure 11.34) with, at its centre, the tilapia fish farm. The nutrients in the effluent water are used as fertiliser in the adjacent fruit plantation and for biogas to operate the pumps. From here, the water is led through a rice field into settlement ponds, where 'Nile cabbage' is grown for use in clearing the fish ponds. A crocodile farm is attached to the water system, as crocodile waste, which is rich in

Figure 11.31

The Bamburi Quarry

Figure 11.32

Casuarina trees planted in coral rubble, Bamburi Quarry

Figure 11.33

The re-created rain-forest ecosystem, Bamburi Quarry

phosphate and nitrogen, is a valuable fertiliser. The crocodiles are part of a planned food chain. Not only are they fed on surplus tilapia, but their eggs are eaten by monitor lizards that help to control the snake population which in turn controls the rodent population. Tourism has become a recent source of income. Baobab Farm, the name of the restored area, is open to school parties each morning and to other visitors in the afternoon. In 1992 it received over 100 000 visitors, making it the largest attraction in the Mombasa area. In brief, the once-barren quarry is now an ecologically and economically sound enterprise (Figure 11.35).

Dr Haller also believes that his intensive aquaculture and agroforestry techniques, geared to maximum yield of food, fuel and income from minimum land area and inputs, offer significant hope for small-scale African farmers who may be short of fertile land in a continent with an explosive population growth and which is ravaged by environmental and human-created disasters. He suggests that these methods could easily be adapted by Africans since their genesis lies in tribal techniques taught to him by local farmers.

Figure 11.34

The Baobab Farm integrated aquaculture system

figure 11.35

From the Bamburi Quarry Nature Trail leaflet

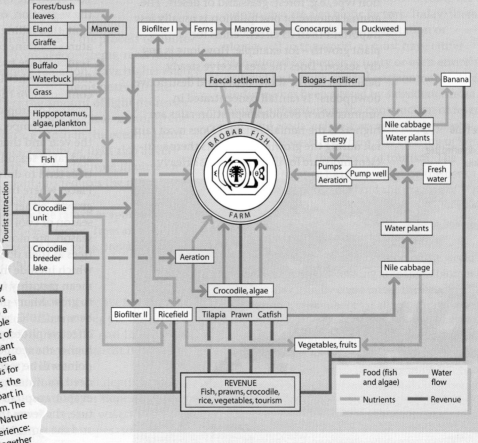

'In 1971 the Bamburi Portland Cement Company embarked on a unique project to re-create a living environment in the vast lunar landscape of its quarry. Today an extraordinary change is evident. Behind the glamorous façade of the tropical paradise regained, a completely balanced and commercially viable aquaculture complex is an important part of the established ecosystem. In a kind of giant jig-saw, casuarina trees, millipedes and bacteria work together, to provide the fertile basis for reclamation. Of particular interest is the surprising variety of wildlife playing a part in the natural balance of the whole system. The living creation of the Bamburi Quarry Nature Trail holds more than just a visual experience: a walk around will enable you to piece together the giant jig-saw puzzle of wasteland rehabilitation ...'

Figure 11.38

World biomes

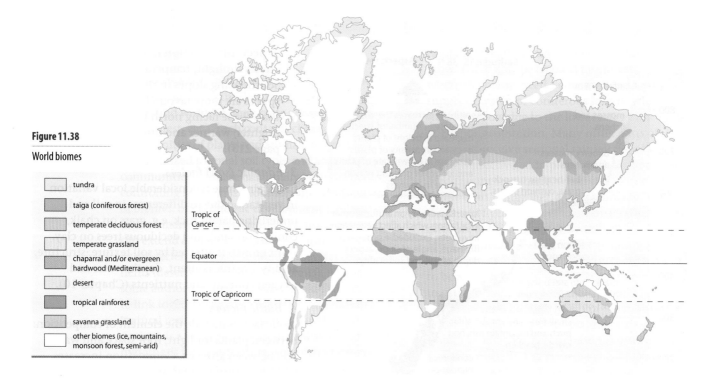

tundra

taiga (coniferous forest)

temperate deciduous forest

temperate grassland

chaparral and/or evergreen hardwood (Mediterranean)

desert

tropical rainforest

savanna grassland

other biomes (ice, mountains, monsoon forest, semi-arid)

Tropic of Cancer

Equator

Tropic of Capricorn

The eight major biomes, as shown in Figure 11.38, can be determined using a variety of criteria; two examples are discussed below and summarised in Figure 11.39.

1 **The traditional method** This links the type and global distribution of vegetation with that of the major world climatic types and zonal soils. This method was based on the understanding that it is climate that exerts the major influence and control over both vegetation and soils. The interactions between climate, soils and vegetation are described and explained in Chapter 12.

2 **The modern method** This is based upon differentiating between relative rates of producing organic matter – i.e. the speed at which vegetation grows. The rate at which organic matter is produced is known as the **net primary production** or NPP, expressed in grams of dry organic matter per square metre per year ($g/m^2/yr$). As shown in Figure 11.40, it is the tropical rainforests, with their large biomass resulting from constant high temperatures, heavy rainfall and year-round growing season, that produce on average the greatest amount of organic matter annually. The tundra (too cold) and the deserts (too dry) produce the least. It may be noted that the average NPP for arable land is 650, lakes and rivers 400 and oceans 125.

Figure 11.40

Net primary production (NPP) of eight major biomes

Net primary production (NPP) ($g/m^2/yr$)

Biome	NPP
Tropical rainforests	2200
Deciduous forests	1200
Tropical grasslands	900
Coniferous forests	800
Mediterranean	700
Temperate grasslands	600
Tundra	140
Deserts	90

1 Traditional method (vegetation, climate and soils subjectively linked)			2 Modern method (scientifically based on net primary production)		
Tropical	1	Rainforests	**High energy**	1	Rainforests
	2	Tropical grasslands		2	Deciduous forest
	3	Deserts	**Average energy**	3	Tropical grasslands
Warm temperate	4	Mediterranean		4	Coniferous forest
Cool temperate	5	Deciduous forest		5	Mediterranean
	6	Temperate grasslands		6	Temperate grasslands
Cold	7	Coniferous forest	**Low energy**	7	Tundra
	8	Tundra		8	Deserts

Figure 11.39

Two methods of clarifying the major biomes (*after* I. Simmonds)

The forests of south-west Australia

Figure 11.41

South-western Australia

Western Australia is ten times the size of the UK, and about 2 per cent of the state was forested before white settlement began in 1829. The forested area stretches from Gingin, 75 km north of Perth, to Walpole, 400 km to the south (Figure 11.41). The Darling and Stirling ranges form the edge of the Darling Plateau and consist mainly of ancient igneous and metamorphic rocks. A number of river valleys cut into the plateau edge. These have broad, flat valley floors.

East of the plateau the old river valleys (now largely dry) are very broad and flat. At the western edge of the scarp, the drainage has been rejuvenated and recaptured by newer fast-flowing streams.

The Blackwood River is an exception. It has maintained enough flow to continue erosion of its bed as the plateau was uplifted. Therefore it has an old meandering course within which is a new cross-sectional V-shaped profile.

The climate of this region is Mediterranean in type, with most rainfall in winter

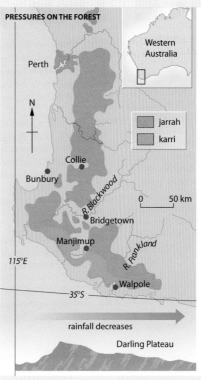

PRESSURES ON THE FOREST

Agricultural clearing
Up to 500 m to allow sheep rearing; wheat grown on well-drained soils to east of forest area; Forest now half extent of 165 years ago.

Settlement
Small towns expanding; most densely populated area of state outside Perth; infrastructure damages forest.

Commercial logging
280 000 m³ p.a. sawn timber for building; timber for woodchips; originally used waste offcuts and damaged timber; 150 000 tonnes jarrah sent to Kemerton for charcoal in silicon manufacture; clear felling now extensive. Greatest pressures in the south, but jarrah forest ecosystem under threat.

Deforestation
Leading to soil erosion, higher water table and salinisation.

Quarrying
Limestone, sand, gravel.

Mining
Bauxite, gold, tin and tantalite; 800 ha forest lost each year; little rehabilitation.

Dieback
Soil-borne fungal disease *Phytophthora cinnamomi* affects 14% of forests; spreading because of winter logging and other human disturbance.

Prescribed burning
Frequent burns in spring reduce flora species and damage food supply for breeding birds; jarrah forest not adapted to short intervals between burns.

Pest infestations
Jarrah leaf miner, gumleaf skeletoniser, affects 62 000 ha. Thinning forest canopy (logging and spring burning) stimulates young foliage, attracts pests.

Loss of habitat
Affects flora and fauna, 26 species of plants and animals in jarrah forests lost or in need of protection; 5 fauna species extinct in karri forests.

Western Australia

Perth

N

jarrah
karri

Collie
Bunbury
R. Blackwood
0   50 km
Bridgetown
Manjimup
R. Frankland
115°E
Walpole
35°S
rainfall decreases
Darling Plateau

Figure 11.42

Jarrah forest

from May to October (700 mm); rainfall is highest (1100 mm) on the western edge of the plateau and decreases rapidly to the east. Temperatures are high in the summer (18–27°C) and lower in the winter (7–15°C). Snow has been known to fall in the Stirling Range!

These conditions allowed the development of high forests, unique to Western Australia, of hardwood trees: varieties of eucalyptus known as karri, jarrah and marri. Jarrah forest is the only tall forest in the world to grow in a truly Mediterranean climate.

The great karri trees, which grow to over 80 m in height, are found in the south-west where the soils have a higher moisture content and are more fertile. The quality of the forest deteriorates to the east, with a variety of eucalypts reflecting lower rainfall. The jarrah forest is more extensive and has a very high species diversity (Figure 11.42). The forests provide important wildlife habitats for birds and animals—over 50 species live in the hollows of the trees.

Since the coming of the white settlers in 1829, half the tall forest cover has been removed (nearly 2 million ha). Much of the

early clearance was for agriculture, with pastures of clover and grasses for sheep and cattle replacing the 500-year-old trees. The timber provided a valuable secondary source of income. Farmers themselves were never able to sell the timber. The state sold it for 'royalty' to timber industry firms or it was burnt as the land was cleared. Much of it was wasted.

The commercial value of the tall forests was soon realised, and large-scale lumbering began. The Western Australian Government controls 1260 ha of native forest in State forests managed for timber production. The Department of Conservation and Land Management (CALM) states that there is 139 000 ha of 'old growth' or unlogged virgin forest left, with 1 121 000 ha regrowth or forest which has been logged in the past one hundred years. The annual cut has been increased to over 1 500 000 m³ in spite of opposition from conservation groups, including the Western Australian Forest Alliance. This has allowed the large timber companies to produce woodchips, saw-logs and poles and use the 'residue' which is sent to the silicon smelter in large

HERITAGE FORESTS FACE THE AXE

IRREPLACEABLE FORESTS TO BE CLEAR FELLED FOR WOODCHIPS

In March 1994, the Australian Heritage Commission (AHC) officially included 40 areas of WA's world-unique native forests on the interim list of the Registers of the National Estate, the highest national recognition of the ecological, aesthetic, scientific or cultural value of an area. Once an area has been interim-listed, it is considered to be on the Register and entitled to protection. The Federal Minister for the environment is legally bound to prevent logging in these areas until an examination has shown that there are no 'prudent and feasible alternative log sources'.

In spite of this protection, CALM (Department of Conservation and Land Management) plans to clear-fell many interim-listed forests such as Rocky,

Sharpe and Hawke, mainly to produce export woodchips. Some of the listed areas are already being clear-felled, roaded and burnt with the full knowledge of the AHC and the Federal government.

In addition, there is supposed to be a moratorium on logging in all high-conservation-value forests. Now that at least some of the best of WA's remaining native forests have been given official recognition, each of these agencies must back up their self-congratulations with action.

The only action they can reasonably take is to stop all roading and logging in WA's heritage forests immediately.

The Real Forest News, April 1994, published by Western Australian Forest Alliance, Perth, WA.

Figure 11.43

of removal, all 'old' forest will have disappeared by 2030.

The forests are an attraction to tourists, providing a further source of income for the region.

CALM has total responsibility for logging and regeneration of cut areas within State forests. CALM calls for tenders for cutting and hauling, then sells the logs to sawmillers and woodchippers. The chief market for West Australian timber is in the form of wood-chips, sold to Japan. Since 1976 over 10 million tonnes of karri have been exported as woodchips, through the port of Bunbury.

The main method of removal is by clear felling, in compartments, or coupes, which vary from 60 ha in karri forests, to 10 ha in the jarrah (Figure 11.45). The size of the coupes has been reduced. Every tree in a coupe is felled, then the logged area is burnt. This is mainly 'old growth': native forest which has not been touched previously. These trees are the largest in the forest. A 1993 report predicted that all the valuable old-growth forest would have been removed within 50 years.

'Occasionally it takes a double trailer to carry just one log. These must be coming from the last of the best virgin karri. A regular traveller on the road passes six trucks in a half-hour journey' (M. Frith, Bridgetown WA).

There is a programme of regeneration in the karri forests. Karri seedlings, which grow more rapidly than jarrah, are planted by hand – sometimes in the jarrah forests, in

quantities and used as industrial charcoal. The timber mills provide employment for approximately 2000 people. This is important in an area where sheep farming has been affected by low world prices for wool. However, it is hoped that the introduction of plantation agriculture (agroforestry) may help the situation.

Conservationists are attempting to stop this rapid logging increase in the virgin forests (Figure 11.43). The rate of loss of forests with many unique wild species of small mammals, birds and flora, raises questions of sustainability of the forests. It is known that already some species are extinct (Figure 11.44). At the present rate

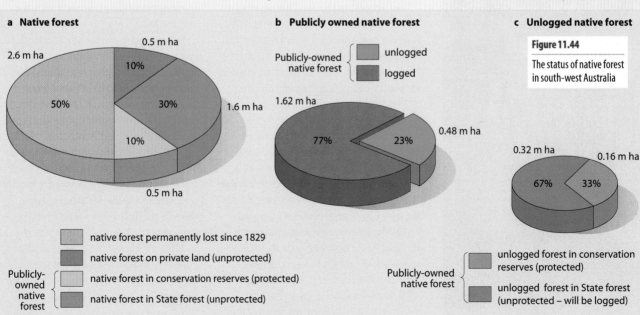

a Native forest

0.5 m ha — 10%
2.6 m ha — 50%
30%
10%
1.6 m ha
0.5 m ha

b Publicly owned native forest

Publicly-owned native forest: unlogged / logged

1.62 m ha — 77%
23% — 0.48 m ha

c Unlogged native forest

Figure 11.44

The status of native forest in south-west Australia

0.32 m ha — 67%
33% — 0.16 m ha

native forest permanently lost since 1829

native forest on private land (unprotected)

Publicly-owned native forest {
native forest in conservation reserves (protected)

native forest in State forest (unprotected)
}

Publicly-owned native forest {
unlogged forest in conservation reserves (protected)

unlogged forest in State forest (unprotected – will be logged)
}

an attempt to extend the range of the karri. However, in dry years, many die of drought. Also, in some of the clear-felled karri areas, three karri trees per hectare are left, to drop their seeds, then they too are felled.

Clear felling has now been reintroduced to the jarrah forests. It was abandoned in the 1940s as the jarrah trees did not regenerate successfully. Jarrah timber is commercially valuable for its dark-red colour, hardness and durability. It grows very slowly, hence the difficulty in sustainable production. The Western Australian Government at present allows 490 000 m³ of jarrah to be removed each year from the forests. The timber now being taken by this method is largely for charcoal and for chipboard. Much of the jarrah forest is mixed jarrah/marri. The marri is used widely for woodchips. With the increased demand for this product, much of the jarrah forest is being felled either by selection logging (removing all saleable trees) or by clear felling.

Bridgetown

Bridgetown district lies in the steep, winding valley of the Blackwood River (Figure 11.46). The lower slopes were cleared for farming between 1835 and 1950. The upper slopes are covered with relatively undisturbed jarrah/marri forest. This includes the Hester State forest on the undeveloped watersheds north-east of Bridgetown. This is an area of difficult access, with few major stands of commercial timber as the valuable trees are widely dispersed. There are already a number of gazetted nature reserves and tourist trails here.

CALM proposes to take 16 000 m³ of jarrah from the area in one year (1999/2000) – a very rapid rate of clearance. In May 1999 the

Shire council and forest conservation bodies managed to prevent clearance of over 20 000 m³ from land close to Bridgetown of high-amenity value. New roads have to be driven into the forests, adding to the destruction of the habitats of birds and small forest animals. The edges of the forest suffer damage from strong winds, sun and encroachment of weeds when roads are widened and extended into the forests.

Impacts of deforestation

- **Salinisation of streams (page 496)** This is a serious problem in the south-west. It increases rapidly when the cover of tall forest is removed for agriculture or for timber production. Salts accumulate in the laterite soil and can be moved relatively easily by an increase in groundwater. Removal of trees causes the groundwater level to rise and eventually the saline water enters the drainage system. Clearance of trees for agriculture caused salts to flow into the Hester Brook and then into the Blackwood. Forest forms 80 per cent of the catchment cover in this area, and CALM is being urged to use less intensive logging to help to reduce the effects of salinity.
- **Eutrophication** In the western forest areas, with over 1100 mm of rain, agricultural and horticultural land uses and housing developments have caused eutrophication (page 281) of coastal wetlands and inlets. Why?
- **Visual and physical degradation of the landscape** This is especially bad in clear-felled areas. Where slopes are steep, tree removal reduces transpira-

tion, increases infiltration and removes nutrients already loosened by burning. In the rainy season, soil is removed downslope with problems of rill and sheet erosion. The sediment load in streams and rivers increases, causing flooding.
- **Loss of native flora and fauna** Western Australia is famous for its wild flowers. Groundcover in the forests is being removed on an extensive scale. In the jarrah forest, 27 native species are gazetted as rare. These include the chuditch, a rare marsupial, the Western ringtailed possum, and the water rat. Birds such as the whipbird are disappearing with the destruction of habitat through frequent fires.

What can be done?

1. CALM has a management programme which includes replanting logged areas with young karri and jarrah trees. Jarrah are expected to regenerate themselves naturally from seedlings called ground coppice. These may live for years as small shrubs whilst building up a foodstore in below-ground tubers. Under natural conditions these grow rapidly in height once the canopy above is removed when an old tree dies or falls.
2. Straight rows of fast-growing timber such as pines and eucalypts are planted for quick returns on logged land. This plantation timber, with all the trees of similar age, height and type, gives an unnatural uniformity to the landscape. Uniform monoculture

.45

ng of karri, getown

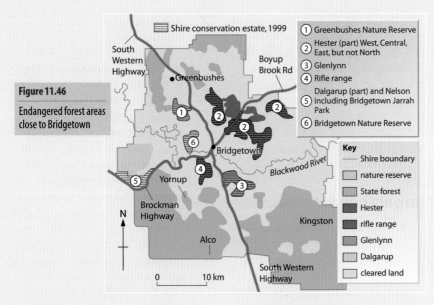

Figure 11.46

Endangered forest areas close to Bridgetown

of plans to place 436 000 ha of native eucalyptus forest in conservation reserves, felling is taking place at an increased rate within some of the areas to be designated.

Discussion

This forest destruction may not be on the scale of the Amazon deforestation, but to the people of south-west Australia it is as damaging. Jim Frith, a farmer and agroforester from near Bridgetown who has seen the rapid destruction of the forests near his farm in the last 35 years, said, 'We are witnessing a tragedy of gigantic proportions – the destruction of an ecosystem.'

What do *you* think could and should be done? Do you consider that preservation of an ecosystem for future generations is necessary? Remember that in Western Europe there are only the smallest remnants of the original native mixed forests left. Should the production of paper have greater importance than protection of trees and wildlife?

If you were a young person looking for work in a town such as Bunbury, Western Australia, your ideas might be different from those of farmers within the forest areas.

It is easy to become emotive on a topic such as this. One needs to gather the facts and draw conclusions by looking into both sides of the question.

plantations may be no substitute for native forest, especially for its biodiversity. However, they are now generally grown on former pasture land in the high-rainfall agricultural areas. They now have a capacity to completely replace native forest products. Pine now has a 40 per cent sawn timber market locally and is increasing. It has 70 per cent in the Eastern States.

3 Landholders are being encouraged to practise agroforestry. Trees are planted on agricultural land in belts separated by grass pasture which can be used for sheep grazing. Hay and other crops can be sown. Little fertiliser is required

– the agroforest will use up surplus agricultural fertiliser and reduce the damage to coastal streams from phosphates. At present the native forests produce 1.3 m³ of timber each year. Five times that amount could be produced using quick-growing varieties of timber for sawlogs and pulp wood.

4 The number of conservation reserves could be increased with the creation of National Parks, conservation parks and nature reserves. There would need to be strong measures to ensure that these were adequately protected – it is already possible to mine within a National Park. In spite

References

Bradbury, I. K. (1998) *The Biosphere*, 2nd edn, John Wiley and Sons.

Bradshaw, M. (1977) *Earth, The Living Planet*, Hodder & Stoughton.

Dobson, A. P. (1996) *Conservation and Biodiversity*, Freeman Scientific American Library.

Huggett, R. J. (1998) *Fundamentals of Biogeography*, Routledge.

Nature at Work: Introducing Ecology (1978) British Natural History Museum Publications.

O'Hare, G. (1998) *Soils, Vegetation and Ecosystems*, Oliver & Boyd.

Roberts, M. (1986) *Biology: A Functional Approach*, Nelson.

Simmons, I. (1982) *Biogeographical Processes*, Allen & Unwin.

Smith, R. L. and Smith, T. M. (1998) *Elements of Ecology*, The Benjamin/Cummings Publishing Company [an imprint of Addison Wesley Longman].

Websites

The Radford University Virtual Geography Department's 'Biome' pages:
http://www.runet.edu/~swoodwar/CLASSES/GEOG235/biomes/main.html#tabcont

The Union of Concerned Scientists' 'Understanding biodiversity' pages:
http://www.ucsusa.org/resources/biodiversity.html

Terrestrial ecozones of Canada:
http://www.cciw.ca/eman-temp/ecozones/ecozones.html

Virginia, USA: conserving Virginia's biodiversity:
http://www.state.va.us/~dcr/vaher.html

Forests of Australia: the National Association of Forest Industries home page:
http://www.nafi.com.au/home.html

Bridgetown – Greenbushes Friends of the Forest:
http://www.wn.com.au/bgff/

See also for more links:
http://www.nelsonthornes.com/gaia

Whitlingham Country Park

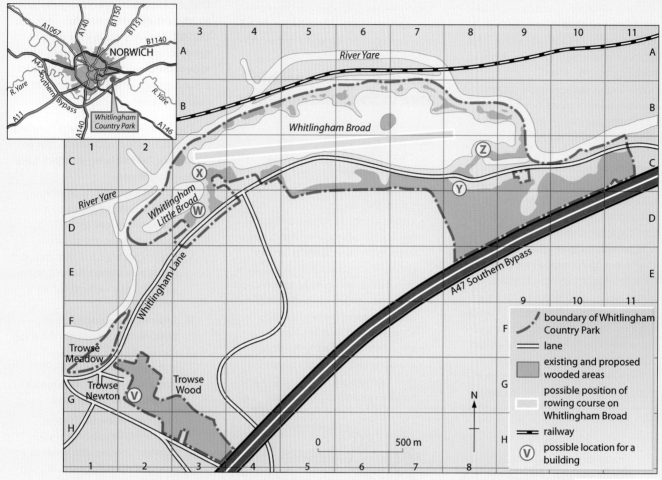

Figure 11.47

Whitlingham Broads and Country Park

Whitlingham Country Park lies on the south-eastern edge of Norwich (see inset, Figure 11.47). It is part of a privately-owned estate which was granted planning approval in 1995 to allow extraction of gravel from the floodplain of the Yare – but on condition that the gravel pits will be flooded and restored to form lakes once working has finished. These will be leased to the Whitlingham Trust, which manages the Country Park.

The new broads, together with a meadow, two areas of woodland and a picnic site, will form an area for quiet recreation, including water-based activities. The broads will be handed over in sections, as gravel extraction and landscaping are completed. The final broad will be handed over in spring 2006.

Trowse Meadow and Trowse Wood are important environments with a wide variety of species. Neither has been much affected by modern land use management techniques. The woodland around the broad is a less interesting ecosystem, but it is a valuable recreational resource for walkers and picnickers.

The shoreline of Whitlingham Little Broad will slope downwards fairly steeply and will be made of gravel. It will be separated from Whitlingham Broad by a stretch of land 70 m wide, containing gas pipes and other utilities. Whitlingham Broad will have gently-sloping shorelines, covered with silt, gravel or sand.

The lease refers to the area being used for 'quiet enjoyment'. The activities that fall within that definition are:

- birdwatching
- canoeing
- field study
- fishing
- rowing
- subaqua sports
- sailing
- windsurfing
- other activities such as picnicking, paddling, walking, cycling, etc.

The Trust has enough funds to cover about 70 per cent of the costs of running the Park and maintaining the environment. It will need to generate the remaining income from users of the site.

Activities in the Country Park

A proposal has been made to build an international standard rowing course in Whitlingham Broad (see Figure 11.47). This will provide a facility that is absent from this part of the country. It will need markers

311

along its whole length but these can be removable. The rowers also need a boathouse, a platform from which to launch boats, a cycle track for use by coaches on the bank, and standing areas for spectators.

Sailing, windsurfing and canoeing can also take place on Whitlingham Broad. A 'time-zoning' plan could allow all of these activities to take place when the Broad is not being used by rowers.

Whitlingham Little Broad is not large enough for water sports, but is an ideal location for training, e.g. for canoeing, windsurfing and subaqua sports. It will be a safe environment, but easy access for emergency vehicles will be essential.

The Trust will provide a boathouse for storage of equipment for the different water sports, along with a club room and a café/bar, to be open to the public.

The woods, the banks of the River Yare and Trowse Meadow provide a variety of ecosystems. These attract a range of bird and animal species which, in turn, attract many naturalists and field study groups. It is hoped that the reclamation of the pits will increase the variety of habitats to make the area an even better resource for these groups.

There are good facilities for anglers in many parts of the Broads, but the best of these are rented by private clubs and companies, and access to these waters is expensive and often very restricted.

The Country Park already attracts many people from the Norwich region for outdoor leisure pursuits. It is an accessible local resource, which should become even more important if it is sensitively developed.

Samara – a comparable area in the Somme valley

Samara lies in the Somme valley in northern France. It has a history of settlement from prehistoric times right up to the present. Since 1990 it has been made into a tourist, conservation and study site. It has three main themes:

- a series of rebuilt Stone- and Bronze-Age houses where people try to reconstruct many aspects of the life and crafts of those times, both as a study of archaeology and as a tourist attraction
- a botanic garden where native plants and trees are conserved as a tourist attraction and wildlife haven
- a marshland area on the floodplain, managed to show stages in the development of a hydrosere. This also provides a wildlife haven (Figure 11.48).

Decision Making

Imagine that you have been employed by the Whitlingham Trust. They wish you to use the knowledge and understanding that you have developed as a geographer to advise them on the development of the Park. They hope that your study of geography will enable you to take a broad view of the physical, environmental and human aspects of the area. Complete the following tasks.

1   The Trust wishes to develop a wetland conservation area, similar to that at Samara. The following locations have been suggested:

   i   Stop all water sports on Whitlingham Little Broad and use it as a conservation area instead.

   ii Separate the northern part of Whitlingham Broad from the water sports area, using a floating barrier, and use it as a wetland conservation area.

   iii Separate the eastern part of Whitlingham Broad from the water sports area as in (ii), and use that as the conservation area.

   iv Flood Trowse Meadow to form a wetland conservation area.

   a   Which of these areas is most suitable? Give reasons for your decision. Consider the views of a variety of potential users of the Park.   **(20 marks)**

   b   Outline the main features of a management plan to ensure that the wetland conservation area would show all the stages in the development of a hydrosere, as happens at Samara. Illustrate your answer with a sketch map.

                                       (15 marks)

2   The Whitlingham Trust needs to build:

   a   a boathouse/clubhouse with a car park covering about 1000 m^2

   b   a field centre for use by school parties and by conservation officers, with a small car park of no more than 100 m^2.

   Suggest which of the sites in Figure 11.48 – V, W, X, Y or Z – is most suitable for **each** of these. Give the reasons for your choice. Take account of your decision in question **1**.   **(15 marks)**

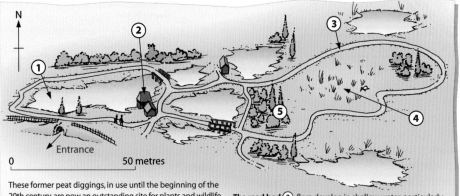

These former peat diggings, in use until the beginning of the 20th century, are now an outstanding site for plants and wildlife. The trail shows the evolution of the marsh and the successive stages of filling in and drying out.

The marsh is full of life. Plants spring up, flourish and die here. In the **tail of the marsh** ①, water, plants, foliage, branches and silt are all intermingled. This vegetable debris gradually turns to peat. People learned to exploit this ecosystem by extracting peat. **The peat digger's hut** ② shows the tools that were used. Other people cleared undergrowth and cut back the willows that grew in the edges of the marsh. Without this human activity the marsh ecosystem would have disappeared, filled in by silting. Marsh passes through several stages of silting, reflected in different plant communities, which follow.

The reed bed ③: flora develop in shallow water, particularly the Reed. Other plants found here are False reed, Bur-reed, Bulrush and Spearwort. The reed bed provides shelter for a wide range of birds and also helps natural purification of the surrounding water.

The megaphorbiae ④: in the ground 'prepared' by the reed bed, a vigorous, dense, herbaceous flora develops as the ground dries out. The spreading network of roots contributes to the firmer ground here.

The willow bed ⑤: this forms where the marsh has virtually dried out, and willows and alder colonise the site. The types of willow most frequently seen at Samara are White, Grey and Eared willows.

11.48

Samara wetland

1 a What are: i herbs ii shrubs iii trees? **(3 marks)**
  b What is plant succession? **(3 marks)**
  c How do herbs and shrubs help to prepare the ground so that trees can grow? **(6 marks)**
  d How would you carry out a field survey to discover the distribution of plants in the area of a playing field? **(5 marks)**
  e What kinds of plants would you expect to find on an abandoned urban railway track? Suggest reasons for your answer. **(4 marks)**
  f Flowers which grow in deciduous woodland are early spring flowers such as bluebell and primrose. Why do these plants flower so early in the year? **(4 marks)**

2 a Study Figure 11.25 (page 298).
    i Explain the roles played by plants in the carbon cycle. **(4 marks)**
    ii Human activity (combustion) releases CO_2 into the air. What is the source of this carbon? **(3 marks)**
  b i Study Figure 11.26 (page 298). Why is nitrogen important for plant life? **(2 marks)**
    ii What is the main source of new nitrogen into the nitrogen system? **(2 marks)**
    iii What is the main cause of loss of nitrogen from the system? **(2 marks)**
  c What is the meaning of the term 'biomass'? **(2 marks)**
  d What is the role of humans in the food chain? **(2 marks)**
  e As CO_2 builds up in the atmosphere, plant growth is increased. Suggest two effects of this on the material cycles. **(4 marks)**
  f Explain the 'greenhouse effect'. **(4 marks)**

3 a Why does temperature change with height on a mountain? **(5 marks)**
  b Draw a diagram to show the changes in vegetation cover with height on a mountainside in a temperate humid area (such as the British Isles). **(6 marks)**
  c i Identify **two** factors, other than height, which could affect the type of vegetation which grows at a particular height on a mountainside. **(2 marks)**
    ii For **one** of these factors describe the effect it would have. **(4 marks)**
  d Study Figure 11.28 (page 300). What is meant by each of the following terms:
    i store ii transfer iii input? **(6 marks)**
  e What, other than the mass of plant tissue, is included in the term 'biomass'? **(2 marks)**

4 Study Case Study 11 (pages 307–310).
  a i What is the extent of deforestation in south-west Australia since white settlement started? **(2 marks)**
    ii Identify the proportion of:
      (i) conserved native forest (ii) public ownership of the forest (iii) forest in danger of being logged. **(3 marks)**
    iii Identify and explain **three** reasons for deforestation in south-west Australia. **(6 marks)**
  b Explain **two** impacts of deforestation on areas such as south-west Australia. **(6 marks)**
  c Describe **two** advantages of the native forest to south west Australia and its people. **(4 marks)**
  d Explain **one** way of protecting the forest lands of south-west Australia. **(4 marks)**

5 a What is meant by:
    i seral change
    ii climatic climax vegetation cover? **(6 marks)**
  b Assume that there has been a landslide on an area of non-calcareous rock in lowland Britain. Describe and explain the sequence of vegetation which would occur so that the area eventually achieved a climatic climax vegetation cover. **(12 marks)**
  c Why is the vegetation cover within an urban area different from the climatic climax vegetation for a similar rural area? **(7 marks)**

6 a i What is the 'polyclimax' theory of development of natural vegetation cover?
    ii Why is the modern 'polyclimax' theory more likely to explain the occurrence of a stable vegetation cover in an area than the climatic climax theory of F. E. Clements? **(10 marks)**
  b Choose one of a *psammosere*, a *halosere*, or a *hydrosere*.
    i Draw an annotated diagram **only** to show the variation in vegetation cover across the environment. **(5 marks)**
    ii Explain the variation in vegetation cover shown on your diagram. **(10 marks)**

7 a Study Figure 11.28 (page 300).
    i Explain the meaning of the term 'litter'.
    ii What do the arrows mean? **(5 marks)**
  b Why are the transfers in the tropical forest (selvas) in Figure 11.29 (page 301) so large? **(5 marks)**
  c Figure 11.29 shows the nutrient cycles in three different environments. Explain the difference between the tropical forest (selvas) and coniferous forest (taiga) ecosystems in terms of their major nutrient stores. **(7 marks)**
  d For the steppes and prairies draw a *nutrient cycle diagram* to include the impact of farming on both the transfers and the stores. Explain the impacts you have shown in your diagram. **(8 marks)**

8 a Identify **two** sources of evidence that climate has changed in the British Isles. For **one** of these sources show how it can be interpreted to show climatic change. **(5 marks)**
  b Why should climatic change have an effect on the pattern of vegetation over the British Isles? **(8 marks)**
  c Identify and suggest reasons for **three** changes of climate which have affected the British Isles over the last 2 million years. **(12 marks)**

Chapter 12

World climate, soils and vegetation

'There was ... an instant in the distant past when the living things, the rocks, the air and the oceans merged to form the new entity, Gaia.'

James Lovelock, *The Ages of Gaia*, 1989

Although it is possible to study climatic phenomena in isolation (Chapter 9), an understanding of the development of soils (Chapter 10) and vegetation (Chapter 11) necessitates an appreciation of the interrelationships between all three (Figure 12.1a). This chapter attempts to show how the integration and interaction of climate, soils and vegetation give the world its major ecosystems, or biomes, and how these have often been modified, in part or almost totally, by human activity (Figure 12.1b).

Soils can be grouped, at the simplest of levels, under zonal, azonal and intrazonal (page 273) with each group, in turn, being subdivided (zonal Figure 12.2, azonal page 273, and intrazonal page 274). Likewise, the major vegetation and fauna groupings (biomes) were listed on page 306 and their generalised global locations and distributions shown in Figure 11.38. In a similar way, geographers seek – despite the difficulties and limitations – to classify different world climates (Framework 7, page 167).

Classification of climates

By studying the weather – the atmospheric conditions prevailing at a given time or times in a specific place or area – it is possible to make generalisations about the climate of that place or area, i.e. the average, or 'normal' conditions over a period of time (usually 35 years). Any area may experience short-term departures from its 'normal' climate, especially if the 35-year mean coincided with an unusually wet/dry or hot/cold period, but, at the same time, it may have long-term similarities with regions in other parts of the world.

In seeking a sense of order, the geographer tries to group together those parts of the world that have similar measurable climatic characteristics (temperature, rainfall distribution, winds, etc.) and to identify and to explain similarities and differences in spatial and temporal distributions and patterns. Areas may then be compared on a global scale – bearing in mind the problems resulting from short-term and long-term climatic change – to help to identify and to explain distributions of soil, vegetation and crops.

Bases for classification

The early Greeks divided the world into three zones based upon a simple temperature description: torrid (tropical), temperate, and frigid (polar); they ignored precipitation.

In 1918, **Köppen** advanced the first modern classification of climate. To support his claim that natural vegetation boundaries were determined by climate, he selected as his basis what he considered were appropriate temperature and seasonal precipitation values. His resultant classification is still used today, although a modification by **Trewartha**, with 23 climatic regions, has become more widely accepted. **Thornthwaite**, in the 1930s and 1940s, suggested

Figure 12.1

Relationship between climate, vegetation and soils

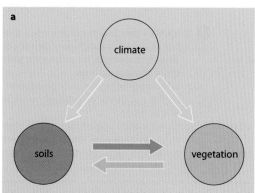

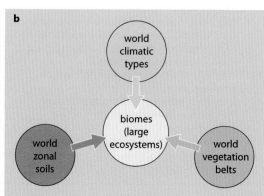

and later modified a classification with a more quantitative basis. He introduced the term 'effectiveness of precipitation' (his P/E index – page 178) which he obtained by dividing the mean monthly precipitation of a place by its mean monthly evapotranspiration, and taking the sum of the 12 months. The difficulty was, and still is, in obtaining accurate evapotranspiration figures. (How can you measure transpiration loss from a forest?) This classification resulted in 32 climatic regions.

In Britain, in the 1930s, **Miller** proposed a relatively simple classification based upon five latitudinal temperature zones which he determined by using just three temperature figures: 21°C (the limit for growth of coconut palms); 10°C (the minimum for tree growth); and 6°C (the minimum for grasses and cereals). He then subdivided these zones longitudinally according to seasonal distributions of precipitation. The advantages of this classification include its ease of use and convenience; and its close relationship to vegetation zones and also, as these are a response on a global scale to climate and vegetation, to zonal soils.

All classifications have weaknesses: none is perfect.

- They do not show transition zones between climates, and often the division lines are purely arbitrary.
- They do not allow for mesoscale variation (the Lake District and London do *not* have exactly the same climate) or microscale (local) variation.
- They can be criticised for being either too simplistic (Miller) or too complex (Thornthwaite).
- They ignore human influence and climatic change, both in the long term and the short term.
- Most tend to be based upon temperature and precipitation figures, and neglect recent studies in heat and water budgets, air-mass movement and the transfer of energy.
- All suffer from the fact that some areas still lack the necessary climatic data to enable them to be categorised accurately.

However, climatic classifications such as those named above are rarely used today. Instead, as we saw in Chapter 11, the relationship between climate, vegetation and soils can best be described and understood at this level through the study of ecosystems, especially the largest of the ecosystems: the **biomes** (Figure 12.1b). Figure 12.2 lists eight of the more important biomes and shows, simplistically, the links between climate, vegetation and soils. These links are described in more detail and explained in the remainder of this chapter, using knowledge and understanding gained from Chapters 9, 10 and 11.

Climate type		Text reference number	Climatic characteristics	Biome (based on NPP)	Soil (zonal type)
arctic		8	very cold all year	tundra	tundra
cold		7	cold all year	coniferous forest (taiga)	podsols
cool temperate	western margin	6	rain all year, winter maximum	temperate deciduous forest	brown earths
	continental	5	summer rainfall maximum	temperate grassland	chernozems prairie chestnut
warm temperate	western margins: Mediterranean	4	winter rain	Mediterranean	Mediterranean
	eastern margins: monsoon	4A	some rain all year, summer maximum	tropical deciduous forest	
tropical	desert	3	little rain	desert (xerophytes)	desert
	continental	2	summer rain	tropical grassland (savanna)	ferruginous
	monsoon	1B		jungle	
	tropical eastern margins	1A	rain all year	rainforest	ferralitic
	equatorial	1			

Figure 12.2

World biomes: the relationship between climate, vegetation and soils at the global scale

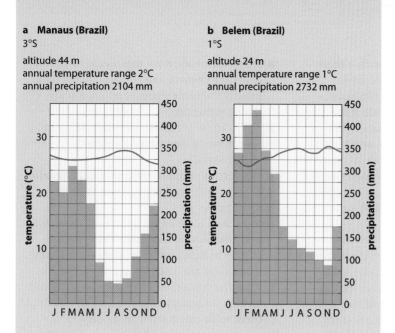

a Manaus (Brazil)
3°S

altitude 44 m
annual temperature range 2°C
annual precipitation 2104 mm

b Belem (Brazil)
1°S

altitude 24 m
annual temperature range 1°C
annual precipitation 2732 mm

Figure 12.3

Climate graphs for the equatorial biome

Figure 12.4

Emergents rising above the rainforest canopy, Borneo

1 Tropical rainforests

The rainforest biome is located in the tropics and principally within the equatorial climate belt, 5° either side of the Equator. It includes the Amazon and Congo basins and the coastal lands of Ecuador, West Africa, and extreme south-east Asia (Figure 11.38).

Equatorial climate

Temperatures are high and constant throughout the year because the sun is always high in the sky. The annual temperature range is under 3°C inland (Manaus) and 1°C on the coast (Belem, Figure 12.3). Mean monthly temperatures, ranging from 26°C to 28°C, reflect the lack of seasonal change. Slightly higher temperatures may occur during any 'drier' season. Insolation is evenly distributed throughout the year, with each day having approximately 12 hours of daylight

and 12 hours of darkness. The diurnal temperature range is also small, about 10°C. Evening temperatures rarely fall below 22°C while, due to the presence of afternoon cloud, daytime temperatures rarely rise above 32°C. It is the high humidity, with its sticky, unhealthy heat, that is least appreciated by Europeans.

Annual rainfall totals usually exceed 2000 mm (Belem, 2732 mm) and most afternoons have a heavy shower (Belem has 243 rainy days per year). This is due to the convergence of the trade winds at the ITCZ and the subsequent enforced ascent of warm, moist, unstable air in strong convection currents (Figure 9.34). Evapotranspiration is rapid from the many rivers, swamps and trees. Most storms are violent, with the heavy rain, accompanied by thunder and lightning, falling from cumulo-nimbus clouds. Some areas may have a drier season when the ITCZ moves a few degrees away from the Equator at the winter and summer solstices (Belem), and others have double maxima when the sun is directly overhead at the spring and autumn equinoxes. The high daytime humidity needs only a little night-time radiation to give condensation in the form of dew. The winds at ground-level at the ITCZ are light and variable (doldrums) allowing land and sea breezes to develop in coastal areas (page 240).

Rainforest vegetation

It is estimated that the rainforests provide 40 per cent of the net primary production of terrestrial energy (NPP, page 306). This is a result of high solar radiation, an all-year growing season, heavy rainfall, a constant moisture budget surplus, the rapid decay of leaf litter and the recycling of nutrients.

Figure 12.5

Buttress roots, Queensland, Australia

Figure 12.6

Rainforest vegetation has to adapt to the wet environment: water lilies, *Victoria regia*, native to the Amazon Basin

are adapted to living in the shade of their taller neighbours.

The climate is at the optimum for photosynthesis. The trees grow tall to try to reach the sunlight, and the tallest have buttress roots which emerge over 3 m above ground-level to give support (Figure 12.5). The trunks are usually slender and branchless. Some, like the cacao, have flowers growing on them, and their bark is thin as there is no need for protection against adverse climatic conditions. Tree trunks also provide support for lianas, vine-like plants, which can grow to 200 m in length. Lianas climb up the trunk and along branches before plunging back down to the forest floor. Leaves are dark green, smooth and often have **drip tips** to shed excess water.

Epiphytes – plants that do not have their roots in the soil – grow on trunks, branches and even on the leaves of trees and shrubs. Epiphytes simply 'hang on' to the tree: they derive no nourishment from the host and are *not* parasites. Less than 5 per cent of insolation reaches the forest floor, with the result that undergrowth is thin except in areas where trees may have been felled by shifting cultivators or where a giant emergent has fallen, dragging with it several of the top canopy trees. Vegetation is also dense along the many river banks, again because sunlight can penetrate the canopy here. Alongside the Amazon, many trees spend several months of the year growing in water as the river and its tributaries rise over 15 m in the rainy season. Huge water lilies with leaves exceeding 2 m in width are found in flooded areas adjacent to rivers (Figure 12.6). Mangrove swamps occur in coastal areas.

Vegetation consists of trees of many different species. In Amazonia, there may be over 300 species in 1 km², including rosewood, mahogany, ebony, greenheart, palm and rubber. The trees, which are mainly hardwoods, have an evergreen appearance for, although deciduous, they can shed their leaves at any time during the continuous growing season. The tallest trees, **emergents**, may reach up to 50 m in height and form the habitat for numerous birds and insects. Below the emergents are three layers, all competing for sunlight (Figure 12.4).

The top layer, or **canopy**, forms an almost continuous cover which absorbs over 70 per cent of the light and intercepts 80 per cent of the rainfall. The crowns of these trees merge some 30 m above ground-level. They shade the underlying species, protect the soil from erosion, and provide a habitat for most of the birds, animals and insects of the rainforest.

The second layer, or **undercanopy**, consists of trees growing up to 20 m (similar in height to deciduous trees in Britain). The lowest, or **shrub layer**, consists of shrubs and small trees which

Figure 12.7

The rainforest nutrient cycle

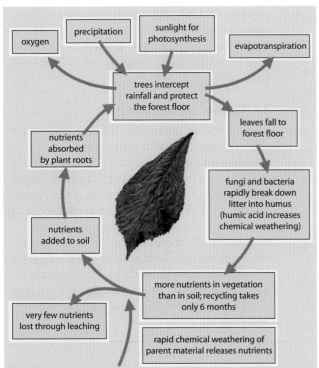

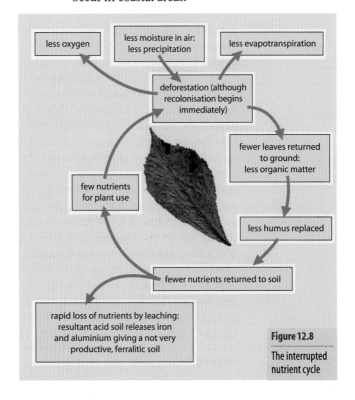

Figure 12.8

The interrupted nutrient cycle

Figure 12.9

A ferralitic soil profile

Although ground animals are relatively few in number, the rainforests of Brazil alone are said to be the habitat for 2000 species of birds, 600 species of insects and mosquitoes, and 1500 species of fish.

The productivity of this biome, upon which the world depends to replace much of its used oxygen, is due largely to the rapid and unbroken recycling of nutrients. Figure 12.7 shows the natural nutrient cycle and Figure 12.8 the conse-quences of breaking the system, e.g. by felling the forest. In areas where the forest has been cleared, the secondary succession differs from that of the original climax vegetation. The new dominants are less tall; the trees are less stratified; there are fewer species and many are intolerant of shade – even though there is more light at ground-level which encourages a dense undergrowth.

Ferralitic soils (latosols)

These soils result from the high annual tempera-ture and rainfall which cause rapid chemical weathering of bedrock and create the optimum conditions for breaking down the luxuriant vege-tation. Continuous leaf fall within the forest gives a thick litter layer, but the underlying humus is thin due to the rapid decomposition and mixing of organic matter by intensive biota activity, e.g. ants and termites. A key feature of these soils is a dense root mat in the top 20–30 cm of the A horizon. According to research, this intercepts and can take up as much as 99.9 per cent of the nutri-ents released by the decomposition of organic matter. The root map helps the rapid recycling of nutrients in the humus cycle (Figure 12.7). Even so, many soils have a low nutrient status (94 per cent of soils in the Amazon Basin have a nutrient deficiency) and fertility is only maintained by the rapid and continuous replacement from the lush vegetation. Where the tree canopy is absent, or is removed, the heavy rainfall causes the release of iron (giving the soil its characteristic red colour – Figure 12.9) and aluminium (most ferralitic soils

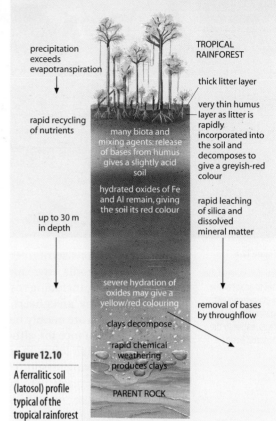

suffer from aluminium toxicity) from the parent material. Leaching results in the removal of silica.

The continual leaching and abundance of mixing agents inhibit the formation of horizons (Figure 12.10). The lower parts of the profile may have a more yellowish-red tint due to the extreme hydration of aluminium and iron oxides. The clay-rich soils are also very deep, often up to 20 m, due to the rapid breakdown of parent material. Ferralitic soils have a loose structure and, if exposed to heavy rainfall, are easily gullied and eroded. Despite their depth, the soils of the rainforest are not agriculturally productive. Once the source of nutrients (the trees) has been removed, the soil rapidly loses its fertility and local farmers, often shifting cultiva-tors, have to move to clear new plots (Places 66, page 480).

1A Tropical eastern margins

Located within the tropics, the eastern coasts of central America, Brazil, Madagascar and Queensland (Australia) receive rain throughout the year. The rain is brought by the trade winds which blow across warm, offshore ocean currents (Figure 9.9) before being forced to rise by coastal mountains. Temperatures are generally very high, although there is a slightly cooler season when the overhead sun appears to have migrated into the opposite hemisphere. The resultant ve-getation and soil types are, therefore, similar to those found in the equatorial belt, i.e. rainforest and ferralitic.

2 Tropical grasslands

These are mainly located between latitudes 5° and 15° north and south of the Equator and within central parts of continents, i.e. the Llanos (Venezuela), the Campos (Brazilian Highlands), most of central Africa surrounding the Congo Basin, and parts of Mexico and northern Australia (Figure 11.38).

Tropical continental climate

Although temperatures are high throughout the year, there is a short, slightly cooler season (in comparison with the equatorial) when the sun is overhead at the tropic in the opposite hemisphere (Figure 12.11). The annual range is also slightly greater (Kano 8°C) due to the sun's slightly reduced angle in the sky for part of the year, the greater distance from the sea, and the less complete cloud and vegetation cover. Temperatures may drop slightly at the onset of the rainy season. For most of the year, cloud amount is limited, allowing diurnal temperatures to exceed 25°C.

The main characteristic of this climate is the alternating wet and dry seasons. The wet season occurs when the sun moves overhead bringing with it the heat equator, the ITCZ, and the equatorial low pressure belt (Figure 12.12). Heavy convectional storms can give 80 per cent of the annual rainfall total in four or five

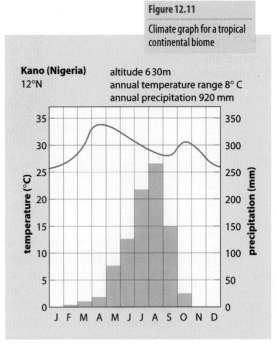

Figure 12.11

Climate graph for a tropical continental biome

months. The dry season corresponds with the moving away of the ITCZ, leaving the area with the strong, steady trade winds. The trades are dry because they are warming as they blow towards the Equator and they will have shed any moisture on distant eastern coasts. Places nearer to the desert margins tend to experience dry, stable conditions (the subtropical high pressure) caused by the migration of the descending limb of the Hadley cell (page 226). Humidity is also low during this season.

Tropical or savanna grassland vegetation

The tropical grasslands are estimated to have a mean NPP of 900 g/m^2/yr (page 306). This is considerably less than the rainforest, partly because of the smaller number of trees, species and layers and partly because, although grasslands have the potential to return organic matter back to the soil, the rate of decomposition is reduced during the winter drought leaving considerable amounts left stored in the litter.

As shown in Figure 12.13, the savanna includes a series of transitions between the rainforest and the desert. At one extreme, the 'closed' savanna is mainly trees with areas of grasses; at the other, the 'open' savanna is vegetated only by scattered tufts of grass. The trees are deciduous and, like those in Britain, lose their leaves to reduce transpiration, but, unlike in Britain, this is due to the winter drought rather than to cold. Trees are xerophytic, or drought-resistant. Even when leaves do appear, they are small, waxy and sometimes thorn-like. Roots are long and extend to tap any underground water. Trunks are gnarled and the bark is usually thick to reduce moisture loss.

Figure 12.12

Causes of seasonal rainfall in places with a tropical continental (savanna) climate

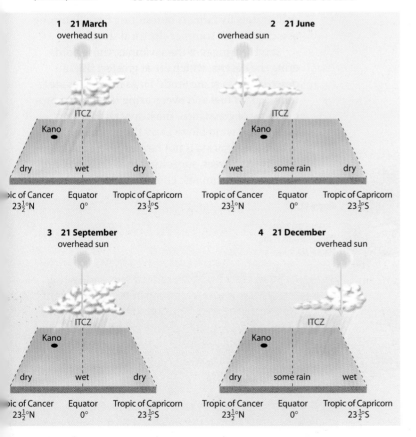

World climate, soils and vegetation

319

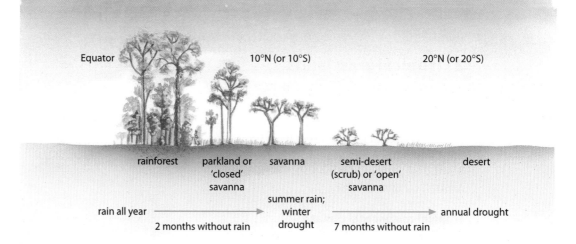

Figure 12.13

Transect across the savanna grasslands

Equator   10°N (or 10°S)   20°N (or 20°S)

rainforest   parkland or   savanna   semi-desert   desert
                 'closed'   (scrub) or 'open'
                 savanna   savanna

                                   summer rain;

rain all year   ⟶   winter   ⟶   annual drought
       2 months without rain   drought   7 months without rain

The baobab tree (also known as the 'upside-down tree') has a trunk of up to 10 m in diameter in which it stores water. Its root-like branches hold only a minimum number of tiny leaves in order to restrict transpiration (Figure 12.14). Some baobabs are estimated to be several thousand years old and, like other savanna trees, are pyrophytic, i.e. their trunks are resistant to the many local fires. Acacias, with their crowns flattened by the trade winds (Figure 12.15), provide welcome though limited shade – as do the eucalyptus in Australia. Savanna trees reach 6–12 m in height. Many have Y-shaped, branching trunks – ideal for the leopard to rest in after its meal! The number of trees increases near to rivers and waterholes. Grasses grow in tufts and tend to have inward-curving blades and silvery spikes. After the onset of the summer rains, they grow very quickly to over 3 m in height: elephant grass reaches 5 m. As the sun dries up the vegetation, it becomes yellow in colour. By early winter, the straw-like grass has died down, leaving seeds dormant on the surface until the following season's rain. By the end of winter, only the roots remain and the surface is exposed to wind and rain.

Over 40 different species of large herbivore graze on the grasslands, including wildebeest, zebra and antelope, and it is the home of several carnivores – both predators, such as lions, and scavengers, such as hyenas. Termites and microbes are the major decomposers. As previously mentioned (page 293), fire is possibly the major determinant of the savanna biome – either caused deliberately by farmers or resulting from lightning associated with summer electrical storms.

It is the fringes of the savannas, those bordering the deserts, which are at greatest risk of desertification (Case Study 7). As more trees are removed for fuel and overgrazing reduces the productivity of grasslands, the heavy rain forms gulleys and wind blows away the surface soil. Where the savanna is not farmed, there are usually more trees, suggesting that grass may not be the natural climatic climax vegetation.

Figure 12.14

A baobab tree, Malawi

Figure 12.15

Savanna grassland during the wet season in the Maasai Mara, Kenya

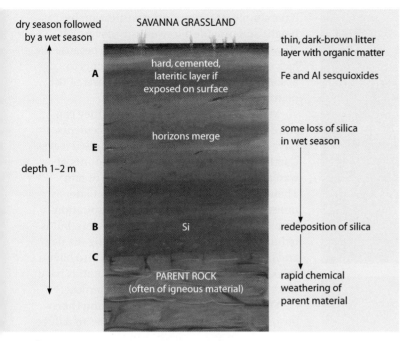

dry season followed by a wet season

SAVANNA GRASSLAND

depth 1–2 m

A — hard, cemented, lateritic layer if exposed on surface

thin, dark-brown litter layer with organic matter

Fe and Al sesquioxides

E — horizons merge

some loss of silica in wet season

B — Si

redeposition of silica

C —

PARENT ROCK (often of igneous material)

rapid chemical weathering of parent material

Figure 12.16

A ferruginous soil profile

Ferruginous soils

As savanna grasses die back during the dry season, they provide organic matter which is readily broken down to give a thin, dark-brown layer of humus (Figure 12.16). During the wet season, rapid leaching removes silica from the upper profile, leaving behind the red-coloured oxides of iron and aluminium. As these soils contain few nutrients, they tend to be acidic and lacking in bases. Although the process of capillary action might be expected to operate during the dry season, in practice it rarely does as the water table invariably falls too low at this time of year.

Ferruginous soils tend to be soft unless exposed at the surface where, being subject to wet and dry seasons, they can harden to form a cemented crust known as **laterite**. The term laterite is derived from the Latin for 'brick'. Indeed this deposit is used as a building material because, being initially soft, it can easily be dug from the soil, shaped into bricks and left to harden by exposure to cycles of wetting and drying. It is only when the laterite crust forms that drainage and plant root penetration is impeded.

As these soils hold few nutrients and tend to dry out during the dry season, they are not particularly suited to agriculture; together with the grassland they support, they are better suited to animal rearing than to arable farming. Where a lateritic crust forms on the surface, or when deep ploughing removes the surface vegetation, the upper soil tends to dry out during the dry season, becoming highly vulnerable to erosion by wind and, when the rains return, by water.

3 Hot deserts

The hot deserts of the Atacama and Kalahari-Namib and those in Mexico and Australia, are all located in the trade wind belt, between 15° and 30° north or south of the Equator, and on the west coasts of continents where there are cold, off-shore, ocean currents (Figures 7.2, 9.9 and 11.38). The exception is the extensive Sahara-Arabian-Thar desert which owes its existence to the size of the Afro-Asian landmass.

Climate

Desert temperatures are characterised by their extremes. The annual range is often 20–30°C and the diurnal range over 50°C (Figure 12.17). During the daytime, especially in summer, there are high levels of insolation from the overhead sun, intensified by the lack of cloud cover and the bare rock or sand ground surface. In contrast, nights may be extremely cold with temperatures likely to fall below 0°C. Coastal areas, however, have much lower monthly temperatures (Arica in the Atacama has a warmest month of only 22°C) due to the presence of offshore, cold, ocean currents (Figure 9.9).

Although all deserts suffer an acute water shortage, none is truly dry. Aridity and extreme aridity have been defined by using Thornthwaite's P/E index (Figure 7.1), and four of the main causes of deserts are described on page 179. Amounts of moisture are usually small and precipitation is extremely unreliable. Death Valley, California, averages 40 mm a year, yet rain may fall only once every two or three years. Whereas mean annual totals vary by less than 20 per cent a year in north-west Europe, the equivalent figure for the Sahel is 80–150 per cent (Figure 9.28). Rain,

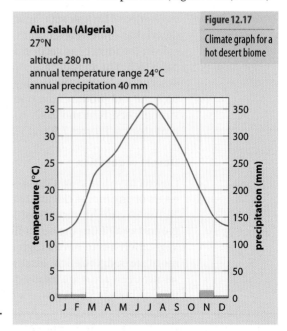

Ain Salah (Algeria)
27°N

altitude 280 m
annual temperature range 24°C
annual precipitation 40 mm

Figure 12.17

Climate graph for a hot desert biome

Figure 12.18

Saguaros cacti in the Arizona desert, USA

have simple structures, no stratification by height and provide a low-density cover. However, plants must be xerophytic because the lack of water hinders the ability of roots to absorb nutrients and of any green parts of the plants to photosynthesise.

Many plants are **succulents**, i.e. they can store water in their tissues. Many succulents have fleshy stems and some have swollen leaves. Cacti (Figure 12.18) absorb large amounts of water during the infrequent periods of rain. Their stems swell up, only to contract later as moisture is slowly lost through transpiration. Transpiration takes place from the stems, but is reduced by the stomata closing during the day and opening nocturnally. The stems also have a thick, waxy cuticle. Australian eucalyptids have thick, protective bark for the same purpose.

Most plants, for example cactus and thorn-bush, have small, spiky or waxy leaves to reduce transpiration and to deter animals. Roots are either very long to tap groundwater supplies – those of the acacia exceed 15 m – or they spread out over wide areas near to the surface to take the maximum advantage of any rain or dew, like those of the creosote bush. Bushes are, therefore, widely spaced to avoid competition for water. Some plants have bulbous roots for storing water. Seeds, which usually have a thick case protecting a pulpy centre, can lie dormant for months or several years until the next rainfall.

Following a storm, the desert blooms (Figure 12.19). Many plants are **ephemerals** and can complete their life-cycles in two or three weeks. Others, like the saltbush, are **halophytic** and can survive in salty depressions; yet others, like the date-palm, survive where the water table is near enough to the surface to form oases. Due to the lack of grass and the limited number of green plants, there are very few food chains: desert biomes have a low capacity to sustain life. There is insufficient plant food to support an abundance of animal life. Food chains (page 296) are simple, often just a single linear sequence (in contrast to the interlocking webs characteristic of, for example, forests). This is why the desert ecosystem is 'fragile': organisms do not have the alternative sources of food which are available in more complex ecosystems. Many animals are small and nocturnal (the camel is an exception) and burrow into the sand during the heat of the day. Reptiles are more adaptable, but bird life is limited. The desert fringes form a delicately balanced ecosystem which is being disturbed by human activity and population growth which are, together, increasing the risk of desertification (Case Study 7).

Figure 12.19

Ephemerals in flower following a desert rainstorm

when it does fall, produces rapid surface runoff which, together with low infiltration and high evaporation rates, minimises its effectiveness for vegetation. The Atacama, an almost rainless desert, has some vegetation as moisture is available in the form of advection fog (Places 24, page 180). The subsiding air, forming the descending limb of the Hadley cell, creates high pressure and produces the trade winds which are strong, persistent and likely to cause localised dust storms (Figures 7.9 and 9.34).

Desert vegetation

Deserts have the lowest organic productivity levels of any biome (Figure 11.40). The average NPP is 90 g/m^2/yr, most of which occurs underground away from the direct heat of the sun. Vegetation has to have a high tolerance to the moisture budget deficit, intense heat and, often, salinity. Few areas are totally devoid of vegetation, although desert plants are few in species,

Desert soils

In desert areas, the climate is too dry and the vegetation too sparse for any significant chemical weathering of bedrock or the accumulation of organic material. In the relatively few places where the water table is near to the surface, soil moisture is likely to be drawn upwards by capillary action. This process causes salts and bases, such as magnesium, sodium and calcium, to be deposited in the upper profile to give a slightly alkaline soil. Many desert soils are grey in colour as the lack of moisture often restricts hydrolysis and, therefore, the release of red-coloured iron (page 42). Soils, which tend to lack both structure and horizons (Figure 12.20), are often thin, although their depth can vary depending upon the origin of the parent material, i.e. *in situ* weathering or the deposition of material by wind or water (Chapter 7). A characteristic of many desert soils is the presence of either a thin crust, 2–3 mm thick, caused by the impact of high-intensity rainfall, and/or a 'desert pavement' (Figure 7.10) which consists of small stones, often ventifacted and covered in desert varnish (page 182), which help stabilise the surface.

Desert soils are unproductive mainly because of the lack of moisture and humus, but potentially they are not particularly infertile. Areas under irrigation are capable of producing high-quality crops, although this farming technique is being threatened by salinisation (Figures 10.22 and 16.53).

4 Mediterranean (warm temperate, western margins)

This type of biome is found on the west coasts of continents between 30° and 40° north and south of the Equator, i.e. in Mediterranean Europe (which is the only area where the climate penetrates far inland), California, central Chile, Cape Province (South Africa) and parts of southern Australia (Figure 11.38).

Climate

The climate is noted for its hot, dry summers and warm, wet winters (Figure 12.21). Summers in southern Europe are hot. The sun is high in the sky, though never directly overhead, and there is little cloud. Winters are mild, partly because the sun's angle is still quite high but mainly due to the moderating influence of the sea. Other 'Mediterranean' areas are less warm in summer and have a smaller annual range due to cold, off-shore currents (compare San Francisco, 8°C in January and 15°C in July, with Malta). Diurnal temperature ranges are often high due to the fact that many days, even in winter, are cloudless.

As the ITCZ moves northwards in the northern summer, the subtropical high pressure areas migrate with it to affect these latitudes. The trade winds bring arid conditions, with the length of the dry season increasing towards the desert margins. In winter, the ITCZ, and subsequently the subtropical jet stream (page 228), move southwards allowing the westerlies, which blow from the sea, to bring moisture. Most areas are backed by coastal mountains and so the combined effects of orographic and frontal precipitation give high seasonal totals. Areas with adjacent, cold, offshore currents experience advection fogs (California). The Mediterranean Sea region is noted for its local winds (Figure 12.22). The **sirocco** and **khamsin** are two of the hot, dry winds that blow from the

Figure 12.20

A desert soil profile

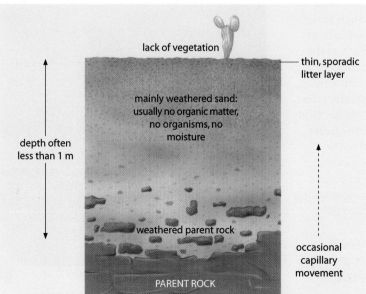

lack of vegetation

thin, sporadic litter layer

mainly weathered sand: usually no organic matter, no organisms, no moisture

depth often less than 1 m

weathered parent rock

occasional capillary movement

PARENT ROCK

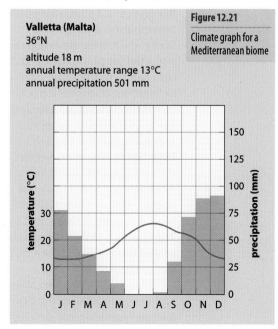

Valletta (Malta)
36°N

altitude 18 m
annual temperature range 13°C
annual precipitation 501 mm

Figure 12.21

Climate graph for a Mediterranean biome

Figure 12.22

Mediterranean winds

Sahara and can raise temperatures to over 40°C. The **mistral** is a cold wind which originates over the Alps and is funnelled at considerable speed down the Rhône valley.

Vegetation

The NPP of Mediterranean ecosystems is about 700 g/m²/yr (Figure 11.40). It is limited by the summer drought and has probably been reduced considerably over the centuries by human activity. Indeed, human activity, together with frequent fires, has left very little of any original climatic climax vegetation. The climax vegetation was believed to have been, in Europe at least, open woodland comprising a mixture of broad-leaved, evergreen trees (e.g. cork oak and holm oak) and conifers (e.g. aleppo pines, cypresses and cedars). The sequoia, or giant redwood, is native in California.

The present vegetation, which is mainly **xerophytic** (drought-resistant), is described as 'woodland and sclerophyllous scrub'. **Sclerophyllous** means 'hard-leaved' and is used to describe those evergreen trees or shrubs that have small, hard, leathery, waxy or even thorn-like leaves and which are efficient at reducing transpiration during the dry summer season. Many of the trees are evergreen, maximising the potential for photosynthesis. Trees such as the cork oak have thick and often gnarled bark to help reduce transpiration. Others, such as the olive and eucalyptus, have long tap roots to reach groundwater supplies and, in some cases, may have bulbous roots in which to store water. High temperatures during the dry summer limit the amount and quality of grass. Citrus fruits, although not indigenous, are suited to the climate as their thick skins preserve moisture. Most trees only grow from 3 to 5 m in height. They provide little shade, as they grow at widely spaced intervals, and they are **pyrophytic** (fire-resistant, page 293).

Where the natural woodland has been replaced, and in areas too dry for tree growth, a scrub vegetation has developed. The scrub is known as *chaparral* in California, *maquis* or *garrigue* in Europe, and *mallee* in Australia. In Mediterranean Europe, the type of scrub depends upon the underlying parent rock. **Maquis** (Figure 12.23), which is taller, denser and more tangled, grows in areas of impermeable rock (granite). It consists of shrubs, such as heathers and broom, which reach a height of 3 m. **Garrigue** (Figure 12.24) grows on drier and more permeable rocks (limestone). It is less tall and less dense than maquis. Apart from gorse, with its prickles, the more common plants include aromatic shrubs such as thyme, lavender and rosemary.

The limited leaf litter tends to decompose slowly during the dry summer, even though temperatures are high enough for year-round bacterial activity. Wildlife and climax vegetation have retreated as human activity has advanced. Arguably, the Mediterranean regions of Europe and California (together with the temperate deciduous forests) form the biome most altered by human activity.

Figure 12.23

Maquis vegetation

Figure 12.24

Garrigue vegetation

Soils

Mediterranean soils are transitional between brown earths on the wetter margins and desert soils at the drier fringes. Initially formed under broad-leaved and coniferous woodland, the soil is partly a relict feature from a previously forested landscape.

There are often sufficient roots and decaying plant material to provide a significant humus layer. Winter rains cause some leaching of bases, sesquioxides of iron and aluminium and the translocation of clays. The *B* horizon is therefore

Figure 12.25

A Mediterranean soil profile

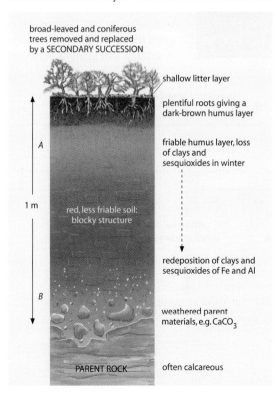

broad-leaved and coniferous trees removed and replaced by a SECONDARY SUCCESSION

shallow litter layer

plentiful roots giving a dark-brown humus layer

friable humus layer, loss of clays and sesquioxides in winter

A

1 m

red, less friable soil: blocky structure

redeposition of clays and sesquioxides of Fe and Al

B

weathered parent materials, e.g. $CaCO_3$

PARENT ROCK   often calcareous

clay-enriched and may be coloured a bright red by the redeposition of iron and aluminium. The soils, which are often thin, are less acid than the brown earths as there is less leaching in the dry season and calcium is often released, especially in limestone areas (Figure 12.25).

In many Mediterranean areas, parent rock is locally a more important factor in soil formation than climate. This leads to the development of intrazonal soils such as rendzina and terra rossa (Figures 10.23 and 10.24).

4A Eastern margin climates in Asia (monsoon)

South-east and eastern Asia are dominated by the monsoon (page 239). Temperature figures and rainfall distributions are similar to those of places having a tropical continental climate with a very warm and dry season from November to May and a hot and very wet season from June to October (Places 32, page 240). The major difference between the two climates is that monsoon areas receive appreciably higher annual amounts of rain. The natural vegetation is jungle (tropical deciduous forest) and the dominant soil type is ferralitic. Both vegetation and soils, therefore, share many similarities with the tropical rainforest.

5 Temperate grasslands

The temperate grassland biome lies in the centre of continents approximately between latitudes 40° and 60° north of the Equator. The two main areas are the North American Prairies and the Russian Steppes (Figure 11.38).

Cool temperate continental climate

The annual range of temperature is high as there is no moderating influence from the sea (38°C at Saskatoon, Figure 12.26). The land warms up rapidly in summer to give maximum mean monthly readings of around 20°C. However, the rapid radiation of heat from mid-continental areas in winter means there are several months when the temperature remains below freezing point. The clear skies also result in a large diurnal temperature range.

In Russia, precipitation decreases rapidly towards the east as distance from the sea – and therefore from the rain-bearing winds – increases; in North America, however, totals are lowest to the west which is directly in the rainshadow of the Rockies. Annual amounts in both areas only average 500 mm and there is a threat of drought, as experienced in North America in 1988. Although, fortuitously, 75 per cent of

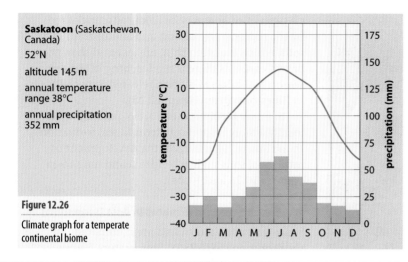

Saskatoon (Saskatchewan, Canada)

52°N

altitude 145 m

annual temperature range 38°C

annual precipitation 352 mm

Figure 12.26

Climate graph for a temperate continental biome

precipitation falls during the summer growing season, it can occur in the form of harmful thunderstorms and hailshowers. The ground can be snow-covered for several months between October and April. Overall, there is a close balance between precipitation and evapotranspiration. In winter, both areas are open to cold blasts of arctic air, although the chinook may bring temporary warmer spells to the Prairies (page 241).

Temperate grassland vegetation

This type of vegetation lies to the south of the coniferous forest belt in the dry interiors of North America and Russia. Temperate grasslands are, however, also found sporadically in parts of the southern hemisphere, where they usually lie between 30° and 40°S. The Pampas (South America) and the Canterbury Plains (New Zealand) are towards the eastern coast, while the Murray–Darling basin (Australia) and the Veld (South Africa) are further inland. The NPP of 600 g/m²/yr is considerably less than that of the tropical grasslands because the vegetation grows neither as rapidly nor as tall (Figure 11.40). Whatever the original climax vegetation of the biome may have been, the ecosystem has been significantly altered by fire and human exploitation to leave, today, grama and buffalo grass as the dominants. There are two main types of grass. Feather grasses grow to 50 cm and form a relatively even coverage, whereas tufted (tussock) grasses, reaching up to 2 m, are found in more compact clumps (Figure 12.27). The grass forms a tightly knit sod which may have restricted tree growth, and certainly made early

Figure 12.27

Tufted grasses on the North American Prairies, USA

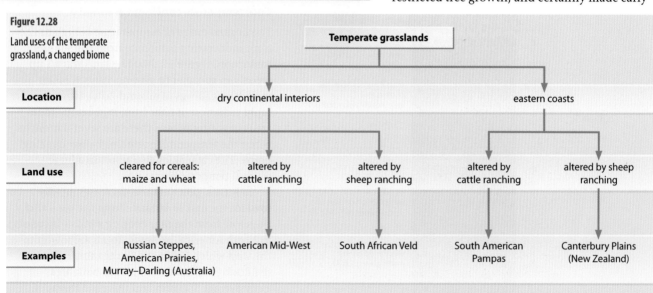

Figure 12.28

Land uses of the temperate grassland, a changed biome

	Temperate grasslands				
Location	dry continental interiors			eastern coasts	
Land use	cleared for cereals: maize and wheat	altered by cattle ranching	altered by sheep ranching	altered by cattle ranching	altered by sheep ranching
Examples	Russian Steppes, American Prairies, Murray–Darling (Australia)	American Mid-West	South African Veld	South American Pampas	Canterbury Plains (New Zealand)

Figure 12.29

A chernozem (black earth) soil profile

Figure 12.30

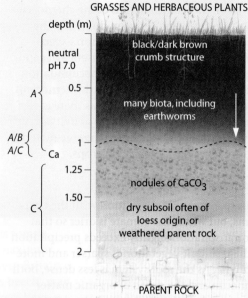

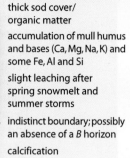

A chernozem (black earth) soil profile typical of the continental grasslands biome

precipitation = evapotranspiration
GRASSES AND HERBACEOUS PLANTS

depth (m)

neutral pH 7.0

A

0.5

black/dark brown crumb structure

many biota, including earthworms

A/B
A/C

Ca

1

C

1.25

1.50

2

nodules of CaCO₃

dry subsoil often of loess origin, or weathered parent rock

PARENT ROCK

thick sod cover/ organic matter

accumulation of mull humus and bases (Ca, Mg, Na, K) and some Fe, Al and Si

slight leaching after spring snowmelt and summer storms

indistinct boundary; possibly an absence of a B horizon

calcification

ploughing difficult. The deep roots, which often extend to a depth of 2 m in order to reach the water table, help to bind the soil together and so reduce erosion. Most of the organic material is in the grass roots and it is the roots and rhizomes that provide the largest store of nutrients (Case Study 12B).

During autumn, the grasses die down to form a turf mat in which seeds lie dormant until the snowmelt, rains and higher temperatures of the following spring. Growth in early summer is rapid and the grasses produce narrow, inward-curving blades to limit transpiration. By the end of summer, their blue-green colour may have turned more parched. Herbaceous plants and some trees (willow) grow along water courses. In response to the windy climate, many prairie and steppe farms are protected by trees planted as windbreaks. The decay of grasses in summer causes a rapid accumulation of humus in the soil, making the area ideal for cereals or, in drier areas, for cattle ranching (Figures 12.27 and 12.28).

The temperate grasslands are a resilient ecosystem. The grasses provide food for burrowing animals such as rabbits and gophers, and for large herbivores such as antelopes, bison and kangaroos. These, in turn, may be consumed by carnivores (wolves and coyotes) or by predatory birds (hawks and eagles).

Chernozems or black earths

The thick grass cover and the importance of roots as a source of organic matter together provide a plentiful supply of mull humus which forms a black, crumbly topsoil (Figure 12.29). While the abundance of biota, especially earthworms, causes the rapid decay and mixing of organic matter during the warm summer,

decomposition is arrested during drier spells and in the long, cold winter. Due to rapid mixing, humus is spread throughout the A horizon, and as a result of rapid decomposition there is effective recycling as the grasses take up and return nutrients to the soil. The late spring snowmelt and early summer storms cause some leaching (Figure 12.30), and bases such as potassium and magnesium may be slowly moved downwards. In late summer, and in places where the water table is near to the surface, capillary water may bring bases nearer to the surface to maintain a neutral or slightly alkaline soil (pH 7–7.5). The grasses have an extensive root system which gives a deep (up to 1 m) dark-brown to black A horizon.

The alternating dry and wet seasons immobilise iron and aluminium sesquioxides and clay within aggregates (peds) in the upper horizon and this, together with the large number of mixing agents, limits the formation of a recognisable B horizon. The subsoil, often of loess origin (page 136), is usually porous and this, together with the capillary moisture movement in summer, means that it remains dry. This upward movement of moisture causes calcium carbonate to be deposited, often in the form of nodules, in the upper C horizon.

Chernozems are regarded as the optimum soil for agriculture as they are deep, rich in organic matter, retain moisture, and have an ideal crumb structure with well-formed peds. After intensive ploughing, chernozems may require the addition of potassium and nitrates.

Brown earths tend to be free-draining as they do not have a hard pan. There is considerable recycling as the deciduous trees take up large amounts of nutrients from the soil in summer, only to return them through leaf-fall the following autumn. Brown earths are usually deeper than podsols, partly because tree roots can penetrate and break up the bedrock (Figure 2.5) and are more fertile, mainly because of the higher content of organic matter and clay (although they often benefit from liming).

7 Coniferous forests

The coniferous forest, or taiga, biome occurs in cold climates to the poleward side of 60°N in Eurasia and North America as well as at high altitudes in more temperate latitudes and in southern Chile (Figure 11.38).

Cold climates

Winters are long and cold. Minimum mean monthly temperatures may be as low as –25°C (–24°C at Fairbanks, Figure 12.35) – there is little moderating influence from the sea and no insolation as, at this time of year, the sun never rises in places north of the Arctic Circle. Strong winds mean there is a high wind-chill factor (frostbite is a hazard to humans); any moisture is rapidly evaporated (or frozen); and snow is frequently blown about in blizzards. Summers are short, but the long hours of daylight and clear skies mean that they are relatively warm. Precipitation is light throughout the year because the air can hold only limited amounts of moisture, and

Figure 12.36

Coniferous forest, British Columbia

most places are a long way from the sea. The slight summer maximum is caused by isolated convectional rainstorms.

Coniferous forest or taiga

The coniferous forest has an average NPP of 800 g/m²/yr (Figure 11.40). The coniferous trees have developed distinctive adaptations which enable them to tolerate long, cold winters; cool summers with a short growing season; limited precipitation; and podsolic soils. The size of the dominant trees and the fact that they are evergreen – giving them the potential for year-round photosynthesis – result in their relatively high NPP. The trees, which are softwoods, rarely number more than two or three species per km². Often there may be extensive stands of a single species, such as spruce, fir or pine. In colder areas, like Siberia, the larch tends to dominate. Although larches are cone-bearing, the European larch is deciduous and sheds its leaves in winter. All trees in the taiga, some of which attain a height of 40 m, are adapted to living in a harsh environment (Figure 12.36).

Figure 12.37

Forest floor in a coniferous forest, Cumbria

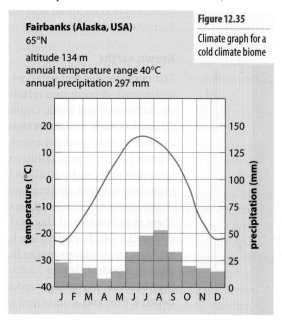

Figure 12.35

Climate graph for a cold climate biome

Fairbanks (Alaska, USA)
65°N

altitude 134 m
annual temperature range 40°C
annual precipitation 297 mm

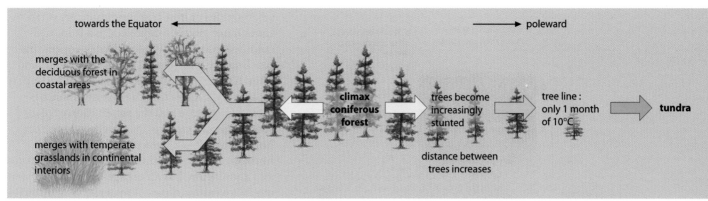

towards the Equator ◄—————— ——————► poleward

merges with the
deciduous forest in
coastal areas

**climax
coniferous
forest**

trees become
increasingly
stunted

tree line :
only 1 month
of 10°C

tundra

merges with temperate
grasslands in continental
interiors

distance between
trees increases

Figure 12.38

The coniferous
forest and its
transition zones

Conditions for photosynthesis become favourable in spring as incoming radiation increases and water becomes available through snowmelt (days in winter are long and dark and soil moisture is frozen). The needle-like leaves are small and the thick cuticles help to reduce transpiration during times of strong winds and during the winter when moisture is in a form unavailable for absorption by tree roots. Cones shield the seeds and thick, resinous bark protects the trunk from the extreme cold of winter and the threat of summer forest fires. The conical shape of the tree and its downward-sloping, springy branches allow the winter snows to slide off without breaking the branches. The conical shape also gives some stability against strong winds as the tree roots are usually shallow. There is usually only one layer of vegetation in the coniferous forest. The amount of ground cover is limited, due partly to the lack of sunlight reaching the forest floor and partly to the deep, acidic layer of non-decomposed needles (Figure 12.37). Plants that can survive on the forest floor include mosses, lichens and wood sorrel. The cold climate and acid soil discourage earthworms and bacteria. Needles decompose very slowly to give an acid mor humus (page 262) with most of the nutrients held within the litter (Figure 11.29a). Evapotranspiration rates are very low and, as they are usually less than precipitation totals, leaching occurs and the few nutrients that are returned to the podsol soil are soon lost. Conifers require few nutrients, taking only 225 kg of plant nutrient annually from each hectare compared with the 430 kg taken by deciduous trees. The limited food supply means that animal life is not abundant. The dark woods are not favoured by bird life, although deer, wolves, brown bears, moose, elk and beavers are found in certain areas.

In North America and Eurasia, the coniferous forest merges into the tundra on its northern fringes (Figure 12.38). The tree line, the point above which trees are unable to grow, is often clearly marked in mountainous areas (Figure 12.36). South of the taiga lie either the deciduous forest or the temperate grassland biomes (Figure 11.38), depending upon whether the location is coastal or inland.

Podsols

Podsols develop in areas where precipitation exceeds evapotranspiration; under coniferous forest, heathland and other vegetation tolerant of low-nutrient-status soils; and where parent materials produce coarse-textured soils. Although podsols usually occur in places with a cool climate, they can be found virtually anywhere between the Equator and the Arctic, providing the required conditions are present.

BSSS WYE 1973
PROFILE   B2
HOTHFIELD   SERIES
HUMO-FERRIC   PODZOL

Figure 12.39

Soil profile of a podsol

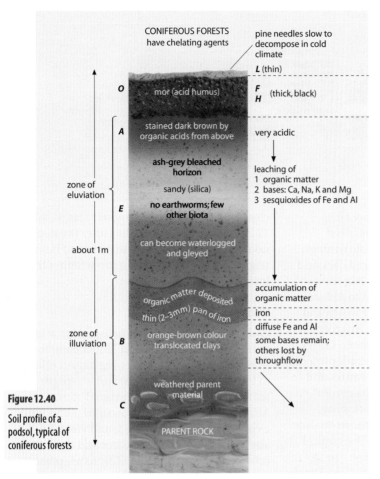

CONIFEROUS FORESTS
have chelating agents

pine needles slow to decompose in cold climate

L (thin)

O   mor (acid humus)

F
H   (thick, black)

A   stained dark brown by organic acids from above

very acidic

ash-grey bleached horizon

sandy (silica)

no earthworms; few other biota

E

leaching of
1 organic matter
2 bases: Ca, Na, K and Mg
3 sesquioxides of Fe and Al

zone of eluviation

can become waterlogged and gleyed

about 1m

organic matter deposited

accumulation of organic matter

thin (2–3mm) pan of iron

iron

diffuse Fe and Al

zone of illuviation

B   orange-brown colour translocated clays

some bases remain; others lost by throughflow

weathered parent material

C

PARENT ROCK

Figure 12.40

Soil profile of a podsol, typical of coniferous forests

Pine needles, with their thick cuticles, provide only a thin leaf litter and inhibit the formation of humus. Any humus formed is very acid (mor) and provides chelating agents and fulvic acid which help to make the iron and aluminium minerals more soluble. The cold climate discourages organisms and the soil is too acidic for earthworms. Consequently, well-defined horizons develop due to the slow decomposition of leaf litter and the lack of mixing agents. The downward percolation of water through the soil, especially following snowmelt, causes the leaching of bases, the translocation of organic matter, and the eluviation of the sesquioxides of iron and aluminium. This leaves an ash-grey, bleached *A* horizon (podsol is Russian for 'ash-like') composed mainly of quartz sand and silica (Figures 12.39 and 12.40).

Pedologists accept that different processes (physical, chemical and biological) can be employed in the translocation of materials, e.g. humus and clay in suspension, bases in solution, sesquioxides by biochemical agents in solution and, perhaps most significantly, movement caused by soil fauna mixing the soil.

The dark-coloured humus is redeposited at the top of the *B* horizon. Beneath this the

sesquioxides of iron and aluminium are often – though not always – deposited as a thin, rust-coloured, hard pan. Where it is developed, this pan is rarely more than 2 or 3 mm in depth and often has a convoluted shape. It acts as an impermeable layer restricting the downward movement of moisture and the penetration of plant roots. This can cause some waterlogging in the *E* horizon to give a gleyed podsol. The lower *B* horizon, an area of diffuse accumulation of iron and aluminium, has an orange-brown colour and overlies weathered parent material. Any throughflow from this horizon is likely to contain bases in solution. Although these soils are not naturally fertile, they can be improved by the addition of lime and fertiliser, or by ripping the iron pan with a deep, single-line plough.

8   The tundra

The tundra, which lies to the north of the taiga, includes the extreme northern parts of Alaska, Canada and Russia, together with all of Greenland (Figure 11.38). The ground, apart from the top few centimetres in summer when temperatures are high enough for some plant growth, remains permanently frozen (the permafrost, Chapter 5).

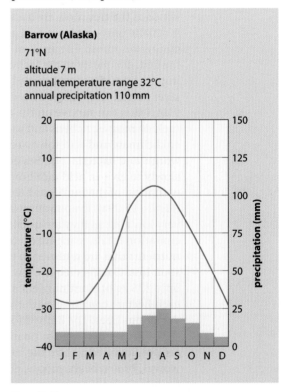

Barrow (Alaska)

71°N

altitude 7 m
annual temperature range 32°C
annual precipitation 110 mm

Figure 12.41

Climate graph for a tundra biome

Arctic climate

Summers may have lengthy periods of continuous daylight but, with the angle of the sun so low in the sky, temperatures struggle to rise above freezing-point (Barrow 3°C, Figure 12.41) and the growing season is exceptionally short. Nearer the poles, the climate is one of perpetual frost. Although winters are long, dark and severe, and the sea freezes, the water has a moderating effect on temperatures, keeping them slightly higher than inland places further south (Siberia). Precipitation, which falls as snow, is light – indeed, Barrow with 110 mm would be classified as a desert if temperatures were high enough for plant growth.

Tundra vegetation

The tundra ecosystem is one with very low organic productivity. The NPP of only 140 g/m²/yr is the second-lowest of the major land biomes (Figure 11.40). In Finnish, *tundra* means a 'barren or treeless land', which accurately describes its winter appearance, and in Russian a 'marshy plain', which is what large areas are in summer. Any vegetation must have a high degree of tolerance of extreme cold and of moisture-deficient conditions – the latter because water is unavailable for most of the year when it is stored as ice or snow. There are fewer species of plants in the tundra than in any other biome. Most are very slow- and low-growing, compact and rounded to gain protection against the wind (plants as well as people are affected by wind-chill), and most have to complete their life-cycles within 50–60 days. There is no stratification of vegetation by height.

The five main dominants, each with its specialised local habitat, are lichens, mosses, grasses, cushion plants, and low shrubs (Figure 12.42). Most have small leaves to limit

Figure 12.43

Waterlogged tundra in the summer season, Alaska

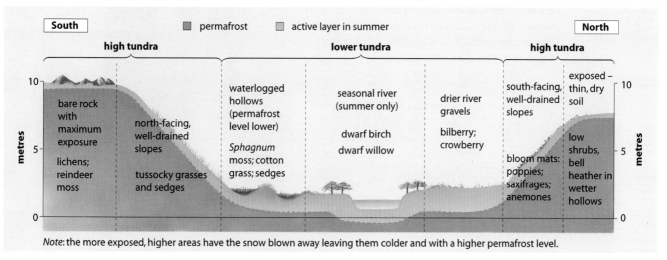

transpiration and short roots to avoid the permafrost. Lichens are pioneer plants in areas where the ice is retreating, and they can help date the chronology of an area following deglaciation (page 288). Much of the tundra is waterlogged in summer (Figure 12.43) due to the impermeable permafrost preventing infiltration. Where relief is gentle and evaporation rates are low, mosses, cotton grass and sedges thrive. On south-facing slopes and in better-drained soils, cushion plants provide a mass of bright colour in summer (Figure 12.44). These 'bloom mats' include arctic poppies, anemones, orchids, pink saxifrages and gentians. Where decaying vegetation accumulates (there is little bacterial action to decompose dead plants), the resultant peat is likely to be covered in heather, whereas on drier

Figure 12.42

Relationship between vegetation and site factors in the tundra

South	permafrost	active layer in summer	North

high tundra — **lower tundra** — **high tundra**

bare rock with maximum exposure

lichens; reindeer moss

north-facing, well-drained slopes

tussocky grasses and sedges

waterlogged hollows (permafrost level lower)

Sphagnum moss; cotton grass; sedges

seasonal river (summer only)

dwarf birch
dwarf willow

drier river gravels

bilberry; crowberry

south-facing, well-drained slopes

bloom mats: poppies; saxifrages; anemones

exposed – thin, dry soil

low shrubs; bell heather in wetter hollows

Note: the more exposed, higher areas have the snow blown away leaving them colder and with a higher permafrost level.

World climate, soils and vegetation

Figure 12.44

'Bloom mats' at
Prudhoe Bay, Alaska

gravels, berried plants (e.g. bilberry and crow-berry) are the dominants. Adjacent to the seasonal snowmelt rivers, dwarf willows, horizontal junipers and stunted birch grow, but only to a maximum of about 30 cm; even so their crowns are often distorted and misshapen by the wind. In winter, the whole biome is covered in snow, which acts as insulation for the plants.

The lack of nitrogen-fixing plants, other than in the pioneer community (page 286), limits fertility, and the cold, wet conditions inhibit the breakdown of plant material. Photosynthesis is hindered by the lack of sunlight and water for most of the year, though the presence of autotrophs, such as lichens and mosses, does provide the basis for a food chain longer than might be expected. Herbivores such as reindeer, caribou and musk-ox survive because plants like reindeer moss have a high sugar content. However, these animals have to migrate in winter to find pasture that is not covered by snow. The major carnivores are wolves and arctic fox; owls are also found here.

The tundra is an extremely fragile ecosystem in a delicate balance. Once it is disturbed by human activity, such as tourism or oil exploration and extraction, it may take many years before it becomes re-established.

Tundra soils

The limited plant growth of this biome only produces a small amount of litter and, as there are few soil biota in the cold soil, organic matter decomposes only very slowly to give a thin peaty layer of humus or mor. There are many sites where there is free drainage. Where this occurs, water is able to percolate downwards, usually as meltwater in late spring, giving limited leaching and, due to the fulvic acid within it (the pH can be under 4.5), allowing the release of iron. Underlying the soil, at a very variable depth but usually under 50 cm, is the permafrost. This, acting as an impermeable layer, severely restricts moisture percolation and causes extreme waterlogging and gleying (Figure 12.45). Few mixing agents can survive in the cold, wet, tundra soils, which are thin and have no developed horizons (an exception is the arctic brown soil which develops on better-drained sites). Where bedrock is near to the surface, the parent material is physically weathered by freeze–thaw action. The shattered angular fragments are raised to the surface by frost-heave, preventing the formation of horizons and creating a range of periglacial land-forms (Figure 5.17).

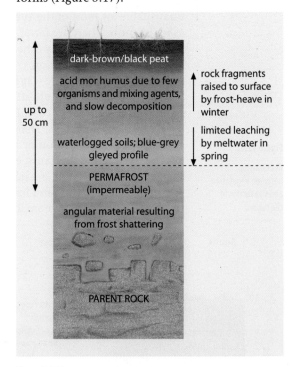

Figure 12.45

Soil profile of a
typical tundra soil

The management of grasslands A Tropical grasslands in Kenya

The main expanses of tropical grassland in Kenya lie within the Rift Valley and on the adjacent plains of the Mara (an extension of the Serengeti) and Loita (Figure 12.48). Their appearance is one of open savanna (Figures 12.13 and 12.46) with small acacia and evergreen trees (Figure 12.15). There is evidence, however, that the original climax vegetation was forest, but that this has been altered by fires, started both naturally and by humans (page 293), by overgrazing (Figure 12.47) and by climatic change.

The climate is very warm and dry for most of the year with, usually, a short season (three months) of fairly reliable and abundant rainfall and an even shorter period known as the 'little rains' (Figure 12.49). Both periods of rainfall follow soon after the ITCZ and the associated overhead sun have passed over the

Figure 12.47

Scattered trees and over-grazed land in the central Rift Valley, Kenya

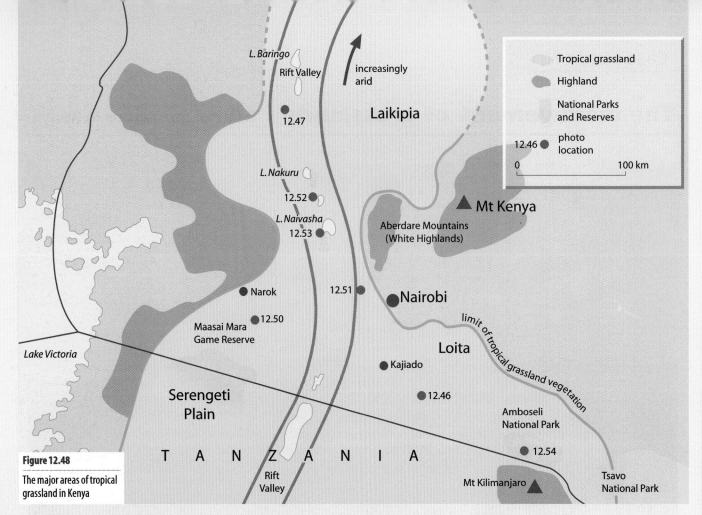

Figure 12.48

The major areas of tropical grassland in Kenya

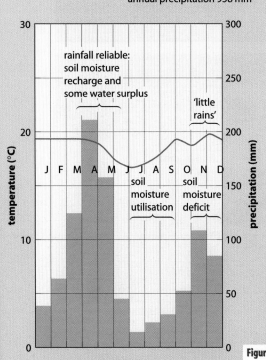

Nairobi (Kenya)

latitude 1°S

altitude 1820 m
annual temperature range 3°C
annual precipitation 958 mm

rainfall reliable: soil moisture recharge and some water surplus

'little rains'

soil moisture utilisation

soil moisture deficit

Equator (Figure 12.12). The annual water balance shows a deficit (Figure 3.3) so that, although there is some leaching during the rainy season, for most of the year capillary action occurs. This has resulted in the development of ferruginous soils with, in places, a lateritic crust (page 321). Water supply is therefore a major management problem in this part of Kenya.

Water is obtained from springs at the foot of Mount Kilimanjaro – the mountain itself is in Tanzania – which are fed by melting snow; from several of the Rift Valley lakes (not all, as some are highly saline); from rivers (many of which are seasonal); and from waterholes. Even so, there have been, in the last 100 years alone, several major droughts when the carrying capacity of the region was exceeded. The **carrying capacity** (page 378) is the maximum number of a population (people, animals, plants, etc.) that can be supported by the resources of the environment in which they live, e.g. the greatest number of cattle that can be fed adequately on the available amount of grassland.

Figure 12.49

Climate graph and water balance for Nairobi (note that, due to its higher altitude, Nairobi is cooler and wetter than the surrounding grasslands)

Human pressure on the natural resources

Maasai pastoralists

Maasai are defined as 'people who speak the Maa language'. Their ancestors were Nilotic, coming from southern Sudan during the first millennium AD. They kept cattle and grew sorghum and millet. The present Maasai may be descendants of the last of several migration waves. Latest evidence suggests that they may have only been in Kenya for 300 years. Over time, they specialised more in cattle and came to see themselves, and to be seen by others, historically and ethnically as 'people of cattle'. Figure 12.50 is a stereotype photo of the Maasai, dressed in their red cloaks and with their humped zebu cattle. While all Maasai are Maa speakers, not all Maa speakers are Maasai – nor, today, are all Maasai pastoralists! The Maasai became semi-nomadic, moving seasonally with their cattle in search of water and pasture (two wet seasons and two dry seasons meant four moves a year; Figure 12.49). Herds had to be large enough to provide sufficient milk and meat for their owners and to reproduce themselves over time, including the ability to recover from drought and disease.

Kikuyu (Bantu) farmers

The Kikuyu were one of several Bantu tribes who arrived in Kenya, from the south, some 2000 years ago. They became subsistence farmers growing crops on the higher land which bounded the eastern side of the Rift Valley. The Kikuyu and Maasai often lived a complementary life-style. For the Maasai, Kikuyu in the highlands were a secure source of foodstuffs and a place of refuge during times of drought and cattle disease. For the Kikuyu, Maasai provided a constant supply of cattle products and wives. The division between them only appeared in early colonial times when the Maasai were forcibly moved from places like Laikipia (Figure 12.48) southwards onto the newly created Maasai reservation (the districts of Narok and Kajiado). The vacuum left was filled by newly arrived European settlers, and by Kikuyu (their rising numbers were causing a land shortage in the highlands). Figure 12.51 shows numerous, small, Kikuyu shambas (smallholdings) on the eastern edge of the Rift Valley to the north-west of Nairobi.

Colonial (European) settlers

While many Europeans settled in the so-called 'White Highlands', others developed huge estates within the Rift Valley. The most famous was Lord Delamere from Cheshire. He introduced, in turn, Australian sheep (they died, as the local grass was mineral-deficient); British sheep and clover (the sheep died, as African bees did not pollinate British clover); British cattle (wiped out by local diseases); wheat (which was more successful unless trampled by wild animals); and, finally and successfully, drought-resistant beef cattle. The present

Figure 12.50
Maasai herdsmen

Figure 12.51
Kikuyu shambas near Nairobi

Delamere estate (Figure 12.52) covers 22 600 hectares (divided into 180-hectare paddocks); it has 10 900 long-horned Boran cattle (the carrying capacity is 12 000) crossed with 300 Friesian bulls; and 280 permanent workers. Although the estate is managed by 'whites', the stockmen are Maasai. More recently, transnational firms have set up large flower farms (Figure 12.53) and vegetable (especially peas and beans) farms in and near the Rift Valley. The closeness to Nairobi airport means that these perishable products can be transported to and sold in European markets, out of season, the day after they are picked.

Figure 12.52
Commercial cattle ranching, Delamere Estate, Kenya

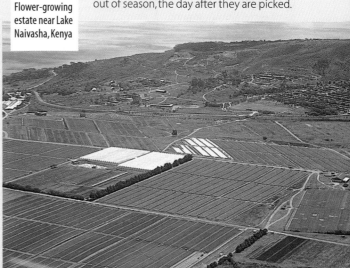

Figure 12.53
Flower-growing estate near Lake Naivasha, Kenya

Population growth and urbanisation

Kenya, an economically less developed country, has one of the world's fastest-growing population rates. This means increased pressure on the land, especially the grasslands, to grow more subsistence crops to feed the growing domestic market; more cash crops to earn needed money from increased exports; and more land lost to urban growth.

Maasai in the late 1990s

The traditional Maasai way of life and their grassland habitat are under constant threat. Figure 12.55 summarises, but does insufficient justice to, some of the present-day problems. Change, as in many societies, is being forced upon the Maasai. While many values and traditions are still known and held, the basis of their economy – the concept of land as territory – has been so transformed that the survival of the herding system is in jeopardy. For some years, many Maasai have tried either to buy individual ranches (IRs) or to amalgamate to create group ranches (GRs), a practice which seems to fail at times of severe drought. The Maasai are also having to come to terms with a sedentary rather than a semi-nomadic lifestyle. Intermediate Technology Development group (ITDG), a British development group (Places 90, page 577), has been working with Maasai people to improve the standard of housing. In response to the main complaint of Maasai women, IT have helped to design a watertight cement skin which can be laid over an old mud roof (it was the women's job to apply more dung and mud onto a leaking roof during a wet night), and have improved ventilation within the house (where all the cooking is done). The government have laid a pipeline from Kilimanjaro to Kajaido, to ensure a more reliable water supply. The quality of Maasai herds has improved, with some cattle being sold for meat in Nairobi. The improvement to herds has been aided by IT who have helped train local villagers to become 'vets' (wasaidizi) capable of vaccinating animals and dealing with common diseases. Some Maasai have begun to grow crops, while others have begun to benefit from tourism. In Amboseli National Park, the Maasai are allowed to sell artefacts from their own shop. They can retain all the income which had, previously, gone to the government.

Figure 12.54

Amboseli National Park, Kenya, watered by melting snow from Mt Kilimanjaro (Tanzania)

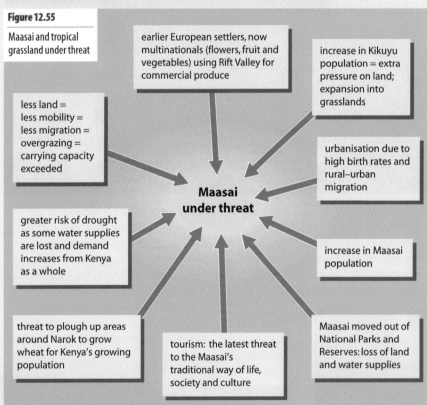

Figure 12.55

Maasai and tropical grassland under threat

earlier European settlers, now multinationals (flowers, fruit and vegetables) using Rift Valley for commercial produce

increase in Kikuyu population = extra pressure on land; expansion into grasslands

less land = less mobility = less migration = overgrazing = carrying capacity exceeded

urbanisation due to high birth rates and rural–urban migration

Maasai under threat

greater risk of drought as some water supplies are lost and demand increases from Kenya as a whole

increase in Maasai population

threat to plough up areas around Narok to grow wheat for Kenya's growing population

tourism: the latest threat to the Maasai's traditional way of life, society and culture

Maasai moved out of National Parks and Reserves: loss of land and water supplies

National parks and reserves

The passing of the National Parks Ordinance in 1945 meant that specific areas were set aside either exclusively for wildlife (no permanent settlement in National Parks other than at tourist lodges) or where other types of land use were permitted only at the discretion of local councils. While wildlife has become a major source of income for Kenya, it has meant less land being available for crops and, to the Maasai, denial of access to important resources of dry-season water and pasture (Amboseli; Figure 12.54). Maasai herds were heavily depleted during the droughts of 1952 and 1972–76.

B The temperate grasslands in North America: the Prairies

Early travellers such as the Spaniard Coronado in the 16th century, who rode into Kansas from Mexico, and later French trappers and explorers in Canada, reported vast extents of waist-high, green grasses sometimes so tall that men on horseback stood in their stirrups to see where they were going. The plains seemed so vast that no limit could be found. Nineteenth-century settlers moving westwards across the Mississippi–Missouri in their wagons or drawing their handcarts, must have wondered if they would ever see woods, forests and mountains again. Today, the extent of the interior grasslands of North America is well known.

The Native Americans, who used the ecosystem, did little to alter the grassland, which remained in its original state of natural balance until the late 19th century.

'It is a wild garden. Each week from April through September, about a dozen new kinds of flowers come into bloom. Once the layer of dead grass gets too thick, though, it starts to choke off the smaller grasses and wild flowers. Meantime, woody plants – they like shade and moisture – can gain a foothold in the sod and spread. If you go long enough without fire, much of this countryside will be covered with trees' (quoted in Chadwick, 1993, p.113).

Grasses such as blue stem and buffalo grass have a network of roots which can extend to considerable depth to absorb water and obtain nutrients. Root systems may make up over 80 per cent of the vegetative biomass in the prairie. These, together with the smaller herbs, have helped to develop a thick sod close to the surface. Plants can survive from year to year because they die back to the ground and lie dormant during the cold winters (page 327).

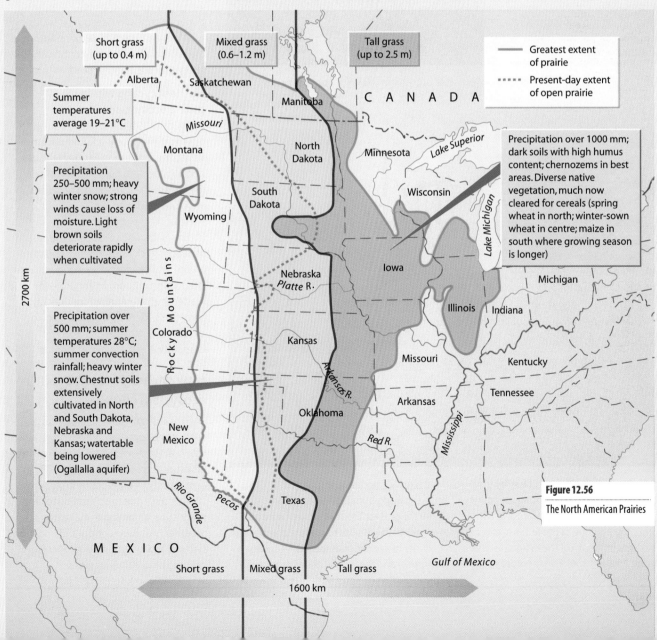

Figure 12.56

The North American Prairies

Figure 12.57

Short-grass prairie with deciduous trees (cottonwoods and aspen) and some conifers: west central Wyoming

Figure 12.58

Long-grass prairie (buffalo grass): Saskatchewan

(bison and humans). The presence of humus also helps to conserve moisture. Due to frequent droughts, the vegetation has developed protective mechanisms, such as leaves that curl up to prevent evaporation loss, and well-developed root fibres which can obtain moisture from deep in the soil.

The rapid spring growth and early maturity of grass allows it to produce seeds early. It then becomes semi-dormant until autumn and can survive heat and drought. Late-growing species may not be able to compete and this has led to an extension of short grass into the mixed-grass zone during a succession of long dry periods.

In the 17th century, there were estimated to be 60–70 million bison roaming the grasslands with 50 million antelope, plus grizzly bears, wolves and prairie dogs, together with many species of birds – hawks, larks, buntings, etc. – and insects and reptiles such as snakes and lizards. Today there are few of the larger mammals left except in wilderness refuges such as National Parks.

Nutrient cycling within the prairie ecosystem

In Figure 11.29b, the small litter store reflects the relatively small amount of vegetative matter and low leaf fall. Litter decomposes into humus and nutrients are released to the soil, giving it good crumb structure. Moderate rainfall reduces loss from runoff. The large soil storage is a result of the weathering of rock and the presence of deep, rich chernozems (in the eastern and central prairies) which have accumulated a high proportion of humus (organic matter) in the temperate continental conditions. There is little or no leaching because the rainfall is exceeded by evaporation in the summer months.

How and why may this ecosystem change?

There is little tall-grass prairie left, and estimates by government agencies indicate that there is less than 34 per cent true mixed-grass prairie and less than 23 per cent true short-grass prairie in existence. This is mainly due to conversion to crop production, damming of major rivers together with flood control and irrigation systems, and in favourable areas the draining of wetlands for crops.

Soils grade in colour and fertility from brown in the western short-grass prairie (Figure 12.57) through chestnut in the mixed-grass zone to the fertile black chernozems (millisols) of the tall-grass eastern zone (Figure 12.58). The chernozems have a high humus content (page 327).

Decaying humus releases minerals slowly for the grasses. The soils are kept light and aerated which helps to prevent compaction under heavy rain (summer convection storms) and the weight of heavy animals

Figure 12.59

Cattle farming in the Prairies

Figure 12.60

Cattle feed lots in Denver

1 Natural conditions

- **Unpredictable rainfall** and **drought** have been a major factor in change in the North American Prairies. Low rainfall in the 1930s allied to bad farming practices led to the creation of the American Dust Bowl; reduced the natural fodder for animals; and permitted an eastward extension of the short-grass prairies into the eastern tall grass.

- **Lightning** is a frequent cause of fire in the grasslands in summer. This destroys the surface vegetation; kills small animals; and damages the food supply. In the years following serious fires, lower bird numbers have been recorded, as many nest on the ground.

- **Bison herds** have been effective in change. They are heavy grazers and reduce the coarser medium grasses, leaving short grasses. In the spring and early summer when mosquitoes hatch they plague the bison, causing them to roll on the ground to reduce the itching! This forms depressions in the prairie surface. These bare soils may be re-colonised later by seeds carried by birds.

2 Human activity

- **Hunting** The earliest inhabitants were the Native Americans who hunted animals for food, using fire and traps to kill unselectively. With the coming of the Europeans and the introduction of the horse and the rifle, they were able to kill large numbers of bison almost to extinction.

- **Trapping** Fur-traders moving west from the Great Lakes and the Mississippi in the 18th and early 19th centuries helped to reduce the population of bison and elk which were slaughtered for winter food and the hides tanned and sold. The herds were forced into remote areas such as the high basins of Montana and Wyoming. Wolves and mountain lions further reduced them. It is only in the late 20th century that the bison have been increasing again as a result of careful conservation in National Parks and refuges such as the Houck Ranch in South Dakota.

- **Cereal farming** The prairie sod proved difficult to remove and cultivate using wooden and iron ploughs, so much of the grassland was used initially for cattle ranching. After the 1840s, when the steel plough was developed, breaking the sod for cultivation became possible. Cereal crops were successfully introduced. Overcultivation in the 1930s led to extensive soil erosion. Soil conservation methods on a large scale have helped to reduce the area that has been permanently damaged. Droughts between 1950 and 1985 damaged the soils even more: 'some plowed fields had lost a metre of their soil. As much as 25 tonnes of soil per hectare had been blown away in two counties' (23 February 1977).

- **Cattle ranching** became the main farming activity on the western prairies (Figure 12.59). The huge herds used the range lands formerly occupied by the bison. Serious problems of overgrazing have occurred in the lower-rainfall regions. Irrigation is normal in Alberta, Montana, the western parts of the Dakotas and Nebraska. This produces fodder crops for cattle which have to be fed outside during the long cold winters. It is not without its problems, for irrigation is lowering the water table.

- **Intensive cattle production** is now taking place on huge feedlots close to railheads such as Denver (Figure 12.60). Young cattle are fattened in stockyards on grain and silage bought in from farmers on the prairie. The animals do not lose condition by being left to roam freely on the open range.

- **Mineral extraction** In the 1970s there were pressures to develop extensive strip mining for coal, as well as demands for oil exploration during the OPEC crises. This meant that new roads, railways and pipelines were constructed across former grazing lands, with little environmental consideration. The 'environmental lobby' is now strong enough to resist such demands.

What can be done to protect this ecosystem?

There are conflicting views about conservation of the grasslands. A proposal to designate 320 000 acres (130 000 hectares) of prairies in Kansas as a National Park has been strongly opposed by local cattle ranchers. Conservationists believe that cattle ranching (Figure 12.61) has already damaged the natural balance of the area. Unless cattle ranching and controlled grazing are held in check, they argue, the prairie will become degraded.

The US Government has established 1 539 478 hectares of prairie to be managed by the Forest Service as National Grasslands, for 'recreation, range, timber, watershed wildlife and fish purposes'. This involves working with groups of graziers who hold long-term leases on public land, as well as private landowners, to provide the best management system for the grasslands. There has been an increase in fencing on the open range, but progress has been made in rehabilitating grassland suffering from overgrazing. Management can be difficult because of the number of private land holdings close to and within the 18 National Grasslands that have already been set up.

Cattle ranchers believe that ...	Conservationists believe that ...
controlled burning to renew pasture should be allowed (this was a Native American custom)	there is too much burning; the prairie does not recover if burns are too frequent
overgrazing can be avoided by careful pasture management	new information gained from research will help both graziers and conservation
soil and water conservation are already practised	the prairie needs restoration to maintain its ecosystem
tourists, picnic sites and more roads will damage the environment	the prairie has already been damaged by cattle grazing

Figure 12.61

Cattle ranchers versus conservationists

The Pawnee National Grassland in north-east Colorado is an area of 79 000 hectares within an area 48 by 96 km. State and private rangeland, part of the remaining short-grass prairie ecosystem, is also included in this area. Rainfall of between 300 and 375 mm falls mainly in summer, and there is considerable snow cover in winter; the area can suffer from summer drought. Strong winds blow across the open grasslands and evaporation rates are high. The common grasses are buffalo and blue grama grass, and the prickly pear and saltbush also grow here. There are few trees except along the intermittently flowing creeks.

Management authorities recognise the interdependence between grasses and grazing herds – originally bison, now cattle. In summer, 8700 cattle graze the Pawnee. Profits from grazing fees go to the local county authorities to pay for roads and schools. Wnters are severe and only 1500 cattle remain on the range. One aim of the Forest Service managers is to maintain vegetation cover in order to reduce the effect of heavy rainfall which causes surface runoff. This is one reason why the cattle numbers are reduced during the winter season.

A large government range research centre was established 60 years ago, and this looks at many aspects of life on the prairies, including the impact of oil and gas leases. Revenue from 50 wells and 112 km of pipelines brings in the equivalent of £200 000 each year for the Forest Service. There are also a number of military sites in the area. A recent development has been the provision of camping grounds, picnic sites and facilities for birdwatching, which are all part of conservation and environmental protection schemes.

References

Bradshaw, M. (1977) *Earth, The Living Planet*, Hodder & Stoughton.

Brady, N. C. and Weil, R. R. (1999) *The Nature and Properties of Soils*, Prentice Hall.

Briggs, D. (1977) *Soils*, Butterworths.

Courtney, F. and Trudgill, S. T. (1984) *The Soil*, Hodder & Stoughton.

FitzPatrick, E. A. (1980) *Soils, Their Formation, Classification and Distribution*, Longman.

Goudie, A. (1993) *The Nature of the Environment*, Blackwell.

King, T. J. (1989) *Ecology*, 2nd edn, Thomas Nelson.

Money, D. C. (1978) *Climate, Soils and Vegetation*, University Tutorial Press.

Simmons, I. (1982) *Biogeographical Processes*, Allen & Unwin.

White, R. E. (1986) *Introduction to the Principles and Practice of Soil Science*, 2nd edn, Blackwell.

Websites
Forest types of the world:
http://hyperion.advanced.org/17456/typesall.html

Biome resource page (links):
http://www.jlhs.nhusd.k12.ca.us/Classes/Science/Net_Lessons/Biomes/Biomelinks.html

A virtual tour through the tropical rainforest in Suriname:
http://www.euronet.nl/users/mbleeker/suriname/suri-eng.html

A virtual field trip to Indian Peaks, Colorado (altitudinal zonation of vegetation from mixed forest to alpine tundra):
http://www.uwsp.edu/acaddept/geog/projects/virtdept/ipvft/ipvftmod.html

See also for more links:
http://www.nelsonthornes.com/gaia

1 a What is the 'climate' of a place? **(3 marks)**
  b What is the reason for wanting to classify climates? **(3 marks)**
  c Why do many geographers use the natural vegetation of a place as an indication of the climate? **(3 marks)**
  d For any **one** world climatic zone:
    i Name the climatic zone and identify **two** places which experience this climate.
    ii Draw and annotate a graph to show the pattern of temperature and precipitation which is typical of the climatic zone.
    iii Name the typical natural vegetation cover of the climatic zone and a typical zonal soil type. **(10 marks)**
  e Explain the causes of one of the climatic characteristics (temperature or precipitation) you have identified in **d ii**. **(6 marks)**

2 a Describe the climate of the tropical rainforest. **(3 marks)**
  b Draw a diagram to show the composition and structure of the characteristic vegetation of the tropical rainforest. **(6 marks)**
  c Explain **one** way in which the vegetation of the tropical rainforest is adapted to the climate of the area. **(4 marks)**
  d Describe **one** zonal soil type of tropical rainforest areas. **(4 marks)**
  e Why is there not much litter on the forest floor in tropical rainforest areas? **(3 marks)**
  f Identify **one** major threat to the natural vegetation of the tropical rainforest, explain its origin and the reason it can be seen as a threat. **(5 marks)**

3 Choose **one** of the world biomes which you have studied.
  a i Describe the main characteristics of the climate. **(3 marks)**
    ii Describe and explain the nutrient cycle in your chosen biome. You should include a diagram of the mineral nutrient cycle in your answer. **(6 marks)**
  b Describe the zonal soil of your chosen biome. **(6 marks)**
  c How is the natural vegetation of the biome adapted to the climatic conditions there? **(6 marks)**
  d How have farmers altered the natural nutrient cycle in your chosen biome? **(4 marks)**

4 a Describe the climate of the areas which have natural temperate deciduous forests. **(3 marks)**
  b Draw a diagram to show the characteristic structure and composition of the vegetation of temperate deciduous forests. **(6 marks)**
  c Explain **one** way in which the vegetation of the temperate deciduous forests is adapted to the climate of the area. **(4 marks)**
  d Describe **one** zonal soil type of the temperate deciduous forests. **(4 marks)**
  e Why is there litter on the forest floor in the temperate deciduous forests? **(3 marks)**
  f Explain what has happened to most of the world's temperate deciduous forests since the settlement of these areas by people. **(5 marks)**

5 a Choose **one** system of climatic classification you have studied.
    i Name the system and describe the factors it uses to group climates. **(5 marks)**
    ii For one climatic zone within your chosen classification describe the significant features of the climate and the relationship to the classification criteria. **(10 marks)**
  b i Compare the system of climatic classification you chose with one other system of climatic classification.
    ii What are the advantages of your chosen classification over the other system? **(10 marks)**

6 a Describe and account for the climatic pattern experienced in areas with a Mediterranean climate. **(8 marks)**
  b Describe the vegetation and explain **two** ways in which it is adapted to the climatic conditions of the Mediterranean. **(9 marks)**
  c How has the presence of people for a long period affected the relationship between climate, soils and vegetation? **(8 marks)**

7 Choose **one** biome and answer the following questions about it.
  a Describe and explain the relationships within the nutrient cycle of the biome. **(10 marks)**
  b Describe one way the natural vegetation of the area is used by people and the effect of this use on the structure and composition of the vegetation. **(10 marks)**
  c How can damage due to past human uses of the biome be reduced? **(5 marks)**

8 Study the information about tropical grasslands in Kenya in Case Study 12A (pages 335–338) and answer the following questions.
  a i Describe the temperature and rainfall pattern of Nairobi (which is typical of many areas of tropical grassland). **(6 marks)**
    ii Describe the structure and composition of vegetation in tropical grassland areas. **(6 marks)**
  b Describe the ways in which the use of the grassland by people can cause a reduction in the variety and ground cover provided by the grass. **(7 marks)**
  c Using examples from the case study and your own knowledge, suggest how the diversity of the natural environment can be maintained while allowing the people who use it to continue their traditional ways of life. **(6 marks)**

AS

A2

Chapter 13

Population

'There is a real danger that in the year 2000 a large part of the world's population will still be living in poverty. The world may become overpopulated and will certainly be overcrowded.'

Willy Brandt, *North–South: A Programme for Survival,* 1980

'In 1999, 600 million children in the world lived in poverty – 50 million more than in 1990.'

United Nations

In demography – the study of human population – it is important to remember that the situation is dynamic, not static. Population numbers, distributions, structures and movements constantly change in time, in space and at different levels (the micro-, meso- and macro-scales in the population system).

Figure 13.1

World distribution of population

Distribution and density

Population distribution describes the way in which people are spread out across the Earth's surface. The distribution is uneven and there are often considerable changes over periods of time.

Population distributions are often shown by means of a dot map, where each dot represents a given number of people. For example, in Figure 13.1 this method effectively shows the concentration of people in the Nile valley in Egypt, where 99 per cent of the country's population live on 4 per cent of the total land area. However, Figure 13.1 is also misleading because it suggests, incorrectly, that areas away from the Nile are totally uninhabited. In fact, parts are populated, but have insufficient numbers to warrant a symbol. When drawing a dot map, therefore, it is important to select the best possible dot value; and when using one, it is necessary to bear in mind its limitations.

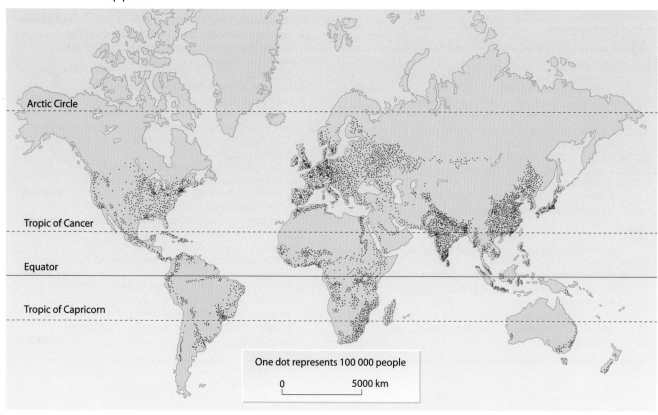

Arctic Circle

Tropic of Cancer

Equator

Tropic of Capricorn

One dot represents 100 000 people

0   5000 km

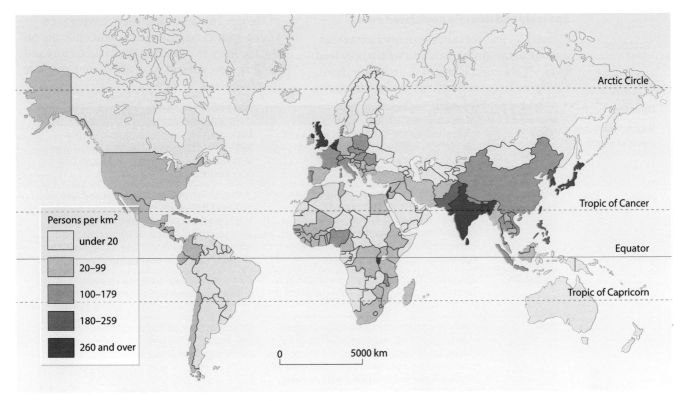

Figure 13.2

World density of
population

Population density describes the number of people living in a given area, usually a square kilometre (km^2). Population densities are often shown by means of a choropleth map, of which Figures 13.2 and 13.5 are examples. Densities are obtained by dividing the total population of a country (or administrative area) by the total area of that country (or area). The densities are then grouped into classes, each of which is coloured lighter or darker to reflect lesser or greater density. Although these maps are easy to read, they hide concentrations of population within each unit area. Figure 13.2, for example, gives the impression that the population of Egypt is equally distributed across the country; it also suggests that there is an abrupt change in population density at the national boundary. A poorly designed system of colouring or shading can make quite small spatial differences seem large – or make huge differences look smaller.

Figures 13.1 and 13.2 both show that there are parts of the world which are sparsely populated and others which are densely populated. One useful generalisation that may be made – remembering the pitfalls of generalisation (Framework 11, page 347) – is that, at the global scale, this distribution is affected mainly by physical opportunities and constraints; whereas, at regional and local scales, it is more likely to be influenced by economic, political and social factors.

Land accounts for about 30 per cent of the Earth's surface (70.9 per cent is water). Of the land area, only about 11 per cent presents no serious limitations to settlement and agriculture (Figure 13.3). Much of the remainder is desert, snow and ice, high or steep-sided mountains, and forest. Usually there are several reasons why an area is sparsely or densely populated.

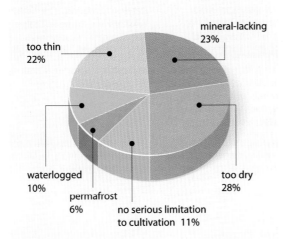

Figure 13.3

The uninhabitable Earth:
how valuable are the
world's soils for food
production?

Sparsely and densely populated areas

Figure 13.4 lists some of the many factors that operate at the global scale and which may lead to an area being sparsely or densely populated. Compare these factors with the patterns shown in Figures 13.1 and 13.2. Then, having read Framework 11 opposite, comment critically on the accuracy and value of the listed factors and suggest, for each factor, an alternative example (or examples).

Figure 13.4

Major factors affecting population density

Factors	Sparsely populated areas	Densely populated areas
Physical	Rugged mountains where temperature and pressure decrease with height; active volcanoes (the Andes); high plateau (Tibet) and worn-down shield lands (the Canadian Shield, Figure 1.9).	Flat, lowland plains are attractive to settlement (the Netherlands and Bangladesh, Places 48 page 377) as are areas surrounding some volcanoes (Mt Pinatubo, Case Study 1 and Mt Etna).
Climate	Areas receiving very low annual rainfall (the Sahara Desert, page 178); areas having a long seasonal drought or unreliable, irregular rainfall (the Sahel countries, page 223); areas suffering high humidity (the Amazon basin, page 316); very cold areas, with a short growing season (northern Canada, page 333).	Areas where the rainfall is reliable and evenly distributed throughout the year; with no temperature extremes and a lengthy growing season (north-west Europe, page 223); where sunshine (the Costa del Sol) or snow (the Alps) is sufficient to attract tourists, (Chapter 20); and areas with a monsoon climate (South-east Asia, page 239).
Vegetation	Areas such as the coniferous forests of northern Eurasia and northern Canada (page 330), and the rainforests of the tropics (page 317).	Areas of grassland tend to have higher population densities than places with dense forest or desert.
Soils	The frozen soils of the Arctic (the permafrost in Siberia, Case Study 5); the thin soils of mountains (Nepal); the leached soils of the tropical rainforest (the Amazon Basin); also, increasingly large areas are experiencing severe soil erosion resulting from deforestation and overgrazing (the Sahel Case Study 7).	Deep, humus-filled soils (the Paris Basin) and, especially, river-deposited silt (the Ganges delta, Places 67 page 481, and Nile delta, Places 73 page 490) both favour farming.
Water supplies	Many areas lack a permanent supply of clean fresh water: mainly due either to insufficient, irregular rainfall or to a lack of money and technology to build reservoirs and wells or lay pipelines (Ethiopia).	Population is more likely to increase in areas with a reliable water supply. This may result from either a reliable, evenly distributed rainfall (northern England) or where there is the wealth and technology to build reservoirs and to provide clean water (California). Places with heavy seasonal rainfall (the monsoon lands of South-east Asia, page 239) also support many people.
Diseases and pests	These may limit the areas in which people can live or may seriously curtail the lives of those who do populate such areas (malaria in central Africa).	Some areas were initially relatively disease- and pest-free; others had the capital and medical expertise to eradicate those which were a problem (the formerly malarial Pontine Marshes, near Rome).
Resources	Areas devoid of minerals and easily obtainable sources of energy rarely attract people or industry (Paraguay).	Areas having, or formerly having, large mineral deposits and/or energy supplies (the Ruhr) often have major concentrations of population; these resources often led to the development of large-scale industry (the Pittsburgh region, USA, Places 105 page 641).
Communications	Areas where it is difficult to construct and maintain transport systems tend to be sparsely populated, e.g. mountains (Bolivia), deserts (the Sahara) and forests (the Amazon Basin and northern Canada).	Areas where it is easier to construct canals, railways, roads and airports have attracted settlements (the North European Plain), as have large natural ports which have been developed for trade (Singapore, Places 97 page 608).
Economic	Areas with less developed, subsistence economies usually need large areas of land to support relatively few people (although this is not applicable to South-east Asia). Such areas tend to fall into three belts: tundra (the Lapps), desert fringes (the Rendille, Places 65 page 479) and tropical rainforests (shifting cultivators, Places 66 page 480).	Regions with intensive farming or industry can support large numbers of people on a small area of land (as in the Netherlands, Places 71 page 487).
Political	Areas where the state fails to invest sufficient money or to encourage development – either economically or socially (the interior of Brazil, Places 38).	Decisions may affect population distribution, e.g. by creating new cities, such as Brasilia; or by opening up 'pioneer' lands for development, as in Israel (page 391).

The study of an environment, whether natural or altered by human activity, involves the study of numerous different and interacting processes. The relative importance of each process may vary according to the scale of the study, i.e. global or **macro-scale**; intermediate or **meso-scale**; and local or **micro-scale**. It may also vary according to the timescale chosen, i.e. whether processes are studied through **geological time**, **historical time**, or **recent time**.

In the study of soils (Chapters 10 and 12), it is clearly climate that tends to impose the greatest influence upon the formation and distribution of the major global (zonal) types (the podsol and chernozem). At the regional level, rock type may be the major influencing factor (Mediterranean areas with their terra rossas and rendzinas). Within a small area, such as a river valley with homogeneous climate and rock type, relief may be dominant (the catena, pages 261 and 276).

In the study of erosion, time is a major variable: a stretch of coastline may be eroded by the sea during a period of several decades or centuries; footpath erosion may occur during a single summer.

A common problem with spatial and time scales, as with models (Framework 12, page 352), is that a chosen level of detail may become inappropriate to all or part of the problem under study: it may become either too large and generalised, or too small and complex. For example, population distributions and densities may be studied at a variety of spatial and time scales. At the world scale (Figures 13.1 and 13.2), the pattern shown is so general and deterministic that it may lead the student into an over-simplified understanding of the processes that produced the apparent distribution and/or density. Such generalised patterns usually break down into something more complex when studied at a more local level or over a period of time.

Although it may often be easier to identify and account for distributions, densities, anomalies and changes at the national level, it is more difficult in the case of a country the size of Brazil (Figure 13.5) than it is for a smaller country such as Uruguay. It is often only when looking at a smaller region (Figure 13.6) or an urban area (Figure 13.7), perhaps over a relatively short time period, that the complexities of the various processes can be readily understood.

Even a quick look at the population density map of Brazil (Figure 13.5) shows a relatively simple, generalised pattern. Over 90 per cent of Brazilians live in a discontinuous strip about 500 km wide, adjacent to the east coast. This strip accounts for less than 25 per cent of the country's total area. The density declines very rapidly towards the north-west, where several remote areas are almost entirely lacking in permanent settlement.

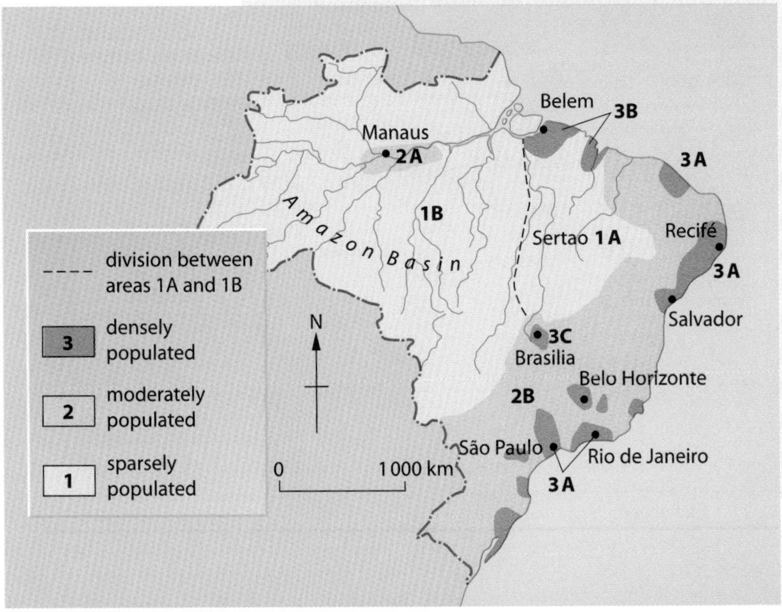

The area marked **1A** on Figure 13.5 is the dry north-east (the Sertao). Here the long and frequent water balance deficit (drought), high temperatures and poor soils combine to make the area unsuitable for growing high-yield crops or rearing good-quality animals. The Sertao also lacks known mineral or energy reserves; communications are poor; and the basic services of health, education, clean water and electricity are lacking. Although birth rates are exceptionally high (many mothers have more than ten babies), there is a rapid outward migration to the urban areas (page 366), a high infant mortality rate (page 354), and a short life expectancy (page 353).

Area **1B** is the tropical rainforest, drained by the River Amazon and its tributaries. Here the climate is hot, wet and humid; rivers flood annually; and there is a high incidence of disease. In the past, the forest has proved difficult to clear, but once the protective trees have gone, soils are rapidly leached and become infertile. Land communications are difficult to build and maintain. The area has suffered, as has **1A**, from a lack of federal investment and can only support subsistence economies.

Figure 13.5

Population density in Brazil: the national scale

0 1 000 km

division between areas 1A and 1B

3 densely populated

2 moderately populated

1 sparsely populated

There are, however, two anomalies in Amazonia. The first is a zone along the River Amazon centred on Manaus (2A on Figure 13.5). Originally a Portuguese trading post, Manaus has had two growth periods. The first was associated with the rubber boom at the turn of the 19th/20th centuries, while the second began in the 1980s with the development of tourism and the granting of its new status as a free port (compare Places 97, page 608). The second anomaly has followed the recent exploit-ation of several minerals (iron ore at Carajas and bauxite at Trombetas) and energy resources (hydro-electricity at Tucuri).

The more easterly parts of the Brazilian Plateau are moderately populated (area 2B). The climate is cooler and it is considerably healthier than on the coast and in the rainforest. The soil, in parts, is a rich terra rossa (page 274) which here is a weathered volcanic soil ideal for the growing of coffee. Several precious minerals have been found. However, rainfall is irregular with a long winter drought; communications are still limited; and federal investment has been insufficient to stimulate much population growth.

Except where the highland reaches the sea, the eastern parts of the plateau around São Paulo and Belo Horizonte and the east coast have the highest population densities (area 3A). Although the coastal area is often hot and humid, the water supply is good. Several natural harbours proved ideal for ports and this encouraged trade and the growth of industry. Salvador, the first capital, was the centre of the slave trade. Rio de Janeiro became the second capital, developing as an economic, cultural and administrative centre. More recently, it has received increasing numbers of tourists from overseas and migrants from the north of Brazil.

One of the fastest-growing cities in the world is São Paulo. The cooler climate and terra rossa soils initially led to the growth of commercial farming based on coffee. Access to minerals such as iron ore and to energy supplies later made it a major industrial centre. The São Paulo region has had high levels of federal investment, leading to the development of a good communications network and the provision of modern services.

Area 3B is a recent growth pole (page 569) based on the discovery and exploitation of vast deposits of iron ore and bauxite, the construction of hydro-electric power stations and the advantages of access along the coastal strip and Amazon corridor. 3C is the new federal capital, Brasilia, built in the early 1960s to try to redress the imbalance in population density and wealth between the south-east of the country and the interior.

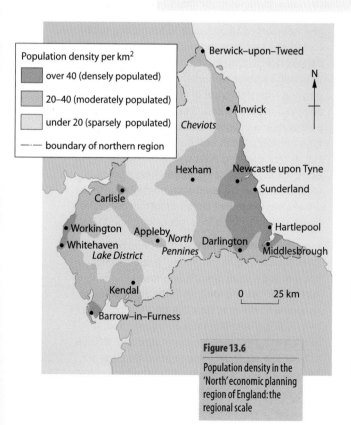

Figure 13.6

Population density in the 'North' economic planning region of England: the regional scale

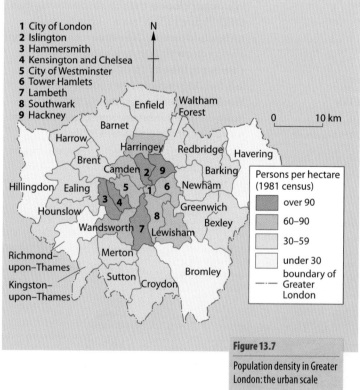

Figure 13.7

Population density in Greater London: the urban scale

Lorenz curves

Lorenz curves

Lorenz curves are used to show inequalities in distributions. Population, industry and land use are three topics of interest to the geographer which show unequal distributions over a given area. Figure 13.8 illustrates the unevenness of population distribution over the world. The dia-gonal line represents a perfectly even distribution, while the concave curve (it may be convex in other examples) illustrates the degree of concentration of population within the various continents. The greater the concavity of the slope, the greater the inequality of population distribution (or industry, land use, etc.).

Figure 13.8

A Lorenz curve: the distribution of world population in 1998

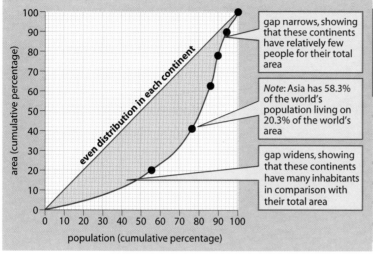

gap narrows, showing that these continents have relatively few people for their total area

Note: Asia has 58.3% of the world's population living on 20.3% of the world's area

gap widens, showing that these continents have many inhabitants in comparison with their total area

even distribution in each continent

area (cumulative percentage)

population (cumulative percentage)

Continents ranked in descending order of population (1998)	Population (%)	Population (cumulative %)	Area (%)	Area (cumulative %)
Asia	58.3	58.3	20.3	20.3
Europe and Russian Federation	14.8	73.1	20.1	40.4
Africa	12.7	85.8	22.3	62.7
Latin America	8.5	94.3	15.2	77.9
North America	5.2	99.5	15.8	93.7
Oceania	0.5	100.0	6.3	100.0

Population changes in time

It has already been stated (page 344) that populations are dynamic, i.e. their numbers, distributions, structure and movement (migration) constantly change over time and space. Population change is another example of an open system (Framework 3, page 45) with inputs, processes and outputs (Figure 13.9).

Birth rates, death rates and natural increase

The total population of an area is the balance between two forces of change: **natural increase** and **migration** (Figure 13.9). The natural increase is the difference between birth rates and death rates. The **crude birth rate** is the number of live births per 1000 people per year and the **crude death rate** is the number of deaths per 1000 people per year. Throughout history, until the last few years in a small number of the economically most developed countries, birth rates have nearly always exceeded death rates. Exceptions have followed major outbreaks of disease (the bubonic plague and AIDS, page 645) or wars (as in Rwanda). Any natural change in the population, either an increase or a decrease, is usually expressed as a percentage and referred to as the **annual growth rate**. Population change is also affected by migration. Although migration does not affect world population totals, it does affect the way people are distributed across the world. Migration leads to *either* an increase in the population – when the number of immigrants exceeds the number of emigrants (as in Hong Kong and Congo) – *or* a decrease in population – when the number of emigrants exceeds the number of immigrants (as in the former Yugoslavia and Rwanda).

Figure 13.9

Simplistic model showing population change as an open system

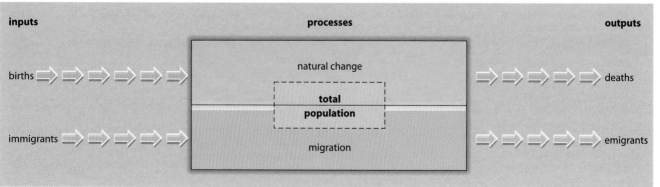

inputs

processes

outputs

births

natural change

total population

immigrants

migration

deaths

emigrants

The demographic transition model

The demographic transition model describes a sequence of changes over a period of time in the relationship between birth and death rates and overall population change. The model, based on population changes in several industrialised countries in western Europe and North America, suggests that *all* countries pass through similar demographic transition stages or **population cycles** – or will do, given time. Figure 13.10 illustrates the model and gives reasons for the changes at each transition stage. It also gives examples of countries that appear to 'fit' the descriptions of each stage.

Figure 13.10

The demographic transition model

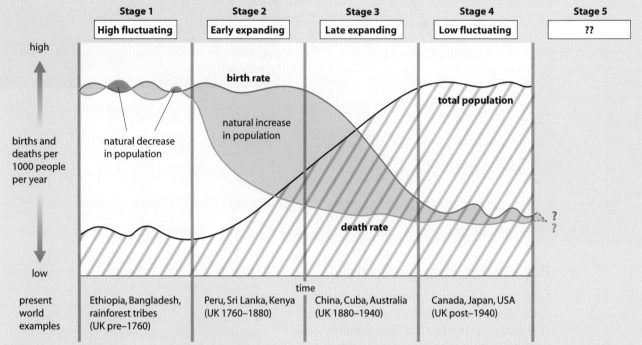

Stage 1	Stage 2	Stage 3	Stage 4	Stage 5
High fluctuating	Early expanding	Late expanding	Low fluctuating	??
Ethiopia, Bangladesh, rainforest tribes (UK pre–1760)	Peru, Sri Lanka, Kenya (UK 1760–1880)	China, Cuba, Australia (UK 1880–1940)	Canada, Japan, USA (UK post–1940)	

present world examples

Stage 1: Here both birth rates and death rates fluctuate at a high level (about 35 per 1000) giving a small population growth.

Birth rates are high because:

- no birth control or family planning
- so many children die in infancy that parents tend to produce more in the hope that several will survive
- many children are needed to work on the land
- children are regarded as a sign of virility
- Some religious beliefs (Roman Catholics, Muslims and Hindus) encourage large families.

High death rates, especially among children, are due to:

- disease and plague (bubonic, cholera, kwashiorkor)
- famine, uncertain food supplies, poor diet
- poor hygiene: no piped, clean water and no sewage disposal
- little medical science: few doctors, hospitals, drugs.

Stage 2: Birth rates remain high, but death rates fall rapidly to about 20 per 1000 people giving a rapid population growth.

The fall in death rates results from:

- improved medical care: vaccinations, hospitals, doctors, new drugs and scientific inventions
- improved sanitation and water supply
- improvements in food production, both quality and quantity
- improved transport to move food, doctors, etc
- a decrease in child mortality.

Stage 3: Birth rates now fall rapidly, to perhaps 20 per 1000 people, while death rates continue to fall slightly (15 per 1000 people) to give a slowly increasing population.

The fall in birth rates may be due to:

- family planning: contraceptives, sterilisation, abortion and government incentives
- a lower infant mortality rate leading to less pressure to have so many children
- increased industrialisation and mechanisation meaning fewer labourers are needed
- increased desire for material possessions (cars, holidays, bigger homes) and less desire for large families
- an increased incentive for smaller families
- emancipation of women, enabling them to follow their own careers rather than being solely child–bearers.

Stage 4: Both birth rates (16 per 1000) and death rates (12 per 1000) remain low, fluctuating slightly to give a steady population.

(Will there be a **Stage 5** where birth rates fall below death rates to give a declining population?

Some evidence suggests that this might be occurring in several western European countries.)

Like all models, the demographic transition model has its limitations (Framework 12, page 352). It failed to consider, or to predict, several factors and events:

1 Birth rates in several of the most economically developed countries have, since the model was put forward, fallen below death rates (Germany, Sweden). This has caused, for the first time, a population decline which suggests that perhaps the model should have a fifth stage added to it.

2 The model, being more or less Eurocentric, assumed that in time all countries would pass through the same four stages. It now seems unlikely, however, that many of the economically less developed countries, especially in Africa, will ever become industrialised.

3 The model assumed that the fall in the death rate in Stage 2 was the consequence of industrialisation. Initially, the death rate in many British cities rose, due to the insanitary conditions which resulted from rapid urban growth, and it only began to fall after advances were made in medicine. The delayed fall in the death rate in many developing countries has been due mainly to their inability to afford medical facilities. In many countries, the fall in the birth rate in Stage 3 has been *less* rapid than the model suggests due to religious and/or political opposition to birth control (Brazil), whereas the fall was much *more* rapid, and came earlier, in China following the government-introduced 'one-child' policy (Case Study 13).

4 The timescale of the model, especially in several South-east Asian countries such as Hong Kong and Malaysia, is being squashed as they develop at a much faster rate than did the early industrialised countries.

The model can be used:

■ to show how the population growth of a country changes over a period of time (the UK, in Figure 13.11)

■ to compare rates of growth between different countries at a given point in time (Figure 13.12).

Figure 13.11 shows that Britain had:

■ a high birth rate and a high death rate between 1700 and 1760, which gave a slow natural increase in population (Stage A)

■ a rapidly falling death rate and a still high birth rate which resulted in a rapid natural increase in population between 1760 and 1880 (stage B)

■ a rapidly falling birth rate and a declining death rate between 1880 and 1940, which led to a slower natural increase (stage C)

■ low, fluctuating birth and death rates since 1940, which has caused only a small natural increase (stage D).

Figure 13.11

The demographic transition cycle: changes in Britain's population, 1700–1998

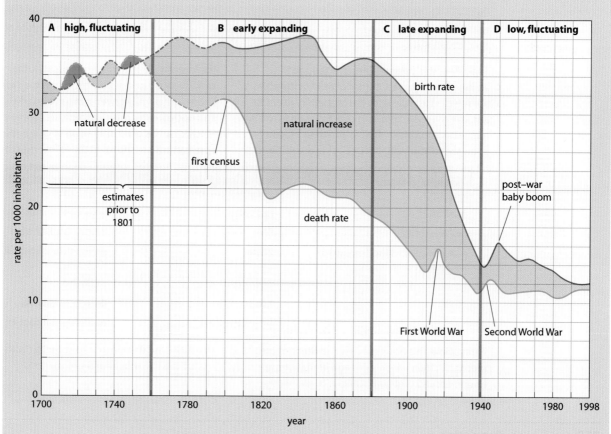

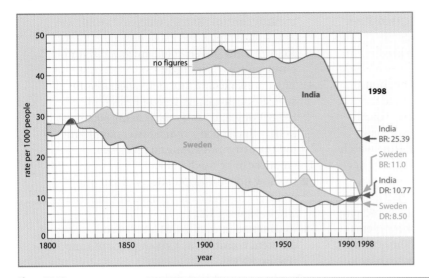

Figure 13.12 shows how Sweden has long since reached Stage 4 of the demographic transition model – a characteristic of most economically more developed countries – whereas India is still at Stage 3 – a characteristic of many economically less developed countries (remember that some of the least economically developed countries are still at Stage 2).

Figure 13.12

A comparison between the demographic cycles of Sweden and India, 1800–1998

Framework 12   Models

Models form an integral and accepted part of present-day geographical thinking and teaching. Nature is highly complex and, in an attempt to understand this complexity, geographers try to develop simplified models of it.

Chorley and Haggett described a model as:

> a simplified structuring of reality which presents supposedly significant features or relationships in a generalised form … as such they are valuable in obscuring incidental detail and in allowing fundamental aspects of reality to appear.

They stated that a model:

> can be a theory or a law, an hypothesis or structured idea, a role, a relationship, or equation, a synthesis of data, a word, a graph, or some other type of hardware arranged for experimental purposes.

A good model will stand up to being tested in the real world and should fall between two extremes:

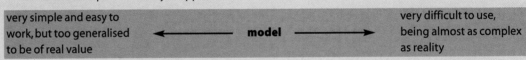

| very simple and easy to work, but too generalised to be of real value | ← model → | very difficult to use, being almost as complex as reality |

To achieve this balance, several – though sometimes only one – critical criteria or variables are selected as a basis for the model. For example, J. H. von Thünen (page 471) chose distance from a market as his critical variable and then tried to show the relationship between this variable and the intensity of land use. If necessary, other variables may be added which, as in the case of von Thünen's navigable river and a rival market, may add both greater reality and greater complexity. Models can be used in all fields of geography. Some applications are shown in the following table.

Physical (landforms)	Climate, soils and vegetation	Human and economic
beach profile	atmospheric circulation	cities (Burgess)
slope form	heat budget	land use (von Thünen)
corrie development	seres	industrial location (Weber)
geomorphological systems (Framework 3), drainage basin, glacier budget	catena	settlement size and distribution (Christaller)
	depression	gravity models
	biome	demographic transition
	soil profiles (podsols)	economic growth (Rostow)

Throughout this book, models and theories are presented; their advantages and limitations are examined; and their applications to real-world situations are demonstrated, together with their usefulness in explaining that situation.

Population structure

The rate of natural increase or decrease, resulting from the difference between the birth and death rates of a country, represents only one aspect of the study of population structure. A second important aspect is population. This is important because the make-up of the population by its age and sex, together with its life expectancy, has implications for the future growth, economic development and social policy of a country. **Life expectancy** is the number of years that the average person born in a given area may expect to live. Differences in language, race, religion, family size, etc. can all affect a country's socio-economic welfare.

Population pyramids

The population structure of a country is best illustrated by a **population** or **age–sex pyramid**. The technique normally divides the population into 5-year age groups (e.g. 0–4, 5–9, 10–14) on the vertical scale, and into males and females on the horizontal scale (Figure 13.13). The number in each age group is given as a percentage of the total population and is shown by horizontal bars, with males located to the left and females to the right of the central axis. As well as showing past changes, the pyramid can predict both short-term and long-term future changes in population.

Whereas the demographic transition model shows only the natural increase or decrease resulting from the balance between births and deaths, the population pyramid shows the effects of migration, the age and sex of migrants (Figure 13.45) and the effects of large-scale wars and major epidemics of disease. Figure 13.13 is the population pyramid for the United Kingdom in mid-1999. Notice the following:

- a narrow pyramid showing approximately equal numbers in each age group
- a low birth rate (meaning fewer school places will be needed) and a low death rate (suggesting a need for more elderly people's homes) which together indicate a steady, almost static, population growth
- the greater number of boys in the younger age groups (a higher birth rate) but more females than males in the older age groups (women having the longer life expectancy)
- a relatively large proportion of the population in the pre- and post-reproductive age groups, and a relatively small number in the 15–64 age groups which produce most of the national wealth (see dependency ratios, page 354).

Figure 13.13

Constructing the population pyramid for the UK, mid-1999

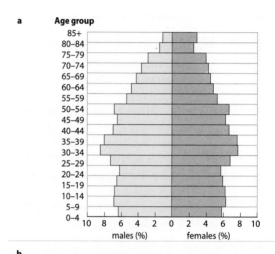

b

Age group	Males Number (000s)	Males Percentage	Females Number (000s)	Females Percentage
0–4	1845	6.35	1753	5.82
5–9	1977	6.88	1897	6.30
10–14	1980	6.81	1872	6.22
15–19	1900	6.54	1797	5.94
20–24	1804	6.21	1716	5.71
25–29	2118	7.29	2045	6.80
30–34	2374	8.18	2318	7.71
35–39	2350	8.02	2292	7.62
40–44	2030	6.99	2006	6.67
45–49	1883	6.48	1884	6.26
50–54	1993	6.86	2005	6.67
55–59	1552	5.35	1589	5.28
60–64	1389	4.78	1452	4.82
65–69	1218	4.19	1349	4.49
70–74	1042	3.59	1275	4.24
75–79	823	2.83	1195	3.97
80–84	420	1.47	761	2.53
85+	318	1.09	870	2.89
Total	29 038		Total 30 076	

Figure 13.14

Population pyramids characteristic of each stage of the demographic transition model

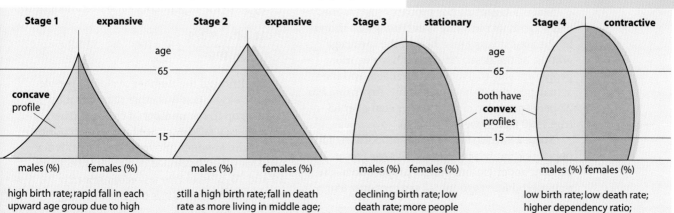

| Stage 1 | expansive | Stage 2 | expansive | Stage 3 | stationary | Stage 4 | contractive |

high birth rate; rapid fall in each upward age group due to high death rates; short life expectancy

still a high birth rate; fall in death rate as more living in middle age; slightly longer life expectancy

declining birth rate; low death rate; more people living to an older age

low birth rate; low death rate; higher dependency ratio; longer life expectancy

economically least developed countries ⟶ economically more developed countries

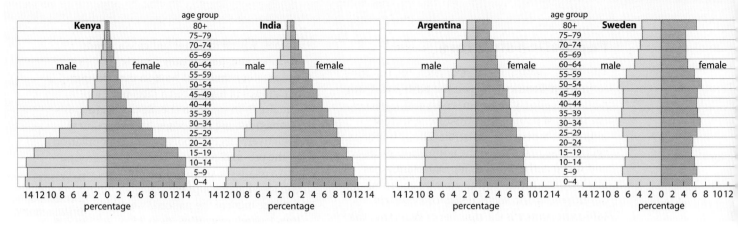

Figure 13.15

Population pyramids for four selected countries, mid-1999

A model has also been produced to try to show the characteristics of four basic types of pyramid (Figure 13.14). As with most models, many countries show a transitional shape which does not fit precisely into any pattern. Figure 13.15 shows the pyramids for *selected* countries – chosen because they *do* conform closely to the model!

Stage 1 Kenya's pyramid has a concave shape, showing that the birth rate is very high. Almost half the inhabitants (43 per cent) are under 15 years old (the corresponding figure for 1990 was 51 per cent); there is a rapid fall upwards in each age group showing a high death rate (including infant mortality) and a low life expectancy, with less than 3 per cent who can expect to live beyond 65. The **infant mortality rate** is the average number of children out of every 1000 born alive who die under the age of one year.

Stage 2 India appears to have reached Stage 2 (shown by the more uniform sides). All pyramids in this stage have a broad base indicating a high birth rate but, as the infant mortality and death rates decline, more people reach middle age and the life expectancy is slightly longer. The result is that although the actual numbers of children may be the same, they form a smaller percentage of the total population (as shown by the narrower base). The large youthful population will soon enter the reproductive period and become economically active. India has 34 per cent under 15; and 4.5 per cent over 65 (the corresponding figures for 1990 were 39 and 3 per cent respectively).

Stage 3 Argentina has probably just reached this stage as its birth rate is declining – as shown by the almost equal numbers in the lower age groups. As the death rate is much lower, more people are able to live to a greater age, and the actual growth rate becomes stable. Argentina has 27 per cent under 15; and 10 per cent over 65 (the 1990 figures were 26 and 6).

Stage 4 Sweden has a smaller proportion of its population in the pre-reproductive age groups (18 per cent under 15) and a larger proportion in the post-reproductive groups (17 per cent over 65),

indicating low birth, infant mort-ality and death rates and a long life expectancy (the equivalent figures for 1990 were 22 and 16). As the numbers entering the reproductive age groups decline there will be, in time, a fall in the total population.

Dependency ratios

The population of a country can be divided into two categories according to their contribution to economic productivity. Those aged 15–65 years are known as the **economically active** or **working population**; those under 15 (the youth dependency ratio) and over 65 (the old age dependency ratio) are known as the **non-economically active population**. (Perhaps in Britain the division should be made at 16, the school-leaving age; in developing countries, however, the cut-off point is much lower as many children have to earn money from a very young age.)

The dependency ratio can be expressed as:

$$\frac{\text{children (0–14)}\ \text{and elderly (65 and over)}}{\text{those of working age}} \times 100$$

e.g. UK 1971 (figures in millions):

$$\frac{13\,387\ +\ 7\,307}{31\,616} \times 100 = 65.45$$

So for every 100 people of working age there were 65.45 people dependent upon them.

By 1999 the dependency ratio had changed to:

$$\frac{11\,345 + 9\,271}{38\,498} \times 100 = 53.55$$

So although the number of elderly people had increased, this was more than offset by the larger drop in the number of children (the dependency ratio does not take into account those who are unemployed). The dependency ratio for most developed countries is between 50 and 70, whereas for less economically developed countries it is often over 100.

Trends in population growth

1 Global trends

In Mother Earth's 46 years (Places 1, page 9), it was only 'in the middle of last week, in Africa, that some manlike apes turned into ape-like men' and the world's human population slowly began to grow. In the absence of any census, this population is estimated to have been about 500 million by 1650. It was only after the Industrial Revolution in western Europe that numbers 'began to multiply prodigiously'. The most rapid increase has taken place over the last 70 years (Figure 13.18).

The United Nations Fund for Population Activities (UNFPA) designated 11 July 1987 as the date of the arrival of the five billionth (5000 millionth) human being on Earth. Of course that 'celebration' was fictitious as nobody knows exactly how many people are living on the Earth at a given moment since, in many areas, census figures are either inaccurate or non-existent. However, although that figure is approximate, it was presumed that in 1999 the world's population was still growing by about 140 people every minute and 78 million each year.

Recent evidence has shown that fertility rates in many economically developing countries have, at last, begun to fall (Figure 13.16). The **fertility rate** is the number of children born to women of child-bearing age. The Population Division of the United Nations claimed that by 1999 the annual growth rate of the world's population, which had been 2.1 per cent between 1964 and 1970, had fallen to 1.3 per cent – a fall mainly credited to China's one child per family policy (Case Study 13). The consequence has been that the world's population was only expected to pass the 6000 million mark in the year 2000 instead of the 6300 million predicted in 1992 and the 7600 million it would have reached had the growth rate of 1950–80 continued. The UN also predicted that the annual growth rate was likely to fall to 1.0 per cent in 2015 and to 0.5 per cent by 2050. By the mid-21st century, world population will be in the range of 7300 million (low estimate) to 10 700 million (high estimate), with a medium variant suggesting 8900 million (Figures 13.17 and 13.18).

2 Regional trends

What these figures fail to show is the marked variation between different areas in the world and especially between the economically developed and economically developing continents. At present, the average population growth rate for developed countries is 0.58 per cent per year compared with 1.92 per cent in those described as developing (Figures 13.17 and 13.18). According to the UN, 97 per cent of the world's population increase is taking place in the less developed regions. Each year the population of Asia is increasing by 50 million, Africa by 17 million and Latin America by nearly 8 million. Africa has the highest annual growth rate with 2.36 per cent compared with Europe whose 0.03 per cent is the lowest. By 2050, one-third of the world's population is likely to live in either India (the highest) or China.

Figure 13.16

Global and regional trends in population growth

In global terms the major trend has been a decline in the rate of population growth from a peak of 2.1% p.a. in 1965–70 to approximately 1.7% in 1992 (World Bank). There continue to be still more persons (over 90 millions p.a.) being added in total to the population than ever before in human history. Whilst the distinction between low (or no) population growth in 'developed' countries and high population growth in 'less developed' countries continues, a major feature of the last four decades has been the wide divergence of the demographic trajectories of the 'less developed' countries. Whilst the detailed pattern is complex, in general terms there has been rapid fertility decline in East Asia, moderate decline in parts of Latin America and to a lesser extent South Asia and very limited decline in Sub-Saharan Africa. However, there have been recent encouraging indications that a number of African countries are moving towards fertility decline. Even those countries which have recently undergone rapid fertility decline continue to grow due to population momentum. Although there is always the possibility of stalling in the process of fertility decline, the main question is not whether, but when and at what total, global population will stabilise. The most authoritative estimates point towards global population rising from 5.5 billions in 1993, to 6.5 billions by 2010 possibly stabilising at 10.1 billions by the middle of the twenty-second century (World Bank 1992) even if population growth follows the rapid fertility decline scenario. Population projections beyond the near future need to be taken into consideration with especial caution.

(N. Ford, at the International Conference on Population and Development, Cairo, September 1994)

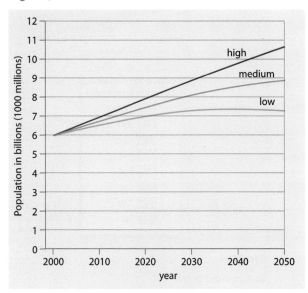

Figure 13.17

Predictions on world population growth

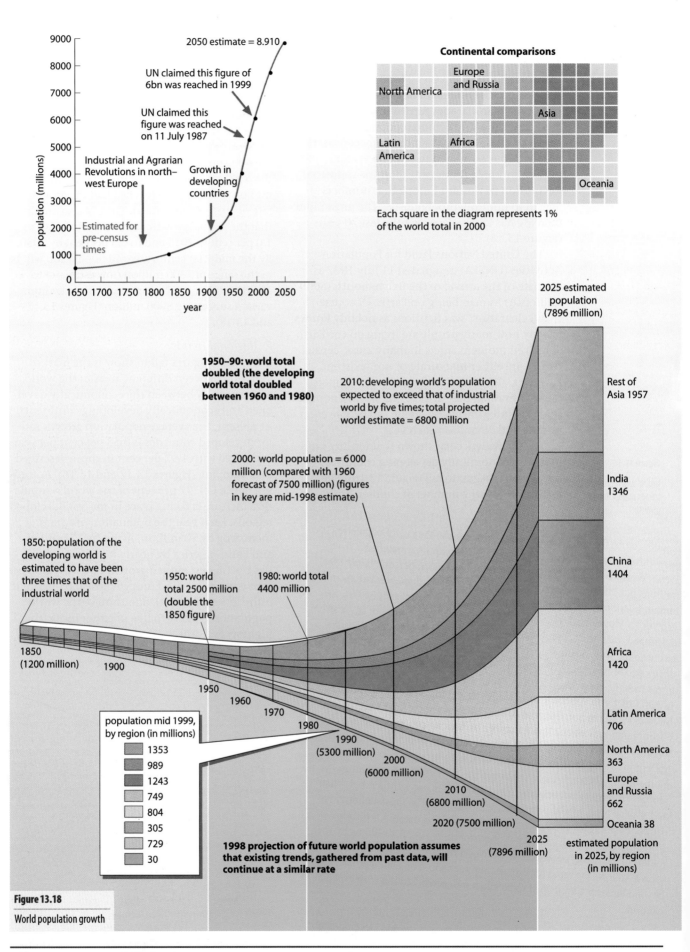

Continental comparisons

Each square in the diagram represents 1% of the world total in 2000

population (millions)

2050 estimate = 8.910

UN claimed this figure of 6bn was reached in 1999

UN claimed this figure was reached on 11 July 1987

Industrial and Agrarian Revolutions in north–west Europe

Growth in developing countries

Estimated for pre-census times

year

North America

Europe and Russia

Asia

Latin America

Africa

Oceania

1950–90: world total doubled (the developing world total doubled between 1960 and 1980)

2010: developing world's population expected to exceed that of industrial world by five times; total projected world estimate = 6800 million

2000: world population = 6 000 million (compared with 1960 forecast of 7500 million) (figures in key are mid-1998 estimate)

1850: population of the developing world is estimated to have been three times that of the industrial world

1950: world total 2500 million (double the 1850 figure)

1980: world total 4400 million

1850 (1200 million)

1900

1950

1960

1970

1980

1990 (5300 million)

2000 (6000 million)

2010 (6800 million)

2020 (7500 million)

2025 (7896 million)

2025 estimated population (7896 million)

Rest of Asia 1957

India 1346

China 1404

Africa 1420

Latin America 706

North America 363

Europe and Russia 662

Oceania 38

estimated population in 2025, by region (in millions)

population mid 1999, by region (in millions)

1353
989
1243
749
804
305
729
30

1998 projection of future world population assumes that existing trends, gathered from past data, will continue at a similar rate

Figure 13.18

World population growth

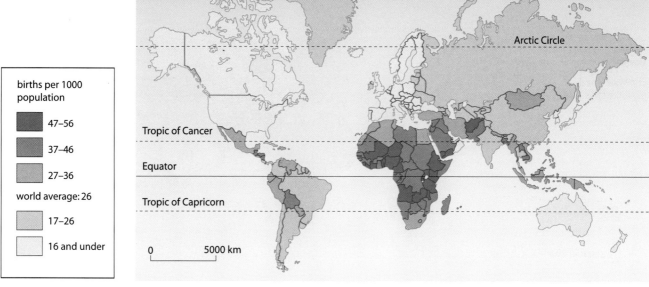

births per 1000 population

- ■ 47–56
- ■ 37–46
- ■ 27–36

world average: 26

- ■ 17–26
- □ 16 and under

0   5000 km

Arctic Circle

Tropic of Cancer

Equator

Tropic of Capricorn

Figure 13.19

World crude birth rates, 1997

3 Birth rates, total fertility rates (TFR) and replacement rates

The world's crude birth rate in 1997 was 24 per 1000. Estonia, Russia and Italy had the lowest with 9 per 1000 while, in contrast, the highest rates were recorded in many African Sahel countries (Figure 7.28) and in several Gulf States in the Middle East where over 50 per 1000 was recorded (Figure 13.19).

The **total fertility rate (TFR)** is the number of children a woman is expected to have during her lifetime, based on current birth rates. The present world average is 3.1. The TFR is one of the best indicators of future population growth. In most economically developed countries the TFR is low and still declining and, while it is still much higher in economically less developed countries, it appears that changing attitudes there will eventually lead to lower TFR in the future. High birth and fertility rates have been considered characteristic of 'underdevelopment'. Indeed although there does seem to be a close correlation between a country's birth rate and its GNP (Framework 19, page 635), the UN have claimed that 'a high birth rate is a consequence, not a cause, of poverty'. Typically, the lower the use of contraceptives, the higher the TFR, and the higher the level of female education, the lower the TFR (Figure 13.20). Government policies also have an enormous impact on the number of children women are likely to have (Places 39).

It is now generally recognised that the three key factors influencing fertility decline are improvements in family planning programmes, in health care, and in women's education and status (arguably in that order, although they are all interrelated). The major world movement is now towards 'children by choice rather than chance', a goal that can only be achieved by giving women reproductive options.

The causes of this unmet need for contraception include lack of knowledge of contraception methods and/or sources of supply; limited access to and low quality of family planning services; lack of education, especially among women; cost of contraception commodities; disapproval of husbands and family members; and opposition by religious groups. Improvements in health care include safer abortions and a reduction in infant mortality – the latter meaning that fewer children need to be born as more of them survive. Improved education raises the status of women and postpones the age of marriage. A survey of younger mothers in Brazil showed that those with secondary education had, on average, 2.5 children whereas those without such education had 6.5. Several governments, especially in South-east Asian countries, have attempted in recent years to 'encourage' couples to have fewer children. Places 39 summarises the degree of success and the problems created by the Singaporean government's attempt to reach zero birth rate while, at the same time, encouraging higher birth rates among selected groups. Some governments, notably China (Case Study 13) have sought to strengthen the effectiveness of their population policies by a range of 'beyond family planning measures'. There is an international consensus which is generally critical of such strategies where they involve coercion.

> ... The most direct way to bring about significant fertility declines is by implementing comprehensive and high-quality family planning programmes. Although past efforts have been substantial, there are still many countries where services are poor or even non-existent. Recent surveys in developing countries reveal that many couples who wish to delay or stop childbearing are not using contraception. About 100 million women – one in six married women in the developing countries (outside China) – have an unmet need for contraception. This unmet need is highest in countries of sub-Saharan Africa – averaging near 25 per cent – but even in Asia and Latin America, where services are more accessible, unmet need levels of around 15 per cent are typical. As a consequence many women bear more children than they want. Approximately one in four births in the developing world (again excluding China) is unwanted, and there are approximately 25 million abortions each year, many of them unsafe. Both unintended pregnancies and unsafe abortions can be prevented if woman are given greater control over their sexual and reproductive lives.
>
> **The Population Council, based in New York, 1994**

Figure 13.20

How may a decrease in fertility be achieved?

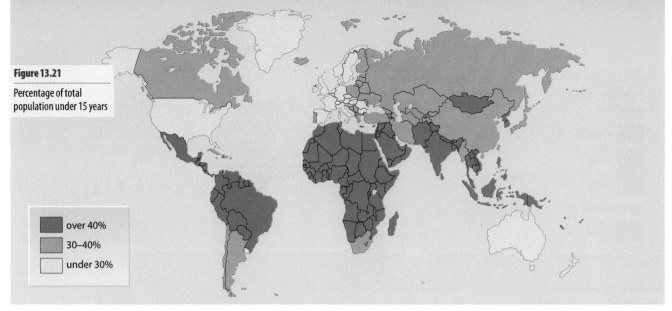

Figure 13.21

Percentage of total population under 15 years

over 40%

30–40%

under 30%

High birth and fertility rates result in a high proportion of the total population being aged 15 or under (see Kenya's population structure in Figure 13.15). Countries where this occurs – many of them in Africa, Latin America and southern Asia (Figure 13.21) – are likely, in the future, to:

■ need greater health care and education – two services many can ill afford

■ have more women reaching child-bearing age. In contrast, many of the more economically developed countries have such low birth and fertility rates that there is a growing problem of 'too few' rather than 'too many' children: is this the possible Stage 5 in the demographic transition model (Figure 13.10)?

The UN stated that, by 1998, 61 countries had a TFR below 2.1 – the figure needed by a country to replace its population (the **replacement rate**, said to be 2.1 children per woman, is when there are just sufficient children born to balance the number of people who die). Throughout history, the replacement rate has always been exceeded (except in times of plague or war) – hence the continual growth in world population. Some countries, including Spain and Italy (TFR 1.2), Germany and China (TFR 1.3) and the UK (TFR 1.7), where the replacement rate is not being met, fear that they will, in time, have too few consumers and skilled workers to keep their economy going; they will see a reduction (especially in the most developed countries) in their competitive advantage in science and technology; will have villages bereft of children and schools closed for lack of students; and are likely to experience a lack of security in providing pensions for an ageing population.

Places 39   Singapore: family planning

The government introduced a massive family planning scheme in the late 1960s. The main objectives of this were:

• to establish family planning clinics and to provide contraceptives at minimal charge

• to advertise through the media the need for and advantages of smaller families – a 'stop at two' policy

• to legislate so that under certain circumstances both abortions and sterilisation could be allowed

• to introduce social and economic incentives such as paid maternity leave, income tax relief, housing priority, cheaper health treatment and free education which would cease as the size of the family grew.

By the early 1990s, this policy had been so successful that the country had an insufficient supply of labour to fill the growing number of job vacancies and fewer young people to support an increasingly ageing population – hence the changed family planning slogan of 'stop at three if you can afford three'. The government were worried that the middle-class elite were having the fewest children. A Social Development Unit was set up to encourage graduates to meet on 'blind dates', hoping that the result would be marriage. Female graduates are encouraged to have three or more children through financial benefits and large tax exemptions. Even so, the average female graduate has only one child so that she can pursue a career and live in a more expensive house/area. Low-income non-graduates only receive housing benefits if they stop at two children. In 1998, Singapore's TFR was 1.9.

Figure 13.23

Percentage of total population aged 60 and over in 1999

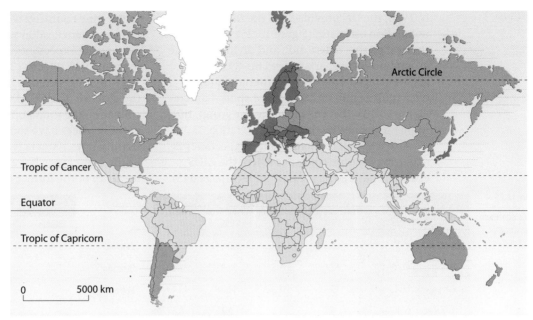

percentage aged 60 and over	
	20 – 24
	10 – 19
	0 – 9
	No data

0 ——— 5000 km

	1970		1998		2025 (estimate)	
	Male	**Female**	**Male**	**Female**	**Male**	**Female**
Japan	71	76	77	83	82	85
Italy	69	75	75	82	79	84
UK	69	75	74	79	76	82
USA	68	75	73	79	76	82
China	63	64	69	72	74	78
India	51	49	60	61	66	68
Bangladesh	46	44	56	56	61	63
Kenya	49	53	54	55	58	59
Cambodia	39	42	49	52	53	56

Figure 13.22

Life expectancies in selected countries

Figure 13.24

Growth in the percentage of population aged 65 and over in selected countries, 1950–2020

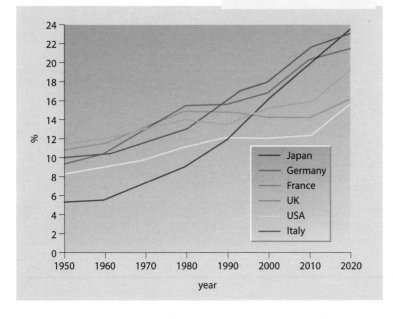

4 Death rates and life expectancy

Death rates, whether it be infant mortality or among children and adults, have, traditionally, declined as a country develops (i.e. Stage 2 onwards in the demographic transition model – Figure 13.10). Due to improvements in medical facilities, hygiene and the increased use of vaccines, the decline in the death rate has led to a sharp increase in life expectancy, initially in the economically more developed countries but, more recently, also in many of the economically less developed countries (Figure 13.22). Already, several of the more developed countries have over 16 per cent of their population aged over 65 (Figures 13.23 and 13.24) and several others have, for the first time in history, more people aged over 65 than they have children aged under 15 (Places 40). In Europe – the major area to be affected by ageing until overtaken recently by eastern Asia – the proportion of children is projected to decline from 18 per cent in 1998 to 14 per cent by 2050, while the proportion of people aged over 65 is expected to rise from 20 per cent to 35 per cent in the same period (the most rapid increases being in Spain and Italy). By 2025 the UN predict that there will be more older people in the world than there will be children. This trend is likely to mean:

■ a greater demand for services (e.g. pensions, medical care and residential homes) which will have to be provided (i.e. paid for) by a smaller percentage of people of working age (i.e. in the economically active age group) in the more developed countries, and

■ a rapid increase in population size with an associated strain on the often already overstretched resources of the less developed countries.

In 1998, the UN provided, for the first time, population estimates (Figure 13.25) for what they called the 'oldest-old', and divided this age group into octogenarian (aged 80–89), nonagenarian (aged 90–99) and centenarian (aged over 100). Figure 13.25 does not show – unlike Figure 13.13 – that the older the age group, the higher its femineity (female:male) ratio.

An exception to increased life expectancy is occurring in those countries where the AIDS epidemic has had its greatest impact (page 645).

In the nine countries with the highest adult HIV prevalence, all in sub-Saharan Africa, the average life expectancy is predicted to be 48 years in 2000 (Figure 13.26) whereas it might have been expected to have reached 58 years in the absence of AIDS. (Even in the worst affected countries, the total population is unlikely to fall as these places have some of the world's highest fertility rates.) Russia is an other anomaly in that its life expectancy has also decreased since the break-up of the former USSR.

Figure 13.25

Age composition of the world's 'oldest-old'

| Age group | Millions | | % | |
	1998	2050	1998	2050
Oldest-old: 80+	66.0	370.4	100	100
Octogenarian: 80–89	58.6	311.3	88.8	84.0
Nonagenarian: 90–99	7.3	56.9	11.0	15.4
Centenarian: 100+	0.1	2.2	0.2	0.6

UN Population Division

| | 1970 | | 1998 | |
	Males	Females	Males	Females
Zambia	46	49	36	36
Uganda	45	49	39	40
Zimbabwe	50	53	36	36

Figure 13.26

The effect of AIDS on life expectancy in selected countries

Places 40   Japan: an ageing population

Japan is developing an ominously top-heavy demographic profile (Figure 13.27). Japanese women, who are marrying and having babies at an increasingly later age, gave birth to 1.38 children on average in 1998 (one of the lowest TFRs in the world) compared with more than 5 in 1928 and 1.66 in 1988, according to the Ministry of Health and Welfare. In contrast, Japan has the world's greatest longevity (Figure 13.22). The Ministry claimed that by September 1999 there would be 10 158 Japanese, 83 per cent of them women, who would be centenarians – 1667 more (over 16 per cent) than a year before. The numbers worry government planners. By 2025, nearly 26 per cent of the Japanese population will be over the age of 65, compared with just 12 per cent in 1990. The potential to the Japanese economy, in terms of care for the elderly and smaller tax revenues, has convinced many officials of the need for welfare reform. While the government is reviewing its elderly care programme and is trying to make child rearing more attractive, several local authorities have already implemented schemes to encourage women to have more children.

Figure 13.27

Changes in the population structure of Japan

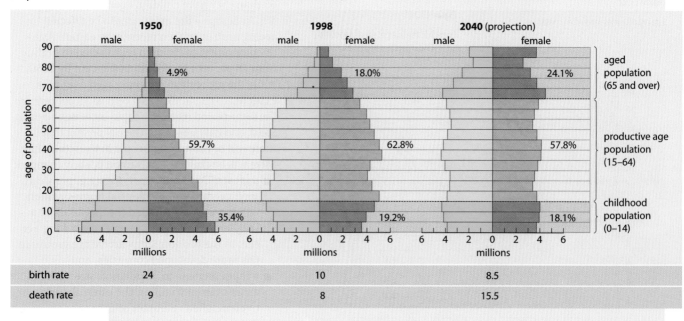

	1950	1998	2040 (projection)
birth rate	24	10	8.5
death rate	9	8	15.5

Migration: change in space and time

Migration is a movement and in human terms usually refers to a permanent change of home. It can also, however, be applied more widely to include temporary changes involving seasonal and daily movements. It includes movements both between countries and within a country. Migration affects the distribution of people over a given area as well as the total population of a region and the population structure of a country or city. The various types of migration are not easy to classify, but one method is given in Figure 13.28.

Internal and external (international) migration

Internal migration refers to population movement within a country, whereas external migration involves a movement across national boundaries and between countries. External migration, unlike internal movement, affects the total population of a country. The **migration balance** is the difference between the number of **emigrants** (people who leave the country) and **immigrants** (newcomers arriving in the country). Countries with a **net migration loss** lose more through emigration than they gain by immigration and, depending upon the balance between birth and death rates, may have a declining population. Countries with a **net migration gain** receive more by immigration than they lose through emigration and so are likely to have an overall population increase (assuming birth and death rates are evenly balanced).

Voluntary and forced migration

Voluntary migration occurs when migrants move from choice, e.g. because they are looking for an improved quality of life or personal freedom. Such movements are usually influenced by **'push and pull' factors** (page 366). **Push** factors are those that cause people to leave because of pressures which make them dissatisfied with their present home, while **pull** factors are those perceived qualities that attract people to a new settlement. When people have virtually no choice but to move from an area due to natural disasters or because of economic, religious or social impositions (Figure 13.29), migration is said to be **forced**.

Times and frequency

Migration patterns include people who may move only once in a lifetime, people who move annually or seasonally, and people who move daily to work or school. Figure 13.28 shows the considerable variations in timescale over which migration processes can operate.

Distance

People may move locally within a city or a country or they may move between countries and continents: migration takes place at a range of spatial scales.

Migration laws and a migration model

In 1885, E. G. Ravenstein put forward seven 'laws of migration' based on his studies of migration within the UK. These laws stated that:

1 Most migrants travel short distances and their numbers decrease as distance increases (distance decay, page 410).
2 Migration occurs in waves and the vacuum left as one group of people moves out will later be filled by a counter-current of people moving in.
3 The process of dispersion (emigration) is the inverse of absorption (immigration).
4 Most migrations show a two-way movement as people move in and out: net migration flows are the balance between the two movements.
5 The longer the journey, the more likely it is that the migrant will end up in a major centre of industry or commerce.
6 Urban dwellers are less likely to move than their rural counterparts.
7 Females migrate more than males within their country of birth, but males are more likely to move further afield.

Permanent	External (international):	between countries
	1 voluntary	West Indians to Britain
	2 forced (refugees)	African slaves to America, Kurds, Rwandans
	Internal:	within a country
	1 rural depopulation	most developing countries
	2 urban depopulation	British conurbations
	3 regional	from north-west to south-east of Britain
Semi-permanent	for several years	migrant workers in France and (former West) Germany
Seasonal	for several months or several weeks	Mexican harvesters in California, holidaymakers, university students
Daily	commuters	south-east England

Figure 13.28

Types of migration

Forced migration	Prevention of voluntary movement

Forced migration

Religious: Jews; Pilgrim Fathers to New England

Wars: Muslims and Hindus in India and Pakistan; Rwanda, Chechnya

Political persecution: Ugandan Asians, Kosovar Albanians

Slaves or forced labour: Africans to south-east USA

Lack of food and famine: Ethiopians into the Sudan

Natural disasters: floods, earthquakes, volcanic eruptions (Mt Pinatubo, Case Study 1)

Overpopulation: Chinese in South-east Asia

Redevelopment: British inner-city slum clearance

Resettlement: Native Americans (USA) and Amerindians (Brazil) into reservations

Environmental: Chernobyl (Ukraine), Bhopal (India)

Voluntary migration

Jobs: Bantus into South Africa, Turks into former West Germany (Places 44), Mexicans into California

Higher salaries: British doctors to the USA

Tax avoidance: British pop/rock and film stars to the USA

Opening up of new areas: American Prairies; Israelis into Negev Desert, Brasilia

Territorial expansion: Roman and Ottoman Empires

Trade and economic expansion: former British colonies

Retirement to a warmer climate: Americans to Florida

Social amenities and services: better schools, hospitals, entertainment

Prevention of voluntary movement

Government restrictions: immigration quotas, Berlin Wall, work permits

Lack of money: unable to afford transport to and housing in new areas

Lack of skills and education

Lack of awareness of opportunities

Illness

Threat of family division and heavy family responsibilities

Reasons for return

Racial tension in new area

Earned sufficient money to return

To be reunited with family

Foreign culture proved unacceptable

Causes of initial migration removed (political or religious persecution)

Barriers to return

Insufficient money to afford transport

Standard of living lower in original area

Racial, religious or political problems in original area

Figure 13.29

Causes of migration, with examples

More recent global migration studies have largely accepted Ravenstein's 'laws', but have demonstrated some additional trends:

8 Most migrants follow a step movement which entails several small movements from the village level to a major city, rather than one traumatic jump.

9 People are leaving rural areas in ever-increasing numbers.

10 People move mainly for economic reasons, e.g. jobs and the opportunity to earn more money. (In many contexts in the early 1990s, many migrants moved to escape from civil war.)

11 Most migrants fall into the 20–34 age range.

12 With the exception of short journeys in developed countries, males are the more mobile. (In many societies, females are still expected to remain at home.)

13 There are increasing numbers of migrants who are unable to find accommodation in the place to which they move; this forces them to live on the streets, in shanties (Places 57, page 443) and in refugee camps (page 367).

The examples in Figure 13.29 help to explain the migration model shown in Figure 13.30.

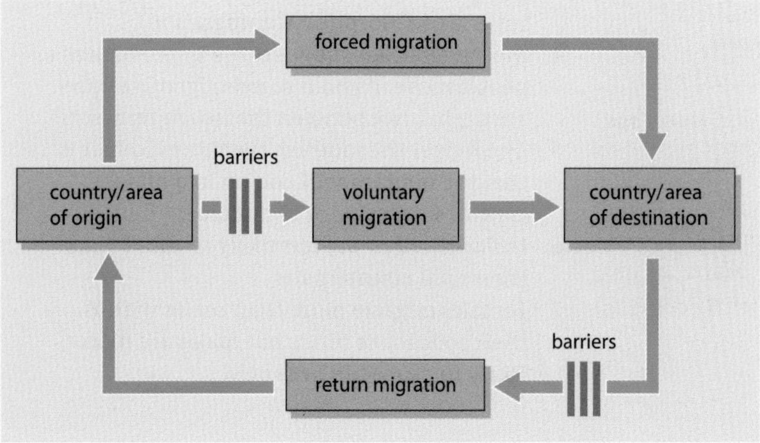

Internal migration in developed countries

Certain patterns of internal migration are more characteristic of developed countries than economically less developed countries. Three examples have been chosen to illustrate this: rural–urban movement; regional movement; and movements within and out of large urban areas.

Rural–urban movement

Although rural depopulation is now a worldwide phenomenon, it has been taking place for much longer in the more developed, industrialised countries. Figure 13.31 and Places 41 describe and explain the changing balance between rural and urban dwellers in the USA since 1870.

Figure 13.30

A migration model (*after* Hornby and Jones)

The 1870s was the decade after the American Civil War and the abolition of slavery. Many black farm-workers, most of whom were share-croppers (page 467), found it impossible either to find vacant land or to afford to buy any that might have become available. Consequently, many began to move to the cities to seek work.

During the 1930s, drought and soil erosion in the Dust Bowl of the Midwest caused many farmers to give up their land. At about the same time, the economic depression lowered prices for farm produce. Farmhands, who had always been poorly paid yet who worked harder and much longer hours compared with their counterparts in industry, were often on yearly or shorter-term contracts and so were paid off by farmers whose profits were falling. A further decline in the farm population occurred in the early 1940s when America's export markets were cut off during the Second World War.

During the 1950s and 1960s, the mechanisation of farming and the availability of cheap oil for fuel saw a big reduction in the number of farmworkers. The consolidation of farms into larger units and the introduction of contract farming, e.g. farmers with combine harvesters who travelled northwards in 'teams' as the cereal ripened, meant employment for even fewer people living in rural areas. Since the 1970s there has been an increase in the number of 'suitcase farmers' – farmers who live in the city and travel out to their farms periodically. Added to this may be the limited number of schools, hospitals, and places of entertainment for members of the rural community, together with fewer job opportunities for females. Improvements in private transport have led to the decline of small service centres. Recent waves of immigrants, mainly Hispanics from Mexico, Puerto Rico and Cuba, and Asians from China, the Philippines, Japan and Korea, have been attracted to urban centres in general and to inner cities in particular.

In spite of their professed love of the 'big outdoors', most Americans seem to prefer living in urban areas – although the percentage of the white population (European origin) living in cities has declined slightly from 72.4 in 1970 to 68.6 in 1998.

Regional movement in Britain

For over a century there has been a drift of people from the north and west of Britain to the south-east of England. The early 19th century was the period of the Industrial Revolution when large numbers of people moved into large urban settlements on the coalfields of northern England, central Scotland and South Wales, and to work in the textile, steel, heavy engineering and shipbuilding industries. However, since the 1920s there has been more than a steady drift of population away from the north of Britain to the south (Figures 13.32 and 13.33). Some of the major reasons for this movement are listed here.

- A decline in the farming workforce and rural population, for reasons similar to those quoted in Places 41 on rural–urban movements.
- The exhaustion of supplies of raw materials (coal and iron ore).
- The decline of the basic heavy industries such as steel, textiles and shipbuilding. Many indus-trial towns had relied not only on one form of industry but, in some cases, on one individual firm. With no alternative employment, those wishing to work had to move south.

- Higher birth rates in the industrial cities meant more potential job-seekers.
- New post-war industries, which included car manufacturing, electrical engineering, food processing and, since the 1980s, micro-elec-tronics and high-tech industries, have tended to be market-oriented. They are said to be footloose, in the sense that they have a free choice of location – unlike the older industries which had to locate near to sources of raw materials and/or energy supplies.

Year	Urban (%)	Rural (%)
1870	24	76
1900	40	60
1920	51	49
1940	56	44
1960	70	30
1980	76	24
1990	77	23
1998	77	23

Figure 13.31

Rural and urban dwellers in the USA as a per-centage of the total population, 1870–1998

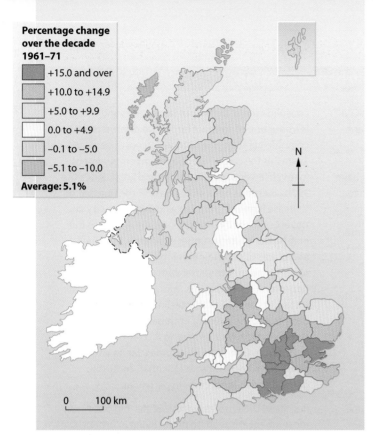

Percentage change over the decade 1961–71

- +15.0 and over
- +10.0 to +14.9
- +5.0 to +9.9
- 0.0 to +4.9
- −0.1 to −5.0
- −5.1 to −10.0

Average: 5.1%

N

0   100 km

Figure 13.32

Population changes in the UK, 1961–71 (boundaries adjusted to the 1974 changes)

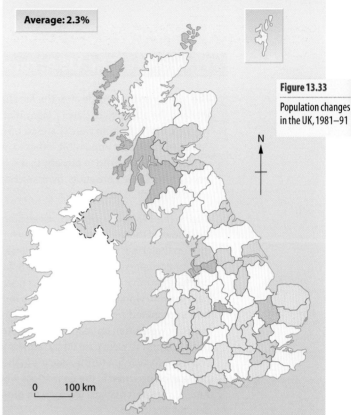

Average: 2.3%

Figure 13.33

Population changes in the UK, 1981–91

N

0   100 km

- The growth of service industries has been mainly in the south-east. This has resulted from the many office firms wanting a prestigious London address, the growth of government offices, the demand for hospitals, schools and shops in a region where one in five British people live, and tourism taking advantage of Britain's warmer south coast.
- Joining the EU meant increased job opportunities in the south and east, while traditional industrial areas and ports such as Glasgow, Liverpool and Bristol, which had links with the Americas, have declined.
- Salaries were higher in the south.
- With so much older housing, derelict land and waste tips, the quality of life is often perceived as being lower in the north, despite the beauty of its natural scenery and slower pace of life.
- There are more social, sporting and cultural amenities in the south.
- Communications were easier to construct in the flatter south. Motorways, railways, international airports, cross-Channel ports and the Channel Tunnel were built and/or improved as this region had the greatest wealth and population size.

Movements within urban areas

Since the 1930s there has been, in Britain, a movement away from the inner cities to the suburbs – a movement accelerated first by improved public transport provision and then by the increase in private car ownership. The former included more bus services, the extension of the London Underground and the construction of the Tyne and Wear Metro. Some of the many stereotyped reasons for this outward movement are summarised in Figure 13.34. The result, in human terms, has been a polarisation of groups of people within society and the accentuation of inequalities (wealth and skills) between them. (Beware, however, of the dangers of stereotyping when discussing these inequalities; Framework 14, page 437).

- The inner cities tended to be left with a higher proportion of low-income families, handicapped people, the elderly, single-parent families, people with few skills and limited qualifications, first-time home-buyers, the unemployed, recent immigrants and ethnic minorities.
- The suburbs tended to attract people moving towards middle age, married with a growing family, possessing higher skills and qualifications, earning higher salaries, in secure jobs and capable of buying their own homes and car. Recently there has been, in part at least, a reversal of this movement and parts of some inner cities have become regenerated and 'fashionable'. This re-urbanisation is partly due to energy conservation, partly to changes in housing markets, partly to planning initiatives such as refurbished waterfronts (London, Places 56, page 440; Baltimore) and partly to new employment growth (leisure, financial services).

	Inner city	Suburbs
Housing	Poor quality; lacking basic amenities; high density; overcrowding	Modern; high quality; with amenities; low density
Traffic	Congestion; noise and air pollution; narrow, unplanned streets; parking problems	Less congestion and pollution; wider, well-planned road system; close to motorways and ring roads
Industry	Decline in older secondary industries; cramped sites with poor access on expensive land	Growth of modern industrial estates; footloose and service industries; hypermarkets and regional shopping centres; new office blocks and hotels on spacious sites
Jobs	High unemployment; lesser-skilled jobs in traditional industries	Lower unemployment; cleaner environment; often more skilled jobs in newer high-tech industries
Open space	Limited parks and gardens	Individual gardens; more, larger parks; nearer countryside
Environment	Noise and air pollution from traffic and industry; derelict land and buildings; higher crime rate; vandalism	Cleaner; less noise and air pollution; lower crime rate; less vandalism
Social factors	Fewer, older services, e.g. schools and hospitals; ethnic and racial problems	Newer and more services; fewer ethnic and racial problems
Planning and investment	Often wholesale redevelopment/clearance; limited planning and investment	Planned, controlled development; public and private investment
Family status/wealth	Low incomes; often elderly and young; large family or none	Improved wealth and family/professional status

Figure 13.34

Some causes of migration from the inner cities to the suburbs

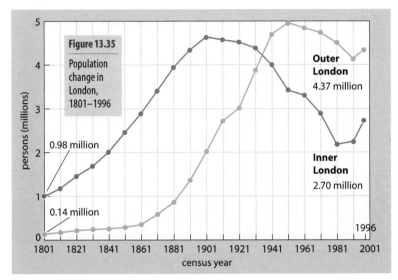

Figure 13.35

Population change in London, 1801–1996

0.98 million

0.14 million

Outer London 4.37 million

Inner London 2.70 million

1996

Movements away from conurbations

After the mid-1950s, large numbers of people moved out of London altogether (Figure 13.35). Initially these were people who were virtually forced to move as large areas of 19th-century inner-city housing were demolished. Many of these people were rehoused in one of the several planned new towns that were created around London. More recently, even the outer suburbs have lost population (Figure 13.36) as people moved, often voluntarily, to smaller towns, or into commuter and suburbanised villages, with a more rural environment. This process of counterurbanisation (page 419) became characteristic of all Britain's conurbations (Figure 13.37) until the 1990s when a reverse movement began, mainly due to regeneration of inner-city areas, especially those with quayside locations (Places 56, page 440).

Conurbation	1961–71	1971–81	1981–91	1991–96
Greater London	–6.8	–11.3	–4.9	+5.9
Inner London	–13.2	–20.0	–6.6	+8.1
Outer London	–1.8	–5.0	–4.1	+7.7
Greater Manchester	+0.3	–5.6	–5.5	+3.0
Merseyside	–3.6	–9.3	–9.1	+1.2
South Yorkshire	+1.5	–2.3	–4.1	+3.3
Tyne and Wear	–2.6	–6.3	–5.4	+2.9
West Midlands	+17.8	–5.9	–5.5	+3.6
West Yorkshire	+3.1	–2.2	–2.7	+4.8
Glasgow City	–13.8	–23.1	–15.5	+2.6

Figure 13.37

Population change in UK conurbations (percentage change per decade)

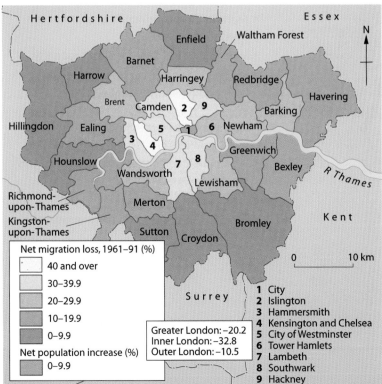

Net migration loss, 1961–91 (%)
- 40 and over
- 30–39.9
- 20–29.9
- 10–19.9
- 0–9.9

Greater London: –20.2
Inner London: –32.8
Outer London: –10.5

Net population increase (%)
- 0–9.9

1 City
2 Islington
3 Hammersmith
4 Kensington and Chelsea
5 City of Westminster
6 Tower Hamlets
7 Lambeth
8 Southwark
9 Hackney

Figure 13.36

Population movement in Greater London, 1961–96

Internal migration in economically less developed countries

There is usually a much greater degree of migration within developing countries than there is in more developed countries. Two examples have been chosen to illustrate this: rural–urban movement and political resettling.

Rural–urban movement

Many large cities in developing countries are growing at a rate of more than 20 per cent every decade. This growth is partly accounted for by rural 'push' and partly by urban 'pull' factors.

Push factors are those that force or encourage people to move – in this case, to leave the countryside. Many families do not own their own land or, where they do, it may have been repeatedly divided by inheritance laws until the plots have become too small to support a family. Food shortages develop if the agricultural output is too low to support the population of an area, or if crops fail. Crop failure may be the result of overcropping and overgrazing (Case Studies 7 and 10), or natural disasters such as drought (the Sahel countries), floods (Bangladesh), hurricanes (the Caribbean) and earth movements (in Andean countries). Elsewhere, farmers are encouraged to produce cash crops for export to help their country's national economy instead of growing sufficient food crops for themselves. Mechanisation reduces the number of farmers needed, while high rates of natural increase may lead to overpopulation (page 376). Some people may move because of a lack of services (schools, hospitals, water supply) or be forced to move by governments or the activities of trans-national companies.

Pull factors are those that encourage people to move – in this case, to the cities. People in many rural communities may have a perception of the city which in reality does not exist. People migrate to cities hoping for better housing, better job prospects, improved lifestyle (aspirational), more reliable sources of food, and better services in health and education. While it is usually true that in most countries more money is spent on the urban areas – where the people who allocate the money themselves live – the present rate of urban growth far exceeds the amount of money available to provide accommodation for all the new arrivals. Recent studies seem to confirm that many migrants make a stepped movement from their rural village first to small towns, then to larger cities and finally to a major city.

Figure 13.38

Migration patterns

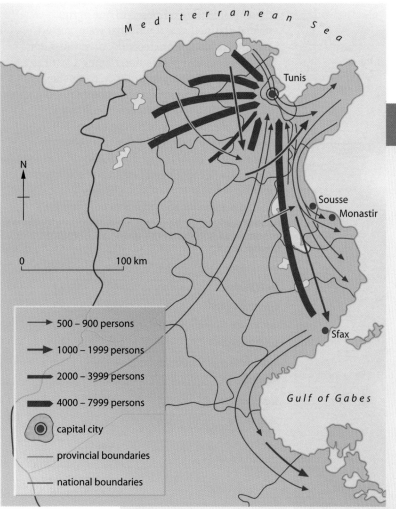

Places 42
Tunisia: migration patterns

Figure 13.38 shows migration patterns in Tunisia. There are several points to notice:

- There is a greater movement of rural than of urban dwellers.
- Most migrants move to Tunis, the capital city.
- Most migrants tend to travel short distances: relatively few make long journeys (distance decay factor, page 410).
- Most move from rural, inland, desert areas to urban, coastal areas.
- A few move from Tunis to coastal towns, such as the holiday resorts of Sousse and Monastir.
- Very few migrants return to rural districts.
- There is evidence of a twofold movement into and out of Tunis and Sfax.

Political resettling

National governments may, for political reasons, direct, control or enforce movement as a result of decisions which they believe to be in the country's (or their own) best interests. Some governments have actively encouraged the development of new community settlements, especially in areas which were, at the time, sparsely populated, e.g. the creation of *kibbutzim* in Israel and of *ujamaa* in Tanzania. Other govern-ments have founded new capital cities in an attempt to develop new growth regions, e.g. Brasilia, Dodoma (Tanzania) and Abuja (Nigeria); while others have built settlements to try to strengthen their claims to an area (e.g. Israeli settlements in the West Bank) or to rehouse people moved for flood control and the production of energy (Three Gorges Project in China – Case Study 3B and Places 82, page 544).

In Brazil and the USA, minority groups of indigenous people – the Amerindians and Native Americans respectively – have been forced off their tribal lands and onto reservations. In South Africa, under *apartheid*, the black population was forced to live either in shanty settlements in urban townships or on homelands in rural areas, which lacked resources (Places 45, page 372). In the last few years, an increasing number of people have been forced to move due to so-called 'ethnic cleansing' policies enforced by several governments, as in the former Yugoslavia.

External migration

Refugees

The United Nations' definition of a **refugee** is 'a person who cannot return to his or her own country because of a well-founded fear of persecution for reasons of race, religion, nationality, political association or social grouping'. The definition is often expanded to include people fleeing from war or other armed conflict. Refugees, who are increasing in number due to internal strife (civil wars) and environmental disasters, e.g. earthquakes and famine, move to other countries hoping to find help and security (Places 43). **Asylum seekers** are people who claim to be refugees. They usually have to undergo a legal procedure in which the host country decides whether they qualify for refugee status (e.g. arrivals in south-east England during 1999). International law recognises the right of individuals to seek asylum, but does not force states to provide it. **Internally displaced persons** are people who have been forced from their homes for refugee-like reasons, but who remain within the borders of their own country. Because they are still under the jurisdiction of a government that might not want international agencies to help them, internally displaced persons may be vulnerable to persecution or violence. Globally, internally displaced persons outnumber refugees.

Before the Second World War, the majority of refugees tended to become assimilated in their new host country but, in the last 50 years, the number of permanent refugees has risen rapidly to approximately 14 million in 1989 and 23 million in 1995 (with another 25 million 'displaced'), according to UN estimates. However, as most refugees are illegal immigrants, the UN admits that these figures could be very inaccurate. The first, apparently insoluble, refugee problem was the setting up of Palestinian Arab camps following the creation of the state of Israel in 1948. The refugee problem has intensified during 30 years of conflict in parts of South-east Asia such as Vietnam, and more recently following food shortages and political unrest in much of sub-Saharan Africa, including Ethiopia, Mozambique and Rwanda (Figure 13.39).

Half of the world's refugees are children of school age; most adult refugees are female; and four-fifths are in the developing countries which have fewest resources to deal with the problem. Refugees usually live in extreme poverty and lack food, shelter, clothing, education and medical care. They rarely have citizenship, and few (if any) civil, legal or basic human rights. There is little prospect of returning home and the long periods spent in camps mean that they often lose their sense of identity and purpose.

Figure 13.39

Part of a displacement camp at Nyaconga, Rwanda

The United Nations High Commission for Refugees (UNHCR) claimed that the world refugee numbers fell from a peak of 17.60 million in 1992 to 13.57 million by the beginning of 1998 (Figure 13.40).

a

World total	13 566 000
Africa	2 944 000
Europe	2 020 000
The Americas and the Caribbean	616 000
East Asia and the Pacific	535 000
Middle East	5 708 000
South and Central Asia	1 743 000

Figure 13.40

World refugees, 1 January 1998

b

Origin of refugees	Number	Origin of refugees	Number
Palestinians	3 721 000	Eritrea	315 600
Afghanistan	2 647 000	Angola	262 700
Iraq	630 700	Azerbaijan	234 200
Somalia	524 400	Armenia	201 000
Burundi	515 800	Congo	165 700
Liberia	486 700	China (Tibet)	119 800
Bosnia/Herzegovina	391 200	Myanmar (Burma)	110 800
Sierra Leone	361 000	Bhutan	108 700
Sudan	351 300	Cambodia	100 000
Croatia	342 000	Sri Lanka	93 600
Vietnam	316 000		

UNHCR, *Statistical Report 1998*

The tables represent long-term refugees such as the Palestinians who are living in Gaza and the West Bank, as well as in Jordan, Kuwait and other Middle East states. The majority of Afghan refugees are in camps in Pakistan (1.2 million) and Iran (1.4 million). It is estimated that there are over 86 000 Saharawis from Western Sahara in camps in Algeria close to the border with Morocco.

1999 was a bad year for refugees. In the spring and early summer almost a million Kosovar Albanians were forced out of their homes, moving into the Federal Republic of Macedonia, Albania and countries further afield. The UNHCR stated (1 September) that while 772 000 Kosovars had returned home, a growing number of Serbs who had lived in Kosova for many years were moving out, while 90 000 Kosovar gypsies had been forcibly ejected by ethnic Albanians. In early September,

reports stated that 'As many as 200 000 people, nearly a quarter of East Timor's population, have been driven from their homes in the last four days' and, a few days later, 'thousands of refugees are pouring into West Timor'. Fighting in Chechnya and Dagestan had displaced 22 000 people and the UNHCR was making provision for aid to the refugees in an already poverty-stricken region of the former USSR. Thousands of Turks were still displaced following the August earthquake.

On a more positive note, the UNHCR reported that 'camps established to help the Rwandan refugees on the Congolese border are being closed as people have returned home; refugee numbers in Tanzania are falling as conditions in the north and west of East Africa seem to be stabilising, and the last Guatemalan refugees had left Mexico after over two decades away from home due to civil war'.

Figure 13.41

Directions of some major international migrations since 1970

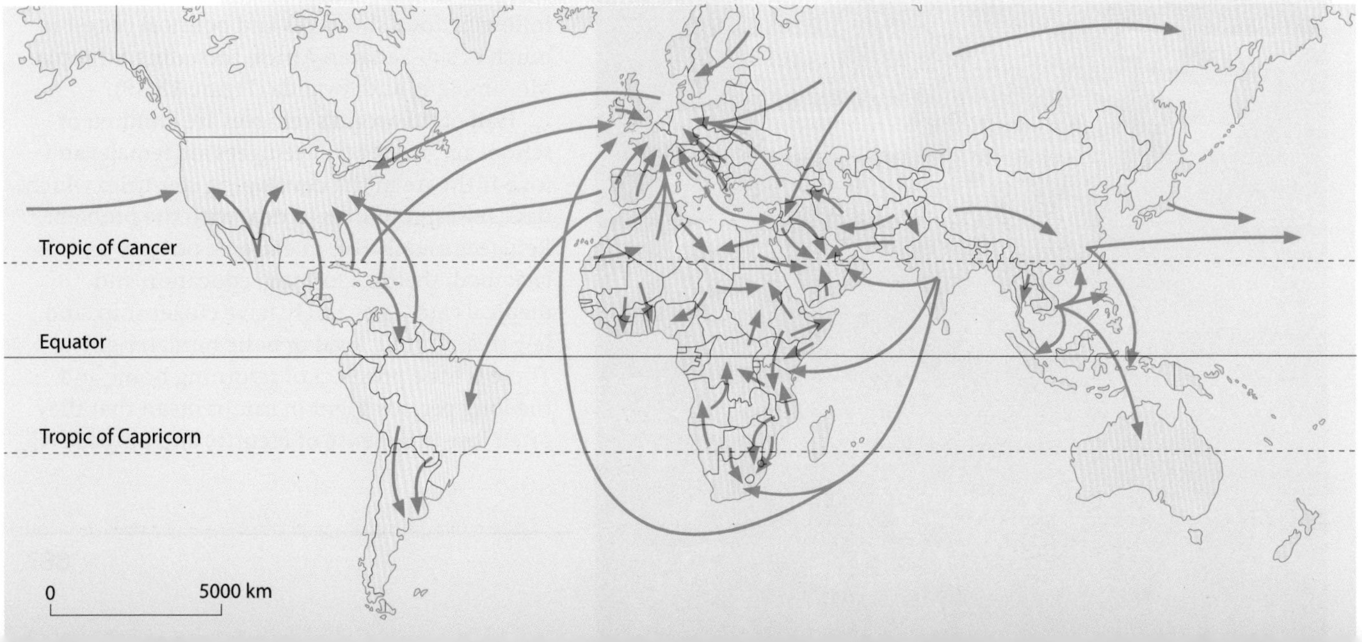

Tropic of Cancer

Equator

Tropic of Capricorn

0   5000 km

Migrant workers

As economic development has taken place at different rates in different countries, supplies of and demand for labour are uneven, and due to improvements in transport, there has been an increase in the number of people who move from one country to another in search of work. Such cross-border movements in search of work can operate at different timescales. For example:

■ **Permanent** For a century and a half, the UK has received Irish workers and, since the 1950s, West Indians and citizens from the Indian subcontinent. Most of these migrants have made Britain their permanent home.

■ **Semi-permanent** After the Second World War, several European countries experienced a severe labour shortage. In order to help rebuild their economies, countries such as France and the then West Germany, accepted cheap, semi-skilled labour from North Africa and Portugal (into France) and from Turkey, Yugoslavia and the Middle East (into Germany, Places 44).

■ **Short-term and seasonal** The South African economy depends largely upon migrant black labour from adjacent nation states. Under apartheid, these workers were not allowed to remain in South Africa for more than six months at a time. In North America, large numbers of Mexicans find seasonal employment picking fruit and vegetables in California.

■ **Daily** With the introduction of free movement for EU nationals within the EU, an increasing number of workers commute daily into adjacent countries.

Places 44   Germany: Turkish migrant workers

1945–89

Sakaltutan is a village of 900 inhabitants in central Turkey (Figure 3.42). Until recently it was a poor, isolated settlement dependent upon agriculture. With a high birth rate and limited resources, the village had become overpopulated (page 376). There were too many males to work on the land (women were not expected to work outside the home) and the demand for craftsmen was limited.

When an all-weather road was built this encouraged the sale of surplus produce in the nearby large towns and led to an increase in mechanisation and a further decline in the need for agricultural labourers. The village school was expanded from one teacher to seven and farmers were able to obtain advice on how to increase output. The result was a growth in the aspirations of the villagers. Outward migration followed, either to the city of Adana, or to the capital Ankara, or overseas to West Germany.

Pforzheim is an industrial town near Stuttgart in Germany. Like other west European towns it had to be rebuilt after 1945 at a time when there were more job vacancies than workers. The extra labour needed was obtained from the poorer parts of southern Europe and the Middle East. Many of these 'guest workers' or *gastarbeiters* initially went into agriculture as most were originally farmers, but they soon turned to the relatively better-paid jobs in factories and the construction industry. These jobs were not taken by the local Germans because they were dirty, unskilled, poorly paid and often demanded long and unsociable hours.

At one stage there were 15 families from Sakaltutan living in Pforzheim. The first arrivals were males in their twenties (Figure 3.45), all of whom had had some education and were skilled at crafts, making them acceptable in the local car factory or in construction. In an attempt to earn as much money as possible, they often took accommodation in poorly equipped company hostels, missed meals and used public transport to reach work.

Although by Turkish standards the migrants were earning high salaries, they found the West German cost of living much higher, and what was intended to be a short working stay became more permanent. In time, the families of *gastarbeiters* arrived and by the mid-1980s over 3 per cent of the West German workforce was Turkish (Figure 13.43).

This movement of labour had advantages and disadvantages to Sakaltutan in Turkey, the losing village/country, and to Pforzheim in West Germany, the receiving town/country. These are summarised in Figure 13.44.

In 1973 the West German government imposed a ban on the recruitment of foreign workers, although Turks still arrived to make family reunions. When in 1980 grants were offered to Turks wishing to return home, very few took advantage of the offer. Of those who remained, fewer than 1 per cent had taken out German citizenship as that meant giving up Turkish nationality.

Figure 13.42

A Turkish village

Figure 13.43

A Turkish worker in Germany

	Advantages	Disadvantages
Sakaltutan	Reduces pressure on jobs and local resources	People of working age leave
	Birth rate may be lowered as people of childbearing age leave	People with skills and education are most likely to leave
	Money may be sent back to the village: estimates suggest that 50 per cent of Sakaltutan's income is from overseas	It is mainly males who migrate and this divides families
	Migrants may develop new skills overseas which they then bring back to village	An elderly population is left, resulting in a higher death rate
		In the long term, creates dependency upon money sent back to home villages
Pforzheim	Labour shortage is solved, especially in dirty, poorly paid, unskilled jobs	Resentment towards Turks when Germany's rate of unemployment rises
	Cheaper labour for less desirable jobs	Turks form an ethnic group which does not mix: they live in poor housing, and retain their own culture
	Cultural advantages of discovering new foods, music, pastimes, etc.	Turks feel discriminated against with racial tension and police harassment
	Migrants tend to be the more economically active and skilled	Migrants may be a drain on local services
		Migrants are mainly male, which can lead to social problems

Figure 13.44

Advantages and disadvantages for Sakaltutan and Pforzheim

Figure 13.45

Population structure of Turks living in Germany and Turkey, 1998

Since 1990

During the early 1990s, when there were still 1.8 million Turks living in Germany, tension between the two communities began to increase, culminating in several arson attacks on Turkish property and the loss of several Turkish lives. The demolition of the Berlin Wall and the reunification of Germany in 1989 brought with it several problems. As large numbers of migrants arrived from the former East Germany and other former communist-bloc countries, the demand for jobs and houses in the former West Germany increased. These new arrivals took the menial, poorer-paid jobs previously held by the Turks. The cost of housing (rents) increased, as did levels of unemployment and inflation; few people anticipated the economic problems of unification. The Turks, being an ethnic group living within their own community, were blamed by the Germans for the decline in the country's economy – even though it was the Turks who suffered first and the most.

Turks from villages like Sakaltutan are now turning to Saudi Arabia and Libya as places in which to work and to live.

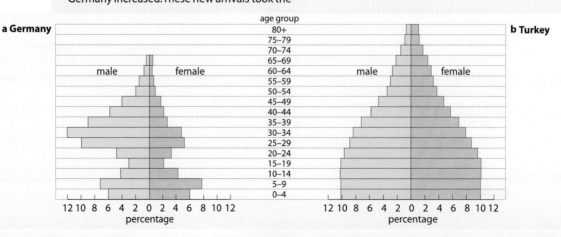

Multicultural societies

This is often a sensitive and emotive issue. Attempts here to explain terms are not intended to cause insult or resentment.

The latest scientific research suggests that humans evolved in central Africa about 200 000 years ago and began, 100 000 years later, to migrate to other parts of the world. This common origin, identified by the study of genes, shows that humans are genetically homogeneous to a degree unparalleled in the animal kingdom.

Previous scientific opinion suggested that the many peoples of the modern world had descended from three main races. These were the Negroid, Mongoloid and Caucasoid. The dictionary definition of *race* is 'a group of people having their own inherited characteristics distinguishing them from people of other races', e.g. colour of skin and physical features. In reality, often because of intermarriage, the distinction between races is now so blurred that the word 'race' has little significant scientific value. Today, while colour still remains the most obvious visible characteristic, groups of people differ from one another in religion, language, nationality and culture. These differences have led to the identification of many ethnic groups.

What criteria do members of various ethnic groups prefer to use when identifying themselves?

- **Colour of skin** Whereas people of 'European' stock have long accepted being called 'white', it is only in more recent years that, in Britain, people from Africa and the Caribbean have preferred to be known collectively as 'black'. The 1971 UK census divided immigrants born in Commonwealth countries into the Old (white) and New (black) Commonwealth. (It made no allowance for children born in the UK of New Commonwealth parentage.)
- **Place of birth (nationality)** *The Annual Abstract of Statistics* for the UK lists immigrants under the heading 'country of last residence' – thus avoiding a reference to colour. Most groups of people, in the USA for example, have been identified by their place of birth, or that of their ancestors, and are known as Chinese, Puerto Rican, etc. There is currently a major movement in the USA (and to a lesser extent in the UK) by blacks, also wishing to be identified by place of origin, to be referred to as African-Americans. Will black people in the UK eventually prefer to be known as African-Caribbean, African-British, or another term not yet invented?
- **Language** At present, the largest group of migrants moving into the USA is the Hispanics, i.e. Spanish speakers. These migrants, mainly from Mexico, Central America and the West Indies, have been identified and grouped together by their common language and higher fertility.
- **Religion** Other ethnic groups prefer to be linked with, and are easily recognised by, their religion, e.g. Jews, Sikhs, Hindus and Muslims.

The 1991 UK census asked respondents, for the first time, to identify themselves by ethnic group. Figure 13.46 lists these groups, and gives the results of this question. These results have been used elsewhere (Robinson, 1994) to describe the spatial distributions of different ethnic groups in England and Wales.

The migrations of different ethnic groups have led to the creation of multicultural societies in many parts of the world. In most countries there is at least one minority group. While such a group may be able to live in peace and harmony with the majority group, unfortunately it is more likely that there will be prejudice and discrimination leading to tensions and conflict. Four multicultural countries with differing levels of integration and ethnic tension are: South Africa (Places 45), the USA and Brazil (Places 46),and Singapore (Places 47). Remember, though, that when we look at these countries from a distance we can rarely appreciate the feelings generated by, or the successes/failures of, different state or government policies.

Figure 13.46

Ethnic groups in Britain, 1991 and 1997

Ethnic groups	Percentage in each group	
	1991	1997
White	94.5	94.0
Black-Caribbean	0.9	1.7
Black-Africa	0.4	
Black-other (please describe)	0.3	
Indian	1.5	1.6
Pakistani	0.9	1.3
Bangladeshi	0.3	
Chinese	0.3	1.4
Any other group (please describe)		
a) Asian	0.4	
b) Other	0.5	
Total non-white	5.5	6.0

If the person is descended from more than one ethnic or racial group, please tick the group to which he/she considers he/she belongs.

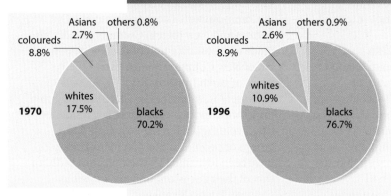

Figure 13.47

Ethnic groups in South Africa, 1970 and 1996

The population census of 1996, the first to be held in South Africa after the first universal suffrage elections of 1994, gave a total population of 40.58 million (the 1999 figure was estimated to be 43.43 million). The 1996 figures also showed a significant change in the balance between blacks and whites since the 1970 census (Figure 13.47).

The first inhabitants of the area were the San (Bushman) people and the Khoi-Khoin people. However, the majority of South Africa's black people are descended from Bantu-speakers who migrated into the area many centuries ago. The white population (most of whom are now known as Afrikaner) are descended from Dutch, German, French, British and Belgian immigrants since 1652. The coloured people are descended from mixed relations among the European settlers, indigenous peoples and people from Madagascar, India, Indonesia and Malaysia. The Asian population arrived after 1860 and came mainly from India.

A policy of segregation between black and white originated in the first Dutch settlement, the Cape, in 1652. This practice became customary, and was established legally as apartheid by the first National Party government in 1948 when some Afrikaners in the Party united to protect their language, culture and heritage from a perceived threat by the black majority and to assert their economic and political independence from British colonial domination.

Statutory apartheid regulated the lives of all groups, but particularly of blacks, coloureds and Indians. The Population Registration Act categorised the nation into White, Black, Indian, Malay and Coloured citizens. Further Acts made mixed marriages illegal, and prescribed segregation in restaurants, transport, schools, places of entertainment and political parties. The Group Areas Act stipulated where and with whom people could live; and the Black Authorities Act established black homelands.

The outcome of all this legislation was the unequal division of rights and resources. This included the disproportionate division of land; the unequal distribution of funding for education; and the general denial of constitutional rights for the majority of South Africans.

Legalised racial discrimination was abolished in the early 1990s. After long negotiations, the first all-party elections, held in 1994, established a multi-party Government of National Unity. The five-year interim constitution ended the existence of the homelands and included a section on human rights; a Reconstruction and Development Programme is addressing housing problems, equality in education and health care, and the encouragement of private home ownership; a Land Court is to address land claims from dispossessed people.

However, the legacy of apartheid will take many years to eradicate. Some aspects of apartheid are described on the next page.

Figure 13.48

Segregated residential areas in two South African cities

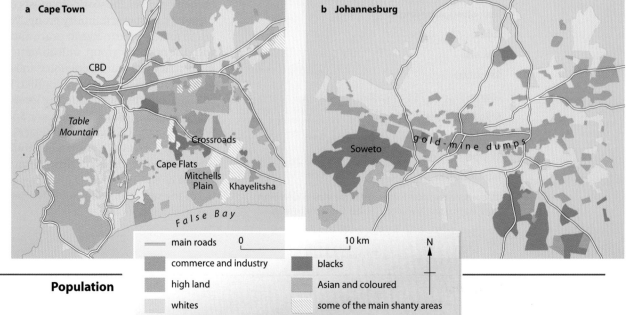

Population

Housing

The Group Areas Act (1950) ensured that white, coloured and Asian communities lived in different parts of the city (Figure 13.48) with the whites having the best residential areas (Figure 13.49). Buffer zones at least 100 m wide, often along main roads or railway lines, were created to try to prevent contact between the three groups. Blacks were treated differently. Those who had lived in the city since birth, or had worked for the same employer for 10 years, were moved to newly created townships on the urban fringes. The remainder were forced away from the cities to live on one of ten designated reserves or homelands, where the environmental advantages were minimal (drought, poor soils and a lack of raw materials). The homelands took up 13 per cent of South Africa's land; held 72 per cent of its total population; and produced 3 per cent of the country's wealth. Most blacks living in the homelands were employed on one-year contracts, to prevent them gaining urban residential rights.

Life in the townships was no less difficult. These were built far away from white residential areas, which meant that those blacks who found jobs in the cities had long and expensive journeys to work. Many of the original shanty towns have been bulldozed and replaced by rows of identical, single-storey houses (Figure 13.50). These have four rooms and a backyard toilet, but only 20 per cent have electricity. Corrugated-iron roofs make the buildings hot in summer and cold in winter. The settlements lack infrastructure and services and, due to rapid population growth (high birth rates and in-migration), are surrounded by vast shanty settlements (Figure 13.51). Two of the better-known townships are Soweto in Johannesburg (an estimated 4 million inhabitants) and Crossroads in Cape Town.

When the African National Congress (ANC) came to power in 1993, it promised to build a million houses during its first term in office to redress the socio-economic imbalances of apartheid. By the 1999 elections, it had built 700 000 houses. Although the ANC was proud of its record, there were still thousands of blacks living either in poverty-stricken and squalid conditions of squatter camps which developed during the apartheid era, or in new, but mainly one-roomed, low-cost housing which, their owners claimed, was often poorly constructed and far too small for the large African families (up to ten people per house).

Employment

Under apartheid, blacks were severely restricted in mobility and type of job. A man had to return to his homeland in order to apply for a job. If successful, he was given a contract to work in 'white' South Africa for 11 months, after which he had to return to his homeland if he wished to renew the contract. This system prevented blacks from becoming permanent residents in the city.

Throughout the late 1990s, unemployment remained the core cause of poverty and social division in South Africa. In 1998, unemployment was 38 per cent among blacks (with another 11 per cent underemployed), 21 per cent among coloureds, 11 per cent among Indians and 4 per cent among whites. The average wage for whites was twice that of Indians, three times that of coloureds and seven times that of blacks.

Education

Under apartheid, schooling was free and compulsory for all whites and Indians, but not for coloureds or blacks (the 1996 census showed that one-quarter of blacks had not received any formal education). Despite attempts by the ANC to improve education, mainly by renovating existing schools, building new classrooms, encouraging school attendance and increasing access to higher education, the pupil–teacher ratio remained, in 1998, 20:1 for whites and 300:1 for blacks.

The United States of America

Americans have long prided themselves that their country is a 'melting pot' in which peoples of all ethnic groups could be assimilated into one nation (Figure 13.52). Yet problems do exist. The indigenous Indian populations have been granted reservations, usually in areas lacking resources, although they are not forced to live on them. It is, however, only on reservations occupied by tribes such as the Navajo and Hopi that traditional Amerindian culture still flourishes. Indians who have drifted to towns often become the focus of social problems.

The African-Americans (blacks), released from slavery after the Civil War, could not find land on which to farm (Places 41). They too moved to large urban centres where they have congregated in inner-city 'ghettos' (Chicago, Places 52 page 421; and Case Study 15B). Despite some improvements following the Civil Rights movement of the 1960s, this ethnic group remains socially and economically deprived in comparison with most whites.

Despite the US claim that it has an 'open-door' policy, strong restrictive laws have frequently been imposed as a barrier to immigration (Figure 13.30). This policy has been noticeable against Chinese in the 1920s, Japanese during the Second World War, Mexicans since the 1980s and, most recently, illegal migrants from Central America and Haiti. Many immigrant groups still identify themselves with their 'home' country and its culture, living and marrying within their own ethnic groups (Puerto Ricans in New York) and congregating to form ethnic areas – Los Angeles, for example, has its Chinatown, Japantown, Koreatown and Filipinotown.

Brazil

Most of the inhabitants of Brazil, having almost every colour of skin conceivable, regard themselves as Brazilians, and the country rightly claims that it has little racial discrimination or prejudice. The Census Department does, however, recognise the following divisions based upon colour:

1 Whites (*Branco*): anyone with a *café au lait* or lighter-coloured skin (the USA and South Africa would regard *café au lait* as black or coloured). This group includes many of the European migrants who came from Portugal (the original colonists), Italy, Germany and Spain.

2 Mulatto (*Pardo*): darker skins but with a discernible trace of European ancestry. They are the result of mixed marriages or 'liaisons' between the early Portuguese male settlers and either female Indians or African slaves. There is pride rather than prejudice in coming from two racial backgrounds.

3 Blacks (*Preto*): those of pure African descent.

4 Orientals (*Amarelo*): those who have emigrated more recently from south and east Asia.

5 Amerindians: a continually declining, yet still distinctive, indigenous group.

All these groups mix freely, especially at football matches, in carnivals and on the beach. Yet despite the lack of racial tension there tends to be a correlation between colour and social status and jobs. Walking into a hotel on arrival in Rio, it is apparent that the baggage-carriers are black, hotel porters a slightly lighter colour and the receptionists and cashiers *café au lait*. In the army, officers are usually white and the ranks black or mulatto. Similarly, the lighter the colour of skin, the more likely it is that a person will become a doctor, bank manager, solicitor or airline pilot. Yet the author has met two Indian tourist guides, one at Manaus in the Amazon rainforest and the other at Iguaçu on the border with Paraguay, who both stated that while blacks and Amerindians had the opportunity of reaching the top in Brazil they preferred to avoid the stresses of modern life.

Figure 13.52

Ethnic groups in the USA, 1998

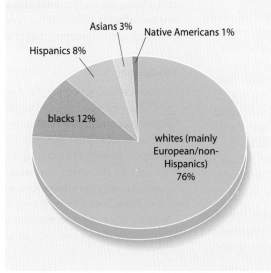

Asians 3%
Native Americans 1%
Hispanics 8%
blacks 12%
whites (mainly European/non-Hispanics) 76%

US Bureau of the Census

The three main races of Singapore have separate religions, yet each is completely tolerant of the others and most people even celebrate all three 'New Years' (Figure 13.53). Although in 1994 there was still a Chinatown (Figure 15.48), restricted to ten streets, Arab Street (four streets) and Little India (six streets), the Singapore government has pulled down most of the old houses in these areas. Ethnic concentrations have been broken up and 87 per cent of Singaporeans now live in modern high-rise flats in new towns (Places 60, page 450). Posters promote racial harmony (Figure 13.54), and all races, religions and income groups live together in what appears to be a most successful attempt to create a national unity – a unity best seen on National Day.

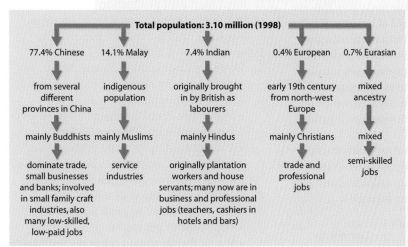

Total population: 3.10 million (1998)

77.4% Chinese	14.1% Malay	7.4% Indian	0.4% European	0.7% Eurasian
from several different provinces in China	indigenous population	originally brought in by British as labourers	early 19th century from north-west Europe	mixed ancestry
mainly Buddhists	mainly Muslims	mainly Hindus	mainly Christians	mixed
dominate trade, small businesses and banks; involved in small family craft industries, also many low-skilled, low-paid jobs	service industries	originally plantation workers and house servants; many now are in business and professional jobs (teachers, cashiers in hotels and bars)	trade and professional jobs	semi-skilled jobs

Figure 13.53

Ethnic and religious groups in Singapore

Figure 13.54

Racial harmony poster, Singapore

Daily migration: commuting

A commuter is a person who lives in one community and works in another. There are two types of commuting:

1 **Rural–urban**, where the commuter lives in a small town or village and travels to work in a larger town or city. There is rarely much movement in the reverse direction. The **commuter village** is sometimes also referred to as a **dormitory village** or a **suburbanised village** (page 398).

2 **Intra-urban**, where people who live in the suburbs travel into the city centre for work. This category now includes inhabitants of inner-city areas who have to make the reverse journey to edge-of-city industrial estates and regional shopping centres.

A **commuter hinterland**, or **urban field**, is the area surrounding a large town or city where the work-force lives. Patterns of commuting are likely to develop where:

■ hinterlands are large, communications are fast and reliable (the London Underground), public transport is highly developed and private car ownership is high (south-east England)

■ modern housing is a long way from either the older inner-city industrial areas or from the CBD (as in the New Towns in central Scotland)

■ there is a nearby city or conurbation with plenty of jobs, especially in service industries (London)

■ there is no rival urban centre within easy reach (Plymouth)

■ salaries are high so that commuters can afford travelling costs

■ people feel that their need to live in a cleaner environment outweighs the disadvantages of time and cost of travel to work (people living in the Peak District and working in Sheffield or Manchester)

■ housing costs are high so that younger people are forced to look for cheaper housing further away from their work (as in south-east England)

■ flexible working hours allow people to travel during non-rush-hour times

■ the more elderly members of the workforce buy homes in the country or near to the coast and commute until they retire (the Sussex coast)

■ there have been severe job losses which force people to look for work in other areas/towns (some of the inhabitants of Cleveland work in south-east England).

Optimum, over- and under-population

Optimum population

The **optimum population** of an area is a theoretical state in which the number of people, when working with all the available resources, will produce the highest per capita economic return, i.e. the highest standard of living and quality of life. If the size of the population increases or decreases from the optimum, the output per capita and standard of living will fall. This concept is of a dynamic situation changing with time as technology improves, as population totals and structure change (age and sex ratios), trade opportunities alter, and as new raw materials are discovered to replace old ones which are exhausted or whose values change over a period of time.

The **standard of living** of an individual or population is determined by the interaction between physical and human resources and can be expressed in the following formula:

$$\text{Standard of living} = \frac{\text{Natural resources} \begin{Bmatrix} \text{minerals,} \\ \text{energy,} \\ \text{soils etc.} \end{Bmatrix} \times \text{Technology}}{\text{Population}}$$

Overpopulation

Overpopulation occurs when there are too many people relative to the resources and technology locally available to maintain an 'adequate' standard of living. Bangladesh, Ethiopia and parts of China, Brazil and India are often said to be overpopulated as they have insufficient food, minerals and energy resources to sustain their populations. They suffer from localised natural disasters such as drought and famine; and are characterised by low incomes, poverty, poor living conditions and often a high level of emigration. In the case of Bangladesh (Places 48), where the population density increased from 282 people per km^2 in 1950, to 704 in 1985, and to 895 in 1998, it is easier to appreciate the problem of 'too many people' than in the case of the north-east of Brazil where the density is less than 2 persons per km^2 (Places 38, page 347).

Underpopulation

Underpopulation occurs when there are far more resources in an area, e.g. of food, energy and minerals, than can be used by the number of people living there. Canada, with a total population of 30 million in the late 1990s, could theoretically double its population and still maintain its standard of living (Places 48). Countries like Canada and Australia can export their surplus food, energy and mineral resources, have high incomes, good living conditions, and high levels of technology and immigration. It is probable that standards of living would rise, through increased production and exploitation of resources, if population were to increase.

However, care is needed when making comparisons on a global scale.

1 There does not seem to be any direct correlation between population density and over-/underpopulation:
   • north-east Brazil is considered to be 'overpopulated' with 2 people per km^2
   • California, despite water problems and pollution, is perceived to be 'underpopulated' with over 550 persons per km^2.
2 Similarly, population density is not necessarily related to gross domestic product (GDP) per capita:
   • the Netherlands and Germany both have a high GDP per capita and a high population density
   • Canada and Australia have a high GDP per capita and a low population density
   • Bangladesh and Puerto Rico have a low GDP per capita and a high population density
   • Sudan and Bolivia have a low GDP per capita and low population density.

The balance of population and resources within a country may also be uneven. For example:
■ a country may have a population that is too great for one resource such as energy, yet too small to use fully a second, such as food supply
■ some parts of a country may be well off, e.g. south-east Brazil, while others may be relatively poor, e.g. north-east Brazil.

The relationships between population and resources are highly complex and the terms 'overpopulation' and 'underpopulation' must therefore be used with extreme care.

Bangladesh and Canada: overpopulation and underpopulation

Is Bangladesh overpopulated?

Bangladesh, with 124 million inhabitants (1998), has a high population density of 953 persons per km^2 (Figure 13.55). It has a high, but falling, birth rate (49 per 1000 in 1970; 30 per 1000 in 1998) and fertility rate (7 per woman in 1970; 4 per woman in 1998), and a declining death rate (28 in 1970; 11 in 1998). This led to a high and accelerating rate of natural increase from 1.6 per cent in 1950 to 2.7 per cent in 1990; although this fell rapidly to 1.0 per cent in 1998 (page 349). Infant mortality is falling, but is still very high (140 per 1000 in 1970; 100 per 1000 in 1998), and over 44 per cent of the population is under 15 years of age. Life expectancy, which was 46 for men and 44 for women in 1970, is now (late 1990s) 56 for both sexes. The GNP of US$ 240 per person is extremely low, with 40 per cent of people living in poverty (defined by the UN as receiving less than 2122 calories per day – page 500).

As most of the country is a flat delta, it is prone to frequent and severe flooding. This results either from flooding by the Ganges and Brahmaputra rivers, due mainly to the monsoon rains and also partly to deforestation in the Himalayas, or from tropical cyclones moving up the Bay of Bengal (Places 19 page 148 and Places 31 page 238). Most of the inhabitants are farmers (59 per cent) who live in rural communities (82 per cent). There is a shortage of industry, services and raw materials (Bangladesh has no energy or mineral resources of any note), and the transport network is limited. The low level of literacy (36 per cent) has led to limited internal innovation, and a lack of capital has meant that the country can ill afford to buy technical skills from overseas (its trade is valued at US$ 54 per person per year). Bangladesh receives $11 per person per year in aid.

Figure 13.55

High population density in Bangladesh

Figure 13.56

Low population density in Canada

Is Canada underpopulated?

Canada, with 30.2 million inhabitants (1998), has a low population density of 3 persons per km^2 (Figure 13.56). It has a low birth rate (16 per 1000 in 1970; 13 per 1000 in 1998), a low fertility rate (2.2 per woman in 1970; 2.0 per woman in 1998), a low death rate (7 per 1000 in both 1970 and 1998), and a low infant mortality rate (16 per 1000 in 1970; 6 per 1000 in 1998). Together, this gives a low, but slightly rising, rate of natural increase (1.0 per cent in 1970; 1.8 per cent in 1998). Only 21 per cent of the population is under 15 years of age. Life expectancy, which was 70 for men and 77 for women in 1970, is now (late 1990s) 76 for men and 83 for women. The GNP of US$ 19 380 per person is very high.

Natural disasters are rare. Relatively few of the inhabitants are farmers (5 per cent) or live in rural areas (23 per cent). Canada has developed industries, services, energy supplies and mineral resources (of which it has considerable reserves) and an efficient transport network. The high level of literacy (99 per cent) and national wealth have enabled it to develop its own technology and to import modern innovations (Canada's trade is valued at US$ 9355 per person per year). Canada donates US$ 73 per person per year of aid.

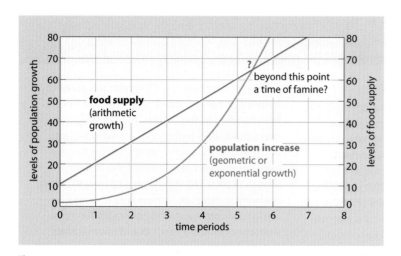

Figure 13.57

Relationships between population growth and food supply (*after* Malthus)

Theories relating to world population and food supply

Malthus

Thomas Malthus was a British demographer who believed that there was a finite optimum population size in relation to food supply and that an increase in population beyond that point would lead to a decline in living standards and to 'war, famine and disease'. He published his views in 1798 and although, fortunately, many of his pessimistic predictions have not come to pass, they form an interesting theory and provide a possible warning for the future. His theory was based on two principles.

1   Human population, if unchecked, grows at a **geometric** or **exponential rate**, i.e.
    $1 \rightarrow 2 \rightarrow 4 \rightarrow 8 \rightarrow 16 \rightarrow 32$, etc.

2   Food supply, at best, only increases at an **arithmetic rate**, i.e. $1 \rightarrow 2 \rightarrow 3 \rightarrow 4 \rightarrow 5 \rightarrow 6$, etc. Malthus considered that this must be so because yields from a given field could not go on increasing for ever, and the amount of land available is finite.

Malthus demonstrated that any rise in population, however small, would mean that eventually population would exceed increases in food supply. This is shown in Figure 13.57, where the exponential curve intersects the arithmetic curve. Malthus therefore suggested that after five years, the ratio of population to food supply would increase to 16:5, and after six years to 32:6. He suggested that once a ceiling had been reached, further growth in population would be curbed by negative (preventive) or by positive checks (Figure 13.59).

Preventive (or **negative**) **checks** were methods of limiting population growth and included abstinence from, or a postponement of, marriage which would lower the fertility rate. Malthus noted a correlation between wheat prices and marriage rates (remember that this was the late 18th century): as food became more expensive, fewer people got married.

Positive checks were ways in which the population would be reduced in size by such events as a famine, disease and war, all of which would increase the mortality rate and reduce life expectancy.

The carrying capacity of the environment

The concept of a population ceiling, first suggested by Malthus, is of a saturation level where the population equals the carrying capacity of the local environment. The **carrying capacity** is the largest population of humans/animals/plants that a particular area/environment/ecosystem can carry or support.

Three models portray what might happen as a population, growing exponentially, approaches the carrying capacity of the land (Figure 13.58).

1   The rate of increase may be unchanged until the ceiling is reached, at which point the increase drops to zero (Figure 13.58a). This highly unlikely situation is unsupported by evidence from either human or animal populations.

2   Here, more realistically, the population increase begins to taper off as the carrying capacity is approached, and then to level off when the ceiling is reached (Figure 13.58b). It is claimed that populations which are large in size, have long lives and low fertility rates, conform to this 'S' curve pattern.

3   In this instance, the rapid rise in population overshoots the carrying capacity, resulting in a sudden check – e.g. famine and reduced birth rates – which causes a dramatic fall in the total population. After this, the population recovers and fluctuates around, and eventually settles down at, the carrying capacity. This 'J' curve appears more applicable to populations that are small in number, and have short lives and high fertility levels (Figure 13.58c).

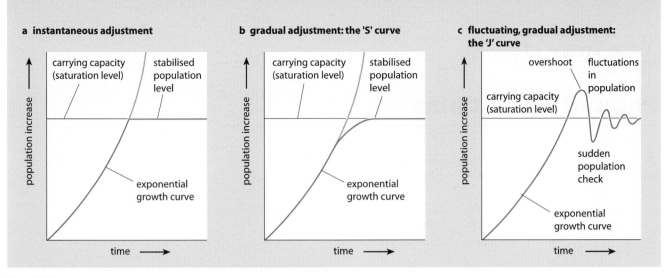

a instantaneous adjustment

carrying capacity (saturation level)

stabilised population level

population increase

exponential growth curve

time →

b gradual adjustment: the 'S' curve

carrying capacity (saturation level)

stabilised population level

population increase

exponential growth curve

time →

c fluctuating, gradual adjustment: the 'J' curve

overshoot

fluctuations in population

carrying capacity (saturation level)

population increase

sudden population check

exponential growth curve

time →

Figure 13.58

Three models illustrating the relationships between an exponentially growing population and an environment with a limited carrying capacity

The links between population growth and economic development

These were summarised by Ford at the International Conference on Population and Development, held in Cairo in 1994.

'It is convenient to view the main lines of thinking about links between population growth and economic development in terms of the three broad perspectives of "Transition theory", "Orthodoxy" (or "Neo-Malthusianism") and "Revisionism". Transition theory provided the first comprehensive explanation of fertility change, which viewed transition from high to low birth and death rates, as being achieved through industrialisation removing the social and structural 'props' which supported high fertility. This view and its policy implications were expressed by some countries at the 1974 World Population Conference (Bucharest) in terms of the slogan "development is the best contraceptive". Neo-Malthusianism (1986), which arose in recognition of the dramatic acceleration in population growth following declines in mortality rates in developing countries, questioned transition theory postulated on

the grounds that rapid population growth could itself impede economic development by exacerbating social and economic problems. Transition theory was further undermined by the empirical findings that fertility decline could often take place prior to appreciable economic development, partly in response to a range of social forces. By the time of the 1984 World Population Conference (Mexico City) an emphasis upon taking positive steps to reduce population growth (largely through family planning programmes) had become the consensus view among both international agencies and the governments of many less developed countries. The general consensus view articulated the need for population policies in integration with other development strategies. The "Revisionist" view (Boserup) argued that economic growth could be stimulated by the search for innovation to increase production in the face of population pressure.'

Boserup was a Danish economist who, in 1965, put forward an alternative theory to that of Malthus. She asserted that in a pre-industrial society an increase in population was likely to stimulate change in agrarian technology and result in increased food production, i.e. 'necessity is the mother of invention'. She suggested that as population increased, farming became more intensive due to innovation and the introduction of new methods and technology.

An international team, which included scientists and administrators and known collectively as the **Club of Rome**, predicted through the use of computer models in 1972, that if the then present trends in population growth and resource utilisation continued, then a sudden decline in economic growth would occur within the next century. Their suggested plans for global equilibrium, few of which have been implemented, included **a** the stabilisation of population growth, use of resources, industrial growth and economic development, and **b** an emphasis on food production and conservation.

Figure 13.59

Population growth and population checks

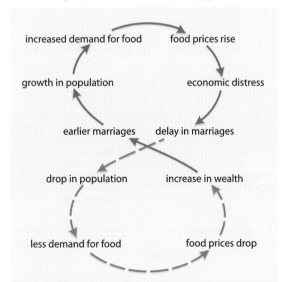

increased demand for food → food prices rise

growth in population → economic distress

earlier marriages → delay in marriages

drop in population → increase in wealth

less demand for food → food prices drop

Population in China

China had, in 1999, an estimated population of 1.24 billion, which was 21 per cent of the world's total. As in other countries, this population was far from evenly distributed (Figure 13.1), with 94 per cent living in the south and east of the country on only 40 per cent of the total land area (Figure 13.60). A population density map (Figure 13.61) shows that the highest densities are either in the coastal provinces or in those in the middle and lower Yangtze Basin. In contrast, densities are extremely low in the mountainous and desert provinces in the north and west of China.

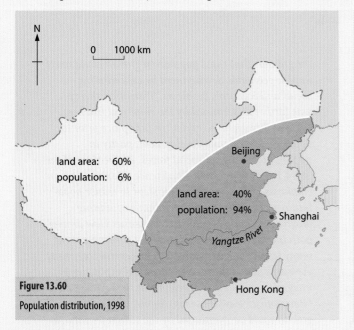

Figure 13.60

Population distribution, 1998

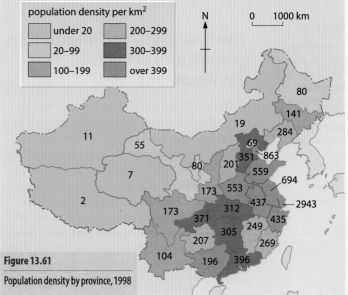

Figure 13.61

Population density by province, 1998

Figure 13.62 shows the demographic cycle for China since the formation of the People's Republic in 1950; and Figure 13.63 China's age structure based on estimates for mid-1999.

The high birth rate of the 1950s was a response to the state philosophy that 'a large population gives a strong nation', and people were encouraged to have as many children as possible. At the same time, death rates were falling, mainly due to improved food supplies and medical care.

The period between 1959 and 1961 coincided with the 'Great Leap Forward'. It was a time when industrial production had to be increased at all costs, and little attention was paid to farming (Places 63, page 468). The result was a catastrophic famine in which an estimated 20 million people died; infant mortality rates rose and birth rates fell.

During the 1960s, attempts to control population growth were thwarted by the Cultural Revolution. Every three years, China's population increased by 55 million – i.e. by the same amount as the total UK population.

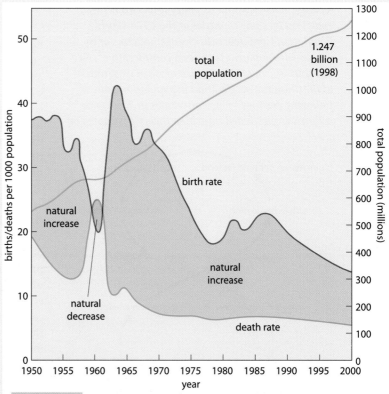

Figure 13.62

The demographic cycle since 1950

State family planning programmes were introduced in the 1970s. By 1975 the average family size had fallen to three children, but this was still regarded by the state as being too many. The state began an advertising campaign for *Wan-xi-shao* – 'later, longer, fewer' (later marriages, longer gaps between children, and fewer children). Even if the family size had been reduced to two, it would still have meant that China's population would, due to the millions of couples in the reproductive age group, have doubled within 50 years. A Chinese demographer, Liu Zeng, calculated that China's optimum population was 700 million (at a time when the total population was approaching 1000 million) and that the state should aim to reduce the population to that figure by 2080. To achieve this, the total fertility rate (TFR) would have to be reduced to a maximum of 1.5. The state decided, in 1979, to play safe and to introduce a rigorous 'one child per family policy' (i.e. a TFR of 1.0). Inducements to have only one child included free education, priority housing, pension and family benefits. Not only were these lost after a second child was born, but fines of up to 15 per cent of the family's income were imposed. The marriageable age for men was set at 22 and for women at 20, with couples having to apply to the state for permission to marry and, later, to have a child (Figure 13.64).

Royle and Philips, writing in the *Geography Review* of September 1997, claimed that the policy, in its early years, 'was operated fiercely. Couples who had more than one child were subject to economic penalties; women pregnant for the second or subsequent time were often coerced into having abortions, sometimes quite late; persistent offenders might be "offered" sterilisation; workplaces and homes were subject to visits from family-planning officials; and the infamous "granny police" would try to ensure that families under their charge did not break the rules. Contraceptive advice and devices were pressed upon the people and over 80 per cent of married women had access to contraception, a tremendous achievement given China's lack of development, huge size and largely rural population.' There were frequent reports of female infanticide where the firstborn was a girl (especially in rural areas where girls were considered less useful for working in the fields), in the hope of a later child being a boy

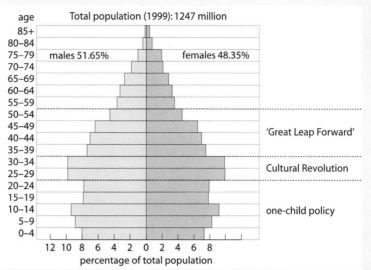

age
Total population (1999): 1247 million

males 51.65%   females 48.35%

'Great Leap Forward'

Cultural Revolution

one-child policy

12 10 8 6 4 2 0 2 4 6 8
percentage of total population

Figure 13.63

Age structure of Chinese population, mid-1999 estimate

Figure 13.64

One-child family, Kunming, Yunnan

Discriminatory abortion is creating a worrying imbalance in China's sex ratio, many couples terminating a pregnancy if they suspect the baby will be a girl. Recognising that ultrasound techniques were being abused for this purpose, authorities have made it illegal to tell parents the sex of their child before birth, but as Chinese doctors are highly underpaid and are used to accepting gifts then...

The state has now banned both forced and late abortions. Termination, even when a pregnancy is 'outside the plan', is no longer compulsory and officials can only fine families who ignore family planning advice. However, there are still reports from rural areas of over-zealous task forces performing forced abortions. Even in the larger cities, abortions as late as sev are common. To reduce these pressures, the authorities do try to encourage, often forcibly in rural areas, women who have had their quota to undergo tubal ligation.

South China Morning Post, 1994

Figure 13.65

The effects of the one-child policy

(Figure 13.65); and of the emergence of a generation of spoilt single children referred to as 'Little Emperors'.

After 1987 the government began to relax its rigid policy, partly in response to intermittent outrage over cases of coercion and brutality and partly because, as a general rule, the policy was working well. During May 1999, the present author spent a month in China. Seeking up-to-date information on the 'one-child policy' he was told that it was 'very complicated'. Complications result mainly from the different needs and differences in enforcement between rural and urban areas, as well as between the Han majority (who account for 92 per cent of the total population) and the 56 other minority groups. Although it is dangerous to generalise, the views usually expressed to the author confirmed that:

- the Han are allowed one child only, unless the first child is mentally or physically handicapped, or dies – under these circumstances, the parents may apply to have a second child
- in many rural areas, farmers can have a second child if their firstborn happens to be a girl; if the second is a girl, then hard luck!
- minority groups, many of whom live in outlying provinces, are allowed two children (Figure 13.66) or, if they live in a very isolated area, possibly up to four children (fewer officials to keep a check)
- Han people living in rural areas who have a second child may be allowed to keep it on the payment of a fine; the fines are imposed by each province and can vary from being relatively small (in less well-off areas where the family income and standard of living are low) to very high (in small urban areas)
- people working for state firms are likely to be made redundant should they have a second child, whereas those working for overseas (transnational) companies are often allowed to keep their job as such policy decisions are made outside China
- in the event of twins, the state will pay the extra costs
- as more people receive a better education, fewer couples are taking up the option to apply for even a single child

- as the first single-family children are now reaching marriageable age, if two 'only' children marry they can have more than one child.

In the author's experience, most people, especially those living in urban areas, seem to accept the necessity of the policy. Those in their early twenties, who were the first generation to be 'only' children, admitted that they would have liked to have had a brother or sister, but none acted like 'Little Emperors', and all seemed to acknowledge that the policy had probably raised their standard of living. With the birth rate having fallen from 44 per 1000 in 1950 to 16 in 1999 (5.8 children per family to 1.9), the authorities are considering further reductions in birth-control restrictions (Figure 13.67), for example by:

- abolishing quotas for child births in 300 trial districts and replacing them with voluntary family planning education programmes which also allow, for the first time, choice between different kinds of contraception
- relaxing penalties for those having larger families
- allowing more exceptions to the one-child rule.

Although the one-child policy risked internal dissent and brought international criticism on grounds of human rights, without it China's population would have risen at a phenomenal rate, leading to severe overpopulation and a depletion of both national and global resources (only 70 million, instead of 300 million, children were born in the first two decades of the

policy's implementation). Figure 13.68 gives three projections, made in 1980, of how China's population structure may have looked in 2040 had the TFR been 2.0 (**a**) or 1.5 (**b**) rather than the enforced 1.0 (**c**).

Figure 13.66

Two-boy family, Lijiang, Yunnan

Figure 13.67

Beijing: new policy on birth control, 1999

By Tang Min

AFTER October 1, married couples in Beijing will be allowed to decide when to have their first baby.

'However, before the wife gets pregnant, the couple will need a special certificate to be issued by the local birth control administration,' said Li Yunli, vice-director of the Beijing Municipal Birth Control Commission, yesterday. It will be the same for married Beijing couples who have already received a special approval for a second child.

This major change in Beijing's birth control policy was laid down officially in an Amendment to the Beijing Municipal Birth Control Regulations, passed on May 14.

Before this amendment takes effect, all married Beijing couples who want a child must apply in compliance with a 'birth quota' regulation.

Since the quota is valid for only one year, the couple must apply again the following year if they fail to fulfil their wish.

According to the amendment, monthly allowances for single-child families will be prolonged for four years.

With the previous regulation, the allowance ended when the child reached 14 years old. The amendment makes it possible for the families to receive the allowance until the

China Daily, 20 May 1999

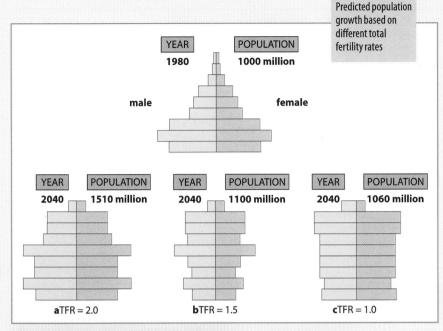

YEAR	POPULATION
1980	**1000 million**

male | female

YEAR	POPULATION
2040	**1510 million**

a TFR = 2.0

YEAR	POPULATION
2040	**1100 million**

b TFR = 1.5

YEAR	POPULATION
2040	**1060 million**

c TFR = 1.0

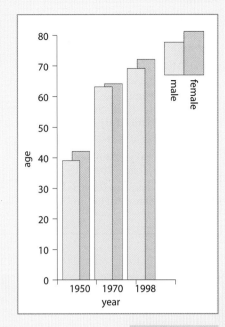

Figure 13.70

Increase in life expectancy, 1950–98

Figure 13.69

Ageing Naxi women, Lijiang, Yunnan

Whereas China's major population concern in the 1970s and 1980s was population growth, at the beginning of the 21st century it has become that of an ageing population (Figure 13.69). Most countries, including China, are bracing themselves for an ageing population due to an increase in life expectancy. Figure 13.70 shows that a person born in China in 1998 can expect to live 30 years longer than a person born in 1950. However, the problem of ageing in China is expected to be more acute than in other countries. This is due to the distortions created by the baby boom encouraged by Chairman Mao in the 1950s and 1960s (notice the 25–34 age groups in

China's population age structure, Figure 13.63) followed by the one-child policy implemented since 1979. Predictions suggest that the proportion of those over 60 will rise from the 10 per cent of 1998 to 22 per cent by 2030, and that the ratio of workers to dependants will fall from 10:1 to 4:1. The World Bank claims that 'China will have the old-age burden of a high-income economy, like that of Japan, with only the resources of a middle-income economy to shoulder it'.

These changes will, according to a report in the *China Business Handbook* (1999), 'have a massive impact on Chinese society and require urgent reform of the provision

of pensions, health care and other benefits. Indeed, there is no pension system at all for the majority of the population, especially in rural areas'. In time, more single-children families will have to support up to two parents and four grandparents – the so-called 4-2-1 pattern. At present most grandparents in rural areas still live, as they have always done, with their family – a situation that has become unusual in large urban areas. Only recently has the state begun to take some responsibility for welfare of the elderly.

The imbalance in the sexes, resulting from the traditional preference for boys, is another concern. Statistics are hard to come by due to sensitivities on the subject, but a 1996 government survey revealed that there were 118 newborn boys reported for every 100 girls, compared with a world average of 105 male births to every 100 female births. More recently, a survey in Hunan Province revealed that there were 2.85 million more males than females in the total local population of 64 million. The distortions were particularly noticeable among the younger generation, prompting one magazine to predict the imminent emergence of 'a large army of bachelors'.

References

Annual Abstract of Statistics 1998 (1998) HMSO

Barke, M. and O'Hare, G. (1984) *The Third World*, Oliver & Boyd

Hornby, W. F. and Jones, M. (1980) *An Introduction to Population Geography*, Cambridge University Press.

Key Data 1993/94 (1994) HMSO

Philips, D.R. (ed.) (2000) *Ageing in the Asia-Pacific Region*, Routledge.

Philip's Geography Digest 1998–1999 (1998) Heinemann-Phillips.

Waugh, D. (1998) *The New Wider World*, Thomas Nelson.

Websites

UN: 1998 Revision of World Population Estimates and Projections:
http://www.popin.org/pop1998/1.htm

US Census Bureau, International Programs Center (see links to World Population Information and International Data Base):
http://www.census.gov/ipc/www/

Population Reference Bureau:
http://www.igc.org/prb/

The Union of Concerned Scientists Population/Environment:
http://www.ucsusa.org/resources/population.html

World Population Slide Show:
http:www.snowcrest.net/geography/slides/pop/index.html

The US High Commissioner for Refugees:
http://www.unhcr.ch/

The China Population Information and Research Center:
http://www.cpirc.org.cn/eindex.htm

See also for more links:
http://www.nelsonthornes.com/gaia

Questions

1
a What do you understand by:
  i birth rate? **(1 mark)**
  ii life expectancy? **(1 mark)**
  iii overpopulation? **(2 marks)**
b Figure 13.71 illustrates Malthus's view of the relationship between population and food supply in a typical country or region.
  i Describe what the diagram shows. **(2 marks)**
  ii Explain what Malthus thought would be the consequences of the changes shown in the model. **(2 marks)**
  iii Suggest why Malthus's predictions did not come true in England following publication of his ideas in the early 18th century. **(4 marks)**
  iv Describe the views of Boserup on the balance between population and resources, and explain how these are different from the views of Malthus. **(5 marks)**
c Choose one country that has attempted to manage its population by introducing laws which it hopes will affect birth rate.
  i Explain how the population policy was intended to operate.
  ii Discuss the consequences of the policy, mentioning both its successes and its failures. **(8 marks)**

Growth of population	1	2	4	8	16	etc.
Growth of food supply	1	2	3	4	5	etc.
Time periods	→	→	→	→	→	→

Figure 13.71
Malthus's view of population and food supply

2
a What do the following terms mean:
  i birth rate? **(2 marks)**
  ii natural increase of population? **(2 marks)**
  iii annual growth rate of population? **(2 marks)**
b Study Figure 13.11 on page 351. What statistical change marks the move from:
  i Stage A (high, fluctuating) to Stage B (early expanding)?
  ii Stage B to Stage C (late expanding)?
  iii Stage C to Stage D (low, fluctuating)? **(3 marks)**
c Explain how social and/or economic changes could have brought about each of the moves described in (b). **(10 marks)**
d Suggest how the total population of the UK will change over the next 50 years. Give reasons for your suggestions. **(6 marks)**

AS

3 Study Figure 13.71 above.
a Outline the theory developed by Malthus to explain the relationship between population increase and the increase in food supply. **(5 marks)**
b Malthus wrote in the early 18th century. He predicted population growth could soon cause widespread famine and other disasters in England. His predictions have not come true. Explain why. **(10 marks)**
c In recent years views described as neo-Malthusian have become common. Explain why these ideas have developed. Contrast the neo-Malthusian view with the more optimistic view of population growth developed by Boserup. **(10 marks)**

4
a The demographic transition model is divided into four, or sometimes five, stages. At what point is the most rapid population growth? Why is the growth so rapid at this point? **(3 marks)**
b Many demographers say that the key to reducing the birth rate in less economically developed countries lies in changing the educational and economic status of women. Discuss this view, with reference to one or more countries that you have studied. **(8 marks)**
c i Name **one** country which has a falling population because its birth rate has fallen very low (which might be described as being in Stage 5 of the demographic transition). Explain what has caused its falling birth rate. **(4 marks)**
  ii Discuss the issues that might arise for countries experiencing a falling population, and explain how some of these issues might be managed. **(10 marks)**

A2

5 a What is meant by:
   i life expectancy? **(1 mark)**
   ii fertility rate? **(3 marks)**
   iii dependency ratio? **(2 marks)**
b Study Figure 13.72.
   i Describe and compare the changes in life expectancy in Japan and the UK shown in Figure 13.72. **(3 marks)**
   ii Suggest why life expectancy is changing in these countries. **(3 marks)**
   iii Apart from changes in life expectancy, name one other factor that might be increasing the proportion of elderly people in the populations of these two countries. Explain your answer. **(4 marks)**
   iv These changes present many challenges to planners in these countries. Explain how they might affect:
   • housing policy
   • economic policy
   • employment policy
   • immigration policy. **(9 marks)**

6 a Study the map of Brazil's population distribution (Figure 13.5 on page 347).
   i Describe the distribution of areas of dense population. **(2 marks)**
   ii Brazil was settled as a colony by Europeans from the 16th century onwards. Suggest how this has influenced the pattern of population density. **(2 marks)**
   iii The area marked 1B is the tropical rainforest. Suggest why this area, or any other area of tropical rainforest that you have studied, has a very sparse population. **(4 marks)**
b Study the map of population density in London (Figure 13.7 on page 348).
   i Describe the distribution of population shown on this map, and suggest why this pattern has developed. **(4 marks)**
   ii During the 20th century there was a large movement of people out from central London into the suburbs and beyond. Explain why people wished to move, and how changing technology allowed them to make the move. **(6 marks)**
c Name an example of an area where population densities have fallen because of the forced migration of large numbers of people. Describe the causes of this migration, and explain how the problems arising from this migration have been managed. **(7 marks)**

Figure 13.72

Consequences of the falling death rate

a the world's ageing population

age	1970 total (millions)	1970 percentage of total world population	2000 (estimate) total (millions)	2000 (estimate) percentage of total world population
over 60	291	7.8	620	9.8
over 85	26	0.8	58	1.9

b life expectancy in Japan and the UK (age in years)

	1960 male	1960 female	1992 male	1992 female	2025 (estimate) male	2025 (estimate) female
UK	68	74	73	79	76	82
Japan	67	71	76	82	82	85

7 Study Figure 13.72.
a Apart from increasing life expectancy, name the other main factor that is leading to a growing proportion of elderly people in the world's population. Explain your answer. **(5 marks)**
b i With reference to one or more countries, explain the socio-economic and geographical consequences that might arise from the growing proportion of elderly people in the population. **(12 marks)**
   ii Suggest how planners in both the public and private sectors of the economy might prepare for the ageing of the population. **(8 marks)**

8 The demographic transition model describes changes in the population which many countries in Europe and North America went through during the 19th and 20th centuries. Many less economically developed countries (LEDCs) went through the early stages of the model in the second half of the 20th century. Are these countries likely to go through the later stages of the model in the 21st century? With reference to **at least two** LEDCs, explain what factors are likely to influence their rates of population change in the 21st century. **(25 marks)**

9 Study Figure 13.41 on page 368.
   i Name **one** major migration of people from a less economically developed country (LEDC) to a more economically developed country (MEDC).
   ii Name **one** major migration of people from one LEDC to a different region within that LEDC, or to another LEDC.
For **each** of your chosen migrations, explain the causes and the consequences of the movement of people. **(12 marks + 13 marks)**

10 Read Case Study 13 on pages 380–383.
a Account for the size of the population in:
   i the 25–34 age cohort (born 1965–74)
   ii the 15–24 age cohort (born 1975–84)
   iii the 0–14 age cohort (born 1985–99). **(10 marks)**
b In 1979 the Chinese authorities made a decision to introduce the one-child policy. As in any decision making exercise, they based their decision on certain criteria. Suggest what were the criteria, on which this decision was based. **(7 marks)**
c Assess the successes and failures of the One-Child Policy. **(8 marks)**

11 Study the two population pyramids below.
  a i What do you understand by the term
         'dependency ratio'?   **(2 marks)**
     ii Calculate the dependency ratios for
         each pyramid.   **(2 marks)**
     iii Suggest, with reasons, what stage of the
         demographic transition is represented by
         each of the pyramids.   **(4 marks)**
  b Choose **one** country which has a population
     structure similar to the one shown in pyramid A.
     i Suggest two problems that are likely to arise in that
        country as a consequence of the large proportion of
        the population in the 0–15 age group.   **(4 marks)**
     ii How is the country attempting to manage
        these problems?   **(5 marks)**

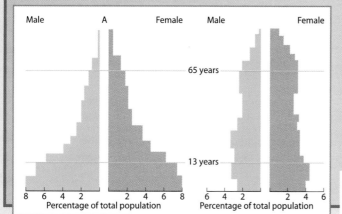

13 Study the two population pyramids in Figure 13.73.
  a i Compare and contrast the shapes of the
         two pyramids.   **(3 marks)**
     ii Account for the differences between the
         two pyramids.   **(4 marks)**
     iii Suggest what population problems are likely
         to be met in these two countries during the
         next 20 years or so.   **(8 marks)**
  b i Name a country which has adopted policies to
         help it to manage its total population and its
         rate of population change.
     ii Describe its population policies and assess
         how successful these policies have been.   **(10 marks)**

14 Study the two tables of data in Figure 13.74.
  a i Describe how the education of women has
         affected fertility rates in Morocco.   **(3 marks)**
     ii Suggest reasons for the changes you have
         described.   **(4 marks)**
  b i Describe the changes to the fertility indices
         of women from the Maghreb, who have
         migrated to France.   **(3 marks)**
     ii Suggest reasons for these changes.   **(4 marks)**
     iii Suggest how the emigration from the Maghreb
         may have affected the fertility rate of women
         who have remained in those countries.   **(5 marks)**
  c What lessons can be learned from these figures
     by development workers in countries suffering
     from pressure caused by rapidly increasing
     population?   **(6 marks)**

Figure 13.74

  c Choose **one** country which has a population
     structure similar to that shown in pyramid B.
     Describe two problems that might arise in future
     as a consequence of the ageing population
     structure, and suggest how these problems
     might be managed.   **(8 marks)**

12 The period following the Second World War saw some
    of the biggest international migrations that the world
    has ever known.
  a Name **one** major international migration that took
     place during this period. Refer to the source and
     the destination of the migrants.   **(1 mark)**
  b Explain the causes of the migration. Refer to
     pushes from the source and pulls to the
     destination.   **(6 marks)**
  c Discuss the consequences of the migration for:
     i the source country
     ii the host country
     iii the migrants themselves.   **(12 marks)**
  d Suggest why large international migrations
     have been so common in the period since the
     Second World War.   **(6 marks)**

Figure 13.73

Population pyramids

AS

15 The proportion of people aged 60 and over is expected
    to increase rapidly in most countries in western Europe
    during the next 25 years. What implications does this
    have for social and economic planners in
    these countries?   **(25 marks)**

a Average number of children per Moroccan women according to age
     and educational level, 1994

Age group	No education	Primary	Secondary	Higher
20–24	1.6	1.1	0.7	0.5
25–29	2.7	1.7	1.3	0.9
30–34	3.9	2.6	1.9	1.3
35–39	5.1	3.4	2.4	2.2
40–44	6.1	4.2	3.2	2.7
45–49	6.7	4.9	3.4	2.5

b Fertility indices (average number of children born to each woman) of
     Maghreb women in country of origin and country of residence*

| | Algerian women in | | Moroccan women in | | Tunisian women in | |
Year	Algeria	France	Morocco	France	Tunisia	France
1977	7.47	4.73	5.93	5.75	5.84	5.05
1981	6.39	4.35	5.92	5.84	5.19	–
1985	6.24	4.24	–	4.47	4.53	4.67
1987	5.29	3.95	4.46	4.09	4.10	4.49
1989	4.72	3.66	3.95	3.71	3.40	4.30
1991	–	3.35	–	3.25	3.34	3.88
1992	–	3.27	3.28	2.99	3.36	3.56

*'The Maghreb' is the western part of North Africa. It was colonised by France, gaining independence in the 1950s and
1960s. These countries still have close ties with France, and there has been much migration from Maghreb to France.

For information on **Key Skills** and the **Key Skills Certificate** see page 101.

At Level 3 the key skills are described as needing to be part of a 'substantial activity'. The activities on this page can be used to provide evidence for the Key Skill of **Application of Number** to Level 3. On page 101 you will find activities to cover the Key Skill of Communication and on page 259 activities to cover Information Technology, also to Level 3.

These activities are a particular way of carrying out tasks that would form a part of a Geography course and its assignments. As you carry out such activities to gather the evidence, your teachers and tutors are expected to monitor the work (as an examination board would require of any coursework to ensure that the work is done by the candidate). Their testimony will provide one source of evidence that *you* did the work. The monitoring also has a teaching role, helping you to strengthen plans and activities to create a final result that achieves all of the outcomes planned for the activities.

The activities suggested below can be used as individual tasks to support Key Skill development; as a completed piece of evidence towards a Key Skills portfolio; or as a learning vehicle to consider the Geography content of the material.

Application of Number – Level 3

This opportunity for access to the Key Skill of Application of Number is based around the **demographic transition model**. Once you understand the requirements of Part A of the specification ('what you need to know') you are required to plan and carry out 'at least one substantial and complex activity' in Part B. This requires you to use *two* different types of source to:
- plan, gather and interpret data including 'a large data set'
- carry out 'multi-stage calculations' covering a range of manipulative skills
- interpret results, present findings and justify methods, using a range of images.

Co-operation in data collection could produce information to allow each student to develop the activity, as well as providing evidence for the Key Skill of 'Working with Others'.

The tasks suggested below use the demographic transition model to allow you to work in a way that will, in total, present the full 'package' of required competences. Although some of the 'range' data manipulation will not be included in these activities, these can be tracked through the use of other activities in this textbook (see the table right) (Note that there is also an external test for this Key Skill).

Task: Research the validity of the demographic transition model

Relate the demographic transition model to the changes in population in *two* chosen countries over a period of 100 years. One country must be a more economically developed country (MEDC), and the other a less economically developed country (LEDC).

In doing this you will need to:
- identify the meanings of statistical data (birth and death rates)
- explain the theory of the demographic transition model
- research sources as necessary to collect data for real countries (an MEDC and an LEDC)
- compare the results with each other and with the model.

Activity One
Plan and document the actions you will need to take to complete the activities involved in the task. This plan should identify the methods you will use (and source suggestions) to collect the information you will need, and the use you will make of it to achieve the results. It is understood that in 'normal' working with numbers you will not always need to plan in detail for others to follow, but for the portfolio for Key Skills it is necessary to prove to an assessor that you *did* plan your work.

Activity Two
Identify and explain the demographic transition model and its relationship to the level of economic development of a country. Your explanation should include an identification of the mathematical terms used, including how they are calculated. You should also explain the reasons for using **rates** rather than raw data (absolute numbers) and include examples to show that you understand.

Activity Three
Identify your sources of information and collect information about birth rates and death rates for the last 100 years for your *two* chosen countries (one MEDC and one LEDC). Explain how the data is collected in each case. Choose appropriate methods to manipulate the data you have gathered to achieve the numerical structures you will need. Compare the patterns found with each other and with the demographic transition model. From the data, calculate the natural change in population over this period in each country.

Activity Four
Present your findings using at least *one* graph and *one* chart, as well as a written description. Then use the data to test the hypothesis that the demographic transition model can be applied to population changes in LEDCs as well as MEDCs.

Activity Five
Evaluate your response, identifying why you chose particular types of image to present your data, and explaining how the results of your mathematical manipulations relate to the purpose of your investigation. Your evaluation should also assess the data to identify the likelihood of your results being an accurate reflection of reality.

Note: The complete range required of the Application of Number Key Skill is not present in this piece of work. The missing items are:

Item	Possible source
Use estimation to plan	
Rounding to 1 significant figure	
Make accurate and reliable observations over time	Focus on Key Skills: Communication (page 101) – river basin study
Use suitable equipment to measure in appropriate units	
Use compound measures (e.g. metres per second)	

Chapter 14

Settlement

*'The largest single step in the ascent of man is the
change from nomad to village agriculture.'*

J. Bronowski, *The Ascent of Man*, 1973

Origins of settlement

About 8000 BC, at the end of the last ice age, the
world's population consisted of small bands of
hunters and collectors living mainly in sub-tropical lands and at a subsistence level (page 478).
These groups of people, who were usually migratory, could only support themselves if the whole
community was involved in the search for food.
At this time two major technological changes,
known as the 'Neolithic revolution', turned the
migratory hunter-collector into a sedentary
farmer. The first was the domestication of animals
(sheep, goats and cattle) and the second the
cultivation of cereals (wheat, rice and maize).
Slow improvements in early farming gradually
led to food surpluses and enabled an increasing
proportion of the community to specialise in
non-farming tasks.

The evolution in farming appears to have
taken place independently, but at about the
same time, in three river basins: the Tigris–
Euphrates (in Mesopotamia), the Nile, and the
Indus (Figure 14.1). These areas had similar
natural advantages:

- hills surrounding the basins provided pasture
  for domestic animals
- flat floodplains next to large rivers
- rich, fertile silt deposited by the rivers during
  times of flood
- a relatively dry – but not too dry – climate
  which maintained soil fertility (i.e. limited
  leaching) and enabled mud from the rivers to
  be used to build houses (climatically, these
  areas were more moist than they are today)
- a warm subtropical climate, and
- a permanent water supply from the rivers
  for domestic use and, as farming developed,
  for irrigation.

By 1500 BC, larger towns and urban centres had
developed with an increasingly wider range of
functions. Administrators were needed to organise the collection of crops and the distribution of
food supplies; traders exchanged surplus goods
with other urban centres; early engineers introduced irrigation systems; and a ruling elite
appropriated taxes from the agricultural and
trading population to support the military, the
priesthood, and 'non-productive' members of
society, such as artists, philosophers and
astronomers. Craftsmen were required to make
farming equipment and household articles – the
oldest-known pottery and woven textiles were
found at Catal Huyuk in present-day Turkey –

Figure 14.1

Civilisations and
cities before 1500 BC

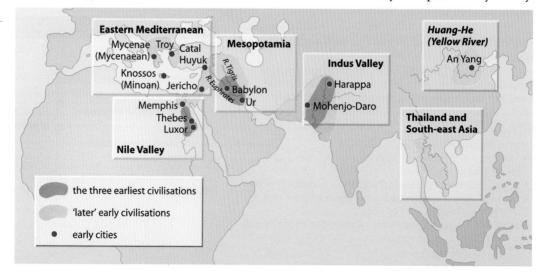

and copper and bronze were being worked by 3000 BC. As towns continued to grow, it became necessary to have a legal system and an army for defence. Although there is divergence of opinion over the exact dates, Figure 14.2 gives a chronological sequence of early settlements.

Figure 14.2

A chronology of early settlement

Approximate date BC		Near East { Tigris–Euphrates / Nile / Indus }	Rest of world
9000		Hunters and collectors	
8000	8500	First domesticated animals and cereals	Northern Europe recovering from the last ice age
	8300	Jericho: first walled city	
7000			
6000	6250	Catal Huyuk: first pottery and woven textiles; became largest city in world	
5000	5500	Growth of villages in Mesopotamia Growth of many villages in Nile and Indus valleys	
	5000	Early methods of irrigation	Rice cultivation in South-east Asia
4000		Bronze casting	
3000	3500	Invention of the wheel and plough in Mesopotamia, and the sail in Egypt	First Chinese city
	3000	Cities in Mesopotamia	First crops grown in central Africa; bronze worked in Thailand
2000	2600	Pyramids	
	2000	Minoan civilisation in Crete	Metal-working in the Andes
1000	1600	Mycenaean civilisation in Greece	

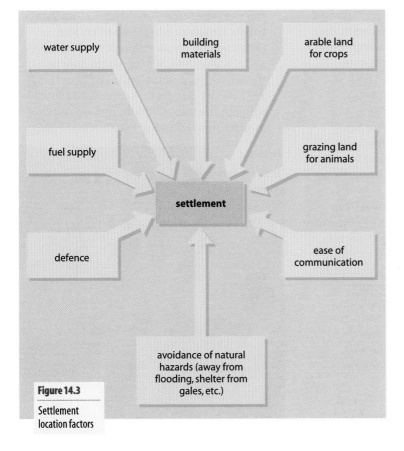

Figure 14.3

Settlement location factors

Site and situation of early settlements

Site describes the characteristics of the actual point at which a settlement is located, and was of major importance in the initial establishment and growth of a village or town. **Situation** describes the location of a place relative to its surroundings (neighbouring settlements, rivers and uplands). Situation, along with human and political factors, determined whether or not a particular settlement remained small or grew into a larger town or city (Figure 14.9).

Early settlements developed in a rural economy which aimed at self-sufficiency, largely because transport systems were limited. While the most significant factors in determining the site of a village include those shown in Figure 14.3 and described below, remember that several factors would usually operate together when a choice in the location of a settlement was being made.

Among the most important factors are:

■ **Water supply** A nearby, guaranteed supply was essential as water is needed daily throughout the year and is heavy to carry any distance. In earlier times, rivers were suffi-ciently clean to give a safe, permanent supply. In lowland Britain, many early villages were located along the spring line at the foot of a

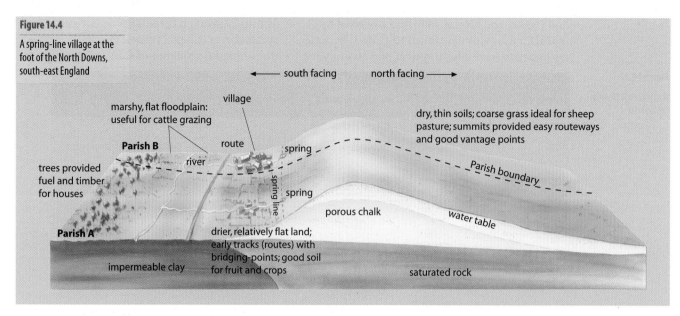

← south facing | north facing →

village

marshy, flat floodplain: useful for cattle grazing

dry, thin soils; coarse grass ideal for sheep pasture; summits provided easy routeways and good vantage points

Parish B

route

spring

trees provided fuel and timber for houses

river

Parish boundary

spring line

spring

porous chalk

water table

Parish A

drier, relatively flat land; early tracks (routes) with bridging-points; good soil for fruit and crops

impermeable clay

saturated rock

chalk or limestone escarpment (Figures 8.10 and 14.4). In regions where rainfall is limited or unreliable, people settled where the water table was near to the surface (a desert oasis, Figure 14.5) enabling shallow wells to be dug. Such settlement sites are known as **wet-point** or **water-seeking sites**.

- **Flood avoidance** Elsewhere, the problem may have been too much water. In the English Fenlands, and on coastal marshes, villages were built on mounds which formed natural islands (Ely). Other settlements were built on river terraces (page 82) which were above the flood level and, in some cases, avoided those diseases associated with stagnant water. Such sites are known as **dry-point** or **water-avoiding sites**.

- **Building materials** Materials were heavy and bulky to move and, as transport was poorly developed, it was important to build settlements close to a supply of stone, wood and/or clay.

- **Food supply** The ideal location was in an area that was suitable for both the rearing of animals and the growing of crops – such as the scarps and vales of south-east England (page 199). The quality, quantity and range of farm produce often depended upon climate and soil fertility and type.

- **Relief** Flat, low-lying land such as the North German Plain was easier to build on than steeper, higher ground such as the Alps. However, the need for defence sometimes overruled this consideration.

- **Defence** Protection against surrounding tribes was often essential. Jericho, built over 10 000 years ago (about 8350 BC), is the oldest city known to have had walls. In

Britain, the two best types of defensive site were those surrounded on three sides by water (Durham, Figure 14.6) or built upon high ground with commanding views over the surrounding countryside (Edinburgh). Hilltop sites may, however, have had problems with water supply (Figure 14.7).

- **Nodal points** Sites where several valleys meet were often occupied by settlements which became **route centres** (Carlisle – Places 49, page 396 – and Paris). **Confluence towns** are found where two rivers join (Khartoum at the junction of the White Nile and the Blue Nile, St Louis at the junction of the Mississippi and the Missouri (Figure 3.59)). Settlements on sites that command routes through the hills or mountains are known as **gap towns** (Dorking and Carcassonne).

Figure 14.5

An oasis: Morocco

Figure 14.6

A settlement within a meander loop: Durham

Figure 14.7

A hilltop defensive site: Andalucia, Spain

■ **Fuel supply** Even tropical areas need fuel for cooking purposes as well as for warmth during colder nights. In most early settlements, firewood was the main source – and still is in many of the least economically developed areas, such as the Sahel.

■ **Bridging-points** Settlements have tended to grow where routes had to cross rivers, initially where the river was shallow enough to be forded (Oxford) and later where the site was suitable for a bridge to be built. Of great significance for trade and transport was the lowest bridging-point before a river entered the sea (Newcastle upon Tyne).

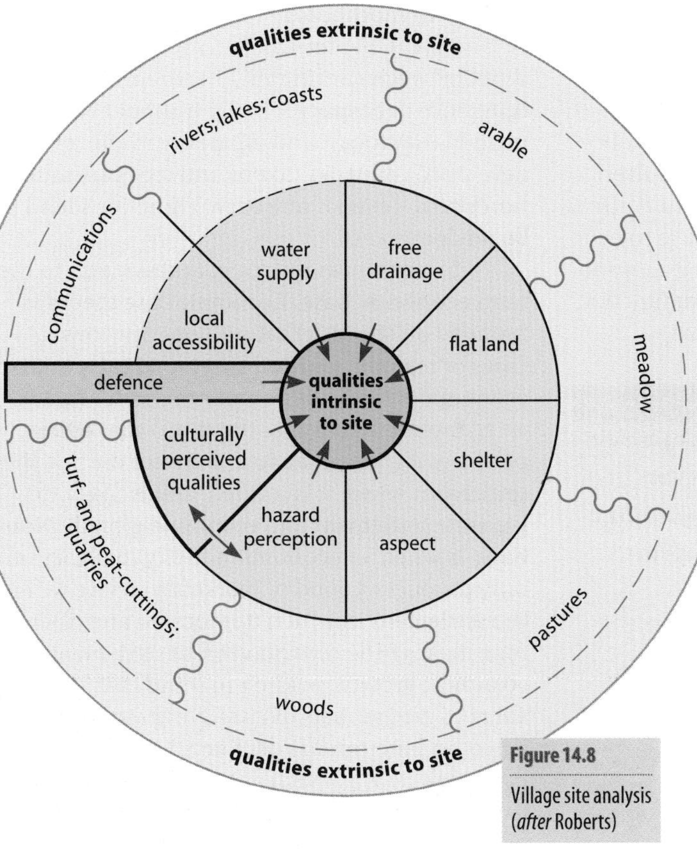

■ **Harbours** Sheltered sea inlets and river estuaries provided suitable sites for the establishment of coastal fishing ports, such as Newquay in Cornwall; later, deep-water harbours were required as ships became larger (Southampton and Singapore, Places 97 page 608). Port sites were also important on many major navigable rivers (Montreal on the St Lawrence) and large lakes (the Great Lakes in North America).

■ **Shelter and aspect** In Britain, south-facing slopes offer favoured settlement sites because they are protected from cold, northerly winds and receive maximum insolation (Torquay).

■ **Resources** Settlements also grew in places with access to specific local resources such as salt (Nantwich, Cheshire), iron ore, coal, etc.

Whereas most of the factors listed above were natural, today the choice of a site for a new settlement is more likely to be **political** (Israeli settlements on the West Bank; Brasilia), **social** (some of Britain's new towns) or **economic** (Blaenau Ffestiniog for its slate – Places 78 page 523 – or, in Brazil, Carajas for its iron ore, and Iguaçu for its hydro-electricity).

Roberts has produced a model (Figure 14.8) which draws together not only site and situation factors, but also the perceptions of different settler groups as to the relative importance of the specific factors – e.g. in a desert, water may be perceived to be the most important; in parts of Mediterranean Europe, it may have been defence. The inner circle in Figure 14.8 is concerned with desirable **site** characteristics (intrinsic qualities) and the outer circle with the general **situation** factors (extrinsic qualities). Roberts stresses that each settlement location represents a complex balancing act of all these factors (London, Figure 14.9), with few sites and situations being ideal. You should be aware that 'extrinsic' factors change over time, and that settlements are dynamic in nature.

Figure 14.8

Village site analysis (*after* Roberts)

Diagram labels:
qualities extrinsic to site
rivers; lakes; coasts
arable
communications
water supply
free drainage
local accessibility
flat land
defence
meadow
qualities intrinsic to site
culturally perceived qualities
shelter
turf- and peat-cuttings; quarries
hazard perception
aspect
woods
pastures
qualities extrinsic to site

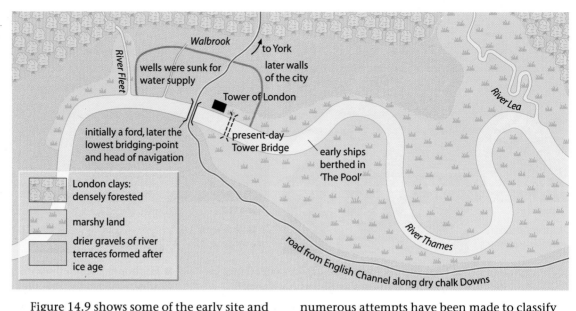

Figure 14.9

The site and situation of early London

Walbrook

to York

River Fleet

wells were sunk for water supply

later walls of the city

Tower of London

River Lea

initially a ford, later the lowest bridging-point and head of navigation

present-day Tower Bridge

early ships berthed in 'The Pool'

London clays: densely forested

marshy land

drier gravels of river terraces formed after ice age

road from English Channel along dry chalk Downs

River Thames

Figure 14.9 shows some of the early site and situation factors that helped determine the original location and early growth of London. As with other settlements, some of these early factors are no longer important, e.g. London now has piped water, has numerous shops to provide food, has bridges and tunnels to cross the river, and no longer needs a castle or city walls for defence.

Functions of settlements

As early settlements grew in size, each one tended to develop a specific function or functions. The **function** of a town relates to its economic and social development and refers to its main activities. There are problems in defining and determining a town's main function and often, due to a lack of data such as employment and/or income figures, subjective decisions have to be made. As settlements are very diverse, it helps to try to group together those with a similar function (Framework 7, page 167). Over the years,

numerous attempts have been made to classify settlements based on function, but these tended to refer to places in industrialised countries and are often no longer applicable to post-industrial societies. Further problems arose when the growth of some settlements was based upon an activity that no longer exists (the former coalmining villages of north-east England and South Wales), or where the original function has changed over **time** (a Cornish fishing village may now be a holiday resort). As functions change in time, this has a direct bearing on settlement morphology (page 394) and patterns of land use (Chapter 15). Functions may also differ between continents – i.e. there is also a difference over **space**. Finally it should be realised that today, and especially in the more developed countries, towns and cities are multi-functional – even if one or two functions tend to be predominant.

It may be worth referring, at this stage, to the term **economic base**. Economic base theory is founded on the idea that settlements (towns, cities or regions) perform two broad categories of economic activity: basic and non-basic. **Basic** is an economic activity (or function) that either produces a good or markets a service outside the settlement where it is located, and is likely to generate settlement and economic growth. **Non-basic** is when an economic activity (or function) only produces a good or markets a service within the settlement in which it is located and, therefore, makes little contribution to settlement or economic growth. Bearing in mind that the value to geographers of classifying settlements based on function has declined, Figure 14.10 has been included, as much as anything, as a check-list should you wish to conduct personal field-work or make an individual study of this topic.

Figure 14.10

Classification of settlement based on function

Rural	Urban	
	Developed countries	**Developing countries**
Market and agricultural	Mining	Administration
Route centre/transport	Manufacturing/industrial	Marketing/agricultural
Small service town	Route centre/transport	Route centre/port
Defensive	Retail/wholesale	Mining
Dormitory/overspill/satellite	Religious/cultural	Commercial
	Trade/commerce/financial	Religious
	Administration	Residential
	Resort/recreation	
	Residential	
	New towns	

Differences between urban and rural settlement

Figure 14.11 shows the commonly accepted types of settlement, but hides the divergence of opinion as to how and where to draw the borders between each type. Several methods have been suggested in trying to define the difference between a village, or rural settlement, and a town, or urban settlement.

- **Population size** There is a wide discrepancy of views over the minimum size of population required to enable a settlement to be termed a town, e.g. in Denmark it is considered to be 250 people, in Ireland 500, in France 2000, in the USA 2500, in Spain 10 000 and in Japan 30 000. In India, where many villages are larger than British towns, a figure of less than 25 per cent engaged in agriculture is taken to be the dividing point.
- **Economic** Rural settlements have traditionally been defined as places where most of the workforce are farmers or are engaged in other primary activities (mining and forestry). In contrast, most of the workforce in urban areas are employed in secondary and service industries. However, many rural areas have now become commuter/dormitory settlements for people working in adjacent urban areas or, even more recently, a location for smaller, footloose industries, such as high-tech industries.
- **Services** The provision of services, such as schools, hospitals, shops, public transport and banks, is usually limited, at times absent, in rural areas (Figure 14.21).
- **Land use** In rural areas, settlements are widely spaced with open land between adjacent villages. Within each village there may be individual farms as well as residential areas and possibly small-scale industry. In urban areas, settlements are often packed closely together and within towns there is a greater mixture of land use with residential, industrial, services and open-space provision.
- **Social** Rural settlements, especially those in more remote areas, tend to have more inhabitants in the over-65 age group, whereas the highest proportion in urban areas lies within the economically active age group (page 354) or those under secondary school age.

It has becoming increasingly more difficult to differentiate between villages and towns, especially where urban areas have spread outwards into the **rural fringe**. The term **rural–urban continuum** (page 516) is used to express the fact that in many highly urbanised countries such as Japan and the UK, there is no longer either physically or socially a simple, clear-cut division between town and country. Instead there is a gradation between the two, with no obvious point where it can be said that the urban way of life ends and the rural way of life begins (Figure 17.1). It is, therefore, more realistic to talk about a transition zone from 'strongly rural' to 'strongly urban'. Cloke (1979) devised an **index of rurality** based upon 16 variables taken largely from census data for England and Wales (Figure 17.2). These variables included people aged over 65; proportion employed in primary, secondary and tertiary sectors; population density; population mobility (those moving home in the previous 5 years); proportion commuting; and distance from a large town (Figure 14.20). Cloke then identified four categories (Figure 17.3):

- extreme rural (parts of south-west England, central Wales, East Anglia and the northern Pennines)
- intermediate rural
- intermediate non-rural, and
- extreme non-rural (mainly suburbanised villages (page 398) around London in Surrey, Cambridgeshire, Hertfordshire and Essex).

Figure 14.11

Type of settlement

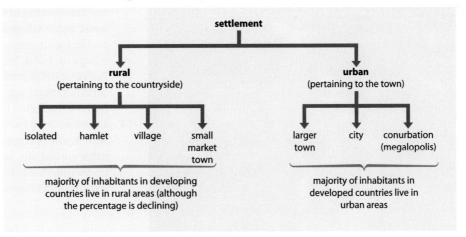

Figure 14.13

Dispersed settlement: in North Yorkshire

Figure 14.12

Isolated settlement: in the Amazon rainforest

Rural settlement

Pattern and morphology

Geographers have become increasingly interested in the **morphology**, i.e. the pattern (numbers 1 and 2 below) and shape (numbers 3–7 below) of settlements. Although village shapes vary spatially in Britain and across the world, it has been – again traditionally – possible to identify seven types (remember that, as in other classifications, some geographers may identify more or fewer categories).

1   **Isolated** This refers to an individual building, usually found in an area of extreme physical difficulty where the natural resources are insufficient to maintain more than a few inhabitants, e.g. the Amazon rainforests where tribes live in a communal home called a *maloca* (Figure 14.12). Isolated houses may also be found in planned pioneer areas such as on the Canadian Prairies where the land was divided into small squares, each with its own farm buildings.

2   **Dispersed** Settlement is described as dispersed when there is a scatter of individual farms and houses across an area; there are either no nucleations present, or they are so small that they consist only of two or three buildings forming a hamlet (Figure 14.13). Each farm or hamlet may be separated from the next by 2 or 3 km of open space or farmland. In the Scottish Highlands and Islands, some communities consist of crofts spaced out alongside a road or raised beach. Hamlets are common in rural areas of northern Britain, on the North German Plain (where their name *urweiler* means 'primeval hamlet') and in sub-Saharan Africa.

3   **Nucleated** Nucleated settlement is common in many rural parts of the world where buildings have been grouped closely together for economic, social or defensive purposes (Figure 14.14). In Britain, where recent evidence suggests that nucleation only took place after the year 1000, villages were surrounded by their farmland, where the inhabitants grew crops and grazed animals in order to be self-sufficient; this led to an unplanned and variable spacing of villages, usually 3–5 km apart. Some villages grew up around crossroads and at T-junctions, as is the case of many villages in India. Many border villages in Britain, hilltop settlements around the Mediterranean Sea, and *kampongs* in Malaysia became nucleated for defensive reasons.

4   **Loose-knit** These are similar to nucleated settlements except that the buildings are more spread out, possibly due to space taken up by individual farms which are still found within the village itself.

Figure 14.14

Nucleated settlement: in Sumatra, Indonesia

5 **Linear**, or **ribbon** Where the buildings are strung out along a main line of communication or along a confined river valley (Figure 14.15), the settlement is described as linear. **Street villages** – planned linear villages – were common in medieval England. Unplanned linear settlements also developed on long, narrow, flood-avoidance sites, e.g. along the raised beaches of western Scotland and on river terraces, as in London. Later, unplanned linear settlement grew up along the floors of the narrow coalmining valleys of South Wales and on main roads leading out of Britain's urban areas following the increase in private car ownership and the development of public transport. In the Netherlands, Malaysia and Thailand, houses have been built along canals and waterways.

6 **Ring and 'green' villages** Ring villages are found in many parts of sub-Saharan Africa and the Amazon rainforest (Figure 14.16). Houses were built around a central area which was left open for tribal meetings and communal life. In Kenya, the Maasai built their houses around an area into which their cattle were driven for protection during most nights. In England, many villages have been built around a central green.

Figure 14.15

Linear (ribbon) settlement: Combe Martin, Devon

7 **Planned** Although many early settlements were planned (Pompeii, York), the apparently random shape of many British villages appears to suggest that they were not. More recently, villages surrounding large urban areas in, for example, Britain and the Netherlands, have expanded and become suburbanised, having small and often crescent-shaped estates (Places 49).

If you study maps of village plans, it is very likely that you will find many settlements with a mixture of the above shapes, e.g. a village may have a nucleated centre, a planned estate on its edges and a linear pattern extending along the road leading to the nearest large town (Figure 14.17).

Roberts (1987) suggested a different basis for classification (Figure 14.18). Even so, he concedes there are difficulties in trying to fit a particular village into a specific category, as when determining if a strip of grass is large enough to be called a green, and concludes that many villages are **composite** (or **polyfocal**), incorporating several plans and phases of development.

Figure 14.16

Ring village: Kraito, in the Amazon rainforest

Basic shape	Plan and morphology		Village green
Linear (in a row)	regular		with
		———————	without
	irregular		with
		— — —	without
Agglomerated (more nucleated)	regular grid	⌗	with
			without
	regular radial	✳	with
			without
	irregular grid	⁚⁚	with
		⊣⊢⊣⊢	without
	irregular agglomerated	●	with
			without

Figure 14.18

A method of classifying village types in Britain (*after* Roberts)

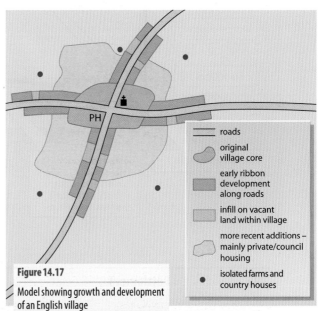

Figure 14.17

Model showing growth and development of an English village

roads

original village core

early ribbon development along roads

infill on vacant land within village

more recent additions – mainly private/council housing

isolated farms and country houses

Settlement

Figure 14.19 is a map of Carlisle in 1810. It shows some of the original **site** factors (some of which still applied), the developing **morphology** (pattern) of the city, and some of its initial and subsequent **functions**.

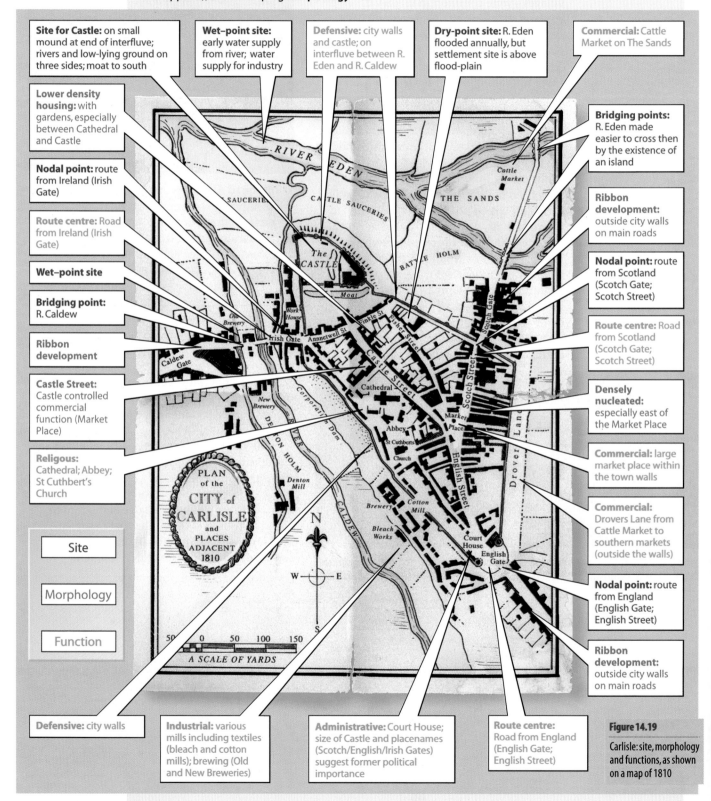

Site for Castle: on small mound at end of interfluve; rivers and low-lying ground on three sides; moat to south

Lower density housing: with gardens, especially between Cathedral and Castle

Nodal point: route from Ireland (Irish Gate)

Route centre: Road from Ireland (Irish Gate)

Wet–point site

Bridging point: R. Caldew

Ribbon development

Castle Street: Castle controlled commercial function (Market Place)

Religous: Cathedral; Abbey; St Cuthbert's Church

Wet–point site: early water supply from river; water supply for industry

Defensive: city walls and castle; on interfluve between R. Eden and R. Caldew

Dry-point site: R. Eden flooded annually, but settlement site is above flood-plain

Commercial: Cattle Market on The Sands

Bridging points: R. Eden made easier to cross then by the existence of an island

Ribbon development: outside city walls on main roads

Nodal point: route from Scotland (Scotch Gate; Scotch Street)

Route centre: Road from Scotland (Scotch Gate; Scotch Street)

Densely nucleated: especially east of the Market Place

Commercial: large market place within the town walls

Commercial: Drovers Lane from Cattle Market to southern markets (outside the walls)

Nodal point: route from England (English Gate; English Street)

Ribbon development: outside city walls on main roads

Site

Morphology

Function

PLAN of the CITY of CARLISLE and PLACES ADJACENT 1810

A SCALE OF YARDS

Defensive: city walls

Industrial: various mills including textiles (bleach and cotton mills); brewing (Old and New Breweries)

Administrative: Court House; size of Castle and placenames (Scotch/English/Irish Gates) suggest former political importance

Route centre: Road from England (English Gate; English Street)

Figure 14.19

Carlisle: site, morphology and functions, as shown on a map of 1810

Dispersed and nucleated rural settlement

Whether settlement is dispersed or nucleated depends upon local physical conditions; economic factors such as the time and distance between places; and social factors which include who owns the land and how the people of the area live and work on it.

Causes of dispersion

The more extreme the physical conditions and possible hardship of an area, the more probable it is that the settlement will be dispersed. Similarly, dispersed settlement develops in areas where natural resources are limited and insufficient to support many people (Figure 13.4). This lack of resources could include a limited water supply (the Carboniferous limestone outcrops of the Pennines); forested areas (the Canadian Shield and the Amazon Basin); and marginal farmland (the Scottish Highlands and the Sahel countries), where pastoral farming is limited by the quality and quantity of available grass. Areas with physical difficulties are also less likely to have good transport networks.

Forms of land tenure can also result in dispersed dwellings, especially in those parts of the world where inheritance laws have meant that the farm is successively divided between several sons. Similar patterns, though with larger farm units, can be found in pioneer areas such as the Canadian Prairies and the Dutch polders.

The 'agrarian revolution' in Britain in the 18th century ended the open-field system, in which strips were owned individually but the crops and animals were controlled by the community. It was replaced by enclosing several fields which were owned by a farmer who became responsible for all the decisions affecting that farm; new farmhouses were sometimes built outside the village.

Two other changes at about the same time increased the incidence of dispersed settlement. The first was the growth of large estates belonging to wealthy landowners. The second was the extension of farming in hilly areas, in the 18th and again in the 19th century, to produce the extra food needed to feed the rapidly growing urban areas. Much moorland in the Pennines was walled; while fenland areas, previously of limited value, were drained and farmed. Areas of downland were also put under the plough. Increased mechanisation reduced labour needs, resulting in overpopulation and, eventually, out-migration.

Finally, settlement was more likely to develop a dispersed pattern where there was less risk of war or civil unrest as there was then less need for people to group together for protection.

Causes of nucleation

The majority of humans have always preferred to live together in groups, as witnessed by the cities of ancient Mesopotamia and Egypt (Figure 14.1), and the present-day conurbations and cities with more than 1 million inhabitants (Figure 15.3). Two major reasons for people to group together have been either a limited or an excess water supply. Settlements have grown up around springs, as at the foot of chalk escarpments in southern England (Figures 14.4 and 8.10), and at waterholes and oases in the desert (Figure 14.5). Settlements have also been built on mounds in marshy fenland regions and on river terraces above the level of flooding (Figures 14.9).

A further cause was the need to group together for defence and protection. Examples of defensive settlements include living in walled cities on relatively flat plains (Jericho and York); behind stockades (African kraals); in hilltop villages in southern Italy and Greece (Figure 14.7); or in meander loops, taking advantage of a natural water barrier (Durham, Figure 14.6).

In Anglo-Saxon England, when many villages had their origin, the feudal open-field system of farming encouraged nucleation: the local lord could better supervise his serfs if they were clustered around him; while the serf, living in the village, was probably equidistant from his fragmented strips of farmland (Places 51, page 400). Today, the more intensive the nature of farming, the more nucleated the settlement tends to be. People like to be as near as possible to services so that the larger and more nucleated the village, the more likely it is to have a wide range of services such as a primary school, shops and a public house (Figure 14.21).

Transport and routeways have always had a major influence on the clustering of dwellings. Buildings tend to be grouped together at crossroads and T-junctions; controlling a gap through hills; at bridging-points (Places 49); and along main roads, waterways and railways. Compact settlement patterns are also found in areas with an important local resource (a Durham coalmining town or a North Wales slate quarry village – Places 78 page 523), or where there was an abundance of building materials. More recently, many governments have encouraged new, nucleated settlements in an attempt to achieve large-scale self-sufficiency. Examples may be found as far afield as the Soviet collective farm, the Chinese commune (Places 63, page 468), the Tanzanian *ujamaa* and the Israeli *kibbutz*.

Changes in rural settlement in Britain

Within the British Isles, there are areas, especially those nearer to urban centres, where the rural population is increasing and others, usually in more remote locations, where the rural population is decreasing (rural depopulation). These population changes affect the size, morphology and functions of villages. Figure 14.20 shows that there is some relationship between the type and rate of change in a rural settlement and its distance from, and accessibility to, a large urban area.

Accessibility to urban centres

As public and private transport improved during the inter-war period (1919–39), British cities expanded into the surrounding countryside at a rapid and uncontrolled rate. In an attempt to prevent this urban sprawl, a **green belt** was created around London following the 1947 Town and Country Planning Act. The concept of a green belt, later applied to most of Britain's conurbations, was to restrict the erection of houses and other buildings and to preserve and conserve areas of countryside for farming and recreational purposes.

Beyond the green belt, **new towns** and **overspill towns** were built, initially to accommodate new arrivals seeking work in the nearby city and, later, those forced to leave it due to various redevelopment schemes. These new settlements, designed to become self-supporting both economically and socially, developed urban characteristics and functions. New towns, overspill and green belts were part of a wider land-use planning process which aimed to manage urban growth (compare Figure 14.22).

Meanwhile, despite the 1968 Town and Country Planning Act, uncontrolled growth also continued in many small villages beyond the green belt. Referred to during the inter-war period as **dormitory** or **commuter villages** (page 375), these settlements have increasingly adopted some of the characteristics of nearby urban areas and have been termed **suburbanised villages**. Figure 14.21 lists some of the changes which occur as a village becomes increasingly suburbanised.

Less accessible settlements

These villages are further in distance from, or have poorer transport links to, the nearest city, i.e. they are beyond commuting range. This makes the journey longer in time, more expensive and less convenient. Though these villages may be relatively stable in size, their social and economic make-up is changing. Many in the younger age groups move out, pushed by a shortage of jobs and social life. They are replaced by retired people seeking quietness and a pleasant environment but who often do not realise that rural areas lack many of the services required by the elderly such as shops, buses, doctors and libraries.

Villages in National Parks and other areas of attractive scenery in upland or coastal areas are being changed by the increased popularity of **second** or **holiday homes** (Figure 14.20 and Places 50). The more wealthy urban dwellers, seeking relaxation away from the stress of their local working and living environment, buy vacant properties in villages. While this may bring trade to the local shop and improve the quality of some buildings, it means that the local inhabitants can no longer afford the inflated house prices, and many properties may stand empty for much of the year.

Figure 14.20

Rural settlements and distance from large urban areas

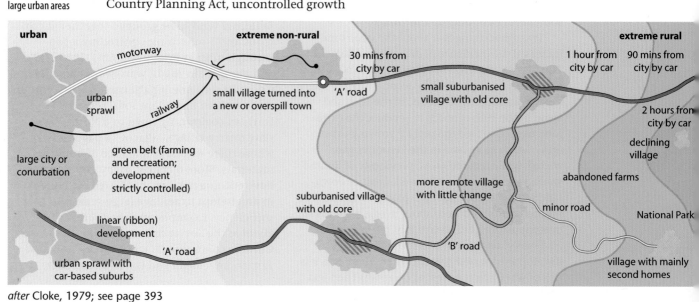

urban   extreme non-rural   extreme rural

motorway

30 mins from city by car

1 hour from city by car   90 mins from city by car

urban sprawl

railway

small village turned into a new or overspill town

'A' road

small suburbanised village with old core

2 hours from city by car

large city or conurbation

green belt (farming and recreation; development strictly controlled)

declining village

abandoned farms

more remote village with little change

suburbanised village with old core

minor road

National Park

linear (ribbon) development

'A' road

'B' road

urban sprawl with car-based suburbs

village with mainly second homes

after Cloke, 1979; see page 393

Characteristic	Extreme non-rural (increasingly suburbanised)	Original village	Extreme rural (increasingly depopulated)
Housing	Many new detached houses, semi-detached houses and bungalows; renovated barns and cottages; expensive estates	Detached, stone-built houses/cottages with slate/thatch roofs; some farms, many over 200 years old; barns	Poor housing lacking basic amenities; old stone houses, some derelict, some converted into holiday/second homes
Population structure	Young/middle-aged married couples with children; very few born in village; professional/executive groups; some wealthy retired people	An ageing population; most born in village; labouring/manual groups	Mainly elderly/retired; born and lived all life locally; labouring/manual groups; younger people have moved away
Employment	New light industry (high-tech and food processing); good salaries; many commuters (well-paid); tourist shops	Farming and other primary activities (forestry, mining); low-paid local jobs	Low-paid; unemployment; farming jobs (declining if in marginal areas) and other primary activities; some tourist-related jobs
Transport	Good bus service (unless reduced by private car); most families have one or two cars; improved roads	Bus service (limited); some cars; narrow/winding roads	No public transport; poor roads
Services	More shops; enlarged school; modern public houses/restaurants; garage	Village shop; small junior school; public house; village hall	Shop and school closed; perhaps a public house
Community/Social	Local community swamped; division between local people and newcomers; may be deserted during day (commuters absent)	Close-knit community (many are related)	A small community; more isolated
Environment	Increase in noise and pollution, especially from traffic; loss of farmland/open space	Quiet, relatively pollution-free	Quiet; increase in conserved areas (National Parks/forestry)

Remote areas

These areas suffer from a population loss which, by leaving houses empty and villages decreasing in size, adds to the problems of rural deprivation (Figure 14.21 and Places 50). Resultant problems include a lack of job opportunities, fewer services and poor transport facilities. Employment is often limited to the shrinking primary industries which are low-paid and lack future prospects. The cost of providing services to remote areas is high, and there is often insufficient demand to keep the local shop or village school open. With fewer inhabitants to use public transport, bus services may decline or stop altogether, forcing people to move to more accessible areas.

Places 50   Hennock, Devon: a village

The following extract, with the title 'Portrait of a village crisis, where the shop has closed and the pub is for sale', was the introduction to an article that appeared in the *Daily Telegraph* on 7 March 1998.

HENNOCK, a Devon village of some 250 people stretched out on the side of a hill 200 m above the Teign Valley, encapsulates many of the problems facing the countryside today. Farmers whose land surrounds the village are facing crisis after crisis as beef, lamb and milk prices have plummeted. Ore mines, which caused the village population to grow to more than 1000 in 1861, have long since closed and there is little local employment. Hennock's location just inside the Dartmoor National Park means property prices are driven up beyond the reach of local people and planning restrictions mean no new affordable housing has been built.

Hennock is served by just two buses a week, making its inhabitants ever more dependent on the car, which has brought the supermarkets of Newton Abbot and Exeter closer and killed off the village shop and post office.

The pub is up for sale after a succession of landlords failed to make it work. There is still some resentment that the church, that other mainstay of village life, has to share a vicar with nearby Bovey Tracey.

The positive signs are to be found in the school, where the roll is growing, and in a determination to revitalise the old village hall. Village opinion on the future is divided. Some say Hennock is dying. Others say there is new hope. They all agree it is a crucial time.

The article, which ran to two sides, continued with views of village life expressed by several residents – the vicar, the village hall secretary, two newcomers who ran a field study centre, the publican, the headmisstress, a couple who were lifetime residents, someone born in the village who had just returned with a family of his own, the shopkeeper, a commuter and a farmer.

Figure 14.22

Pressure on green belts

THE URBAN SPRAWL

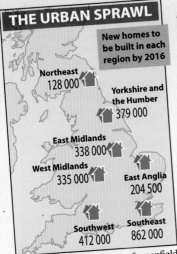

New homes to be built in each region by 2016

Northeast 128 000

Yorkshire and the Humber 379 000

East Midlands 338 000

West Midlands 335 000

East Anglia 204 500

Southwest 412 000

Southeast 862 000

THOUSANDS of hectares of greenfield sites are to be concreted to make way for 20 new towns and dozens of housing estates under plans being drawn up by local authorities.

The plans, which fly in the face of government pledges to save the countryside, will hit the south-east of England particularly hard.

Earlier this year John Prescott, the deputy prime minister who has responsibility for the environment, said no more than 40% of the 5 million new homes needed in Britain by 2016 should be built on greenfield sites, with the rest being built on derelict land (i.e. 2 million in total).

Developers have already proposed a new town outside Cambridge – codenamed C2 – which would provide homes for at least 50000 people. There have also been proposals for two settlements in Devon, a 10 000-home development between Ripon and Thirsk in North Yorkshire, a new town in Worcestershire and a new town with 7000 homes in Hampshire.

'These plans are a disaster for the countryside,' said Tony Burton, of the Council for the Protection of Rural England. 'We don't think there is a need to allocate any more greenfield sites for development. It would undermine the urban renewal which is meant to be the strategy of this government.'

Sunday Times, 5 November 1998

Green Belt homes plan goes ahead

ENVIRONMENTALISTS reacted angrily last night after a council decided to press ahead with plans to build thousands of homes on green belt farmland in Hertfordshire.

Earlier this month thousands of people turned out for a 5 km march and rally through Langley Valley, on the outskirts of Stevenage, which has been earmarked for new homes.

But despite objections, Hertfordshire County Council agreed yesterday to adopt the plan to build 3600 homes on the controversial land. Another 1000 houses will be built on a green belt site in Hemel Hempstead and hundreds of others on smaller areas by the Government's deadline of 2011.

The council said building new homes on the two sites was unavoidable because urban building was reaching saturation point.

Daily Telegraph, 1 May 1998

Places 51   Britain: evolution of settlement

When Britain's first census was taken in 1801, almost 80 per cent of the population still lived in hamlets and villages. (The corresponding figure in the 1991 census was 7 per cent, rising to an estimated 10 per cent in 1998.) Most people have their own mental image of a 'traditional' hamlet, village, or market town. However, in reality, the development of rural settlement has been so dynamic and complex that, due to differences in site, form (morphology) and function (Places 49), there is no such thing as a 'typical' rural settlement (Figure 14.23) – nor is there a 'typical' urban settlement.

Iron Age settlements

Palaeolithic man left behind flint tools, but few marks on the landscape. The first people to alter their natural surroundings were those of the Neolithic period, the Bronze Age and the Iron Age (Figure 11.18). They began, despite limited technology, to clear woodlands and to leave a legacy of stone circles, tumuli, barrows, hillforts (Figure 14.24) and settlement sites. The hillfort built on the volcanic sill at Drumadoon (Figure 1.37) had a fine panorama of an enemy approaching from the sea, while the steep cliffs prevented any frontal attack. Hillforts may, however, have only been settled during times of

Villages in the British landscape

Figure 14.23

There is a tendency to think of country life as stable, conservative and unchanging but this is far from the truth. Settlements, like the people who live in them, are mortal. There is, however, no recognisable expected life-span, and a village can survive for twenty or two thousand years depending on its ability to adapt to changing economic and social conditions. In addition to extant village communities there are in Britain thousands of former occupation sites which have been abandoned.

Rural settlement in the past reflected the ever-changing relationship between man and his environment. Human society is never completely static and the settlements which serve it can never remain absolutely still for very long; and before a well-balanced form of settlement becomes generally established, new forces will be at work altering that form. The forces which created our hamlets and villages have involved factors as varied as the pace of technological change, the nature of local authority, inheritance customs, the presence of arable or pasture, and the availability of building materials. Village history tells a story of fluctuating expansion, decline and movement, sometimes reflecting national factors such as pestilence, economic changes and social development, and sometimes purely local events, such as the silting up of a river estuary or the bankruptcy of a local entrepreneur. Such factors have combined to give each village a unique history and plan.

T. Rowley, 1978

attack, as few had a guaranteed water supply. Not all Iron Age settlements were hillforts; some forts were located in lowland areas, while other settlements may have had a religious or market function as opposed to a military one.

Figure 14.24

Maiden Castle hillfort, Dorset, England

Romano-British settlements

While the Romans preferred to live in well-planned towns or in large rural villas, it is clear that at the same time many nucleated villages existed in lowland Britain, many of which showed evidence of Roman influence by having well-planned streets. One characteristic feature of Romano-British villages was the presence of small-scale industrial activity – usually pottery production and iron-working.

Anglo-Saxon settlements

Although many English village and town names have Anglo-Saxon origins, it does not prove that they existed during those times. Most Anglo-Saxon settlements were sited in clearings in the natural forest, on 'islands' in marshy areas or near to the coast. Archaeological evidence suggests that most settlements were likely to have consisted of several farms grouped together to form self-contained hamlets. The houses, or rather huts, were rectangular in shape and built from local materials – wood for the frame from the forest, mud and wattle (interlaced twigs and branches) for the walls from

the river and forest, and thatch for the roof from local reeds or straw left over after the harvest. The huts, which were shared with the animals in winter, may have been protected by a stone or wooden wind-break. It was only by late Anglo-Saxon times that larger nucleated villages, with their open fields worked in strips by a heavy plough drawn by oxen, became more commonplace.

Medieval settlements

By medieval times, each village was dominated by a large farm, or manor, house in which the lord of the manor lived. The village would have contained several peasant cottages, built with materials similar to those of Anglo-Saxon homes, a church, a house for the priest, a blacksmith's forge and a mill. Surrounding the village were (usually) three large open fields – open because they had neither hedges nor fences as boundaries. Each field was divided into numerous, long, narrow strips, shared between the peasants. Two of the fields were likely to be growing cereals such as wheat, barley and rye (mainly for bread), while the third was left fallow (allowed to rest). The crops were rotated so that each field was left fallow every third year – the three-field system of crop rotation. When the fields were ploughed, a ridge was formed about 0.3 m above an adjacent furrow. Over many years of ploughing, the ridges built up so that they can still be recognised in our present-day landscape (Figure 14.25).

In the scarp-and-vale areas of south-east England (page 199), the villages were often close together along the spring lines. The parish boundaries were laid out between each village and parallel to each other, so that each individual parish had a long, narrow strip of land extending across the clay vale and over the chalk escarpment (Figure 14.4). This allowed each parish to be self-contained by having a permanent water supply together with land suitable for both rearing animals and growing crops. Although individual parishes no longer need to be self-supporting, the old boundaries still remain.

Figure 14.25

Ridge and furrow, south-east Leicestershire

Measuring settlement patterns

Several theories and statistical tests have been put forward to explain and to allow objective comparisons to be made between settlements in different parts of the world, e.g. within a country or between countries.

- **Nearest neighbour analysis** is a statistical test to describe the settlement pattern.
- **The rank–size rule** seeks to find a numerical relationship between the population size of settlements.
- **Central place theory** is concerned with the functional importance of places.
- **Gravity models** seek to determine the interaction (i.e. movement) between places.

Nearest neighbour analysis

Settlements often appear on maps as dots. Dot distributions are commonly used in geography, yet their patterns are often difficult to describe. Sometimes patterns are obvious, such as when settlements are extremely nucleated or dispersed (Figure 14.26). As, in reality, the pattern is likely to lie between these two extremes, then any description will be subjective. One way in which a pattern can be measured objectively is by using nearest neighbour analysis.

This technique was devised by a botanist who wished to describe patterns of plant distributions. It can be used to identify a tendency towards nucleation (clustering) or dispersion for settlements, shops, industry, etc., as well as plants. Nearest neighbour analysis gives a precision that enables one region to be compared with another and allows changes in distribution to be compared over a period of time. It is, however, only a technique and therefore does *not* offer any explanation of patterns.

The formula used in nearest neighbour analysis produces a figure (expressed as *Rn*) which measures the extent to which a particular pattern is clustered (nucleated), random, or regular (uniform) (Figure 14.26).

- **Clustering** occurs when all the dots are very close to the same point. An example of this in Britain is on coalfields where mining villages tended to coalesce. In an extreme case, *Rn* would be 0.
- **Random** distributions occur where there is no pattern at all. *Rn* then equals 1.0. The usual pattern for settlement is one that is predominantly random with a tendency either towards clustering or regularity.
- **Regular** patterns are perfectly uniform. If ever found in reality, they would have an *Rn* value of 2.15 which would mean that each dot (settlement) was equidistant from all its neighbours. The closest example of this in Britain is the distribution of market towns in East Anglia.

Using nearest neighbour analysis
Figure 14.27 shows settlements in part of north-east Warwickshire and south-west Leicestershire, an area of the English Midlands where it might be expected that there would be evidence of regularity in the distribution.

1 The settlements in the study area were located. (The minimum number recommended for a nearest neighbour analysis is 30.) Each settlement was given a number.
2 The nearest neighbour formula was applied. This formula is:

$$Rn = 2\bar{d}\sqrt{\frac{n}{A}}$$

where:

Rn = the description of the distribution
$\bar{d}$ = the mean distance between the nearest neighbours (km)
n = the number of points (settlements) in the study area
A = the area under study (km²).

Figure 14.26

Nearest neighbour values (*Rn*)

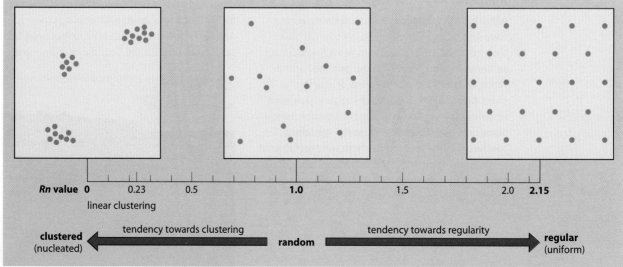

Rn value   0   0.23   0.5   1.0   1.5   2.0 2.15

linear clustering

clustered (nucleated) ← tendency towards clustering — **random** — tendency towards regularity → regular (uniform)

Figure 14.27

Nearest neighbour analysis: a worked example for part of north-east Warwickshire and south-west Leicestershire

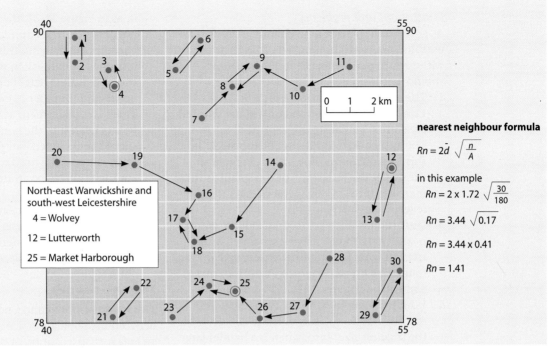

North-east Warwickshire and south-west Leicestershire
- 4 = Wolvey
- 12 = Lutterworth
- 25 = Market Harborough

nearest neighbour formula

$$Rn = 2\bar{d}\sqrt{\frac{n}{A}}$$

in this example

$$Rn = 2 \times 1.72 \sqrt{\frac{30}{180}}$$

$$Rn = 3.44\sqrt{0.17}$$

$$Rn = 3.44 \times 0.41$$

$$Rn = 1.41$$

Settle-ment number	Nearest neigh-bour	Distance (km)
1	2	1.0
2	1	1.0
3	4	0.6
4	3	0.6
5	6	1.6
6	5	1.6
7	8	1.8
8	9	1.3
9	8	1.3
10	9	2.1
11	10	2.2
12	13	2.2
13	12	2.2
14	15	3.3
15	18	1.7
16	17	1.3
17	18	1.0
18	17	1.0
19	16	3.0
20	19	3.2
21	22	1.6
22	21	1.6
23	24	2.1
24	25	1.1
25	24	1.1
26	25	1.5
27	26	1.8
28	27	2.5
29	30	2.2
30	29	2.2
		Σ51.7

3 To find $\bar{d}$, measure the straight-line distance between each settlement and its nearest neighbour, e.g. settlement 1 to 2, settlement 2 to 1, settlement 3 to 4, and so on. One point may have more than one nearest neighbour (settlement 8) and two points may be each other's nearest neighbour (settlements 1 and 2). In this example, the mean distance between all the pairs of nearest neighbours was 1.72 km – i.e. the total distance between each pair (51.7 km) divided by the number of points (30).

4 Find the total area of the map: i.e. 15 km x 12 km = 180 km².

5 Calculate the nearest neighbour statistic, Rn, by substituting the formula. This has already been done in Figure 14.27 and gives an Rn value of 1.41.

6 Using this Rn value, refer back to Figure 14.26 to determine how clustered or regular is the pattern. A value of 1.41 shows that there is a fairly strong tendency towards a regular pattern of settlement.

7 However, there is a possibility that this pattern has occurred by chance. Referring to Figure 14.28, it is apparent that the values of Rn must lie outside the shaded area before a distribution of clustering or regularity can be accepted as significant. Values lying in the shaded area at the 95 per cent probability level show a random distribution. (*Note:* with fewer than 30 settlements, it becomes increasingly difficult to say with any confidence that the distribution is clustered or regular.) The graph confirms that our Rn value of 1.41 has a significant element of regularity.

How can the nearest neighbour statistic be used to compare two or more distributions? Figure 14.28 shows the Rn value for three areas in England, including that for our worked example, the English Midlands. The Rn statistic of 1.57 for part of East Anglia shows that the area has a more pronounced pattern of regularity than the Midlands. An Rn value of 0.61 for part of the Durham coalfield indicates that it has a significant tendency towards a clustered distribution.

Figure 14.28

Interpretation of Rn statistic: significant values

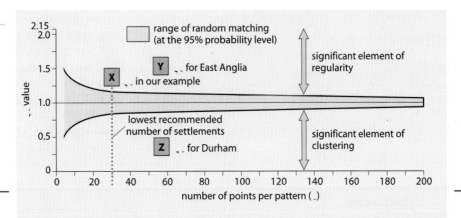

403

Limitations and problems

Limitations and problems

As noted earlier, nearest neighbour analysis is a useful statistical technique but it has to be used with care. In particular, the following points should be considered:

1 The size of the area chosen is critical. Comparisons will be valid only if the selected areas are a similar size.
2 The area chosen should not be too large, as this lowers the *Rn* value (i.e. it exaggerates the degree of clustering), or too small, as this increases the *Rn* value (i.e. it exaggerates the level of regularity).
3 Distortion is likely to occur in valleys, where nearest neighbours may be separated by a river, or where spring-line settlements are found in a linear pattern as at the foot of a scarp slope (Figures 8.10 and 14.4).
4 Which settlement sizes are to be included? Are hamlets acceptable, or is the village to be the smallest size? If so, when is a hamlet large enough to be called a village (page 393)?
5 There may be difficulty in determining the centre of a settlement for measurement purposes, especially if it has a linear or a loose-knit morphology.
6 The boundary of an area is significant. If the area is a small island or lies on an outcrop of a particular rock, there is little problem; but if, as in Figure 14.27, the area is part of a larger region, the boundaries must have been chosen arbitrarily (in this instance by predetermined grid lines). In such a case, it is likely that the nearest neighbour of some of the points (e.g. number 20) will be off the map. There is disagreement as to whether those points nearest to the boundary of the map should be included, but perhaps of more importance is the need to be consistent in approach and to be aware of the problems and limitations.

Despite these problems, nearest neighbour analysis forms a useful basis for further investigation into why any clustering or regularity of settlement has taken place.

The rank–size rule

This is an attempt to find a numerical relationship between the population size of settlements within an area such as a country or county. The rule states that **the size of settlements is inversely proportional to their rank**. Settlements are ranked in descending order of population size, with the largest city placed first. The assumption is that the second-ranked city will have a population one-half that of the first-ranked, the third-ranked city a population one-third of the first-ranked, the fourth-ranked one-quarter of the largest city, and so on.

The rank–size rule is expressed by the formula:

$Pn = Pl \div n$ (or R)

where:

Pn = the population of the city
Pl = the population of the largest (primate) city
n (or R) = the rank–size of the city.

For example, if the largest city has a population of 1 000 000, then:

the second-largest city will be 1 000 000 ÷ 2, i.e. 500 000
the third-largest city will be 1 000 000 ÷ 3, i.e. 333 333
the fourth-largest city will be 1 000 000 ÷ 4, i.e. 250 000.

If such a perfect negative relationship actually occurred (Framework 19, page 635), it would produce a steeply downward-sloping, smooth, concave curve on an arithmetic graph (Figure 14.29a). However, it is more usual to plot the rank–size distribution on a logarithmic scale, in which case the perfect negative relationship would appear as a straight line sloping downwards at an angle of 45° (Figure 14.29b). Figure 14.30 shows the rank–size rule applied to Brazil.

Figure 14.29

The rank–size rule

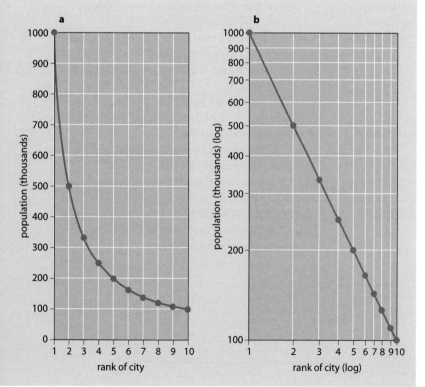

Variations from the rank–size rule

In reality, it is rare to find a close correlation between the city size of a country and the rank–size rule. There are, however, two major variations from the rank–size rule.

1 **Primate distribution** (urban primacy) is found where the largest city, often the capital, completely dominates a country or region (in terms of population size, economic development, wealth, services and cultural activities). In such a case, the primate city will have a population size many times greater than that of the second-largest city (Buenos Aires in Figure 14.31). Montevideo in Uruguay is 17 times larger than the second-largest city, and Lima in Peru is 11 times larger.

2 **Binary distribution** occurs where there are two very large cities of almost equal size within the same country: one may be the capital and the other the chief port or major industrial centre. Examples of binary distribution include Madrid and Barcelona in Spain, and Quito and Guayaquil in Ecuador. It has been suggested (though there are many exceptions) that the rank–size rule is more likely to operate if the country is developed; has been urbanised for a long time; is large in size; and has a complex and stable economic and political organisation. In contrast, primate distribution is more likely to be found (also with exceptions, including France and Austria) in countries which are small in size; less developed; former colonies of European countries; only recently urbanised; and which have experienced recent changes in political organisation and/or boundaries.

Two schools of thought exist concerning the causes of variation in urban primacy. One suggests that as a city begins to dominate a country it attracts people, trade, industry and services at an increasingly rapid rate and at the expense of rival cities (arguably this is more applicable to economically less developed countries). The other claims that as a country becomes more urbanised and industrialised, the growth of several cities tends to be stimulated, thus reducing the importance of the primate city (arguably more applicable to economically more developed countries where some of the largest cities are now experiencing urban depopulation, page 365).

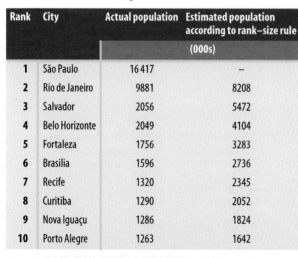

Rank	City	Actual population	Estimated population according to rank–size rule
		(000s)	
1	São Paulo	16 417	–
2	Rio de Janeiro	9881	8208
3	Salvador	2056	5472
4	Belo Horizonte	2049	4104
5	Fortaleza	1756	3283
6	Brasilia	1596	2736
7	Recife	1320	2345
8	Curitiba	1290	2052
9	Nova Iguaçu	1286	1824
10	Porto Alegre	1263	1642

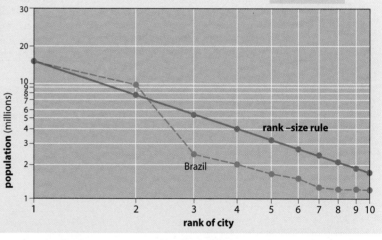

Figure 14.30

The rank–size rule applied to Brazil

Rank	USA 1994 actual population (000s)		ITALY 1994 actual population (000s)		ARGENTINA 1991 actual population (000s)		JAPAN 1994 actual population (000s)	
1	New York	16 329	Rome	2688	Buenos Aires	10 990	Tokyo	26 836
2	Los Angeles	12 410	Milan	1334	Cordoba	1198	Osaka	10 601
3	Chicago	7688	Naples	1062	Rosario	1096	Yokohama	3228
4	San Francisco	6410	Turin	946	Mendoza	775	Nagoya	2159
5	Philadelphia	4949	Palermo	695	La Plata	640	Sapporo	1732
6	Washington DC	4466	Genoa	660	San Miguel de Tucuman	622	Kobe	1509
7	Detroit	4307	Bologna	395	Mar de Plata	521	Kyoto	1452
8	Houston	3653	Florence	393	Santa Fé	395	Fukuoka	1269
9	Atlanta	3331	Bari	339	Salta	370	Kawasaki	1200
10	Boston	3240	Catania	327	San Juan	353	Hiroshima	1102

Settlement

Figure 14.31

Largest cities in four selected countries

Figure 14.32

Size, spacing and functions of settlements

Central place	Population	Distance apart (km)	Sphere of influence (km²)	Functions (services)
Hamlet	10–20	2	—	probably none
Village	1 000	7	45	church, post office, shop, junior school
Small town	20 000	21	415	shops, churches, senior school, bank, doctor
Large town	100 000	35	1 200	shopping centre, small hospital, banks, senior schools
City	500 000	100	12 000	shopping complex, cathedral, large hospital, football team, large bus and rail station, cinemas, theatre
Conurbation	1 million	200	35 000	shopping complexes, several CBDs
Capital or primate city	several million	—	whole country	government offices, all other functions

Notes: The distances and service areas have been taken from Christaller's work in southern Germany (1933) with, in some cases, a rounding-off of figures for simplicity. The population figures and functions are more applicable to the UK and the present time. Populations, distances and service areas vary between and within countries and should be taken as comparative and approximate rather than absolute. All places in the hierarchy have all the services of the settlements below them.

Central place theory

A **central place** is a settlement that provides goods and services. It may vary in size from a small village to a conurbation or primate city (Figures 14.32 and 14.33) and forms a link in a hierarchy. The area around each settlement which comes under its economic, social and political influence is referred to as its **sphere of influence**, **urban field** or **hinterland**. The extent of the sphere of influence will depend upon the spacing, size and functions of the surrounding central places.

Functional hierarchies

Four generalisations may be made regarding the spacing, size and functions of settlements.

1 The larger the settlements are in size, the fewer in number they will be, i.e. there are many small villages, but relatively few large cities.
2 The larger the settlements grow in size, the greater the distance between them, i.e. villages are usually found close together, while cities are spaced much further apart.
3 As a settlement increases in size, the range and number of its functions will increase (Figure 14.33).
4 As a settlement increases in size, the number of higher-order services will also increase, i.e. a greater degree of specialisation occurs in the services (Figure 14.32).

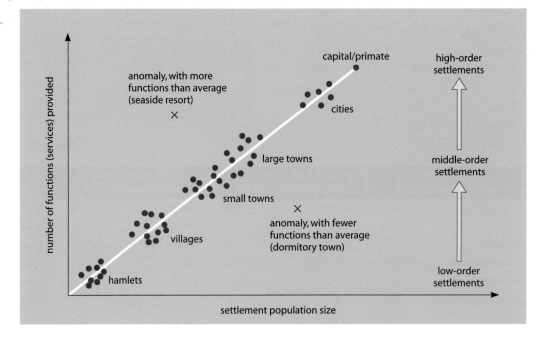

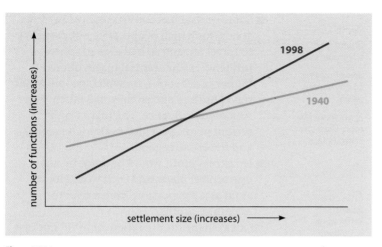

Figure 14.34

Relationship between the number of functions and settlement size in the UK, 1940 and 1998

The range and threshold of central place functions

Central place functions are activities, mainly within the tertiary sector, that market goods and services from central places for the benefit of local customers and clients drawn from a wider hinterland. The **range** of a good or service is the maximum distance that people are prepared to travel to obtain it. It is dependent upon the value of the good, the length of the journey, and the frequency that the service is needed. People are not prepared to travel as far to buy a newspaper (a low-order item), which they need daily, as they are to buy furniture (a high-order item), which they might purchase only once every several years. Low-order functions, such as corner shops and primary schools, need to be spaced closely together as people are less willing and less able to travel far to use them. High-order functions, such as regional shopping centres and hospitals, are likely to be widely spaced as people are more prepared to travel considerable distances to them (page 432).

The **threshold** of a good or service is the minimum number of people required to support it. It is assumed, incorrectly in practice, that people will always use the service located nearest to them (the nearest superstore). As a rule, the more specialised the service, the greater the number of people needed to make it profitable or viable. It has been suggested that, in the UK, about 300 people are necessary for a village shop, 500 for a primary school, 2500 for a doctor, 10 000 for a senior school or a small chemist's shop, 25 000 for a shoe shop, 50 000 for a small department store, 60 000 for a large supermarket, 100 000 for a large department store, and over 1 million for a university. Services locate where they can maximise the number of people in their catchment area and maximise the distance from their nearest rival. Threshold analysis was used by planners of British new towns who equated, for example, 20 000 people with a cinema, 10 000 people with a swimming pool and 100 000 people with a theatre.

Changes in population size and number of functions

Figure 14.34 shows that over the last 50 years in the UK there has been a decrease in the number of services available in small settlements and an increase in the number of functions provided by large settlements. This may be due to many factors, for example:

■ Small villages are no longer able to support their former functions (village shop) as the greater wealth and mobility (car ownership) of some rural populations enable them to travel further to larger centres where they can obtain, in a single visit, both high- and low-order goods (Places 50, page 329).

■ Domestic changes (deep freezers, convenience foods) mean that rural householders need no longer make use of daily, low-order services previously available in their village.

■ As larger settlements attract an increasingly larger threshold population, they can increase the variety and number of functions and, by reducing costs (supermarkets), are likely to attract even more customers.

■ In areas experiencing rural depopulation, villages may no longer have a population large enough to maintain existing services.

Christaller's model of central places

Walter Christaller was a German who, in 1933, published a book in which he attempted to demonstrate a sense of order in the spacing and function of settlements. He suggested that there was a pattern in the distribution and location of settlements of different sizes and also in the ways in which they provided services to the inhabitants living within their sphere of influence. Regardless of the level of service provided, he termed each settlement a **central place**. Although Christaller's **central place theory** was based upon investigations in southern Germany, and it was not translated into English until 1966, his work has contributed a great deal to the search for order in the study of settlements.

The two principles underlying Christaller's theory were the **range** and the **threshold** of goods and services. He made a set of assumptions which were similar to those of two earlier German economists, von Thünen (agricultural land use model, page 471) and Weber (industrial location theory, page 557).

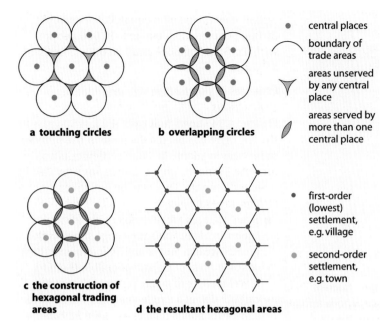

a **touching circles**

b **overlapping circles**

- central places

- boundary of trade areas

- areas unserved by any central place

- areas served by more than one central place

c **the construction of hexagonal trading areas**

d **the resultant hexagonal areas**

- first-order (lowest) settlement, e.g. village

- second-order settlement, e.g. town

Figure 14.35

Constructing spheres of influence around settlements (*after* Christaller)

These assumptions were:
- There was unbounded flat land so that transport was equally easy and cheap in all directions. Transport costs were proportional to distance from the central place and there was only one form of transport.
- Population was evenly distributed across the plain.
- Resources were evenly distributed across the plain.
- Goods and services were always obtained from the nearest central place so as to minimise distance travelled, i.e. the assumed rational behaviour that all consumers will minimise their travel in the pursuit of goods and services.
- All customers had the same purchasing power (income) and made similar demands for goods.

Figure 14.36

Christaller's central places and spheres of influence

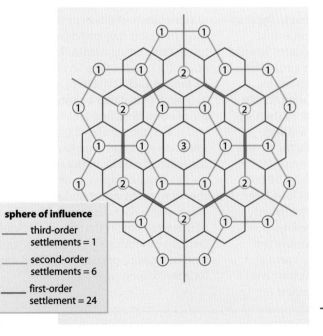

central place	sphere of influence
③	—— third-order settlements = 1
②	—— second-order settlements = 6
①	—— first-order settlement = 24

- Some central places offered only low-order goods, for which people were not prepared to travel far, and so had a small sphere of influence. Other central places offered higher-order goods, for which people would travel further, and so they had much larger spheres of influence. The higher-order central places provided both higher-order and lower-order goods.
- No excess profit would be made by any one central place, and each would locate as far away as possible from a rival to maximise profits.

The ideal shape for the sphere of influence of a central place is circular, as then the distances from it to all points on the boundary are equal. If the circles touch at their circumferences, they leave gaps which are unserved by any central place (Figure 14.35a); if the circles are drawn so that there are no gaps, they necessarily overlap (Figure 14.35b) – which also violates the basic assumptions of the model. To overcome this problem, the overlapping circles are modified to become touching hexagons (Figure 14.35c). A hexagon is almost as efficient as a circle in terms of accessibility from all points of the plain and is considerably more efficient than a square or triangle (Figure 14.35d). A hexagonal pattern also produces the ideal shape for superimposing the trading areas of central places with different levels of function – the village, town and city of Christaller's hierarchy. Figure 14.36 shows a large trade area for a third-order central place, a smaller trade area for the six second-order central places, and even smaller trade areas for the 24 first-order central places.

By arranging the hexagons in different ways, Christaller was able to produce three different patterns of service or trading areas. He called these $k = 3$, $k = 4$ and $k = 7$, where k is the number of places dependent upon the next-highest-order central place.

The following should be noted at this point.
- Where $k = 3$, the trade area of the third-order (i.e. highest) central place is three times the area of the second-order central place, which in turn is three times larger than the trade area of the first-order (lowest) central place.
- Where $k = 4$, the trade area of the third-order central place is four times the area of the second-order central place, which is four times larger than the trade area of the first-order central place.
- Where $k = 7$, the trade area of each order is seven times greater than the order beneath it.

Figure 14.37

Christaller's $k = 3$

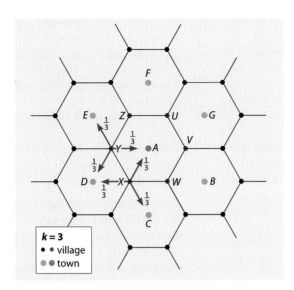

Figure 14.38

Christaller's $k = 4$

$k = 3$

The arrangement of the hexagons in this case is the same as given in Figure 14.36 and the explanation of how $k = 3$ is reached is shown in Figure 14.37, where:

> A is the central place or third-order settlement
> B, C, D, E, F and G are 6 second-order settlements surrounding A
> U, V, W, X, Y and Z are some of the 24 first-order settlements which lie between A and the second-order settlements.

It is assumed that one-third of the inhabitants of Y will go to A to shop, one-third to D and one-third to E. Similarly, one-third of people living at X will shop at A, one-third at D and one-third at C. This means that A will take one-third of the customers from each of U, V, W, X, Y and Z ($6 \times \frac{1}{3} = 2$) plus all of its own customers (1). In total, A therefore serves the equivalent of three central places (2 + 1).

Christaller based the $k = 3$ pattern on a **marketing principle** which maximises the number of central places and thus brings the supply of higher-order goods and services as close as possible to all the dependent settlements and therefore to the inhabitants of the trade area.

$k = 4$

In this case, the size of the hexagon is slightly larger and it has been re-oriented (Figure 14.38). The first-order settlements, again labelled U, V, W, X, Y and Z, are now located at the mid-points of the sides of the hexagon instead of at the apexes as in k = 3. Customers from Y now have a choice of only two markets, A and N, and it is assumed that half of those customers will go to A and half to N. Similarly, half of the customers from X will go to A and the other half to M. A will therefore take half of the customers from each of the six settlements at U, V, W, X, Y and Z ($6 \times \frac{1}{2} = 3$) plus all of its own customers (1) to serve the equivalent of four central places (3 + 1). This pattern is based on a **traffic principle**, whereby travel between two centres is made as easy and as cheap as possible. The central places are located so that the maximum number may lie on routes between the larger settlements.

$k = 7$

Here the pattern shows the same high-order central place, A, but all the lower-order settlements, U, V, W, X, Y and Z, lie within the hexagon or trade area (Figure 14.39). In this case, all of the customers from the six smaller settlements will go to A ($6 \times 1 = 6$), together with all of the inhabitants of A (1). This means that A serves seven central places (6 + 1). As this system makes it efficient to organise or control several places, and as the loyalties of the inhabitants of the lower-order settlements to a higher one are not divided, it is referred to as the **administrative principle**.

Figure 14.39

Christaller's $k = 7$

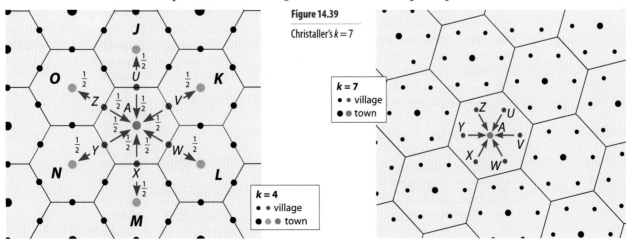

Why, with the possible exception of the reclaimed Dutch polders, can no perfect example of Christaller's model be found in the real world? The answer lies mainly in the basic assumptions of the model.

■ Large areas of flat land rarely exist and the presence of relief barriers or routes along valleys means that transport is channelled in certain directions. There is more than one form of transport; costs are not proportional to distance; and both systems and types of transport have changed since Christaller's day.

■ People and wealth are not evenly distributed.

■ People do not always go to the nearest central place – for example, they may choose to travel much further to a new edge-of-city hypermarket.

■ People do not all have the same purchasing power, or needs.

■ Governments often have control over the location of industry and of new towns.

■ Perfect competition is unreal and some firms make greater profits than others.

■ Christaller saw each central place as having a particular function whereas, in reality, places may have several functions which can change over time.

■ The model does not seem to fit industrial areas, although there is some correlation with flat farming areas in East Anglia, the Netherlands and the Canadian Prairies.

Christaller has, however, provided us with an objective model with which we can test the real world. His theories have helped geographers and planners to locate new services such as retail outlets and roads.

Interaction or gravity models

These models, derived from Newton's law of gravity, seek to predict the degree of interaction between two places. Newton's law states that:

'Any two bodies attract one another with a force that is proportional to the product of their masses and inversely proportional to the square of the distance between them.'

When used geographically, the words 'bodies' and 'masses' are replaced by 'towns' and 'population' respectively.

The interaction model in geography is therefore based upon the idea that as the size of one or both of the towns increases, there will also be an increase in movement between them. The further apart the two towns are, however, the less will be the movement between them. This phenomenon is known as **distance decay**.

This model can be used to estimate:

1 traffic flows (page 411)

2 migration between two areas

3 the number of people likely to use one central place, e.g. a shopping area, in preference to a rival central place (page 618).

It can also be used to determine the sphere of influence of each central place by estimating where the **breaking point** between two settlements will be, i.e. the point at which customers find it preferable, because of distance, time and expense considerations, to travel to one centre rather than the other.

Reilly's law of retail gravitation (1931)

Reilly's interaction breaking-point is a method used to draw boundary lines showing the limits of the trading areas of two adjacent towns or shopping centres. His law states that:

'Two centres attract trade from intermediate places in direct proportion to the size of the centres and in inverse proportion to the square of the distances from the two centres to the intermediate place.'

Unlike Christaller, Reilly suggested that there were no fixed trade areas, that these areas could vary in size and shape, and that they could overlap.

This can be expressed by the formula:

$$Db = \frac{Dab}{1 + \sqrt{\dfrac{Pa}{Pb}}}$$

or similarly

$$djk = \frac{dij}{1 + \sqrt{\dfrac{Pi}{Pj}}}$$

where:

Db (or djk) = the breaking-point between towns A and B

Dab (or dij) = the distance (or time) between towns A and B

Pa (or Pi) = the population of town A (the larger town)

Pb (or Pj) = the population of town B (the smaller town).

Taking as an example Grimsby–Cleethorpes which has a population of 131 000 and Lincoln, 71 km away, with a population of 75 000, the formula can be written as:

$$Db = \frac{71}{1 + \sqrt{\dfrac{131\,000}{75\,000}}}$$

which means that

$$Db = \frac{71}{1 + 1.32}$$
$$\therefore Db = 30.58$$

Thus the breaking-point is 30.58 km from Lincoln (town B) and 40.42 km from Grimsby–Cleethorpes (town A). This is shown in Figure 14.40.

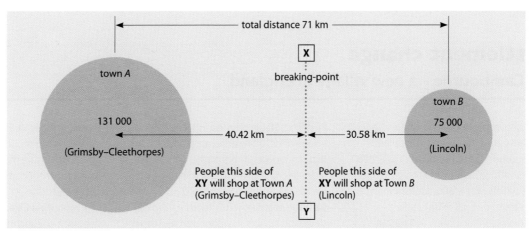

Figure 14.40

Reilly's breaking-point between settlements of different sizes, applied to north Lincolnshire

Limitations of Reilly's model

As with other models, Reilly's model is based on assumptions which are not always applicable to the real world. In this case, the assumptions are that:

- the larger the town, the stronger its attraction
- people shop in a logical way, seeking the centre which is nearest to them in terms of time and distance.

These assumptions may not always be true. For example:

- there may be traffic congestion on the way to the larger town and, once there, car parking may be more difficult and expensive
- the smaller town may have fewer but better-quality shops
- the smaller centre may be cleaner, more modern, safer and less congested, and
- the smaller town may advertise its services more effectively.

A variation on Reilly's law of retail gravitation

Like central place theory, Reilly's law seems to fit rural areas better than closely packed, densely populated urban areas. One of several variations on Reilly's law of retail gravitation is based upon the drawing power of shopping centres (i.e. the number and type of shops in each) rather than distance between the two towns. (Other variations include retail floorspace and retail sales.)

The version based on the drawing power of shopping centres has the formula:

$$Db = \frac{Dab}{1 + \sqrt{\dfrac{Sa}{Sb}}}$$

where:

Sa = the number of shops in town A
Sb = the number of shops in town B.

Referring to our original example, suppose Grimsby–Cleethorpes has 800 shops and Lincoln has 300 shops. The formula could then be written:

$$Db = \frac{71}{1 + \sqrt{\dfrac{800}{300}}}$$

$$\therefore Db = 27$$

This means that out of every 71 shoppers, 44 would go to Grimsby–Cleethorpes and 27 to Lincoln.

In reality, the competitive commercial relationships between urban centres can change over a period of time. On Humberside, for example, there have been the effects of the opening of the Humber Bridge on places either side of the estuary, the construction of the M62 and M180 motorways, and the development of new out-of-town shopping centres (pages 433 and 458).

Measuring settlement patterns: conclusion

Nearest neighbour analysis, the rank–size rule, Christaller's central place theory and the inter-action models are all difficult to observe in the real world. Their value lies in the fact that they form hypotheses against which reality can be tested – provided you do not seek to *make* reality fit them (Framework 10, page 299)! Also, they offer objective methods of measuring differences between real-world places. When theory and reality diverge, the geographer can search for an explanation for the differences. An important shared characteristic of these approaches is that they aim to find order in spatial distributions.

Settlement change

A Cambourne – a new village in England

Work began in late 1998 on a new village in South Cambridgeshire to be called Cambourne (Figure 14.41). Eventually 8000 people will live here, in 3300 houses (up to 900 of which will be 'affordable homes'), which are to be built over 12 years. Cambourne, which covers 400 hectares, will be laid out as three distinct villages (Figure 14.43), each with its own central green (Figure 14.42). There will also be a church, two primary schools, a library, 18 hectares of playing fields, a multi-purpose sports centre, a health centre, police and fire stations. The developers have agreed to provide funds for a park and ride scheme, cycle tracks and a bus service. The development aims to enhance the environment by including 69 hectares of planted woodland, 56 hectares for a new Country Park, and the construction of a series of lakes. It is hoped that a new 20 hectare business park will eventually create up to 3000 new jobs, many of which, as the village is so close to Cambridge, are likely to be high-tech (Places 86, page 566). In time, the A428 arterial route linking Cambourne to Cambridge will become a dual carriageway.

Speaking on what has been described as an exciting design which uses modern ideas on sustainability and which is set to become a showpiece for new developments of this kind nationwide, the District Planning Officer claimed, in 1998: 'A new village offers many advantages over increased development in our existing villages. Because of the economies of scale we have been able to negotiate a £32 million package of public infrastructure from the developers. If this development had been scattered over many villages, this infrastructure would either not have been built or would have had to be paid for by the taxpayer. By creating 3000 jobs we enable people to live and work in the community, reducing the need for commuting.'

Naturally not everyone shares the views of the District Planning Officer. Some, like local farmer Robin Page, writing in the *Daily Telegraph* on 5 June 1999, point out that it is another 400 hectares of rural land lost to the developers. He, among others, questions why Britain needs so many more new homes – an estimated 5 million by 2016 (Figure 14.22), when the total population is only rising by 0.2 per cent per year. He also questions the site of Cambourne which, he claims: 'is on the highest part of west Cambridgeshire, guaranteeing immense drainage problems and a terrible eyesore on the skyline. The 400 hectares form the watershed of the Bourn Brook, a tributary of the River Cam, on which the Countryside Restoration Trust is a major landowner and where otters are supposed to be a protected species. The whole project was delayed for 12 months until the developers agreed to ensure that water drainage and sewage did not damage the Bourn Valley or the River Cam.'

In reply, Ross Clark wrote in the *Daily Telegraph* of 3 July 1999 that, if anything, Cambourne was being built too far from Cambridge (15 km from the city centre). To him, the area was not especially attractive and, while 25 000 jobs had been created in the area between 1961 and 1966, only 16 000 new homes had been built – many of them scattered in villages some distance from Cambridge where the economic growth is concentrated. The result had been the rocketing of house prices (some have doubled in a decade), the turning of village roads and rural lanes into racetracks, and the threat that new high-tech industries, if they cannot find space here, will be forced to locate overseas. Mr Clark also pointed out that it was not 'the high-tech industries that have been poisoning the watercourses with nitrates and exterminating the wild flowers with pesticides: it was good old farming'.

Cambourne is being built. Several similar new villages are planned countrywide. The argument for and against their development will continue.

Figure 14.41

In early 2000, much of Cambourne remained an empty site, with only basic roads and street lighting

Figure 14.42

The brand-new village green early in 2000

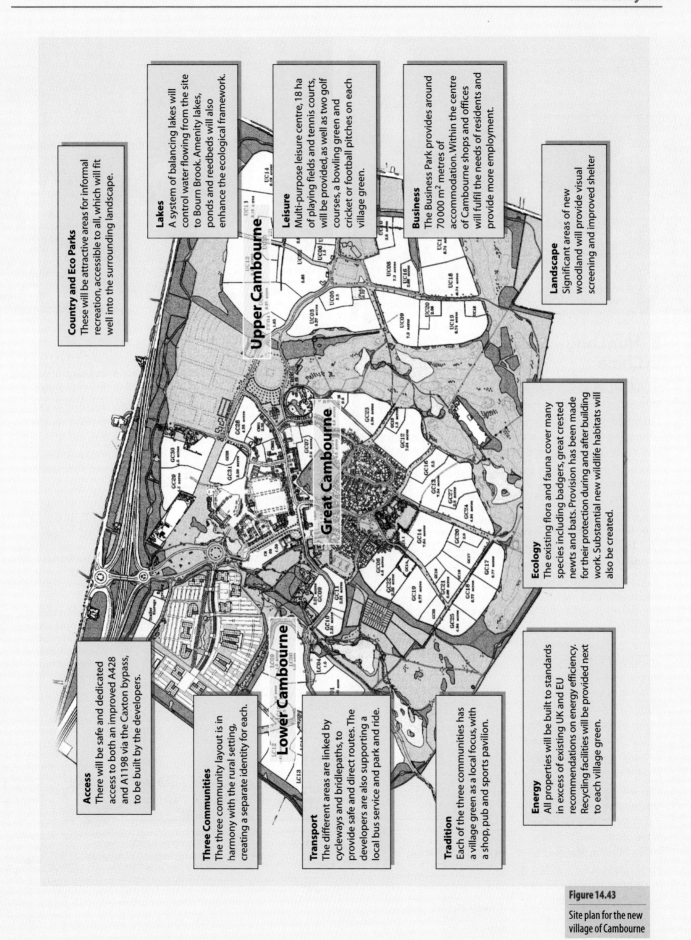

Country and Eco Parks
These will be attractive areas for informal recreation, accessible to all, which will fit well into the surrounding landscape.

Lakes
A system of balancing lakes will control water flowing from the site to Bourn Brook. Amenity lakes, ponds and reedbeds will also enhance the ecological framework.

Leisure
Multi-purpose leisure centre, 18 ha of playing fields and tennis courts, will be provided, as well as two golf courses, a bowling green and cricket or football pitches on each village green.

Business
The Business Park provides around 70000 m² metres of accommodation. Within the centre of Cambourne shops and offices will fulfil the needs of residents and provide more employment.

Landscape
Significant areas of new woodland will provide visual screening and improved shelter

Ecology
The existing flora and fauna cover many species including badgers, great crested newts and bats. Provision has been made for their protection during and after building work. Substantial new wildlife habitats will also be created.

Access
There will be safe and dedicated access to both an improved A428 and A1198 via the Caxton bypass, to be built by the developers.

Three Communities
The three community layout is in harmony with the rural setting, creating a separate identity for each.

Transport
The different areas are linked by cycleways and bridlepaths, to provide safe and direct routes. The developers are also supporting a local bus service and park and ride.

Tradition
Each of the three communities has a village green as a local focus, with a shop, pub and sports pavilion.

Energy
All properties will be built to standards in excess of existing UK and EU recommendations on energy efficiency. Recycling facilities will be provided next to each village green.

Upper Cambourne

Great Cambourne

Lower Cambourne

Figure 14.43

Site plan for the new village of Cambourne

Figure 14.44

Farmhouses around a central courtyard

Figure 14.46

Farmer within the courtyard

B Hua Long – a village in China

Hua Long is situated in the province of Sichuan, 280 km from Chengdu and 180 km from the Yangtze port of Chongqing. Like many other villages in the area it dates from the later Ming period (1550–1644). Between that time and the 1990s, little changed. Today, some 2000 people live in the village, which is fairly small by Chinese standards. Hua Long is linear in shape with most of its buildings strung out along the wide, but poorly maintained, 'main' road which passes through it (Figure 14.47).

Most families in Hua Long are farmers (Sichuan is known as the 'rice bowl of China'), working long hours at little more than a subsistence level (page 477). Many live in farmhouses which are usually grouped together, in typical Chinese fashion, around a central courtyard. Around the courtyard shown in Figure 14.44 are 13 doors, signifying 13 families (Figure 14.45). The address of this group of families – a legacy of the commune days of the 1960 and 1970s (Places 63, page 468), is Group 4 Team 1. Most of the families (Figure 14.46) have lived in these one-roomed

houses for several generations. Despite the lack of running water and sewerage, and the presence of several pigs, there is no smell. The wooden or mud-bricked houses have tiled roofs and shutters, or iron bars, across openings that served as windows. There are no chimneys. Central to the courtyard is an area for collecting household waste that can be fed to the pigs. The

remainder of the area is used for drying and storing crops, or as a social meeting-place. The houses are soon to be pulled down (the families visited by the author in 1999 wanted copies of the photos shown here as mementoes), and although many of the occupants will be sad to lose their ancestral home, they are looking forward to living in modern, brick-built houses with water and electricity.

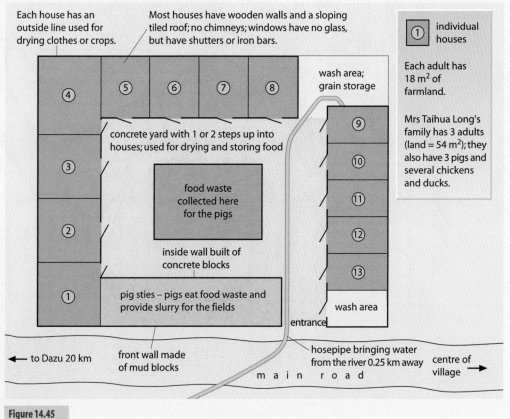

Figure 14.45

Courtyard plan

Each house has an outside line used for drying clothes or crops.

Most houses have wooden walls and a sloping tiled roof; no chimneys; windows have no glass, but have shutters or iron bars.

① individual houses

Each adult has 18 m² of farmland.

Mrs Taihua Long's family has 3 adults (land = 54 m²); they also have 3 pigs and several chickens and ducks.

concrete yard with 1 or 2 steps up into houses; used for drying and storing food

food waste collected here for the pigs

inside wall built of concrete blocks

pig sties – pigs eat food waste and provide slurry for the fields

wash area; grain storage

wash area

entrance

← to Dazu 20 km

front wall made of mud blocks

hosepipe bringing water from the river 0.25 km away

centre of village →

m a i n   r o a d

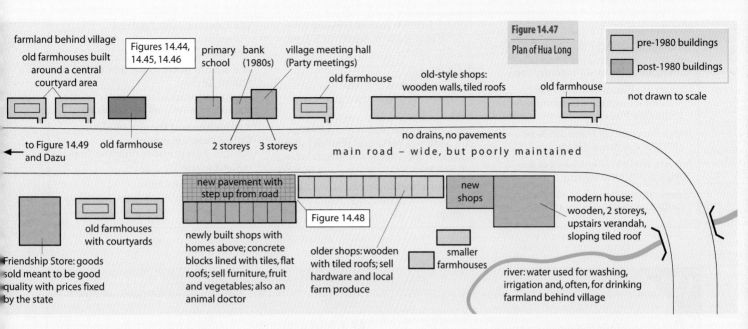

Figure 14.47

Plan of Hua Long

farmland behind village

old farmhouses built around a central courtyard area

Figures 14.44, 14.45, 14.46

primary school

bank (1980s)

village meeting hall (Party meetings)

old farmhouse

old-style shops: wooden walls, tiled roofs

old farmhouse

pre-1980 buildings

post-1980 buildings

not drawn to scale

to Figure 14.49 and Dazu

old farmhouse

2 storeys   3 storeys

no drains, no pavements

main road – wide, but poorly maintained

new pavement with step up from road

Figure 14.48

new shops

modern house: wooden, 2 storeys, upstairs verandah, sloping tiled roof

old farmhouses with courtyards

Friendship Store: goods sold meant to be good quality with prices fixed by the state

newly built shops with homes above; concrete blocks lined with tiles, flat roofs; sell furniture, fruit and vegetables; also an animal doctor

older shops: wooden with tiled roofs; sell hardware and local farm produce

smaller farmhouses

river: water used for washing, irrigation and, often, for drinking farmland behind village

Several changes have taken place in Hua Long in the last 20 years (Figure 14.47). These include the building of a bank (not needed before 1980 as people were not allowed to earn money), an improved primary school for children aged 6 to 12 (funded by the voluntary Hope Project which aims to improve education in the poorest parts of China) and a Friendship Store. Along the main road are several tile-faced, double-storey buildings (Figure 14.48) where newly rehoused people live above shops and small workshops, and several new detached houses (signs of increasing wealth among a few of the inhabitants).

In contrast, the Yangs live, with their two children, in a large, two-year-old brick farmhouse built on the outskirts of the village (Figure 14.49). Mr Yang is a farmer, but he also operates a trishaw 'taxi' in the nearby town Dazu. The family saved enough money, and borrowed the rest from Mr Yang's cousin, to replace their old wooden farm with a seven-roomed, double-storeyed house (though some rooms are only used for storing crops, and furniture is sparse). The Yangs claim that most people in the village are better off and much happier than they were 20 years ago (Figure 16.8).

Figure 14.48

New shops and houses

Figure 14.49

A new farmhouse

References

Bradford, M. G. and Kent, W. A. (1977) *Human Geography: Theories and Their Applications*, Oxford University Press.

Hornby, W. F. and Jones, M. (1991) *Settlement Geography*, Cambridge University Press.

Philips Geographical Digest 1998–99 (1998) Heinemann-Philips.

Roberts, B. (1987) *The Making of the English Village*, Longman.

Waugh, D. and Rowley, C. (2000) *Images of Change: China*, CD Rom, Geopix.

Wilson, J. (1984) *Statistics in Geography for 'A' level Students*, Schofield & Sims.

Websites

Early civilisation in Crete:
http://www.dilos.com/region/crete/kn_01.html

The Countryside Agency's links page (UK National Parks and regional sites):
http://www.countryside.gov.uk/who/f_link.htm

Gretton: a Northamptonshire village:
http://www.skynet.co.uk/maurice/gretton/

The future of rural England:
http://www.ruralnet.org.uk/rwp/bground.htm

The official guide to the new town of Milton Keynes:
http://ourworld.compuserve.com/homepages/gareth_lewis/ofguida.htm

A guided tour of the city of Al Ain, United Arab Emirates:
http://www.geocities.com/The Tropics/Canbana/8632/

See also for more links:
http://www.nelsonthornes.com/gaia

Questions

Q

1 a What is the meaning of:
   i the 'site' of a settlement? **(1 mark)**
   ii the 'situation' of a settlement? **(1 mark)**

 b In the past various factors had to be considered by people seeking a settlement site. Explain what each of the following terms means, and why each type of site was sometimes chosen for settlements:
   i a 'wet point site'
   ii a 'dry point site'
   iii a 'nodal point'. **(6 marks)**

 c Many towns and cities in the UK have changed their functions many times since they were first built. This has often caused serious planning problems because the original sites are not suitable for the modern functions of the settlement. Name one town or city in the UK which has problems caused by its original site.
   i Describe the site, and explain why it was originally chosen. **(4 marks)**
   ii Explain why that site causes problems now. **(4 marks)**
   iii Describe how the planners are attempting to tackle the problems caused by the site. **(6 marks)**
   iv To what extent have the planners been successful in tackling the problems? **(3 marks)**

2 a What is meant by:
   i the morphology of a settlement?
   ii a nucleated settlement?
   iii dispersed settlement? **(3 marks)**

 b Name an example of each of the settlement types listed below. Describe the main features of each of the settlements that you name. Explain why each of the named settlements developed at that location.
   i linear settlement **(4 marks)**
   ii ring or green village **(4 marks)**
   iii commuter village **(4 marks)**

 c Study Figure 14.50. It shows the development of second homes in a remote area of rural North Wales.
   i Suggest why such a high proportion of houses have become second homes for people who have their main homes elsewhere. **(5 marks)**
   ii Explain why the growth of second home ownership can create problems in areas such as that shown on the map. **(5 marks)**

AS

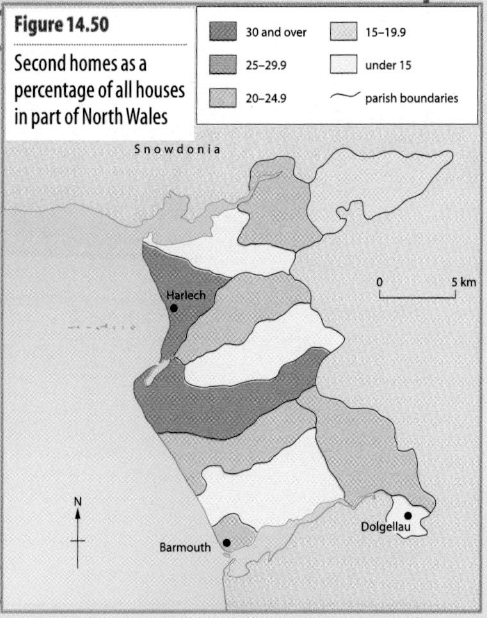

3 a What do the following phrases mean?
   i the 'range' of a good or service **(2 marks)**
   ii the 'threshold population' for a good or service. **(2 marks)**

 b When geographers develop models they always make a set of assumptions before they start to describe the model. Explain **three** of the assumptions that Christaller made before he developed his central place model. **(6 marks)**

 c Why did settlements have market areas shaped like hexagons in Christaller's central place model? **(4 marks)**

 d In a $k = 3$ version of the model there was one third-order settlement with a market area of 1000 square km². How large would the market area be of:
   i each second-order settlement?
   ii each first-order settlement? **(2 marks)**

 e Why is the $k = 3$ version of the model known as the 'market principle' version of the model? **(3 marks)**

 f How useful is Christaller's central place model for modern geographers? **(6 marks)**

Figure 14.50

Second homes as a percentage of all houses in part of North Wales

Legend:
- 30 and over
- 25–29.9
- 20–24.9
- 15–19.9
- under 15
- parish boundaries

Snowdonia

Harlech

Barmouth

Dolgellau

0   5 km

N

4 Study the section 'Hennock, Devon: a village' on page 399.
   a i Give two main reasons why the income of people working in the local area has fallen. **(2 marks)**
      ii Give two main reasons why local people cannot buy houses in the village. **(2 marks)**
      iii Give two examples (other than the closure of shops) to show how the village is losing services, and explain the economic reason for the loss of these services. **(4 marks)**
      iv Why has the village shop closed? You should refer to changing conditions in the village, and also to changing patterns in the nature of shopping in the surrounding region. **(4 marks)**
   b Imagine that 'Hennock Action for the Village Environment' (HeAVE) has been set up by a group of local residents. Their aim is to keep the village as an attractive place to live for everyone – long-time residents and newcomers. They have suggested that if the village can raise £10 000 they can apply for a National Lottery grant to rebuild the village hall. They feel that this would be an important step towards revitalising the village.
      i Outline the views of HeAVE, explaining why their proposal could be important for the village. **(4 marks)**
      ii Suggest why some people might be opposed to this plan. **(4 marks)**
      iii Imagine that the people mentioned in ii developed an alternative plan to help to revitalise the village. Describe and justify an alternative plan. You should outline your aims for the village and suggest how your plan will fulfil those aims. **(5 marks)**

6 Study Case Study 14A on pages 412–413.
   a Explain why the area's planning officers think that a new village is needed in Cambourne, so close to Cambridge. **(6 marks)**
   b Imagine that you work for an estate agent which is trying to sell houses in Cambourne. Write a description of the village to go in a property magazine aimed at young, graduate professional people living in East Anglia. **(6 marks)**
   c The third paragraph of the case study states that concern was expressed that the new village might cause problems with the Rivers Cam and Bourn Brook. Describe some of the potential problems, and suggest how good management of the development could reduce one of these problems. **(6 marks)**
   d A primary school will be needed in the village. Refer to Figure 14.43 on page 413 to see where it would be located. Taking account of access, safety and the environmental concerns of children, parents and other people without children, justify why this location was chosen for the school. **(7 marks)**

7 Study Figure 14.20 on page 398.
   a Six settlements are shown outside the main conurbation. Explain why these settlements have developed in different ways. **(12 marks)**
   b Choose a region in which rural settlement changes in nature with distance away from a large urban area. Discuss the extent to which this model helps to explain variations in the form of settlements in your chosen region. **(13 marks)**

5 a Study Figure 14.51 of the MetroCentre on Tyneside.
      i What evidence supports the view that this site was chosen because it was:
         • easily accessible to a large number of people? **(5 marks)**
         • built on comparatively cheap land? **(5 marks)**
      ii What evidence shows that the MetroCentre has been carefully designed to allow customers to have the easiest possible access to all parts of the complex? **(5 marks)**
   b Some modern offices are built as close as possible to city centres, whilst others are located on the rural-urban fringe. Compare the advantages of these two types of locations for office locations. Refer to specific examples. **(10 marks)**

Figure 14.51

MetroCentre, Gateshead

8 Choose a town in the UK which shows evidence of its evolution through different periods of history.
   a Describe how the present settlement shows evidence of the form of the settlement in previous periods. **(12 marks)**
   b Discuss the problems and benefits which the historical development presents for today's inhabitants of the town. **(13 marks)**

9 Study Figure 14.51. Referring to evidence from the photograph, and to your knowledge of urban structures, explain the links between:
   • the morphology of cities,
   • the distribution of offices and shops, and
   • the price of land. **(25 marks)**

10 Name one town or city that you have studied. Explain how the growth and development of the town have been influenced by the physical geography of its site. **(25 marks)**

Chapter 15

Urbanisation

*'The invasion from the countryside . . . is over-
whelming the ability of city planners and governments
to provide affordable land, water, sanitation, transport, building materials and food for the urban poor.
Cities such as Bangkok, Bogota, Bombay, Cairo,
Delhi, Lagos and Manila each have over one million
people living in illegally developed squatter settlements or shanty towns.'*

L. Timberlake, *Only One Earth*, 1987

Urban growth – trends and distribution

Urbanisation is defined as the process by which
an increasing proportion of the total population,
usually that of a country, lives in towns and cities.
Although the process began at least as far back as
the fourth millennium BC (Figure 14.2), the
number of people living in urban areas formed,
until fairly recently, only a small proportion of a
country's population. One estimate suggests that
in 1800 only 3 per cent of the world's population
were urban dwellers, a figure that has risen,
according to latest UN estimates, to 48 per cent
(1998) and which is predicted to rise to 60 per
cent before 2025.

Rapid urbanisation has occurred twice in
time and space.

1 During the 19th century, in what are now
   referred to as the economically more developed countries, industrialisation led to a
   huge demand for labour in mining and
   manufacturing centres. Urbanisation was,
   in these parts of the world, a consequence
   of economic development.

2 Since the 1950s, in the economically less
   developed countries, the twin processes of
   migration from rural areas (page 366) and the
   high rate of natural increase in population
   (resulting from high birth rates and falling
   death rates, Figure 13.10) have resulted in
   the uncontrolled growth of many cities.
   Urbanisation is, in the developing countries,
   a consequence of population movement and
   growth and is not, as was previously
   believed, an integral part of development.
In 1998, the UN estimated that 75 per cent
of people in the more economically developed
countries, and 40 per cent in the less economically
developed countries, lived in urban areas (the
corresponding prediction for 2020 is 77 and
53 per cent respectively – Figure 15.1). Figure 15.2
shows national levels of urbanisation.

Simultaneous with urbanisation has been the
growth of very large cities. Whereas the only
cities in the world with a population exceeding
1 million in 1900 were London and Paris, there
were, according to the *Philip's Geographical Digest*,
70 by 1950 and 296 in the mid-1990s. Of these,
34 cities had a population in excess of 5 million,
including 6 with over 15 million (Figure 15.3).
Previous editions of this book have ranked, over a
period of time, the world's largest cities. Although
the largest cities in the mid-1990s are named on
Figure 15.3, their population size is not given due
to uncertainties arising from:
- the use of different criteria by different countries to define the size of an urban area
- problems in collecting accurate census
  figures (e.g. within shanty towns and illegal
  settlements) and in estimating natural
  increase changes which are made annually
  between each census
- not all countries take their census at the same
  time (Framework 15, page 448)
- difficulties in obtaining accurate migration
  figures, especially where refugees and illegal
  immigrants are involved.

There have been several noticeable trends
in the growth of the largest cities since the
mid-1980s.

Figure 15.1

Urban population
growth

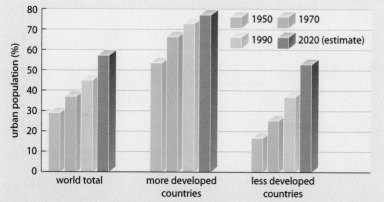

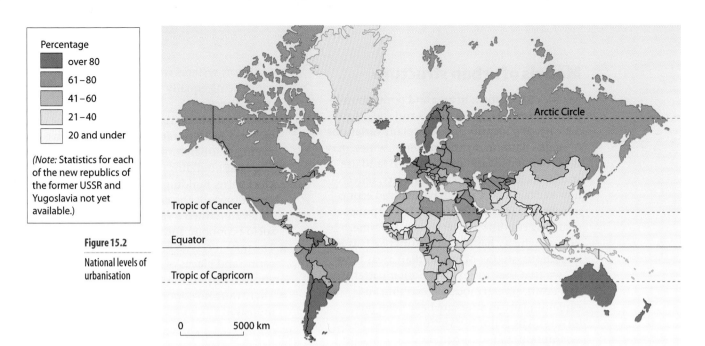

Figure 15.2

National levels of urbanisation

Percentage
- over 80
- 61–80
- 41–60
- 21–40
- 20 and under

(Note: Statistics for each of the new republics of the former USSR and Yugoslavia not yet available.)

0   5000 km

- The fastest-growing cities are in Latin America and in the Japan–Korea region.
- Most of the fastest-growing cities are located in less developed countries, with cities in the least developed countries often having the greatest increase.
- The rate of growth in most of the developing regions was less rapid than had previously been predicted.
- Although there are over 70 Chinese cities in excess of 1 million inhabitants, these have grown less quickly than predicted due to state policies restricting family size (Case Study 14B) and movement from rural areas.
- Cities in North America and western Europe are showing a decrease in overall size – the process of **counter-urbanisation** (page 365).

What affects most people who live in large urban areas is not the actual population size of the city but rather its density. Of the 85 largest cities in the world, the 10 with the lowest population density are in developed countries (9 are in North America) while the 10 with the highest density are in developing countries (headed, in rank order, by Hong Kong, Lagos, Jakarta, Bombay and Ho Chi Minh City). Despite this popular image that the largest cities in developing countries are growing so rapidly, it should be remembered that over one-third of the country's inhabitants, especially in India and China, still live in smaller towns of 20 000–100 000 inhabitants. **Over-urbanisation** occurs when migrants are driven from rural areas to large cities where slow economic growth does not allow the provision of sufficient jobs, or shelter (page 446).

Figure 15.3

Distribution of world cities with populations over 1 million, 1998

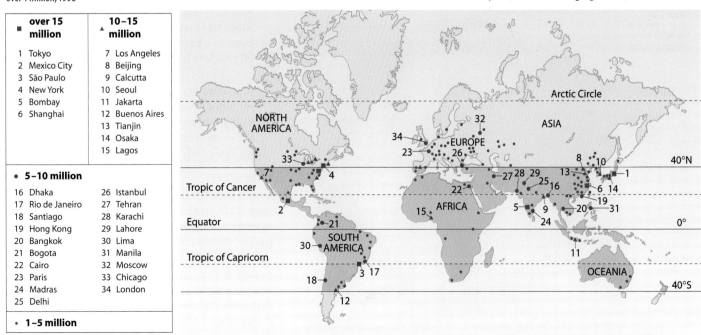

■ over 15 million	▲ 10–15 million
1 Tokyo	7 Los Angeles
2 Mexico City	8 Beijing
3 São Paulo	9 Calcutta
4 New York	10 Seoul
5 Bombay	11 Jakarta
6 Shanghai	12 Buenos Aires
	13 Tianjin
	14 Osaka
	15 Lagos

● 5–10 million

16 Dhaka	26 Istanbul
17 Rio de Janeiro	27 Tehran
18 Santiago	28 Karachi
19 Hong Kong	29 Lahore
20 Bangkok	30 Lima
21 Bogota	31 Manila
22 Cairo	32 Moscow
23 Paris	33 Chicago
24 Madras	34 London
25 Delhi	

● 1–5 million

Urbanisation

Models of urban structure

As cities have grown in area and population in the 20th century, geographers and sociologists have tried to identify and to explain variations in spatial patterns. Spatial patterns, which may show differences and similarities in land use and/or social groupings within a city, reflect how various urban areas have evolved economically and socially (culturally) in response to changing conditions over a period of time. While each city has its own distinctive pattern, or patterns, studies of other urban areas have shown that they too exhibit similar patterns. As a result, several models describing and explaining urban structure have been put forward. The following section lists the basic assumptions of four urban models, describes the theory behind them, applies them to the real world, and gives their limitations.

1 Burgess, 1924

Burgess attempted to identify areas within Chicago based upon the outward expansion of the city and the socio-economic groupings of its inhabitants (Places 52).

Basic assumptions
Although the main aim of his model was to describe residential structures and to show processes at work in a city, geographers have subsequently presumed that Burgess made certain assumptions:

- The city was built on flat land which therefore gave equal advantages in all directions, i.e. morphological features such as river valleys were removed.
- Transport systems were of limited significance being equally easy, rapid and cheap in every direction.
- Land values were highest in the centre of the city and declined rapidly outwards to give a zoning of urban functions and land use.
- The oldest buildings were in, or close to, the city centre. Buildings became progressively newer towards the city boundary.
- Cities contained a variety of well-defined socio-economic and ethnic areas.
- The poorer classes had to live near to the city centre and places of work as they could not afford transport or expensive housing.
- There were no concentrations of heavy industry.

Burgess's concentric zones
The resultant model (Figure 15.4) shows five concentric zones:

1 The **central business district** (CBD) contains the major shops and offices; it is the centre for commerce and entertainment, and the focus for transport routes.
2 The **transition** or **twilight zone** is where the oldest housing is either deteriorating into slum property or being 'invaded' by light industry. The inhabitants tend to be of poorer social groups and first-generation immigrants.
3 Areas of **low-class housing** are occupied by those who have 'escaped' from zone 2, or by second-generation immigrants who work in nearby factories. They are compelled to live near to their place of work to reduce travelling costs and rent. In modern Britain, these zones are equated with the inner cities.
4 **Medium-class housing** of higher quality which, in present-day Britain, would include inter-war private semi-detached houses and council estates.
5 **High-class housing** occupied by people who can afford the expensive properties and the high cost of commuting. This zone also includes the commuter (suburbanised) villages beyond the city boundary – although there were very few of these when Burgess produced his model in 1924.

Limitations
Urban models, like all models (Framework 12, page 352), have limitations and are therefore open to criticism. Despite the advantage of simplicity, and his own admission that it was specific to one place (Chicago, Places 52) and one period of time (the 1920s), the Burgess model has been criticised (Figure 15.15) – in some instances on grounds that did not exist when it was put forward.

Figure 15.4

The Burgess concentric model

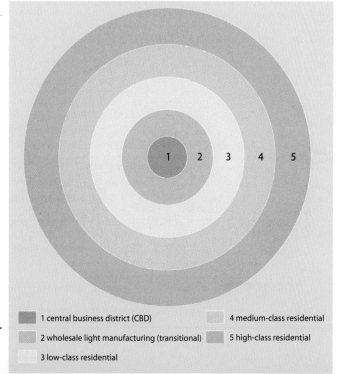

1 central business district (CBD)

2 wholesale light manufacturing (transitional)

3 low-class residential

4 medium-class residential

5 high-class residential

Burgess, in producing his model, was influenced by the emerging science of plant ecology at the University of Chicago. He made analogies with such ecological processes as the **invasion** of an area by competing groups, **competition** between the invaders and the natural groups, and the eventual **dominance** of the area by the invaders which allowed them to **succeed** the natural groups.

Relating this to urban geography, Burgess suggested that people living in the inner zone were **invaded** by newcomers and, in face of this **competition** by immigrants who became **dominant** there, **succeeded** to the next outer zone – a process also referred to as **centrifugal movement**. The energy to maintain this dynamic system came from a continual supply of immigrants to the centre, and existing groups being forced (or choosing) to move towards the periphery.

Chicago lies on the shores of Lake Michigan, with its CBD, known as the 'Loop', facing the lake. Surrounding the CBD, the city's housing developed a distinctive pattern (Figure 15.5). The initial migrants, from north-western Europe, settled around the CBD. In time, they were replaced by

newer immigrants from southern Europe (especially Italy) and by Jews who were, in turn, replaced by blacks from the American south (Figure 15.6). This led to the creation of a series of income, social and ethnic zones radiating outwards from the centre. These zones showed:

1 That wealth, as seen by the quality of housing, increased towards the outskirts of the city. People with the highest incomes lived in the newest property (on the north-west fringe) while those with the lowest incomes occupied the poorest housing next to the CBD.

2 That people in their early twenties or over 60 tended to live close to the CBD, while middle-aged people and families with young children tended to live nearer to the city boundary.

3 That areas of ethnic segregation existed, with the early white immigrants – whose wealth had tended to increase in relation to the length of time they had lived in the city – living towards the outskirts, and non-white groups living nearer to the city centre, e.g. in China Town and the black belt.

Figure 15.5

Urban areas of Chicago (*after* Burgess)

Figure 15.6

Centrifugal movement in Chicago

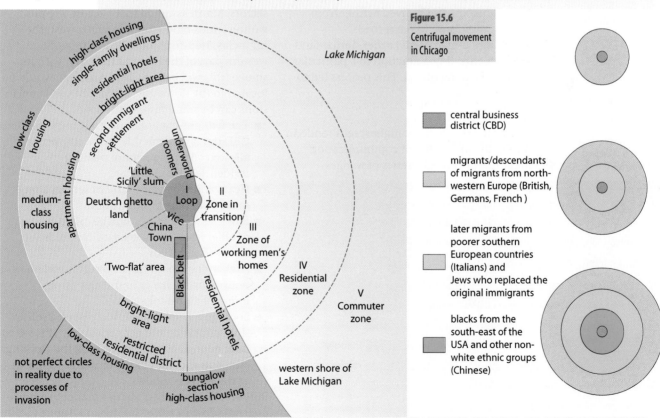

1 CBD (central business district)

2 wholesale light manufacturing (transitional)

3 low-class residential

4 medium-class residential

5 high-class residential

Figure 15.7

The Hoyt sector model

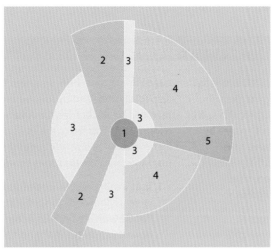

2 Hoyt, 1939

Hoyt's model was based on the mapping of eight housing variables for 142 cities in the USA. He tried to account for changes in, and the distribution of, residential patterns.

Basic assumptions

Hoyt made the same implicit assumptions as had Burgess, with the addition of three new factors:

- Wealthy people, who could afford the highest rates, chose the best sites, i.e. competition based on 'ability to pay' resolved land use conflicts.
- Wealthy residents could afford private cars or public transport and so lived further from industry and nearer to main roads.
- Similar land uses attracted other similar land uses, concentrating a function in a particular area and repelling others. This process led to a 'sector' development.

Hoyt's sector model

Hoyt suggested that areas of highest rent tended to be alongside main lines of communication and that the city grew in a series of wedges

Figure 15.8

Urban areas of Calgary, 1961 (*after* Hoyt)

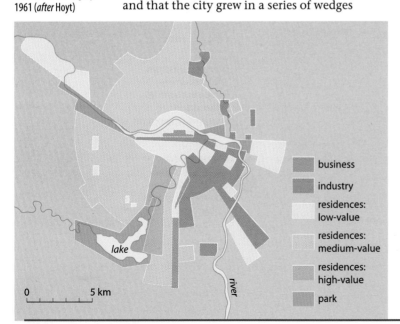

business

industry

residences: low-value

residences: medium-value

residences: high-value

park

0   5 km

lake

river

(Figure 15.7). He also claimed that once an area had developed a distinctive land use, or function, it tended to retain that land use as the city extended outwards, e.g. if an area north of the CBD was one of low-class housing in the 19th century, then the northern suburbs of the late 20th century would also be likely to consist of low-class estates. Calgary, in Canada, is the standard example of a city exhibiting the characteristics of Hoyt's model (Figure 15.8).

Limitations

Many criticisms are similar to those made of Burgess's model and have been summarised in Figure 15.15. It should be remembered that this model was put forward before the rapid growth of the car-based suburb, the swallowing-up of small villages by urban growth, the redevelopment of inner-city areas and the relocation of shopping, industry and office accommodation on edge-of-city sites.

3 Mann, 1965

Mann tried to apply the Burgess and Hoyt models to three industrial towns in England: Huddersfield, Nottingham and Sheffield. His compromise model (Figure 15.9) combined the ideas of Burgess's concentric zones and Hoyt's sectors. Mann assumed that because the prevailing winds blow from the south-west, the high-class housing would be in the south-western part of the city and industry, with its smoke (this was before Clean Air Acts), would be located to the north-east of the CBD. His conclusions can be summarised as follows.

- The twilight zone was not concentric to the CBD but lay to one side of the city which allowed, elsewhere, more wealthy residential areas.
- Heavy industry was found in sectors along main lines of communication.
- Low-class housing should be called the 'zone of older housing' (age-based classification, rather than social).
- Higher-class or, in Hoyt's terms, 'modern' housing was usually found away from industry and smoke.
- Local government (politics) played a role in slum clearance and gentrification. This led to large council estates which took the working class/low incomes to the city edge (opposite of the Burgess model).

Robson (1975) applied Mann's model to a north-eastern industrial town, Sunderland (Figure 15.10), and to Belfast. Mann's model does show, despite its small sample, that a variety of approaches are possible to the study of urban structures.

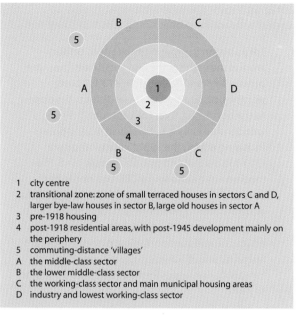

1 city centre
2 transitional zone: zone of small terraced houses in sectors C and D, larger bye-law houses in sector B, large old houses in sector A
3 pre-1918 housing
4 post-1918 residential areas, with post-1945 development mainly on the periphery
5 commuting-distance 'villages'
A the middle-class sector
B the lower middle-class sector
C the working-class sector and main municipal housing areas
D industry and lowest working-class sector

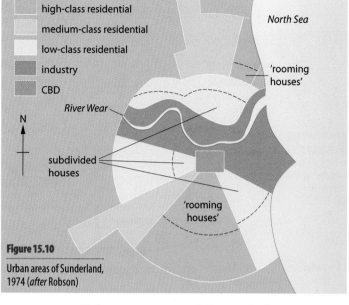

Figure 15.10

Urban areas of Sunderland, 1974 (*after* Robson)

Figure 15.9

Mann's model of urban structure

4 Ullman and Harris, 1945

Ullman and Harris set out to produce a more realistic model than those of Burgess and Hoyt but consequently ended with one that was more complex (Figure 15.11) – and more complex models may become descriptive rather than predictive if they match reality too closely in a specific example (Framework 12, page 352).

Basic assumptions

■ Modern cities have a more complex structure than that suggested by Burgess and Hoyt.

■ Cities do not grow from one CBD, but from several independent nuclei.

■ Each nucleus acts as a growth point, and probably has a function different from other nuclei within that city. (In London, the City is financial; Westminster is government and administration; the West End is retailing and entertainment; and Dockland was industrial.)

■ In time, there will be an outward growth from each nucleus until they merge as one large urban centre (Barnet and Croydon now form part of Greater London; Figure 13.7).

■ If the city becomes too large and congested, some functions may be dispersed to new nuclei. (In Greater London, edge-of-city retailing takes place at Brent Cross and new industry has developed close to Heathrow Airport/M25/M4.)

Multiple nuclei developed as a response to the need for maximum accessibility to a centre, to keep certain types of land use apart, for differences in land values and, more recently, to decentralise (Places 53).

Urban structure models: conclusions

The four models described were put forward to try to explain differences in structure within cities in the developed world. It must be remembered that:

■ each model will have its limitations (Figure 15.15)

■ if you make a study of your local town or city, you must avoid the temptation of saying that it *fits* one of the models – at best it will show characteristics of one or possibly two; each city is unique and will have its own structure – a pattern not necessarily derived according to any existing model (Framework 12, page 352).

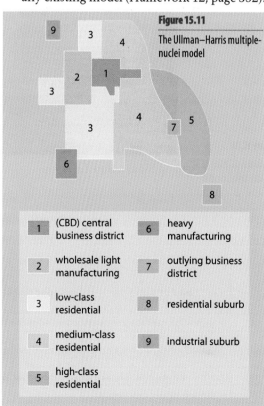

Figure 15.11

The Ullman–Harris multiple-nuclei model

1	(CBD) central business district	6	heavy manufacturing
2	wholesale light manufacturing	7	outlying business district
3	low-class residential	8	residential suburb
4	medium-class residential	9	industrial suburb
5	high-class residential		

Tokyo began to grow in the late 16th century around the castle of the Edo Shogunate (near the present Imperial Palace, Figure 15.12). Later religious, cultural and financial districts developed to the north-east. Over the centuries, the mainly wooden-built city was destroyed several times, including during the 1923 Kanto earthquake (140 000 deaths) and by US aircraft in 1945. The modern city has no single CBD but, rather, has several nuclei each with its own specialist land use and functions – government offices (Figure 15.13), shopping (Figure 15.14), finance, entertainment, education and transport. Most of these nuclei are linked by one of Tokyo's many railways, the Yamanote line, which forms a circle with a diameter of 7 km.

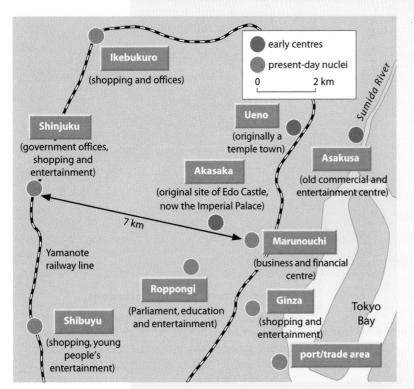

Figure 15.12

Multiple nuclei in Tokyo, 1994

Map labels:
- early centres
- present-day nuclei
- 0   2 km
- **Ikebukuro** (shopping and offices)
- **Shinjuku** (government offices, shopping and entertainment)
- **Ueno** (originally a temple town)
- **Asakusa** (old commercial and entertainment centre)
- **Akasaka** (original site of Edo Castle, now the Imperial Palace)
- Sumida River
- 7 km
- Yamanote railway line
- **Marunouchi** (business and financial centre)
- **Roppongi** (Parliament, education and entertainment)
- **Ginza** (shopping and entertainment)
- **Shibuyu** (shopping, young people's entertainment)
- Tokyo Bay
- port/trade area

Figure 15.14

The Ginza shopping district

Figure 15.13

The Shinjuko business district

Figure 15.15

Limitations/criticisms of the four urban models

		Burgess	Hoyt	Mann	Ullman–Harris
1		zones, in reality, are never as clear-cut as shown on each model			
2		each zone usually contains more than one type of land use/housing			
3		no consideration of characteristics of cities outside USA and north-west Europe			
		based on 1 USA city	based on 142 USA cities	based on 3 English cities (in north and Midlands)	based on cities in economically more developed world
4		redevelopment schemes and modern edge-of-city developments are not included (most of the models pre-date these developments)			
5		based mainly on housing: other types of land use neglected		industry not always to north-east of British cities	
6		cities not always built upon flat plains			
7		tended to ignore transport			

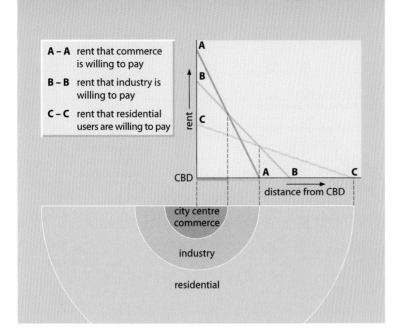

A – A rent that commerce is willing to pay

B – B rent that industry is willing to pay

C – C rent that residential users are willing to pay

Figure 15.16

Bid–rent curves

The land value model or bid–rent theory

This model is the urban equivalent of von Thünen's rural land use model (page 471) in that both are based upon locational rent. The main assumption is that in a free market the highest bidder will obtain the use of the land. The highest bidder is likely to be the one who can obtain the maximum profit from that site and so can pay the highest rent. Competition for land is keenest in the city centre. Figure 15.16 shows the locational rent that three different land users are prepared to pay for land at various distances from the city centre.

The most expensive or 'prime' sites in most cities are in the CBD, mainly because of its accessibility and the shortage of space there. Shops, especially department stores, conduct their business using a relatively small amount of ground-space, and due to their high rate of sales and turnover they can bid a high price for the land (for which they try to compensate by building upwards and by using the land intensively). The

most valuable site within the CBD is called the **peak land value intersection** or PLVI – a site often occupied by a Marks and Spencer store! Competing with retailers are offices which also rely upon good transport systems and, traditionally, proximity to other commercial buildings (this concept does not have the same relevance in centrally planned economies).

Away from the CBD, land rapidly becomes less attractive for commercial activities – as indicated by the steep angle of the bid–rent curve (**A–A**) in Figure 15.16. Industry, partly because it takes up more space and uses it less intensively, bids for land that is less valuable than that prized by shops and offices. Residential land, which has the flattest of the three bid–rent curves (**C–C**), is found further out from the city centre where the land values have decreased due to less competition. Individual householders cannot afford to pay the same rents as shopkeepers and industrialists.

The model helps to explain housing (and population) density. People who cannot afford to commute have to live near to the CBD where, due to higher land values, they can only obtain small plots which results in high housing densities. People who can afford to commute are able to live nearer the city boundary where, due to lower land values, they can buy much larger plots of land, which creates areas of low housing density. Figure 15.17 shows the predicted land use pattern when land values decrease rapidly and at a constant rate from the city centre. The resultant pattern is similar to that suggested by Burgess (Figure 15.4).

One basis of this model is 'the more accessible the site, the higher its land value'. Rents will therefore be greater along main routes leading out of the city and along outer ring roads. Where two of these routes cross, there may be a secondary or subsidiary land value peak (Figure 15.18). Here the land use is likely to be a small suburban shopping parade or a small industrial estate. The 'retail revolution' of the 1980s (page 432), which led to the development of large edge-of-city shopping complexes (MetroCentre in Gateshead, Places 55, page 433, Bluewater, page 458 and Brent Cross in north London), has altered this pattern. Similarly, large industrial estates and science parks (Places 86, page 566) have been located near to motorway interchanges.

Figure 15.17

Urban land use patterns based on land values

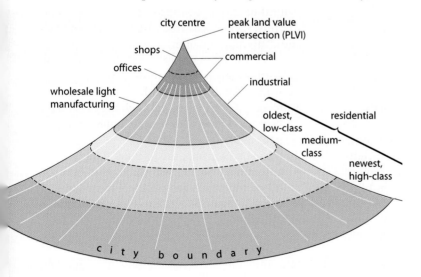

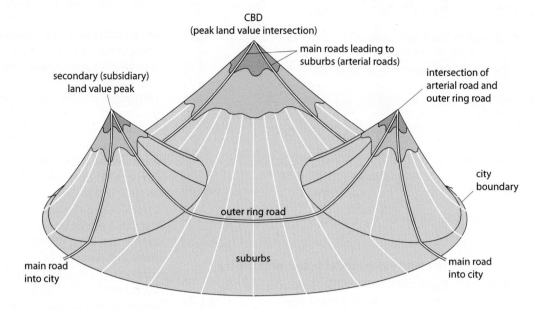

Figure 15.18

Secondary land value peaks

Diagram labels:
- CBD (peak land value intersection)
- main roads leading to suburbs (arterial roads)
- secondary (subsidiary) land value peak
- intersection of arterial road and outer ring road
- city boundary
- outer ring road
- suburbs
- main road into city
- main road into city

Functional zones within a city

Different parts of a city usually have their own specific functions (Figure 15.12). These functions may depend upon:

- the age of the area: buildings usually get older towards the city centre except that most CBDs and many old inner-city areas have been redeveloped and modernised
- land values: these increase rapidly from the city boundary in towards the CBD (Figure 15.16)
- accessibility: some functions are more dependent upon transport than others.

While each urban area will have its own unique pattern of functional zones and land use, most British cities exhibit similar characteristics. These characteristics have been summarised and simplified in Figure 15.19 where:

Zone A = the CBD (shops and offices)

Zone B = old inner city (including, before re-development, 19th-century/low-cost/low-class housing, industry and ware-housing and, after redevelopment/regeneration, modern low-cost housing and small industrial units

Zone C = inter-war (medium-class housing)

Zone D = suburbs (modern/high-cost/high-class housing, open space, new industrial estates/science and business parks, shopping complexes and office blocks).

The central business district (CBD)

The CBD is regarded as the centre for retailing, office location and service activities (banking and finance). It contains the principal commercial streets and main public buildings and forms the **core** of a city's business and commercial activities. Some large cities, such as London and Tokyo (Figure 15.12) may have more than one CBD. Other types of city-centre land use, such as government and public buildings, churches and educational establishments, are classed as non-CBD functional elements.

The delimitation of the CBD

Most of you are likely to have relatively easy access to a town or city centre. If so, your geography group may be able to make one or more visits to that CBD with the aim of trying to delimit its extent. Bearing in mind possible dangers, such as from moving traffic, your group could attempt one or more of several methods, based on the pioneer work of Murphy and Vance in North America, and described in Places 54 page 430. Ideally you should:

1 formulate one (or more) hypothesis before you begin your fieldwork (Framework 10, page 299)
2 collect, as a group, the relevant data
3 determine how you will record that data (i.e. using which geographical techniques)
4 discuss – again as a group – your findings.

One of several dangers that may result from putting forward geographical models and from making generalisations is that of creating stereotypes. For example:

a Urban models suggest that all housing in inner-city areas is low-class/low-income (Burgess) and that all elderly and single-parent families live in this zone while families with two or three children only live in the outer zone.

b Different groups of people tend to develop their own customs and ways of life. By putting such characteristics together, we make mental pictures and develop preconceptions of different groups of people, i.e. we create stereotypes.

The following unsupported, emotive statements may not only be grossly inaccurate, they may also be considered, by many, to be offensive.

- The Germans, on holiday, are always first to the swimming pool and dining room.
- All Italians drive cars dangerously.
- All Chinese and Japanese are small.
- *Favelas* are shanty settlements whose inhabitants have no chance of improving their living conditions.
- The inhabitants of a *favela* can only survive by a life of crime (see below).
- The Amazon Amerindian way of life remains undeveloped as the people are lazy and unintelligent (see below).

The following accounts are based on the author's experiences in Brazil.

Example One

'According to books which I had read in Britain and advice given to me by guides in São Paulo, *favelas* were to be avoided at all costs (Places 57, page 443). Any stranger entering one was sure to lose his watch, jewellery and money and was likely to be a victim of physical violence.

With this in mind, I set off in a taxi to take photographs of several *favelas*. On reaching the first *favela*, to my horror the driver turned into the settlement and we bumped along an unmade track. He kept stopping and indicating that I should take photographs. Expecting at each stop that the car would be attacked and my camera stolen, I

hastily took pictures – which turned out to be over-exposed because, not daring to open windows, I took them through the windscreen and looking into the sun!

Suddenly the taxi spluttered and stopped. In one movement, I had hidden my camera and was outside trying to push the car. I raised my eyes to find three well-built males helping me to push the car. Which one would hit me first? I smiled and they smiled. I pointed to each one in turn and called him after one of Brazil's football players and then referred to myself as "Lineker". Huge smiles, big pats on the back and comments like " *Ingleesh amigo* " were only halted by the car re-starting. As we drove away, I began to question my original stereotyped view of a *favela* inhabitant.'

Example Two

'I was surprised to find, on landing at Manaus airport in the middle of the Amazon rainforest, that our courier was an Amerindian. He dashed around quickly getting our party organised and our luggage collected. (He certainly did not seem to be slow or lazy.) He later admitted, and proved, that he could speak in seven languages (hardly the sign of someone unintelligent – how many can *you* speak?). I asked him why so few Amerindians appeared to have good jobs and why he kept talking about returning to the jungle. His reply was simple: 'to avoid hassle'. He considered that the Indian lifestyle was preferable to the Western one with its quest for material possessions. Had he returned to the jungle, he would have rejoined his family and become a shifting cultivator living in harmony with the environment (Places 66, page 480). Is that traditional way of life really less demanding of intelligence than that imposed by invading timber and beefburger transnationals engaged in the destruction of large tracts of rainforest?'

From these examples, we can see how easy it is to accept stereotypes without realising we are doing so, and also how seeing a situation for ourselves may lead us to question our original picture. Should geographers take a role in overcoming the problems of stereotyped images (on the basis of which planning decisions, for example, may be made) by helping to provide relatively unbiased information to improve knowledge and understanding?

A CBD

A1

Figure 15.19

Functional zones in a British city

A1 Indoor shopping mall (St Enoch's Centre, Glasgow)
A2 High-rise office development (the City of London)

B1 An inner-city corner shop (Leeds)
B2 19th-century terraced housing (Lancashire)
B3 Inner-city redevelopment (London)
B4 19th-century industry and transport (Manchester)

B Inner city

B1

B2

C Inter-war areas

C1

C2

D Edge of city

D1

D2

1   Shopping types

2   Residential styles

C1 A suburban shopping parade
C2 Inter-war semi-detached private housing (Enfield)
C3 Inter-war council housing estate (Carlisle)
C4 Public open space (Brockwell Park, London)

D1 Edge-of-city shopping complex (Lakeside shopping centre, Dartford)
D2 A modern private housing estate (Wirral)
D3 Post-war edge-of-city council housing estate (Kenton, Newcastle upon Tyne)
D4 Business/science park (Guildford)

2 Residential styles

3 Other land uses

The main characteristics of the CBD

1 The CBD contains the major retailing outlets. The principal department stores and specialist shops with the highest turnover and requiring largest threshold populations compete for the prime sites (Figure 15.19 A1).

2 It contains a high proportion of the city's main offices (Figure 15.19 A2).

3 It contains the tallest buildings in the city (more typical in North America), mainly due to the high rents which result from the competition for land (Figure 15.16).

4 It has the greatest number and concentration of pedestrians.

5 It has the greatest volume and concentration of traffic. The city centre grew at the meeting point of the major lines of communication into the city and therefore had the greatest accessibility.

6 It has the highest land values in the city (Figure 15.17).

7 It is constantly undergoing change, with new shopping centres, taller office blocks and traffic schemes. Some of the grandiose schemes of the early 1960s are now viewed as out of date and unattractive (Birmingham's Bull Ring; London's Paternoster Square at St Paul's). Many have since been demolished and rebuilt.

Recent studies have shown that the CBD of many cities is advancing in some directions (**zone of assimilation**) and retreating in others (**zone of discard**). The zone of assimilation is usually towards the higher-status residential districts whereas the zone of discard tends to be nearer the industrial and poorer-quality residential areas (Figure 15.20). There has also been a trend in many CBDs for retailing to be static, or even declining (due to competition from out-of-town developments), while offices, banks and insurance companies are increasing in terms of space taken and income generated.

Mapping the characteristics of the CBD

The following fieldwork methods may be used to evaluate the seven characteristics described above.

1 **Land use mapping of shops**

   a Plot the location of all the shops. Where the ratio of shops to other properties is more than 1:3, count that area as being within the CBD (based on evidence that over 33 per cent of buildings in the CBD are connected with retailing).

   b An alternative method is to include within the CBD all shops that are within 100 m (or any agreed distance) of adjacent shops. This may produce a central 'core' and several smaller groupings.

   c A third possibility is to take the mean frontage (in metres) of, for example, the middle five buildings or shop units in a block. Shop frontages are likely to be greatest near to the PLVI where most department stores are located.

2 **Land use mapping of offices** Method **1a** above could be repeated using offices instead of shops, and a ratio of 1:10. This recognises that, at ground-floor level, offices are less numerous than shops. Include banks and building societies in your count.

3 **Height of buildings** Plot the height (i.e. the number of storeys) of individual buildings, or the mean of a group of buildings in the centre of a block. Most cities tend to have a sharp decline in building height at the edge of the CBD.

4 **Number of pedestrians** This is a group activity – the more groups the better! Each group counts the number of pedestrians passing a given point at a given time (e.g. 1100–1115 hours). The greater the number of sites (ideally chosen by using random numbers, Framework 6, page 159), the greater the accuracy of the survey. Define a pedestrian as someone of school age and over, walking into, out of or past a shop on your side of the street. These criteria may be altered as long as they are applied by all the groups.

5 **Accessibility to traffic** This is similar to the previous survey except that here vehicles are counted. Make sure all groups have the same

Figure 15.20

The core and frame concept for the CBD

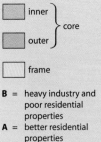

- inner ⎱
- outer ⎰ core
- frame

B = heavy industry and poor residential properties

A = better residential properties

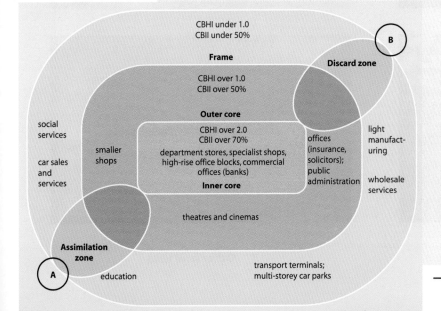

definition of a vehicle, e.g. do you include a bicycle and/or pram?

6 **Land values** These might be expected to decline outwards at a fairly uniform rate. Providing rateable values can be obtained (try the rates office) and there is the time to process them (or a sample of the total), this is often a good indicator of the CBD. It may be useful to take the PLVI point and call this 100 per cent, and then convert the rateable index for all other properties as a percentage of the PLVI. It has been suggested that a figure of 20 per cent delimits the CBD for a British city.

7 **Changing land use and functions** This is a mapwork exercise using old maps of the central area (shopping maps are produced by GOAD plans) and superimposing onto them present-day land uses. Look for evidence of zones of assimilation and discard (Figure 15.20).

• **Central business index** This is probably the best method as it involves a combination of land use characteristics, building height and land values. The problem is in obtaining the necessary data, i.e.

  a the total floor area of all central or CBD functions

  b the total ground floor area (central and non-central functions)

  c the total floor area (upstairs floor area as well as the ground floor).

You may laboriously work this out from a large-scale plan, or choose to compromise by taking the mean of a sample of buildings in each block. From these data, two indices can be derived:

• The **central business height index**, or **CBHI**, which is expressed as:

$$CBHI = \frac{\text{total floor area of all CBD functions}}{\text{total ground floor area}}$$

• The **central business intensity index**, or **CBII**, which is expressed as:

$$CBII = \frac{\text{total floor area of all CBD functions}}{\text{total floor area}} \times \frac{100}{1}$$

To be considered part of the CBD, the CBHI of a plot should be over 1.0 and the CBII over 50 per cent (Figure 15.21).

Plotting the data

Careful consideration should be given as to which cartographic technique is best applied to each set of collected data. You may wish to use one or several of the following: land use maps, isolines, choropleths, flow graphs, histograms, bar graphs, scattergraphs and transects. Alternatively, you may be able to devise a technique of your own. You may save time and produce results that are easier to compare by using tracing overlays and/or a computer.

Delimiting the CBD: conclusions

If you have carried out your own survey, your report might include comments on the following questions:

1 What problems did you encounter in collecting and refining the data?

2 Which of the methods used in collecting the data appeared to give the most, and the least, accurate delimitation of the CBD?

3 In your town, was there an obvious CBD; did you find an inner and an outer core (Figure 15.20)? Was there evidence of zones of assimilation and discard? Were there any specific functional zones other than shops and offices? Was the area of the CBD similar to your mental map (your preconceived picture) of its limits?

4 What refinements would you make to the techniques used if you had to repeat this task in a different urban area?

Figure 15.21

The central business index (CBI)

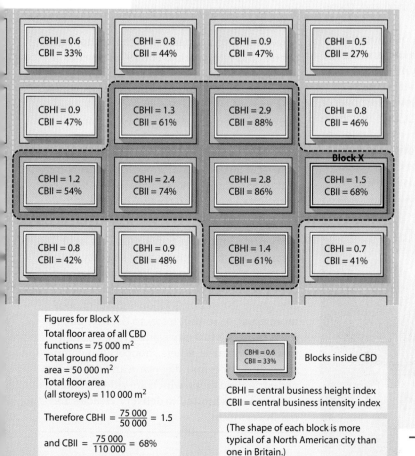

Figures for Block X
Total floor area of all CBD functions = 75 000 m²
Total ground floor area = 50 000 m²
Total floor area (all storeys) = 110 000 m²

Therefore CBHI = $\frac{75\,000}{50\,000}$ = 1.5

and CBII = $\frac{75\,000}{110\,000}$ = 68%

CBHI = central business height index
CBII = central business intensity index

(The shape of each block is more typical of a North American city than one in Britain.)

Retailing

Traditional shopping patterns

Traditionally, as neatly summarised by Prosser, 'Retailing in British cities has been based upon a well-established hierarchy, from the CBD or "High Street" at the top, through major district centres, local suburban centres, to neighbourhood parades and the local corner shop. Using numbers of outlets, floor space, type and range of goods, for example, as measures of size or "mass", Christaller's central place (page 407) and gravity models (page 410) have been applied to the hierarchical structure, relating mass to spatial distribution of shopping centres and their spheres of influence.'

Within this hierarchy were two main types of shop:

1   Those selling **convenience** or **low-order goods** which are bought frequently, usually daily, and are not sufficiently high in value to attract customers from further than the immediate catchment area, e.g. newsagents and small chain stores.

2   Those selling **comparison** or **high-order goods** which are purchased less frequently but which need a much higher threshold population, e.g. goods found in department stores and specialist shops.

The preferred location of these two types of shop was usually determined by the frequency of visit, their accessibility and the cost of land and, therefore, rent (page 425).

Convenience shops are commonly located in housing estates, both in the inner city and the suburbs, and in neighbourhood units so as to be within easy reach of their customers – often within walking distance. With a lower turnover of goods than retail units in the CBD, they may have to charge higher prices but their rent and rates are lower. Ideally, they are located along suburban arterial roads or at a crossroads for easier access and, possibly, to encourage impulse buying by

motorists driving into the CBD (Figure 15.18). Convenience shops are also located in inner cities where the corner shop (Figure 15.19 B1) caters for a population that cannot afford high transport costs; in suburban shopping parades (Figure 15.19 C1) where the inhabitants live a long way from the central shopping area; and along side-streets in the CBD where they take advantage of lower rents to provide daily essentials for those who work in the city centre.

Comparison shops need a large threshold population (page 407) and therefore have to attract people from the whole urban area and beyond. As they bid for a central location, they must have a high turnover in order to pay the high rents. This central area has traditionally afforded the greatest accessibility for shoppers, with public transport competing with the private motorist. Large department stores and specialist shops usually locate within the CBD (Figure 15.19 A1), although comparison shops may also locate in the more affluent suburbs.

The retailing revolution

There have been so many significant changes in retailing since the early 1970s that, collectively, they have been referred to as a 'revolution'. These changes include the shopping behaviour of customers, the organisation of the retail industry, the character of the shopping environment, the planning policies of local and central government and, most important of all, the location and introduction of new retail outlets (superstores, hypermarkets, regional shopping centres and retail parks). The 1970s saw the emergence of superstores, often within existing shopping centres, and hypermarkets, usually on new edge-of-city sites (Figure 15.19 D1). The early 1980s saw the development of non-food retail parks which included furniture, DIY, carpet stores and garden centres. The late 1980s saw the growth of large regional shopping centres (Places 55). Since 1980, the numbers employed in the retail industry have nearly trebled.

Town centres

Congestion and other traffic problems have meant that urban access is now less easy than in the past. Many city centres have had to undergo constant change either to try to attract more customers (Newcastle's Eldon Square) or, as is more usual, to restrict losses to the new out-of-town centres. Most city centres now contain under-cover malls, where shoppers can compare styles and prices and stay warm and dry, and pedestrianised areas, which are either traffic-free or have access limited to delivery vehicles and public transport.

Figure 15.22

An environmentally improved city centre shopping area in Sheffield

Many local councils are now encouraging city centres to try to win back or replace custom lost to the edge-of-city shopping area. This is being achieved by varying the land use of the 'High Street' and extending the general function of the CBD from mainly retailing towards leisure by increasing the number of eating, drinking and entertainment establishments and by improving the environment (Figure 15.22). Nottingham, for example, refurbished St Peter's Square by making the area more attractive: planting shrubs and other plants, giving permission for open-air cafés, and employing staff whose early morning job is to ensure that the place is clean and undamaged.

Out-of-town shopping centres

Between 1980 and 1990 over 80 per cent of new retail floorspace was built on out-of-town green-field sites. Shopping outlets locating here took advantage of economies of scale, lower rents and a pleasant, planned shopping environment. Hypermarkets and superstores were built on cheap land at, or beyond, the city margins, allowing them to buy space for their immediate use, for possible future expansion and for the necessary large car parking areas. Their aim was to capture a high threshold population, especially one that is wealthy and mobile. The ideal location is often adjacent to a main road leading directly to a motorway interchange, as this facilitates access for both customers and delivery drivers. The early outlets were mainly food stores, and included the present 'big five' of Sainsbury's, Tesco, ASDA, Argyll and Gateway. They all offer a full range of goods, convenience and comparison, under one roof so that shoppers can complete their purchases in the warmth, at times suitable to the working family, and in a cleaner, less congested environment.

Later developments have included retail parks and regional shopping centres. Retail parks, which have also been attracted to redeveloped inner-city areas, concentrate mainly upon the sale of non-food, non-convenience items (e.g. B & Q, MFI, Do-it-All, Comet, Toys Я Us). Regional shopping centres not only sell 'everything' under one roof, but often include restaurants, children's play areas and cinemas. The five largest – Gateshead MetroCentre (Places 55), Sheffield Meadowhall, Dudley Merry Hill, Thurrock Lakeside and Dartford Bluewater (page 458) – each cover over 100 000 m^2 (1.2 million square feet). These centres are highly controversial as not only do they take a large amount of business away from the local city centre, they also attract, literally, coach-loads of shoppers from places over 150 km away. Research suggests that, in 1998, almost 40 per cent of retail sales were from out-of-town locations, compared with 5 per cent in 1980. In 1994 the government announced, to the annoyance of property developers, its intention to discourage further out-of-town developments where they were 'likely to make an unacceptable impact on the vitality of traditional town centres'. At present it is difficult to obtain planning permission for such large-scale schemes. In 1999, Carlisle City Council rejected an application by Tesco to expand their out-of-town location, in order to protect shops in the city centre, yet at the same time were both actively discouraging cars from entering the CBD and failing to provide public transport for shoppers living outside the city boundary.

Figure 15.23

Aerial view of the MetroCentre site, Gateshead: to the left is the A1 (Western by-pass) and to the right the Newcastle–Carlisle railway, with station, and the River Tyne

Places 55   Gateshead: the MetroCentre

Sir John Hall, whose brainchild the MetroCentre is, was a pioneer among modern entrepreneurs in recognising that shopping could become part of a family activity linked to the wider leisure experience as people's mobility, disposable income and leisure time (for those with jobs) increased. The MetroCentre was the prototype for a new concept in retailing in Britain – the regional shopping centre. Its success was assured after Marks & Spencer opted to make the MetroCentre its first out-of-town location.

Location

The site, before development, was marshland. This meant, together with its edge-of-Gateshead location, that a large area of land was both available and relatively cheap to buy.

433

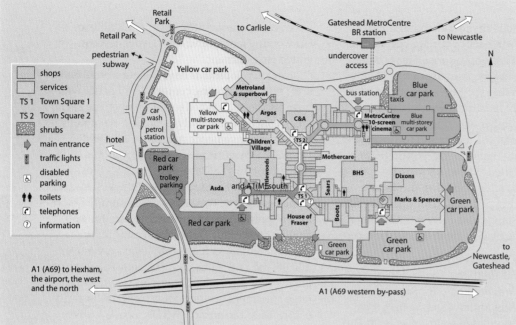

Figure 15.24

Planned layout of the MetroCentre

Shopping environment

There are now over 300 retail units set in a pleasant shopping environment (Figure 15.25) with wide tree-lined malls, air conditioning, 1 km^2 of glazed roof to let in natural light (supplemented by modern lighting in 'old world' lamps), numerous seats for relaxing, window boxes, escalators, and lifts for the disabled. A street market atmosphere has been created by traders selling from barrows, and there are over 40 eating places offering all kinds of cuisine from Mexican to Chinese.

Access

The site is adjacent to the western by-pass which now forms part of the main north–south trunk road, the A1, which avoids central Newcastle and Gateshead. It has 10 000 free car parking spaces with special facilities for the disabled motorist, 100 buses per hour, and 69 trains daily. The centre has had both a bus and a railway station purpose-built (Figures 15.23 and 15.24).

Local and central government planning

The MetroCentre took advantage of tax concessions by locating in a government Enterprise Zone (page 439). The local council were highly co-operative, seeing the scheme as a major source of revenue and of employment – there are over 6000 employees. There are plans (1999) to enlarge the site (Figure 15.26).

Additional amenities

Leisure is a vital part of the scheme. There is a 'children's village' and crèche, a 10-screen cinema, a 'space city' for computer and space enthusiasts, a covered 'fantasy-land' with all the attractions of the fair without the worries of the British climate, and theme areas such as 'Antiques Village' and 'The Roman Forum'. The complex also has a 150-bedroom hotel and leisure centre, an office block, an exhibition centre and a petrol station (Figure 15.24).

Visitors

MetroCentre's sphere of influence extends from Carlisle to Scandinavia, from York to Scotland. Since its opening in 1987, the centre has shown a steady increase in both volume of traffic (from 4.8 million cars in 1987 to 13.1 million in 1998), and the number of visitors (from 12.2 million in 1987 to 29.7 million in 1998).

Figure 15.25

The internal environment of the Metro-Centre

Figure 15.26

Reactions to the proposed expansion of the MetroCentre, 1999

North councils object to MetroCentre expansic

By Brian Nicholls

A DATE is now set for an inquiry into the MetroCentre's multi-million-pound expansion plans, which include 27 new shop units and a new car park, and would hoist the profitability of Tyne and Wear county even higher. But there are objections, so a public inquiry will be held next February 2.

Donald Gordon, chairman of the MetroCentre's main stakeholder, Capital Shopping Centres, said: 'CSC has every confidence that this inquiry will endorse the natural evolution

of this successful shopping centre – a major local employer with an economic multiplier effect [page 569] throughout the region.'

When the MetroCentre opened in the mid-1980s some 12 million shoppers used it in the first year. Now 29.7 million a year use it, and that figure does not include many other visitors to the neighbouring retail park, Ikea and the new ASDA store. The new £50 m development proposed for the MetroCentre would occupy the 'red' quadrant where ASDA previously stood [Figure 15.24]. However, objectors to

this plan include the cou of Newcastle, Sunderl Durham, North Tynes Chester-le-Street, Derwen and Tynedale.

Both Sunderland Durham have their own retail developments in har a bid to keep shoppers' m in their own cities, an Gateshead itself two comp planning applications are going through the proces factory shops on land we the new ASDA store.

MetroCentre was rece ranked top in a national su of shoppers, retailers investors.

Newcastle Journal, January

Financial institutions and offices

Financial institutions employ large numbers of people, especially in world centres such as New York, Tokyo, Hong Kong and London. These institutions, which include banking, insurance and accountancy, operate within offices. Traditionally offices have vied with shops for city centre locations regardless of the country's level of development (compare Tokyo, Figure 15.13; London, Figure 15.19 A2; Hong Kong, Figure 15.27; and Nairobi, Figure 15.36). However, whereas shops offer assistance to local individuals, offices form part of an agglomeration of businesses usually served by, and in close association with, a myriad of consultants, media, hospitality and recreational establishments.

Company head offices and major institutions such as the stock exchange locate in the capital city. As offices use land intensively, they compete with shops for prime sites within city centres (Figure 15.17). Increasingly, due to high land values, they have had to locate in ever-taller office blocks. Elsewhere in city centres, offices may locate above shops in the main street or on ground-floor sites in side-streets running off the main shopping thoroughfares. Banks can afford prime corner sites, while building societies and estate agents vie for high-visibility locations. A city centre office location may have been desirable for prestige reasons, for ease of access for clients and staff, for proximity to other

functional links (banks, insurance and entertainment) and sources of data and information, and for face-to-face contact.

The recent trend, in London for example, has been for offices to seek sites away from the CBD. Initially firms moved their offices to the new towns which were created around London after the Second World War. Later the decentralisation of government offices saw a movement to more distant regions (DHSS to Newcastle, car licences to Swansea, Giro Bank to Bootle). More recent movements from 'the City' have been to environmentally more attractive sites with either an out-of-town or a small rural town location (Cambourne, Case Study 14A), or to the Isle of Dogs Enterprise Zone (Places 56, page 440). These firms may have chosen, or been forced, to move due to the high rents, congestion and pollution of the city centre. They seek sites – often for the same reasons as new retailing outlets – where land values are lower; there is more space for car parking and possible future expansion; the working environment is more pleasant; and there is a greater proximity to motorways and, for overseas clients, airports.

New technology has allowed the easier transfer of data and has reduced the need for face-to-face contact, while increased computerisation has often led to a reduced workforce (banking) but one that is more highly skilled. Many new office locations are on purpose-built business, office or science parks (Figure 15.19 D4).

Figure 15.27

Office development on Hong Kong Island

Industrial zones

Industry within urban areas has changed its location over time. In the early 19th century, it was usually sited within city centres, e.g. textile firms, slaughter houses and food processing. However, as the Industrial Revolution saw the growth in size and number of factories, and later when shops began to compete for space in the city centre, industry moved centrifugally outwards into what today is the inner city (Places 52, page 421). Inner-city areas could provide the large quantity of unskilled labour needed for textile mills, steelworks and heavy engineering. The land was cheaper and had not yet been built upon. Factories were also located next to main lines of communication: originally, rivers and canals, then railways and finally roads (Figure 15.19 B4). Firms including bakeries, dairies, printing (newspapers) and furniture, which have strong links with the city centre, are still found here.

Between the 1950s and the 1980s this zone increasingly suffered from industrial decline as older, traditional industries closed down and others moved to edge-of-city sites. In Britain, recent changes in government policy have led to attempts to regenerate industry in these areas through initiatives such as Enterprise Zones, derelict land grants and Urban Development Corporations (page 439). Even so, the replacement industries are often on a small scale and compete for space with warehouses and DIY shops.

Most modern industry is 'light' and clean in comparison to that of the last century and has moved to greenfield sites near to the city boundary (Figure 15.19 D4). Industrial estates and modern business and science parks are located on large areas of relatively cheap land where firms have built new premises and use up-to-date equipment (Places 86, page 566). The internal road systems are purpose-built for cars and lorries and are linked to nearby motorway interchanges and other main roads. The wide range of skills and increasing demand for female labour are satisfied by local housing estates. Most of the industries are 'footloose' and include high-tech, electronics, assembling, food processing and distributive firms.

Residential zones

The Industrial Revolution also led to the rapid growth in urban population and the outward expansion of towns. Long, straight rows of terraced houses (Figure 15.19 B2) were constructed as close as possible to the nearby factories where most of the occupants worked. The closeness was essential as neither private nor public transport was yet available. Houses and factories competed for space. As a result, houses were small, sometimes with only one room upstairs and one downstairs or they were built 'back to back'. The absence of gardens and public open space added to the high housing density.

By the 1950s, many of these inner-city areas, the low-class/low-income houses of the urban models, had become slums. Wholesale clearances saw large areas flattened by bulldozers and redeveloped with high-rise blocks of flats (Figure 15.19 B3). Within 20 years, the previously unforeseen social problems of these flats led to a change in policy. Under urban renewal, older housing was improved, rather than replaced, by adding bathrooms, kitchens, hot water and indoor toilets.

Some inner-city areas have undergone a process known as **gentrification**. This is where old, substandard housing is bought, modernised and occupied by more wealthy families. In some Inner London districts, like Chelsea, Fulham and Islington, such properties are much sought-after and have become very expensive. The process is partly triggered by the proximity of employment and services in the city centre and partly through the availability of improvement grants. Once begun, it is often maintained by the perception of social prestige derived from living in such areas. More recently, inner-city areas with a waterfront location, as in London, Bristol, Manchester, Liverpool and Newcastle, have undergone a renaissance which has also seen them becoming fashionable and expensive (Places 56, page 440).

The outward growth of the city continued both during the inter-war period when, aided by the development of private and public transport, large estates of semi-detached houses were built (the medium-class houses of the urban models, Figure 15.19 C2 and C3), and after the 1950s. Many of the present edge-of-city estates consist of low-density private housing. Due to low land values (Figure 15.17), the houses are large, and have gardens and access to open space (Figure 15.19 D2). Other estates were created by local councils in an attempt to rehouse those people forced to move during the inner-city clearances. These estates, a mixture of high-rise and low-rise buildings (Figure 15.19 D3), have a high density and, like some older inner-city areas, are now experiencing extreme social and economic problems (page 441).

In the past, some teachers, perhaps because of their religious or political opinions, felt it necessary to pass on their own values to students, while the majority tended to avoid considering the role of values in geography in an attempt to remain neutral and non-doctrinaire. Today it is being accepted, perhaps among some with reservations, that in order to understand the character of places and the behaviour of people in relation to their environments, it is necessary to consider the motivations, values and emotions of those people involved.

The present author has tried, rightly or wrongly, to maintain a 'neutral' stance. Some would claim that what has been included in this book has been influenced by the author's own values and attitudes, e.g. a belief in the fundamental role of physical geography in an understanding of environmental problems; a preference for living in a semi-rural area rather than an inner city. Criticism could also be levelled for using personal experiences as exemplars in some **Places** and **Case Study** sections. What the author has tried to do is to present readers with information in the hope that they may become more aware of their own values in relation to the behaviour of others, and to enable them to discuss, with fewer prejudices and preconceptions, the foundations of their own values.

This may be illustrated with reference to the following section on inner cities which is structured as follows:

1 **The problem of inner cities** Will these issues be seen differently by the inhabitant of an inner-city area and a person living in a rural environment?

2 **The image of an inner-city area** Will a description of inner-city problems give a negative picture of the quality of life in those environments and in doing so perpetuate the problems, or could it help in the understanding and tackling of them?

3 **Possible solutions to the inner-city problem** Would solutions proposed by inner-city residents be similar to those suggested and implemented by the government or the local authority?

4 **What successes have government schemes had?** Your answer to this may depend upon your own political views. Before the 1997 general election, Conservatives pointed out the many achievements of the period 1980–97; Labour, the Liberal Democrats and other opposition parties claimed that little had been done. Who, if either, was correct? Presumably since that election, which led to a reversal of roles, the two main parties will be changing their attitudes!

Issues in Britain's inner cities

The widest definition of an inner city is 'an area found in older cities, surrounding the CBD, where the prevailing economic, social and environmental conditions pose severe problems'. While much activity in inner cities may be positive, such as London's Notting Hill Carnival and the lively multicultural society of Brixton, and despite the increasing number that are experiencing a renaissance (Places 56, page 440), the more likely perception of such areas by non-residents is usually negative (Framework 13, page 427). To them, inner cities have an image of poverty, dirt, crime, overcrowding, unemployment, poor housing conditions and racial tension. While a description of these problems *may* present a negative picture (Framework 14), it is necessary to identify them if people are to begin to understand the difficulties and to offer workable solutions.

■ **Economic problems** Inner cities have long suffered from a lack of investment – especially after 1945 when much money was channelled into the 'new towns'. Traditional industries declined, while those remaining shed much of their labour force due to improved technology and falling demand (shipbuilding and textiles). A subsequent decline and shift in trade affected ports such as Liverpool and London. Relatively few new industries have sought an inner-city location partly due to the environment and partly because the former labour force often lacked the relevant skills needed in the new service and high-tech industries.

■ **Social inequalities** Burgess, in his urban model, accepted that there were well-defined socio-economic and ethnic areas within a city (Figure 15.4). The segregation of different groups of people in British cities may result from differences in wealth, class, colour, religion, education and the quality of the environment. Much inner-city housing remains high density (although this should not automatically mean poor quality) and is pre-1914 in age. Relatively few occupants can afford to buy their homes and the houses may be in a poor state of repair.

The small-area census statistics (1991) reveal the following characteristics.

1 A lack of basic amenities with over 1 million dwellings still lacking either a bathroom, WC or hot water.

2 Overcrowding as a result of either large families living in small houses or, where slum clearances have taken place, large numbers living in poorly built high-rise flats. (The census figures do not include the increasing number of homeless.)

3 Higher death and infant mortality rates, a lower life expectancy and a greater incidence of illness than in other residential environments.

4 A predominance of low-income, semi-skilled and manual workers.

5 A higher incidence of single-parent families, elderly people, children in care and free school meals than other districts.

6 Concentrations of ethnic minorities which may cause tension. The Scarman Report, following the 1981 riots, concluded that racial discrimination was a major issue.

■ **Environmental factors** Inner-city areas may suffer from noise and air pollution caused by heavy traffic and the few remaining factories; visual pollution in the form of derelict factories and houses, waste land, rubbish, vandalism and graffiti; and water pollution in rivers and old canals. Although some open space has been created by the clearance of slum property, there is likely to be a general absence of trees and grass.

Indicators of welfare and deprivation
Deprivation is defined by the Department of the Environment, as when 'an individual's well-being falls below a level generally regarded as a reasonable minimum for Britain today' (Figures 15.28 and 15.29). In late 1999, of 2000 children born in Britain each day, one-third will be born into poverty and almost 11 million people, including 3 million children, live in households where the annual income is less than half the national average. Holterman (1975) used 18 chosen indicators from the 1971 census figures. He noted that while Inner London ranked highest for sub-standard housing, Clydeside scored as the worst area for nearly all of the other indicators (which included aspects of education, health, employment, housing and delinquency). Smith (1987) used five indicators to highlight economic and social inequalities between whites and blacks in Atlanta, USA. These were wealth (mean family income); house prices (median prices of owner-occupied houses); overcrowding (houses with over 1 person per room); education (median of completed school years); and health (infant mortality rates)(Case Study 15B).

Figure 15.28

Deprivation index by wards, Newcastle upon Tyne, 1991 census

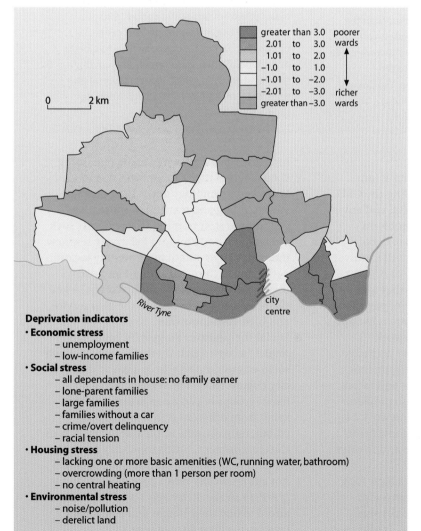

greater than 3.0	poorer wards
2.01 to 3.0	
1.01 to 2.0	
−1.0 to 1.0	
−1.01 to −2.0	
−2.01 to −3.0	richer
greater than −3.0	wards

0   2 km

River Tyne

city centre

Deprivation indicators
• **Economic stress**
  – unemployment
  – low-income families
• **Social stress**
  – all dependants in house: no family earner
  – lone-parent families
  – large families
  – families without a car
  – crime/overt delinquency
  – racial tension
• **Housing stress**
  – lacking one or more basic amenities (WC, running water, bathroom)
  – overcrowding (more than 1 person per room)
  – no central heating
• **Environmental stress**
  – noise/pollution
  – derelict land

Cycle of poverty, or deprivation

This is a concept that is largely, though not exclusively, linked to inner-city problems. It offers some explanation of how the problems have arisen. The cycle of poverty involves a continuous process which transmits relative poverty from one generation to another and which makes escape from deprivation very difficult. Certain occupational groups earn very low incomes, which makes for a low standard of living, including poor housing (since they cannot afford any better). The poor environment may produce stresses and strains in the household, and poor health amongst household members. In turn this affects the educational and other prospects of younger members in the family. The school and neighbourhood may lack the resources and skilled people needed to improve conditions for the young who are caught in this cycle of poverty. They tend to leave school early with insecure job prospects.

Poor conditions and poor prospects encourage criminal activity and lack of interest in the neighbourhood environment, discouraging outside investment and incentives to improve it. On the contrary, the neighbourhood becomes even more run-down and an adverse image of it is created, discouraging inward movement of all but the desperate households who have nowhere else to go.

The cycle of poverty is thus characteristic of the underclasses and is also increasingly concentrated in particular areas of the city and in certain housing estates on the edge of some cities.

Source: Department of the Environment

Figure 15.29

Cycle of poverty, or deprivation

Government policies for the inner cities since 1980

An increasing number of inner-city initiatives have been introduced since the 1980s. These, in total if not individually, have sought to:

- enhance the job prospects and the ability of local residents to compete for them
- bring derelict land and buildings back into use
- improve housing conditions
- encourage private sector investment
- encourage self-help and to improve the social fabric
- improve the quality of the environment.

1 **Urban Development Corporations (UDCs)** UDCs were introduced in 1980 to spearhead the then government's attempts to regenerate areas that contained large amounts of derelict, unused or underused land and buildings. The UDCs had the power to acquire, reclaim and service land and to restore buildings to effective use; to promote new industrial activity and housing developments; and to support local community facilities. They were to be financed by private-sector investment. The first two, the London Docklands (LDDC, Places 56) and the Merseyside Development Corporation (MDC), were set up in 1981. A further 11 were established between 1986 and 1993: the Black Country, Teesside, Trafford Park, Tyne and Wear, Bristol, Leeds, Sheffield, Central Manchester, Birmingham Heartlands, Plymouth and Cardiff. The English UDCs were all wound up by the end of March 1998, and Cardiff Bay in 2000. In his evaluation of the UDCs, Hudson wrote (*Geography Review*, January 1999):

   'It is clearly difficult to reach a simple judgement about the "success" or "failure" of the UDCs and their property-led policies towards inner-city regeneration. There is certainly evidence of significant changes in these areas. However, there are also questions about the desirability of some of the changes, and concerns about the processes through which regeneration was pursued. At times these seemed to ride "rough shod" over the interests of local people, raising questions of equity and democratic accountability.'
   The former UDCs are now the responsibility of local authorities.

2 **Enterprise Zones (EZs)** The first EZs were created in 1981 to try to stimulate economic activity in areas of high unemployment by lifting certain tax burdens, e.g. exemption from paying rates for 10 years after their designation and from land tax; 100 per cent grants for machinery and buildings; and relaxing or speeding up applications for planning permission (Places 55 and Figure 19.5).

3 **Land registers (1981)** These listed unused and underused land.

4 **Urban Development Grants (1982)** These tried, by bridging the gap between costs and value on completion and by linking the public and private sectors, to enable investors to make a reasonable capital return.

5 **Derelict Land Grants (1983)** These were paid to the voluntary sector to improve derelict land, e.g. converting derelict mills, utilising former railway sidings and property, and included sites used for the National Garden Festivals. This was, perhaps, the most successful of the schemes.

6 **Inner-city task force (1987)** This was a temporary scheme which developed training opportunities and created 50 000 new jobs.

7 **City Challenge (1991)** This was, in effect, a competition between 15 selected inner cities to see which could suggest the best 'high-quality schemes aimed at revamping squalid housing and industrial estates and rebuilding derelict sites', and who could raise money totally from private enterprise (no more was made available by the government).

8 **Single Generation Budgets (SGB – 1997)** This was similar to the City Challenge, but it followed a change in government, with local authorities bidding against each other for project funding.

It is difficult to generalise on the effectiveness of such a wide range of schemes introduced over a long period of time (Framework 11, page 347). There have been many positive achievements, especially with environmental improvements such as in the Lower Swansea Valley, but many social and economic problems remain, especially with high rates of unemployment and crime, and poor standard of health. It has also been argued that several schemes have not benefited local people, e.g. new housing has often been too expensive, and new jobs have been inappropriate for local people – as in London's Docklands, Places 56.

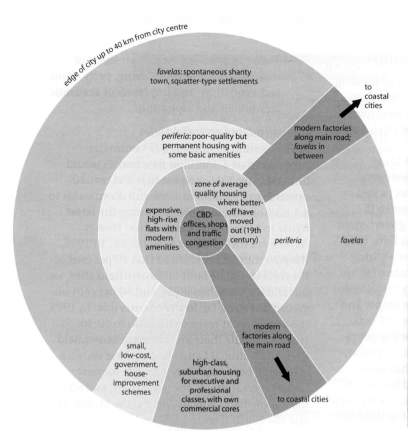

edge of city up to 40 km from city centre

favelas: spontaneous shanty town, squatter-type settlements

periferia: poor-quality but permanent housing with some basic amenities

zone of average quality housing where better-off have moved out (19th century)

expensive, high-rise flats with modern amenities

CBD: offices, shops and traffic congestion

modern factories along main road; *favelas* in between

to coastal cities

periferia

favelas

small, low-cost, government, house-improvement schemes

high-class, suburban housing for executive and professional classes, with own commercial cores

modern factories along the main road

to coastal cities

Figure 15.32

Model showing land use and residential areas in Brazilian cities (excluding Brasilia): the zoning of housing, with the more affluent living near to the CBD and the poorest further from the centre, is typical of cities in developing countries (*after* Waugh, 1983)

Figure 15.33

Living conditions in Howrah, Calcutta

Cities in developing countries

Cities in economically less developed countries, which have grown rapidly in the last few decades (page 419), have developed different structures from those of older settlements in developed countries. Despite some observed similarities between most developing cities, few attempts have been made to produce models to explain them. Clarke has proposed a model for West African cities, McGhee for South-east Asia, and the present author (based on two television programmes on São Paulo and Belo Horizonte together with some limited fieldwork) for Brazil (Figure 15.32).

Functional zones in developing cities

The CBD is similar to those of 'Western' cities except that congestion and competition for space are even greater (São Paulo, Cairo, and Nairobi, Places 58).

Inner zone In pre-industrial and/or colonial times, the wealthy landowners, merchants and administrators built large and luxurious homes around the CBD. While the condition of some of these houses may have deteriorated with time, the well-off have continued to live in this inner zone – often in high-security, modern, high-rise apartments, sometimes in well-guarded, detached houses.

Middle zone This is similar to that in a developed city in that it provides the 'in between' housing, except that here it is of much poorer quality. In many cases, it consists of self-constructed homes to which the authorities *may* have added some of the basic infrastructure amenities such as running water, sewerage and electricity (the *periferia* in Figure 15.32 and the 'site and service' schemes on page 449 and in Figure 15.41).

Outer zone Unlike that in the developed city, the location of the 'lower-class zone' is reversed as the quality of housing decreases rapidly with distance from the city centre. This is where migrants from the rural areas live, usually in shanty towns (the *favelas* of Brazil and *bustees* in Calcutta, Places 57 and Figure 15.33) which lack basic amenities. Where groups of better-off inhabitants have moved to the suburbs, possibly to avoid the congestion and pollution of central areas, they live together in well-guarded communities with their own commercial cores.

Industry This has either been planned within the inner zone or has grown spontaneously along main lines of communication leading out of the city.

Calcutta's *bustees*

Although over 100 000 people live and sleep on Calcutta's streets, one in three inhabitants of the city lives in a *bustee* (Figure 15.33). These dwellings are built from wattle, with tiled roofs and mud floors – materials that are not particularly effective in combating the heavy monsoon rains. The houses, packed closely together, are separated by narrow alleys. Inside, there is often only one room, no bigger than an average British bathroom. In this room the family, often up to eight in number, live, eat and sleep. Yet, despite this overcrowding, the interiors of the dwellings are clean and tidy. The houses are owned by landlords who readily evict those *bustee* families who cannot pay the rent.

Rio de Janeiro's *favelas*

A *favela* is a wildflower that grew on the steep *morros*, or hillsides, which surround and are found within Rio de Janeiro. Today, these same *morros* are covered in *favelas* or shanty settlements (Figure 15.34). A *favela* is officially defined as a residential area where 60 or more families live in accommodation that lacks basic amenities. The *favelados*, the inhabitants, are squatters who have no legal right to the land they live on. They live in houses constructed from any materials available – wood, corrugated iron, and even cardboard. Some houses may have two rooms, one for living in and the other for sleeping. There is no running water, sewerage or electricity, and very few local jobs, schools, health facilities or forms of public transport. The land upon which the *favelas* are built is too steep for normal houses. The most favoured sites are at the foot of the hills near to the main roads and water supply, although these may receive sewage running in open drains downhill from more recently built homes above them. Often there is only one water pump for hundreds of people and those living at the top of the hill (with fine views over the tourist beaches of Copacabana and Ipanema!) need to carry water in cans several times a day. When it rains, mudslides and flash floods occur on the unstable slopes (page 55). These can carry away the flimsy houses (over 200 people were killed in this way in February 1988).

Several attempts have been made to clean up the *favelas* or to remove them altogether. In some cases, new homes have been built for the *favelados*, but over 40 km away, in areas lacking jobs and transport. In other examples, the evicted inhabitants have simply moved back again as they had nowhere else to go. In 1998 more than 1 million people lived in *favelas* – about 10 per cent of the total population of Rio. Roçinha and Morro de Alemao, with populations in excess of 100 000 each, are the largest of Rio's 600 *favelas*.

Figure 15.34

A *favela* in Rio de Janeiro

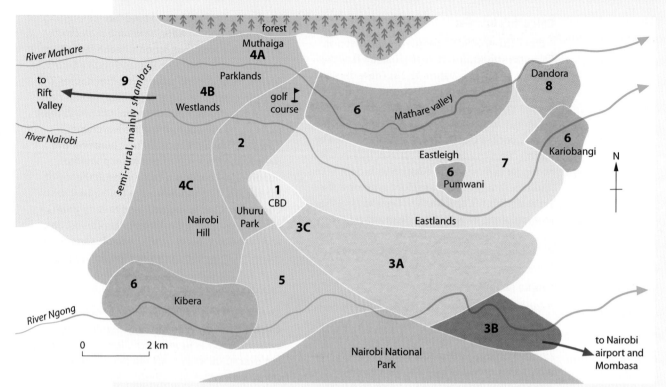

forest

River Mathare

Muthaiga
4A

to
Rift
Valley

9

Parklands

Dandora
8

Semi-rural, mainly shambas

4B
Westlands

golf
course

6

Mathare valley

6
Kariobangi

River Nairobi

2

Eastleigh

7

N

4C

1
CBD

6
Pumwani

Nairobi
Hill

Uhuru
Park

3C

Eastlands

Nairobi
National
Park

6
Kibera

5

3A

River Ngong

3B

to Nairobi
airport and
Mombasa

0   2 km

Figure 15.35

Functional zones
and residential
areas in Nairobi

In 1899 a railway, being built between Mombasa, on the coast, and Lake Victoria, reached a small river which the Maasai called *enairobi* (meaning 'cool'). The land that surrounded the river was swampy, malarial and uninhabited. Despite these seemingly unfavourable conditions, a railway station was built and, less than a century later, the settlement at Nairobi had grown to over 1.5 million people. The present-day functional zones (Figure 15.35) show the early legacy of Nairobi as a colonial settlement and the more recent characteristics associated with a rapidly growing city in an economically developing country.

1 CBD This is the centre for administration; it includes the Parliament Buildings, the prestigious Kenyatta International Conference Centre, commerce and shopping (Figure 15.36). Also located here are large hotels and, in the north, the University and the National Theatre.

2 Open space Immediately to the west and north of Nairobi's CBD (unlike in developed cities), are several large areas of open space. These include Uhuru (Freedom) Park and several other parks, sports grounds and a golf course. Other areas of open space, notably the Nairobi National Park to the south and the Karura Forest to the north, lie outside the city boundary.

Figure 15.36

The CBD (zone 1)

Figure 15.37

Higher-income
housing (zone 5)

Figure 15.38

Shanty settlement, Mathare Valley (zone 6)

Figure 15.39

Inside a shanty settlement, Kibera (zone 6)

3 Industrial zone Early industry, much of which is formal, grew up in a sector that borders the railway linking Nairobi with the port of Mombasa (**3A** in Figure 15.35). The main industries, most of which are formal (Figure 19.00), include engineering, chemicals, clothing and food processing. A modern industrial area (**3B**) extends alongside the airport road and contains many well-known transnational firms. This zone includes (**3C**) the *Jua kali* workshops (Places 89, page 575).

4 High-income residential Wealthy European colonists and, later, immigrant Asians lived on ridges of highland to the north and west of the CBD where they built large houses above the malarial swamps (Figure 15.37). Today,

Figure 15.40

Low-income, council-built housing (zone 7)

Figure 15.41

Dandora 'site and services scheme' (zone 8)

Europeans tend to concentrate in Muthaiga (**4A**) and the Asians and more wealthy Africans in Parklands and Westlands (**4B**). Westlands, with its shops and restaurants, forms a small secondary core while several large hotels are located on Nairobi Hill (**4C**). Many of the largest private properties have their own security guards.

5 Middle-income residential The southern sector was originally built for Asians who worked in the adjacent industrial zone. The estates, which were planned, are now mainly occupied by those Africans who have found full-time employment.

6 Shanty settlements As in other developing cities, shanty settlements have grown up away from the CBD on land that had previously been considered unusable – in Nairobi, this was on the narrow, swampy floodplains of the Rivers Mathare and Ngong. The two largest settlements are those that extend for several kilometres along the Mathare valley (Figure 15.38) and in Kibera (Figure 15.39). Estimates suggest that over 100 000 people, almost exclusively African, live in each area. They find work in informal industries (page 574).

7 Low-income residential These areas include flats, 3–5 storeys in height and council-built (Figure 15.40), and former shanty settlements to which the council has added a water supply, sewerage and electricity.

8 Self-help housing Under this scheme (page 449), the council provided basic amenities and, at a cheap price, building materials. In Dandora (Figure 15.41), which has over 120 000 residents, relatively wealthy people bought plots of land and built up to six houses around a central courtyard. The council then installed a tap and a toilet in each courtyard and added electricity and roads to the estate. The 'owner' is able to sell or rent the houses that are not needed by his/her own family.

In 1993, an article in Nairobi's daily newspaper *The Nation* stated that 'Kenya has been hailed as Africa's leading example of multi-racial harmony, yet one has only to tour its residential districts to see a form of "apartheid". Despite a façade of racial harmony, people live according to colour and status and, unlike in the UK or USA, do not feel they have to mix with each other. We, whether African, Asian or European, may live and work together and may even like each other a bit, but we like our space and keep ourselves to ourselves.'

Relocation housing and new towns

Some of the more wealthy developing countries, such as Venezuela with its oil revenue and Hong Kong and Singapore with their income from trade and finance, have made considerable efforts to provide new homes to replace squatter settlements. In most cases, high-rise blocks of flats have been built on sites as close as possible to the CBD or in new towns beyond the city boundary (Places 60).

Places 60   Singapore: a housing success story

Faced with a large and rapidly increasing number of slum dwellers, and an overcrowded, unplanned, central area, the Singapore government set up, in 1960, the Housing and Development Board (HDB). The HDB cleared old property near to the CBD, especially in the Chinese, Arab and Indian ethnic areas (Figure 15.48), and created purpose-built estates (with 10 000–30 000 people) within a series of new towns (each of up to 250 000 people). By 1994 there were 14 new towns, all within 12 km of the CBD.

In both cases, the HDB constructed housing units of 1–3 rooms in closely packed high-rise flats (Figure 15.49). The flats were initially for low-income families and rents were kept to a minimum. However, one-quarter of every wage-earner's salary is automatically deducted and individually credited by the government into a central pension fund (CPF). Western-style welfare benefits are regarded as an anti-work ethic, but Singaporeans can use their CPF capital to buy their own apartment or flat. Since 1974 the HDB have built many 4-room and 5-room units for the average and higher-income groups who have then been expected to buy their own property. In 1998, 86 per cent of Singaporeans lived in government-built housing and 80 per cent owned their own homes (compared with 9 per cent in 1960).

The large estates are functional in design and were developed on the neighbourhood concept of British new towns. Each estate contains much greenery and is well provided with amenities such as shops, schools, banks, medical and community centres. Where several estates are in close proximity, better services are provided such as department stores and entertainment facilities. All the new towns have been linked to, and are within half an hour of, the city centre by the MRT (mass rapid transport railway). Each estate has its own light industries producing, usually, clothing, food

products and high-tech goods. As everywhere else in Singapore, the estates are models of cleanliness with lawns trimmed and even the oldest apartments being constantly repainted. They are free of litter and graffiti: the minimum fine for each is $500 Singapore (about £130). The lifts are clean and almost always work. As in many other countries with high-rise flats, there was the initial problem of the lifts being used as toilets. The HDB put sensors in the floors which were activated by salt in the urine. This locked the doors automatically and set off an alarm. The offenders were *very* heavily fined and the problem was quickly solved!

By 1999 the HDB had built over 825 000 flats. Its aim now is to provide every householder with a minimum flat size of 3 rooms. This is being achieved by pulling down and replacing some of the earliest apartment blocks, merging adjacent flats to make them bigger, and building even more estates. To ensure that all Singaporeans have a home, the HDB has been buying 3-roomed flats on the open market and selling them at discount prices to low-income families. The newest estates (Figures 15.50 and 15.51) have been architect-designed and are set within quite large areas of open space.

In 1995, the government announced its Selective Enbloc Redevelopment Scheme (SERS). Under the scheme, estates that became 17 years old would be extensively modernised (providing that 75 per cent of the residents agreed). This included allowing owners to apply for a larger flat and/or to relocate to a newly built estate, refurbishing the interior and decorating the exterior of existing flats and constructing new public and private housing. A corresponding improvement in public utilities and services included the addition, or upgrading, of communal facilities such as community centres, modernising the centre of the estate and improving the road and transportation network. By 1999, when 31 estates had been improved, there was little evidence of high-rise decay typical of cities in many other parts of the world.

In 1999, the HDB introduced a new scheme aimed at enabling more low-income families to own a 3-room flat. Called the 'Rent and Purchase Scheme', it allowed families with a low income (and who had previously only been eligible for a 1-room or a 2-room flat) to first rent a 3-room flat from the HDB and subsequently to purchase it. The scheme, whose objective is to assist low-income families to become householders, was available to applicants over the age of 21, who were first-time buyers, were non-property owners, who had a monthly income of under $1500 (Singaporean), and whose household had a minimum of four people.

Figure 15.50

An early 1990s estate in Bishan

Figure 15.51

A late 1990s estate

Los Angeles A Physical hazards

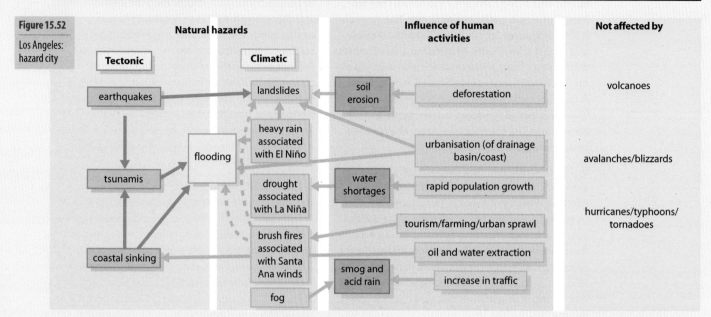

Figure 15.52

Los Angeles: hazard city

For several generations, southern California was seen as America's promised land. Now it seems that the 'sunshine state' is cursed by natural disasters such as earthquake, fire, drought and flood (by rivers and tsunamis) – disasters which, in part, are created or exacerbated by the life-style and economic activities of its inhabitants (Figure 15.52). Los Angeles, with a population in excess of 13 million, has become known as 'hazard city'.

Earthquakes

Not only does the San Andreas Fault, marking the conservative boundary between the Pacific and North American Plates, cross southern California (Places 6, page 21), but Los Angeles itself has been built over a myriad transform faults (Figure 15.53). Although the most violent earthquakes are predicted to occur at any point along the San Andreas Fault between Los Angeles and San Francisco, earth move-

ments frequently occur along most of the lesser-known faults. The most recent of 11 earthquakes to affect Los Angeles since 1970 occurred in January 1994. It registered 6.7 on the Richter scale, lasted for 30 seconds, and was followed by aftershocks lasting several days. The quake killed 60 people, injured several thousand, caused buildings and sections of freeways to collapse, ignited fires following a gas explosion, and left 500 000 homes without power and 200 000 without water (Figure 15.54).

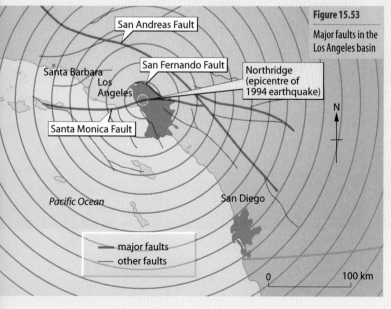

Figure 15.53

Major faults in the Los Angeles basin

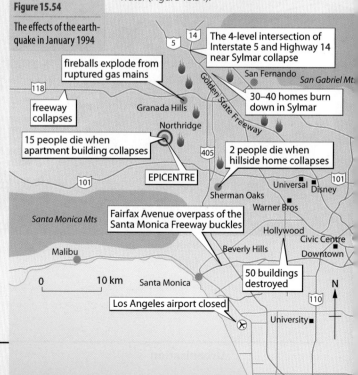

Figure 15.54

The effects of the earthquake in January 1994

Tsunamis

Tsunamis are large tidal waves triggered by submarine earthquakes which can travel across oceans at great speed. The 1964 Alaskan earthquake caused considerable damage in several Californian coastal regions. Although Los Angeles has escaped so far, it is considered to be a tsunami hazard-prone area.

Sinking coastline

The threat of coastal flooding has increased due to crustal subsidence. Although this may, in part, be due to tectonic processes, the main cause has been the extraction of oil and, to a much lesser extent, subterranean water. Parts of Long Beach have sunk by up to 10 m since 1926. Although this sinking has now been checked, parts of the harbour area lie below sea-level and are protected from flooding by a large sea wall.

Landslides and mudflows

Landslides and mudflows occur almost annually during the winter rainfall season within the city boundary of Los Angeles. They have increased in number and frequency due to effects of urbanisation such as the removal of vegetation from, and the cutting of roads through, steep hillsides and by channelling rivers (Figure 3.8). Figure 15.55 describes the effects of one mudflow (February 1994). Landslides are frequent along coastal cliffs, and the 1994 earthquake caused several thousand of them in the hills surrounding the city.

Heavy rain

Winter storms bring rain and strong winds. These are especially severe during an El Niño event (Case Study 9). Although most rivers in the Los Angeles basin are short in length and seasonal, they can transport large volumes of water during times of flood. Deforestation and brush fires on the steep surrounding hillsides, and rapid urbanisation (page 63), have increased surface runoff. Large dams have been built to try to hold back floodwater but even so the flood risk remains. In February 1992 (during an El Niño event) eight people died and dozens of cars and caravans were swept out to sea when, following two days of torrential rain, floodwaters poured through a caravan park to the south of Malibu. Heavy rain also triggers landslides and mudflows.

A fierce winter storm has brought more misery and destruction to areas of southern California already devastated by brushfires and earthquakes. Torrents of mud have trapped residents and washed away cars. More than a dozen expensive beachfront homes were swamped by landslides in the star colony of Malibu, 30 km west of Los Angeles, as rain-soaked hillsides, stripped of their vegetation by last year's fires, suddenly gave way. Some people became trapped in their homes and had to be rescued by city council workers who picked them up in the scoops of bulldozers and earthmovers. Hundreds of people were forced to evacuate their homes.

Exclusive oceanside homes of stars and entertainment industry executives in the Malibu colony faced double jeopardy – mudslides from the mountains and 2.5 m waves crashing in from the Pacific. Police reported walls of mud sweeping across the Pacific Coast Highway, through the front doors of homes and carrying furniture out of ocean-facing doors and onto the beach. Parts of the highway were buried in over 1 m of mud and water, trapping some drivers in their cars. More than a dozen parked cars were swept away. People who lost power in last month's earthquake found themselves in the dark again as the storms knocked out electricity to more than 3000 homes in Malibu.

Figure 15.55

The effect of a mudflow in 1994

El Niño and La Niña events

El Niño events seem to coincide with years of above-average rainfall, and La Niña events with periods of drought, though to a lesser extent (Figure 15.56). In February 1998, parts of southern California were declared a disaster area. El Niño was blamed for the serious floods, mudflows, landslides, storms and, in the mountains, heavy snowfalls.

Figure 15.56

Los Angeles rainfall, 1950–98

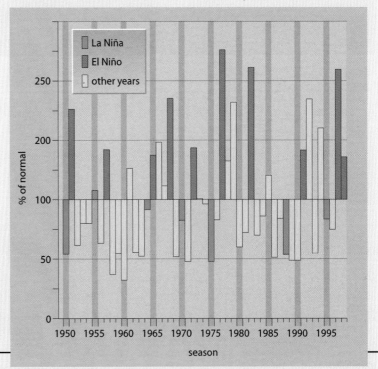

Drought

The long, dry summers associated with the Mediterranean climate may be ideal for tourists but, as the population of Los Angeles continues to grow, they put tremendous pressure on the limited water resources. Much of the city's water comes, via the Colorado aqueduct, from the River Colorado 400 km to the east. So much water is now extracted from the river that, in very dry years, it almost dries up before reaching the sea (Case Study 3C).

Brush fires

Much of the Los Angeles basin is covered in drought-resistant (xerophytic) chapparal, or brush vegetation (page 324). By the autumn, after six months without rain, this vegetation becomes tinder-dry. The Santa Ana is a hot, dry wind that owes its high temperature to adiabatic heating as it descends from the mountains. The heat and extreme low humidity of Santa Ana winds cause discomfort to humans and increase the dryness in vegetation. A careless spark or an electrical storm can prove sufficient to set off serious fires. In September 1970 a fire, 56 km in width, swept down from the Santa Monica Mountains (Figure 15.54) into Malibu. Some 72 000 hectares of brush and 295 houses were destroyed, and three people died. In November 1993 the homes of several American film and recording stars were among over 1000 destroyed in another severe brush fire (Figure 15.57).

Fog and smog

Advection fog (page 222) occurs when cool air from the cold offshore Californian current drifts inland where it meets warm air. Fog can form most afternoons between May and October as the strength of the sea-breeze increases (page 240). This event can cause a temperature inversion (page 217), where warm air becomes trapped under cold air. When many pollutants from Los Angeles' traffic, power stations and industry are released into the air, the result is smog (Figure 9.25) and, when they return to Earth, acid rain. Smog in Los Angeles can be a major health problem (Figure 15.58). In 1999, a health maintenance organisation confirmed a correlation between smog and hospital admissions. For each 10 microgram increase in airborne particulate concentrations, admissions jumped 7 per cent for chronic respiratory patients and 3.5 per cent for cardiovascular disease patients. According to another recent study reported in the *Los Angeles Times*, local residents show lung damage that might be expected of someone who smoked half a pack of cigarettes every day.

Figure 15.57

Brush fire near Malibu

Figure 15.58

Smog over Los Angeles

B Social contrasts

Living in Los Angeles presents great contrasts in lifestyle and opportunity. The census data for Watts (see Figure 15.62), an area 2 km from the CBD, indicates one side of life in the city (Figures 15.59 and 15.61). This is in contrast to the idealised picture given by films and TV of the expensive lifestyle enjoyed by the wealthy in Beverly Hills. In the past decade the picture has altered, with emphasis being placed on aspects of a city with serious problems of poor housing, crime, traffic and pollution. In 1998 Los Angeles had over 2 million people living below the poverty level (defined as $19,680 a year for a five-person household.)

Figure 15.59

Contrasts in Los Angeles

	Beverly Hills	Mission Viejo	Watts
Average household income	$55 463	$24 160	$15 639
Households below poverty level	7%	2%	17%
Households earning over $150 000 per annum	19.8%	4.5%	0.7%
Unemployment rate (% of labour force)	3.95%	2.89%	11.0%

Source: Census 1990 and estimates 1997 and 1998

Immigrant issues

Metropolitan Los Angeles has grown rapidly in the past 50 years, both in area and population, attracting many immigrants from other parts of the USA and from overseas (Figure 15.60). It covers five counties, with an estimated total population of 15.7 million (1998). Los Angeles county (9 million) has a population of 2 million Hispanics, 1.25 million of these being men of working age, including some who arrived as illegal immigrants from Mexico and Central America. Most are young, have

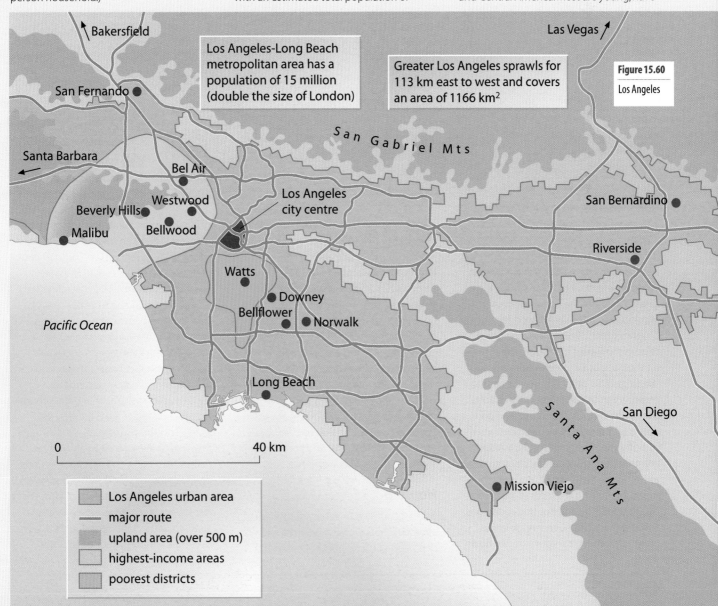

Bakersfield

Las Vegas

Los Angeles-Long Beach metropolitan area has a population of 15 million (double the size of London)

Greater Los Angeles sprawls for 113 km east to west and covers an area of 1166 km²

Figure 15.60

Los Angeles

San Fernando

San Gabriel Mts

Santa Barbara

Bel Air

Westwood

Los Angeles city centre

San Bernardino

Beverly Hills

Bellwood

Malibu

Riverside

Watts

Downey

Bellflower

Norwalk

Pacific Ocean

Long Beach

San Diego

Santa Ana Mts

Mission Viejo

0   40 km

Los Angeles urban area
major route
upland area (over 500 m)
highest-income areas
poorest districts

Figure 15.61

Contrasting data for Beverly Hills (an affluent city in western suburbs), Mission Viejo (a new city in south-Orange County), and Watts (a deprived inner city area)

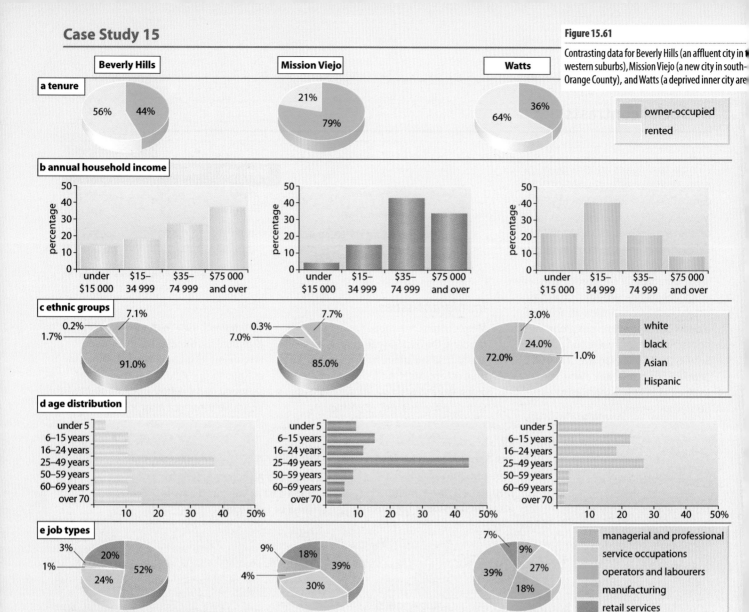

Source: US Census 1990 and estimates 1997 and 1998

little money, and few qualifications. They are attracted by California's wealthy image (stereotype – see Framework 13, page 427), but the reality they face on arrival is often very different. Until they can obtain a Green Card from the Department of Immigration they may not work legally; nor can they obtain welfare. They are forced to take very low-paid jobs, often in the informal sector. Even for legal immigrants language difficulties may make it hard to obtain a permanent job. Low educational standards, lack of qualifications and poor health and housing have been characteristic of some immigrant and Afro-American communities, e.g. Watts. The city authorities as well as the State and Federal governments are making attempts to improve health, education and housing conditions.

The growing number of Asian immigrants are helping to redevelop run-down areas of the city with help from the government-funded Community Development Association. Japanese and Korean immigrants, mainly highly educated professionals and business persons, are developing low- and medium-cost housing, creating jobs and helping to provide some social services for their own communities. In districts where they have settled, such as Norwalk, neighbourhood schools have improved, there is less street violence, and house prices have risen in response to demand. Unfortunately there have been tensions between Afro-American communities and Koreans, sometimes resulting in violence.

Housing

Increased immigration has led to a lack of affordable housing. Regeneration and demolition of older housing means a lack of family housing units. Many immigrants earn less than $4 an hour, so are unable to afford the high rents asked. Forty per cent of Asians and 55 per cent of Hispanics in Los Angeles County live in overcrowded conditions, paying up to 34 per cent of their income for their homes; it is estimated that 200 000 are living in garages. Once they establish themselves with regular jobs, the movement out of these conditions may become possible, starting the succession of movement into the suburbs. In these immigrant areas there is now a range of ethnic supermarkets with Hispanic, Chinese, Mexican and Korean names. Mini-malls on major intersections provide places for small family businesses. In Mexican areas the Taco-Trucks (mobile fast-food outlets) are to be seen outside factories and schools. Street vendors have become common, although they are officially illegal.

Figure 15.62

Watts: a poor-quality environment

Figure 15.63

Beverly Hills: a high-quality environment

Urban sprawl

The built-up urban area of Los Angeles stretches for over 115 km from east to west. This urban sprawl has taken over much of the former farmland in order to create a number of 'edge cities' or 'exurbs', e.g. Mission Viejo in Orange County. People are moving out of the concentrated built-up area of Los Angeles in search of the 'good life', attracted by the pleasant environment, good schools and clean, safe streets. In 1998 the estimated population of Mission Viejo was 80 470, an increase of 10.4 per cent since 1990. However, this brings other problems. Commuting is time-consuming, as the number of cars increases. Car-sharing schemes are now obligatory in most companies, to help reduce the impact on the environment. In spite of this, 82 per cent of workers from Mission Viejo drive alone to their places of employment. There is little public transport available from the outlying districts into the city centre.

Social workers and police are reporting the emergence of youth gangs and latch-key kids among well-to-do children when one or more parents is away for 12–14 hours a day. Insufficient community recreational provision has been made in some of these new communities, which are largely made up of professional people, with 38 per cent of households earning $60–100 000 a year, and only 11 per cent less than $30 000 a year. Most houses are detached and less than 25 years old; many have swimming pools. This should provide an ideal environment, but the risk of job losses means that in many families both parents are wage earners, commuting long distances each day.

References

Barke, M. and O'Hare, G. (1991) *The Third World*, Oliver & Boyd.

Bradford, M. G. and Kent, W. A. (1982) *Human Geography: Theory and Applications*, Hodder & Stoughton.

Hornby, W. F. and Jones, M. (1991) *Settlement Geography*, Cambridge University Press.

Philip's Geographical Digest 1998–1999 (1998) Heinemann-Philips.

Prosser, R. (1992) *Human Systems and the Environment*, Thomas Nelson.

Waugh, D. (1998) *The New Wider World*, Thomas Nelson.

Websites

World urbanisation:
http://cities.canberra.edu.au/publications/OECDpaper/World_urbanisation.htm

Social geography of Chicago:
http://uwec.edu/Academic/Geography/Ivogeler/w188/ch.html

Real Brazil, Lessons from Recife (favelas):
http://www.lanic.utexas.edu/project/ppb/rb/favelas.html

Los Angeles, Northridge earthquake page:
http://www.scecdc.scec.org/northreq.html

USGS Response to an Urban Earthquake:
http://geohazards.cr.usgs.gov/northridge/

Air pollution in Los Angeles:
http://www.doc.mmu.ac.uk/aric/langeles.html

US Census Bureau:
http://www.census.gov/hhes/www/poverty.html

See also for more links:
http://www.nelsonthornes.com/gaia

Bluewater

Figure 15.64

Location of Bluewater shopping centre

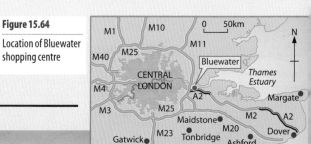

Figure 15.65

Bluewater shopping centre

Bluewater, the largest out-of-town shopping centre in England, opened in March 1999. It was built by Lend Lease, an Australian company, on a 97 ha site in a former chalk quarry, close to the M25, M20 and A2, and near to the proposed Ebbsfleet Channel Tunnel rail link (Figure 15.64). Costing £1.2 billion, it has already become a major attraction for shoppers in the south-east of England and further afield, including northern France.

The shopping centre has been constructed in a triangular shape with three 'anchor' stores – John Lewis, Marks and Spencer, and House of Fraser – at the corners joined by streets or malls on two levels with 320 shops and 3 leisure villages catering for different age groups and tastes. There are two cinemas, one with twelve screens. Ten million people live within an hour's drive (1.5 million are within 20 minutes' drive) and spend a total of £5.5 billion in the economy a year. Research indicated that they wanted something more 'upmarket' than the Lakeside centre, 15 km to the north across the River Thames in Essex. There are therefore no supermarkets or convenience stores in Bluewater.

Environmental impact

A centre of this size is certain to have major impacts on the environment and people of the area. However, attempts have been made to minimise this through the design: the white buildings with extensive glass lie below the steep quarry face, and the former quarry has been carefully landscaped to have seven small lakes with walkways and picnic areas (Figure 15.65). Over a million trees have been planted, and they will in time soften the present stark quarry environment.

The Council for the Preservation of Rural England (CPRE) calculated that 90 per cent of the projected 30 million visitors a year would travel to the centre by car. On the first day of the Christmas 1999 sales, 140 000 people visited Bluewater; most travelled by car, causing serious pressure on the approach roads. The 13 000 parking places

help to draw customers (described as 'guests' at Bluewater) from other shopping areas. Parking is carefully controlled in the busiest periods, but the CPRE believe there is a strong case for compulsory car parking charges to reduce the number of cars.

Regular bus services are provided for shoppers and workers at the centre. Frequent services run from towns and villages within a 10 mile radius and from large towns in the region, helping to improve rural transport links. Buses also run to Lakeside and to Tilbury and Romford north of the Thames. There are also regular shuttle buses from the renamed Bluewater/Greenhithe train station to the centre.

Residential areas to the north and west include new estates of modern housing close to the edge of the old quarries. Bean and Stonewood are close to the major new road junction on the A2 and B255. Local residents near the approach roads complain of increased noise from traffic and the impact of car headlights in the evening when the leisure facilities are open.

Economic impact

The south-east of England is a region of economic growth and increasing population in the suburbs of London and smaller towns in the Thames corridor. The area along the north Kent coast has been declining because of the closure of tradi-

tional industries and Chatham dockyard. Unemployment is high but efforts have been made to attract new industries, some linked to tourism such as the development of the dockyard as a historic site and the new focus on Rochester's links with Charles Dickens. There is a major depot planned for the Channel Tunnel rail link at Ebbsfleet. Bluewater has brought 5000 new jobs to the area in the tertiary sector.

The impact of the centre on other town shopping centres such as Dartford, Bexleyheath, Bromley and Gravesend is still being assessed. The owner of a local chain of fashion shops said that his Bexleyheath store had lost 50 per cent of its trade in the first month after Bluewater opened. His stores in Bromley, Gravesend, Maidstone and Chatham lost between 15 per cent and 40 per cent. A sports store in Bexleyheath commented that takings were down by £3000 per week in the first month. However, by the end of September 1999 the situation appeared to be improving in Chatham, Dartford and Gravesend. Investment in these centres by the local authorities and developers to increase car parking, street lighting and to improve their general appearance is helping to keep the more traditional town centres from declining.

Lakeside appeared almost deserted in the first week after Bluewater opened – perhaps due to the curiosity factor. It was possible to park close to the main entrances and the numbers of shoppers was small. However, customers are returning, according to the Lakeside centre management; 120 000 people visited the centre on the opening day of January 2000.

'The two centres have different catchment areas. Bluewater attracts most of its customers from south of the Thames, while only a small proportion of our (Lakeside) customers – around 10 per cent – come from south of the river.'

Richard Best, Lakeside Manager

Lakeside now benefits from increased multistorey parking and an improved link to the railway station.

The future

Following the opening of Bluewater, developers revealed further plans for the area. It is envisaged as a new township to the east of London to help ease the pressure on development in central London and along the M4 corridor to the west of London. The Bluewater architect calls it a 'linear city of the future'. This would extend Bluewater with housing, leisure facilities, offices and industry on 800 ha of former chalk quarry to the east beyond the B255 (Figure 15.66). This would link up with projects associated with the Ebbsfleet Channel Tunnel station. Already 7000 houses are planned in the quarry; the first stage of 450 homes is to be built on 10

The report on Bluewater, Dartford (*The Guardian*, 11 March 1999) repeated the propaganda of local planners and developers about the adjoining areas they plan to develop. It is not true that these areas are mainly 'old quarries'. In the main they are agricultural land, naturally regenerated areas and marshland.

Take the 'adjoining quarry' where 7000 houses are planned. This 260 ha site was green-belt land until last year. It is a condition of the quarrying there that the land is restored to agricultural use, and much of it already has been. Many parts of the site are agricultural land that was never quarried.

Further to the east, the Ebbsfleet valley (175 ha) is to be extensively developed, but this is mainly agricultural land. The other area under threat – Swanscombe Peninsula, 170 ha – is labelled as a 'brownfield site' when most of it is agricultural land and marshes. Skylarks abound on the rough grassland here, and there is also a heronry.

The motivation for landowners and developers to rubbish open land as 'brownfield' is the huge profits to be made from developing it.

Dartford is set to lose these large areas of recreational and environmentally important open land and to become an extension of London's urban sprawl. Many may accept this as simply development of 'urban fringe', but remember that when this land is developed, the countryside next to it then becomes urban fringe.

Local resident in a letter to
The Guardian, March 1999

Figure 15.67

ha within walking distance of the centre. The development includes one-bedroom flats as well as four-bedroom homes, and is intended for white-collar staff working in the local area. It will cost £60 million. In total 30 000 homes could be built housing 70 000 people by 2020. This would involve the development of new infrastructure, such as primary schools and health centres, and possibly small convenience shopping areas.

Industrial development will be linked to the developments at Ebbsfleet but the location alongside the A2 would suggest a possible site for modern high-tech industry and manufacturing of consumer goods.

There is local opposition to the new proposals, including local environmental groups and residents in the villages (Figure 15.67).

Decision Making

Imagine that you are a member of the local authority planning team for North Kent. You have been asked as a geographer to provide a report on the possible developments suggested for the Bluewater valley, and to make suggestions on the possible use of the area, including housing and industry. You will need to consider any changes in transport and infrastructure that will be required, together with possible environmental impacts.

You should use the maps, photograph and text here to develop your report. It may also help you to consult the wider area around Bluewater on the OS map (Landranger Sheet 177).

Complete the following tasks:

1 Using Figure 15.66, make a reasoned proposal for development of the eastern valley. **(10 marks)**
2 Identify the different groups involved in any planning enquiry, giving reasons why they may wish to be involved. Develop each case clearly in tabular form. **(10 marks)**
3 Assess the environmental, social and economic costs and benefits to the area of the proposal to build 7000 homes. Use a bi-polar matrix or force field scaling to help your assessment. **(15 marks)**
4 Provide a written evaluation of your assessment together with your advice on the way forward. **(15 marks)**

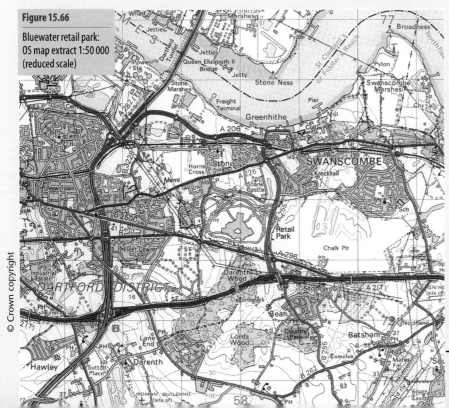

Figure 15.66

Bluewater retail park: OS map extract 1:50 000 (reduced scale)

© Crown copyright

1 Study Figure 15.68.
a Describe and compare the rates of urbanisation shown in the table for:
  i North America
  ii Latin America
  iii Africa. **(8 marks)**
b Choose **one** of the less economically developed regions shown in the diagram. Explain why that region is experiencing rapid urbanisation. **(5 marks)**
c Choose **one** of the more economically developed regions shown in the diagram. Explain why the rate of urbanisation was comparatively slow in the last 30 years of the 20th century. **(5 marks)**
d Describe and explain the problems that can be caused for the planning authorities when:
  either i cities in less economically developed countries grow very quickly
  or ii cities in more economically developed countries lose population. **(7 marks)**

2 Study Figure 15.69 below.
a i Describe and account for the differences between Zone A and Zone B, which are both described as part of the CBD. **(5 marks)**
  ii Which two types of land use occupy the most area in Zone C? Explain why this is so. **(4 marks)**
  iii How would you expect the appearance of the housing areas in Zone D to be different from those in Zone E? **(4 marks)**
  iv Explain why Zone F has more industry and warehouses, offices and comparison shops than Zones D and E. **(5 marks)**
b Name a city in the UK that you have studied. Assess how closely it matches the idealised city shown in the diagram. Make specific reference to named areas within your chosen city. **(7 marks)**

Urban population (percentage)				
Area	**1950**	**1970**	**1990**	**2020 (estimate)**
World	29.2	37.1	45.2	57.4
More developed regions	53.8	66.6	73.0	77.2
Less developed regions	17.0	25.4	37.1	53.1
Europe and CIS	56.3	66.7	73.4	76.7
North America	63.9	73.8	74.3	78.9
Oceania	61.3	70.8	71.3	75.1
Latin America	41.0	57.4	75.1	83.0
Asia (excluding CIS)	16.4	24.1	28.2	49.3
Africa	15.7	22.5	33.9	52.2

Figure 15.68
The proportion of world population living in urban areas

4 a Describe the main features of the Burgess model of urban structure, and explain why the model is useful to geographers. **(5 marks)**
b Select one of the following models of urban development:
  • the Hoyt model   • the Mann model
  • the Ullman and Harris model.
  i Describe your chosen model, and explain how it is different from the Burgess model of urban development. **(5 marks)**
  ii Discuss the limitations of the model. **(5 marks)**
c With reference to a named city, describe the structure of the city and discuss the extent to which any of the models of urban structure fit that city. **(10 marks)**

5 Study Figure 15.69.
a Describe and explain the changes in land use along the transect. **(13 marks)**
b Draw an idealised transect from the CBD to the city boundary for a typical city in a less economically developed country. Add notes below your transect to explain some of the key features of your diagram. **(12 marks)**

3 Study Figure 15.68 above. Compare and contrast the rates of urbanisation in the more economically developed regions and the less economically developed regions of the world. Suggest reasons for the differences that you have observed. **(25 marks)**

Land use
- residential
- open space
- public buildings
- industry and warehouses
- offices
- convenience shops
- comparison shops

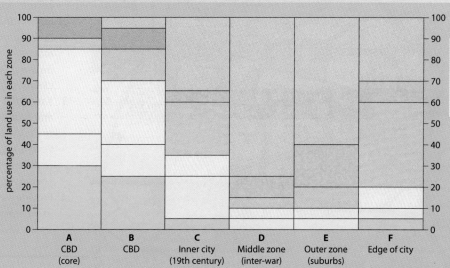

Figure 15.69
Land use in a British city: an idealised transect from the CBD to the city boundary

percentage of land use in each zone

A CBD (core)
B CBD
C Inner city (19th century)
D Middle zone (inter-war)
E Outer zone (suburbs)
F Edge of city

AS

A2

6 There are many factors which 'push' people away from rural areas and make them migrate to cities in less economically developed countries. These include poverty, shortage of land, famine and natural disasters, and the lack of opportunity.

a Explain what the main 'pull' factors are which attract people to move to the cities. **(4 marks)**

b Many of the newcomers in the cities find themselves living in 'squatter settlements' on the outskirts of the city.

  i Why do many newcomers end up living in such settlements? **(2 marks)**

  ii Why are such settlements often found on the edges of cities? **(2 marks)**

  iii Describe the main features of a squatter settlement in a named city which you have studied. **(5 marks)**

c With reference to a named example, explain why traffic congestion can be a problem in cities in less economically developed countries. **(4 marks)**

d Name a city in a less economically developed country. Explain how that city is tackling the problem of housing its growing population, and show how successful it has been. **(8 marks)**

7 a What is the meaning of:

  i urbanisation? **(2 marks)**

  ii gentrification? **(2 marks)**

  iii brownfield development? **(2 marks)**

b With reference to one or more inner city areas in the UK, explain what is meant by the 'cycle of deprivation'. **(5 marks)**

c Choose **two** of the following policies for inner city redevelopment which have been tried in the UK:
  • Urban Development Corporations (UDCs)
  • Enterprise Zones (EZs)
  • derelict land grants
  • Single Regeneration Budgets (SRBs)
  • English Partnership (EP) agreements.
For each of your chosen schemes:

  i describe how the scheme has affected one area in which it has been tried **(10 marks)**

  ii comment on how successful the scheme has been. **(4 marks)**

AS

c With reference to an example from a rapidly growing city in a less economically developed country, explain how the authorities and the people are managing the rapid growth of the city's population. **(10 marks)**

8 Several different schemes have been developed by UK governments since 1979 to improve conditions in declining inner city areas.

Choose any **two** of these schemes. Describe the aims and methods of both of the schemes. Assess the successes and failures of each scheme, with reference to one or more cities where the schemes were put into practice. **(25 marks)**

9 The tables in Figure 15.70 are taken from a questionnaire survey carried out by a team from Newcastle upon Tyne University into Home Based Enterprises (HBEs) in a squatter settlement in Delhi, India. They interviewed 50 householders with HBEs.

a Referring to tables **a** and **b**, comment on the suggestion that cities in many less economically developed countries are overcrowded as a result of the high birth rate. **(5 marks)**

b In this settlement approximately 50 per cent of all households had some kind of Home Based Enterprise. Discuss the importance of these HBEs to the economy of the settlement and the world beyond it, and to the development of the buildings in the settlement. **(10 marks)**

a Average household size and composition

People in household	5
Adults in household	2.4
Children in household	2.6
Dependency ratio (children / adult)	1

b Average dwelling characteristics

Number of rooms	2
Floor space (m²)	16
Living area (m²)	8.32
Living area / person (m²)	1.66

(NB 18 out of 50 dwellings have an upper floor)

c Structural modifications to houses for HBE (percentages)

No modification	49	Wall / roof modified	5
Shelf(s) added	13	Room added	4
Upper floor added	7	Roof added	4
Shelter added	5	Repairs to wall/roof	4
Wall added	5	Built with HBE in mind	4

Many of the buildings had substantial shelves high up in the room to enable floor space to be used below the storage. The shelves are used to store household goods while working, and working goods at other times. One householder has built over a narrow access lane, and uses that space as storage. Another has placed a large tin trunk with one edge on his roof and the other edge on his neighbour's roof – for storage. The coat-hanger polisher has poles erected below the ceiling to hang up his finished products. Others prop their beds outside in the street, during the day, to make space for their HBE.

d Income characteristics of HBEs by type

	Sturdy goods manufacture [1]	Light manu-facture [2]	Retail [3]	Services	Packaging goods	Keeping live-stock
	(n = 5)	(n = 21)	(n = 15)	(n = 6)	(n = 2)	(n = 1)
% of sample	10	42	30	12	4	2
Av. workers / HBE	2	1	2	1	2	1.5
Income / worker / mth (Rs) [4]	1288	850	750	700	800	n.a.
Hours worked / person / week	72	70	70	40	75	30

Figure 15.70

Tables from survey into HBEs in Delhi, India

[1] includes metal workers: motor winding, making metal straps, repairing metal drums and repairing gas appliances

[2] includes tailoring, assembling plastic toys, making rubber gaskets for TATA truck manufacturer, polishing hard wood coat-hangers for export

[3] Services: creche, ironing, doctor

[4] On average, earnings from the HBE contributed about 50 per cent of the income for the household in which they were based

A2

Chapter 16

Farming and food supply

'But of all the occupations by which gain is secured, none is better than agriculture, none more profitable, none more delightful, none more becoming to a free man.'

Cicero, *De Officiis*, 1.51

'Behold, there shall come seven years of great plenty throughout all the land of Egypt: and there shall arise after them seven years of famine; and all the plenty shall be forgotten in the land of Egypt; and the famine shall consume the land …'

The Bible, Genesis 41:29, 30

'He who slaughters his cows today shall thirst for milk tomorrow.'

Muslim proverb

The location of different types of agriculture at all scales depends upon the interaction of physical, cultural and economic factors (Figure 16.25). Where individual farmers in a market economy (capitalist system) or the state in a centrally planned economy have a knowledge, or understanding, of these three influences, then decisions may be made. How these decisions are reached involves a fourth factor: the behavioural element.

Environmental factors affecting farming

Although there has been a movement away from the view that agriculture is controlled solely by physical conditions, it must be accepted that environmental factors do exert a major influence in determining the type of farming practised in any particular area. Increasingly, the environment is seen to be an input converted into monetary terms, e.g. yields and slopes.

In 1966 McCarty and Lindberg produced their optima and limits model, an adaptation of which appears in Figure 16.1. They suggested that there was an optimum or ideal location for each specific type of farming based on climate, soils, slopes and altitude. The optimum is defined as where the total cost of production per unit output (TCP) is minimised for that crop or live-stock. As distance increases from this optimum, conditions become less than ideal, i.e. too wet or dry; too steep or high; too hot or cold; or a less suitable soil. Consequently, the profitability of producing the crop or rearing animals is reduced, and the **law of diminishing returns** operates when either the output decreases or the cost of maintaining high yields becomes prohibitive. Eventually a point is reached where physical conditions are too extreme to permit production on an economically viable scale, and later at even a subsistence level (page 477). McCarty and Lindberg applied their model to the cotton belt of the USA (Figure 16.2), but it can equally be adapted to account for the growth of spring wheat on the Canadian Prairies (Figure 16.3).

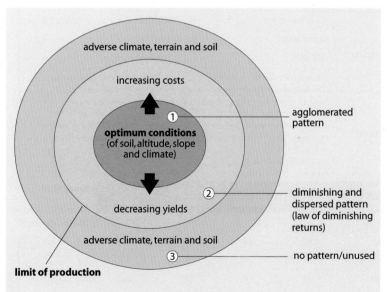

adverse climate, terrain and soil

increasing costs

① **optimum conditions** (of soil, altitude, slope and climate)

decreasing yields

adverse climate, terrain and soil

② ③

limit of production

agglomerated pattern

diminishing and dispersed pattern (law of diminishing returns)

no pattern/unused

Figure 16.1

The optima and limits model (*after* McCarty and Lindberg)

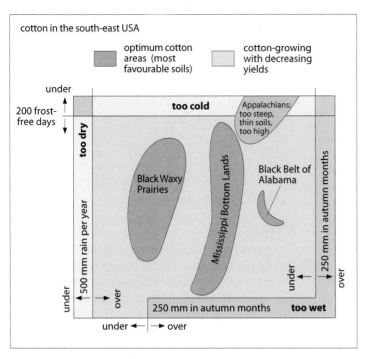

cotton in the south-east USA

optimum cotton areas (most favourable soils)

cotton-growing with decreasing yields

Figure 16.2

Optima and limits model applied to the former cotton belt in south-eastern USA

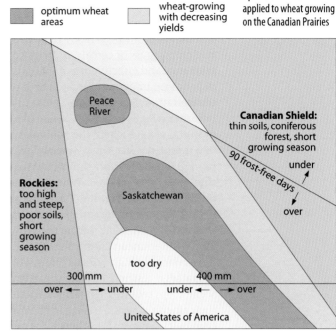

optimum wheat areas

wheat-growing with decreasing yields

Figure 16.3

Optima and limits model applied to wheat growing on the Canadian Prairies

Temperature

This is critical for plant growth because each plant or crop type requires a minimum growing temperature and a minimum growing season. In temperate latitudes, the critical temperature is 6°C. Below this figure, members of the grass family, which include most cereals, cannot grow – an exception is rye, a hardy cereal, which may be grown in more northerly latitudes.

In Britain, wheat, barley and grass begin to grow only when the average temperature rises above 6°C, which coincides with the beginning of the growing season. The growing season is defined as the number of days between the last severe frost of spring and the first of autumn. It is therefore synonymous with the number of frost-free days that are required for plant growth. Figures 16.2 and 16.3 show that cotton

needs a minimum of 200, and spring wheat 90. Barley can be grown further north in Britain than wheat, and oats further north than barley because wheat requires the longest growing season of the three and oats the shortest. Frost is more likely to occur in hollows and valleys. It has beneficial effects as it breaks up the soil and kills pests in winter, but it may also damage plants and destroy fruit blossom in spring.

Within the tropics there is a continuous growing season, provided moisture is available. As well as decreasing with distance from the Equator, both temperatures and the length of the growing season decrease with height above sea-level. This produces a succession of natural vegetation types according to altitude, although many have been modified for farming purposes (Figure 16.4).

Figure 16.4

The effect of altitude on farming and vegetation

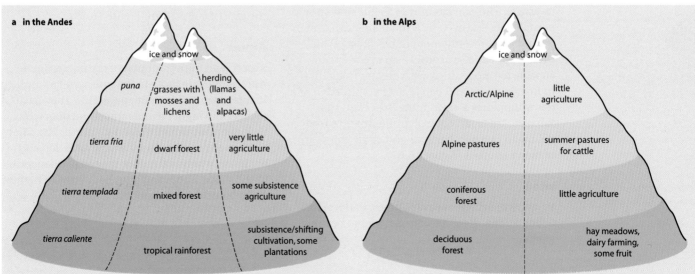

Farming and food supply

Precipitation and water supply

The mean annual rainfall for an area determines whether its farming is likely to be based upon tree crops, grass or cereals, or irrigation. The relevance and effectiveness of this annual total depends on temperatures and the rate of evapotranspiration. Few crops can grow in temperate latitudes where there is less than 250 mm a year or in the tropics where the equivalent figure is 500 mm. However, the seasonal distribution of rainfall is usually more significant for agriculture than is the annual total. Wheat is able to grow on the Canadian Prairies (Places 70, page 486) because the summer rainfall maximum means that water is available during the growing season. The Mediterranean lands of southern Europe have relatively high annual totals, yet the growth of grasses is restricted by the summer drought. Some crops require high rainfall totals during their ripening period (maize in the American corn belt), whereas for others a dry period before and during harvesting is vital (coffee).

The type of precipitation is also important (page 62). Long, steady periods of rain allow the water to infiltrate into the soil, making moisture available for plant use. Short, heavy downpours can lead to surface runoff and soil erosion and so are less effective for plants. Hail, falling during heavy convectional storms in summer in places such as the Canadian Prairies, can destroy crops. Snow, in comparison, can be beneficial as its insulates the ground from extreme cold in winter and provides moisture on melting in spring. In Britain we tend to take rain for granted, forgetting that in many parts of the world amounts and occurrence are very unreliable (Figure 9.28). India depends upon the monsoon; if this fails, there is drought and a risk of famine (page 502). Even in the best of years, the Sahel countries receive a barely adequate amount of moisture. The ecosystem is so fragile that should rainfall decrease even by a small amount (and in several years recently no rain has fallen at all), then crops fail disastrously – an event which appears to be occurring with greater frequency. In Britain, we would barely notice a shortfall of a few millimetres a year: in the Sahel and sub-Saharan Africa, an equivalent fluctuation from the mean can ruin harvests and cause the deaths of many animals (Figure 16.61).

Wind

Strong winds increase evapotranspiration rates which allows the soil to dry out and to become vulnerable to erosion. Several localised winds have harmful effects on farming: the *mistral* brings cold air to the south of France (Figure 12.22); the *khamsin* is a dry, dust-laden wind found in Egypt; Santa Ana winds can cause brush fires in California (Case Study 15A); and hurricanes, typhoons and tornadoes can all destroy crops by their sheer strength. Other winds are beneficial to agriculture: the *föhn* and *chinook* (page 241) melt snows in the Alps and on the Prairies respectively, so increasing the length of the growing season.

Altitude

The growth of various crops is controlled by the decrease in temperature with height. In Britain few grasses, including those grown for hay, can give commercial yields at heights exceeding 300 m, whereas in the Himalayas, in a lower (warmer) latitude, wheat can ripen at 3000 m. As height increases, so too does exposure to wind and the amounts of cloud, snow and rain, while the length of the growing season decreases. Soils take longer to develop as there are fewer mixing agents; humus takes longer to break down and leaching is more likely to occur. Those high-altitude areas where soils have developed are prone to erosion (Case Study 10).

Angle of slope (gradient)

Slope (see catena, page 276) affects the depth of soil, its moisture content and its pH (acidity, page 269), and therefore the type of crop that can be grown on it. It influences erosion and is a limitation on the use of machinery. Until recently, a 5° slope was the maximum for mechanised ploughing but technological improvements have increased this to 11°. Many steep slopes in South-east Asia have been terraced to overcome some of the problems of a steep gradient and to increase the area of cultivation (Figure 16.29).

Aspect

Aspect is an important part of the microclimate. **Adret** slopes are those in the northern hemisphere that face south (Places 28, page 213). They have appreciably higher temperatures and drier soils than the **ubac** slopes which face north. The adret receives the maximum incoming radiation and sunshine, whereas the ubac may be permanently in the shade. Crops and trees both grow to higher altitudes on the adret slopes.

Soils (edaphic factors)

Farming depends upon the depth, stoniness, water-retention capacity, aeration, texture, structure, pH, leaching and mineral content of the soil (Chapter 10). Three examples help to show the extent of the soil's influence on farming:

1 Clay soils tend to be heavy, acidic, poorly drained, cold, and give higher economic returns under permanent grass.
2 Sandy soils tend to be lighter, less acidic, perhaps too well-drained, warmer and more suited to vegetables and fruit.
3 Lime soils (chalk) are light in texture, alkaline, dry, and give high cereal yields.

Although soils can be improved, e.g. by adding lime to clay and clay to sands, and by applying fertiliser, there is a limit to the increase in their productivity – i.e. the law of diminishing returns operates.

Global warming

Despite uncertainty as to the exact effects of global warming, scientists agree that the greenhouse effect will not only lead to an increase in temperature but also to changes in rainfall patterns. The global increase in temperature will allow many parts of the world to grow crops which at present are too cold for them: wheat will grow in more northerly latitudes in Canada and Russia, while maize, vines, oranges and peaches may flourish in southern England (Case Study 9B). Of greater significance will be the changes in precipitation, with some places becoming wetter and more stormy (Australia and South-east Asia) while others are likely to become drier (the wheat-growing areas of the American Prairies and the Russian Steppes).

Places 61   Northern Kenya: precipitation and water supply

The Rendille tribe live on a flat, rocky plain in northern Kenya where the only obvious vegetation is a few small trees and thorn bushes. Their traditional way of life has been to herd sheep, goats and camels, moving about constantly in search of water. (See Places 65, page 497 and Figure 16.5.)

'On the government map of Kenya, the realities of the Rendille's land are summarised in a few words:

Figure 16.5

Rendille herders at a shallow hand-dug well

"Koroli Desert", it says, and just above this is the warning "Liable to Flood". There are two rainy seasons here: the long rains in April and May and the short rains in November. But the word 'season' suggests that the rains are much more predictable and steady than they are in reality. Add together rainfall from the long and the short rains and you arrive at only 150 mm on the Rendille's central plains in an average year. But the word average means nothing here, because 'normal variation' from that average can bring only 35 mm of rain one year and 450 mm the next. Variation from place to place is even more erratic than variation from year to year. Rains can be heavy when they do come, and water often rushes off the baked ground in flash floods; thus the apparent contradiction of a flood-prone desert.

It may suddenly rain in a valley for the first time in ten years; and it may not rain there for another decade. Therefore, the Rendille do not so much follow the rains as chase them, rushing to get their animals on to new grasses, which are more easily digested and converted into milk than are the drier, older shoots.'

L. Timberlake, *Only One Earth*, p. 92

Although the former Soviet Union is the largest country in the world, physical controls of climate, relief and soils have restricted farming to relatively small parts of the country. Of the land area of 22.27 million km², only 27 per cent was farmed in 1989 (10 per cent arable and 17 per cent pastoral), mainly in the deciduous forest belt, where the land had been cleared, and on the Steppes. The remaining 73 per cent (non-farmed) consisted of forest (42 per cent), tundra, desert and semi-desert (Figure 16.6).

After the Second World War, farmers were offered incentives to exceed their production targets. This task was most difficult for those farmers who were 'encouraged' by state directives rather than by financial incentives to develop the 'virgin lands' (Figure 16.6), in such states as Kazakhstan, by ploughing up the natural grassland in order to grow wheat and other cereals. Unfortunately, the unreliable rainfall, with totals often less than 500 mm a year, did not guarantee reliable crop yields. Later, to help cereal production, irrigation schemes were begun. These have since been extended into semi-desert areas where cotton is now grown. This necessitated the Soviets constructing large-scale transfer schemes by which water from rivers in the wetter parts of the country was diverted to areas suffering a deficiency.

Future water-transfer schemes are even more ambitious and may never reach fruition, as they involve diverting water from the northward-flowing Pechora, Ob and Yenisei rivers towards the south. Apart from the cost, environmentalists fear that this could result in the saline Arctic Ocean receiving less cold river water and then being warmed up sufficiently to cause the pack ice to melt and sea-levels to rise.

Figure 16.6

Physical controls on farming in the former Soviet Union

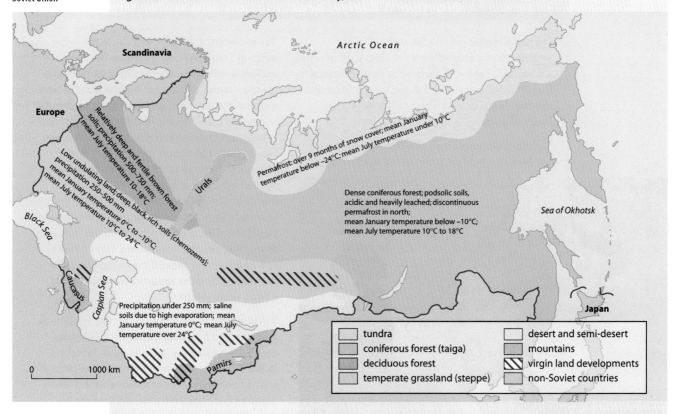

Cultural (human) factors affecting farming

Land tenure

Farmers may be owner-occupiers, tenants, landless labourers or state employees on the land which they farm. The **latifundia** system is still common to most Latin American countries.

The land here is organised into large, centrally managed estates worked by peasants who are semi-serfs. Even in the mid-1980s it was estimated that in Brazil 70 per cent of the land belonged to 3 per cent of the landowners. Land is worked by the landless labourers among the peasantry who sell their labour, when conditions permit, for substandard wages on the large estates or commercial plantations.

Other peasant farmers in Latin America have some land of their own held under insecure tenure arrangements. This land may be owned by the farmer, but it is more likely to have been rented from a local landowner or pawned to a moneylender. This latter type of tenancy takes two forms: cash-tenancy and share-cropping. **Cash-tenancy** is when farmers have to give as much as 80 per cent of their income or a fixed pre-arranged rent to the landowner. If the farmer has a short-term lease, he tends to overcrop the area and cannot afford to use fertiliser or to maintain farm buildings. If the lease is long-term, the farmer may try to invest but this often leads to serious debt. **Share-cropping** is when farmers have to pay, as a form of 'rent-in-kind' for occupying the land, part of their crop or animal produce to the landowner. As this fraction is usually a large one, the farmer works hard with little incentive and remains poor. This system operated in the cotton belt of the USA following the abolition of slavery and still persists in places. Both forms of tenancy, together with that of latifundia, resemble feudal systems found in earlier times in western Europe. The **plantation** is a variant form of the large estate system in that it is usually operated commercially, producing crops for the world market rather than for local use as in latifundia. On some plantations (oil palm in Malaysia – Places 68, page 483), the labourers are landless but are given a fixed wage; on others (sugar in Fiji), they are smallholders as well as receiving a payment.

In economically more developed, capitalist countries, many farmers are owner-occupiers, i.e. they own, or have a mortgage on, the farm where they live and work. Such a system should, in theory, provide maximum incentives for the farmer to become more efficient and to improve his land and buildings. Tenant farmers have been and still are, albeit in reduced numbers, an important part of land tenure in developed countries as well as in developing countries.

In sharp contrast to the neo-feudalist (latifundia, cash-tenancy and share-cropping) and capitalist systems of land tenure is the socialist system. In the former USSR, the individual farmer and the company-run estates were replaced by the *kolkhoz* (collective farm) and *sovkhoz* (state farm) system of organisation and management. Other forms of socialist tenure include the **commune** system which operated during the early years of communism in China (Places 63) and the *kibbutz*, which is a form of communal farming in Israel.

Inheritance laws and the fragmentation of holdings

In several countries, inheritance laws have meant that on the death of a farmer the land is divided equally between all his sons (rarely between daughters). Also, dowry customs may include the giving of land with a daughter on marriage. Such traditions have led to the sub-division of farms into numerous scattered and small fields. In Britain, fragmentation of land parcels may also result from the legacy of the open-field system (Places 51, page 400) or, more recently, from farmers buying up individual fields as they come onto the market. Fragmentation results in much time being wasted in moving from one distant field to another, and may cause problems of access. It may, however, be of benefit as it can enable a wider range of crops to be grown on land of different qualities.

Farm size

Inheritance laws, as described above, tend to reduce the size of individual farms so that, often, they can operate only at subsistence level or below. In most of the EU and North America, the trend is for farm sizes to increase as competitive market capitalism leads to the demise of small farms, and their land being purchased for enlargement by larger and more efficient, economically successful farms (page 493). Capital-intensive farms use much machinery, fertiliser, etc. and have a wide choice in types of production.

In South-east Asia and parts of Latin America and Africa, the rapid expansion of population is having the reverse effect. Farms, already inadequate in size, are being further divided and fragmented, making them too small for mechanisation (even if the farmers could afford machines). They are increasingly limited in the types of production possible, and output in certain areas, such as sub-Saharan Africa, is falling (Figure 16.57). Although farms of only 1 ha can support families in parts of South-east Asia where intensive rice production occurs and several croppings a year are possible, the average plot size in many parts of Taiwan, Nepal and South Korea has fallen to under 0.5 ha (about the size of a football pitch). In comparison, farms of several hundred hectares are needed to support a single family in those parts of the world where farming is marginal (upland sheep farming in Britain, cattle ranching in northern Australia).

Pre-1949

Before the establishment of the People's Republic in 1949, farming in China was typical of South-east Asia, i.e. it was mostly intensive subsistence (page 481). Farms were extremely small and fragmented, with the many tenants having to pay up to half of their limited produce to rich, often absentee landlords. Cultivation was manual or using oxen. Despite long hours of intensive work, the output per worker was very low. The need for food meant that most farmland was arable, with livestock restricted to those kept for working purposes or which could live on farm waste (chickens and pigs).

People's communes, 1958

After taking power in 1949, the communists confiscated land from the large landowners and divided it amongst the peasants. However, most plots proved too small to support individual farmers. After several interim experiments, the government created the 'people's communes'. The communes, which were meant to become self-sufficient units, were organised into a three-tier hierarchy with communist officials directing all aspects of life and work (Figure 16.7). Members of the commune elected a people's council, who elected a subcommittee to ensure that production targets, set by the Central Planning Committee (the government) in a series of Five Year Plans, were met. The committee was also responsible for providing an adequate food supply to make the unit self-supporting (crops, livestock, fruit and fish), for providing small-scale industry (mainly food processing and making farm implements), organising housing and services (hospital, schools) and for flood control and irrigation systems. Most communes had a research centre which trained workers to use new forms of machinery, fertiliser and strains of seed correctly (Green Revolution, page 504). By pooling their resources, farmers were able to increase yields per hectare.

Responsibility system, 1979

The introduction in 1979 of this more flexible approach, which encouraged farming families to become more 'responsible', preceded the abolition of the commune system in 1982. Under it, individual farmers were given rent-free land in their own village or district. They then had to take out contracts with the government, initially for 3 years but now extended to 15, to deliver a fixed amount of produce. To help meet their quota, individual farmers were given tools and seed. Once farmers had fulfilled their quotas, they could sell the remainder of their produce on the open market for their own profit. The immediate effect, due to farmers working much harder, was an increase in yields by an average of 6 per cent per year throughout the 1980s. Rural markets thrived and some farmers have become quite wealthy. Profits were used to buy better seed and machinery and to create village industry. Although most farmers have improved their standard of living, admittedly from an extremely low base, those living near to large cities (large nearby market) and in the south of the country (climatic advantages) have benefited the most.

1999

Hua Long (Case Study 14B) was one of several villages where the residents claimed that both their standard of living and quality of life had improved considerably over the last 20 years (Figure 16.8). Even the more rural villages were showing signs of an improvement in services and amenities (Figure 14.45), while the more efficient and prosperous farmers were able to save money and to invest it in new homes (Figure 14.47) and machinery.

50 families	= 1 **production** team	(300 people, 20 ha)	Responsible for own finances and payment of taxes for welfare services
10 production teams	= 1 **brigade**	(3000 people, 200 ha)	Responsible for overall planning, although they left the details to the production team
5 brigades	= 1 **commune**	(15 000 people, 1000 ha)	Responsible for ensuring that production targets set by the state were met

Figure 16.7

The structure of a former Chinese commune

Figure 16.8

Group 4 Team 1 in Hua Long village

Large farms are often ...	extensive on more marginal land	commercial in the EU and North America	animal grazing (sheep, cattle ranching); plantations; and temperate cereals (wheat)	further from large cities	areas of low population density and/or under-populated	increasing in size and efficiency due to amalgamation and mechanisation
Small farms are often ...	intensive on flat, fertile land	subsistence in Asia, Latin America and Africa	tropical crops (rice); and market gardening	nearer large cities	areas of high population density and/or overpopulated	decreasing in size and efficiency due to fragmention and hand labour

Figure 16.9

Reasons for spatial variations in farm size

Bearing in mind the dangers of making generalisations (Framework 11, page 347), Figure 16.9 gives some of the spatial variations, and reasons for these variations, between large and small farms. Differences in farm size also affect other types of land use and the landscape.

Economic factors affecting farming

However favourable the physical environment may be, it is of limited value until human resources are added to it. Economic man – a term used by von Thünen – applies resources to maximise profits. Yet these resources are often available only in developed countries or where farming is carried out on a commercial scale.

Transport

This includes the types of transport available, the time taken and the cost of moving raw materials to the farm and produce to the market. For perishable commodities, like milk and fresh fruit, the need for speedy transport to the market demands an efficient transport network, while for bulky goods, like potatoes, transport costs must be lower for output to be profitable. In both cases, the items should ideally be grown as near to their market as possible.

Markets

The role of markets is closely linked with transport (perishable and bulky goods). Market demand depends upon the size and affluence of the market population, its religious and cultural beliefs (fish consumed in Catholic countries, abstinence from pork by Jews), its preferred diet, changes in taste and fashion over time (vegetarianism) and health scares (BSE and GM foods).

Capital

Most economically developed countries, with their supporting banking systems, private investment and government subsidies, have large reserves of readily available finance, which over time have been used to build up **capital-intensive** types of farming (Figure 16.24) such as dairying, market gardening and mechanised

cereal growing. Capital is often obtained at relatively low interest rates but remains subject to the law of diminishing returns. In other words, the increase in input ceases to give a corresponding increase in output, whether that output is measured in fertiliser, capital investment in machinery, or hours of work expended.

Farmers in developing countries, often lacking support from financial institutions and having limited capital resources of their own, have to resort to **labour-intensive** methods of farming (Figure 16.24). A farmer wishing to borrow money may have to pay exorbitant interest rates and may easily become caught up in a spiral of debt. The purchase of a tractor or harvester can prove a liability rather than a safe investment in areas of uncertain environmental, economic and political conditions.

Technology

Technological developments such as new strains of seed, cross-breeding of animals, improved machinery and irrigation may extend the area of optimal conditions and the limits of production (Green Revolution, page 504). Lacking in capital and expertise, developing countries are rarely able to take advantage of these advances and so the gap between them and the economically developed world continues to increase.

The state

We have already seen that in centrally planned economies it is the state, not the individual, that makes the major farming decisions (Places 62 and 63). In the UK, farmers have been helped by government subsidies. Initially, organisations such as the Milk and Egg Marketing Boards ensured that British farmers got a guaranteed price for their products. Today, most decisions affecting British farmers are made by the EU. Sometimes EU policy benefits British farmers (support grants to hill farmers) and sometimes it reduces their income (reduction in milk quotas). Certainly countries in the EU have improved yields, evident by their food surpluses (pages 487 and 493), and have adapted farming types to suit demand. Governments are also responsible for agricultural training schemes and for giving advice on new methods.

Figure 16.10

Farming in China: the relationship between precipitation and farming type

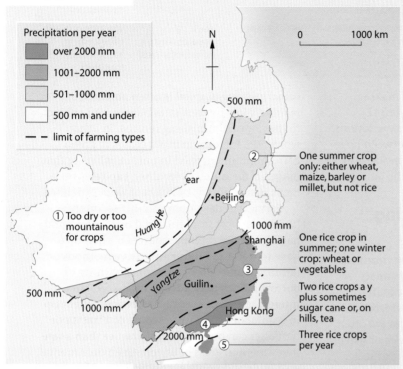

Precipitation per year
- over 2000 mm
- 1001–2000 mm
- 501–1000 mm
- 500 mm and under
- — — limit of farming types

N

0   1000 km

500 mm

ear

•Beijing

Huang He

① Too dry or too mountainous for crops

② One summer crop only: either wheat, maize, barley or millet, but not rice

1000 mm

Shanghai

③ One rice crop in summer; one winter crop: wheat or vegetables

Yangtze

Guilin •

500 mm

1000 mm

Hong Kong

④ Two rice crops a y plus sometimes sugar cane or, on hills, tea

2000 mm

⑤ Three rice crops per year

It is very difficult to generalise (Framework 11, page 347) about farming in a country that is the world's third largest in terms of area (40 times that of the UK) and largest in population. An atlas will show more accurately that, in general, the height of the land decreases, while temperatures and rainfall together with the length of the growing and rainy season (the monsoon, page 239), increase from the deserts and mountains of north and west China to the subtropics of the south-east. The type of farming – i.e. the type of crop grown and the number of croppings per year – shows a close correlation with such physical factors as the length of the growing season and the amount and distribution of annual rainfall (Figure 16.10).

Although there has been a population movement towards the towns, increasingly since 1979, especially to those near to the coast, and an increase in employment in the manufacturing and service sectors, 71 per cent of Chinese still live in rural areas and 73 per cent are farmers. Despite many improvements both in farming and in rural settlements (Case Study 14B), most farmers still have a very hard life and live at, or only a little above, subsistence level (page 477). Many work in their fields from daylight to dusk and have to rely upon hand labour (Figure 16.11). Although machinery is increasingly being used on the larger, flatter fields and the bigger farms of north-east China (Figure 16.12), animals such as the water buffalo are better suited to the smaller fields and farms found towards the south of the country where every conceivable piece of land is intensively used (Figure 16.13). Pastoral farming is practised in the higher, drier lands to the north and west (Figure 16.14).

Most farmers are still short of capital, although since the introduction of the responsibility system (Places 63) they now have the freedom to grow those crops or rear the animals they choose, together with the incentive to produce more and to diversify, as they can now sell any surplus. (The creation of wealth was not allowed during the first 30 years of the People's Republic, which coincided with a time when food shortages caused the deaths of millions of people.) As a result farmers across the country now claim that their standard of living, their quality of life and the country's food supply are better than they have been in living memory (Figure 16.8).

Figure 16.11

Intensive farming: planting rice near Kunming

Figure 16.12

Extensive farming: wheat and oilseed rape near Xi'an

Figure 16.14

Pastoral farming: northern China

Figure 16.13

Use of animals: water buffalo near Dazu

Von Thünen's model of rural land use

Heinrich von Thünen, who lived during the early 19th century, owned a large estate near to the town of Rostock (on the Baltic coast of present-day Germany). He became interested in how and why agricultural land use varied with distance from a market, and published his ideas in a book entitled *The Isolated State* (1826). To simplify his ideas, he produced a model in which he recognised that the patterns of land use around a market resulted from competition with other land uses. Like other models, von Thünen's makes several simplifying assumptions. These include:

■ The existence of an isolated state, cut off from the rest of the world (transport was poorly developed in the early 19th century).

■ In this state, one large urban market (or central place) was dominant. All farmers received the same price for a particular product at any one time.

■ The state occupied a broad, flat, featureless plain which was uniform in soil fertility and climate and over which transport was equally easy in all directions.

■ There was only one form of transport available. (In 1826 this was the horse and cart.)

■ The cost of transport was directly proportional to distance.

■ The farmers acted as 'economic men' wishing to maximise their profits and all having equal knowledge of the needs of the market.

In his model, von Thünen tried to show that with increasing distance from the market:

a the intensity of production decreased, and
b the type of land use varied.

Both concepts were based upon **locational rent** (*LR*) which von Thünen referred to as **economic rent**. Locational rent is the difference between the revenue received by a farmer for a crop grown on a particular piece of land and the total cost of producing and transporting that crop. Locational rent is therefore the profit from a unit of land, and should not be confused with **actual rent**, which is that paid by a tenant to a landlord.

Since von Thünen assumed that all farmers got the same price (revenue) for their crops and that costs of production were equal for all farmers, the only variable was the cost of transport, which increased proportionately with distance from the market. Locational rent can be expressed by the formula:

$$LR = Y(m - c - td)$$

where:

LR = locational rent
Y = yield per unit of land (hectares)
m = market price per unit of commodity
c = production cost per unit of land (ha)
t = transport cost per unit of commodity
d = distance from the market.

Since Y, m, c and t are constants, it is possible to work out by how much the LR for a commodity decreases as the distance from the market increases. Figure 16.15 shows that LR (profit) will be at its maximum at **M** (the market), where there are no transport costs. LR decreases from **M** to **X** with diminishing returns, until at **X** (the **margin of cultivation**) the farmer ceases production because revenue and costs are the same – i.e. there is no profit.

Farming and food supply

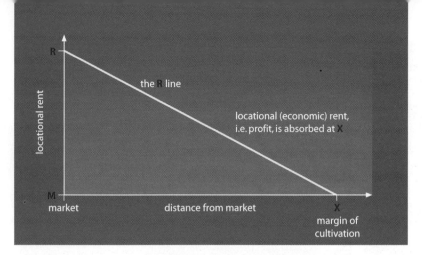

Figure 16.15

The relationship between locational (economic) rent and distance from the market

Details of von Thünen's theory

Von Thünen tried to account for the location of several crops in relation to the market. He suggested that:

a bulky crops, such as potatoes, should be grown close to the market as their extra weight would increase transport costs

b perishable goods, such as vegetables and dairy produce, should also be produced as near as possible to the market (he wrote before refrigeration had been introduced)

c intensive crops should be grown nearer to the market than extensive crops (Figure 16.16).

Consequently, bulky, perishable and intensive crops (or commodities) will have steep *R* lines (Figures 16.15, 16.16 and 16.17).

Figure 16.16 shows the result of two crops, potatoes and wheat, grown in competition. The two *R* lines, showing the locational rent or profit for each crop, intersect at **Y**. If a perpendicular is drawn from **Y** to **Z**, locational rent can be trans-

Figure 16.16

Locational rents for two crops grown in competition

lated into land use. Potatoes, an intensive, bulky crop, are grown near to the market (between **M** and **Z**) as their transport costs are high. Wheat, a more extensively farmed and less bulky crop, is grown further away (between **Z** and **X**) because it incurs lower transport costs.

What happens if three crops are grown in competition? This is the combination of von Thünen's two concepts: variation of intensity and type of land use, with distance from market. Let us suppose that wool is produced in addition to potatoes and wheat (Figure 16.17).

Potatoes give the greatest profit if grown at the market, and wool the least. However, as potatoes cost £10 to transport every kilometre, after 7 km their profit will have been absorbed in these costs (£70 profit – £70 transport = £0). This has been plotted in Figure 16.18a which is a **net profit graph**. Wheat costs £3/km to transport and so can be moved 15 km before it becomes unprofitable (£45 profit – £45 transport = £0). Wool, costing only £1/km to transport, can be taken 30 km before it, too, becomes unprofitable. Figure 16.18 also shows that although potatoes can be grown profitably for up to 7 km from the market, at point **A**, only 3.5 km from the market, wheat farming becomes equally profitable and that, beyond that point, wheat farming is more lucrative. Similarly, wheat can be grown up to 15 km from the market, but beyond 7.5 km it is less profitable than, and is therefore replaced by, wool. The point at which one type of land use is replaced by another is called the **margin of transference**.

The types of land use can now be plotted spatially. Figure 16.18b shows three concentric circles, with the market as the common central point. As on the graph, potatoes will be grown within 3.5 km of the market. This is because competition for land, and consequently land values, are greatest here so only the most intensive farming is likely to make a profit. The plan also shows that wheat is grown between 3.5 and 7.5 km from the market, while between 7.5 and 30 km, where the land is cheaper, farming is extensive and wool becomes the main product. Von Thünen's land use model is therefore based on a series of concentric circles around a central market.

The formula for locational rent (page 471) assumed that market prices (*m*), production costs (*c*) and transport costs (*t*) were all constant. What would happen to a crop's area of production if each of these in turn were to alter?

If the market price falls or the cost of production increases, there is a decrease in both the

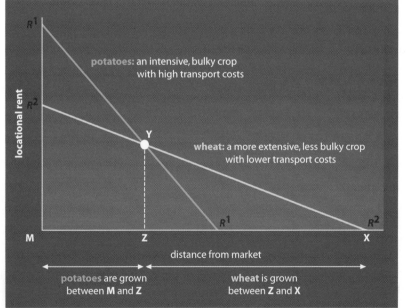

Farm product	Market price per unit of commodity	Production costs per unit of land (ha)	Transport costs per unit of commodity	Profit if grown at market
Potatoes	100	30	10	70
Wheat	65	20	3	45
Wool	45	15	1	30

Figure 16.17

Locational rents for three commodities in competition

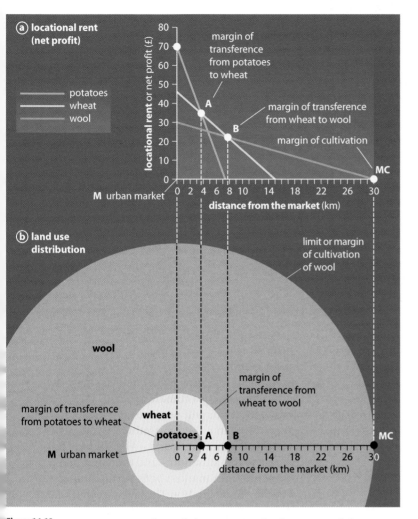

Figure 16.18

Locational rent (net profit) and land use of three commodities produced in competition

margin of cultivation. Changes in transport costs will not affect any farm at the market (Figure 16.19c) but an increase in transport costs reduces profits for distant farms, causing a decrease in the margin of cultivation. Conversely, a fall in transport costs makes those distant farms more profitable and enables them to extend their margin of cultivation.

Von Thünen's land use model

Von Thünen combined his conclusions on how the intensity of production decreased and the type of land use varied with distance from the market to create his model (Figure 16.20a). He suggested six types of land use which were located by concentric circles.

1 Market gardening (horticulture) and dairying were practised nearest to the city, due to the perishability of the produce. Cattle were kept indoors for most of the year and provided manure for the fields.
2 Wood was a bulky product much in demand as a source of fuel and as a building material within the town (there was no electricity when von Thünen was writing). It was also expensive to transport.
3 An area with a 6-year crop rotation was based on the intensive cultivation of crops (rye, potatoes, clover, rye, barley and vetch) with no fallow period.
4 Cereal farming was less intensive as the 7-year rotation system relied increasingly on animal grazing (pasture, rye, pasture, barley, pasture, oats and fallow).

profit and the margin of cultivation of that crop (Figure 16.19a and b). Conversely, if the market price rises or the costs of production decrease, profits would rise, leading to an extension in the

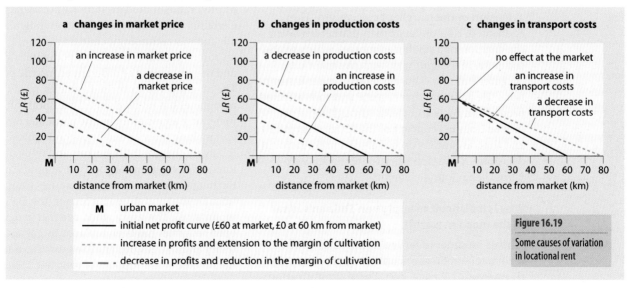

Figure 16.19

Some causes of variation in locational rent

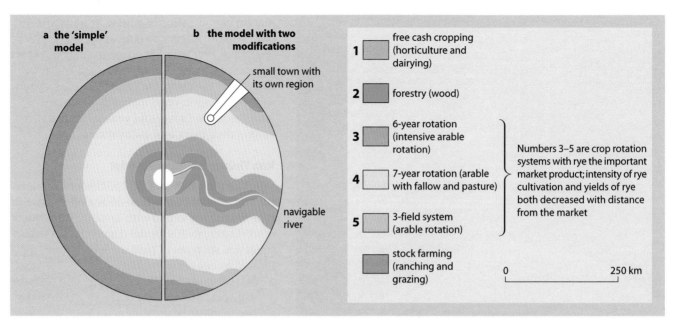

a the 'simple' model

b the model with two modifications

small town with its own region

navigable river

1 free cash cropping (horticulture and dairying)

2 forestry (wood)

3 6-year rotation (intensive arable rotation)

4 7-year rotation (arable with fallow and pasture)

5 3-field system (arable rotation)

Numbers 3–5 are crop rotation systems with rye the important market product; intensity of rye cultivation and yields of rye both decreased with distance from the market

stock farming (ranching and grazing)

0   250 km

Figure 16.20

The von Thünen land use model

5 Extensive farming based on a 3-field crop rotation (rye, pasture and fallow). Products were less bulky and perishable to transport and could bear the high transport costs.

6 Ranching with some rye for on-farm consumption. This zone extended to the margins of cultivation, beyond which was wasteland.

Modifications to the model

Later, von Thünen added two modifications in an attempt to make the model more realistic (Figure 16.20b). This immediately distorted the land use pattern and made it more complex. The inclusion of a navigable river allowed an alternative, cheaper and faster form of transport than his original horse and cart. The result was a linear, rather than a circular, pattern and an extension of the margin of cultivation. The addition of a secondary urban market involved the creation of a small trading area which would compete, in a minor way, with the main city.

Later still, von Thünen relaxed other assumptions. He accepted that climate and soils affected production costs and yields (though he never moved from his concept of the featureless plain) and that, as farmers do not always make rational decisions, it was necessary to introduce individual behavioural elements.

Why is it difficult to apply von Thünen's ideas to the modern world?

Models, in order to represent the totality of reality, rely upon the simplifying of assumptions (Framework 12, page 352). These simplifications can, in turn, be subject to criticisms which in the case of von Thünen's model can be grouped under four headings:

a Oversimplification There are very few places with flat, featureless plains, and where such landscapes do occur they are likely to contain several markets rather than one. As large areas with homogeneous climate and soils rarely exist, certain locations will be more favourable than others. Similarly, the 'isolated state' is rarely found in the modern world – Albania may be nearest to this situation – and there is much competition for markets both within and between countries. Von Thünen accepted that while his model simplified real-world situations, the addition of two variables immediately made it more complex (Figure 16.20b).

b Outdatedness As the model was produced 170 years ago, critics claim it is out-dated and of limited value in modern farming economics. Certainly since 1826 there have been significant advances in technology, changing uses of resources, pressures created by population growth, and the emergence of different economic policies. The invention of motorised vehicles, trains and aeroplanes has revolutionised transport, often increasing accessibility in one particular direction and making the movement of goods quicker and relatively cheaper. Milk tankers and refrigerated lorries allow perishable goods to be produced further from the market (London uses fresh milk from Devon) and stored for longer (the EU's food mountains). The use of wood as a fuel in developed countries has been replaced by gas and electricity and so trees need not be grown so near to the market, while supplies of timber in developing countries are being rapidly consumed and not always replaced. Improved farming techniques using fertilisers and irrigation have improved yields and extended the margins of cultivation.

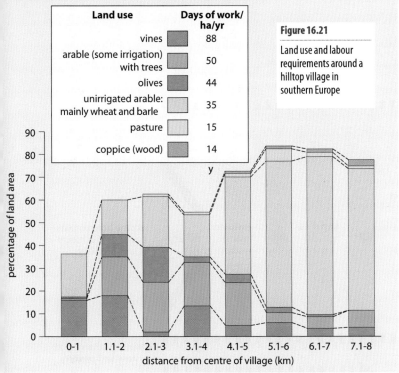

Distribution of land uses around the village together with labour requirements in days of work per hectare per year

Land use	Days of work/ha/yr
vines	88
arable (some irrigation) with trees	50
olives	44
unirrigated arable: mainly wheat and barle	35
pasture	15
coppice (wood)	14

Figure 16.21

Land use and labour requirements around a hilltop village in southern Europe

y-axis: percentage of land area

x-axis: distance from centre of village (km)
0-1, 1.1-2, 2.1-3, 3.1-4, 4.1-5, 5.1-6, 6.1-7, 7.1-8

Elsewhere, farmland has been taken over by urban growth or used by competitors who obtain higher economic rents.

c Failure to recognise the role of government
Governments can alter land use by granting/reducing subsidies and imposing/removing quotas. The EU (page 493) has recently reduced milk quotas and paid farmers to take land out of production (set-aside). Centrally planned economies, as in the former USSR and in the early years of the People's Republic of China (Places 63, page 468), directly control the types and amount of production rather than manipulating market mechanisms.

	urban areas
	market gardening (orchards and vines)
	dairying
	intensive cereals (arable)
	cereals with livestock
	extensive sheep grazing
	extensive cattle ranching (beef)

Figure 16.22

Land use patterns in Uruguay

BRAZIL

ARGENTINA

30°S

Rio Uruguay

Rio Negro

URUGUAY

Fray Bentos

Rio de la Plata

Montevideo

0   100 km

N

d Failure to include behavioural factors Von Thünen has been criticised for assuming that farmers are 'rational economic men'. Farmers do not possess full knowledge, may not always make rational or consistent decisions, may prefer to enjoy increased leisure time rather than seeking to maximise profits and may be reluctant to adopt new methods. Farmers, as human beings, may have different levels of ability, ambition, capital and experience and none can predict changes in the weather, government policies or demand for their product.

How relevant is von Thünen's theory to the modern world?
Chapter 14 concluded with the observation that although theories are difficult to observe in the real world, they *are* useful because reality can be measured and compared against them. In the case of von Thünen's model:

a Figure 16.21 takes, at a **local** level, a relatively remote, present-day hill village in the Mediterranean lands of Europe. Many villages in southern Italy, Spain and Greece have hilltop sites (in contrast to von Thünen's featureless plain) where, usually, transport links are poor, affluence is limited and the village provides the main – perhaps the only – market (Figure 14.7). As the distance from the village increases, the amount of farmland used, and the yields from it, decrease. Two critical local factors are the distance which farmers are prepared to travel to their fields and the amount of time, or intensity of attention, needed to cultivate each crop.

b Figure 16.22 shows, at the **national** level, the spatial pattern of land use in Uruguay. The capital city, Montevideo, is located on the coast, and Fray Bentos is on the navigable Rio Uruguay: a situation similar in some respects to von Thünen's modified model (Figures 16.20b and 16.35).

Conclusions
Von Thünen's land use model still has some modern relevance, particularly at the local level, provided its limitations are understood and accepted. His concept of locational rent, which is useful in studying urban as well as rural land use (page 425) is still applicable today, as conceptually the land use providing the greatest locational rent will be the one farmed. Concentric circles of land use may not exist around every urban centre, yet many areas, perhaps particularly at the local scale, do have patterns which show a similarity to the von Thünen land use model, and can be partially explained by it.

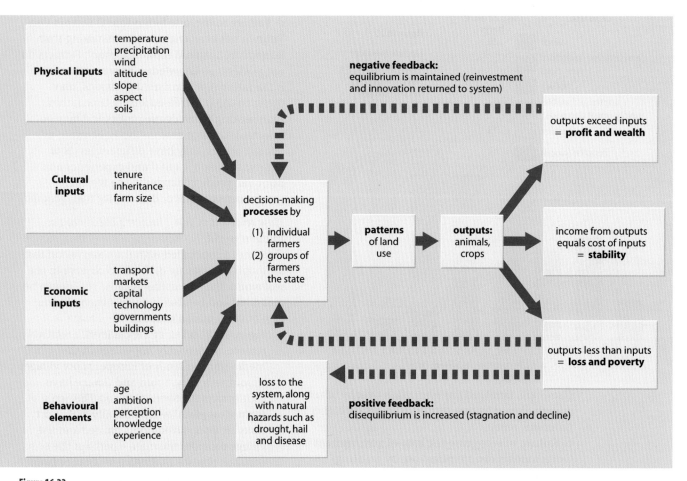

Figure 16.23

The farming system

The farming system

Farming is another example of a system, and one which you may have studied already (Framework 3, page 45). The system diagram (Figure 16.23) shows how physical, cultural, economic and behavioural factors form the inputs. In areas where farming is less developed, physical factors are usually more important but as human inputs increase, these physical controls become less significant. This system model can be applied to all types of farming, regardless of scale or location. It is the variations in inputs that are responsible for the different types and patterns of farming.

Types of agricultural economy

The simplest classifications show the contrasts between different types of farming.

1 Arable, pastoral and mixed farming
Arable farming is the growing of crops, usually on flatter land where soils are of a higher quality. It was the development of new strains of cereals which led to the first permanent settlements in

the Tigris–Euphrates, Nile and Indus valleys (Figure 14.1). Much later, in the mid-19th century, the building of the railways across the Prairies, Pampas and parts of Australia led to a rapid increase in the global area 'under the plough' (page 485). Today, there are few areas left with a potential for arable farming. This fact, coupled with the rapid increase in global population, has led to continued concern over the world's ability to feed its present and future inhabitants, a fear first voiced by Malthus (page 378). Already, there has been a decrease in the amount of arable land in some parts of the world, especially those parts of Africa affected by drought and soil erosion (Places 75, page 503).

Pastoral farming is the raising of animals, usually on land which is less favourable to arable farming (i.e. colder, wetter, steeper and higher land). However, if the grazed area has too many animals on it, its carrying capacity is exceeded or the quality of the soil and grass is not maintained, and then erosion and desertification may result (Case Study 7).

Mixed farming is the growing of crops and the rearing of animals together. It is practised on a commercial scale in developed countries, where it reduces the financial risks of relying upon a single crop or animal (monoculture), and at a subsistence level in developing countries, where it reduces the risks of food shortage.

2 Subsistence and commercial farming
Subsistence farming is the provision of food by farmers only for their own family or the local community – there is no surplus (Places 67, page 481). The main priority of subsistence farmers is self-survival which they try to achieve, whenever possible, by growing/rearing a wide range of crops/animals. The fact that subsistence farmers are rarely able to improve their output is due to a lack of capital, land and technology, and not to a lack of effort or ability. They are the most vulnerable to food shortages.

Commercial farming takes place on a large, profit-making scale. Commercial farmers, or the companies for whom they work, seek to maximise yields per hectare. This is often achieved – especially within the tropics – by growing a single crop or rearing one type of animal (Places 68, page 483). Cash-cropping operates successfully where transport is well developed, domestic markets are large and expanding, and there are opportunities for international trade (Places 69 page 484, and 70 page 486).

3 Shifting and sedentary farming
Many of the earliest farmers moved to new land every few years, due to a reduction in yields and also reduced success in hunting and gathering supplementary foods. Shifting cultivation is now limited to a few places where there are low population densities and a limited demand for food; where soils are poor and become exhausted after three or four years of cultivation (Places 66, page 480); or where there is a seasonal movement of animals in search of pasture (Places 65, page 479). However, farming over most of the world is now sedentary, i.e. farmers remain in one place to look after their crops or to rear their animals.

4 Extensive and intensive cultivation
These terms have already been used in describing von Thünen's model (Figure 16.16). Extensive farming is carried out on a large scale, whereas intensive farming is usually relatively small-scale. Farming is extensive or intensive depending upon the relationship between three factors of production: labour, capital and land (Figure 16.24). Extensive farming occurs when:

■ Amounts of labour and capital are small in relation to the area being farmed. In the Amazon Basin (Places 66, page 480), for example, the yields per hectare and the output per farmer are both low (Figure 16.24a).
■ The amount of labour is still limited but the input of capital may be high. In the Canadian Prairies (Places 70, page 486), for example, the yields per hectare are often low but the output per farmer is high (Figure 16.24b).

Intensive farming occurs when:

■ The amount of labour is high, even if the input of capital is low in relation to the area farmed. In the Ganges valley (Places 67, page 481), for example, the yields per hectare may be high although the output per farmer is often low (Figure 16.24c).
■ The amount of capital is high, but the input of labour is low. In the Netherlands (Places 71, page 487), for example, both the yields per hectare and the output per farmer are high (Figure 16.24d).

Figure 16.24

Extensive and intensive farming (*after* Briggs)

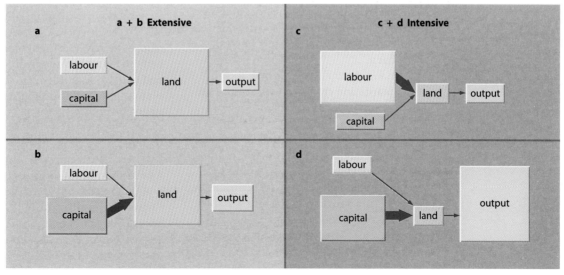

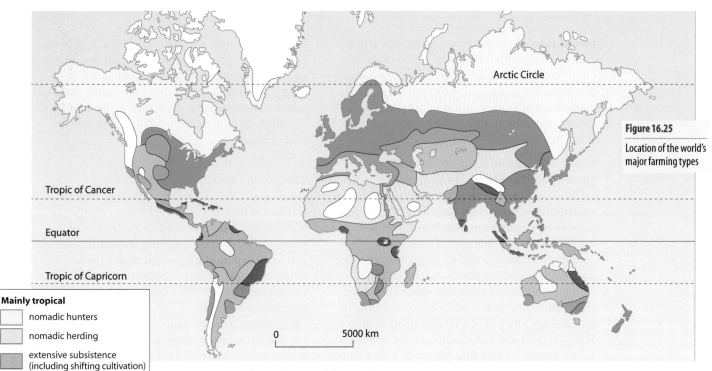

Figure 16.25

Location of the world's major farming types

Mainly tropical

1 nomadic hunters

2 nomadic herding

3 extensive subsistence (including shifting cultivation)

4 intensive subsistence agriculture

5 commercial plantation agriculture

Mainly temperate

6 livestock ranching (commercial pastoral)

7 cereal cultivation (commercial grain)

8 intensive commercial (mixed)

9 Mediterranean agriculture

10 irrigation

11 unsuitable for agriculture

World distribution of farming types

There is no widely accepted consensus as to how the major types of world farming should be classified or recognised (Framework 7, page 167). There is disagreement over the basis used in attempting a classification (intensity, land use, tropical or temperate, level of human input, the degree of commercialisation); the actual number and nomenclature of farming types; and the exact distribution and location of the major types.

You should be aware that:

1 Boundaries between farming types, as drawn on a map, are usually very arbitrary.

2 One type of farming merges gradually with a neighbouring type; there are few rigid boundaries.

3 Several types of farming may occur within each broad area, e.g. in West Africa, sedentary cultivators live alongside nomadic herdsmen.

4 A specialised crop may be grown locally, e.g. a plantation crop in an area otherwise used by subsistence farmers.

5 Types of farming alter over time with changes in economies, rainfall, soil characteristics, behavioural patterns and politics.

Figure 16.25 suggests one classification and shows the generalised location and distribution of farming types based upon the four variables described in the previous section. On a continental scale, this map demonstrates a close relationship between farming types and the physical environment/pattern of biomes

(page 306). It disguises, however, the important human–economic factors that operate at a more local level.

The following section describes the main characteristics of each of these categories of farming together with the conditions favouring their development. A specific example is used in each case (which should be supplemented by wider reading) together with an account of recent changes or problems within that agricultural economy.

1 Hunters and gatherers

Some classifications ignore this group on the grounds that it is considered to be a relict way of life, with the original lifestyle now largely, or totally, destroyed by contact with the outside world. Others feel that even if it did exist then it does not constitute a 'true' farming type, as no crops or domesticated animals are involved. It is included here as, before the advent of sedentary farming, all early societies had to rely upon hunting birds and animals, catching fish, and collecting berries, nuts and fruit in order to survive … which is surely why we rely upon farming today. There are now very few hunter-gatherer societies remaining – the Bushman of the Kalahari, the Pygmies of central Africa, several Amerindian tribes in the Brazilian rainforest, and the Australian Aborigines. All have a varied diet resulting from their intimate knowledge of the environment, but each group need an extensive area from which to obtain their basic needs.

2 Nomadic herding

In areas where the climate is too extreme to support permanent settled agriculture, farmers become **nomadic pastoralists**. They live in inhospitable environments where vegetation is sparse and the climate is arid or cold. The movement of most present-day nomads is determined by the seasonal nature of rainfall and the need to find new sources of grass for their animals, e.g. the Bedouin and Tuareg in the Sahara and the Rendille and Maasai in Kenya (Places 65 and Case Study 12A). The indigenous Sami of northern Scandinavia have to move when their pastures become snow-covered in winter, while the Fulani in West Africa may migrate to avoid the tsetse fly.

There are two forms of nomadism. **Total nomadism** is where the nomad has no permanent home, while **semi-nomads** may live seasonally in a village. There is no ownership of land and the nomads may travel extensive distances, even across national frontiers, in search of fresh pasture. There may be no clear migratory pattern, but migration routes increase in size under adverse conditions, e.g. during droughts in the Sahel. The animals are the source of life. Depending upon the area, they may provide milk, meat and blood as food for the tribe; wool and skins for family shelter and clothing; dung for fuel; mounts and pack animals for transport; and products for barter. Just as sedentary farmers will not sell their land unless they are in dire economic difficulty, similarly pastoralists will not part with their animals, retaining them to regenerate the herd when conditions improve.

Places 65   Northern Kenya: nomadic herders

Rainfall is too low and unreliable in northern Kenya to support settled agriculture (Places 61, page 465). Over the years, the Rendille have learned how to survive in an extreme environment (Figure 16.26). All they need are their animals (camels, goats and a few cattle): all their animals need is water and grass. The tribe are constantly on the lookout for rain, which usually comes in the form of heavy, localised downpours. Once the rain has been observed or reported, the tribe pack their limited possessions onto camels (a job organised by the women) and head off, perhaps on a journey of several days, to an area of new grass growth. In the past, this movement prevented overgrazing, as grazed areas were given time to recover. Camels, and to a much lesser extent goats, can survive long periods without water by storing it within their bodies or by absorbing it from edible plants – food supply is as important as water. Humans, who can go longer than animals without food but much less long without water, rely upon the camels for milk and blood, and the goats for milk and occasional meat. Indeed, the main diet of blood and milk avoids the necessity of cooking and the need to find firewood.

But the Rendille way of life is changing. Land is becoming overpopulated and resources overstretched as the numbers of people and animals increase and as water supplies and vegetation become scarcer. Consequently, as the droughts of recent years continue, pastoralists are forced to move to small towns, such as Korr. Here there is a school, health centre, better housing, jobs, a food supply and a permanent supply of water from a deep well (Figure 16.26). The deep well waters hundreds of animals, many of which are brought considerable distances each day. However, the increase in animal numbers has resulted in overgrazing, and the increase in townspeople has led to the clearance of all nearby trees for firewood. This has resulted in an increase in soil erosion, creating a desert area extending 150 km around the town (desertification, Case Study 7). Although attempts are being made to dig more wells to disperse the population, travelling shops now take provisions to the pastoralists, and the tribespeople have been shown how to sell their animals at fairer prices, many Rendille are still moving to Korr to live. There the children, having been educated, remain, looking for jobs, with the result that there are fewer pastoralists left to herd the animals.

Figure 16.26

Rendille camels and goats at a waterhole

3 Shifting cultivation (extensive subsistence agriculture)

Subsistence farming was the traditional type of agriculture in most tropical countries before the arrival of Europeans, and remains so in many of the less economically developed countries and in more isolated regions. The inputs to this system are extremely limited. Relatively few labourers are needed (although they may have to work intensively), technology is limited (possibly to axes), and capital is not involved. Over a period of years, extensive areas of land may be used as the tribes have to move on to new sites. Outputs are also very low with, often, only sufficient being grown for the immediate needs of the family, tribe or local community.

The most extensive form of subsistence farming is shifting cultivation which is still practised in the tropical rainforests (the *milpa* of Latin America and *ladang* of South-east Asia) and, occasionally, in the wooded savannas (the *chitimene* of central Africa). The areas covered are becoming smaller, due to forest clearances, and are mainly limited to less accessible places within the Amazon Basin (Places 66), Central America, Congo and parts of Indonesia. Shifting cultivation, where it still exists, is the most energy-efficient of all farming systems as well as operating in close harmony with its environment.

Places 66 Amazon Basin: shifting cultivation

With the help of stone axes and machetes, the Amerindians clear a small area of about 1 ha in the forest (Figures 16.27 and 16.28). Sometimes the largest trees are left standing to protect young crops from the sun's heat and the heavy rain; so also are those which provide food, such as the banana and kola nut. After being allowed to dry, the felled trees and undergrowth are burnt – hence the alternative name of **'slash and burn'** cultivation. While burning has the advantage of removing weeds and providing ash for use as a fertiliser, it has the disadvantage of destroying useful organic material and bacteria. The main crop, manioc, is planted along with yams (which need a richer soil), pumpkins, beans, tobacco and coca. The Amerindian diet is supplemented by hunting, mainly for tapirs and monkeys, fishing and collecting fruit.

The productivity of the rainforest depends upon the rapid and unbroken recycling of nutrients (Figure 12.7). Once the forest has been cleared, this cycle is broken (Figure 12.8). The heavy, afternoon, convectional rainstorms hit the unprotected earth causing erosion and leaching. With the source of humus removed, the loss of nutrients within the harvested crop, and in the absence of fertiliser and animal manure, the soil rapidly loses its fertility. Within four or five years, the decline in crop yields and the re-infestation of the area by weeds force the tribe to shift to another part of the forest. Although shifting cultivation appears to be a wasteful use of land, it has no long-term adverse effect upon the environment as, in most places, nutrients and organic matter can build up sufficiently to allow the land to be re-used, often within 25 years.

The traditional Amerindian way of life is being threatened by the destruction of the rainforest. As land is being cleared for highways, cattle ranches, commercial timber, hydro-electric schemes, reservoirs and mineral exploitation, the Amerindians are pushed further into the forest or forced to live on reservations. Recent government policy of encouraging the in-migration of landless farmers from other parts of the country, together with the development of extensive commercial cattle ranching, has meant that sedentary farming is rapidly replacing shifting cultivation. After just a few years, as should have been foreseen, large tracts of some cattle ranches and many individual farms have already been abandoned as their soils have become infertile and eroded.

Figure 16.27

'Slash and burn': a shifting cultivator clearing the rainforest

Figure 16.28

Crops gown in *chagras* (fields) around the *maloca* (communal house)

4 Intensive subsistence farming

This involves the maximum use of the land with neither fallow nor any wasted space. Yields, especially in South-east Asia, are high enough to support a high population density – up to 2000 per km² in parts of Java and Bangladesh. The highest-yielding crop is rice which is grown chiefly on river floodplains (the Ganges) and in river deltas (the Mekong and Irrawaddy). In both cases, the peak river flow, which follows the monsoon rains, is trapped behind bunds, or walls (Places 67). Where flat land is limited, rice is grown on terraces cut into steep hillsides, especially those where soils have formed from weathered volcanic rock as in Indonesia and the Philippines (Figure 16.29). Upland rice, or dry padi, is easier to grow but, as it gives lower yields, it can support fewer people. Rice requires a growing season of only 100 days, which means that the constant high temperatures of South-east Asia enable two, and sometimes even three, croppings a year (Figure 16.10).

The high population density, rapid population growth and large family size in many South-east Asian countries mean that, despite the high yields, there is little surplus rice for sale. The farms, due to population pressure and inheritance laws (page 467), are often as small as 1 hectare. Many farmers are tenants and have to pay a proportion of their crops to a landlord. Labour is intensive and it has been estimated that it takes 2000 hours per year to farm each 1 hectare plot. Most tasks, due to a lack of capital, have to be done by hand or with the help of water buffalo. The buffalo are often overworked and their manure is frequently used as a fuel rather than being returned to the land as fertiliser. Poor transport systems hinder the marketing of any surplus crops after a good harvest and can delay food relief during the times of food shortage which may result from the extremes of the monsoon climate: drought and flood.

Figure 16.29

Rice cultivation on terraced hillsides, Bali

Places 67   The Ganges valley: intensive subsistence agriculture

Rice, with its high nutritional value, can form up to 90 per cent of the total diet in some parts of the flat Ganges valley in northern India and western Bangladesh. Padi, or wet rice, needs a rich soil and is grown in silt which is deposited annually by the river during the time of the monsoon floods. The monsoon climate (page 239) has an all-year growing season but, although 'winters' are warm enough for an extra crop of rice to be grown, water supply is often a problem. During the rainy season from July to October, the *kharif* crops of rice, millet and maize are grown. Rice is planted as soon as the monsoon rains have flooded the padi fields and is harvested in October when the rains have stopped and the land has dried out. During the dry season from November to April, the *rabi* crops of wheat, barley and peas are grown and harvested. Where water is available for longer periods, a second rice crop may be grown.

Figure 16.30

Rice harvesting on the floodplain of the River Ganges

Rice growing is labour intensive with much manual effort needed to construct the bunds (embankments); to build irrigation channels; to prepare the fields; and to plant, weed and harvest the crop (Figure 16.30). The bunds between the fields are stabilised by tree crops. The tall coconut palm is not only a source of food, drink and sugar, but also acts as a cover crop protecting the smaller banana and other trees which have been planted on the bunds. The flooded padi fields may be stocked with fish which add protein to the human diet and fertiliser to the soil.

5 Tropical commercial (plantation) agriculture

Plantations were developed in tropical areas, usually where rainfall was sufficient for trees to be the natural vegetation, by European and North American merchants in the 18th and 19th centuries. Large areas of forest were cleared and a single bush or tree crop was planted in rows (Figure 16.31 and 16.32) – hence the term

In 1964, many Indian farmers and their families were short of food, lacked a balanced diet and had an extremely low standard of living. The government, with limited resources, made a conscious decision to try to improve farm technology and crop yields by implementing Western-type farming techniques and introducing new hybrid varieties of rice and wheat – the so-called Green Revolution (page 504). Although yields have increased and food shortages have been lessened, the 'Green Revolution' is not considered to be, in this part of the world, a social, environmental or political success (Figure 16.63).

monoculture (page 280). This so-called **cash crop** was grown for export and was not used or consumed locally (Places 68).

Plantations needed a high capital input to clear, drain and irrigate the land; to build estate roads, schools, hospitals and houses; and to bridge the several years before the crop could be harvested. Although plantations were often located in areas of low population density, they needed much manual labour. The owners and managers were invariably white. Black and Asian workers, obtained locally or brought in as slaves or indentured labour from other countries, were engaged as they were prepared, or forced, to work for minimum wages. They were also capable of working in the hot, humid climate. Today, many plantations, producing most of the world's rubber, coffee, tea, cocoa, palm oil, bananas, sugar cane and tobacco, are owned and operated by large transnational companies (Figure 16.32).

Plantations, large estates and even small farmers are being increasingly drawn into making commercial contracts to supply fruit and vegetables to consumers in the developed world. Although such contracts may help some developing countries to provide jobs and to pay off their international debts, it also means they have to import greater volumes of staple foods to make up for the land switched from staples to export crops (page 501).

Figure 16.31

A rubber plantation in Malaysia

Figure 16.32

The advantages and disadvantages of plantation agriculture

Advantages	Disadvantages
Higher standards of living for the local workforce	Exploitation of local workforce, minimal wages
Capital for machines, fertiliser and transport provided initially by colonial power, now the transnational corporations	Cash crops grown instead of food crops: local population have to import foodstuffs
Use of fertilisers and pesticides improves output	Most produce is sent overseas to the parent country
Increases local employment	Most profit returns to Europe and North America
Housing, schools, health service and transport provided, also often electricity and a water supply	Dangers of relying on monoculture: fluctuations in world prices and demand
	Overuse of land has led, in places, to soil exhaustion and erosion

A plantation is defined in Malaysia as an estate exceeding 40 ha in size. Many extend over several thousand hectares. The first plantations were of coffee, but these were replaced at the end of the 19th century by rubber. Rubber is indigenous to the Amazon Basin, but some seeds were smuggled out of Brazil in 1877, brought to Kew Gardens in London to germinate and then sent out to what is now Malaysia. The trees thrive in a hot, wet climate, growing best on the gentle lower slopes of the mountains forming the spine of the Malay peninsula. Rubber tends not to be grown on the coasts where the land is swampy, but near to the relatively few railway lines and the main ports. The 'cheap' labour needed to clear the forest, work in the nurseries, plant new trees and tap the mature trees was provided by the poorer Malays and immigrants from India, (Figure 16.31).

The Malaysian government has now taken over all the large estates, formerly run by such trans-nationals as Dunlop and Guthries, having seen them as a relict of colonialism. Publicly owned companies now only account for 20 per cent of the land devoted to rubber: the remainder is in the form of smallholdings. In the early 1970s, the Federal Land Development Authority (FELDA) was set up. Under FELDA, the forest is cleared and land divided into 5 ha plots which are allocated to farmers. For the first four years, the government does all the work, supervising planting and caring for the young trees. The farmer is then put in charge, but is still provided with free fertiliser and pesticide as the trees are too young to provide any income. Once the crop is ready, it is bought and marketed by the government.

Since the Second World War, the world demand for rubber has steadily declined, mainly due to competition from synthetic rubber. Apart from the years immediately after the AIDS scare in 1988 which saw an increased demand for family planning, the price of rubber has continued to fall. By 1999, when the price was less than half that of 1995, it was estimated that the income of one-

Figure 16.33

Oil palm

quarter of the 400 000 smallholders was below the poverty line. Half of Malaysia's smallholders are totally dependent upon rubber, with each having to support an average of four dependents.

The Malaysian plantation industry is now heavily dependent on just one crop – oil palm (Figure 16.33). Oil palm, which now covers 80 per cent of the country's plantations compared with rubber's 20 per cent, has many advantages (Figure 16.34) including higher yields, higher prices, lower production costs and a less intensive use of labour. A director of the Malayan Agricultural Producers Association claimed in 1999 that 'We are concerned about our heavy dependence upon oil palm, but we have no choice. Palm oil is not like rubber. It is more versatile, as it can be used in the food industry. As the world's population continues to grow it will need edible oils. Researchers are also coming up with new uses for palm oil, like vitamins and oleo-chemicals.'

Although oil palm fruits have still to be harvested manually (the fronds get in the way of machines) and the fruits have to be harvested within a short period of time (otherwise the oil is lost), the spraying of herbicides, the application of fertiliser and transportation have all been mechanised.

Figure 16.34

The changing importance of rubber and oil palm

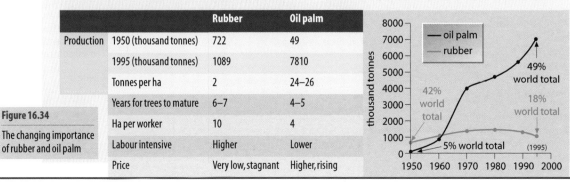

		Rubber	Oil palm
Production	1950 (thousand tonnes)	722	49
	1995 (thousand tonnes)	1089	7810
	Tonnes per ha	2	24–26
	Years for trees to mature	6–7	4–5
	Ha per worker	10	4
	Labour intensive	Higher	Lower
	Price	Very low, stagnant	Higher, rising

oil palm
rubber

42% world total

49% world total

18% world total

5% world total

(1995)

thousand tonnes

8000 7000 6000 5000 4000 3000 2000 1000 0

1950 1960 1970 1980 1990 2000

6 Extensive commercial pastoralism (livestock ranching)

Livestock ranching returns the lowest net profit per hectare of any commercial type of farming. It is practised in more remote areas where other forms of land use are limited and where there are extensive areas of cheaper land with sufficient grass to support large numbers of animals. It is found mainly in areas with a low population density and aims to give the maximum output from minimum inputs – i.e. there is a relatively small capital investment in comparison with the size of the farm or ranch, but output per farm-worker is high. This type of farming includes commercial sheep farming (in central Australia, Canterbury Plains in New Zealand, Patagonia, upland Britain) and commercial cattle ranching (Places 69), mainly for beef (in the Pampas, American Midwest, northern Australia and, more recently, Amazonia and Central America). It corresponds, therefore, to the outer land use zone of von Thünen's model (Figure 16.20) and does not include commercial dairying which, being more intensive, is found nearer to the urban market (Places 71, page 487).

The raising of beef cattle is causing considerable environmental concern. It is a cause of deforestation (uses 40 per cent of the cleared forest in Amazonia), desertification and soil erosion (overgrazing) and global warming (release of methane). It also takes more water and feed to produce one pound of beef than the equivalent amount of any other food or animal product.

Places 69
The Pampas, South America: extensive commercial pastoralism

The Pampas covers Uruguay and northern Argentina. The area receives 500–1200 mm of rainfall a year – enough to support a temperate grassland vegetation. During the warmer summer months the water supply has to be supplemented from underground sources, while in the cooler, drier winter much of the grass dies down. Temperatures are never too high to dry up the grass in summer, nor low enough to prevent its growth in winter. The relief is flat and soils are often of deep, rich alluvium, deposited by rivers such as the Parana which cross the plain (Figure 16.35). The grasses help to maintain fertility by providing humus when they die back (Figure 11.29b).

Many ranches, or *estancias*, exceed 100 km² and keep over 20 000 head of cattle. Most are owned by businessmen or large companies based in the larger cities, and are run by a manager with the help of cowboys or gauchos. Several economic improvements have been added to the natural physical advantages. Alfalfa, a leguminous, moisture-retaining crop, is grown to feed the cattle

Figure 16.35

Land use on the South American Pampas, an area with a zonation similar to that suggested by von Thünen

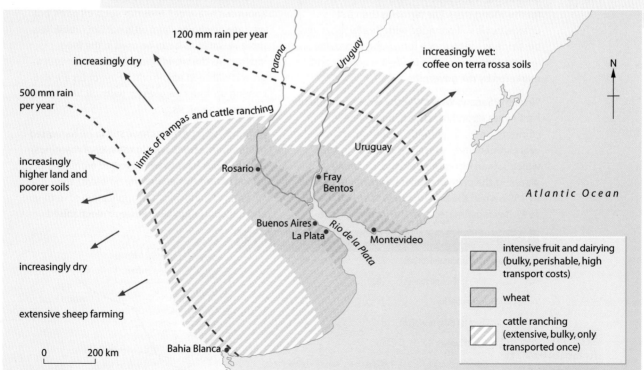

when the natural grasses die down in winter. Barbed wire, for field boundaries, was essential where rainfall was insufficient for the growth of hedges. Pedigree bulls were brought from Europe to improve the local breeds and later British Hereford cattle were crossed with Asian Brahmin bulls to give a beef cow capable of living in warm and drier conditions. Initially, due to distances from world markets, cattle were reared for their hides. It was only after the construction of a railway network, linking places on the Pampas to the stockyards (*frigorificos*) at the chief ports of Rosario, Buenos Aires, La Plata and Montevideo on the Rio de la Plata (Figure 16.35), that canned products such as corned beef became important. Later still, the introduction of refrigerated wagons and ships meant that frozen beef could be exported to the more industrialised countries.

7 Extensive commercial grain farming

As shown on the map of the Pampas (Figure 16.35) and in the von Thünen model (Figure 16.20), cereals utilise the land use zone closer to the urban market than commercial ranching. Grain is grown commercially on the American Prairies (Places 70), the Russian Steppes (Figure 16.6) and parts of Australia, Argentina and north-west Europe (Figure 16.25). In most of these areas, productivity per hectare is low but per farmworker it is high.

It was the introduction and cultivation of new strains of cereals that led to the first permanent settlements (Figure 14.1) and, later, it was a reliance upon these cereals to provide a staple diet which allowed steady population growth in Europe, Russia and South-east Asia (Figure 16.36). A demand for increased cereal production came, in the mid-19th century, from those countries experiencing rapid industrialisation and urban growth. This demand was met following the building of railways in Argentina, Australia and across North America (Figure 16.36). More recent demands have, so far, been met by the 'green revolution' in South-east Asia (page 504) and increases in irrigation and mechanisation.

Figure 16.36

Changes in the world's arable areas, 1870–1990

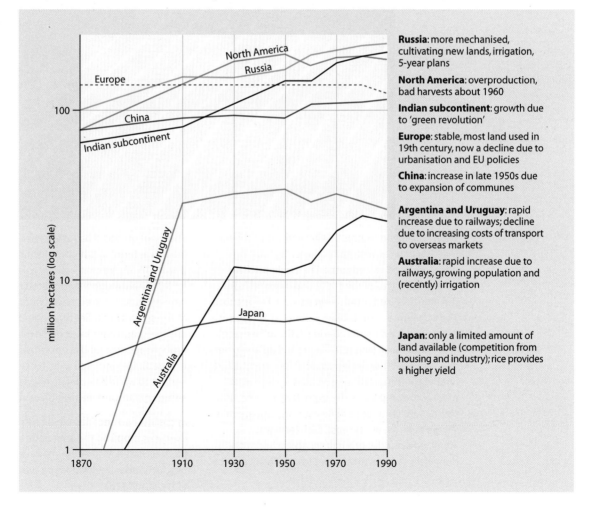

Russia: more mechanised, cultivating new lands, irrigation, 5-year plans

North America: overproduction, bad harvests about 1960

Indian subcontinent: growth due to 'green revolution'

Europe: stable, most land used in 19th century, now a decline due to urbanisation and EU policies

China: increase in late 1950s due to expansion of communes

Argentina and Uruguay: rapid increase due to railways; decline due to increasing costs of transport to overseas markets

Australia: rapid increase due to railways, growing population and (recently) irrigation

Japan: only a limited amount of land available (competition from housing and industry); rice provides a higher yield

The Prairies have already been referred to in the optima and limits model (Figure 16.3). Although this area has many favourable physical character-istics, it also has disadvantages (Case Study 12B). Wheat, the major crop, ripens well during the long, sunny, summer days, while the winter frosts help to break up the soil. However, the growing season is short and in the north falls below the minimum requirement of 90 days. Precipitation is low, about 500 mm, but though most of this falls during the growing season there is a danger of hail ruining the crop, and droughts occur periodically. The winter snows may come as blizzards but they do insulate the ground from severe cold and provide moisture on melting in spring. The chinook wind (page 241) melts the snow in spring and helps to extend the growing season, but tornadoes in summer can damage the crop. The relief is gently undulating, which aids machinery and transport. The grassland vegetation has decayed over the centuries to give a black (chernozem – page 327) or very dark brown (prairie) soil (page 328). However, if the natural vegetation is totally removed, the soil becomes vulnerable to erosion by wind and convectional rainstorms.

Figure 16.37

Extensive commercial cereal farming on the Canadian Prairies

When European settlers first arrived, they drove out the local Indians, who had survived by hunting bison, and introduced cattle. The world price for cereals increased in the 1860s and demand from the industrialised countries in western Europe rose. The trans-American railways were built in response to the increased demand (and profits to be made) and vast areas of land were ploughed up and given over to wheat. The flat terrain enabled straight, fast lines of communication to be built (essential as most of the crop had to be exported) and the land was divided into sections measuring 1 square mile (1.6 km²). In the wetter east, each farm was allocated a quarter or a half section; while in the drier west, farmers received at least one full section.

The input of capital has always been high in the Prairies as farming is highly mechanised (Figure 16.37). Mechanisation has reduced the need for labour although a migrant force, with combine harvesters, now travels northwards in late summer as the cereals ripen. Seed varieties have been improved, and have been made disease-resistant, drought-resistant and faster-growing. Fertilisers and pesticides are used to increase yields and the harvested wheat is stored in huge elevators while awaiting transport via the adjacent railway.

In the last two decades, wheat has become less of a specialist crop and the area upon which it is grown has decreased. Many farms have diversified to produce sugar beet, flax, dairy produce and beef (Case Study 12B).

8 Intensive commercial (mixed) agriculture

This corresponds with von Thünen's inner zone where dairying, market gardening (horticulture) and fruit all compete for land closest to the market. All three have high transport costs, are perishable, bulky, and are in daily demand by the urban population. Similarly, all three require frequent attention, particularly dairy cows which need milking twice daily, and market gardening. Although this type of land use is most common in the eastern USA and north-west Europe (Places 71), it can also be found around every large city in the world. Intensive commercial farming needs considerable amounts of capital to invest in high technology and numerous workers: it is labour intensive. The average farm size used to be under 10 ha but recently this has been found to be uneconomic and amalgamations have been encouraged by the American government and the EU in order to maximise profits. This type of farming gives the highest output per hectare and the highest productivity per farmworker.

Food surpluses

As farming in the more developed countries of North America and the EU continued to become more efficient, output increased. Farmers were paid subsidies, or a guaranteed minimum price, for their produce. The result was the overproduction of certain commodities for the American and European markets. Although food surpluses are needed for trade to take place, the issue is the unit cost at which the surpuses are produced. The problem arises when surpluses are produced at costs and subsidised prices that are above world market prices. This means that the commodities can only be sold on at a further subsidised price which then either distorts world markets or makes them inaccessible to developing countries. The EU has been trying to bring down 'internal' prices to meet world prices since the early 1990s, and attempts to slim down the 'mountains' and 'lakes' have met with some success (page 493). The EU has spent most money on dairy products. In 1986 it cost almost £1 million a day to store butter, and when some of it was sold to the then USSR at 7p/lb there were political repercussions. This action, together with some limited sales at reduced prices to the poor and the elderly within the Community, used up large amounts of the EU budget. Although the decision in 1986 to reduce milk output by 9.5 per cent by 1989 may have appeared sensible, it should be remembered that this has forced many dairy farmers to sell up and to slaughter some 5 million cattle. In 1988, attention was turned to cereals. It was agreed that £35/ha should be paid to arable farmers if they 'set-aside' their land (i.e. took it out of production) instead of growing crops on it (page 493). The beef mountain and wine lake were tackled after 1989. In the late 1990s the EU, in an attempt to keep land in production and farmers in employment without generating food surpluses, encouraged the cultivation of non-food crops such as fibres and oleochemicals.

Places 71   The western Netherlands: intensive commercial farming

Most of the western Netherlands, stretching from Rotterdam to beyond Amsterdam, lies 2–6 m below sea-level. Reclaimed several centuries ago from the sea, peat lakes or areas regularly flooded by rivers, this land is referred to as the **old polders**. Today, they form a flat area drained by canals which run above the general level of the land. Excess water from the fields is pumped (originally by windmills) by diesel and electric pumps into the canals. With 469 persons per km² in 1998 (compared with only 360 in 1975), the Netherlands has the highest population density in Europe. Consequently, with farmland at a premium, the cost of reclamation so high and the proximity of a large domestic urban market, intensive demands are made on the use of the land (Figure 16.38).

There are three major types of farming on the old polders.

- Dairying is most intensive to the north of Amsterdam, in the 'Green Heart' and in the south-west of Friesland. It is favoured by mild winters, which allow grass to grow for most of the year; the evenly distributed rainfall, which provides lush grass; the flat land; and the proximity of the *Randstad* conurbation. Most of the cattle are Friesians. Some of the milk is used fresh but most is turned into cheeses (the well-known Gouda and Edam) and butter. Most farms have installed computer systems to control animal feeding.

- The land between The Hague and Rotterdam (Figure 16.38) is a mass of glasshouses where **horticulture** is practised on individual holdings averaging only 1 ha. The cost of production is exceptionally high. Oil and natural gas-fired central heating maintain high temperatures

---

and sprinklers provide water. Heating, moisture and ventilation are all controlled by computerised systems. Machinery is used for weeding and removing dead flowers, and the soils are heavily fertilised and manured. Sometimes plants are grown through a black plastic mulch (heat-absorbing) which has the effect of advancing their growth and thus extending the cropping season to meet market demand for fresh produce. Several crops a year can be grown in the glasshouses, i.e. cut flowers in spring, tomatoes and cucumbers in summer, and lettuce in autumn and winter.

- The sandier soils between Leiden and Haarlem are used to grow **bulbs**. Tulips, hyacinths and daffodils, protected from the prevailing winds by the coastal sand dunes, are grown on farms averaging 8 ha. The flowers form a tourist attraction, especially in spring (Figure 16.39) and bulbs are exported all over Europe from nearby Schipol Airport.

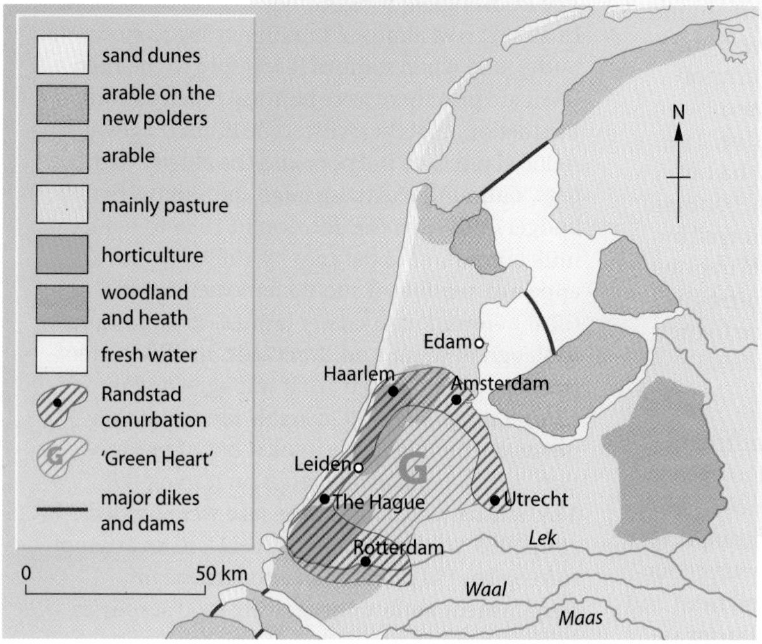

Figure 16.38

Agricultural land use in the western Netherlands

Key:
- sand dunes
- arable on the new polders
- arable
- mainly pasture
- horticulture
- woodland and heath
- fresh water
- Randstad conurbation
- 'Green Heart'
- major dikes and dams

0   50 km

Edam
Haarlem   Amsterdam
Leiden   G
● The Hague   ● Utrecht
Rotterdam   Lek
Waal
Maas

Figure 16.39

Intensive farming on reclaimed polder land in the western Netherlands

9 Mediterranean agriculture

A distinctive type of farming has developed in areas surrounding the Mediterranean Sea. Winters are mild and wet, allowing the growth of cereals and the production of early spring vegetables or *primeurs*. Summers are hot, enabling fruit to ripen, but tend to be too dry for the growth of cereals and grass. As rainfall amounts decrease and the length of the dry season increases from west to east and from north to south, irrigation becomes more important. River valleys and their deltas (the Po, Rhône and Guadalquivir) provide rich alluvium, but many parts of the Mediterranean are mountainous with steep slopes and thin rendzina soils (page 274). Due to earlier deforestation, many of these slopes have suffered from soil erosion. Frosts are rare at lower levels, though the cold mistral and bora winds may damage crops (Figure 12.22).

Farming tends to be labour intensive but with limited capital. There are still many absentee landlords (latifundia, page 466) and outputs per hectare and per farmworker are usually low. Most farms tend to be small in size. Land use (Figure 16.21) shows the importance of tree crops such as olives, citrus and nuts, while land use frequently illustrates that crops which need most attention are grown nearest to the farmhouse or village, and that land use is more closely linked to the physical environment than controlled by human inputs (Places 72). Many village gardens and surrounding fields are devoted to citrus fruits, such as oranges, lemons and grapefruit, as these have thick waxy skins to protect the seeds and to reduce moisture loss. These fruits are also grown commercially where water supply is more reliable, e.g. oranges in Spain around Seville and on *huertas* (irrigated farms) near Valencia, lemons in Sicily and grapefruit in Israel. Recently there has been a rapid increase in the use of polythene, especially in south-east Spain, where the area around Almeria has become known as the 'Costa del Polythene', and in Israel. The polythene, which is stretched across 3 m high poles, creates a hothouse environment suitable for the growth of tomatoes and other crops such as melons, green beans, peppers and courgettes. The crops are harvested twice yearly, usually when they are out of season in more northerly parts of Europe.

Vines, another labour-intensive crop, and olives, the 'yardstick' of the Mediterranean climate, are both adapted to the physical conditions. They tolerate thin, poor, dry soils and hot, dry summers by having long roots and protective bark. Wheat may be grown in the wetter winter period in fields further from the village as it needs less attention, while sheep and goats are reared on the scrub and poorer-quality grass of the steeper hillsides. Grass becomes too dry in summer to support cattle and so milk and beef are scarce in the local diet.

Apart from central Chile, other areas experiencing a Mediterranean climate have developed a more commercialised type of farming based upon irrigation and mechanisation. Central California supports agribusiness based on a large, affluent, domestic market which is, in terms of scale, organisation and productivity, the ultimate in the capitalist system. Southern Australia produces dried fruit to overcome the problem of distance from world markets. All Mediterranean areas have now become important wine producers.

Places 72 The Peloponnese, Greece: Mediterranean farming

Figure 16.40

Land use and farming types in the Peloponnese (not to scale)

Figure 16.40 is a transect, typical of the Peloponnese and many other Mediterranean areas, showing how relief, soils and climate affect land use and farming types. The area next to the coast, unless taken over by tourism, is farmed intensively and commercially (Figure 16.41). As distance from the coast increases, farming becomes more extensive and eventually, before the limit of cultivation, at a subsistence level (Figure 16.42).

A Intensive commercial farming

Coastal plain: flat with deep, often alluvial, soil (washed down from hills by seasonal rivers)

Hot, dry summers; mild, wet winters with no frost

Some mechanisation; irrigation needed in summer

Citrus fruits (oranges, clementines and mandarins); peaches and some figs

B Extensive commercial farming

Undulating land with small hills: soils quite deep and relatively fertile (terra rossa)

Larger farms (villages on hills originally for defence, now above best farmland)

Similar climate, with a slight risk of frost in winter

Some mechanisation, but donkeys still used

Olives and vines with, occasionally, tobacco

C Extensive, subsistence farming

Steeper hillsides covered in scrub: thin, poor soils (rendzina)

Warm, dry summers; cool, wet winters with increasing risk of frost

No mechanisation

Sheep and goats

D Virtually no farming/some rough grazing

Steep hillsides, mountainous, with poor, discontinuous scrub: very little soil (much erosion)

Cool, dry summers; cold, wet and windy winters with a risk of snow

Sheep and goats (mainly in summer)

Figure 16.42

Orange and olive groves in foreground, rough grazing on hillsides beyond: near Mycenae in the Peloponnese

Figure 16.41

Intensive fruit, mainly oranges, next to the coast, settlement on higher, less fertile land in mid-distance, and deforested hills in the background: near Navplion in the Peloponnese

10 Irrigation

Irrigation is the provision of a supply of water from a river, lake or underground source to enable an area of land to be cultivated (Figure 16.43). It may be needed where:

1 rainfall is limited and where evapotranspiration exceeds precipitation, i.e. in semi-arid and arid lands such as the Atacama Desert in Peru (Places 24, page 180) and the Nile valley (Places 73)
2 there is a seasonal water shortage due to drought, as in southern California with its Mediterranean climate (Case Study 15A)
3 amounts of rainfall are unreliable, as in the Sahel countries (Figure 9.28)
4 farming is intensive, either subsistence or commercial, despite high annual rainfall totals, e.g. the rice-growing areas of Southeast Asia.

In economically more developed countries, large dams may be built from which pipelines and canals may transport water many kilometres to a dense network of field channels (Case Study 3C). The flow of water is likely to be computer-controlled. Unfortunately, it is the economically less developed countries, lacking in capital and technology, that suffer most severely from water deficiencies. Unless they can obtain funds from overseas, most of their schemes are extremely labour intensive as they have to be constructed and operated by hand.

Figure 16.43

Boom irrigation in Saudi Arabia

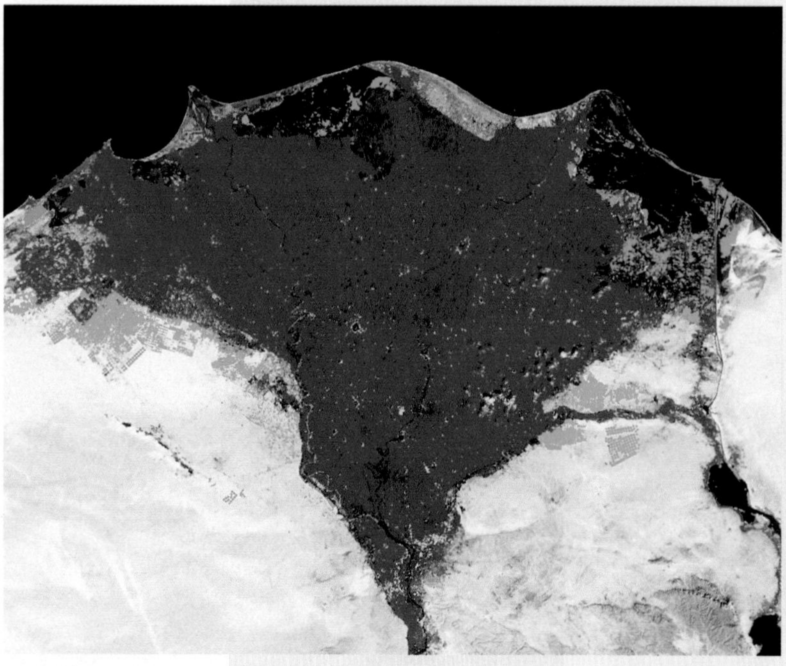

Figure 16.44

Landsat photo of the Nile delta: the River Nile, Suez Canal and Mediterranean Sea are shown in black, the irrigated lands in magenta, and Cairo and other settlements in pale blue

Places 73   The Nile valley: irrigation

From the time of the Pharaohs until very recently, water for irrigation was obtained from the River Nile by two methods. First, each autumn the annual floodwater was allowed to cover the land, where it remained trapped behind small bunds until it had deposited its silt. Secondly, during the rest of the year when river levels were low, water could be lifted 1–2 m by a *shaduf*, *saquia* (*sakia*) **wheel** or **Archimedes screw**. However, the Egyptians had long wished to control the Nile so that its level would remain relatively constant throughout the year. Although barrages of increasing size had been built during the early 20th century, it was the rapid increase in population (which doubled from 25 to 50 million between 1960 and 1987) and the accompanying demand for food that led to the building of the Aswan High Dam (opened 1971) and several new schemes to irrigate the desert near Cairo (late 1980s).

The main purpose of the High Dam was to hold back the annual floodwaters generated by the summer rains in the Ethiopian Highlands. Some water is released throughout the year, allowing an extra crop to be grown, while any surplus is saved as an insurance against a failure of the rains. The river regime below Aswan is now more constant, allowing trade and cruise ships to travel on it at all times. Two and sometimes three crops can now be grown annually in the lower Nile valley (Figure 16.44). Yields have increased and extra income is gained from cash crops of cotton, maize, sugar cane, potatoes and citrus fruits. The dam incorporates a hydro-electric power station which provides Egypt with almost a third of its energy needs for domestic and industrial purposes. Lake Nasser is important for fishing and tourism.

Following the construction of the Aswan High Dam (Figure 16.45), Egypt has modernised its methods of irrigation. Electricity is now used to power pumps which, by raising water to higher levels, allow a strip of land up to 12 km wide on both sides of the Nile to be irrigated. **Drip irrigation** utilises plastic pipes in which small holes have been made; these are laid over the ground and water drips onto the plants in a much less wasteful manner, as less evaporates or drains away. Between the Nile and the Suez Canal, **boom irrigation** has been introduced (Figure 16.43 shows this method in use in Saudi Arabia), creating fields several hectares in diameter.

However, the Dam has created several problems. Environmentally, the cessation of the Nile flood has also meant ending the annual deposition of fertile silt on the fields, which means fertiliser now has to be added; without the silt, the delta has begun to retreat – having lost its supply of sediment. The number of bilharzia snails has increased as a consequence of the greater number of irrigation channels. Economically and socially, development has encouraged farmers to grow cash crops instead of providing a better diet for themselves, and costs have increased due to the need to buy fertiliser. Clay is no longer available for making bricks in the traditional manner. The drought during the 1980s in Ethiopia meant a dramatic decrease in floodwater brought down by the Blue Nile and River Atbara (Figure 16.45). By 1988 Lake Nasser was only 40 per cent full; ships were having difficulty travelling up the shallow Nile; and there was talk of having to close the turbines which generated the hydro-electricity (since then lake levels have risen). Large areas of land reclaimed after 1971 have become saline and have been allowed to return to desert. The UN Food and Agriculture Organisation (FAO) claims that between 1974 and 1984 there was an actual *decrease* of 12 per cent in land under irrigation and that salinisation was affecting 30–40 per cent of the remaining irrigated land.

Figure 16.45

The Nile: sources and uses of water

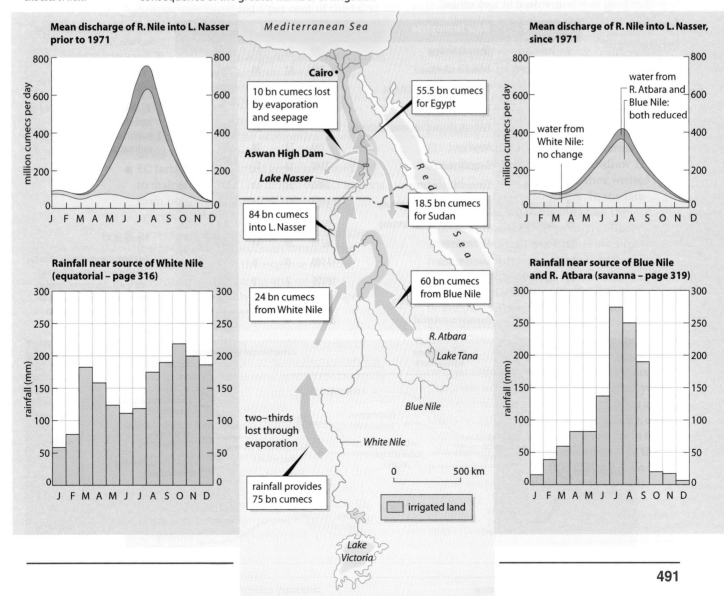

A further attempt, made in March 1999, to reform the CAP was limited in success by the divergent interests of the main EU farming countries (Figure 16.48). At the same time, many farmers in Britain were facing their worst ever financial crisis following the BSE scare, a dramatic fall in the price of farm animals and their products and an average reduction of 70 per cent in two years in their overall income (over 400 per cent drop for cattle and sheep farmers).

Figure 16.48

A judgement of the CAP reforms of early 1999

FUDGED CAP – Vested interests win again

Yesterday's package fails on all counts. It won't liberate enough money to sublimate the CAP into a fund to preserve the environment of the countryside; it won't help the agricultural problems of aspirant east European countries like Poland; and it won't create money to boost the EU's competitive position in other sectors. The CAP was spectacularly successful in turning the post-war food shortage into a surplus but it has queered the pitch for any successor schemes – like, say, a Common Technology Fund. France digs in its heels for peasant farmers and that's that. Britain says No to giving up its hard-earned rebate and that's that as well. Meanwhile the Germans are fed up paying by far the most into EU funds.

The Guardian, 17 March 1999

Figure 16.49

A rural landscape with trees and hedges, Dorset

Figure 16.50

How eutrophication can upset the ecosystem

Farming and the environment

Numerous pressure groups are claiming that the traditional British countryside is being spoilt, yet the countryside of today is not 'traditional' – it has always been changing. The primeval forests, regarded as Britain's climatic climax vegetation (page 286), were largely cleared, initially for sheep farming and later for the cultivation of cereals. Although there is evidence that hedges were used as field boundaries by the Anglo-Saxons, it was much later that land was 'enclosed' by planting hedges and building dry stone walls (page 397). It is this 18th- and 19th-century landscape which has become looked upon, incorrectly, as the traditional or natural environment (Figure 16.49). However, the rate of change has never been faster than in recent years. Estimates by the Nature Conservancy Council suggest that, between 1949 and 1990, 40 per cent of the remaining ancient broadleaved woodlands, 25 per cent of hedgerows, 30 per cent of heaths, 60 per cent of wetlands and 30 per cent of moors have 'disappeared'. While most accusing fingers point to the intensification of agriculture, together with afforestation and building programmes, as the major causes, it should be remembered that farmland too is under threat from rival land users (Figure 17.4).

Farming as a threat to the environment

a The use of chemicals

Fertiliser, slurry and pesticides all contribute to the pollution of the environmental system. Fertiliser, in the form of mineral compounds which contain elements essential for plant growth, is widely used to produce a healthy crop and increase yields. If too much nitrogenous fertiliser or animal waste (manure) is added to the soil, some remains unabsorbed by the plants and may be leached to contaminate underground water supplies and rivers. Where chemical fertiliser accumulates in lakes and rivers, the water becomes enriched with nutrients (eutrophication) and the ecosystem is upset (Figure 16.50). In parts of north-west Europe, levels of nitrates in groundwater are above EU safety limits.

In Britain, the Water Authorities claim that slurry (farmyard effluent) is now the major pollutant of, and killer of life in, rivers. After several decades in which the quality of river water had improved, the last few years have seen levels of pollution again increasing, especially in farming areas. Even so, the Water Authorities themselves asked permission in 1989 to *increase* the discharge of sewage into rivers.

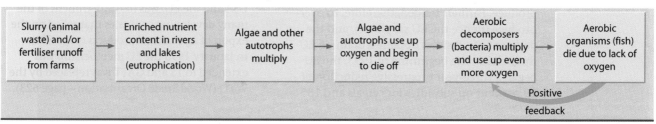

Slurry (animal waste) and/or fertiliser runoff from farms → Enriched nutrient content in rivers and lakes (eutrophication) → Algae and other autotrophs multiply → Algae and autotrophs use up oxygen and begin to die off → Aerobic decomposers (bacteria) multiply and use up even more oxygen → Aerobic organisms (fish) die due to lack of oxygen

Positive feedback

Figure 16.51

The case for and against hedgerows and ponds in a rural area

For	Against
■ Hedgerows	
Form part of the attractive, traditional British landscape	Are not traditional and were initially planted by farmers
Form a habitat for wildlife: birds, insects and plants (Large Blue butterfly is extinct, 10 other species are endangered)	Harbour pests and weeds
Act as windbreaks (and snowbreaks)	Costly and time-consuming to maintain
Roots bind soil together, reducing erosion by water and wind	Take up space which could be used for crops
	Limit size of field machinery (combine harvesters need an 8-m turning circle)
■ Ponds	
Form a habitat for wildlife: birds, fish and plants	Take up land that could be used more profitably
Add to the attractiveness of the natural environment	Stagnant water may harbour disease
i.e. Concern is environmental	**i.e. Concern is economic**

Pesticides and herbicides are applied to crops to control pests, diseases and weeds. Estimates suggest that, without pesticides, cereal yields would be reduced by 25 per cent after one year and 45 per cent after three. The Friends of the Earth claim that pesticides are injurious to health and, although there have been no human fatalities reported in Britain in the last 15 years, there are many incidents in developing countries resulting from a lack of instruction, fewer safety regulations and faulty equipment. A UN report (1994) claims that 25 million agricultural workers in developing countries (3 per cent of the total workforce) experience pesticide poisoning each year. Pesticides are blamed for the rapid decrease in Britain's bee and butterfly populations, and an up to 80 per cent reduction in 800 species of fauna in the Paris basin. Pesticides can dissipate in the air as vapour, in water as runoff, or in soil by leaching to the groundwater.

b The loss of natural habitats
The most emotive outcries against farmers have been at their clearances of hedges, ponds and wetlands. These clearances mean a loss of habitat for wildlife and a destruction of ecosystems, some of which may have taken centuries to develop and, being fragile, may never recover or be replaced. As stated earlier, over 25 per cent of British hedgerows were removed between 1949 and 1990 – in Norfolk, the figure was over 40 per cent. Figure 16.51 lists some of the arguments for and against the removal of hedgerows and the drainage of ponds/wetlands. Figures 16.49 and 16.52 show the contrast between a landscape with trees and hedges, and one where they have been removed.

Farming can increase soil erosion. The rate of erosion is determined by climate, topography, soil type and vegetation cover (Case Study 10), but it is accelerated by poor farming practices (overcropping and overgrazing) and deforestation. (The conditions favouring wind erosion are given on page 182 and the processes of wind transportation are shown in Figure 7.8.) In Britain, wind erosion tends to be restricted to parts of East Anglia and the Fens where the natural vegetation cover, including hedges, has been removed and where soils are light or peaty. Water erosion is most likely to occur after periods of prolonged and heavy rainfall, on soils with less than 35 per cent clay content, in large and steeply-sloping fields and where deep ploughing has exposed the soil.

Arable farming, especially when ploughing is done in the autumn, removes the protective vegetation cover. The intensification of farming, and overcropping, in areas of highly erodible soils in the USA have led to a decrease in yields and an estimated loss of one-third of the country's topsoil – much of it from the Dust Bowl during the 1930s. Deforestation in tropical rainforests, mountainous and semi-arid areas – Brazil, Nepal and the Sahel, respectively – also accelerates soil erosion.

Figure 16.52

An agricultural landscape without trees or hedges, Cambridgeshire

Organic farming has its problems. If it replaces a conventional farming system, yields can drop considerably in the first two years, when artificial fertiliser is no longer used, although they soon rise again as the quality of the soil improves. Also, during the transition period, farmers cannot market any goods as 'organic': they must wait until they meet the Soil Association's standards before receiving its label guaranteeing the authenticity of their produce. Weeds can increase, without herbicides, and may have to be controlled by hand labour or by being covered with either mulch or polythene. This means that, although organic farming is helpful to the environment and, arguably, less harmful to human health, its produce is more expensive to buy.

Britain's Agriculture Minister announced, in 1999, a scheme that would see the doubling in payment rates to farmers wanting to convert to organic farming, that the Ministry's (MAFF) support to farmers wanting to convert would increase six-fold in that year, and that the amount of farmland in organic production in the UK had increased five-fold over the last year. What he failed to point out was that the Ministry had insufficient funds to support all those farmers wishing to convert, that Britain had the lowest percentage of arable land in western Europe under organic crops, and that the rate of increase was less than its EU neighbours. About the same time it was announced that one of Britain's leading organic farmers had won a deal to supply Sainsbury's and was close to signing a contract with Marks & Spencer – two deals demonstrating the growth in demand for organic food, a demand perhaps accelerated by fears over GM foods.

Places 74   South-west England: organic farming

Colin Hutchin farms near Taunton in Somerset. His 130 ha mixed farm is hilly and well-wooded, with thick hedges and hay meadows rich in herbs. He keeps a herd of white Charolais cows, as well as other beef breeds and a flock of sheep. He therefore has no need to buy fertiliser. He devotes 40 hectares of the farm to growing vegetables which he is then able to sell to local wholesalers. The fertility of his land is maintained by adding lime, slag and animal manure. When parts of the farm are left fallow, to allow them to rest and be replenished, large areas are given over to clover. Colin has a rotation pattern of two straw crops followed by a root crop and then either grass and clover (clover, a leguminous crop, returns nitrogen to the soil) or peas or beans (also leguminous crops). He believes that this rotation not only maintains a natural soil fertility, but also reduces pests and diseases by changing the crop before they can take hold. He also relies on ladybirds to do the job of pesticides as they keep down the number of harmful blackfly and aphids.

Step Farm in Gloucestershire, which belongs to the National Trust, receives visitors from all walks of life, but increasingly from groups of farmers interested in seeing the practicalities of organic farming for themselves (Figure 16.55). The farm was converted to being totally organic in the late 1980s and is now benefiting from the current upsurge of interest in organic produce. The centre of the enterprise are Friesian milking cows, beef cattle and sheep, together with 130 ha of cultivated crops which include wheat, barley and beans.

Figure 16.55

Step Farm, Gloucestershire

The concept of sustainable development dominated the environmental agenda during the 1990s and, following the 1992 Earth Summit at Rio de Janeiro, has been embraced by governments at all levels of development. The term is not, however, easy to explain; Dobson, in 1996, claimed that there were over 300 different definitions and interpretations. Of these, the most widely used is that taken from the Brundtland Report (The World Commission on Environment and Development, 1987) which claims that sustainable development 'meets the needs of the present without compromising the ability of future generations to meet their own needs'. This definition, according to Munton and Collins (*Geography*, 1998), 'highlights the socio-economic rather than the environmental basis of sustainable development and, unlike earlier understandings of the term 'environmental sustainability', it gives absolute primacy to improving human conditions and not to environmental limits'.

Put more simply, sustainable development should lead to an improvement in people's:

- quality of life, allowing them to become more content with their way of life and the environment in which they live
- standard of living, enabling them, and future generations, to become better off economically.

This may be achieved in a variety of ways:

- by encouraging economic development at a pace that a country can both afford and manage so as to avoid that country falling into debt
- by developing technology that is appropriate to the skills, wealth and needs of local people irrespective of the country's level of development, and developing local skills so that they may be handed down to future generations
- by using natural resources without spoiling the environment, developing materials that will use fewer resources, and using materials that will last for longer – ideally, once a resource is used, it should either be renewed, recycled or replaced.

Sustainable development needs careful planning and, increasingly as it involves a commitment to conservation, the co-operation of groups of countries and, under extreme conditions, global agreement.

Sustainable development is a theme that keeps re-appearing throughout this book. It is a concept that, from a geographer's point of view, can be studied:

- through a selection of physical and human environments
- at a variety of levels of development

in the context of people and food supply, resources, and natural and human created/adapted environments. Examples referred to in this book, with chapter numbers in brackets, include the following:

- **People and environments**
  - world biomes and fragile environments such as the tropical rainforest (11 and 12) and the tundra (5)
  - smaller-scale ecosystems including wetlands (3 and 16) and sand dunes and saltmarshes (11)
  - effects of economic development on scenic areas and the wildlife of coastal and mountainous areas (4, 6 and 20)
- **People and resources**
  - finite resources of fossil fuels (18) and minerals (17)
  - renewable resources, providing that they are carefully managed, including soils (10); fresh and reliable water supply (3 and 15); forests (11 and 17); crops and food supply (16); energy (18); recycled materials (19); and the atmosphere (9).
- **Socio-economic**
  - population growth and family planning (13)
  - urban growth/loss of countryside (15)
  - housing materials (15 and 19)
  - development of skills and levels of education (22).

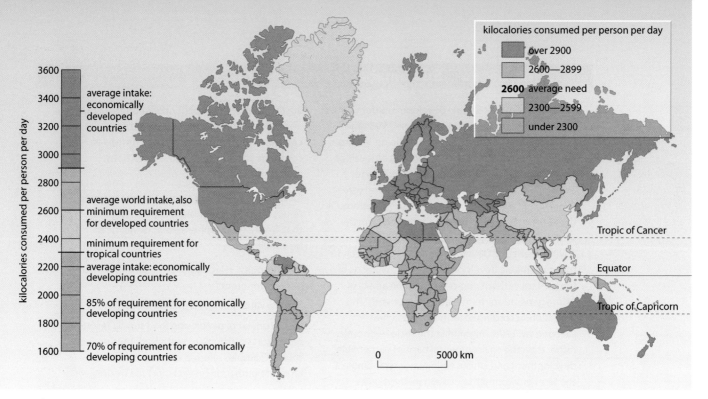

kilocalories consumed per person per day

3600 — 3400 — 3200 — 3000 — 2900 — 2800 — 2600 — 2400 — 2200 — 2000 — 1800 — 1600

average intake: economically developed countries

average world intake, also minimum requirement for developed countries

minimum requirement for tropical countries

average intake: economically developing countries

85% of requirement for economically developing countries

70% of requirement for economically developing countries

kilocalories consumed per person per day

over 2900
2600—2899
2600 average need
2300—2599
under 2300

Tropic of Cancer

Equator

Tropic of Capricorn

0   5000 km

Figure 16.56

World food supply in the late 1990s: average kilocalorie consumption per person/day, by country

Food supplies

Diet and health

It is 200 years since Malthus expressed his fears that world population would outstrip food supply (page 378). Today, despite assurances from various international bodies such as the Food and Agriculture Organisation (FAO) that there is still sufficient food for everyone, it is estimated that three-quarters of the world's population is inadequately fed, and that the majority of these live in less economically developed countries. The problem is, therefore, the unevenness in the distribution of food supplies: massive surpluses exist in North America and the EU; and there are shortages in many developing countries.

This uneven distribution is reflected in Figure 16.56 which shows variations in kilocalorie intake throughout the world. Dieticians calculate that the average adult in temperate latitudes requires 2600 kilocalories a day, compared with 2300 kilocalories for someone living within the tropics. The FAO reports that the actual average intake for the economically more developed world is 3300 kilocalories, but only 2200 kilocalories in less developed countries. However, the quantity of food consumed is not always as important as the quality and balance of the diet. A good diet should contain different types of food to build and maintain the body, and to provide energy to allow the body to work. A balanced diet should contain:

- **proteins**, such as meat, eggs and milk, to build and renew body tissues
- **carbohydrates**, which include cereals, sugar, fats, meat and potatoes, to provide energy, and

- **vitamins** and **minerals**, as found in dairy produce, fruit, fish and vegetables, which prevent many diseases.

Malnutrition and undernutrition, often caused by poverty, affect many people including even a surprisingly high number in developed countries. Malnutrition may not be a primary cause of death, but by reducing the ability of the body to function properly, it reduces the capacity to work and means that people, and especially children, become less resistant to disease and more likely to fall ill. Nutritional diseases, which include rickets (vitamin D deficiency), beri-beri (vitamin B1 deficiency and common in rice-dependent China), kwashiorkor (protein deficiency) and marasmus (shortage of protein and calories), can reduce resistance to intestinal parasitic diseases, malaria and typhoid. In contrast, people in developed countries are at risk from over-eating and from an unbalanced diet which often contains too many animal fats which can cause heart disease.

Trends in food supply

Between the early 1950s and 1986 world food output increased more rapidly than did world population. The increase in output was more rapid in the developing countries, albeit from a much lower base, than in developed countries. The main exceptions to this generalisation were several African countries where food output per person actually fell (Figure 16.57). In 1986 the FAO announced that although there was sufficient food to feed everyone in the world 1 kg of food per day, this could not be achieved due to its uneven distribution and its rapidly rising cost. Disparities

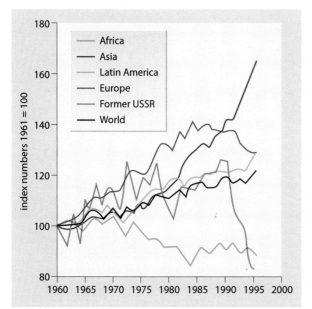

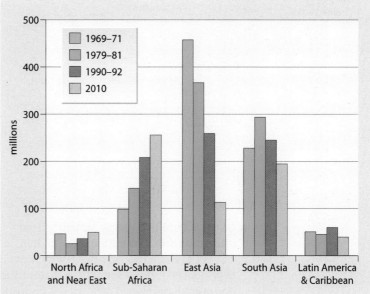

Figure 16.57

Trends in per capita
food production,
1961–96

on a continental scale meant that while there was 5 kg per person for North America, 3.5 kg for Oceania and 2 kg for western Europe, there was only 1 kg for Latin America and South-east Asia and less than 0.5 kg for Africa.

Between 1986 and 1988, mainly due to below-average global rainfall totals, it was estimated that the world's food reserves fell from 101 days' supply to 54 days, and that by 1993 over 1000 million people (35 per cent of the world's total) lived at, or below, starvation level. At that time there was much pessimism about future prospects of feeding an increasing population.

Despite the gloomy forecasts, food production did continue to grow in the early 1990s at a faster rate than did population growth. This, together with the prospects of applying present and new farming techniques (compare Boserup, page 379), made for a more encouraging future. Indeed by 1999, when admittedly world food reserves were put at only 49 days, the WHO (World Health Organisation) claimed that 'nutrition seems to be improving. Life expectancy is growing almost worldwide [page 359], with better nutrition one of the key factors behind this rise. On the surface at least, there is plenty of food. Global supplies are in relatively good shape, with surpluses in many areas of the world.'

However, the same WHO report also referred to the term 'food insecurity' which, defined in the most basic terms, means a lack of nutritious food needed to keep people alive and healthy. Although numbers had dropped from the 1993 peak, 800 million (20 per cent of the world's total) still suffered from chronic undernutrition (Figure 16.58). The WHO report highlighted several areas of continued concern:

■ Throughout history, whenever extra food was needed, people simply cleared more land for crops. Today, most high-quality agricultural land is already in use, or has been built upon. The environmental costs of converting the remaining forest, grassland and wetland habitats into farmland are well known, and much of the remaining soil is less productive and more fragile, i.e. less sustainable (Framework 16, page 499).

■ Although total yields continue to increase on a global scale (Figure 16.59a), there is a disturbing decline in the rate of yield growth of wheat, maize and padi rice (Figure 16.59b).

■ Food production in Africa continues to decline (Places 75), and there was also, in the 1990s, a rapid decrease in the former USSR (Figure 16.57).

■ There has been a fall in global cereal prices, causing farmers who might in the past have justified the expense of additional inputs (e.g. fertiliser and water) to keep yields high, to move away from cereals to more profitable crops.

■ There is increasing globalisation of food production, with transnational corporations and large supermarkets in developed countries sourcing more of their food from developing countries. Plantations, large commercial estates and, increasingly, small farmers are being drawn into contracts to supply fruit, vegetables and wine to consumers in the developed world. The result is a decline in the growth of staple foods within those developing countries.

Figure 16.59

Global changes in
cereal production,
1961–96

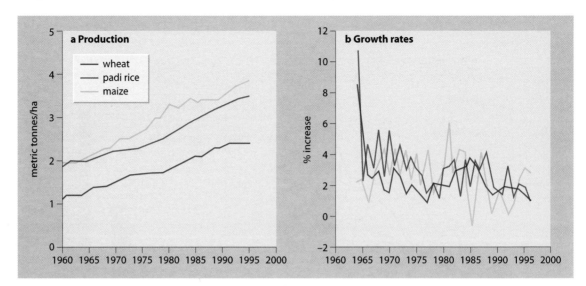

Of the 800 million people suffering from undernutrition, 200 million are children. The WHO claim that one-third of children in developing countries suffer from malnutrition, and even in developed countries malnutrition is associated with more than half the deaths among children under the age of 5. The percentage of under-5s who were malnourished (1998) was 67 in Bangladesh, 53 in India and over 40 in each of Nepal, Vietnam, Eritrea, Pakistan, Laos and Cambodia. Malnutrition in children is generally determined by weight – the percentage under the age of 5 who weigh considerably less than the general population. Low-birthweight babies are children born weighing less than 2500 g, their low weight being attributed to maternal malnutrition. Low-birthweight children are often prone to a shortened lifetime full of health problems, including retarded development and susceptibility to disease.

One concern not expressed but which hit the headlines in the UK in 1999 was that of genetically modified (GM) foods. Whereas it was hoped during the late 1990s that biotechnology might be the major solution to food shortages in certain parts of the world, feelings grew in some developed countries as to the possible risks of such 'untested' methods of crop production, both to the environment and to people's health.

The most widely recognised cause of malnutrition is poverty – the lack of money to buy food, or the land, resources and knowledge needed to grow crops. Other factors include a shortage of clean drinking water – a shortage felt by one-quarter of the world's population – as water is more difficult and expensive to trade than agricultural products.

Famine

Famines caused by continent-wide droughts were once considered inevitable occurrences.

Dry spells continue but, aided by improved planning and early-warning systems, their effects have been reduced even if not eliminated. The 'natural' famines of the past have been replaced with those often created by localised wars and the consequent displacement of people (Places 75). The notion that famine means a total food shortage (as implied in the introductory Biblical quote to this chapter) has been challenged as recent studies suggest that famine only affects certain groups in society (the poorest, least skilled and unemployed). Even during the worst times of famine, some food still appears in local markets – but at a price beyond the reach of most people.

Amartya Sen, an economist, says that 'the idea that the causation of famines can be neatly split into "natural" and "man-made" ones would appear to be a bit of a non-starter'. Indeed, it is now more widely accepted that most famines result from a combination of natural events and human mismanagement. Sen talked about the 'entitlement' of workers. He considered that workers have 'endowments', which may be goods, resources, skills, food and the ability to work, which are used to obtain items which they want or need. 'Entitlement failure' occurs when the endowment is insufficient to avoid starvation, i.e. certain groups in the community cannot afford to buy food, rather than there being no food available. Famine is therefore a decline in access to food (due to cost of transport and/or food supplies) rather than a decline in the available food supply. For example, the devastating 1974 flood in Bangladesh did not totally destroy the rice crop but led to fears that, as the harvest would be poor, rice prices would rise. This, combined with rising economic inflation, resulted in a decrease in rural jobs and wages and a collapse in the entitlement of rural workers.

A UN report issued in mid-1999 stated that nearly 10 million people were in need of emergency food assistance in sub-Saharan Africa. Severe drought and persistent civil strife and insecurity in many countries have displaced large numbers of people and disrupted food production (Figure 16.60). The report listed 16 countries that were facing exceptional food shortages (Figure 16.61), with Angola, Somalia and Ethiopia the worse affected. However, it was not all bad news as food supply in the Sahel (Case Study 7) remained adequate.

The population of sub-Saharan Africa, with its high birth rates and falling death rates, is growing faster than anywhere else in the world. With 71 per cent of the labour force in agriculture and 77 per cent of the population living in rural areas, the income, nutrition and health of most Africans is closely tied to farming. In an area where, due to limited capital and technology, the use of new seeds, fertilisers, pesticides, machinery and irrigation is the lowest in the world, agriculture is almost wholly reliant upon an environment that is not naturally favourable. The soils in many areas have fertility constraints, low water-holding capacity and are vulnerable to erosion. High evapotranspiration rates harm crops, as do the unreliable rains which may cause flooding one year and then fail for several years. While the periods of water budget deficiency (drought) are getting longer and more frequent, experts argue as to whether this is part of a natural climatic cycle, a reduction of moisture in the air as a consequence of deforestation, or the effects of global warming.

With increases in population, fallow periods have been reduced and land has been overgrazed or overcropped. This, together with the destruction of forests for fuelwood, has allowed accelerated erosion and desertification. Efforts to increase food production have been impeded by a lack of money for fertiliser, seeds and tools. Even when overseas financial aid has been given, it has often had to be channelled towards unsuitable projects such as promoting monoculture, increasing cattle herds on marginal land, and ploughing soils that would be better left with a protective vegetation cover. Financial aid schemes from overseas may create problems as the recipient country is likely to fall into debt, while the donor expects cash crops to be grown for export rather than food crops for local consumption. The diet often lacks protein and, during times of shortage, people cannot afford to buy food. Animals are likely to be attacked by the tsetse fly, crops in the field by locusts, and crops in storage by rats and fungi. To add to these difficulties, several countries are torn by civil war and administrative corruption, both of which interrupt farming and the distribution of relief supplies.

Figure 16.60

Children awaiting food aid: Somalia

Civil war	Drought	Population displacement	Agricultural production		Countries facing exceptional food emergencies
•		•		**Angola**	Civil strife and population displacement
•				**Burundi**	Civil strife and insecurity
•		•		**Congo, Dem. Rep.**	Civil strife, IDPs and refugees
•				**Congo, Rep.**	Civil strife
		•		**Eritrea**	IDPs and refugees
	•	•		**Ethiopia**	Drought, large number of vulnerable people and IDPs
•		•		**Guinea-Bissau**	Civil strife, population displacement
	•			**Kenya**	Weather adversity in parts
•			•	**Liberia**	Impact of past civil strife, shortage of farm inputs
			•	**Mauritania**	Localised conflicts
•				**Rwanda**	Insecurity in parts
•		•		**Sierra Leone**	Civil strife, population displacement
•	•			**Somalia**	Drought and civil strife
•				**Sudan**	Civil strife in the south
		•		**Tanzania**	Food deficits in several regions
•	•			**Uganda**	Civil strife in parts and drought

IDP = internally displaced person

Figure 16.61

Sub-Saharan countries in need of emergency food supplies, 1999

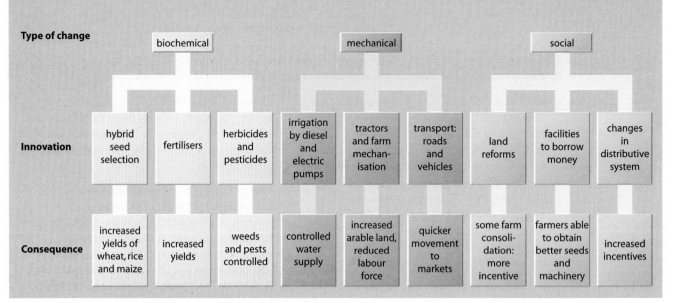

Type of change	biochemical			mechanical			social		
Innovation	hybrid seed selection	fertilisers	herbicides and pesticides	irrigation by diesel and electric pumps	tractors and farm mechan-isation	transport: roads and vehicles	land reforms	facilities to borrow money	changes in distributive system
Consequence	increased yields of wheat, rice and maize	increased yields	weeds and pests controlled	controlled water supply	increased arable land, reduced labour force	quicker movement to markets	some farm consoli-dation: more incentive	farmers able to obtain better seeds and machinery	increased incentives

Figure 16.62

The Green Revolution

What might be done to improve food supplies in developing countries?

As most areas with an average or high agricultural potential have already been used, future extension of cropland can only take place on marginal land where the threats of soil erosion and desertification are greatest. The solution is not, therefore, to extend the cultivated area but to make better use of those areas already farmed.

Land reform can help to overcome some inefficiencies in the use of land and labour. The redistribution of land has been tackled by such methods as the expropriation of large estates and plantations and distributing the land to individual farmers, landless labourers or communal groups; the consolidation of small, fragmented farms; increasing security of tenure; attempting new land colonisation projects; and state ownership. The success of these schemes has been mixed. Not all have increased food production, although many farms in China have seen an increase in yields since the transference of farming decisions to individual farmers under the responsibility system (Places 63, page 468; Places 64, page 470).

The **Green Revolution** refers to the application of modern, Western-type farming techniques to developing countries. Its beginnings were in Mexico when, in the two decades after the Second World War, new varieties or hybrids of wheat and maize were developed in an attempt to solve the country's domestic food problem. At that time, there was no intention of trying to transform the agriculture of other developing countries. The new strains of wheat produced dwarf plants capable of withstanding strong winds, heavy rain and diseases (especially the 'rusts' which had attacked large areas). Yields of wheat and maize tripled and doubled respectively, and the new seeds were taken to the Indian sub-continent. Later, new varieties of improved rice were developed in the Philippines. The most famous, the IR-8 variety, increased yields sixfold at its first harvest. Another 'super rice' increased yields by a further 25 per cent (1994). Further improvements have shortened the growing season required, allowing an extra rice crop to be grown, and new strains have been developed that are tolerant of a less than optimum climate.

In 1964 many farmers in India were short of food, lacked a balanced diet and had an extremely low standard of living. The government, with limited resources, was faced with the choice (Figure 16.62) of attempting a land reform programme (redistributing land to landless farmers) or trying to improve farm technology. It opted for the latter. Some 18 000 tonnes of Mexican HYV (high-yielding varieties) wheat seeds and large amounts of fertiliser were imported. Tractors were introduced in the hope that they would replace water buffalo; communications were improved; and there was some land consolidation. The successes and failures of the Green Revolution in India are summarised in Figure 16.63. In general, it has improved food supplies in many parts of the country, but it has also created adverse social, environmental and political conditions. Certainly the biochemical and mechanical innovations (Figure 16.62) have been far more successful than the social changes as, in the early 1990s, rural areas were still over-populated, experienced out-migration and had a low standard of living. As one television commentator said: 'The poorer families can probably now afford two meals a day instead of one, while the rich can educate their children.'

Successes	Failures
Wheat and rice yields have doubled	HYV seeds need heavy application of fertiliser and pesticides, which has increased costs, encouraged weed growth and polluted water supplies
Often an extra crop per year	Extra irrigation is not always possible and can cause salinisation
Rice, wheat and maize have varied the diet	HYVs not suited to waterlogged soils
Dwarf plants can withstand heavy rain and wind	Farmers unable to afford tractors, seed and fertiliser have become relatively poorer
Farmers able to afford tractors, seed and fertiliser now have a higher standard of living	Farmers with less than 1 ha of land have usually become poorer
Farmers with more than 1 ha of land have usually become more wealthy	Farmers who have to borrow money to buy seed and fertiliser are likely to get into debt
The need for fertiliser has created new industries and local jobs	Still only a few tractors, partly due to cost and shortage of fuel
Some road improvements	Mechanisation has increased rural unemployment
Area under irrigation has increased	Some HYV crops are less palatable to eat
Some land consolidation	
Conclusions	
A production and economic success which has lessened but not eliminated the threat of food shortages	Social, environmental and political failure: bigger gap between rich and poor

Figure 16.63

An appraisal of the Green Revolution in the Indian sub-continent

Appropriate technology (Case Study 18) is needed to replace the many, often well-intentioned schemes that involved importing capital and technology from the more developed countries. Appropriate technology, often funded by non-governmental organisations such as the British-based Intermediate Technology Development Group (Places 90, page 577), seeks to develop small-scale, sustainable projects which are appropriate to the local climate and environment, and the wealth, skills and needs of local people. This means:

- *Not* large dams and irrigation schemes, but more wells so that people do not migrate to the few existing ones, drip irrigation as this wastes less water, stone lines (Figures 10.40 and 16.64) and check-dams (Figure 10.43). For stone lines, stones are laid down, following the contours, even on gentle slopes in Burkina Faso, while small dams built of loess are constructed across gulleys in northern China. In both cases, surface runoff is trapped giving water time to infiltrate into the soil and allowing silt to be deposited behind the barriers. These simple methods, taking up only 5 per cent of farmland, have increased crop yields by over 50 per cent.

- *Not* chemical fertiliser, but cheaper organic fertiliser from local animals (which can also provide meat and milk in the diet). Unfortunately, in many parts of Africa dung is needed as fuel instead of being returned to the fields.

- *Not* tractors, but simple, reliable, agricultural tools made, and maintained, locally.

- *Not* cash crops (often monoculture) on large estates, but smallholdings where both cash crops (income) and subsistence crops (food supply) can be grown. Mixed farming and crop rotation are less likely to cause soil erosion and exhaustion. Intercropping can protect crops and increase yields (smaller plants protected by tree crops).

Figure 16.64

Stone lines in Burkina Faso

Farming   A   Farming in western Normandy

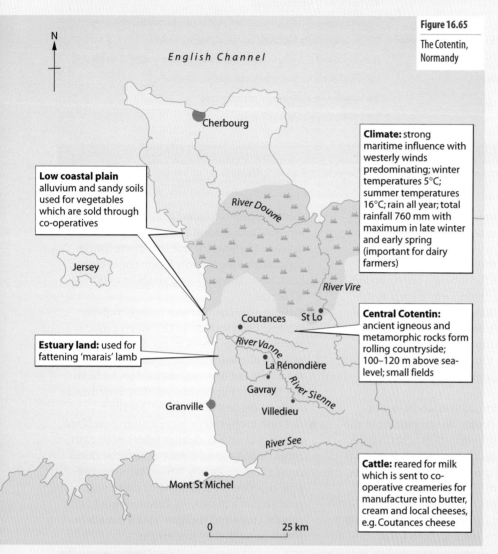

Figure 16.65

The Cotentin, Normandy

Low coastal plain alluvium and sandy soils used for vegetables which are sold through co-operatives

Climate: strong maritime influence with westerly winds predominating; winter temperatures 5°C; summer temperatures 16°C; rain all year; total rainfall 760 mm with maximum in late winter and early spring (important for dairy farmers)

Estuary land: used for fattening 'marais' lamb

Central Cotentin: ancient igneous and metamorphic rocks form rolling countryside; 100–120 m above sea-level; small fields

Cattle: reared for milk which is sent to co-operative creameries for manufacture into butter, cream and local cheeses, e.g. Coutances cheese

0   25 km

used for producing fodder for the animals. The present herd consists of 52 cattle – mainly Friesian, with some traditional Normandy cows. The black-and-white Friesians have high milk yields, but the Normandy cows have better-quality milk with a high cream content. They are kept outdoors all year round, with some protection in the winter. The cattle in milk are brought to the dairy twice a day and they produce on average 116 litres per cow per day (Figure 16.67). The small milking parlour is similar to many in the region. It holds eight cows at a time, and is simpler than large dairies in the English Midlands or on dairy farms close to Paris. The milk is kept under refrigeration on the farm until it is collected by the creamery lorry – each day in summer, but every two days at other times of the year (Figure 16.68).

The cows are artificially inseminated and produce one calf a year. Bull calves are sold in Gavray market for veal, and female calves are sold or used to replenish the herd. Carefully checked for yield and as this drops off they are replaced. They are kept as long as possible, as the return from cull cows is not high.

The present farmer has been on the farm for nearly 20 years, but it was farmed earlier by his parents and grandparents. All the work is done by the farmer, his wife (she is in charge of the dairy) and his father. Neighbours help during silage making. There is a strong tradition of dairy farming in the region.

On the western side of the Cotentin, there is a low-lying plain approximately 15–60 m above sea-level. It contains areas of sandy soils which are important for producing vegetables, including carrots, leeks, sweet corn, lettuce and tomatoes. These vegetables are marketed through co-operatives in the larger towns of the region, as well as in Paris and the UK.

The lowlands along the estuary of the Sienne and the Vanne are used as grazing land for the 'marais lamb'; large flocks of sheep are fattened on the marshes, providing yet another income for the farmers of the region.

The Cotentin lies between the Vire estuary and Mont St Michel Bay (Figure 16.65). It is mainly an agricultural region, although tourism is also important. The maritime climate, with rain (760 mm per year) occurring at all seasons and reaching a maximum in the late winter and spring months, is important for the farming. The maximum occurs just as temperatures are rising and the grass is starting to grow. This has been the basis of the successful dairy farming industry. Cattle are reared for their milk from which Normandy butter is made in addition to many local cheeses and cream. Most farms also produce fodder for their cattle, either in the form of silage in the late spring or as crops of corn in the late summer.

La Rénondière is a typical Cotentin dairy farm (Figure 16.66). It lies at 71 m above sea-level in a small valley whose stream flows into the River Vanne 0.75 km to the north. The land slopes very gently; fields are small and bounded by dense hedges; and most of the farm can be ploughed except for a small area in the valley bottom which becomes very wet. The Normandy-style farmhouse of grey stone covered in creeper, with white shutters, faces south. It is sheltered from the westerly winds, as are most of the buildings grouped around it.

The farm is 44 hectares in area. This is large for Normandy, where the average size is between 15 and 24 ha. Cattle are kept on 4 ha close to the farm; the rest of the land is

Figure 16.66

A typical Normandy farmhouse

Figure 16.67

A small milking parlour

Figure 16.68

A co-operative creamery in Normandy

In addition to their regular enterprises, many Normandy farmers breed and train trotting ponies – making regular visits to the long open sandy beaches to train them at low tide. As in Britain, bed and breakfast accommodation during the short tourist season from June to the end of August provides an additional source of income.

A major issue facing farmers in this part of France is the steady loss of people from the land. Many small farmers are going out of business, leaving houses empty. As in other peripheral regions of Europe, young people are moving to the cities. There is evidence that one or two wealthier large farmers are buying up vacant land. Some of the villages contain summer homes, owned by Parisians, with a number of British residents both in holiday and permanent homes. Prices for some houses without land have been low, encouraging overseas buyers. Villages still contain their bakery and shop, often with a butcher, but children are being forced to travel increasing distances to school. These features of rural life are common to many remoter areas within the European Union.

The impact of EU regulations can be seen. Milk quotas in line with EU rulings have been set by the government (page 493). They are generally higher than in the UK, perhaps due to the political strength of the farmers, and are an established part of the farm economy. However, they are generally unpopular with local farmers. This attitude was clearly conveyed by Mme Leboisellier at La Rénondière, who threw up her hands in a typical Gallic gesture when interviewed on the subject.

Subsidies for lamb encourage the producer to maintain flocks. Demand for lamb is high, as is shown by the high prices in the supermarkets.

The issue of 'set-aside' has become important, with farmers leaving more of their land uncropped as they gain additional subsidies from the EU (page 493).

B Farming in Gwent

Great Llwgy is an intensive sheep farm of 59.7 ha, with an additional 20ha rented. The farming year is organised around the flock of 450 breeding ewes and 100 ewe lambs (Figure 16.69). The main flock is now North Country Mule sheep cross-bred from Swaledale ewes and Blue-faced Leicesters. This produces high-quality lambs intended to be sold for meat. Originally there was a small herd of beef cattle grazing on the hill, but these were sold at the start of the BSE crisis and it is no longer economic to provide feed for these animals.

The availability of good-quality feed is basic to the farming system. The ewes must be in good condition when put to the rams in October and November. This dictates the number and quality of the following year's lamb crop. A very dry summer can reduce lamb numbers by 17–20 per cent. A high percentage of twin and triplet lambs is required. Sheep remain out on grass in the autumn as late as possible, but additional silage may be fed to them if the weather is poor. Housing the flock in winter not only protects them before and during the lambing season, but allows the land to recover from heavy use. The building has now been extended to house more animals. The impact of the animals on the wet ground can severely damage future grazing.

Lambing takes place in late March and April.

An important part of the feed is silage produced on the farm. Fertiliser is added to the grass in spring and early summer when the temperature rises and the grass begins to grow. The amount required has to be adjusted according to the number of sheep being carried on the land. Some 15 tonnes were used in 1994, but this will have to be increased if the number of sheep is raised. Fields that will produce silage are shut off early to protect the grass. This means better silage if it is produced by mid-May; the quality drops off rapidly after that time. Contractors are employed to cut and bale the silage in large black plastic bags, for storage until required. Hay is made by the farmer using his own baler and turner, when there is maximum daylight towards the end of June. Unreliable weather makes it difficult to predict dates for a contractor. It needs a week of warm dry weather.

High technology is used on the farm in January, when the ewe flock is scanned to check the number of ewes bearing triplets, twins or single lambs. These are separated out into groups and those with multiple lambs are fed additional concentrates, as silage does not provide sufficient energy for these ewes. This is important as it does not waste expensive extra supplement on ewes that do not require it.

The main output of the farm is lamb for meat; there is little return now from the wool clip. Shearing takes place twice a year. The pedigree sheep are sheared in December/January. This enables more sheep to be kept inside the shed. The ewes will eat more to keep warm and hence the lambs born in March and April will be larger. The general condition of the ewes is seen to be better. When lambing can be delayed until April, costs are reduced by lambing out in the fields. Hill ewes and ewe lambs are shorn in May and early June. Although approximately 2 kg wool is produced per animal, there is at present little market for it. Cull ewes (those that are kept as long as they can produce good-quality lambs) are now worth very little. Slaughter costs are £9 per animal, which must be paid by the farmer. Government regulations now insist that carcasses of animals over 12 months old being sold for human consumption must be split and the spinal cord removed. This is an additional cost and a precaution against the possible spread of BSE through the sheep disease called scrapie.

The flock has to be watched carefully throughout the year, and the lambs are wormed at three-weekly intervals from

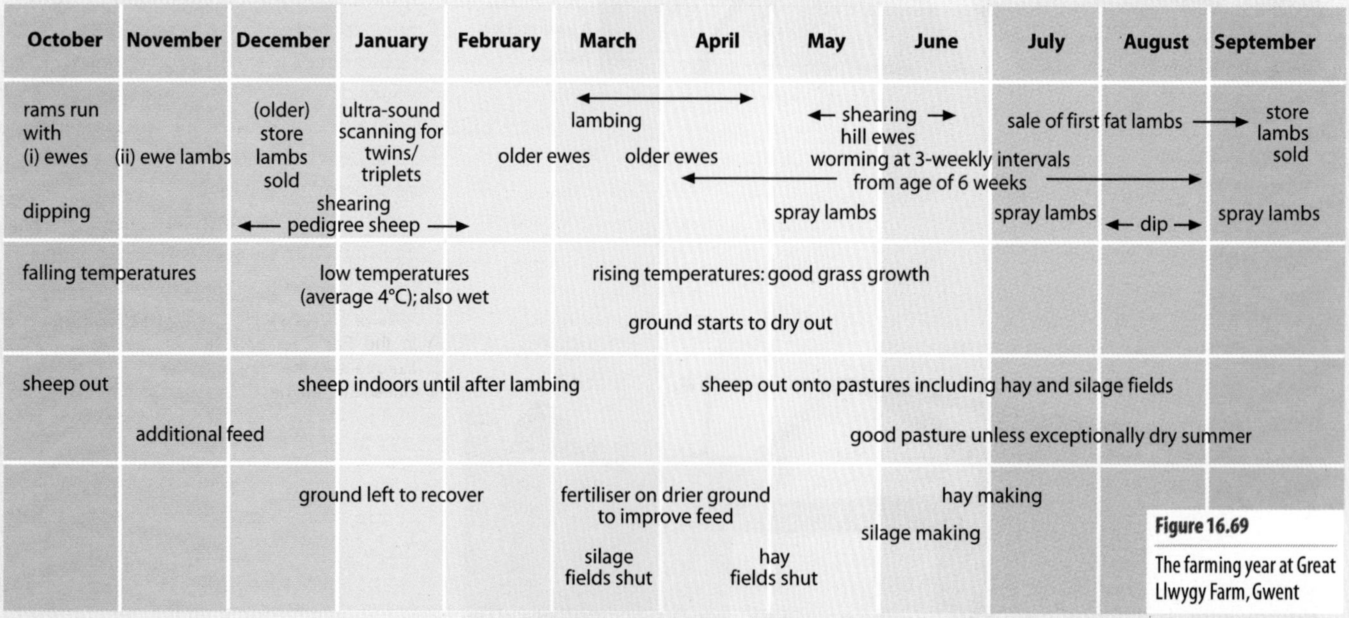

October	November	December	January	February	March	April	May	June	July	August	September
rams run with (i) ewes	(ii) ewe lambs	(older) store lambs sold	ultra-sound scanning for twins/ triplets		lambing		← shearing → hill ewes		sale of first fat lambs →		store lambs sold
					older ewes	older ewes	worming at 3-weekly intervals from age of 6 weeks				
dipping		shearing ← pedigree sheep →					spray lambs		spray lambs	← dip →	spray lambs
falling temperatures		low temperatures (average 4°C); also wet			rising temperatures: good grass growth						
					ground starts to dry out						
sheep out		sheep indoors until after lambing				sheep out onto pastures including hay and silage fields					
	additional feed						good pasture unless exceptionally dry summer				
		ground left to recover			fertiliser on drier ground to improve feed			hay making			
								silage making			
					silage fields shut	hay fields shut					

Figure 16.69

The farming year at Great Llwygy Farm, Gwent

May to August. Chemical sprays are used to prevent flies attacking the lambs in May, July and September. The main sheep dip is now carried out in August, with the minimum of help. Two students are employed for the lambing season, which usually coincides with the university spring break in late March and April. The farm is small in comparison with many extensive high moorland farms (Figure 16.70), but this enables Joe Binns to work it himself with the minimum of help. Contractors are used where it is uneconomic to keep expensive machinery on the farm.

Costs of additional feed, fertiliser, veterinary care, fuel, scanning, shearing and silage-making have to be set against the main output of lamb production. During recent years sheep farmers have been suffering from a general reduction in lamb prices, in 1999 averaging £25 per head as against £41 in 1998. Production costs including transport, fertiliser and foodstuffs continue to increase. Returns from the sale of lambs halved between 1997 and 1999, with hill lambs fetching on £10–£12 each. Hill farmers receive an EU subsidy of £17 per animals. Farmers with 2000 ewes on extensive upland production systems appear to be in a better position than small intensive farms on the margins of the uplands, as in south-east Wales.

Farmers may sell their lambs under contract to supermarkets, on average obtaining only £30 per animal. Increased costs of slaughter have to be borne by the farmer.

Joe Binns has been forced to diversify his farm activities. A cottage on the farm has been modernised and is let to visitors. He also provides an educational visit for students. The National Park planning authorities, who have some control over developments on farms within the boundaries of the Brecon Beacons National Park, have been sympathetic to changes on the farm, even allowing a telephone mast to be erected behind the farm (Figure 16.70), and this also provides income. In February 1999, 3 ha of land adjoining Strawberry Cottage Wood, a Site of Special Scientific Interest, was planted with broadleaved deciduous trees, with the aid of the National Park Challenge Fund; a further 10 ha is to be planted in early 2000. This means that 25 per cent of the productive land on the farm will be out of production for 15 years. This

has been done as a balance against the rapidly falling income from sheep; returns from planting trees appear to be better than farming sheep. Tracks and trails are planned for horses and mountain bikes in the woodland. A major problem in the early stages of agroforestry is the elimination of weeds and gorse. The present economic situation does not encourage an increase in the flock size, which would require additional grazing land and silage. Land on the farm that was formerly used by cattle is now available for sheep, and replaces some of the rougher grazing that has been used for tree production.

The present farming difficulties have been caused by a combination of circumstances.

1 The Russian economic collapse and financial problems in Asia have meant that there is no longer a market there for sheepskins and cattle hides.

2 BSE costs affect both cattle and sheep farmers. Strict regulations require that cattle over 30 months old may not be used for meat. Sheep have to be destroyed and the costs borne by the farmers. Abbatoir costs have increased and regulations and inspections are rigid. There are fewer abbatoirs working, which means additional transport costs for the farmers. Before the BSE crisis, surplus calves were sold to Europe for veal; this market is no longer available.

3 The strength of the pound ensures that prices for UK produce overseas are high. Farmers are at a disadvantage as foreign goods may be imported and sold cheaply. In 1998 farm incomes fell by 75 per cent; the average agricultural wage was just £8000.

4 The value of food produced has slumped 'at the farm gate', i.e. when it leaves the farm. It does not necessarily mean that it is cheaper for the consumer. Costs of production for pig meat (£20 per pig in 1999) and lowland lamb are high so that it is hard for a farmer to make a profit. Sheep farmers are shooting their animals rather than selling at a loss.

5 In a recent survey, 64 per cent of hill farmers were aged over 50 and 11 per cent were over 60.years old. Young people are reluctant to follow the

Figure 16.70

Great Llwygy Farm, Gwent: the shed has been extended, and a telephone mast erected behind the farm

family tradition of entering farming, because of the low level of income.

6 Associated industries such as machinery manufacturers have been forced to lay off staff. The effect on the rural economy is felt in village shops, schools and public transport.

In September 1999 the government announced an aid package of £150 million for the farming industry. This included further subsidies for hill farmers, a reduction in abbatoir costs, and easing of veterinary regulations. There was still no assistance with the disposal of 500 000 surplus bull calves.

Direct sales of farm produce to customers in urban areas is increasing through a system of 'Farmers' markets'. These are similar to markets found on the fringes of large US cities such as Washington. A non-profit making farmers co-operative started in south-east Wales organises the Farmers Ferry, which ships 40 000 lambs each week from Ramsgate to the Continent. This is helping to steady the fall in lamb prices paid to the farmers.

Figure 16.71

The UK farming situation in 1999

CRISES are nothing new in British farming. The current one, however, is of a different order to anything seen since the war. This morning, we report how two Exmoor farmers yesterday had 70 of their ewes shot and buried in mass graves. Two years ago, the ewes might have fetched more than £20 apiece. This year they are worth almost nothing. For the farmers, even transporting their animals to market would have been uneconomical.

Over the past year, the effects of financial collapse in the Far East and Russia have hurt farmers in Europe and the United States, as well as here. But British farmers have also been hard hit by the strength of the pound against the Euro, which has boosted imports across the board. At the same time, they have had to cope with a series of other problems and disadvantages, unique to this country and often the direct result of government action. As a result, farmgate prices have fallen in all sectors – milk by 17 per cent over the past two years, fruit by 25 per cent and pigs by 31 per cent – while costs have risen. Coming, as all this does, hard on the heels of the BSE fiasco, an increasing number of farms are unlikely to survive.

Daily Telegraph Editorial, 16 September 1999

C Banana cultivation

World trade in bananas is dominated by two groups of producers. These are the ACP (African, Caribbean and Pacific) producers and the 'dollar producers', of the Central American republics and Colombia. Production in the latter group is controlled by large American transnationals. Eight per cent of the bananas entering the European market originate from the Caribbean and are produced on small family-owned farms, by people who rely on the income for their livelihood. Until mid-1998 these producers had a guaranteed market in the European Union. Bananas produced on large plantations in the Ivory Coast and Cameroon are also within the ACP countries. Each country had its own quota based on the amount it exported. In June 1998 the EU imposed a single block quota on all ACP countries, as the plantation producers of West Africa were at an advantage. This could have a damaging effect on the economies of the smaller producers in the West Indies.

The EU also decided that there should be discrimination against Latin American bananas distributed by American transnational companies. To counter this, the USA threatened to impose duties on European products entering the USA.

Bananas are cultivated under tropical conditions where the temperatures are high and rainfall exceeds 120 mm per month. In some tropical plantation conditions where evapotranspiration is high, irrigation may be used . Drip irrigation is more effective and produces a better bunch weight of bananas than basin irrigation. In order to meet the demands of the marketing companies, the bunches (or hands) of bananas must be over 270 g in weight. Bananas grown for local consumption are mainly cultivated on small landholdings, whilst those produced for export are grown on large plantations (Figure 16.74a). In most Caribbean countries, bananas are grown on small family-run plots. The crop requires a high labour input, which in the Caribbean islands is mainly provided by the smallholder's family. Suckers taken from a mother plant are rooted and grow well in the deep volcanic soils. Weeds growing between the plants need to be kept down until the plant is tall enough to outgrow them. It is common to see plants being supported by props so that the weight of the bunch does not pull the plant over. Fruit has to be protected from bruising and scarring. Each bunch may be covered by a large plastic bag until it is ready for harvest. This takes place about 10 months after the plant is established. The fruit is cut when it is still green and hard, and then it is taken to the processing plant. Here it is packed and refrigerated before being sold or shipped overseas (Figure 16.74b).

On the Caribbean islands marketing is done through transnationals such as Fyffes. The small farmers rely on the banana industry to provide their basic needs of food, shelter and education. The effect of the EU new quota regulations would be to limit the returns to these smallholders. These small farmers are also the ones who suffer the most from hurricane damage, as for example when the Jamaican crop was destroyed by Hurricane Gilbert in 1989. In October 1998 Hurricane Mitch destroyed much of the plantation area of the eastern lowlands of Nicaragua and Honduras.

The influence of the large transnational companies is strong in the Central American countries where the bananas are grown on the rich alluvial soils found on the coastal lowlands, providing high yields per hectare for large plantations owned by transnationals. Labour is hired and often low-paid. Land is carefully cultivated and more mechanisation is used than on smaller farms. There is intensive use of fertilisers and pesticides which is having cumulative environmental effects. One of the most serious of these is the damage to the coral reefs off the Costa Rican coast, where 90 per cent are now dead as a result of pesticide runoff from banana plantations.

Following years of expansion because of increased demand for the fruit, there is now a problem of oversupply. Economies such as those of St Vincent and St Lucia depend on the crop for survival. There is a need to diversify into food crops and other cash crops to reduce the dependency upon one major export.

Country	Production 1998 (tonnes)	Export 1998 (tonnes)	% total exported	Value (£ '000)
Ecuador	4 563 442	3 989 338	87	668 828
Colombia	2 200 000	1 508 487	68	291 938
Costa Rica	2 101 449	1 800 000	86	367 518
Mexico	1 647 000	244 992	15	45 302
Honduras	947 000	467 281	49	72 358
Guatemala	880 000	794 240	90	119 607
Panama	650 000	462 415	71	86 717
Dominican Republic	359 016	65 391	18	8 391
Jamaica	130 000	62 000	48	22 500
St Lucia	90 000	53 727	60	20 310
Belize	81 000	53 000	65	13 125
Nicaragua	68 830	16 912	25	6 975
St Vincent/Grenadines	43 000	38 890	90	12 841
Dominica	30 000	28 135	93	8 563
Trinidad & Tobago	6 000	93	1.5	18.125
Grenada	4 400	94	2.1	18.75
World	56 404 895	13 714 586	24	4 933 711

Source: FAOSTAT

Figure 16.72

Banana production in the Caribbean and Central America

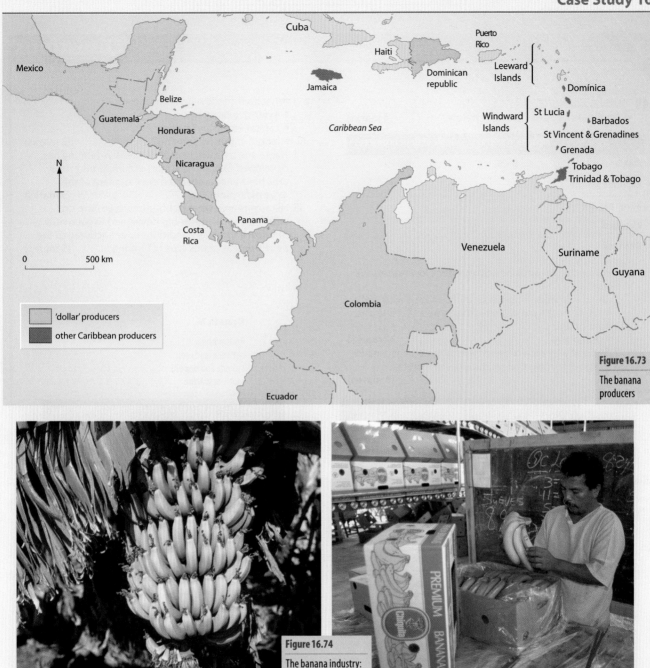

Figure 16.73

The banana producers

Legend:
- 'dollar' producers
- other Caribbean producers

0   500 km

Figure 16.74

The banana industry:
a Bananas on the tree
b Preparing for export

References

Barke, M. and O'Hare, G. (1991) *The Third World*, Oliver & Boyd.

Carr, M. (1997) *New Patterns: Process and Change in Human Geography*, Thomas Nelson.

Philip's Geographical Digest 1998–1999 (1998) Heinemann-Philips.

Timberlake, L. (1987) *Only One Earth*, Earthscan/BBC Books.

Waugh, D. (1998) *The New Wider World*, Thomas Nelson.

Websites

UN Food and Agriculture Organisation:
http://www.fao.org/

UK Ministry of Agriculture, Fisheries and Food (MAFF):
http://www.maff.gov.uk/

US Department of Agriculture (USDA):
http://www.usda.gov/usda.htm

Union of Concerned Scientists (UCS):
http://www.ucsusa.org/warming/index.html

UN World Food Programme (WFP):
http://www.wfp.org/index.htm

World Resources Institute: Feeding the World page:
http://www.igc.org/wri/wri/wri/wr-98-99/feeding.htm

See also for more links:
http://www.nelsonthornes.com/gaia

1 Study Figure 16.75.

   a Complete a copy of the table below. **(4 marks)**

	Altitude in metres	Angle of slope in degrees
Arable	0–20	0–3
Improved pasture		
Rough pasture		
Woodland		
Moorland		

   b Moorland and woodland both produce low returns for farmers.

     i Using information from your table, suggest what is the main physical type of land in this sample that is left as:
- moorland
- woodland. **(2 marks)**

     ii Suggest why each of these types of land is not used for a type of farming that produces better returns. **(6 marks)**

   c i What do the following terms mean:
- extensive farming
- capital-intensive farming
- labour-intensive farming? **(3 marks)**

     ii Name one area where capital-intensive farming has developed. Explain how market conditions in that area have encouraged the development of this type of farming. **(5 marks)**

     iii Name **one** area where labour-intensive farming with low capital inputs has developed. Explain how physical and social conditions have encouraged the development of this type of farming. **(5 marks)**

Figure 16.75

Relationships between land use, altitude and slope in south-east Arran

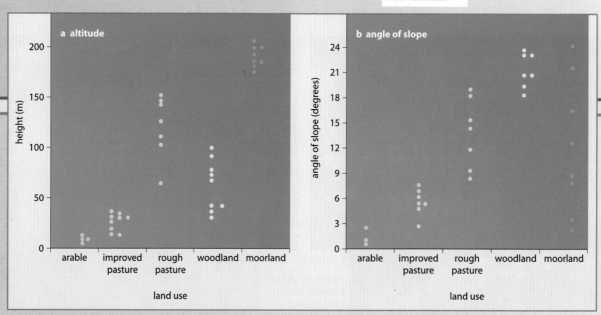

2 a Study the data in Figure 16.75.

     i Which type of farming in this area is strongly influenced by altitude but not strongly influenced by angle of slope? Justify your answer. **(3 marks)**

     ii Describe the type of land where woodland is most likely to be found, and explain why it is found there. **(8 marks)**

   b Name a country or region where variations in physical geography have led to the development of distinct agricultural regions.

     i Describe the location of **four** distinct agricultural regions. **(4 marks)**

     ii Explain why the pattern described in i has developed. **(10 marks)**

3 Two of the biggest causes of problems of food supply in less economically developed countries (LEDCs) are:
- the need for land reform
- the need for access to improved technology.

   i Explain why each of these presents problems for farmers in LEDCs. Refer to one or more examples that you have studied. **(9 marks)**

   ii Describe a scheme to improve land tenure in a named LEDC, and assess how successful that scheme has been. **(8 marks)**

   iii Describe a scheme to improve the level of technology available to farmers in a named LEDC, and assess how successful that scheme has been. **(8 marks)**

4 a Physical controls have an important effect on the type of farming in most agricultural areas. Choose **two** of the following physical factors. For each of your chosen factors, explain how it influences farming. Illustrate each part of your answer with reference to a named area.
  i temperature
  ii precipitation
  iii soil. **(8 marks)**
b The use of technology can reduce the farmer's dependence on physical factors. Explain how this has happened in:
  i a named farming region in a more economically developed country **(5 marks)**
  ii a less economically developed country where intermediate technology has been used. **(5 marks)**
c Figure 16.76 shows how the type of farming varies around a 'typical' isolated town. Physical conditions are almost completely uniform throughout this area. Suggest how distance from the market has influenced the type of farming in this area. **(7 marks)**

5 Study the map in Figure 16.77. It shows the general pattern of intensity of farming in Europe.
a i Describe the location of the areas where average intensity of farming is 75 per cent of the average, or lower. **(2 marks)**
  ii Choose a named location within the area described in **i** and explain why physical geography makes farming difficult in that area. **(3 marks)**
b i Describe the location of the area with average intensity 50 per cent or more above average. **(2 marks)**
  ii Explain how market forces have affected the development of the area of intensive farming you have described in **i**. **(4 marks)**
c Name **one** area of intensive farming that is found within the peripheral area of Europe.
  i Describe the type of farming.
  ii Explain why this area of intensive farming has developed there. **(7 marks)**
d Name **one** area of low-intensity farming found within the farming core.
  i Describe the type of farming.
  ii Explain why this area of low-intensity farming has developed, despite the favourable market conditions. **(7 marks)**

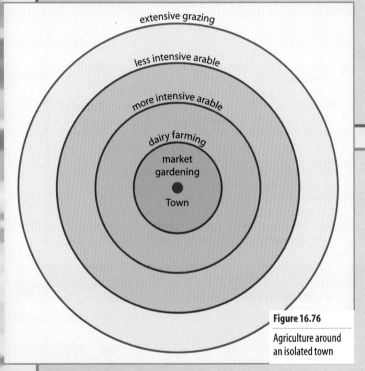

Figure 16.76

Agriculture around an isolated town

6 Study the map in Figure 16.77. It shows the general pattern of intensity of farming in Europe.
a Explain how market forces have influenced the development of this pattern. **(9 marks)**
b Explain why intensive farming can take place in parts of the peripheral region. Refer to one or more case studies. **(8 marks)**
c There are areas of less intensive farming, and even abandoned land, near many towns in the core. Explain why this is so, with reference to one or more case studies. **(8 marks)**

100 = average intensity

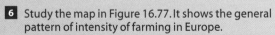

0   1000 km

Figure 16.77

Intensity of agriculture in Europe (*after* van Valkenburg and Held, 1952)

7 a Modern farming practices threaten the environment in many ways. Describe **one** problem that can result from each of the following practices:

   i increasing use of chemicals on the land   **(4 marks)**

   ii increasing the size of fields   **(4 marks)**

   iii draining wetlands.   **(4 marks)**

  b Choose **one** of the problems that you described in a. Explain how changes in the management of the land can reduce this problem.   **(6 marks)**

  c 'I would like to manage my farm in a more eco-friendly way, but I feel that I must farm as intensively as modern scientific techniques will allow. Farmers like me must produce maximum possible yields in order to feed the starving millions in poor countries throughout the world.'

   Imagine that a farmer who ran a very intensive farm in East Anglia made the statement above. How might you reply if you wanted to convince him that he ought to consider a less intensive form of farming?   **(7 marks)**

8 a Name a less economically developed country (LEDC) which has suffered / is suffering from famine. Explain the causes of the famine. You should refer to both natural and human causes.   **(10 marks)**

  b 'Famine and food shortage are likely to increase in future.' Give two reasons why this is likely.   **(5 marks)**

  c i With reference to one or more named case studies, explain how land reform can improve total food production in LEDCs.   **(5 marks)**

   ii With reference to one or more case studies, explain how appropriate technology (intermediate technology) can help increase agricultural yields in LEDCs.   **(5 marks)**

AS

9 a On a copy of Figure 16.78 add:

   i net profit curves for dairying and wheat when locational rent for:

    • dairying is £120 at the market and £0 at 60 km

    • wheat is £80 at the market and £0 at 80 km.   **(6 marks)**

   ii labels to show:

    • the margin of transference from market gardening to dairying

    • the margin of transference from dairying to wheat

    • the margin of cultivation for wheat.   **(3 marks)**

  b Explain why land use changes at the margins of transference.   **(4 marks)**

  c i Explain why von Thünen's model is difficult to apply to agricultural patterns in the modern world.

   ii In what ways is von Thünen's model still useful to an understanding of modern agricultural geography?   **(12 marks)**

10 Study the figures in Figure 16.79.

  a Explain how the Common Agricultural Policy of the European Community (now the European Union) led to the development of surpluses like those shown on the table.   **(8 marks)**

  b Explain how these surpluses have been reduced during the period since 1986.   **(8 marks)**

  c Increasing intensification of farming in the UK and other parts of the European Union has damaged the environment in several ways. Choose one way in which intensification has damaged the environment and:

   i explain how intensification has caused the damage

   ii explain how changes in management could reduce the damage caused by intensification.   **(9 marks)**

Commodity	January 1986	January 1992
	(figures in thousand tonnes unless otherwise stated)	
Butter	1400	300
Skimmed milk powder	800	0
Beef	500	800
Cereals	15 000	7000
Wine/alcohol	4 000 (hectolitres)	2 500 (hectolitres)

Figure 16.79

EU food surpluses

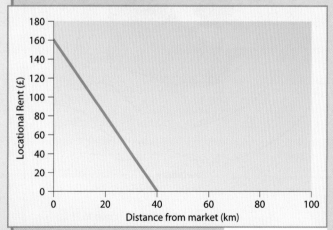

Figure 16.78

Net profit curve for market gardening around a town on a uniform plain

11 'As farming becomes more modernised the influence of economic factors increases while the influence of physical factors decreases.'

Discuss this statement with reference to farming in regions at varying levels of development.   **(25 marks)**

A2

12 Study the photographs Figure 16.80. They were both taken near Guilin in China.

a i Describe evidence in photograph B which shows that farming is intensive in this area. **(2 marks)**

ii Two crops per year can be taken from farmland in photograph A. Suggest how the land is kept fertile, even though the people cannot afford inputs of artificial fertiliser. **(2 marks)**

iii Land in the background of photograph A is not farmed. Suggest why not. **(2 marks)**

b • Before the revolution in 1949, farming in this part of China was mostly subsistence farming. Farms were small and fragmented and tenants had to give up to half their produce to absentee landlords.

• After the revolution, land was divided amongst the peasants, but most plots were too small to support the families which worked them.

• After several experiments the government created 'people's communes' in which around 15 000 people pooled their land and labour to run the farm.

• Since 1979 individual farmers have been given more responsibility, and now they are allowed to sell surplus crops at local markets, and to keep the profits.

Suggest why yields are higher under the present system than they have been under any of the previous systems. **(9 marks)**

c Changes in the Common Agricultural Policy (CAP) of the European Union have caused big changes in the way farmers operate in Europe during the last 45 years. Choose **two** of the following policies:

• payment of subsidies
• the quota system
• set-aside
• encouragement of environmentally friendly farming.

For **each** of your chosen policies:

i explain how it works
ii explain how it has affected farming practices.

(5 marks + 5 marks)

A

B

Figure 16.80

Agriculture in the Li Valley, near Guilin

13 Study the photographs in Figure 16.80. They were both taken near Guilin in China.

a Describe the evidence which shows that farming in this area is intensive. **(5 marks)**

b i Describe how farming organisation has changed in this area since the introduction of the 'Responsibility System' in 1979. **(5 marks)**

ii Explain why these changes have led to increased productivity from China's farms **(5 marks)**

c Name a less economically developed country where the land tenure system discourages farmers from developing their land in the best way possible. Explain why the land tenure system causes these problems, and suggest how changes might improve productivity. **(10 marks)**

14 'Modern agri-business is not a sustainable form of farming.'

Discuss this statement using the following headings:

• The nature of modern agri-business
• Is modern agri-business sustainable?
• Can agri-business be made less damaging to the environment? **(25 marks)**

A2

Chapter 17

Rural land use

'Nor rural sights alone, but rural sounds, exhilarate the spirit'

William Cowper

'I see the rural virtues leave the land'

Oliver Goldsmith

The term **rural** refers to those less densely populated parts of a country which are recognised by their visual 'countryside' components. Areas defined by this perception will depend upon whether attention is directed to economic criteria (a high dependence upon agriculture for income), social and demographic factors (the 'rural way of life' and low population density) or spatial criteria (remoteness from urban centres). Usually it is impossible to give a single, clear definition of rural areas as, in reality, they often merge into urban centres (the rural–urban fringe) and differ between countries. Although generalisations may lead to over-simplifications (Framework 11, page 347), it is useful to identify three main types of rural area.

1 Where there is relatively little demand for land, certain rural activities can be carried out on an extensive scale, e.g. arable farming in the Canadian Prairies and forestry on the Canadian Shield.

2 In many areas, especially in economically developing countries, there is considerable pressure upon the land which results in its intensive use. Where human competition for land use becomes too great to sustain everyone, the area is said to be overpopulated (page 376). This often leads to rural depopulation, e.g. the movement to urban centres in Latin American countries (page 366).

3 In many economically developed countries, competition for land is greater in urban than in rural areas. The resultant high land values and declining quality of life are leading to a repopulation of the countryside (urban depopulation), e.g. migration out of New York and London (page 365).

The urban–rural continuum

It is now unusual to find a clear distinction between where urban settlements and land use end and rural settlements and land use begin. Instead, there is usually a gradual gradation showing a decrease in urban characteristics with increasing distance from the city centre Figure 17.1). This is known as the **urban–rural continuum** (page 393).

Figure 17.1

The urban–rural continuum

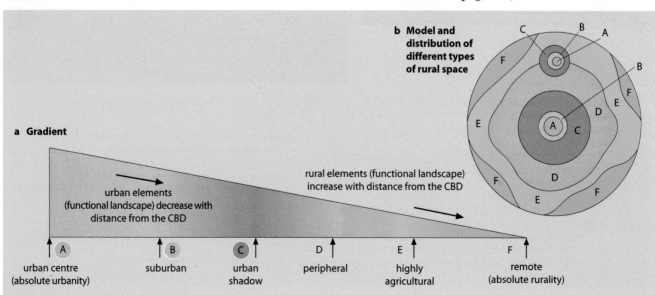

b **Model and distribution of different types of rural space**

a **Gradient**

urban elements (functional landscape) decrease with distance from the CBD

rural elements (functional landscape) increase with distance from the CBD

A — urban centre (absolute urbanity)
B — suburban
C — urban shadow
D — peripheral
E — highly agricultural
F — remote (absolute rurality)

The urban–rural continuum includes the rate at which rural settlements expand or decrease as people move out of or into nearby cities; changes in the socio-economic base as services and other functions are transferred to the countryside; and changes in land use resulting from increased pressure exerted on rural areas by nearby urban areas.

Figure 17.3

Rurality in England and Wales

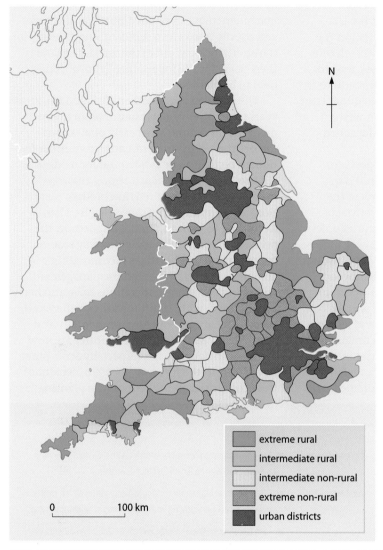

extreme rural
intermediate rural
intermediate non-rural
extreme non-rural
urban districts

0   100 km

Figure 17.2

An index of rurality for England and Wales (*after* Cloke, 1977)

Indices	Characteristics in rural areas
Population per ha	Low
% change in population	Decrease
% total population: over 65 years	High
% total population: male 15–45 years	Low
% total population: female 15–45 years	Low
Occupancy rate: % population at 1.5 per room	Low
Households per dwelling	Low
% households with exclusive use of (a) hot water (b) fixed bath (c) inside WC	High
% in socio-economic groups: 13/14 farmers	High
% in socio-economic group: 15 farmworkers	High
% residents in employment working outside the rural district	Low
% population resident < 5 years	Low
% population moved out in last year	Low
% in-/out-migrants	Low
Distance from nearest urban centre of 50 000	High
Distance from nearest urban centre of 100 000	High
Distance from nearest urban centre of 200 000	High

There are a number of measures of the intensity of change over distance, of which the best known is Cloke's **index of rurality** (Figure 17.2). The index is obtained by combining a range of socio-economic measures or variables, with absolute urbanity at one extreme and absolute rurality at the other. Using his index of rurality, Cloke then produced a map with a five-fold classification to show rurality in England and Wales (Figure 17.3).

Figure 17.4 shows some of the major competitors for land in a rural area. In many parts of the world, farming takes up the majority of the land and, especially in developing countries, employs most of the population.

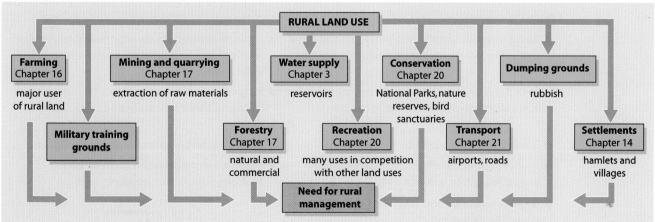

RURAL LAND USE

Farming
Chapter 16
major user of rural land

Mining and quarrying
Chapter 17
extraction of raw materials

Water supply
Chapter 3
reservoirs

Conservation
Chapter 20
National Parks, nature reserves, bird sanctuaries

Dumping grounds
rubbish

Military training grounds

Forestry
Chapter 17
natural and commercial

Recreation
Chapter 20
many uses in competition with other land uses

Transport
Chapter 21
airports, roads

Settlements
Chapter 14
hamlets and villages

Need for rural management

Rural land use

Figure 17.4

Competition for rural land use and the need for management of rural resources

Forestry

In Britain

Neolithic farmers began the clearance of Britain's primeval forests about 3000 years ago. Aided by the development of axes, some clearances may have been on a scale not dissimilar to that in parts of the tropical rainforests of today. In 1919, with less than 4 per cent of the UK covered in trees, the Forestry Commission was set up to begin a controlled replanting scheme. Since then the policy has been to look towards an economic profit over the long term and to try to protect the environment. The area under trees is increasing, as in other countries in western Europe and North America, and had reached 10 per cent in the early 1990s.

Commercial forestry is easier to operate in Britain, Scandinavia and Canada than it is in developing countries as they are nearer to the major world markets, although it is not viable economically in most developed countries unless subsidised by the state or supported by tax incentives for private investment (page 493). Softwoods are in greater demand than tropical hardwoods and, regardless of whether they are grown naturally or have been replanted, as there are few species within a large area, selection and felling is made easy. Afforestation in northern Britain begins by growing young trees in nurseries. These, on reaching a height of 50 cm, are planted in rows on land which has been cleared and drained. The young trees are regularly sprayed with fertiliser and pesticide. The lower and dead branches are removed, while the poorer-quality trees are felled. This carefully controlled process of thinning gives more light and room for the healthier trees to grow – although overthinning can increase the risk of wind damage. The foresters fell trees with modern lightweight chainsaws and use machinery that can remove branches and bark, and cut the trunks into standard lengths ready for transport, by lorry, out of the forest. The land is then cleared ready for another 'crop' of young trees to be planted.

The softwoods of northern Britain take 40–60 years to mature, so afforestation is an investment for the future. Most early plantations (before the 1980s) were not attractive either for human recreation or as a habitat for wildlife (Figure 17.5). Since then, a strong environmental lobby has ensured that modern plantations are carefully landscaped (Figure 17.6). According to Warren, 'In the last decade, the primacy of economics has given way to a broader range of objectives incorporating amenity, conservation, wildlife management and recreation. Forest operations have also been transformed, with a shift towards smaller-scale practices that are more environmentally and aesthetically sensitive. Monocultures are now avoided, antique species (especially Scots pine) have come to the fore and, by 1996, deciduous trees comprised 40 per cent of new planting.'

Figure 17.5

The case for and against forestry in Britain (*after* Warren, *Geography Review*, March 1998)

Advantages	Disadvantages
Socio-economic	**Landscape**
■ National timber needs – the UK supplies only 13% of its own timber and has a large annual import bill for wood products.	■ Early plantations were visually intrusive with their rigid geometric patterns, and with no regard for natural features.
■ Provides employment, especially as located in those rural areas where jobs are in short supply.	■ Often a 'blanket afforestation', using just one species of tree, created a monoculture (Case Study 10) with a uniformity of height and colour.
■ A positive method of using set-aside land.	■ They transformed the landscape and obliterated views. Concern over the speed and scale of replanting.
Non-market/environmental	**Environmental**
■ Trees are a renewable resource if carefully managed and, by planting in the UK, reduces pressures on tropical forests (sustainable development).	■ Introduction of non-native species, such as the North American Sitka spruce and lodgepole pine, as they were faster-growing than indigenous species.
■ Trees replace oxygen in the atmosphere and so help counterbalance the increase of carbon dioxide and its effects on global warming.	■ Destruction of valued environments such as the Flow Country wetlands of Caithness and Sutherland, and moorlands elsewhere in upland Britain.
■ Forests reduce water runoff (page 63).	■ Adverse impacts on flora and fauna, e.g. moorland birds and plants.
■ Forests contribute to biodiversity, providing habitats for a range of fauna and flora, e.g. red deer and red squirrels.	■ Concerns over water quality as afforestation led to increased acidification of lakes and rivers, and disrupted runoff.
■ Forests offer opportunities for recreation, and trees make an aesthetic contribution to the countryside.	
■ Some people argue that forests are part of Britain's traditional landscape.	

summits left clear for heather moorlands which provide a habitat for grouse and golden eagles

mature woodland forms a habitat for tawny owls and provides food for short-eared owls

trees planted at different times as differences in height are scenically more attractive

a variety of species and a lower density of trees replanted: helps to encourage more bird life which feeds on insects and so reduces the need for pesticide spraying

only small areas cleared at one time to reduce 'scars'

grassland provides a habitat for short-eared owls and food for tawny owls

winding forest road

land beside roads/tracks cleared to a width of 100 m and left as grass or planted with attractive deciduous trees

ponds created

cleared forest: branches left to rot; it takes 10 years for the nutrients to be returned to the soil

land next to river left clear for migrating animals such as deer

Figure 17.6

Managing an upland British forest (Kielder) in the 1990s

In developing countries

Commercial forestry is a relatively new venture in the tropics. It is usually controlled by trans-nationals based overseas which look for an immediate economic profit and have little thought for the long-term future or the environment. The forested areas in Latin America, tropical Africa and South-east Asia are rapidly decreasing. Figures giving the rate of forest clearances are diverse and unreliable, but the UN suggests that for every 29 ha deforested in Africa, only 1 ha is replanted (the ratio is 11:1 in tropical Latin America). The underlying causes of deforestation in developing countries are varied. Key issues, according to the World Wide Fund for Nature, include unsustainable levels of consumption; the effects of national debt; pressure for increased trade and development; poverty; patterns of land ownership; and growing populations and social relationships. It is also usual to blame forest destruction on the poor farmers of these countries rather than upon the resource-consuming developed countries.

During commercial operations the forest is totally cleared by chainsaw, bulldozer and fire: there is no selection of trees to be felled. The secondary succession (page 318) is of poorer-quality trees, as little restocking is undertaken. Where afforestation of hardwoods does take place, there is often insufficient money for fertiliser and pesticide. The hope for the future may lie in **agroforestry**, where trees and food crops are grown alongside each other. Forest soils, normally rated unsuitable for crops, can be improved by growing leguminous tree species. Commercial forestry is more difficult to operate in developing countries as they are distant from world markets, the demand for hardwood is less than for softwood, and although there are several hundred species in a small area only a few are of economic value.

The threat of the destruction of the rain-forests has become a major global concern. Some of the consequences of deforestation are described in Places 76 and Case Study 11.

Ethiopia, Amazonia and Malaysia: forestry in developing countries

Ethiopia

Early in the 20th century, 40 per cent of Ethiopia was forested. Today the figure is 13 per cent. In 1901, a traveller described part of Ethiopia as being 'most fertile and in the heights of commercial prosperity with the whole of the valleys and lower slopes of the mountains one vast grain field. The neighbouring mountains are still well-wooded. The numerous springs and small rivers give ample water for domestic and irrigation purposes, and the water meadows produce an inexhaustible supply of good grass the whole year.' In 1985, the same area was described as 'a vast barren plain with eddies of spiralling dust that was once topsoil. The mountains were bare of vegetation and the river courses dry.' As the trees and bushes were cleared, less rainfall was intercepted and surface runoff increased, resulting in less water for the soil, animals and plants.

Amazonia

The clearance of the rainforests means a loss of habitat to many Indian tribes, birds, insects, reptiles and animals. Over half of our drugs, including one from a species of periwinkle which is used to treat leukaemia in children, come from this region. It is possible that we are clearing away a possible cure for AIDS and other as yet incurable diseases. (Despite the rainforests being the world's richest repository of medical plants, only 2 per cent have so far been studied for potential health properties.)

Without tree cover, the fragile soils are rapidly leached of their minerals, making them useless for crops and vulnerable to erosion (Figure 12.8). The Amazon forest supplies one-third of the world's oxygen and stores one-quarter of the world's fresh water – both would be lost if the region was totally deforested. The burning of the forest not only reduces the amount of oxygen given off, but increases the release of carbon dioxide (a contributory cause of global warming). It has also been suggested that the decrease in evapotranspiration, and subsequently rainfall, caused by deforestation could also have serious climatic repercussions – could the Amazon basin become another Ethiopia?

Malaysia: a model for the future?

Malaysia has several thousand species of tree, mainly hardwoods, with timber and logs being the country's third-largest export. However, the government has imposed strict controls, and the Forestry Department 'manages the nation's forest areas to ensure sufficient supply of wood and other forest produce and manages and implements forest activities that would help to sustain and increase the productivity of the forest' (*Malaysia Official Year Book*, 1998). The Department insists that trees reach a specific height, age and girth before they can be felled (Figures 17.7 and 17.8). Logging companies are given contracts only on agreement that they will replant the same number of trees as they remove. Many newly planted hardwoods are ready for harvesting within 20–25 years due to the favourable local growing conditions. Further experiments are being made with acacias and rattan, both of which grow even faster. Consequently, half of Malaysia is still forested and as most of the remaining third is under tree crops such as rubber, oil palm and coca (Places 68, page 483) stocks are being successfully maintained. Even so, Malaysia's rapid industrialisation (Places 91, page 578) is causing increased deforestation, especially around the capital of Kuala Lumpur.

Figure 17.7

Logging operations in Malaysia

Figure 17.8

Timber lorries in the Malaysian rainforest

From September 1997 to June 1998, much of South-east Asia was blanketed by a thick smoke haze, in reality smog, caused by thousands of uncontrolled forest fires, mainly in Sumatra and Borneo (Figure 17.9). At its peak the smoke haze covered an area the size of western Europe and caused visibility to be reduced to 50 m. Its effects were various:

- **Human** The Air Pollution Index on Sarawak reached 851 (300 is considered 'hazardous' for human life), children and high-risk groups already suffering from respiratory or cardio-vascular diseases (Case Study 22) were prone to major health problems, and schools on Sumatra were closed.

- **Economic** Airports throughout the region were closed (an airline crash in Sumatra and a ship collision in the Strait of Malacca were both attributed to the haze), logging operations were suspended and farm crops destroyed.

- **Environmental** An estimated 90 per cent of canopy trees were lost in Sumatra and Borneo, and the rate of secondary succession would be slow; soils were seriously degraded; and wildlife habitats were lost (including those for such endangered species as the orang-utan, Sumatran rhinoceros and Sumatran tiger, and an irreparable loss in biodiversity).

Many Indonesians, accustomed to the humid climate and with little experience of dry weather, still adhere to fire-using traditions. Fire has long been used as a quick and cheap method of land clearance by farmers, and by plantation and forestry-concession owners. In 1997 the monsoon rains failed and the resultant prolonged drought, believed to have been triggered by the El Niño event (Case Study 9A), together with the prevailing land use and land management conditions, proved ideal conditions for the spread of forest fires on an unprecedented scale. The remoteness of the fires and the lack of resources, organisation and expertise combined to make fire-control impossible. Satellite imagery suggested that, although the blame for most of the fires was apportioned to the many small farmers, 80 per cent of the fires were due to large companies. By the time the rains did come, in May 1998, 10 million ha of forest had been burnt. Lessons were not learned, however. The fires and smoke haze returned in 1998, when another 5 million ha were lost, and again in 1999 – two years when there was *no* El Niño!

Figure 17.9

Maximum extent of smoke haze, late 1997

Mining and quarrying

Even since the Neolithic (when flint was excavated from chalk pits), Bronze and Iron Ages, quarrying and mining have been an integral part of civilisation. It was through the extraction and processing of minerals that many of today's 'developed' countries first became industrialised, while to some 'developing' countries the export of their mineral wealth provides the only hope of raising their standard of living. The modern world depends upon 80 major minerals, of which 18 are in relatively short supply, including lead, sulphur, tin, tungsten and zinc.

Minerals are a finite, non-renewable resource which means that, although no essential mineral is expected to run out in the immediate future, their reserves are continually in decline. **Resources** are the total amount of a mineral in the Earth's crust. The quantity and quality are determined by geology. **Reserves** are the amount of a mineral that can be economically recovered.

Although many items in our daily lives originated as minerals extracted from the ground, no mineral can be quarried or mined without some cost to local communities and the environment. Extractive industries provide local jobs and create national wealth, but they also cause inconvenience, landscape scars, waste tips, loss of natural habitats, and various forms and levels of pollution.

The most convenient methods of mining are **open-cast** and **quarrying**. In open-cast mining, all the vegetation and topsoil are removed, thus destroying wildlife habitats and preventing other types of economic activity such as farming (Places 79). Sand and gravel are extracted from depressions which, although shallow, often reach down to the water table, as in the Lea valley in north-east London. Coal and iron ore are often obtained from deeper depressions using drag-line excavators which are capable of removing 1500 tonnes per hour (Figure 17.10). Often, the worst scars (eyesores) result from quarrying into hillsides to extract 'hard rocks' such as limestone and slate (Figure 17.11 and Places 78). There is usually greater economic and political pressure for open-cast coalmining than to quarry any other resource: it is the cheapest method of obtaining a strategic energy resource, but none generates greater social and environmental opposition. The increased demand for aggregates for road building and cement manufacture has led to the go-ahead being given for superquarries to be developed in the Scottish Highlands. One prospective quarry on Harris may see the eventual removal of a mountain and the creation of a new loch (Case Study 8).

Mining involves the construction of either horizontal **adit mines**, where the mineral is exposed on valley sides, or vertical **shaft mines**, where seams or veins are deeper.

Deep mining still affects local communities and the environment either by the piling up of rock waste to form tips – of coal in South Wales valleys (Aberfan, Case Study 2B) and china clay in Cornwall, for example – or by causing surface subsidence – as in some Cheshire saltworkings. Waste can also be carried into rivers where it can cause flooding by blocking channels and, when it contains poisonous substances, can kill fish and plants and contaminate drinking water supplies. This was highlighted in early 1992 when floodwaters from Cornwall's last working tin mine, Wheal Jane, flowed into rivers and to the coast, carrying with them arsenic and cadmium.

Figure 17.10

Opencast mining for coal, West Virginia, USA

Figure 17.11

The Dinorwic slate quarries, Gwynedd, North Wales

The Oakley slate quarries were first worked in 1818. By the 1840s, the most easily obtained slate had been won and mining began. The introduction of steam power and the building of the Ffestiniog railway led to the export of 52 million slates from Porthmadog in 1873. At the quarry's peak productivity, 2000 men and boys were employed on seven different levels. Each level was steeply inclined into the hillsides and was worked to a depth of 500 m. Apart from farming, the slate mines were the sole providers of employment, and Blaenau Ffestiniog's population peaked at 12 000. Working in candlelight in damp and dusty conditions for up to 12 hours a day, and with rock falls common (pressure release, page 41) the life expectancy of miners was short. By the turn of the century, the manufacture of clay roof tiles heralded the beginning of the industry's decline and in 1971 the mine at Blaenau closed. A decade later, renamed Gloddfa Ganol, the underground galleries were re-opened to tourists, some of whom arrive via the narrow-gauge Ffestiniog railway.

As the mines closed, many people became either unemployed or were forced to move to seek work – the present population of Blaenau is under 500. Even so, as a tourist attraction, the slate mines are still the largest single employer. As in many British mining towns, the older houses are rows of cottages coated grey with the dust of earlier economic activity. Dust still blows into the cottages, but the inhabitants no longer die from silicosis. Above the town rise the large and unsightly spoil heaps (Figure 17.12), though they appear more stable than the coal tips which affected Aberfan (Case Study 2B). On average, 10 tonnes of waste were created for every tonne of usable slate. Discarded machinery and unused buildings add to the scene of dereliction, though some are being restored as tourist attractions. Fortunately, there are few signs of the subsidence that is common to other mining areas.

Figure 17.12

Spoil heaps above Blaenau Ffestiniog, Gwynedd

Malaysia (1998) is the world's third major producer of tin (first until 1995), but the mineral is no longer one of the country's major exports. Early tin mining was typical of the colonial trade: British settlers brought in the capital, machinery and technology; supervised the mining; and exported the tin for refining. Malaysia itself received few advantages. Most tin was obtained by open-cast methods together with the use of hydraulic jets.

Since independence, the mines have been nationalised and operated under the Malaysian government. However, the total output, and the number of working mines and tin-workers, have all fallen rapidly due to the depletion of reserves and the low world market price for tin. Many of the old mines now form large tracts of land either covered in spoil (Figure 17.13) or left flooded. Disused buildings, overhead 'railways' and machinery often still remain.

One of the largest abandoned mines lies 15 km south of Kuala Lumpur in an area of rapidly growing housing and industry. It has been converted into a theme park complete with the world's longest aerial ropeway as well as water slides, musical fountains, a pirate ship and various water sports (Figure 17.14).

Figure 17.13

Disused tin mine, Malaysia

Figure 17.14

Sunway Lagoon Theme Park, near Kuala Lumpur, Malaysia

Having completed any sampling exercise (Framework 6, page 159), it is important to remember that patterns exhibited may not necessarily reflect the parent population. In other words, the results may have been obtained purely by chance. Having determined the mean of the sample size, it is possible to calculate the difference between it and the mean of the parent population by assuming that the parent population will conform to the normal distribution curve (Figure 6.37). However, while the sample mean must be liable to some error as it was based on a sample, it is possible to estimate this error by using a formula which calculates the **standard error of the mean** (*SE*).

$$SE\bar{x} = \frac{\sigma}{\sqrt{n}}$$

where: $\bar{x}$ = mean of the parent population
σ = standard deviation of parent population
$\sqrt{n}$ = square root of number of samples

We can then state the reliability of the relationship between the sample mean and the parent mean within the three confidence levels of 68, 95 and 99 per cent (Framework 6). Unfortunately, when sampling, the standard deviation of the parent population is not available and so to get the standard error we have to use the standard deviation of the sample, i.e. using *s* rather than. Although this introduces a margin of error, it will be small if *n* is large (*n* should be at least 30).

For example: a sample of 50 pebbles was taken from a spit off the coast of eastern England. The mean pebble diameter was found to be 2.7 cm and the standard deviation 0.4 cm. What would be the mean diameter of the total population (all the pebbles) at that point on the spit?

$$SE = \frac{0.4}{\sqrt{50}} = \frac{0.4}{7.07} = 0.06 \text{ (to two decimal places)}$$

This means we can say:

1 with 68 per cent confidence, that the mean diameter will lie between 2.7 cm ± 0.06 cm, i.e. 2.64 to 2.76 cm

2 with 95 per cent confidence, that the mean diameter will lie between 2.7 cm ± 2 x *SE* (2 x 0.06 = 0.12 cm), i.e. 2.58 to 2.82 cm

3 with 99 per cent confidence, that the mean diameter will lie between 2.7 cm ± 3 x *SE* (3 x 0.06 = 0.18 cm), i.e. 2.52 to 2.88 cm.

If we wanted to be more accurate, or to reduce the range of error, then we would need to take a larger sample. Had we taken 100 values in the above example we would have had

$$SE = \frac{0.4}{\sqrt{100}} = \frac{0.4}{10.00} = 0.04 \text{ (to two decimal places)}$$

which means we can now say with 68 per cent confidence that the mean diameter size will lie between 2.7 cm ± 0.04 cm (i.e. 2.66 to 2.74 cm). Of course, this also means there is a 32 per cent chance that the mean of the parent population is *not* within these values. This is why most statistical techniques in geography require answers at the 95 per cent confidence level.

This standard error formula is applicable only when sampling actual values (**interval** or **measured data**). If we wish to make a count to discover the frequency of occurrence where the data are **binomial** (i.e. they could be placed into one of two categories), we have to use the **binomial standard error**. For example, we may wish to determine how much of an area of sand dune is covered in vegetation and how much is *not* covered in vegetation. When using binary data, the sample population estimates are given as percentages, not actual quantities – i.e. *x* per cent of points on the sand dune *were* covered by vegetation; *x* per cent of points on the sand dune *were not* covered by vegetation.

The formula for calculating standard error using binomial data is

$$SE = \sqrt{\frac{p \times q}{n}}$$

where: p = the percentage of occurrence of points in one category
q = the percentage of points not in that category
n = the number of points in the sample.

A random sample of 50 points was taken over an area of sand dunes similar to those found at Morfa Harlech (Figure 6.33). Of the 50 points, 32 lay on vegetation and 18 on non-vegetation (sand) which, expressed as a percentage, was 64 per cent and 36 per cent respectively. How confident can we be about the accuracy of the sample?

$$SE = \sqrt{\frac{64 \times 36}{50}} = \sqrt{46.08} = 6.79$$

As the sample found 64 per cent of the sand dunes to be covered in vegetation and knowing the standard error to be ± 6.79, we can say:

1 with 68 per cent confidence, that the vegetated area will lie between 64 per cent ± 6.79, i.e. between 57.21 and 70.79 per cent

2 with 95 per cent confidence, that it will lie between 64 per cent ± 2 x SE (2 x 6.79 = 13.58), i.e. between 50.42 and 77.58 per cent will be vegetated

3 with 99 per cent confidence, that it will lie between 64 per cent ± 3 x SE (3 x 6.79 = 20.37), i.e. between 43.63 and 84.37 per cent.

Minimum sample size

It seems obvious that the larger the size of the sample, the greater is the probability that it accurately reflects the distribution of the parent population. It is equally obvious that the larger the sample, the more costly and time-consuming it is likely to be to obtain. There is, however, a method to determine the minimum sample size needed to get a satisfactory degree of accuracy for a specific task, e.g. to find the mean diameter of pebbles on a spit, or the amount of vegetation cover on sand dunes. This is achieved by reversing the two standard error calculations.

For measured data: Imagine you wish to know the mean diameter of pebbles at a given point on a spit to within ± 0.1 cm at the 99 per cent confidence level.

The 99 per cent confidence level is 3 x SE.

$$3\,SE = \frac{3s}{\sqrt{n}}$$

i.e. $3s = 0.1\sqrt{n}$

i.e. $\dfrac{3s}{0.1} = \sqrt{n}$

We determined earlier that s (standard deviation of the sample) for the pebble size was 0.4, and so by substitution we get:

$$\frac{1.2}{0.1} = \sqrt{n}$$

i.e. $12 = \sqrt{n}$

$$n = 12^2$$

$$n = 144$$

We would need, therefore, to measure the diameter of 144 pebbles to get an estimate of the parent population at the 99 per cent confidence level.

For binomial data: How many sample values are needed to estimate the area of sand dunes which is vegetated, with an accuracy which would be within 5 per cent of the actual area (i.e. at the 95 per cent confidence level)?

$$n = \frac{p \times q}{(SE)^2}$$

Again by substitution we get:

$$n = \frac{64 \times 36}{(5)^2}$$

$$n = 92.16$$

We would therefore have to take a sample of 93 values to achieve results within 5 per cent of the parent population.

The need for rural management

As was shown on Figure 17.4, there is often considerable competition for land in most rural areas and, therefore, there is a need, in most people's opinion, for careful management. In Britain, this management may be the task of national, local or voluntary organisations such as the Department of the Environment, the various National Parks Planning Boards (Places 92, page 592) and the Council for the Protection of Rural England (CPRE). Pressure on the land may be even greater in economically less developed countries where the need to improve people's basic standard of living is likely to take preference over management schemes.

One attempted management scheme in a developing country is described in Places 80. It draws together several topics discussed in this book, i.e. an island (Chapter 6) with interrelated ecosystems (Chapter 11) offering alternative, rural land use possibilities (Chapter 17), where the population is increasing (Chapter 13) and wishing to improve its standard of living (Chapter 22), thus putting pressure on natural resources (Chapter 17).

Over two-thirds of Tanzania's 900 km long coastline consists of three fragile ecosystems – a fringing coral reef, separated from mangrove swamps on the mainland by a lagoon. Mafia Island, where the coral reaches above sea-level, is a national marine park.

An island management plan was put forward in the 1990s to try to maintain economic development, to conserve resources for future generations and to avoid conflict between different land uses and users. Some of the various economic activities threatening the fragile island ecosystems include the following:

- **Coral mining** The removal of live coral for the tourist and curio trade, and of fossilised coral rock for building purposes (Places 37, page 302). For lime, the rock is burnt over fires made from locally collected wood.
- **Fisheries** At all scales from subsistence to commercial, taking fin-fish, octopus, crayfish and edible shellfish.
- **Dynamite fishing** The illegal use of dynamite to stun and kill fish. This destroys the physical structure of the reef and kills virtually every organism within 15 m of the blast.
- **Seaweed farming** Important as a means of diversifying income but suffers from the problems associated with cash crops and could lead to biodiversity loss through the creation of monoculture (page 501).
- **Salt production** By evaporation: hyper-saline seawater is boiled using local mangrove wood for fuel, a crude process that can cause the denudation of large areas of mature trees.
- **Tourism** A rapidly growing industry and one that the government is keen to promote. Coastal tourism includes game-fishing, 'sea-safaris', diving, snorkelling and beach activities. Tourists, per capita, are major consumers of resources (drinking water, fuel and foods), can damage the natural environment (new hotels, destruction of the reef) and can cause cultural conflicts (dress code in a Muslim country).
- **Off-shore gas extraction** From the small Songo-Songo gasfield.

A new management approach

This aims to satisfy economic, social and environmental objectives in order to ensure:

- the maximum sustainable economic benefit from the long-term use of natural resources
- the maintenance of the conditions and productivity of the natural environment

- the allocation of resources between competing uses and users.

These aims are often seen as contradictory, and the main problem is how to cope with the diverse requirements of the different user-groups, especially those who utilise finite resources.

Developing a management model

To achieve an understanding of the nature and conditions of the resources in a proposed management area, the following considerations should be explored:

- **Political factors** What is the scale and structure of the area? Is it stable? Who will pay? Can it provide finance or secure funding? Who will advise? Are there powerful interest groups either for or against?
- **Physical factors** What are the main physical features? Are these stable? Are there any natural hazards?
- **Biological factors** What biological communities exist? In what condition are they? Are there records of change or overuse over a period of time? Are there species of endangered, cultural or commercial importance?
- **Socio-economic** What are the current uses of the area? Who uses it? Are they traditional or local uses? Have they a commercial interest? How are the resources exploited? Are these practices sustainable or destructive?

Once an area has been chosen, four stages can lead to a practical plan for its creation as a multi-user management scheme:

1 The definition of management goals – normally including conservation, sustainable resource use and economic development (Framework 16, page 499).

2 The establishment of an administrative authority – the process of human representation.

3 The formulation of a management strategy and objectives – an assessment of the physical and human characteristics of the whole area and, within it, sub-zones.

4 The development of legislation – to achieve the objectives.

But remember – no plan should be considered as final – it is simply an improvement on what was done before.

Rural conflicts in south western USA

Figure 17.15

National Parks and Recreation Areas in the western USA

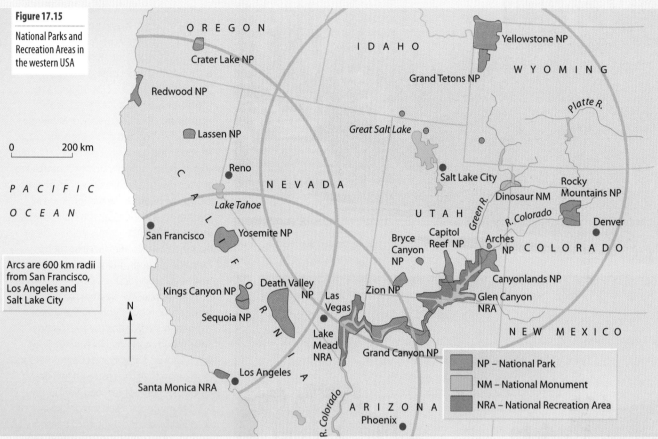

0   200 km

PACIFIC

OCEAN

Arcs are 600 km radii from San Francisco, Los Angeles and Salt Lake City

N

Map labels:

OREGON

Crater Lake NP

Redwood NP

Lassen NP

Reno

NEVADA

Lake Tahoe

San Francisco

Yosemite NP

CALIFORNIA

Kings Canyon NP

Death Valley NP

Sequoia NP

Las Vegas

Lake Mead NRA

Santa Monica NRA

Los Angeles

R. Colorado

Phoenix

IDAHO

Yellowstone NP

WYOMING

Grand Tetons NP

Platte R.

Great Salt Lake

Salt Lake City

Rocky Mountains NP

UTAH

Dinosaur NM

Green R.

R. Colorado

Denver

Bryce Canyon NP

Capitol Reef NP

Arches NP

COLORADO

Canyonlands NP

Zion NP

Glen Canyon NRA

NEW MEXICO

Grand Canyon NP

ARIZONA

Legend:

NP – National Park

NM – National Monument

NRA – National Recreation Area

The scenery of the mountain states of western USA is spectacular and varied. It includes some of the country's highest peaks, as well as extensive desert and wild river scenery. Yellowstone was set up in 1872, and is arguably the world's best known National Park with its variety of mountain and volcanic scenery, geysers, hot springs, canyons, lava flows and wildlife including bear, elk and buffalo (bison). Tourists from all over the world now flock to the region to visit the large number of designated National Parks, Monuments and Recreation Areas. Figure 17.15 indicates the accessibility of the most popular attractions for visitors from major cities and international airports. Over 30 million people live within 500 km of the major National Parks and Recreation Areas.

The National Parks were set up to preserve and protect the environment for future generations. Visitors are encouraged to stay, but apart from workers associated with the National Park there are no permanent residents within the parks. (This is a major difference from National Parks in the UK where farmers and other residents live on the land throughout the year.) Lodges, hotels, tourist villages and regulated camping grounds are provided, together with well-made roads and tracks to the different scenic attractions. This contributes to visitor pressure on 'honeypot' sites such as Old Faithful in Yellowstone where people wait for the geyser to blow once every 85 minutes (Figure 17.16). There are traffic jams as cars stop to watch animals such as bison grazing or herds of elk close to the road (Figure 17.17). Roads are closed by rangers if pressure is considered too great.

Visitor numbers to the Parks and Recreation Areas in the mountain states have continued to increase. Many come in private cars, camper vans and buses. Park authorities are working to provide better traffic management, which includes vehicle

Figure 17.16

Old Faithful, Yellowstone National Park

Figure 17.17

Yellowstone National Park, 1991: bison grazing on the verge; the vegetation is recovering from the 1988 fire

Figure 17.18

Arches National Park: the Three Gossips

Figure 17.19

Grand Tetons National Park: Mt Moran

restriction. The new Management Plan for Grand Canyon has considered the possibility of shuttle bus services from the Park boundaries to the South Rim and Grand Canyon Village. This would mean provision of large parking areas close to Park entrances, which would be an additional environmental pressure. At present most visitors go to the South Rim tourist area, but the North Rim is becoming increasingly popular, although it is closed during the winter months because the roads are snowbound. The Grand Canyon continues to attract rafting and canoeing enthusiasts, although the number of permits to journey through the Canyon is strictly controlled to protect the natural habitat along the river banks. The National Recreation Areas at Lake Mead and Glen Canyon (Lake Powell) received 10 million visitors in 1998. These large numbers require facilities such as camping grounds, marinas, drinking water, sewerage provision, fuel and power, as well as food supplies. All-weather roads have made traffic movements relatively easy.

The numbers visiting the five National Parks within Utah – Arches (Figure 17.18), Bryce Canyon, Canyonlands, Capitol Reef and Zion – with their diverse desert and river-eroded scenery approached 7 million in 1998. Many people used motor caravans, cars or tourist buses, and using such transport it is possible to visit all these areas within a two-week stay. Small lodges and bed and breakfast accommodation are found close to the Parks. Three new hotels have been built in the past five years near the town of Torrey just outside Canyonlands National Park.

Tourism is a major source of income for the states in the western Rockies. Rainfall is generally low in the Basin and Range province between the Sierra Nevada and the Rocky Mountains to the east. The rural economy depends on extensive ranching, irrigation farming and mining. Small settlements within driving distance of the Parks are flourishing. These include St George in southern Utah, which has attracted many 'retirees' wishing to move away from the city areas to a pleasant environment. West Yellowstone in Montana was once a frontier town and railhead for north-west Wyoming but is now a fast-growing resort with many second homes, smart boutiques and supermarkets serving the summer and winter tourist trade. Jackson Hole in

Wyoming, close to the magnificent Grand Tetons National Park (Figure 17.19) and to Yellowstone National Park, has become a busy tourist centre all year. Smart 'condos' (blocks of flats) fringe the lower slopes of the hills, which become ski slopes in the winter. There are a number of large ranches providing extensive grazing for beef cattle on the floor of the valley. In summer, irrigation is used to improve the pasture. A number of the ranches also supplement their income by providing bed and breakfast accommodation for tourists.

Further south, in areas that rely on water from the Colorado and its tributaries, there are conflicts over the growth of settlements. Rapid population increase since 1990 has meant that large cities such as Phoenix in Arizona and Las Vegas in Nevada rely on expensive water brought by canal and pipeline. Many new homes have swimming pools and there is much use of fountains and water as an environmental improvement. Less is available for irrigation on the farms that produce fresh vegetables and fruit for the towns and cities. Small towns are expanding as more people decide to move into rural areas to work from home using computers and electronic communication systems. Apart from California, the south-west has the highest annual percentage increase in population in the USA, particularly to smaller settlements and to the rural/urban fringe of cities such as Salt Lake City, Phoenix and Las Vegas.

Agriculture relies on water from streams and rivers which flow off the mountain slopes. Mormon settlers who trekked westwards in the latter part of the 19th century established many small farms in relatively remote valleys in the high plateaus between the mountains. Irrigation farming transformed the region (Figure 17.20). Winters are severe, with heavy snow and strong winds, and the growing season is restricted to summer months. Close to the larger settlements such as Salt Lake City, vegetables and fruit are produced, along with alfalfa for cattle fodder. Dairy cattle graze on accessible pasture. In remoter areas, as in the Uinta mountains, irrigation is used to improve pasture and provide silage for animals. There are conflicts between the need for local water provision and the overall water allocation for individual states, and these still have to be solved.

Much of the high rangeland in the west

is still owned by the Federal Government, which leases out grazing rights to the ranchers with properties in the high valleys. This has minimised the conflicts between ranchers. Forecasts of the number of animals that can be supported on the ranges (the stocking rate) are made and then permits are granted. Some ranchers overstock, as they claim they cannot make a living unless they increase the number of animals. As costs of ranching increase there is less traditional movement of sheep or cattle to the highest pastures in summer and down to the rangeland for the winter.

Traditionally the mountain states were important mineral producers, especially of copper, silver and gold. World prices and difficulties in extraction have forced the closure of all but the most productive mines. Copper is still produced in large quantities at Kennecott-Bingham mine, west of Salt Lake City. Elsewhere in the region there are many abandoned mines and quarries, forming scars on the landscape. These may be opened up again if and

when the world price for the mineral rises. In southern Utah there is evidence of the rush to produce uranium in the 1980s when mines were permitted in the remote wilderness areas – these are no longer in use.

Glen Canyon National Recreation Area: Lake Powell

Glen Canyon (Figure 17.21) received 2.5 million overnight visitors in 1998.

There has been a strong emphasis on the importance of recreational use of land and water in the management of Lake Powell since its creation. Provision has been made for many different recreational opportunities – boating, water skiing, wind surfing, sailing, fishing, swimming, and camping on the shoreline (Figure 17.22). There are wilderness trails, and back-country roads which can only be used by four-wheel-drive vehicles, but which give access to remote historical and archaeological sites. Houseboats are popular, providing both transport and accommodation.

Figure 17.20

Irrigation on rangelands near Salt Lake City

Figure 17.21

Glen Canyon National Recreation Area: Bullfrog Ferry, Lake Powell

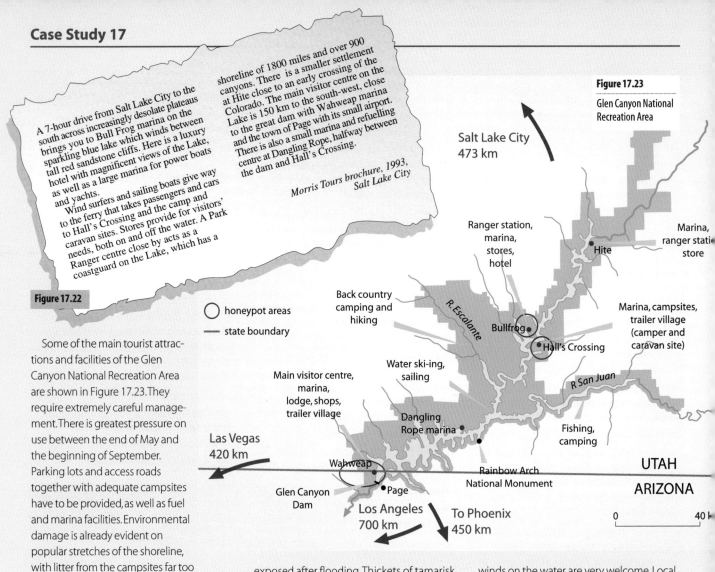

Figure 17.23

Glen Canyon National
Recreation Area

A 7-hour drive from Salt Lake City to the south across increasingly desolate plateaus brings you to Bull Frog marina on the sparkling blue lake which winds between tall red sandstone cliffs. Here is a luxury hotel with magnificent views of the Lake, as well as a large marina for power boats and yachts.

Wind surfers and sailing boats give way to the ferry that takes passengers and cars to Hall's Crossing and the camp and caravan sites. Stores provide for visitors' needs, both on and off the water. A Park Ranger centre close by acts as a coastguard on the Lake, which has a

shoreline of 1800 miles and over 900 canyons. There is a smaller settlement at Hite close to an early crossing of the Colorado. The main visitor centre on the Lake is 150 km to the south-west, close to the great dam with Wahweap marina and the town of Page with its small airport. There is also a small marina and refuelling centre at Dangling Rope, halfway between the dam and Hall's Crossing.

Morris Tours brochure, 1993,
Salt Lake City

Figure 17.22

Some of the main tourist attractions and facilities of the Glen Canyon National Recreation Area are shown in Figure 17.23. They require extremely careful management. There is greatest pressure on use between the end of May and the beginning of September. Parking lots and access roads together with adequate campsites have to be provided, as well as fuel and marina facilities. Environmental damage is already evident on popular stretches of the shoreline, with litter from the campsites far too common. 'Adopt-a-Canyon' has become a powerful slogan, encouraging visitors to boat out everything that they take in.

Water quality and lake use need careful monitoring. When the wind is strong and the wash of pleasure boats is reflected off the canyon walls, it is impossible to water ski and the conditions can lead to a very rough boat ride. Speed limits are enforced and there are now designated areas for some activities.

The lake level fluctuates with the amount of power and irrigation water required. This means that some surfaces are

exposed after flooding. Thickets of tamarisk are invading these areas; they provide leaf litter, attract flies and produce hayfever-causing pollen. When the lake levels rise, this brings problems for water sports as the vegetation is partly submerged.

During the winter months, the main recreational use is by fishermen and there is little activity at the marinas. Day temperatures average 8°C from November to February, with plenty of sunshine. From May to September, temperatures rarely fall below 27°C and often exceed 35°C – so the plateau

winds on the water are very welcome. Local thunderstorms over the lake can be a problem for water-sports enthusiasts.

This is a remote area from the viewpoint of British students, although more tourists from the UK are now including it in their itinerary. To Americans it has become a national recreation centre in most senses of the phrase. People in the trailer parks come from all the different states, even including Alaska, and several different European languages may be heard in and around the marinas and campsites.

References

Pickering, K. and Owen, L. (1994) *Global Environmental Issues*, Routledge.

Waugh, D. and Rowley, C. (2000) *Images of Change: China*, CD Rom, Geopix.

Wilson, J. (1984) *Statistics in Geography for 'A' Level Students*, Schofield & Sims.

Websites

UN Food and Agriculture Organisation Forestry:
http://www.fao.org/fo/forestry.htm

Forestry Commission of Great Britain:
http://www.forestry.gov.uk/

ForestWorld:
http://forestworld.com/index2.html

UK Countryside Agency:
http://www.countryside.gov.uk/

Council for the Protection of Rural England:
http://www.greenchannel.com/cpre/main.htm

Finnish Forest Association:
http://www.metla.fi/forestfin/intro/eng/index.htm

See also for more links:
http://www.nelsonthornes.com/gaia

1 Study Figure 17.24 below. It shows how conflicts may arise in a rural area where recreation and tourism are important.

Figure 17.24

a Name a rural area in a more economically developed country where recreation and tourism are important, and where their development has caused conflicts with local people and conservationists. Describe conflicts in that area between:
  i tourists and the local community
  ii tourists and conservation
  iii the local community and conservation. **(8 marks)**
b Explain how management of the area is attempting to reduce the conflicts described in **a** above. **(7 marks)**
c Can tourism ever lead to sustainable development in rural areas in less economically developed countries? Illustrate your answer with reference to one or more case studies. **(10 marks)**

2 a i What is a mineral resource? **(1 mark)**
    ii What is a mineral reserve? **(1 mark)**
    iii How might the total quantity of reserves of a mineral be affected by:
      ■ increases in the market price of the mineral
      ■ improved technology for mining the mineral? **(4 marks)**
  b Explain the differences between open cast mining or quarrying, and deep mining. Refer to the techniques employed and their effect on the environment. **(7 marks)**
  c The development of mining in rural areas usually goes through several different stages. These include:
    ■ discovery
    ■ exploitation
    ■ exhaustion of the reserves.
    Explain how each of these stages affects local communities and local economies.
    Make reference to one or more case studies. **(12 marks)**

3 a 'Forestry is not usually economically viable in developed countries unless supported by the state with subsidies.' Explain the advantages of forestry in rural areas of the United Kingdom, giving:
    i two socio-economic advantages **(4 marks)**
    ii two environmental advantages. **(4 marks)**
  b i Explain why some people think that commercial forestry plantations have caused environmental damage in some parts of the United Kingdom. **(4 marks)**
    ii 'In the last decade … forest operations have been transformed, with a shift towards smaller-scale practices which are more environmentally and aesthetically sensitive.' Explain how the changes in forest management referred to above have improved the rural environment in parts of the United Kingdom. **(5 marks)**
  c With reference to a named tropical country, explain how commercial forestry in the rainforest can be a form of sustainable development. **(8 marks)**

---

4 Study Figure 17.24.
'The development of the tourist industry can bring both benefits and problems for communities and the environment of rural areas.'
Discuss this statement with reference to examples from more economically developed and less economically developed countries. **(25 marks)**

5 a 'In the last decade … forest operations have been transformed, with a shift towards smaller-scale practices which are more environmentally and aesthetically sensitive.'
    With reference to a named area of forestry in the United Kingdom explain how the changes referred to above have altered forest management practices. Explain how this has benefited the environment. **(10 marks)**
  b i How has commercial forestry caused environmental damage in tropical regions? **(7 marks)**
    ii Suggest how commercial forestry in tropical regions can be managed so that it is a sustainable form of development. **(8 marks)**

6 'Agriculture has become too efficient in many more economically developed countries, so huge surpluses result.'
Discuss the view that farmers may, in future, be paid to conserve the countryside and make it attractive for leisure and tourism rather than to maximise the yield from agriculture. **(25 marks)**

7 The development of mining in rural areas usually goes through several different stages. These include:
  ■ discovery
  ■ exploitation
  ■ exhaustion of the reserves.
Explain, with reference to one or more case studies, how each of these stages affects:
  i local communities and local economies **(13 marks)**
  ii the environment. **(12 marks)**

Chapter 18

Energy resources

'If each Indian were to start consuming the amount of commercial energy a Briton does, that would mean the world finding the equivalent of an extra 3190 million tonnes of oil each year. Imagine what consuming that would do to the greenhouse effect, not to mention its effect on oil and other reserves.'

Mark Tully, *No Full Stops in India*, 1991

What are resources?

Resources have been defined as features that are needed and used by people. Although the term is often taken to be synonymous with **natural resources**, geographers and others often broaden this definition to include **human resources** (Figure 18.1). Natural resources can include raw materials, climate and soils. Human resources may be subdivided into people and capital. A further distinction can be made between **non-renewable resources**, which are finite as their exploitation can lead to the exhaustion of supplies (oil), and **renewable resources**, which, being a 'flow' of nature, can be used over and over again (solar energy). As in any classification, there are 'grey' areas. For example, forests and soils are, if left to nature, renewable; but, if used carelessly by people, they can be destroyed (deforestation, soil erosion).

Figure 18.1

A classification of resources

Reserves are known resources which are considered exploitable under current economic and technological conditions. For example, North Sea oil and gas needed a new technology and high global prices before they could be brought ashore; in contrast, tidal power still lacks the technology, and often the accessibility to markets, that are needed to allow it to be developed on a widespread, commercial scale.

Energy resources

The sun is the primary source of the Earth's energy. Without energy, nothing can live and no work can be done. Coal, oil and natural gas, which account for an estimated 85.5 per cent of the global energy consumed in 1996 (Figure 18.2), are forms of stored solar energy produced, over thousands of years, by photosynthesis in green plants. As these three types of energy, which are referred to as **fossil fuels**, take long periods of time to form and to be replenished, they are classified as non-renewable. As will be seen later, these fuels have been relatively easy to develop and cheap to use, but they have become major polluters of the environment. Nuclear energy is a fourth non-renewable source but, as it uses uranium, it is not a fossil fuel.

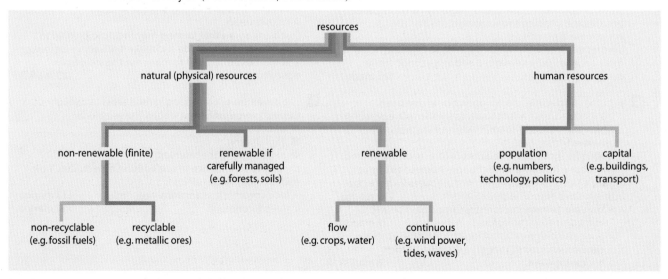

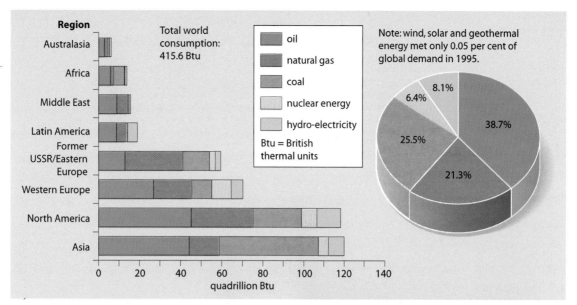

Figure 18.2

World energy consumption: by region, 1995

Total world consumption: 415.6 Btu

oil
natural gas
coal
nuclear energy
hydro-electricity

Btu = British thermal units

Note: wind, solar and geothermal energy met only 0.05 per cent of global demand in 1995.

Renewable sources of energy are mainly forces of nature which can be used continually, are sustainable and cause minimal environmental pollution. They include running water, waves, tides, wind, the sun, geothermal, biogas and biomass. At present, with the exception of running water (hydro-electricity), the wind and biomass, there are economic and technical problems in converting their potential into forms which can be used.

World energy producers and consumers

It has been estimated that, annually, the world consumes an amount of fossil fuel that took nature about 1 million years to produce, and that the rate of consumption increased fourfold between 1950 and 1990. This consumption of energy is not evenly distributed over the globe (Figure 18.2). At present, the 79 per cent of people living in the 'developing' countries consume only 30 per cent of the total energy supply (see table above).

Consumption in:	1990	1995	2000	2015
Developed countries	74.9	69.2	67.0	58.9
Developing countries	25.1	30.8	33.0	41.1

Although recently the consumption of energy in 'developed' countries has begun to slow down, due partly to industrial decline and environmental concerns, it has been increasing more rapidly in 'developing' countries with their rapid population growth and aspirations to raise their standard of living. This led to a conflict of interest between groups of countries at the 1992 Rio Earth Summit conference. The 'industrialised' countries, with only 21 per cent of the world's population yet consuming almost 70 per cent of the total energy, wished to see resources conserved and, belatedly, the environment protected. The 'developing' countries, which blame the industrialised countries for most of the world's pollution and depletion of resources, considered that it was now their turn to use energy resources, often regardless of the environment, in order to develop economically and to improve their way of life.

The world's reliance upon fossil fuels (Figures 18.2 and 18.3) is likely to continue well into the 21st century. However, while the economically recoverable reserves of coal remain high (Figure 18.4), the similar life expectancies of oil and natural gas are much shorter (coal: about 200–400 years; oil: about 50 years; natural gas: about 120 years). The distribution of recoverable fossil fuels is spread very unevenly across the globe, with the former USSR being well endowed with coal and natural gas; North America and parts of Asia with coal; and the Middle East with oil and natural gas (Figure 18.5). As these producers are not always major consumers, there is a considerable world movement of, and trade in, fossil fuels (Figure 18.6).

Figure 18.3

World energy consumption: by type, 1973–98

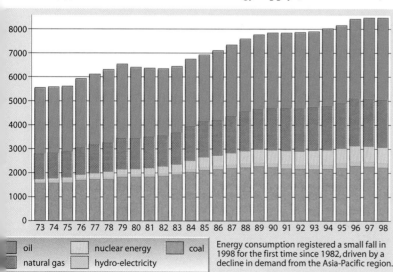

oil
natural gas
nuclear energy
hydro-electricity
coal

Energy consumption registered a small fall in 1998 for the first time since 1982, driven by a decline in demand from the Asia-Pacific region.

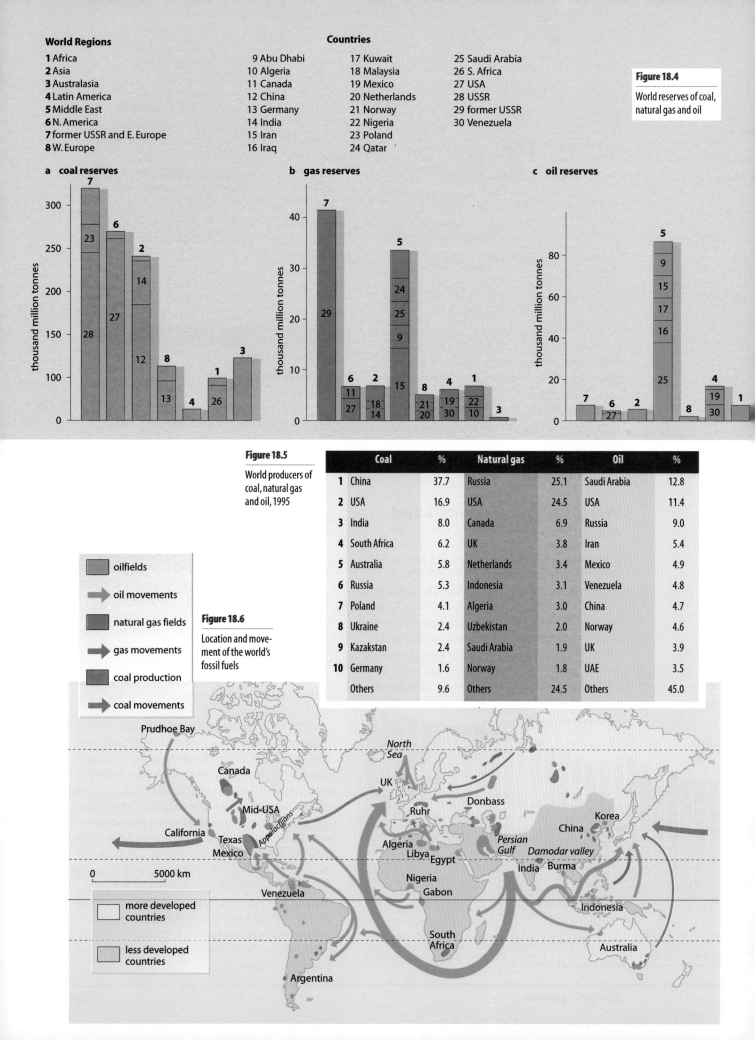

World Regions

1 Africa
2 Asia
3 Australasia
4 Latin America
5 Middle East
6 N. America
7 former USSR and E. Europe
8 W. Europe

Countries

9 Abu Dhabi
10 Algeria
11 Canada
12 China
13 Germany
14 India
15 Iran
16 Iraq
17 Kuwait
18 Malaysia
19 Mexico
20 Netherlands
21 Norway
22 Nigeria
23 Poland
24 Qatar
25 Saudi Arabia
26 S. Africa
27 USA
28 USSR
29 former USSR
30 Venezuela

Figure 18.4

World reserves of coal, natural gas and oil

a coal reserves

b gas reserves

c oil reserves

Figure 18.5

World producers of coal, natural gas and oil, 1995

	Coal	%	Natural gas	%	Oil	%
1	China	37.7	Russia	25.1	Saudi Arabia	12.8
2	USA	16.9	USA	24.5	USA	11.4
3	India	8.0	Canada	6.9	Russia	9.0
4	South Africa	6.2	UK	3.8	Iran	5.4
5	Australia	5.8	Netherlands	3.4	Mexico	4.9
6	Russia	5.3	Indonesia	3.1	Venezuela	4.8
7	Poland	4.1	Algeria	3.0	China	4.7
8	Ukraine	2.4	Uzbekistan	2.0	Norway	4.6
9	Kazakstan	2.4	Saudi Arabia	1.9	UK	3.9
10	Germany	1.6	Norway	1.8	UAE	3.5
	Others	9.6	Others	24.5	Others	45.0

oilfields
oil movements
natural gas fields
gas movements
coal production
coal movements

Figure 18.6

Location and movement of the world's fossil fuels

more developed countries

less developed countries

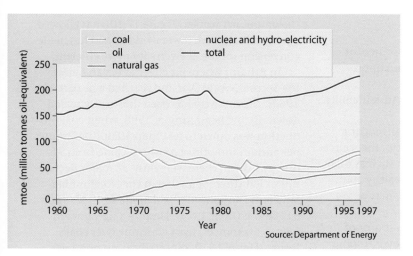

Source: Department of Energy

Figure 18.7

UK primary energy consumption, 1960–97

Recent trends

Energy consumption fell by 0.1 per cent in 1998, the first decline since 1982, during which time it had averaged an annual growth of 2 per cent. Consumption of gas, nuclear and hydro-electric power all rose by more than 1 per cent, but this was insufficient to offset the lack of growth in oil and the accelerated decline in the use of coal. Consumption fell most rapidly in the former USSR and eastern Europe as well as in recession-hit Asia. Although rapid growth continued in Africa and South America, often from an admittedly low base, growth in North America and Europe was slow, despite favourable economic conditions, mainly due to a mild winter.

Figure 18.8

The UK's changing sources of energy, 1950–97

UK energy consumption

The UK has always been fortunate in having abundant energy sources. In the Middle Ages, fast-flowing rivers were used to turn water-wheels while, in the early 19th century, the use of steam, from coal, enabled Britain to become the world's first industrialised country. Just when the most accessible and cheapest supplies of coal began to run short, natural gas (1965) and oil (1970) were discovered in the North Sea, and improvements in technology enabled the controversial production of nuclear power. Looking ahead to a time when the UK's fossil fuels do become less available (300 years for coal and 40 years for oil and natural gas), Britain's seas and weather have the potential to provide renewable sources of energy using the wind, waves and tides.

The total energy consumption in the UK rose from 152.3 toe (million tonnes of oil equivalent, a standard measure for comparing energy consumption) in 1960 to 228.0 toe in 1997 (Figure 18.7). Of that latter figure, 87.8 per cent was from fossil fuels (Figure 18.8) and 11.3 per cent from nuclear energy, leaving only 0.9 per cent from renewable sources (Figure 18.1).

Energy consumption by final user has also changed over time. Between 1972 and 1997 there was a decline in the amount of energy consumed by industry, from 42 per cent of the total to 22 per cent, but an increase in the amount used by transport from 22 per cent to 36 per cent. Domestic demand has fallen slightly due to energy-efficiency schemes (e.g. double-glazing and roof and cavity wall insulation – page 544).

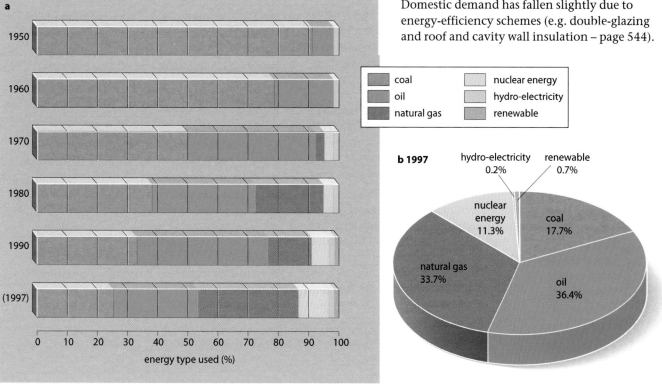

Sources of energy

Decisions by countries as to which source, or sources, of energy to use may depend upon several factors. These include:

- Availability, quality, lifetime and sustainability of the resource.
- Cost of harnessing, as well as transporting (importing or within the country), the source of energy: some types of energy, such as oil, may be too expensive for less wealthy countries; while others, such as tides, may as yet be uneconomical to use.
- Technology needed to harness a source of energy: like costs, this may be beyond some of the less developed countries (nuclear energy), or may yet have to be developed (wave energy).
- Demands of the final user: in less developed countries, energy may be needed mainly for domestic purposes; in more developed countries, it is needed for transport, agriculture and industry.
- Size, as well as the affluence, of the local market.
- Accessibility of the local market to the source.
- Political decisions: for example, which type of energy to utilise or to develop (nuclear), or whether to deny its sale to rival countries (embargoes).
- Competition from other forms of available energy.
- Environment: this may be adversely affected by the use of specific types of energy, such as coal and nuclear; it may only be protected if there are strongly organised local or international conservation pressure groups such as Friends of the Earth and Greenpeace.

These factors will be considered in the next section, which discusses the relative advantages and disadvantages of each available or potential source of energy.

Non-renewable energy

Coal

Coal provided the basis for the Industrial Revolution in Britain, western Europe and the USA. Despite its exploitation for almost two centuries, it still has far more economically recoverable reserves than any of the other fossil fuels (Figure 18.4). Improved technology has increased the output per worker (Figure 17.10), has allowed deeper mining with fewer workers, and has made conversion for use as electricity more efficient.

In Britain, where pit closures began before the Second World War and reached a peak in the 1960s, the industry has been drastically scaled down. By the beginning of 1998, when coal production was down to less than 50 million tonnes per annum (one-third of which came from 91 opencast sites) there were only 22 major deep mines left (there were 958 in the Northumberland and Durham coalfield alone in 1947, and 310 in the UK in 1973). Despite a new government's assurances, fears continue over coal's future (Figure 18.9), especially as no new coal-fired power stations are being built. The social and economic consequences, especially in former one-industry coal-mining villages, have been devastating. Similar problems exist in other old mining communities in Belgium, the Ruhr (Germany) and the Appalachians (USA) (Figure 18.6).

There are many reasons for this decline. The most easily accessible deposits have been used up, and many of the remaining seams are dangerous, due to faulting, and uneconomic to work. Costs have risen due to expensive machinery and increased wages. The demand for coal has fallen for industrial use (the decline of such heavy industries as steel), domestic use (oil- and gas-fired central heating) and power stations (now preferring gas). British coal has had to face increased competition from cheaper imports (USA and Australia), alternative methods of generating electricity (gas-fired) and cleaner forms of energy. Political decisions have seen subsidies paid to the nuclear power industry and a greater investment in gas rather than coal-fired power stations (the 'dash for gas' policy – page 538). Green pressures have also led to a decline in coal mining, which creates dust and leaves spoil tips; and in the use of coal to produce electricity, as this releases sulphur dioxide and carbon dioxide which are blamed for acid rain and for global warming (Case Study 9B). In places such as China and India, which have large reserves and a rapidly growing demand for energy, it is essential to introduce 'clean coal technologies', but these will depend upon their development, and possibly their funding, by the industrialised countries (Places 82, page 544).

Fears raised over coal's future as pit closes

Mining MPs and industry executives are pressing ministers for renewed guarantees that coal has a future in Britain's energy mix as the country's biggest producer comes under pressure.

RJB Mining yesterday announced the closure of its North Selby pit with the loss of up to 300 jobs in order to improve efficiency at its loss-making deep-mined business.

The closure, part of a £100 million cost-cutting programme, is the latest in a series undertaken by RJB since it bought the bulk of the state-owned coal industry in 1994. It reduces the group's deep mines to 12.

North Selby, which merged with the nearby Stillingfleet pit two years ago, is part of the huge Selby complex, the jewel in the crown of the UK coal industry which employs 2300 and produces 7 million tonnes a year. But, despite signing new long-term contracts with generators, RJB is suffering a fresh squeeze on prices and sales as power-producers cut coal-fired output and new entrants switch to cheaper, imported coal.

Coal UK, an industry newsletter, says that Edison Mission Energy, new owners of Ferrybridge and Fiddler's Ferry power stations sold by PowerGen, has started test imports and plans to increase these substantially next year – up to 5 million tonnes a year. It says National Power plans to cut 2 million tonnes of consumption at its 'non-green' plants, and Eastern Electricity is 'awash with coal'.

A new sense of crisis has propelled mining MPs to enter into talks with ministers on coal's future, supposed to have been settled by last year's energy review. RJB, which lost £4.7 million on deep-mined operations in the second half of last year and is only profitable because of its opencast business, said it was concentrating production in low-cost pits with fewer geological problems.

The Guardian, 29 July 1999

Figure 18.9

Fears for the future of coal

Oil

Oil is the world's largest business, with commercial and political influence transcending national boundaries. Indeed, several of the largest transnational enterprises are oil companies. Oil, like other fossil fuels, is not even in its distribution (Figure 18.6), and is often found in areas that are either distant from world markets or have a hostile environment, e.g. the Arctic (Alaska), tropical rainforests (Nigeria and Indonesia), deserts (Algeria and the Middle East) and under stormy seas (North Sea). This means that oil exploration and exploitation is expensive, as is the cost of its transport by pipeline or tanker to world markets. Oil, with its fluctuating prices, has been a major drain on the financial reserves of many developed countries and has been beyond the reach of most developing countries. Countries where oil is at present exploited can only expect a short 'economic boom' as, apart from several states in the Middle East where production may continue for 100 years, most world reserves are predicted to become exhausted within 30 years.

New technology has had to be developed to tap less accessible reserves. Before oil could be recovered from under the North Sea, large concrete platforms, capable of withstanding severe winter storms, had to be designed and constructed. Each platform, supported by four towers, had to be large enough to accommodate a drilling rig, process plant, power plant, helicopter landing-pad and living and sleeping quarters for its crew. The towers may either be used to store oil or may be filled with ballast to provide extra anchorage and stability. Two 90 cm trunk pipelines were laid, by a specially designed pipe-laying barge, over an uneven sea-bed to Sullom Voe on Shetland. Since then, production has spread northwards to even deeper and stormier waters west of Shetland, where the Foinaven and Schiehallion fields are being developed. In 1999 there were 129 offshore oilfields and despite the fall of world oil prices to levels no higher than their 1973 low, production reached a record high.

On a global scale, oil production and distribution are affected by political and military decisions. OPEC (Figure 21.28) is a major influence in fixing oil prices and determining production – although even it is helpless in the face of international conflicts such as Suez (1956), the Iran–Iraq War (early 1980s) and the Gulf War (1991). Closer to home, recent British and EU fuel policies have favoured the gas and nuclear industries at the expense of oil. Oil is used in power stations, by

Figure 18.10

Milford Haven oil refinery

industry, for central heating and by transport. Although it is considered less harmful to the environment than coal, it still poses many threats. Oil tankers can run aground during bad weather (*Braer*, 1993) releasing their contents which pollute beaches and kill wildlife (*Exxon Valdez*, 1989), while explosions can cause the loss of human life (Alpha Piper rig, 1988). To try to reduce the dangers of possible spillages and explosions, oil refineries have often been built on low-value land adjacent to deep, sheltered tidal estuaries, well away from large centres of population (Figures 18.10 and 18.12). During 1991 there were, in England and Wales alone, 5288 reported cases of inland water pollution by oil and petroleum products. Road transport, using petrol, is a major contributor to global warming.

Natural gas

Natural gas has become the fastest-growing energy resource. It provides an alternative to coal and oil and, in 1998, it comprised almost a quarter of the world's primary energy consumption. Latest estimates suggest that global reserves will last another 120 years. Gas is often found in close proximity to oilfields (Figure 18.6) and therefore experiences similar problems in terms of production and transport costs and requirements for new technology. In the UK, as a result of the so-called 'dash for gas', the electricity industry had by 1998 over 20 new gas-fired power stations in operation, several more under construction, and many more at the planning stage. These stations run on gas from 90 gasfields in the North Sea and the Irish Sea. Interconnector pipelines now supply gas to Northern Ireland (1997) and Belgium (1998).

At present gas is considered to be the cheapest and cleanest of the fossil fuels, which is why the British government has favoured new power stations being gas-fired and older ones being converted to gas (mainly at the expense of coal).

As gas causes less atmospheric pollution, it also helps the British government to achieve its promised 60 per cent reduction of greenhouse-gas emissions from power stations by the year 2000.

Nuclear energy

During the early 1950s, nuclear energy, with its slogan 'atoms for peace', was seen by many to be a sustainable, inexpensive and clean energy resource and by others as a dangerous military aid and a threat to the environment.

Nuclear energy uses uranium as its raw material. Compared with fossil fuels, uranium is only needed in relatively small amounts (50 tonnes of uranium a year, compared with 500 tonnes of coal per hour for coal-fired power stations). Uranium has a much longer lifetime than coal, oil or gas and can be moved more easily and cheaply. However, the development of nuclear energy, with its new technology, specially designed power stations and essential safety measures, has been very expensive. As a result, it has generally been adopted by the more wealthy countries, and even then only by those (Figure 18.11) lacking fossil fuels (Japan) or with significant energy deficits (UK, USA and France). As a source of electricity, it is fed into the National Grid; but, as a source of energy, it cannot be used by transport or for heating. The decision to develop nuclear power has been, universally, a political decision. In Britain, for example, the first nuclear programme was initiated in 1955 by a government which felt that the scheme would be uneconomic. Later, two further programmes were announced, respectively, by Labour in 1964 and the Conservatives in 1979. At its peak in the early 1990s, nuclear energy provided the UK with 30 per cent of its energy needs from 22 power stations (Figure 18.12). Since then plans for new stations have been dropped, and the plan for a fast-breeder reactor has been abandoned. The Dounreay reprocessing plant, in Caithness, and several older UK stations that had come to the end of their life, have closed. However, while in 1998 an all-party House of Commons Committee saw no immediate need for public spending on new nuclear capacity, it did not argue against the construction of future nuclear plants. Indeed, no country which at present relies heavily upon nuclear power can afford to close down its plants, as there are few economically viable alternatives. Also, as the threat of global warming becomes more a reality, the industry is projecting itself as the solution to the world's excessive use of fossil fuels.

Figure 18.11

Major users of nuclear energy, 1995

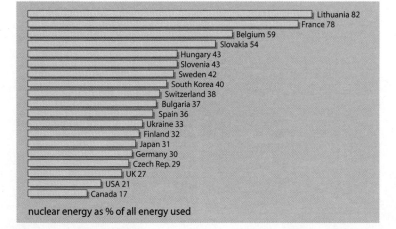

nuclear energy as % of all energy used

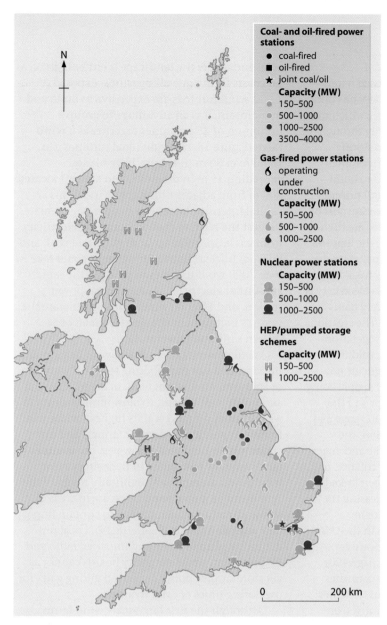

Figure 18.12

Power stations in the UK with 50 MW or more capacity, 1998

emissions of radioactivity, waste disposal, and radioactive contamination accidents. Routine emissions have been linked – without proven evidence – with clusters of increased leukaemia, especially in children, around several power stations (notably Sellafield and Dounreay). Radioactive waste has to be stored safely, either deep underground or at Sellafield. Every radio-active substance has a 'half-life', i.e. the time it takes for half of its radioactivity to die away. Iodine, with a half-life of 8 days, becomes 'safe' relatively quickly. In contrast, plutonium 239, produced by nuclear reactors, has a half-life of 250 000 years and may still be dangerous after 500 000 years. The two worst radioactive acci-dents resulted from the melt-down of reactor cores at Three Mile Island in the USA (1979) and at Chernobyl in the Ukraine (1986). It was mainly for economic and safety reasons that British nuclear power stations (Figure 18.12) were built on coasts and estuaries where there is water for cooling and cheap, easily reclaimable land well away from major centres of popula-tion. However, the British government had to agree in 1998, following renewed calls from several EU countries, to make a large reduction in discharges into the Irish Sea from Sellafield.

Renewable energy

With the depletion of oil and gas reserves during the early years of the 21st century and the unfavourable publicity given to all types of fossil fuels, especially regarding their contribution towards global warming, renewable energy resources are likely to become increasingly more attractive. Renewable sources of energy are likely to become more cost-competitive, offer greater energy diversity, and allow for a cleaner environ-ment. As shown in Figure 18.1, there are two types of renewable energy:

- **Continuous sources** are recurrent and will never run out. They include running water (for hydro-electricity), wind, the sun (solar), tides, waves and geothermal.
- **Flow sources** are sustainable providing that they are carefully managed and maintained (Framework 16, page 499). Biomass, including the use of fuelwood, is sustainable in that it has a maximum yield beyond which it will begin to become depleted.

Since the 1950s, most money spent on devel-oping energy in the UK has been spent on nuclear power whereas, according to conservation groups, it should have been spent on non-polluting, renewable resources. Nuclear power is a major issue in Japan, where the government plans to build more power stations. Japan is desperately short of its own sources of power, but experienced first-hand the worst side of nuclear power when two atomic bombs were dropped on Hiroshima and Nagasaki in 1945.

Although nuclear power stations produce fewer greenhouse gases than thermal (coal-, oil- and gas-fired) power stations, they do present potential risks in three main areas: routine

Hydro-electricity

Hydro-electricity is the most widely used renewable energy resource. Its availability depends upon an assured supply of fast-flowing water which may be obtained from rainfall spread evenly throughout the year, or by building dams and storing water in large reservoirs. The initial investment costs and levels of technology needed to build new dams and power stations, to install turbines and to erect pylons and cables for the transport of the electricity to often-distant markets, are high. However, once a scheme is operative, the 'natural, continual, renewable' flow of water makes its electricity cheaper than that produced by fossil fuels.

Although the production of hydro-electricity is perceived as 'clean', it can still have very damaging effects upon the environment. The creation of reservoirs can mean large areas of vegetation being cleared (Tucurui in Amazonia), wildlife habitats (Kariba in Zimbabwe) and agricultural land (Volta in Ghana) being lost, and people being forced to move home (Aswan in Egypt and the Three Gorges Dam in China – Places 82, page 544). Where new reservoirs drown vegetation, the resultant lake is likely to become acidic and anaerobic. Dams can be a flood risk if they collapse or overflow (Case Study 2B) and have been linked, without any proven evidence, to increasing the risk of earthquake activity (Nurek Dam in Tajikistan, and in the Colorado Basin in the USA, Case Study 3B), and can trap silt previously spread over farmland (Nile valley, Places 73, page 490). Despite these negative aspects, many countries rely upon large, sometimes prestigious, schemes or, increasingly in less developed countries, on smaller projects using more appropriate levels of technology (Case Study 18).

Wind

Wind is an underused energy resource, especially in places where winds are strong, steady and reliable, and where the landscape is either high or, as on coasts facing prevailing winds, exposed.

As wind turbines are expensive to build and to maintain, it is an advantage to group a minimum of 25 machines together as a **wind farm** (Figure 18.13). Individual turbines can stand over 30 m in height, have blades exceeding 35 m in diameter, and are best located 200 m from adjacent turbines. Wind farms are capital intensive but cheap to operate. Estimates suggest the relative cost of producing electricity from gas is 3p/kWh, 4p from wind and coal, and 8p from nuclear power. However, it could take over 7000 wind turbines to produce the same amount of electricity as one nuclear power station, and 100 000 wind farms to generate the same as is at present obtained from all of Britain's nuclear power stations.

Wind power has the advantage of being pollution free and not contributing to global warming or acid rain. Also, winds tend, in both California and north-west Europe, to blow more strongly in winter when demand is at its highest, while wind farms can provide income for farmers and can create jobs in rural areas. However, environmentalists are now less united in their support for wind power as, especially in Britain, many of the actual and proposed wind farms are in areas of scenic attraction which often also contain important wildlife habitats (Figure 18.14). Local residents complain of noise and impaired radio and TV reception, while electricity boards cannot as yet store surplus power generated during gales for use during times of calm weather.

Although the first large-scale wind farms were located in California (Places 81), by 1999 Europe produced over one-third more energy from the wind than did the USA. Denmark, which relies more on wind power than any other country, intends to install 2000 turbines offshore by 2009.

Places 81   California and the UK: wind farms

California

Most wind farms in the USA have been developed by private companies. The developers, who use either their own or leased land, sell electricity to electric utilities. At present, 90 per cent of the USA's capacity comes from California. California's wind farms are in an ideal location mainly because peak winds occur about the same time of year as does peak demand for electricity in the large cities nearby.

Approximately 16 000 turbines within the state produce enough electricity to supply a city the size of San Francisco. The three largest wind farms are at Altamont Pass (east of San Francisco), Tehachapi (between the San Joaquin Valley and the Mojave Desert) and San Gorgonio (north of Palm Springs). The Altamont Pass, with 7000 turbines, is one of the largest wind farms in the world (Figure 18.13). The average wind speed averages between 20 and 37 km/hr. The land is still used for cattle grazing as there is only one turbine for every 1.5–2 ha.

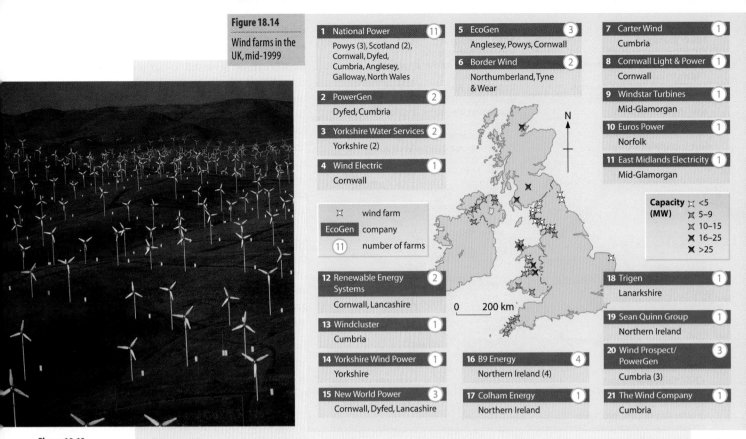

Figure 18.14

Wind farms in the UK, mid-1999

| 1 National Power ⑪ |
| Powys (3), Scotland (2), Cornwall, Dyfed, Cumbria, Anglesey, Galloway, North Wales |

| 2 PowerGen ② |
| Dyfed, Cumbria |

| 3 Yorkshire Water Services ② |
| Yorkshire (2) |

| 4 Wind Electric ① |
| Cornwall |

| 5 EcoGen ③ |
| Anglesey, Powys, Cornwall |

| 6 Border Wind ② |
| Northumberland, Tyne & Wear |

| 7 Carter Wind ① |
| Cumbria |

| 8 Cornwall Light & Power ① |
| Cornwall |

| 9 Windstar Turbines ① |
| Mid-Glamorgan |

| 10 Euros Power ① |
| Norfolk |

| 11 East Midlands Electricity ① |
| Mid-Glamorgan |

☆ wind farm
EcoGen company
⑪ number of farms

Capacity (MW): <5, 5–9, 10–15, 16–25, >25

0   200 km

| 12 Renewable Energy Systems ② |
| Cornwall, Lancashire |

| 13 Windcluster ① |
| Cumbria |

| 14 Yorkshire Wind Power ① |
| Yorkshire |

| 15 New World Power ③ |
| Cornwall, Dyfed, Lancashire |

| 16 B9 Energy ④ |
| Northern Ireland (4) |

| 17 Colham Energy ① |
| Northern Ireland |

| 18 Trigen ① |
| Lanarkshire |

| 19 Sean Quinn Group ① |
| Northern Ireland |

| 20 Wind Prospect/ PowerGen ③ |
| Cumbria (3) |

| 21 The Wind Company ① |
| Cumbria |

Figure 18.13

Wind farm at Altamont Pass, California

The UK

Apart from hydro-electricity, wind is the only renewable option being developed commercially in the UK. Britain's first wind farm was opened in 1991 near Camelford in Cornwall (Figure 18.14). The farm, on moorland 250 m above sea-level and where average wind speeds are 27 km/hr, generates enough electricity for 3000 homes. By mid-1999 there were 21 companies, 44 wind farms and 773 turbines producing 0.37 per cent of the UK's total annual needs – enough to supply a city slightly larger than Aberdeen, Norwich or Swansea. Future proposals include sites with an offshore location.

Solar energy

The sun, as stated earlier, is the primary source of the Earth's energy. Estimates suggest that the annual energy received from the sun (insolation), is 15 000 times greater than the current global energy supply. Solar energy is safe, pollution-free, efficient and of limitless supply. Unfortunately, it is expensive to construct solar 'stations', although many individual homes have had solar panels added, especially in climates that are warmer and sunnier than in Britain. It is hoped, globally, that future improvements in technology will result in reduced production costs. This would enable many developing countries, especially those lying within the tropics, to rely increasingly upon solar energy. In Britain, the solar energy option is less favourable partly due to the greater amount of cloud cover and partly to the long hours of darkness in winter when demand for energy is at its highest.

In 1999 Shell announced plans to establish the first large-scale rural electrification scheme for India, organised purely on a commercial basis, with plans to 'light up' 50 000 homes by 2003. The project will use solar power, and claims to protect the environment while eliminating safety hazards associated with traditional light sources such as paraffin and candles.

Wave power

Waves are created by the transfer of energy from winds which blow over them (page 140). In western Europe, winter storm waves from the Atlantic Ocean transfer large amounts of energy towards the coast where it may, in the future, be used to generate electricity. During the 1990s, experiments with harnessing the power of waves were conducted by Norwegians, at two small 'shore-line' commercial wave power stations. The hope was that similar offshore devices could be designed, to harness greater levels of potential energy, and still withstand the power of the waves. However, after several years of trials, technical problems forced the schemes to be abandoned.

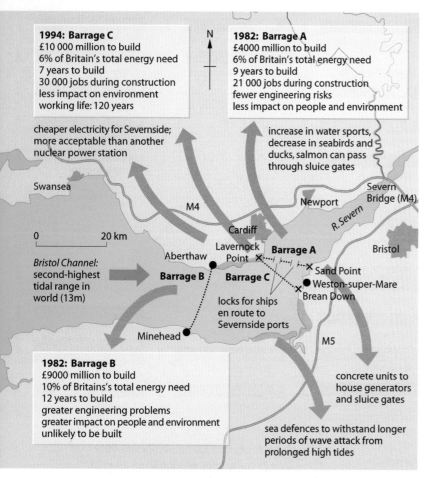

1994: Barrage C
£10 000 million to build
6% of Britain's total energy need
7 years to build
30 000 jobs during construction
less impact on environment
working life: 120 years

1982: Barrage A
£4000 million to build
6% of Britain's total energy need
9 years to build
21 000 jobs during construction
fewer engineering risks
less impact on people and environment

cheaper electricity for Severnside; more acceptable than another nuclear power station

increase in water sports, decrease in seabirds and ducks, salmon can pass through sluice gates

Swansea

N

Severn Bridge (M4)

Newport

M4

Cardiff

R. Severn

Bristol

0   20 km

Aberthaw

Lavernock Point ✕

Barrage A

Sand Point ✕

Bristol Channel: second-highest tidal range in world (13m)

Barrage B

Barrage C

Weston-super-Mare
Brean Down ✕

locks for ships en route to Severnside ports

Minehead

M5

1982: Barrage B
£9000 million to build
10% of Britains's total energy need
12 years to build
greater engineering problems
greater impact on people and environment
unlikely to be built

concrete units to house generators and sluice gates

sea defences to withstand longer periods of wave attack from prolonged high tides

Figure 18.15

Proposals for tidal barrages across the Severn estuary

Tidal energy

Of all the renewable resources, tidal energy is the most reliable and predictable but, to date, only places with the maximum tidal range offer potential sites. Incoming tides in four schemes – the Rance estuary in north-west France, the Bay of Fundy in Canada, at Kislaya in Russia, and Jiangxia in China – turn turbines, the blades of which can be reversed to harness the receding tide. Although the technology is now proven, economic and environmental costs are high. In Britain, the Severn barrage has been discussed for several decades. If built, it would provide the energy of five nuclear power stations but would cost the equivalent of six to construct (Figure 18.15). Other estuaries with large tidal ranges include the Mersey, Morecambe Bay and the Solway Firth.

Tidal barrages do not contribute to global warming nor to acid rain, but they are not favoured by environmentalists because they result in the permanent flooding of marshland and wetland habitats, favoured as feeding grounds by migratory birds, and have an adverse affect on spawning fish.

Geothermal energy

Several countries, especially those located in active volcanic areas, obtain energy from heated rocks and molten magma at depth under the Earth's surface, e.g. Iceland, New Zealand, Kenya, and several countries in central America (Figure 18.16). Cold water (Figure 18.17) is pumped downwards, is heated naturally and is then returned to the surface as steam which can generate electricity. Geothermal energy does pose environmental problems as carbon dioxide and hydrogen sulphide emissions may be high, the water supply can become saline, and earth movements can damage power stations.

Ocean thermal energy conversion

In tropical seas, there is a considerable temperature difference between surface water and that found at depth. The American government is funding an experimental and highly ambitious scheme off the coast of India aimed at converting this heat difference into energy.

Figure 18.16

Wairakei geothermal power station, New Zealand

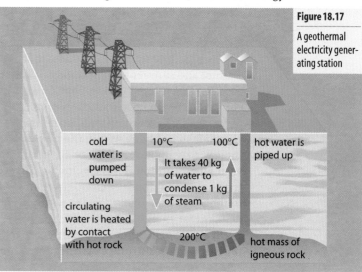

Figure 18.17

A geothermal electricity generating station

cold water is pumped down

10°C   100°C

It takes 40 kg of water to condense 1 kg of steam

hot water is piped up

circulating water is heated by contact with hot rock

200°C

hot mass of igneous rock

Biomass

Biomass refers to energy obtained from organic matter, i.e. crops, plants and animal waste. Brazil uses ethanol, a type of alcohol obtained from sugar cane, instead of petrol as a cheaper car fuel. The EU is at present conducting similar experiments using oilseed rape. In poorer parts of Africa, animal dung is allowed to ferment to produce methane gas. Although the methane provides a vital domestic fuel, it is a greenhouse gas, and if the dung is used for fuel it can no longer be used as a fertiliser. In more developed countries, it is now possible to convert the methane gas released at landfill 'waste' sites into energy. The present production of biomass energy in the USA is equivalent to 10 nuclear power stations. Some Scandinavian countries derive 10–15 per cent of their national energy consumption from biomass, most of it woody waste from the pulp and paper industries (Places 83, page 562).

Figure 18.18

Collecting fuelwood, Sub-Saharan Africa

Fuelwood

Trees are a sustainable resource, provided that those cut down are replaced, or that sufficient time is allowed for natural regeneration. The major source of energy in many of the least developed countries, especially those in Africa, is fuelwood. In these parts of the world, fuelwood – a fossil fuel – is needed daily for cooking and heating. As nearby supplies are used up, collecting fuelwood becomes an increasingly time-consuming task; in extreme cases, it may take all day (Figure 18.18 and Case Study 18). Many of these countries have a rapid population growth, which adds greater pressure to their often meagre resources, and lack the capital and technology to develop or buy alternative resources. In places where the demand for fuelwood outstrips the supply, and where there is neither the money to replant nor the time for regeneration, the risk of desertification and irreversible damage to the environment increases – i.e. the cycle of environmental deprivation (Case Study 7 and Figure 18.19).

Scientists in Wales are planning a trial to turn willow wood into a fuel of the future. Researchers from the School of Biosciences at Cardiff University, using grants from the EU, say that high-yielding varieties of willow can be grown on short-term coppices of three to four years, producing 10–30 tonnes of fresh wood per ha. The willow can then be harnessed for domestic use or turned into chips for electricity generation. If successful, it could transform farms on marginal hill land in upland Wales into forests of alternative energy (page 494). It could also be used to decontaminate heavy metal industrial sites, treating seepages from landfill sites, and de-watering sewage beds. The project runs until the end of 2001.

Energy conservation through greater efficiency

The UN 'Earth Summit' at Rio de Janeiro in 1992 was convened to initiate global action in cleaning up the environment. Energy was on the agenda, mainly in relation to global warming. One of the main objectives was to set global limits and timescales in reducing harmful emissions from vehicles, factories and power stations. Although all 106 participating countries agreed to reduce the amount of greenhouse gases released into the atmosphere and to seek methods of greater energy efficiency, little has been achieved due to opposition from vested interests and a lack of political will.

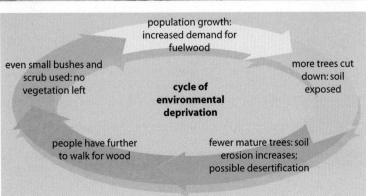

Figure 18.19

The cycle of environmental deprivation

Energy resources

There have been some achievements, however. Several industries, such as steel, have improved their techniques, which has led to a reduction in the amount of energy needed. Many factories have made savings by reducing to a single source electricity needed for heating, lighting and operating machinery. At home, heat loss has been reduced through roof and wall insulation and double glazing. The 'dash for gas' has meant, in the UK, a reduction in the number of coal- and oil-fired power stations (although gas, despite being cleaner, is still a source of greenhouse gases). An EU directive aims to reduce, by 2003, sulphur dioxide emissions to 60 per cent of their 1980 levels. The British government has stated that the UK needs to rely 'more on renewable energy and to reduce total energy consumption'. Whether this is achievable may depend upon interested opposition groups, e.g. national groups such as Friends of the Earth and Greenpeace (proposals for new power stations), RSPB (tidal barrages) and local residents (wind farms and plants to burn industrial waste).

Figure 18.20

Energy production in China, 1980–2005

Places 82   China: changes in energy production

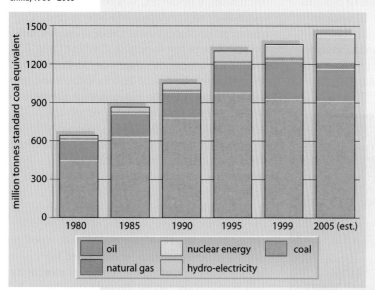

In the mid-1990s, China's energy industry was dominated by coal (Figure 18.20), which was not surprising since the country was producing nearly two-fifths of the world's total (Figure 18.5). Coal is mined in most parts of the country, although production is lower to the south of the Yangtze River and least in the mountains to the west (Figure 18.21). With industry, transport and homes all so reliant upon the burning of coal in one form or another, many Chinese cities experienced severe atmospheric pollution (Figure 18.22), and China was blamed for releasing annually 10 per cent of the world's greenhouse gases.

At that time, the remainder of China's energy budget was made up from oil, in which it was self-sufficient, and hydro-electricity. By 1995 China was ranked second in the world for generated energy and generating capacity. The country had 12 power grids, each characterised by large-scale generating units and power plants, high voltage and increased automation. The country had 35 large power stations – 26 thermal, 7 hydro-electric (the largest being at Gezhouba on the Yangtze), and 2 nuclear (Figure 18.21).

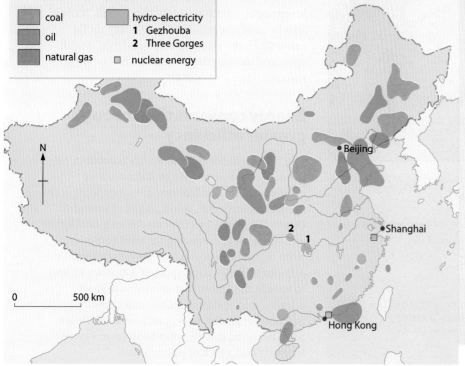

Figure 18.21

Energy supplies in China

Figure 18.22

Atmospheric pollution over Chengdu

By the turn of the century, several major changes were taking place. The most significant is China's attempt to close small coal mines and, by reducing production, also reducing various forms of pollution (Figure 18.23a). The use of oil had increased and there were plans to use more natural gas (Figure 18.23b). Hydro-electricity is expected to become an even more important energy resource once the Three Gorges Dam and its two power stations are completed in 2003 (Figure 18.24). Finally, nuclear power is expected to treble from its present figure of 1 per cent of China's energy output, by 2003 (Figure 18.23c).

a

DESPITE mounting difficulties, China will continue to close small coal mines and reduce coal production.

The work at the current stage is to eliminate the irrationally located coal mines within the large state-owned coal mines' areas of operation and the mines failing to meet the requirements for re-opening following upgrading.

In the first four months this year, the nation shut down 20 125 small coal mines, thereby reducing output by 81.36 million tonnes, according to the State Coal Industry Bureau (SCIB).

The strict policies issued by the State Council call for the closure of 25 800 small coal mines and to reduce output by 250 million tonnes this year because of oversupply in the coal market as well as environmental and safety concerns.

China Daily, 13 May 1999

b

CHINA will launch into its ambitious programme of development and use of clean energy early in the 21st century.

After five years of consideration, China is to shift its energy strategy towards the use of more natural gas and away from coal as pressure for environmental protection mounts. Annual use of natural gas in the country will be increased from last year's 22 billion m^3 to 80 billion in 2010, said the administration's deputy director, Chen Geng.

Natural gas development and transmission in high-pressure pipelines is usually risky. In China, major reserves are all verified in the remote west while most energy-thirsty cities are concentrated in the booming east, thousands of kilometres apart.

China Daily, 18 May 1999

c

NUCLEAR power is expected to contribute 3 per cent of China's power output by 2006, 2 percentage points more than at present, a senior official said yesterday.

China has two nuclear power stations under commercial operation, the Qinshan nuclear power station in East China's Zhejiang Province, and the Daya Bay nuclear power station in South China's Guangdong Province.

Construction of the second and third phases of the Qinshan and Ling'ao nuclear power stations, and the Lianyungang nuclear power plant, is under way.

More than a dozen provinces – including Shandong, Fujian, Hainan, Jiangxi, Hunan, Hubei and Sichuan – are preparing to build nuclear power stations, the official said.

China Daily, 15 May 1999

Figure 18.23

Changes in coal, natural gas and nuclear power production

Figure 18.24

The Three Gorges Dam on the Yangtze River

Development and energy consumption

To many people, especially in developed countries, economic development is linked to the wealth of a country, with wealth being measured by GNP per capita (page 631). Of several other variables that can be used to measure development, one is energy consumption per capita – i.e. how much energy, often given in tonnes of coal or oil equivalent, that each person in a country uses per year. Consequently a correlation between the wealth of a country and energy consumption might be expected (Framework 19, page 635).

The log-log graph in Figure 18.25 seems to show that there is a good, positive correlation between the two variables, i.e. as the wealth of a country increases, so too does its energy consumption. The huge gap in energy consumption between the developed and the developing world is shown in Figure 18.26. Note also that those countries above the line in Figure 18.25 tend to have more natural energy reserves (USA, Russia, Saudi Arabia) than those below the line (Japan, Italy, Peru).

Energy is the driving force behind most human activities, so it is fundamental to development. Energy allows people to make greater use of the resources that they have. According to Intermediate Technology (Case Study 18 and Places 90, page 577):

'reliable, accessible and affordable energy supplies can play an important role in improving living conditions in the developing world. They provide light and heat for homes, and power workshops that create jobs and generate wealth. Poor people in developing countries face particular problems in securing energy for their daily needs. This is a pressing issue in rural areas where most people live. More than half the world's population relies upon biomass fuel (usually wood but also charcoal, crop residues or animal dung). Poor people cannot afford alternative fuels such as gas or kerosene. National grids mainly serve urban areas, or large industrial operations; it is prohibitively expensive to extend them far into the countryside. In places where there are renewable energy resources such as the sun, wind and water, communities often lack the knowledge, expertise and capital needed to install the most appropriate system.'

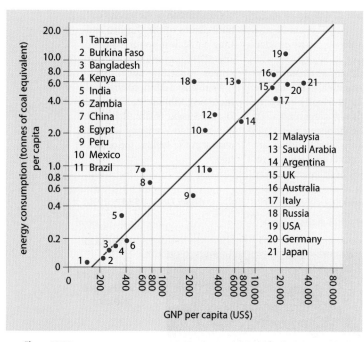

Figure 18.25

Correlation between GNP (US$) and energy consumption, 1995

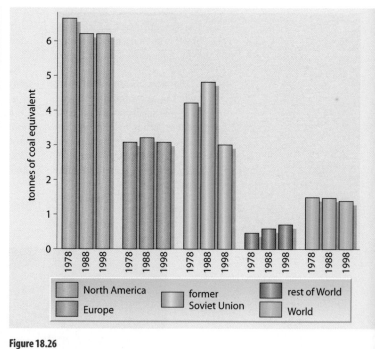

Figure 18.26

Energy consumption per capita in the developed and developing worlds

Appropriate technology: micro-hydro in Nepal

'An Appropriate Technology is exactly what it says – a technology appropriate or suitable to the situation in which it is used (page 576). If that situation is a highly industrialised urban centre the appropriate technology may well be "high tech". If, however, the situation is a remote Nepalese village "appropriateness" will be measured in the following terms:

- Is it culturally acceptable?
- Is it what people really want?
- Is it affordable?
- Is it cheaper or better than alternatives?
- Can it be made and repaired with local material, by local people?
- Does it create new jobs or protect existing ones?
- Is it environmentally sound?

For many decades "Aid" meant sending out the same large-scale, expensive, labour-saving technologies that we use: huge hydro-electric schemes, coal-fired power stations, diesel-powered generators. In some cases, for example towns and industrial areas, these have been appropriate. But such schemes do not reach the poorer communities in the rural areas. What was needed was some way of using local resources appropriately, and best of all some way of using renewable resources to decrease the need for reliance on outside help. Wind, solar and biogas energy are possibilities, but another resource widely available and already in use for thousands of years is water. Water is attracting much attention in the search for renewable sources of energy.

However, despite continuing public outrage at the devastating impact of large hydro-electric schemes on people's livelihoods and the environment (page 540 and Places 82), vast sums of money continue to be pumped into big dams and other inappropriate power generation plans. On the other hand, the intermediate approach, through small-scale hydro, has no negative impact on the environment, offers positive benefits to the local community, and uses local resources and skills.'

Intermediate Technology Development Group, 1990

Intermediate Technology Development Group and micro-hydro in Nepal

'The small Himalayan kingdom of Nepal ranks as one of the ten poorest countries in the world. Around 90 per cent of its 19 million people earn their living from farming, often at a subsistence level. The Himalayas offer Nepal one vast resource – the thousands of streams which pour down from the mountains all year round. Nepali people have harnessed the power in these rivers for centuries, albeit on a small scale (Figure 18.27).

About, 20 years ago, two local engineering workshops began to build small, steel, hydro power schemes for remote villages. These turbines have the advantage of producing more power than the traditional mills, as well as being able to run a range of agricultural processing machines (Figure 18.28). ITDG first became involved in Nepal's micro-hydro sector in the late 1970s when the local manufacturers asked for help in using their micro-hydro schemes to generate electricity.

In the mid-1980s, ITDG ran two training courses on micro-hydro power aimed at improving the technical ability of the nine new water turbine manufacturers that had been established in Nepal. These courses were very successful and prompted an agreement between ITDG and the Agricultural Development Bank (the agency which funds micro-hydro power in Nepal) to collaborate on the development of small water turbines for rural areas. This work not only improved and extended the range and number of micro-hydro schemes in Nepal, but also established ITDG as a leader in the field. In 1990 ITDG was included in a government task force investigating the whole area of rural electrification; and in 1992 ITDG was asked by the government to help establish an independent agency to promote all types of appropriate energy in rural areas of the country.'

Intermediate Technology, 1990

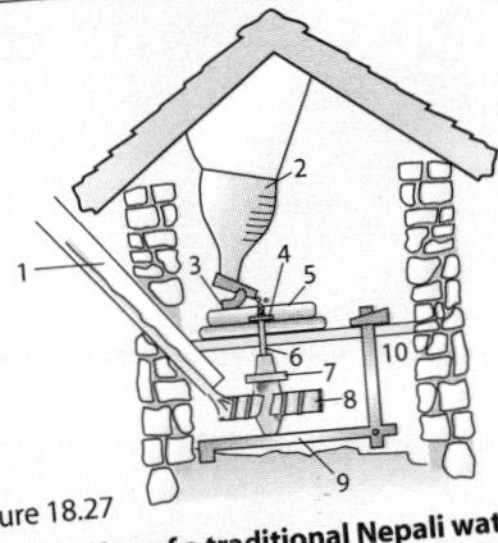

Figure 18.27
Cross-section of a traditional Nepali water mill

1 chute delivering the water to the paddles of the wheel
2 grain hopper (basket)
3 device to keep the grain moving
4 metal piece to lock top of shaft in upper millstone
5 grinding stones
6 metal shaft
7 thick wooden hub
8 wooden horizontal wheel, with obliquely set paddles attached to hub
9 metal pin and bottom piece
10 lifting device to adjust gap between millstones

Figure 18.28
Cross-section of a modern Nepali water turbine

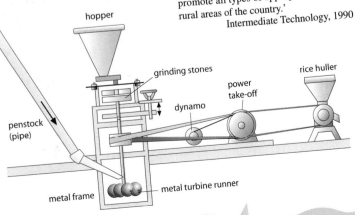

hopper

grinding stones

power take-off

rice huller

dynamo

penstock (pipe)

metal frame

metal turbine runner

Life before power

'Water power, harnessed using water wheels or *ghattas*, has been used for centuries for grinding corn. The micro-hydro system the village now has improved the efficiency of milling, so that what would take a woman four hours to grind by *ghatta* can be done in fifteen minutes. The power can also be used for dehusking rice and extracting oil from sunflower seeds (Figure 18.29).

The mechanical power produced by Ghandruk's micro-hydro system is also converted to electric power, which is distributed to every house in the village. Apart from the obvious benefit of lighting, many households are starting to use electric cookers or *bijuli dekchis*, which work like slow cookers (Figure 18.29).

Women are turning to *bijuli dekchis* because they reduce smoke levels in the kitchen, they save time by reducing the amount of firewood the family needs to collect, and they are more convenient and cook faster than traditional stoves. In a country ravaged by deforestation – villagers spend up to 12 hours on a round trip to collect wood – fuel saving is becoming more and more important.

Micro-hydro schemes like the one in Ghandruk work because the community has "ownership" of the scheme by participating in its planning, installation and management; because the machinery needed can be made and maintained by local manufacturers using local materials available in the country; and because production and consumption are linked within a community.

The lives of villagers all over Nepal are literally being lit up by micro-hydro schemes, and the country could serve as a model for decentralised, sustainable energy production. Already, 700 mechanical and 100 electrical schemes have been installed.

Much of the impetus for the development of hydro in Nepal initially stemmed from the absence of fossil fuel reserves to exploit. However, if the Government can resist the temptations of big dam schemes and the dollars being thrown at them by the big, international donor agencies, it could have the last laugh watching the rest of the world scrabble for the last of fossil fuel reserves.'

Intermediate Technology
Development Group

Crop processing

A woman grinds corn for the family meal.

An elderly Nepalese woman and young girl hull rice with a traditional footpowdered dhiki.

Cooking

Cooking on an open fire.

Living

Light comes from a single kerosene lamp.

Power for life

Figure 18.29

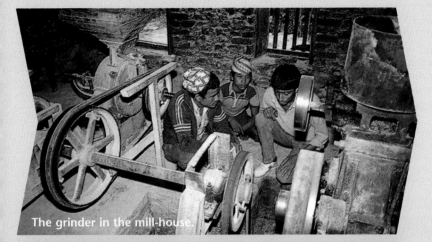

The grinder in the mill-house.

Grinding enough corn to feed a family for just 3 days takes 15 hours when it is done by hand.

By taking corn to the grinder in the mill-house – usually a popular meeting place for villagers – 3 days' worth of corn can be ground in just 15 minutes.

Rice being hulled mechanically in a 3kW mill.

For thousands of women, the supply of power releases them from the many labour-intensive and time-consuming tasks they previously had to carry out by hand.

Villagers can now hull their rice mechanically with this 3 kW mill, driven by a micro-hydro turbine. Time is saved and quality and productivity increased.

Low-wattage electric cooker

Cooking on an open fire burns up a great deal of wood (which is becoming increasingly scarce) and gives off a lot of thick smoke. As a result, the villagers not only have to walk long distances to collect their fuel, but many women and children suffer from serious lung disorders.

IT is helping to develop two low-wattage electric cookers which have been specifically designed to make use of 'off-peak' electricity. The *bijuli dekchi* heats water during off-peak times for use in cooking later on, while the heat storage cooker stores the energy available during off-peak periods and releases it at mealtimes for cooking. Both save fuelwood and help to reduce deforestation.

A nepalese family enjoying the benefits of electic light.

Kerosene lamps are costly to run, and those who can afford them have to collect fuel in cans from towns which are usually several days' walk away.

With electric light, children and adults can improve their education by learning to read and write in the evenings. Electric light is also cheaper, cleaner and brighter than kerosene.

References

Department of the Environment (1992)
The UK Environment, HMSO.

Intermediate Technology (various dates)
Mirco-hydro in Nepal, IT.

Middleton, N. (1999) *The Global Casino:
An Introduction to Environmental Issues*
2nd edition, Arnold.

Philip's Geographical Digest, 1998–99
(1998) Heinemann/Philips.

Pickering, K. and Owen, L. (1994)
Global Environmental Issues, Routledge.

Websites
**US Energy Information
Administration:**
http://www.eia.doe.gov/

**BP Amoco Statistical Review of World
Energy:**
http://www.bpamoco.com/worldenergy/
oil/index.htm

**Royal Dutch/Shell Group,
Discovery Zone:**
http://www.shell.com/zone/directory/
0,1387, 1057, 00.html

British Wind Energy Association:
http://www.bwea.com/contents.html

China Environment, Energy page:
http://www.chinaenvironment.com/
energy/index.html

See also for more links:
http://www.nelsonthornes.com/gaia

Questions

Q

1 a i What are 'natural resources'? **(1 mark)**
  ii What is the difference between renewable and
   non-renewable resources? **(2 marks)**
  iii Name a renewable source of energy which is used
   commercially. State where it is produced and explain
   why conditions in that area are suitable. **(3 marks)**
  iv Explain what will happen to the amount of reserves
   of a fuel such as natural gas if:
   ■ the market price of gas goes up;
   ■ new technology is developed, allowing deeper
    wells to be drilled. **(4 marks)**
 b Study Figure 18.30.
  i Describe the main trends shown by the graph. **(4 marks)**
  ii During the 1990s the use of energy resources by the
   more economically developed countries did not
   increase, and may even have fallen. At the same time
   the amount used by less economically developed
   countries increased.
   Explain this situation. **(4 marks)**
 c Describe the main features of the world trade
  in any one fuel. **(7 marks)**

2 a In many less economically developed countries
  fuelwood is the main source of energy for heating and
  cooking. Explain how this can cause:
  i damage to the environment
  ii social problems. **(10 marks)**
 b i What does 'Appropriate Technology' mean? **(2 marks)**
  ii Appropriate technology can be used by poor people
   in remote areas to harness energy supplies. Describe
   one such scheme in a named region of the world.
    (5 marks)
  iii Explain how the scheme described in ii brings social and
   economic benefits to the people who use it. **(8 marks)**

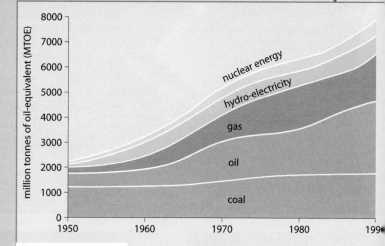

Figure 18.30

World energy use, 1950–90

3 a What is meant by the term 'fossil fuel'? **(2 marks)**
 b Choose **one** country which has important reserves of coal.
  i Briefly describe the distribution of coal reserves in
   that country. **(2 marks)**
  ii Explain the economic factors which are influencing
   decisions about whether those reserves should be
   exploited at the present time. **(6 marks)**
  iii Name **one** environmental problem caused by the use of
   coal as a fuel. Describe the problem. Explain how good
   management can reduce the problem. **(10 marks)**
 c 'Natural gas has become the fastest-growing energy
  resource, providing an alternative to coal and oil.'
  Explain why this is so. **(5 marks)**

AS

4 a Choose **one** country which has important reserves of coal.
  Briefly describe the distribution of coal reserves in
  that country. **(2 marks)**
  Discuss the main factors which influence decisions
  about whether that country's coal reserves should
  be exploited. **(15 marks)**
 b 'Natural gas has become the fastest-growing energy
  resource, providing an alternative to coal and oil.'
  Explain why this is so. **(10 marks)**

5 Lack of a suitable power supply is holding back
 development in many remote areas of the world. For a
 named area:
 a explain how shortage of power has caused economic
  and social problems **(12 marks)**
 b explain how the problems are being reduced by
  provision of an appropriate power supply. **(13 marks)**

A2

6 Study Figure 18.31.
  a 'Wealthier countries usually have a higher IFD than poor countries.'
    i To what extent is this statement supported by Figure 18.31? **(5 marks)**
    ii Suggest why the relationship that you have described in i exists. **(7 marks)**
    iii Choose one country that does not fit the general pattern and suggest why it is an anomaly. **(4 marks)**
  b Choose a renewable source of energy that can be used for generating electricity.
    i Describe the environmental conditions needed for this source to be developed successfully. **(4 marks)**
    ii Explain why the governments of many countries are investing money to subsidise the development of renewable sources of energy. **(5 marks)**

7 **a** In the mid-1990s about 75 per cent of China's energy came from coal.
    i Explain why coal became so important as a source of energy in China. **(5 marks)**
    ii Describe **two** problems caused by China's reliance on coal, and explain what is being done to rationalise China's coal-mining industry. **(8 marks)**
  b As coal production is reduced, China intends to increase its use of:
    ■ natural gas
    ■ hydro-electric power
    ■ nuclear power.
  Choose two of these forms of power. For each of your choices:
    i explain the choices of locations for production of the power;
    ii discuss the environmental consequences of the development of the power source. **(6+6 marks)**

The map shows an Index of Fuel Diversification (IFD). This is based on countries' use of four different sources of energy – coal, oil, natural gas and others (including nuclear, hydro-electric, solar, wind and fuelwood). The IFD varies between 0%, where only one fuel is used, and 100% when the four fuels are used in equal amounts.

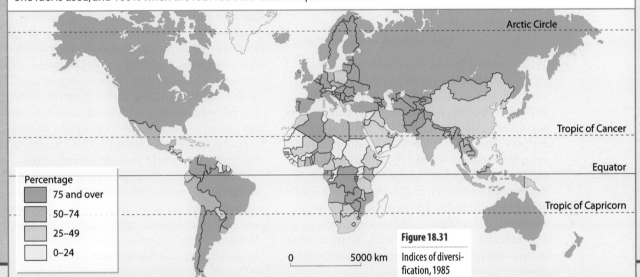

Arctic Circle

Tropic of Cancer

Equator

Tropic of Capricorn

Percentage
  75 and over
  50–74
  25–49
  0–24

0   5000 km

Figure 18.31

Indices of diversification, 1985

8 Study Figure 18.31.
  a To what extent does the map support the statement: 'More economically developed countries usually rely on a single source of energy.' **(17 marks)**
  b Choose:
    i **one** country with a high IFD
    ii **one** country with a low IFD.
    For each of your chosen countries explain how socio-economic and physical conditions have combined to produce this IFD. **(18 marks)**

9 Some of the most heavily populated countries in the less economically developed world are undergoing rapid economic development at the start of the 21st century. This means that their demand for energy is likely to grow quickly. What consequences might this have for the global environment and the global economy? **(25 marks)**

10 **a** Fuel wood is an important source of power in many remote regions in less economically developed countries (LEDCs). Name an example of a region where fuel wood is widely used and:
    i explain why people in that region rely on fuel wood. **(3 marks)**
    ii describe some of the problems caused for the economy and the environment by the reliance on fuel wood. **(5 marks)**
  b Large hydro-electric power schemes have been seen as the solution to the energy shortages of many LEDCs.
    i suggest why some people see the scheme as being a welcome development for that country. **(5 marks)**
    ii suggest why other people see the scheme as being unwelcome. **(5 marks)**
  c Recent conferences on global warming have concluded that more economically developed countries should share their technological knowledge with the LEDCs. How might such sharing help to reduce global warming in future? **(7 marks)**

Chapter 19

Manufacturing industries

'Science finds, industry applies, man confirms.'

Anon, Chicago World Fair 1933

'We need methods and equipment which are cheap enough so that they are accessible to virtually everyone; suitable for small-scale production; and compatible with man's need for creativity. Out of these three characteristics is born non-violence and a relationship of man to nature which guarantees permanence. If one of these three is neglected, things are bound to go wrong.'

E. F. Schumacher, *Small is Beautiful*, 1974

What is meant by industry? In its widest and more traditional sense, the word industry is used to cover all forms of economic activity: **primary** (farming, fishing, mining and forestry); **secondary** (manufacturing and construction); **tertiary** (back-up services such as administration, retailing and transport); and **quaternary** (high-technology and information services/ knowledge economy). In this chapter, the use of the term 'industry' has been confined to its narrowest definition, i.e. manufacturing. Manufacturing industry includes the processing of raw materials (iron ore, timber) and of semi-processed materials (steel, pulp).

It needs to be pointed out, however, that while this definition may be convenient, it does create several major problems. At present, only some 19 per cent of the UK's working population are employed in manufacturing, a trend that is repeated across most of the developed market economies. This shift from an industrial to a post-industrial society is shown in Figure 19.1. In reality, it is also unrealistic to draw boundaries between 'manufacturing' and 'services'. Not only are the two integrated in reality through linkages (page 568 and Figure 19.2), buyer–supplier relations, etc., but many people who are officially classified as working in the manufacturing sector also have occupations that are service based (salespeople, administrators) within 'manufacturing' sector firms. It can be argued, with much justification, that it is conceptually (and empirically) unrealistic to sever manufacturing from services. This distinction becomes particularly problematic when discussing, for example, high-tech developments along the M4 (Places 86, page 566) as, by their nature, many firms are 'information-intensive' and knowledge based rather than production or materials based; or when describing the differences between the 'formal' and 'informal' sectors in less economically developed, less industrialised countries (page 574).

Figure 19.1

Towards a post-industrial economy: employment structure in the UK, 1841–1997

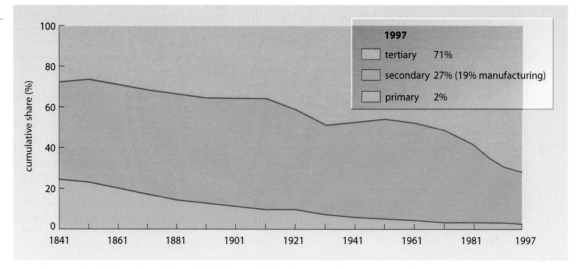

1997

☐ tertiary   71%
☐ secondary 27% (19% manufacturing)
☐ primary   2%

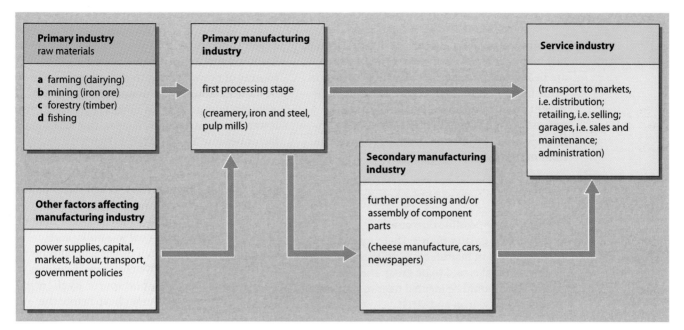

Figure 19.2

Linkages between various sections and types of industry

The location of industry

The processes which contribute to determine the location and distribution of industry are more complex and dynamic than those affecting agriculture. This means that the making of generalisations becomes less easy and the dangers of stereotyping increase. Some reasons for this complexity include:

■ Some locations were chosen before the Industrial Revolution and many more during it. Initial factors favouring a location may no longer apply today. For example, the original raw materials may now be exhausted (iron ore and coal in west Cumbria), or replaced by new innovations (cotton replaced by synthetic fibres) and sources of energy (water power replaced by electricity).

■ New locational factors which were not applicable last century include cheaper and more efficient transport systems, the movement of energy in the form of electricity, new techniques and automation.

■ Some industries have developed from older industries and are linked to these former patterns of production even when the modern product is different (the Mazda Car Corporation began as a cork-making and then a machine-tools firm).

■ Before the 20th century, industry was usually financed and organised by individual **entrepreneurs** who initiated and organised, usually for a profit, an enterprise or business; this included risk-taking, deciding what goods would be produced or services provided, the scale of production, and marketing. Nowadays decisions concerning modern industry are often taken far away from the location of a particular factory, either by the state or by **transnationals** (**multinationals**, page 573).

■ Many factories now produce a single component and therefore are a part of a much larger organisation which they supply.

■ The sites of many early factories were chosen by individual preference or by chance, i.e. the founder of a firm just happened to live at, or to like, a particular location (Unilever at Port Sunlight and Rowntree at York). Other industries have found their present locations as a result of 'trial and error'.

Factors affecting the location of manufacturing industry

Raw materials

Industry in 19th-century Britain was often located close to raw materials (ironworks near iron ore), sources of power (coalfields) or ports (to process imports), mainly due to the immobility of the raw materials which were heavy and costly to move when transport was expensive and inefficient. In contrast, today's industries are rarely tied to the location of raw materials and so are described as **footloose**. There is now a greater efficiency in the use of raw materials; power is more mobile; transport of raw materials, finished products and the workforce is more efficient and relatively cheaper; components for many modern, and especially high-tech, industries are relatively small in size and light in weight; and some firms may simply rely on assembling component parts made elsewhere. A location close to markets, labour supply or other linked firms has become increasingly important.

Industries that still need to be located near to raw materials are those using materials which are heavy, bulky or perishable; which are low in value in relation to their weight; or which lose weight or bulk during the manufacturing process. **Alfred Weber**, whose theory of industrial location is referred to later, introduced the term **material index** or **MI**.

$$MI = \frac{\text{total weight of raw materials}}{\text{total weight of finished product}}$$

There are three possible outcomes.

1 If the MI is greater than 1, there must be a weight loss in manufacture. In this case, the raw material is said to be **gross** and the industry should be located near to that raw material, e.g. iron and steel:

$$MI = \frac{6 \text{ tonnes raw material}}{1 \text{ tonne finished steel}} = 6.0$$

2 If the MI is less than 1, there must be a gain in weight during manufacture. This time the industry should be located near to the market, e.g. brewing:

$$M1 = \frac{1 \text{ tonne raw material}}{5 \text{ tonnes beer}} = 0.2$$

3 Where the MI is exactly 1, the raw material must be **pure** as it does not lose or gain weight during manufacture. This type of industry could therefore be located at the raw material, the market or any intermediate point.

Industries that lose weight during manufacture include food processing (butter has only one-fifth the weight of milk, refined sugar is only one-eighth the weight of the cane), smelting of ores (copper ore is less than 1 per cent pure copper, iron ore has a 30–60 per cent iron content; Places 84, page 563) and forestry (paper has much less mass than trees; Places 83, page 562). Industries that gain weight in manufacture include those adding water (brewing and cement), and those assembling component parts (cars; Places 85, page 565; and electrical goods; Places 86, page 566). In these cases, the end product is more bulky and expensive to move than its many smaller constituent parts.

Power supplies

Early industry tended to be located near to sources of power, which in those days could not be moved. However, as newer forms of power were introduced and the means of transporting it were made easier and cheaper, this locational factor became less important (Figure 19.3).

During the medieval period, when water was a prime source of power, mills had to be built alongside fast-flowing rivers. When steam power took over at the beginning of the Industrial Revolution in Britain, factories had to be built on or near to coalfields, as coal was bulky and expensive to move. When canals and railways were constructed to move coal, new industries were located along transport routes. By the mid-20th century, oil (relatively cheap before the 1973 Middle East War) was being increasingly used as it could be transported easily by tanker or pipeline. This began to free industry from the coalfields and to offer it a wider choice of location (except for such oil-based industries as petrochemicals). Today, oil, coal, natural gas, nuclear and hydro-electric power can all be used to produce electricity to feed the National Grid. Electricity, in addition to its cleanliness and flexibility, has the advantage that it can be transferred economically over considerable distances either to the long-established industrial areas, where activity is maintained by **geographical inertia**, or to new areas of growth. (In 1900, electricity could be transmitted economically only 59 km; today, the distance is over 1500 km.)

Transport

Transport costs were once a major consideration when locating an industry. Weber based his industrial location theory on the premise that transport costs were directly related to distance (compare von Thünen's assumptions, page 471). Since then, new forms of transport have been introduced, including lorries (for door-to-door delivery), railways (preferable for bulky goods) and air (where speed is essential). Meanwhile, transport networks have improved, with the building of motorways, and methods of handling goods have become more efficient through containerisation. For the average British firm, transport costs are now only 2–3 per cent of their total expenditure. Consequently, raw materials can be transported further and finished goods sold in more distant markets without any considerable increase in costs. Firms previously tied to coastal sites relying on overseas materials or markets, or earlier nodal points, now have a freer choice of location – although many still favour growth-point axes such as the M4 corridor (Places 86, page 566).

Period	Source of power	Examples of location
early iron industry	charcoal	wooded areas (the Weald, the Forest of Dean)
later iron industry	waterwheels	fast-flowing rivers (R. Don, Sheffield)
early steel industry	coal	coalfields (South Wales, north-east England)
present-day steel industry	electricity	coastal (Port Talbot, Newport)

Figure 19.3

Power supply and the location of iron and steelworks

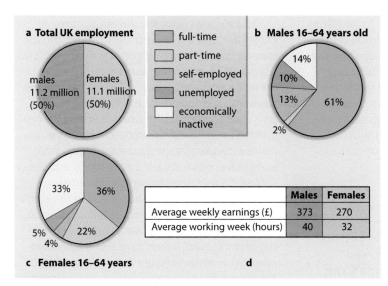

a Total UK employment

full-time
part-time
self-employed
unemployed
economically inactive

males
11.2 million
(50%)

females
11.1 million
(50%)

b Males 16–64 years old

14%
10%
13%
2%
61%

33%
36%
5%
4%
22%

c Females 16–64 years

d

	Males	Females
Average weekly earnings (£)	373	270
Average working week (hours)	40	32

Figure 19.4

UK employment data, 1996

There are two types of transport costs from an economic point of view. **Terminal costs** refer to the time and equipment needed to handle and store goods (cranes and warehouses) and the costs of providing the transport system (building railway tracks, docks and motorways). **Long-haul costs** refer to the cost of actually moving the goods (lorries, petrol and salaries). These costs and the advantages and disadvantages of different forms of transport are summarised in Figure 21.1.

Markets

Today, the pull of a large market is more important than the location of raw materials and power supplies; indeed, it has been suggested that flexibility and rapid response to changing market signals are perhaps the most important determinants of location.

Industries will locate near to markets if:

■ the product becomes more bulky with manufacture or there are many linkage industries involved (the assembling of motor vehicles)

■ the product becomes more perishable after processing (bread is more perishable than flour); it is sensitive to changing fashion (clothes); or it has a short life-span (daily newspapers)

■ the market is very large (north-eastern states of the USA, or south-east England)

■ the market is wealthy

■ prestige is important (publishing).

Labour supply

Labour varies spatially in its cost, availability and quality.

In the 19th century, a huge force of semi-skilled, mainly male, workers operated in large-scale 'heavy' industries doing manual jobs in steelworks, shipyards and textile mills. Today,

there are fewer semi-skilled and more highly skilled workers operating in small-scale 'light' industries which increasingly rely upon machines, computers and robots. The cost of labour, especially in EU countries, can be high, accounting for 10–40 per cent of total production costs. Three consequences of this have been the introduction of mechanisation to reduce human inputs, the exploitation of female labour, and the use of 'cheap' labour in developing countries.

Traditionally, labour has been relatively immobile. Although there was a drift to the towns during the Industrial Revolution, since the First World War British people have expected respective governments to bring jobs to them rather than they themselves having to move for jobs. Certain industries need specialist training (cutlery, furniture-making and electronics) and so are difficult to develop in a new area with an untrained labour force. If you wished, for example, to open a cutlery business you would probably choose Sheffield in preference to Exeter or Aberdeen. However, as industries become more mechanised and transport improvements allow greater mobility, firms can locate, and their workforce can travel, more freely. A problem in the mid-1990s in Britain was that, despite a high level of unemployment, there was a shortage of highly skilled people, partly due to a lack of training.

Similarly, the roles of women and trade unions have both changed. At the turn of the 21st century, half of Britain's workforce were women (Figure 19.4a), with an increasing number either seeking career jobs or prepared to work part-time (Figure 19.4c), even flexi-time, although many still have to accept a lower salary than males (Figure 19.4d). The role of trade unions has declined significantly as their membership numbers have fallen with the decline of the large 'heavy' industries.

Capital

Capital may be in three forms.

1 **Working capital** (money) which is acquired from a firm's profits, shareholders or financial institutions such as banks. Money is mobile and can be used within and exchanged between countries. Location is rarely constrained by working capital unless money is to be borrowed from the government which might direct industry to certain areas (see below). In Britain, capital is more readily available in the City of London, where most of the financial institutions are based.

2 **Physical** or **fixed capital** refers to buildings and equipment. This form of capital is not mobile, i.e. it was invested for a specific use.

3 **Social capital** and cultural amenities are linked to the workforce's out-of-work needs rather than to the factory or office itself. Houses, hospitals, schools, shops and recreational amenities are social capital which may attract a firm, particularly its management, to an area.

Government policies

These should aim to even out differences in employment, income levels and investment within a country. In 1934, regional assistance was granted in Britain to Clydeside, west Cumberland, North-East England and South Wales (compare these areas with the 1993 assisted areas in Figure 19.5). This was followed by the setting up of trading/industrial estates in areas of high unemployment such as the Team Valley (Gateshead), Trafford Park (Manchester) and Treforest (north of Cardiff). Since then successive governments have tried various schemes to try to dissuade firms from locating in certain areas (mainly the South East) and giving inducements if they locate in others.

In the early 1980s, the government set up Enterprise Zones where unemployment and the state of the environment posed serious problems (Figure 19.5). Firms that located in EZs were exempt from rates, received 100 per cent capital allowances on industrial and commercial property, were subject to simplified planning procedures (so long as health, safety and pollution control standards were guaranteed), and were exempt from industrial training levies. Further policies were introduced under the Urban Development Corporations (UDCs) in the 1980s to try to help badly hit docklands (London and Liverpool, Places 56 page 440) and other inner-city areas. Governments have also made direct attempts to attract foreign firms (Nissan and Toyota) and to encourage private–public partnerships. The latest government initiative was the launch, in April 1999, of the Regional Development Agencies (RDAs). These agencies act on behalf of their region to ensure economic development, social cohesion and sustainability. At a time of limited inward investment, the competition between the various UK regions to secure investment is likely to become intense, and may well shape the UK's future economic geography.

Land

In the 19th century, extensive areas of flat land were needed for the large factory units. Today, although modern industry is usually smaller in terms of land area occupied, it prefers cheaper land, less congested and cramped sites and improved accessibility, as are to be found on greenfield sites on the edges of cities and in smaller towns. Government policies during the 1990s, under mounting pressure from environmental and local pressure groups, were aimed at attracting new industry to derelict and underused brownfield sites and former industrial premises, especially those in the inner cities (page 439). Such policies are aimed at utilising existing infrastructure and protecting the green belt.

Environment

The 1980s saw an increasing demand by both managers and workforces to live and work in a more pleasant environment. This has led to firms seeking locations in smaller towns within easy reach of open countryside and away from the commuting problems and high land values of south-east England. There is still, however, a common perception (stereotype) that 'the north' is all Coronation Streets and 'the south' all quiet rural villages. Consequently, potential employers have to work hard to 'sell' any social or environmental advantages which their area possess.

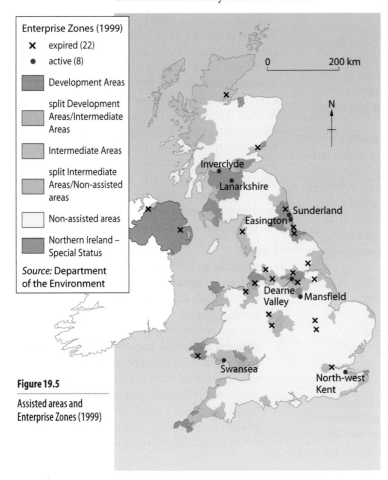

Enterprise Zones (1999)

✕ expired (22)

● active (8)

Development Areas

split Development Areas/Intermediate Areas

Intermediate Areas

split Intermediate Areas/Non-assisted areas

Non-assisted areas

Northern Ireland – Special Status

Source: Department of the Environment

0   200 km

N

Inverclyde

Lanarkshire

Sunderland

Easington

Dearne Valley   Mansfield

Swansea

North-west Kent

Figure 19.5

Assisted areas and Enterprise Zones (1999)

Theories of industrial location

We have already seen that 'models form an integral and accepted part of present-day geographical thinking' (Framework 12, page 352). Models, as in many branches of geography, have been formulated in an attempt to try to explain, in a generalised, simplified way, some of the complexities affecting industrial location. Two of the more commonly quoted models are those based upon:

1 the industrialist who seeks the lowest-cost location (LCL) – Weber, 1909
2 the industrialist who seeks the area that will give the highest profit – Smith, 1971.

Before looking more closely at these two models, it must be remembered that, as models, they form theoretical frameworks which may be difficult to observe in the real world, but against which reality can be tested. Mainly due to constant changes in the world economy, the validity of both 'least cost' and 'profit maximisation' have come under question. What is important, however, is that you should:

a recognise that there is no single body of theory, i.e. model, to explain the real world but, rather, there exists a variety of approaches; and
b begin to appreciate the complexity of industrial organisation in the real world and not to assume that conceptual models actually work.

Weber's model of industrial location

Alfred Weber was a German 'spatial economist' who, in 1909, devised a model to try to explain and predict the location of industry. Like von Thünen before him and Christaller later, Weber tried to find a sense of order in apparent chaos, and made assumptions to simplify the real world in order to produce his model. These assumptions were as follows:

■ There was an isolated state with flat relief, a uniform transport system in all directions, a uniform climate, and a uniform cultural, political and economic system.
■ Most of the raw materials were not evenly distributed across the plain (this differs from von Thünen). Those that were evenly distributed (water, clay) he called **ubiquitous materials**. As these did not have to be transported, firms using them could locate as near to the market as was possible. Those raw materials that were not evenly distributed he called **localised materials**. He divided these into two types: gross and pure (page 554).

■ The size and location of markets were fixed.
■ Transport costs were a function of the mass (weight) of the raw material and the distance it had to be moved. This was expressed in tonnes per kilometre (t/km).
■ Labour was found in several fixed locations on the plain. At each point it was paid the same rates, had equivalent skills, was immobile and in large supply. Similarly, entrepreneurs had equal knowledge, related to their industry, and motivation.
■ Perfect competition existed over the plain (i.e. markets and raw materials were unlimited) which meant that no single manufacturer could influence prices (i.e. there was no monopoly). As revenue would therefore be similar across the plain, the best site would be the one with the minimal production costs (i.e. the least-cost location or LCL).

Possible least-cost locations

Weber produced two types of locational diagram. A straight line was sufficient to show examples where only one of the raw materials was localised (it could be pure or gross). However, when two localised raw materials were involved, he introduced the idea of the **locational triangle**. Figure 19.6 summarises the nine possible variations based on the type of raw material involved.

1 One gross localised raw material. As there is weight loss during manufacture (the material index for a gross raw material is more than 1) then it is cheaper to locate the factory at the source of the raw material – there is no point in paying transport costs if some of the material will be left as waste after production (Figure 19.7a).
2 a One ubiquitous raw material or b one pure localised raw material gaining weight on manufacture (MI less than 1). If the raw material is found all over the plain (ubiquitous) then transport is unnecessary as it is already found at the market. If a pure material gains mass on manufacture then it is cheaper to move it rather than the finished product and so again the LCL will be at the market (Figure 19.7b).
3 One pure localised raw material. If this neither gains nor loses weight during manufacture (MI = 1), the LCL can be either at the market, at the location of the raw material, or at any intermediate point (Figure 19.7c).

4 Two ubiquitous (gross or pure) raw materials. As these are found everywhere, they do not have to be transported and so the LCL is at the market.

5 Two raw materials: one ubiquitous and one pure and localised. The LCL is at the market because the ubiquitous material is already there and so only the pure localised material has to be transported (Figure 19.8a). It will be cheaper to move one raw material than the more cumbersome final product.

6 Two raw materials: one ubiquitous and the other gross and localised. The ubiquitous material is available at every location. As the gross material loses weight, the LCL could theoretically be at any intermediate point between its source and the market. However, if the mass of the product is greater than that of the localised raw material, the LCL is at the market; if it is less, the LCL is at the location of the raw material; and if it is the same, the LCL is at the mid-point (Figure 19.8b).

7 Two raw materials: both localised and pure. In the unlikely event of the two raw mater-ials lying to the same side of and in line with the market, the LCL will be at the market. If the materials do not conform with this arrangement but form a triangle with the market (Figure 19.9), the LCL is at an inter-mediate point near to the market. This is because the weight and therefore the trans-port costs of the raw material are the same as, or less than, those of the product.

8 Two localised raw materials: one pure and one gross. In this case, the industry will locate at an intermediate point (Figure 19.10a). The greater the loss of weight during production, the nearer the LCL will be to the source of the gross material.

9 Two raw materials: both localised and gross. If both raw materials have an equal loss of weight, the LCL will be equidistant between these two sources but closer to them than to the market (Figure 19.10b1). However, if one raw material loses more mass than the other, the industry is more likely to be located closer to it (Figure 19.10b2).

Figure 19.6

Least-cost locations dependent upon types of raw material

Type(s) of raw material (RM) MI = material index	LCL at raw material	LCL at any intermediate point	LCL at market
1 one gross localised RM MI >1	■		
2 one RM gaining weight or one ubiquitous RM MI <1			■
3 one pure localised RM MI = 1		■	
4 two ubiquitous RMs (pure or gross)			■
5 two RMs (one ubiquitous and one pure)			■
6 two RMs (one ubiquitous, one gross)		(could be any site, according to amount of weight loss)	
7 two RMs (both pure)			■
8 two RMs (one pure, one gross)	■ (if big weight loss)	■ (if a small weight loss)	
9 two RMs (both gross)	■ (at RM with greatest weight loss)	■ (equal weight loss)	

Figure 19.7

Least-cost locations with one raw material

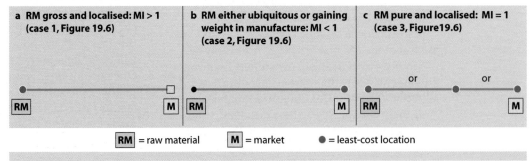

a RM gross and localised: MI > 1 (case 1, Figure 19.6)

b RM either ubiquitous or gaining weight in manufacture: MI < 1 (case 2, Figure 19.6)

c RM pure and localised: MI = 1 (case 3, Figure19.6)

RM = raw material   M = market   ● = least-cost location

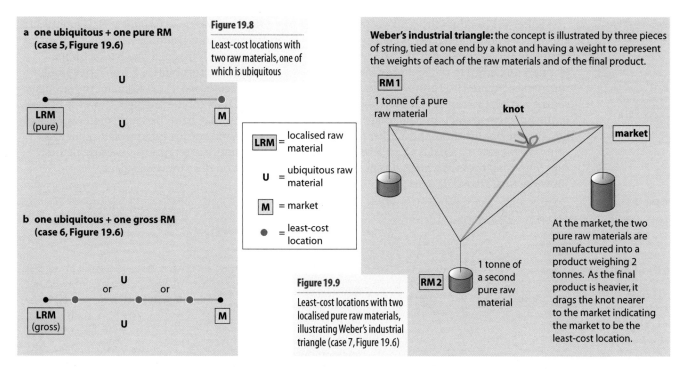

a one ubiquitous + one pure RM (case 5, Figure 19.6)

U

U

LRM (pure) ——————— M

b one ubiquitous + one gross RM (case 6, Figure 19.6)

U

or   or

LRM (gross)   U   M

Figure 19.8

Least-cost locations with two raw materials, one of which is ubiquitous

LRM = localised raw material

U = ubiquitous raw material

M = market

● = least-cost location

Figure 19.9

Least-cost locations with two localised pure raw materials, illustrating Weber's industrial triangle (case 7, Figure 19.6)

Weber's industrial triangle: the concept is illustrated by three pieces of string, tied at one end by a knot and having a weight to represent the weights of each of the raw materials and of the final product.

RM 1

1 tonne of a pure raw material

knot

market

RM 2

1 tonne of a second pure raw material

At the market, the two pure raw materials are manufactured into a product weighing 2 tonnes. As the final product is heavier, it drags the knot nearer to the market indicating the market to be the least-cost location.

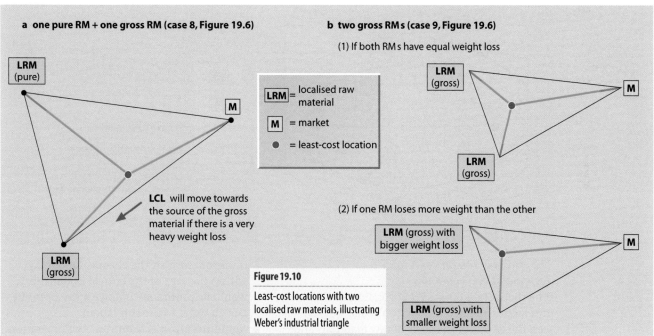

a one pure RM + one gross RM (case 8, Figure 19.6)

LRM (pure)

M

LCL will move towards the source of the gross material if there is a very heavy weight loss

LRM (gross)

LRM = localised raw material

M = market

● = least-cost location

b two gross RMs (case 9, Figure 19.6)

(1) If both RMs have equal weight loss

LRM (gross)

M

LRM (gross)

(2) If one RM loses more weight than the other

LRM (gross) with bigger weight loss

M

LRM (gross) with smaller weight loss

Figure 19.10

Least-cost locations with two localised raw materials, illustrating Weber's industrial triangle

Weber claimed that four factors affected production costs: the cost of raw materials and the cost of transporting them and the finished product, together with labour costs and agglomeration/deglomeration economies (pages 559–561).

Spatial distribution of transport costs
As transport costs lay at the heart of his model, Weber had to devise a technique that could both measure and map the spatial differences in these costs in order to find the LCL. His solution was to produce a map with two types of contour-type lines which he called isotims and isodapanes. An **isotim** is a line joining all places with equal transport costs for moving either the raw material (Figure 9.11a) or the product (Figure 9.11b). An **isodapane** is a line joining all places with equal total transport costs, i.e. the sum of the costs of transporting the raw material and the product (Figure 19.11c).

Figure 19.11a shows the costs of transporting 1 tonne of a raw material (R) as concentric circles. In this example, it will cost 5 t/km (tonne/kilometres) to transport the material to the market. Figure 19.11b shows, also by concentric circles, the cost of transporting 1 tonne of the finished product (P). The total cost of moving the product from the market to the source of the raw material is again 5 t/km. By superimposing these two maps it is possible to show the total transport costs (Figure 19.11c).

If a factory were to be built at X (Figure 19.11c), its transport costs would be 7 t/km (i.e. 2 t/km for moving the raw material plus 5 t/km for the product). A factory built at Y would have lower transport costs of 6 t/km (4t/km for the raw material plus 2 t/km for the product). However, the LCL in this case may be at the source of the raw material, the market or any intermediate point in a straight line between the two because all these points lie on the 5 t/km isodapane.

Figure 19.11

Isotims and isodapanes

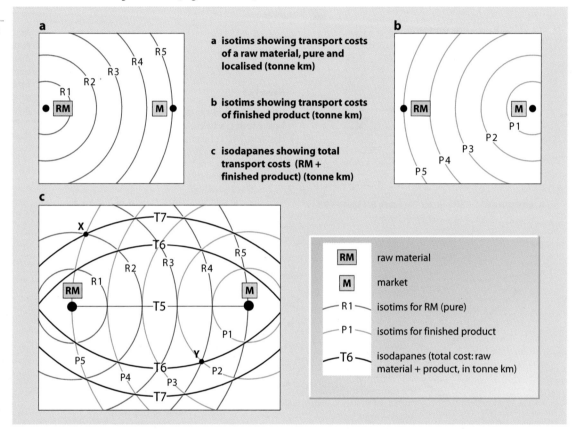

a isotims showing transport costs of a raw material, pure and localised (tonne km)

b isotims showing transport costs of finished product (tonne km)

c isodapanes showing total transport costs (RM + finished product) (tonne km)

RM	raw material
M	market
R1	isotims for RM (pure)
P1	isotims for finished product
T6	isodapanes (total cost: raw material + product, in tonne km)

The effects of labour costs and agglomeration economies
It has been stated that Weber considered that four factors affected production costs: we have seen the effects of the costs of raw materials and transport – let us now look at labour costs and agglomeration economies.

■ **Labour costs** Weber considered the question of whether any savings made by moving to an area of cheaper or more efficient labour would offset the increase in transport costs incurred by moving away from the LCL. He plotted isodapanes showing the increase in transport costs resulting from such a move. He then introduced the idea of the **critical isodapane** as being the point at which savings made by reduced labour costs equalled the losses brought about by extra

transport costs. If the cheaper labour lay within the area of the critical isodapane, it would be profitable to move away from the LCL in order to use this labour.

■ **Agglomeration economies** Agglomeration is when several firms choose the same area for their location in order to minimise their costs. This can be achieved by linkages between firms (where several join together to buy in bulk or to train a specialist workforce), within firms (individual car component units) and between firms and supporting services (banks and the utilities of gas, water and electricity). Deglomeration, in contrast, is when firms disperse from a site or area, possibly due to increased land prices or labour costs or a declining market.

Figure 19.12

Critical isodapanes
and agglomeration
economies

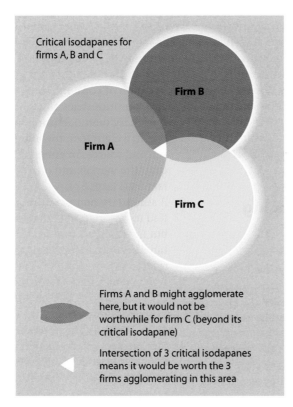

Critical isodapanes for
firms A, B and C

Firm B

Firm A

Firm C

Firms A and B might agglomerate
here, but it would not be
worthwhile for firm C (beyond its
critical isodapane)

Intersection of 3 critical isodapanes
means it would be worth the 3
firms agglomerating in this area

Figure 19.12 shows the **critical isodapane** for
three firms. It would become profitable for all
the firms to locate within the central area
formed by the overlapping of all three critical
isodapanes. It may be slightly more profitable
for firms **A** and **B**, but less profitable for firm **C**,
to locate within the purple area. However, it
would not be additionally profitable for any firm
to move if none of the isodapanes overlapped.
Agglomeration is now considered by many to be
probably the most important single factor in the
location of a firm or industry.

Figure 19.13

Space–cost curve to
show the area of
maximum profit
(*after* Smith, 1971)

Criticisms of Weber's model

The point has already been made with previous
examples and on page 557 that no model is
perfect and all have their critics. Criticisms of

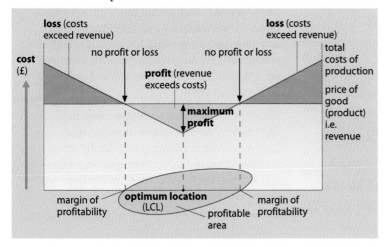

Weber's industrial location model include:

- It no longer relates to modern conditions –
  such as the present extent of government
  intervention (grants, aid to Enterprise Zones),
  improvements in and reduced costs of transport, technological advances in processing raw
  materials, the development of new types of
  industry other than those directly involved in
  the processing of raw materials, the increased
  mobility of labour and the increased complexity of industrial organisation (transnationals instead of single-product firms).
- Each country evolves its own industrial
  patterns and may be in different stages
  of economic development (pages 630–632).
- There are basic misconceptions in his original assumptions. For example, there are
  changes over time and space in demand and
  price; there are variations in transport
  systems; perfect competition is unreal as
  markets vary in size and change over a period
  of time; and decisions made by industrialists
  (who do not all have the same knowledge)
  may not always be rational (von Thünen's
  'economic man', page 471).
- Weber's material index was a crude measure
  and applicable only to primary processing
  or to industries with a very high or very
  low index.
- Dr L. Crewe (1999) claimed that 'traditional
  locational theories, such as that of Weber,
  are becoming increasingly less significant.
  Although labour costs and agglomeration
  factors *are* important in determining the
  location of an economic activity, they are
  handled far too simplistically in our ever
  increasingly complex world. Traditional
  models often cannot cope with the volatile
  global system in which we now live, and in
  which technological change is both rapid
  and endemic. A whole range of organisa-
  tional and institutional forces shape economic change within a global economy. The
  real problem is the interconnectedness and
  complexity of the various processes at work.'

Smith's area of maximum profit

An alternative approach to that of Weber was put
forward by **David Smith** in 1971. He suggested
that, as profits could be made anywhere where
total revenue exceeds total costs, there would be a
wider area where production was still profitable
(Figure 19.13) – although there would be a point
of **maximum profit**. To introduce his **space–cost
curves**, Smith also used isodapanes. By taking a
cross-section he was able to identify, spatially,

margins of profitability. He reasoned that firms rarely located at the ideal, or LCL, site (which had probably already been taken), but somewhere between the two profit margins. In other words, firms choose a **sub-optimal location** because they have imperfect knowledge about production and market demand; they have imperfect decision-makers who do not always act rationally; and they may be tempted or encouraged to locate in areas of high unemployment.

Industrial location: changing patterns

Four different types of industry have been selected as exemplars to try to demonstrate how the importance of different factors affecting the location of industry have changed through time. Their choice may reinforce the generalisation, by no means true in every case, that the more important locational factors in the 19th century were physical, while in modern industry they tend to be human and economic. They also show that while Weber's theory may have had some relevance in accounting for the location of older industries (remembering that it was put forward in 1907), it does not go far in explaining the location of contemporary industry.

The four industries are:

1. A primary manufacturing industry where, due to weight loss, the presence of raw materials and sources of energy is more important than the market and other economic factors (Places 83).

2. A secondary manufacturing industry initially tied to raw materials and sources of energy but in which economic and political factors have become increasingly more important (Places 84).

3. A secondary manufacturing industry where the nearness of a market and labour supply is more important than the presence of raw materials and sources of energy (Places 85).

4. Modern secondary (quaternary) manufacturing industries where human and economic factors are the most important (Places 86).

Places 83   Sweden: wood pulp and paper

There are three stages in this industry: the felling of trees, the processing of wood pulp (primary processing), and the manufacture of paper (secondary processing). In Sweden, most pulp and paper mills (Figure 19.14) are located at river mouths on the Gulf of Bothnia (Figure 19.15). Timber is a gross raw material which loses much of its weight during processing; it is bulky to transport; and it requires much water to turn it into pulp. Towns such

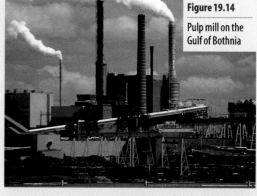

Figure 19.14

Pulp mill on the Gulf of Bothnia

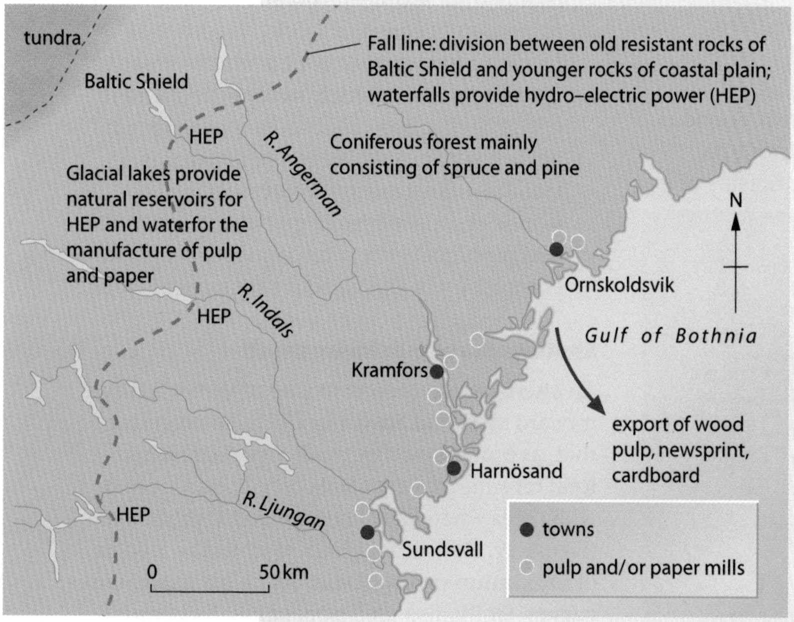

as Sundsvall and Kramfors are ideally situated (Figure 19.15): the natural coniferous forests provide the timber; the fast-flowing Rivers Ljungan, Indals and Angerman which initially provided cheap water transport for the logs are a source of the necessary and cheap hydro-electricity; and the Gulf of Bothnia provides an easy export route. Paper has a higher value than pulp and it is convenient and cheaper to have integrated mills.

Weber's agglomeration economies seem to operate with the clustering of so many mills. Smith's concept of an area of profitability rather than an individual least-cost location also seems applicable as the forest is extensive and there are several processing centres.

Figure 19.15

Location of wood pulp and paper factories in central Sweden

tundra

Baltic Shield

Fall line: division between old resistant rocks of Baltic Shield and younger rocks of coastal plain; waterfalls provide hydro–electric power (HEP)

HEP

R. Angerman

Coniferous forest mainly consisting of spruce and pine

Glacial lakes provide natural reservoirs for HEP and water for the manufacture of pulp and paper

HEP

R. Indals

N

HEP

Ornskoldsvik

Gulf of Bothnia

Kramfors

export of wood pulp, newsprint, cardboard

Harnösand

HEP

R. Ljungan

Sundsvall

● towns
○ pulp and/or paper mills

0   50km

Although the early iron and later steel industries were tied to raw materials, modern integrated iron and steelworks have adopted new locations as the sources of both ore and energy have changed.

- **Before AD 1600**   Iron-making was originally sited where there were surface outcrops of iron ore and abundant wood for use as charcoal (the Weald, the Forest of Dean, Figure 19.16a). Locations were at the source of these two raw materials as they had a high material index, were bulky and expensive to transport, had a limited market and could not be moved far owing to the poor transport system.

- **Before AD 1700**   Local ores in the Sheffield area were turned into iron by using fast-flowing rivers to turn waterwheels as water provided a cheaper source of energy.

- **After AD 1700**   In 1709, Abraham Derby discovered that coke could be used to smelt iron ore efficiently. At this time, it took 8 tonnes of coal and 4 t of ore to produce 1 t of iron, and so new furnaces were located on coalfields. One of the first areas to develop was South Wales where bands of iron ore (blackband ores) were found between seams of coal. The advantages possessed by South Wales at that time are shown in Figure 19.17a. Later, the industry extended into other British coalfields. When local ores became exhausted, the industry continued in the same locations because of geographical inertia, a pool of local skilled labour, a local market using iron as a raw material, improved techniques reducing the amount of coal needed (2 t per 1 t of final product), improved and cheaper transport systems (rail and canal) which brought distant mined iron ore, and the beginnings of agglomeration economies.

- **After 1850**   Until the 1880s, the low ore and high phosphorus content of deposits found in the Jurassic limestone, extending from the Cleveland Hills to Oxfordshire, had not been touched. After 1879, the **Gilchrist–Thomas process** allowed this ore to be smelted economically. As iron ore now had a higher material index than coal it was more expensive to move. As a result, new steelworks were opened on Teesside, near to the Cleveland Hills deposits, and at Scunthorpe and Corby, on the ore fields. However, the major markets remained on the coalfields.

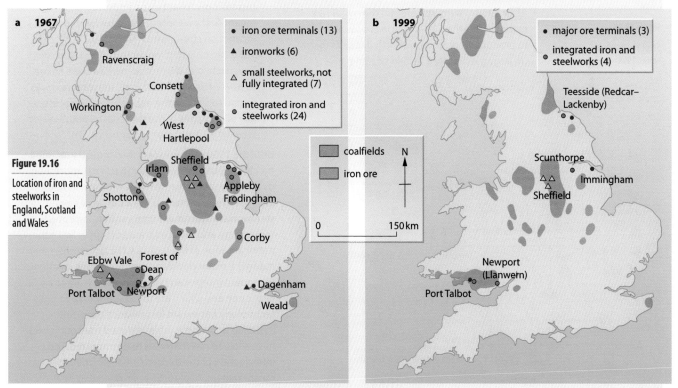

Figure 19.16

Location of iron and steelworks in England, Scotland and Wales

a   1967

- iron ore terminals (13)
- ▲ ironworks (6)
- △ small steelworks, not fully integrated (7)
- ⊙ integrated iron and steelworks (24)

coalfields
iron ore
N
0   150 km

Ravenscraig
Consett
Workington
West Hartlepool
Sheffield
Irlam
Shotton
Appleby Frodingham
Corby
Ebbw Vale
Forest of Dean
Port Talbot   Newport
Dagenham
Weald

b   1999

- major ore terminals (3)
- ⊙ integrated iron and steelworks (4)

Teesside (Redcar–Lackenby)
Scunthorpe
Sheffield   Immingham
Newport (Llanwern)
Port Talbot

Figure 19.17

Growth, decline and changing location of iron and steelworks in South Wales

Period of time			a Location of early 19th-century iron foundries in South Wales (e.g. Ebbw Vale)	b Disadvantages of these early locations by 1960 (e.g. Ebbw Vale)	c Location of integrated steelworks of the early 1990s (Port Talbot and Llanwern, Newport)
Physical	Raw materials	Coal	mined locally in valleys	older mines closing	little now needed; imported
		Iron ore	found within the Coal Measures	had to be imported: long way from coast	imported from N Africa and N America
		Limestone	found locally	found locally	found locally
		Water	for power and effluent: local rivers	insufficient for cooling	for cooling: coastal sites
	Energy/fuel		charcoal for early smelting, later rivers to drive machinery; then coal	electricity from National Grid	electricity from National Grid using coal, oil, natural gas and nuclear power
	Natural routes		materials local; export routes via the valleys	poor; restricted by narrow valleys	coastal sites
	Site and land		narrow valley floor locations	cramped sites; little flat land	large areas of flat, low-potential farmland
Human and economic	Labour		large quantities of semi-skilled labour	still large numbers of semi-skilled workers	still relatively large numbers but with a higher level of skill; fewer due to high-tech/mechanisation
	Capital		local entrepreneurs	no investment	government and EU incentives
	Markets		local	difficult to reach Midlands and ports	tin plate industry (Llanelli) and the Midland car industry
	Transport		little needed; some canals; low costs	poor; old-fashioned; isolated	M4; purpose-built ports
	Geographical inertia		not applicable	not strong enough	tradition of high-quality goods
	Economies of scale		not applicable	worked against the inland sites	two large steelworks more economical than numerous small iron foundries
	Government policy		not applicable	Ebbw Vale kept open by government help	having the capital, governments can determine locations and closures and provide heavy investment
	Technology		small scale: mainly manual	out-of-date	high-technology: computers, lasers, etc.

Figure 19.18

Steel and finishing works, 1998

- **After 1950** With iron ore still the major raw material (less than 1 t of coal was now needed to produce 1 t of steel), but with deposits in the UK largely exhausted, Britain became increasingly reliant upon imported ores. This meant that new **integrated steelworks** were located on coastal sites (Figure 19.16b). By 1980, the only two remaining inland sites, at Ravenscraig and Scunthorpe, had been linked to new, nearby ore terminals. (Ravenscraig was forced to close in the early 1990s.) The present-day advantages of the two coastal South Wales steelworks are shown in Figure 19.17c.

Until the 1950s, the iron and steel industry satisfied much of Weber's theory. After that, and unforeseen by him 40 years earlier, three new elements became increasingly important in the location of new steelworks: government intervention, improved technology, and reduced transport costs. Since the nationalisation of the industry, the government has made the major decisions concerning the location of modern integrated plants, and which are to be kept open and which are to close. Improved technology has meant less reliance upon raw materials and labour, while reduced transport costs have allowed raw materials to be imported. The introduction of the oxygen furnace in the 1950s has meant that relatively little energy (coal) is now needed, while the reduction in the workforce has cut production costs but increased unemployment – an economic gain but a social loss.

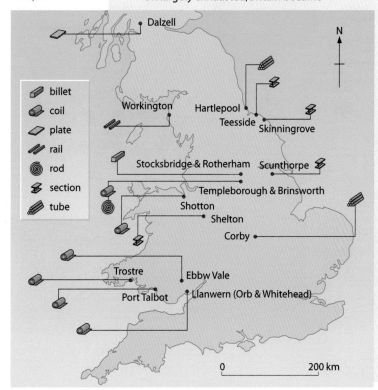

- billet
- coil
- plate
- rail
- rod
- section
- tube

Dalzell

Workington

Hartlepool

Teesside

Skinningrove

Stocksbridge & Rotherham

Scunthorpe

Templeborough & Brinsworth

Shotton

Shelton

Corby

Trostre

Ebbw Vale

Port Talbot

Llanwern (Orb & Whitehead)

N

0   200 km

Japan's production of 9.9 million cars in 1992, which was 24.6 per cent of the world's total, made it the world leader ahead of the USA (6.0 m) and Germany (4.4 m). This has been achieved despite a lack of basic raw materials.

Japan has very limited energy resources for, although it produces hydro-electricity and nuclear energy, it has to import virtually all its coal, oil and natural gas requirements. Similarly, most of the iron ore and coking coal needed to manufacture steel also has to be imported. The result has been the location of the major steelworks on tidal sites found around the country's many deep and sheltered natural harbours. As only 17 per cent of the country is flat enough for economic development (for homes, industry and agriculture), most of the population also has to live in coastal areas and around the harbours. The five major conurbations, linked by modern communications, provide both the workforce and the large, affluent, local markets needed for such steel-based products as cars (Figure 19.19). Within these conurbations, especially Keihin, Chukyo and Setouchi, are numerous firms engaged in making car component parts. This agglomeration of firms limits transport costs and conforms with Weber's concept that industries gaining weight through processing (car assembly) are best located at the market. As many of the smaller, older and original firms have amalgamated into large-scale companies, the extra space required for their factories has had to come from land reclaimed from

the sea (Figure 19.20). These new locations, despite the high costs of reclamation, make excellent sites from which to export finished cars to all parts of the world. The large local labour force contains both skilled and semi-skilled workers. It is well-educated and industrious, with workers who are very loyal to their firms. The car industry, which has received considerable government financial assistance, has an organisation which centres around teamworking, worker involvement, total-quality management, and 'just-in-time' production (this is when various component parts arrive just as they are needed on the assembly line, thus avoiding the need to store or to overproduce). The Japanese car industry has a high level of automation and uses the most modern technology: it produces three times the number of cars per worker as does western Europe. The assembled cars are reliable and universally acceptable in design which means, together with the shift from mass production to lean production, that the Japanese have gained strong footholds in world markets. To expand further into these markets, the Japanese have either built overseas assembly plants or have amalgamated with local companies so that more cars can be produced close to the large urban markets within western Europe and the USA. However, the recession that affected eastern Asia in the late 1990s, causing a drop in both the domestic and overseas markets, led to the closure of several assembly plants within Japan itself.

Figure 19.19

Major industrial areas in Japan

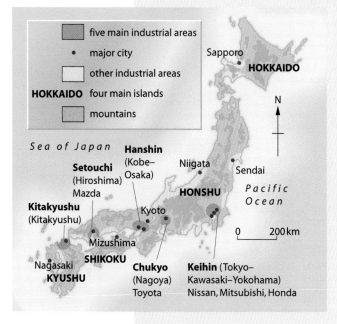

Figure 19.20

Mazda's new Hofu car plant, built upon land reclaimed from the sea

---

Manufacturing industries

The term **high-technology** refers to industries that have developed within the last 25 years and whose processing techniques often involve micro-electronics, but may include medical instruments, biotechnology and pharmaceuticals. These industries, which collectively fit into the **quaternary** sector (page 552), usually demand high inputs of information, expertise and research and development (R&D). They are also said to be **footloose** in that, not being tied to raw materials, they have a free choice of location. However, they do tend to occur in clusters in particular areas, forming what Weber would have called 'agglomerated economies', and they locate, according to Smith's terminology, in areas of maximum profit such as along the M4 and M11 corridors in England (also Silicon Glen in Scotland, Silicon Valley in California and Grenoble and the Côte d'Azur in France). By locating close together, high-tech firms can exchange ideas and information and share basic amenities such as connecting motorways.

Two of the major concentrations of high-tech industries in Britain are along the M4 westwards (Sunrise Valley) from London to Reading, Newbury ('Video Valley'), Bristol (Aztec West) and into South Wales; and the M11 northwards to Cambridge (Figure 19.21). Transport is convenient due to the proximity of several motorways and mainline railways, together with the four main London airports. Transport costs are, in any case, relatively insignificant as the raw materials (silicon chips) are lightweight and the final products (computers) are high in value and small in bulk. Even so, it has been argued that two of the main reasons for high-tech development in this part of Britain were:

1 the presence of government-sponsored research establishments at Harwell and Aldermaston and of government aerospace contractors in the Bristol area

2 its attractive environment, e.g. the valley of the Thames and the nearby upland areas of the Cotswolds, Chilterns and Marlborough Downs (Figure 19.21), and its proximity to cultural centres, e.g. London, Oxford and Cardiff.

Figure 19.21

The M4 and M11 Corridors

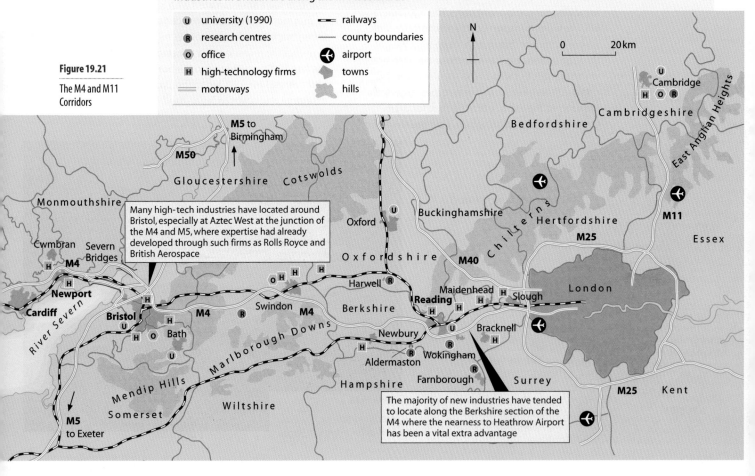

Most firms that have located here claim that the major factor affecting their decision was the availability of two types of labour:

- Highly skilled and inventive research scientists and engineers, the majority of whom were university graduates or qualified technicians. These specialists, whose abilities were in short supply, could often dictate areas where they wanted to live and work, i.e. areas of high environmental, social and cultural quality. The proximity of several universities (Figure 19.21) provided a pool of skilled labour and facilities for R&D.
- Female workers who either tended to be plentiful as an increasing number of career-minded women were among those who had recently moved out of London and into new towns and suburbanised villages (page 398), or were prepared to accept part-time/flexi-time jobs (Figure 19.4).

Science parks are often joint ventures between universities and local authorities. They are usually located adjacent to universities on edge-of-town greenfield sites where, because the land is of lower value, there is plenty of space for car parking, landscaping (ornamental gardens and lakes) and possible future expansion.

The Cambridge Science Park (Figure 19.22) has been developed in conjunction with Trinity College, Cambridge. Opened in 1972, the success of early firms soon attracted more (agglomeration economies), so that by 1999 there were almost 100 companies employing over 2500 people. Existing companies can be divided into those making scientific instruments (38 per cent), electronics (30 per cent), and drugs and pharmaceuticals (22 per cent). Only selected firms, using the high-quality, flexible buildings for specific purposes, are allowed to locate in the business park. Almost one-quarter of these firms are medium-sized, each employing between 20 and 49 workers. Some 70 per cent of the park, which covers 52 hectares, is left as open space with trees, grass and ornamental gardens with lakes (Figure 19.23). As this, and other business and science parks in the Cambridge area, continue to develop, new housing has to be provided, e.g. at Cambourne (Case Study 14A), and building pressure increases on the surrounding countryside (Figure 14.22).

Figure 19.22

The Cambridge Science Park

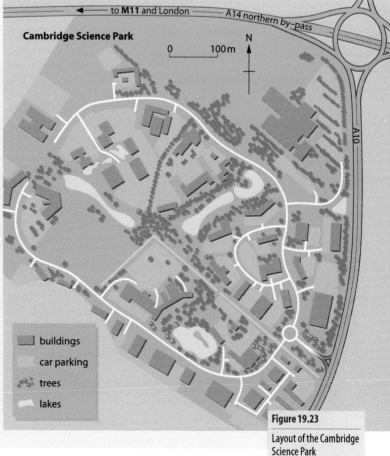

Figure 19.23

Layout of the Cambridge Science Park

Manufacturing industries

Industrial linkages and the multiplier

When Weber introduced the term 'agglomeration economies', he acknowledged that many firms made financial savings by locating close to, and linking with, other industries. The success of one firm may attract a range of associated or similar-type industries (cutlery in Sheffield), or several small firms may combine to produce component parts for a larger product (car manufacture in Coventry). **Industrial linkages** may be divided into **backward linkages** and **forward linkages**:

backward linkages to firms providing raw materials or component parts	← FACTORY →	**forward linkages** to firms further processing the product or using it as a component part

A more detailed classification of industrial linkages is given in Figure 19.24. The more industrially advanced a region or country, the greater is the number of its linkages. Developing countries have few linkages, partly because of their limited number of industries and partly because few industries go beyond the first stage in processing – the simple chain in Figure 19.24a. Industrial linkages may result in:

- energy savings
- reduced transport costs
- waste products from one industry forming a raw material for another
- energy given off by one process being used elsewhere
- economies of scale where several firms buy in bulk or share distribution costs
- improved communications, services and financial investment
- higher levels of skill and further research
- a stronger political bargaining position for government aid (the securing of EU funding now depends upon having a network of linked organisations).

Crewe (1999) stressed the 'increasingly critical importance of local linkages in ensuring competitive success, and the need to emphasise how agglomeration is becoming an increasingly important factor in explaining industrial location. In the fashion quarter of Nottingham's Lace Market, for example, 85 per cent of all firms are linked to others, e.g. supplier links, manufacturers, retailers, local intelligence, and so on. Other examples of linkages and industrial location include the Motor Sport valley in Oxfordshire and car assembly in the West Midlands, together with both the fashion and jewellery agglomerations and the semiconductor clusters in California and the UK' (Places 86).

Figure 19.24

Types of industrial linkage

a Vertical (or simple chain) linkages:

the raw material goes through several successive processes

newsprint

pulp

mill logging

b Horizontal (or multi–origin) linkages:

an industry relies on several other industries to provide its component parts

brakes
gearboxes
electrical equipment → car assembly plant
tyres
radiators

c Diagonal (or multi–destination) linkages:

an industry makes a component which can be used subsequently in several industries

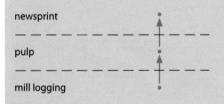

washers, nuts and bolts → watches and clocks / car industry / domestic appliances / repair workshops/garages

d Technological linkages:

a product from one industry is used subsequently as a raw material by other industries

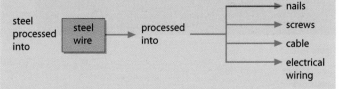

steel processed into → steel wire → processed into → nails / screws / cable / electrical wiring

The multiplier effect and Myrdal's model of cumulative causation

If a large firm, or a specialised type of industry, is successful in an area, it may generate a **multiplier effect**. Its success will attract other forms of economic development creating jobs, services and wealth – a case of 'success breeds success'. This circular and cumulative process was used by **Gunnar Myrdal**, a Swedish economist writing in the mid-1950s, to explain why inequalities were likely to develop between regions and countries. Figure 19.25 is a simplified version of his model.

Myrdal suggested that a new or expanding industry in an area would create more jobs and so increase the spending power of the local population. If, for example, a firm employed a further 200 workers and each worker came from a family of four, there would be 800 people demanding housing, schools, shops and hospitals. This would create more jobs in the service and construction industries as well as attracting more firms linked to the original industry. As **growth poles**, or points, develop there will be an influx of migrants, entrepreneurs and capital,

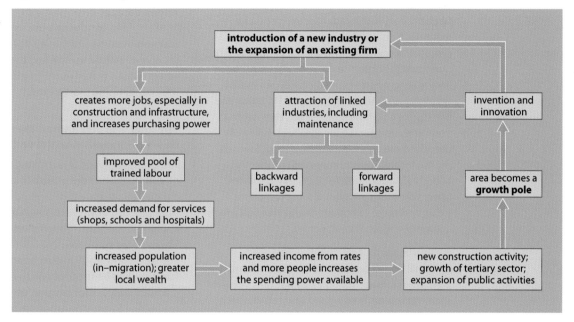

together with new ideas and technology. Myrdal's multiplier model may be used to explain a number of patterns.

1   The growth of 19th-century industrial regions (South Wales and the Ruhr) and districts (cutlery in Sheffield, guns and jewellery in Birmingham and clothing in Nottingham).
2   The development of growth poles (page 640) in developing countries (São Paulo in Brazil and the Damodar Valley in India), where increased economic activity led, in turn, to multiplier effects, agglomeration economies and an upward spiral resulting in core regions (Places 87 and Places 105, page 641). At the same time, cumulative causation worked against regions near the **periphery** where Myrdal's **backwash effects** included a lack of investment and job opportunities.
3   The creation of modern government regional policies which encourage the siting of new, large, key industries in either peripheral, less developed (Trombetas and Carajas in Amazonia) or high unemployment (Nissan and Toyota in England) areas in the hope of

stimulating economic growth. This policy is more likely to succeed if the industries are labour intensive.

Industrial regions

Much of Britain's early industrial success stemmed from the presence of basic raw materials and sources of energy for the early iron, and the later iron and steel, industries; the mass production of materials using the processed iron and steel; and the development of overseas markets. During the 19th century it was the coalfields, especially those in South Wales, northern England and central Scotland, which became the core industrial regions. However, as the initial advantages of raw materials (which became exhausted), specialised skills and technology (no longer needed as the traditional heavy industries declined) and the ability to export manufactured goods (in the face of growing overseas competition) were lost, these early industrial regions have become more peripheral. Recent attempts to revive their economic fortunes have met with varying success (Places 87).

Pre-1920: industrial growth creating a core region

The growth of industry in South Wales was based upon readily obtainable supplies of raw materials (Figure 19.17a). Coking coal and blackband iron ore were frequently found together, exposed as horizontal seams outcropping on steep valley sides. Their proximity to each other meant that the area around Merthyr Tydfil and Ebbw Vale (Figure 19.26) was ideally suited for industrial development (Weber's least-cost location for two gross raw materials, Figure 19.10b). Added to this was the presence of limestone only a few kilometres to the north, and the expertise of the local population in iron-making where water wheels, driven by fast-flowing rivers, had earlier been used to power the blast furnace bellows. By the 1860s there were 35 iron foundries operating in the Welsh valleys. By the time the more accessible coal had been used up, mining techniques had improved sufficiently to allow shafts to be sunk vertically into the valley floors. When local supplies of iron ore became exhausted, there were ports nearby through which substitute ore could be imported.

'… Thus began the spread of the well-known industrial landscape of the Valleys. Pits crammed themselves into the narrow valley bottoms, vying for space with canals, housing and, later, railways and roads. Housing began to trail up the valley sides, line upon line of terraces pressed against the steep slopes [Figure 19.27]. The opening-up of the underground coal seams resulted in massive immigration, much of it from rural areas. Working conditions, living conditions and wages were deplorable while health and safety standards underground were poor. Housing was overcrowded as the provision of homes, financed by the local entrepreneur ironmasters, lagged far behind the supply of jobs.'

The rapid increase in coalmining and iron-working partly resulted from the growth of large overseas markets as both products were mainly exported. Transport to the Welsh ports first involved simply allowing trucks to run downhill under gravity. Later, canals and then railways were used to move the bulky materials. While Barry, Cardiff and Newport developed as exporting ports, Swansea and Neath grew as 'break of bulk' ports smelting the imported ores of copper, nickel and zinc. **Break of bulk** is when a transported product has to be transferred from one form of transport to another – a process that involves time and money. It was easier and cheaper, therefore, to have had the smelting works where the raw materials were unloaded, rather than transporting them inland.

The inter-war and immediate post-war years: depression and industrial decline

Just as the existence of raw materials and overseas markets had led to the growth of local industry, so did their loss hasten its decline. Iron ore had long since been exhausted and it increasingly became the turn of coal, even though there were still over 500 collieries employing 260 000 miners in 1925. The steelworks which had replaced the iron foundries had been built upon the same inland, cramped sites; as they became less competitive mainly due to rising transport costs, so they became increasingly dependent upon government support (Figure 19.17b). Overseas markets were lost as rival industrial regions with lower costs and more up-to-date technology were developed overseas. The difficulties of an economy reliant on a narrow industrial base, dependent upon an increasingly out-of-date infrastructure, and unable to compete with overseas competition, led to major economic, social and environmental problems.

Figure 19.26

Early industrial development in South Wales

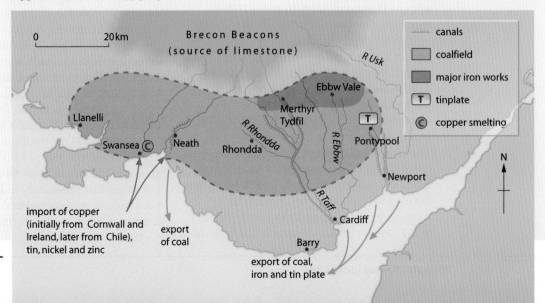

Figure 19.27

Industry, communications and terraced housing strung along the valley floor and lower valley sides: Rhondda Fawr, looking towards Treorchy, mid-Glamorgan

Towards the present: industrial diversification in a peripheral area

Steel-making and non-ferrous metal smelting have been maintained, partly due to geographical inertia, despite a significant fall in output and workers. As the centre of gravity for steel-making has moved to coastal sites, so too has the location of the two South Wales integrated works, to Llanwern (Newport) and Port Talbot (Figures 19.17c and 19.28). Tin plate, using local steel, is produced at Trostre near Llanelli (the Felindre works near Swansea closed in 1989), while the Mond nickel works near Swansea is the world's largest (Figure 19.28).

The major factor to have affected industry in the region in the last 40 years has been government intervention (or lack of it, depending upon your

Figure 19.28

Recent industrial development in South Wales

political views). The Special Areas Act of 1934 saw the first government assistance which set up industrial estates at Treforest, Merthyr Tydfil and Rhondda (Figure 19.28), while Cwmbran became one of Britain's first new towns (1949). Much of the former coalfield remains a Development Area (Figure 19.5). The last NCB colliery closed in 1994, although the Tower Colliery re-opened privately in 1995 employing 350 people. At present 75 per cent of the mined coal comes, despite environmental concerns (page 522) from opencast mining, and the industry's workforce of 2000 contrasts with the 270 000 employed at its peak in 1913.

Two local areas of exceptionally high unemployment, Swansea and Milford Haven Waterway, were designated two of Britain's 27 Enterprise Zones (pages 439 and 556). The Swansea EZ includes five parks – the Enterprise (commerce and light industry), Leisure (recreation facilities), Riverside (heritage and environmental schemes), City (retailing) and Maritime (housing and cultural) Parks. The Ford Motor Company took advantage of government incentives to build two plants in the region, one of which, at Bridgend, has been expanded. It was government policy that built an integrated steelworks at Ebbw Vale in 1938, and which closed it in 1979. The future of Port Talbot (highly modernised and hopefully secure) and Newport is also in government hands. A policy to decentralise some government departments has seen vehicle licensing moved to Swansea and the Royal Mint to Llantrisant (Figure 19.28). Improvements in communications include the M4, the Heads of Valleys Road, the InterCity rail link, and Cardiff international airport – some of which were financed by EU funds.

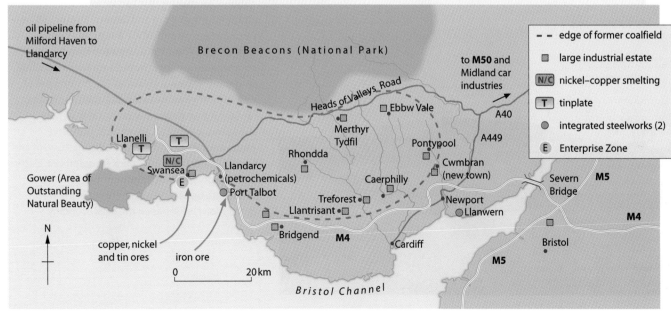

Figure 19.29

Swansea Enterprise Park, west Glamorgan

The Welsh Development Agency (WDA) was set up in 1976 'to attract high-quality investment, to help the growth of Welsh businesses and to improve the environment' (Figure 19.29). It saw as its main advertising points:

- a workforce that was skilled (although it needed retraining for the new-style high-tech industries)
- low labour costs, high productivity and good labour relations
- a well-developed transport infrastructure with modern road, rail and air links
- the availability of advanced factory sites with quality buildings at competitive rates
- a local market, and access to a national and the international market
- low rates and rents for firms wishing to locate in either the Development or Intermediate Areas (Figure 19.5)
- lower house prices and cost of living than south-east England
- the University of Wales with its five separate colleges
- the Welsh countryside, including the Pembrokeshire Coast and Brecon Beacons National Parks and 500 km of Heritage Coastline (including the Gower Peninsula)
- the Welsh culture, including music, the performing arts and sport.

Money has also been spent on landscaping old industrial areas which had been scarred either by metal-smelting industries (lower Swansea Valley) or by slag (Ebbw Vale) and colliery waste tips (Aberfan – Case Study 2B). The Ebbw Vale Garden Festival (1992), sited on part of the former steelworks, was part of a larger scheme aimed at creating new jobs, improving housing, renovating old properties and improving the local environment (page 439). Other schemes, some funded by the WDA (see below), include tourist and cultural facilities such as the Welsh Industrial and Maritime Museum in Cardiff's newly created Marina area.

The Cardiff Bay project, environmentally controversial, was aimed at improving transport and housing as well as providing jobs and retailing and leisure opportunities.

By the end of the 20th century, South Wales had a much more varied and broad economic base than it had ever had before. Both manufacturing and inward investment were growing at a faster rate than anywhere else in the UK. Of over 380 international companies that had located here, 140 were North American (Ford and General Electric), 50 were German (Bosch), and 40 were Japanese (Sony – Figure 19.30, Aiwa, Matsushita and Hitachi). Other companies have come from South Korea (LG), Italy, Singapore, France and Taiwan. The major types of new industry include (alphabetically) aerospace (at Cardiff international airport), car assembly (Bridgend), chemicals, pharmaceuticals and electronics (Bridgend, Hirwaun, Blaenau Gwent and Cardiff), medical devices, optical equipment, plastics and components, and telecommunications. It remains to be seen what difference, if any, Welsh devolution (1999) has on the local economy.

Figure 19.30

Sony's CTV European headquarters at Pencoed, Bridgend, occupies a 25 ha site

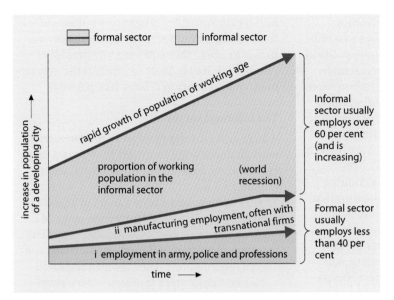

formal sector   informal sector

increase in population of a developing city ↑

rapid growth of population of working age

proportion of working population in the informal sector

(world recession)

ii manufacturing employment, often with transnational firms

i employment in army, police and professions

Informal sector usually employs over 60 per cent (and is increasing)

Formal sector usually employs less than 40 per cent

time →

Figure 19.31

Growth in the informal sector

Advantages to the country	Disadvantages to the country
Brings work to the country and uses local labour	Numbers employed small in comparison with amount of investment
Local workforce receives a guaranteed income	Local labour force usually poorly paid and have to work long hours
Improves the levels of education and technical skills of local people	Very few local skilled workers employed
Brings inward investment and foreign currency to the country	Most of the profits go overseas (outflow of wealth)
Companies provide expensive machinery and introduce modern technology	Mechanisation reduces the size of the labour force
Increased gross national product/ personal income can lead to an increased demand for consumer goods and the growth of new industries and services	GNP grows less quickly than that of the parent company's headquarters, widening the gap between developed and developing countries
Leads to the development of mineral wealth and new energy resources	Raw materials are usually exported rather than manufactured locally, and energy costs may lead to a national debt
Improvements in roads, airports and services	Money possibly better spent on improving housing, diet and sanitation
Prestige value (e.g. Volta Project)	Big schemes can increase national debt (e.g. Brazil)
Widens economic base of country	Decisions are made outside the country, and the firm could pull out at any time
Some improvement in standards of production, health control, and recently in environmental control	Insufficient attention to safety and health factors and the protection of the environment

Figure 19.32

Advantages and disadvantages of transnational (multinational) corporations

Industry in economically less developed countries

In cities in economically less developed countries, the number of people seeking work far outweighs the number of jobs available. As these cities continue to grow, either through natural increase or in-migration, the job situation gets continually worse. In the late 1990s, the UN estimated that there were over 550 million people unemployed in developing countries and only 20 million in economically more developed countries. The UN also estimates that in developing countries, on average, only about 40 per cent of those people with jobs work in the **formal sector** (Figure 19.31). These jobs, which are permanent and relatively well paid, include those offered by the state (police, army and civil service) or by overseas-run **transnational (multinational) corporations**. The remaining 60 per cent – a figure which the UN claims is rising – have to seek work in the **informal sector**. The main differences between the formal and informal sectors are listed in Figure 19.33.

Transnational (multinational) corporations

A transnational, or multinational, corporation is one that operates in many different countries regardless of national boundaries. The headquarters and main factory are usually located in an economically more developed country. Although, at first, many branch factories were in economically less developed countries, increasingly there has been a global shift to the more affluent markets of Europe, North America and Japan. Transnationals (TNCs) are believed to directly employ some 40 million people worldwide, to indirectly influence an even greater

number, and to control over 75 per cent of world trade (compared with only 20 per cent in 1960). The largest TNCs have long been car manufacturers and oil corporations but these have, more recently, been joined by electronic and high-tech firms. Several of the largest TNCs have a higher turnover than all of Africa's GNP in total.

Transnationals, with their capital and technology, have the 'power' to choose what they consider to be the ideal locations for their factories. This choice will be made at two levels: the most suitable country, and the most suitable place within that country. The choice of a country usually depends upon political factors. Most governments, regardless of the level of economic development within their country, are prepared to offer financial inducements to attract transnationals which they see as providers of jobs and a means of increasing exports. (Sony, Figure 19.30, were reputed to have been offered better inducements to locate at Bridgend rather than in Barcelona.) Many governments of economically less developed countries, due to a greater economic need, are prepared to impose fewer restrictions on transnationals because they often have to rely upon them to develop natural

Manufacturing industries

573

resources, to provide capital and technology (machinery, skills, transport), to create jobs and to gain access to world markets. Despite political independence, many poorer countries remain economically dependent (neo-colonialism) upon the large transnationals (together with international banks and foreign aid). Some of the advantages and disadvantages of transnational corporations to developing countries are listed in Figure 19.32.

Transnationals, having selected a country, then have to decide where to locate within that country. If the country is economically developed, the location is likely to be where financial inducements are greatest, land values are low, transport is well developed, and levels of skill and unemployment are high (Japanese companies in South Wales, page 572; a Korean company, Samsung, in north-east England). If the country is economically less developed, the location is more likely to be in the primate city (page 405), especially if that city is also the capital or the chief port. A capital city location, with an international airport, allows quick access to the companies' overseas headquarters; and a

port location enables easier export of manufactured goods. Should several transnational companies locate in the same area, the multiplier effect (page 569) is likely to result in the development of a core region (Places 105, page 64).

The informal sector

A large and growing number of people with work in developing countries have found or created their own jobs in the informal sector (Figure 19.33).

'This sector covers a wide variety of activities meeting local demands for a wide range of goods and services. It contains sole proprietors, cottage industries, self-employed artisans and even moonlighters. They are manufacturers, traders, transporters, builders, tailors, shoemakers, mechanics, electricians [Places 88], plumbers, flower-sellers and many other activities. The [Kenyan] government have recognised the importance of these small-scale *Jua kali* enterprises [Places 89] and a few commercial banks are beginning to extend loans to these new entrepreneurs who are themselves forming co-operatives. There are many advantages in developing these concerns. They use less capital per worker than larger firms; they tend to use and re-cycle materials that would otherwise be waste; they provide low-cost, practical on-the-job training which can be of great value later in more formal employment; and, as they are flexible, they can react quickly to market changes. Their enterprising spirit is a very important national human resource.'

Central Bank of Kenya

The governments of several developing countries now recognise the importance of such local ventures as Kenya's *Jua kali* which, apart from creating employment, provide goods at affordable prices. India, for example, encourages the growth of co-operatives to help family concerns, under the 'Small Industries Development Organisation', by setting up district offices that offer technical and financial advice. Under its Development Plans, the manufacture of 600 products will be exclusively reserved for small firms and family enterprises.

Children, many of whom may be under the age of 10, form a significant proportion of the informal-sector workers. Very few of them have schools to go to and, from an early age, they go onto the streets to try to supplement the often meagre family income. They may try to earn money by shining shoes or selling items such as sweets, flowers, fruit and vegetables (Places 104, page 634).

Figure 19.33

Differences between 'formal' and 'informal' sectors

	Formal	Informal
Description	Employee of a large firm	Self-employed
	Often a transnational	Small-scale/family enterprise
	Much capital involved	Little capital involved
	Capital-intensive with relatively few workers; mechanised	Labour-intensive with the use of very few tools
	Expensive raw materials	Using cheap or recycled waste materials
	A guaranteed standard in the final product	Often a low standard in quality of goods
	Regular hours (often long) and wages (often low)	Irregular hours and uncertain wages
	Fixed prices	Prices rarely fixed and so negotiable (bartering)
	Jobs done in factories	Jobs often done in the home (cottage industry) or on the streets
	Government and transnational help	No government assistance
	Legal	Often outside the law (illegal)
	Usually males	Often children and females
Type of job	Manufacturing: both local and transnational companies	Distributive (street peddlers and small stalls)
	Government-created jobs such as the police, army and civil service	Services (shoe cleaners, selling clothes and fruit)
		Small-scale industry (food processing, dressmaking and furniture repair)
Advantages	Uses some skilled and many low-skilled workers	Employs many thousands of low-skilled workers
	Provides permanent jobs and regular wages	Jobs may provide some training and skills which might lead to better jobs in the future
	Produces goods for the more wealthy (food, cars) within their own country so that profits may remain within the country	Any profit will be used within the city: the products will be for local use by the lower-paid people
	Waste materials provide raw materials for the informal sector	Uses local and waste materials

Don Mariano, a refugee from Nicaragua, lives with his family in a shanty settlement in San Jose, Guatemala. His hut, which was built with boards and roofed with corrugated iron, also contains his 'workshop'. It is a separate room with a worktable and an electric connection. In this workshop he repairs all kinds of electrical appliances: heating plates, electric heaters, radios and television sets. His equipment is totally inadequate – two screwdrivers and one pair of pliers. For a voltmeter he uses an old lamp – he can tell the voltage from its brightness. He does not own a blowtorch. Don Mariano's primary problem is not a lack of skill but a lack of equipment. If he had the money he could buy a blowtorch and a voltmeter. If he had those he could do more repairs and earn more money. If …

(adapted from Timberlake, *Only One Earth*, 1987)

Jua kali means 'under the hot sun'. Although there are many smaller *Jua kali* in Nairobi, the largest is near to the bus station where, it is estimated, over 1000 workers create jobs for themselves (Figure 19.34). The plot of land upon which the metal workshops have been built measures about 300 metres by 100 metres. The first workshops were spontaneous and built illegally as their owners did not seek permission to use the land, which did not belong to them. As more workshops were set up and the site developed, the government was faced with the option of either bulldozing the temporary buildings, as governments had done to shanty settlements in other developing countries, or encouraging and supporting local initiative. Realising that the informal workshops created jobs in a city where work was hard to find, the government opted to help. The Prime Minister himself became personally involved by organising the erection of huge metal sheds which protected the workers from the hot sun and occasional heavy rain.

Groups of people are employed touring the city collecting scrap. The scrap is melted down, in charcoal stoves, and then hammered into various shapes including metal boxes and drums, stoves and other cooking utensils, locks and water barrels, lamps and poultry water troughs (Figure 19.35). Most of the workers are under 25 and have had at least some primary education. The technology they use is appropriate and sustainable, suited to their skills and the availability of raw materials and capital. Most of the products are sold locally and at affordable prices.

It is estimated that there are approximately 600 000 people engaged in 350 000 small-scale *Jua kali* enterprise units in Kenya. This figure needs to be compared with the 180 000 recorded as employed in large-scale manufacturing and the 2.2 million total in all areas of the non-agricultural economy. *Jua kali* form, therefore, a most significant part of the total employment picture.

Figure 19.34

Jua kali workshops

Figure 19.35

Jua kali end products

Intermediate (appropriate) technology

Dr E. F. Schumacher developed the concept of **intermediate technology** as an alternative course for development for poor people in the 1960s. He founded the Intermediate Technology Development Group (ITDG) in 1966 and published his ideas in a book, *Small is Beautiful* (1973). Schumacher himself wrote:

'If you want to go places, start from where you are.

If you are poor, start with something cheap.

If you are uneducated, start with something relatively simple.

If you live in a poor environment, and poverty makes markets small, start with something small.

If you are unemployed, start using labour power, because any productive use of it is better than letting it lie idle.

In other words, we must learn to recognise boundaries of poverty.

A project that does not fit, educationally and organisationally, into the environment, will be an economic failure and a cause for disruption.'

In 1988 the ITDG stated that:

'Essentially, this alternative course for development is based on a local, small-scale rather than the national, large-scale approach. It is based on millions of low-cost workplaces where people live – in the rural areas – using technologies that can be made and controlled by the people who use them and which enable those people to be more productive and earn money.'

These ideas challenged the conventional views of the time on aid. Schumacher said:

'The best aid to give is intellectual aid, a gift of useful knowledge … The gift of material goods makes people dependent, but the gift of knowledge makes them free – provided it is the right kind of knowledge, of course.'

To illustrate this he quoted an old proverb:

'Give a man a fish and you feed him for a day; teach him how to fish and he can feed himself for life.'

The first part of this might be seen as the traditional view of aid where 'giving' leads to dependency. The second part, 'teaching', is a move in the direction of self-sufficiency and self-respect. Schumacher added a further dimension to the proverb by saying: 'teach him to make his own fishing tackle and you have helped him to become not only self-supporting but also self-reliant and independent'.

In most developing countries, not only are high-tech industries too expensive to develop, they are also usually inappropriate to the needs of local people and the environment in which they live. Examples of intermediate, or **appropriate technology** as it is now known (Places 90), include:

- labour-intensive projects; since, with so many people already being either unemployed or underemployed, it is of little value to replace workers by machines
- projects encouraging technology that is sustainable and the use of tools and techniques designed to take advantage of local resources of knowledge and skills
- the development of local, low-cost schemes using technologies which local people can afford, manage and control rather than expensive, imported techniques
- developing projects that are in harmony with the environment.

Newly industrialised countries (NICs)

Newly industrialised countries (NICs) are those which, although still perceived as being 'developing', have achieved a considerable level of industrialisation. Although their number includes Brazil, Mexico and Argentina, the majority are located in eastern and south-eastern Asia. Encouraged by Japan's success, other governments in Asia's Pacific Rim set out to improve their standards of living. They achieved this by:

- encouraging the processing of primary products, as this added value to their exports
- investing in manufacturing industry, initially by developing heavy industries such as steel and shipbuilding, and later by concentrating on high-tech products
- encouraging transnational firms to locate within their boundaries (many countries now have their own TNCs)
- grouping together to form ASEAN (Figure 21.28) to promote, among other aims, economic growth
- having a dedicated workforce that was reliable and, initially, prepared to work long hours for relatively little pay
- long-term industrial planning.

The term **'tiger economies'** was first given to Hong Kong, Singapore, South Korea and Taiwan because of their ferocious growth after 1970. This growth continued during the 1980s at a time when economic growth in the developed world was slowing down. Since the 1980s, Malaysia (the most successful, Places 91), Thailand and, to a lesser extent, Indonesia and the Philippines, have emerged as the newest NICs. The latest, potentially the largest and at an increasingly faster rate, is China (Case Study 19).

Intermediate Technology Development Group (ITDG) is a British charitable organisation which works with people in developing countries, especially those living in rural areas, by helping them to acquire the tools and techniques needed if they are to work themselves out of poverty. ITDG helps people to meet their basic needs of food, clothes, housing, energy and employment. ITDG uses, and adds to, local knowledge by providing technical advice, training, equipment and financial support so that people can become, in Schumacher's words, 'more self-sufficient and independent'. At present, ITDG is working in response to requests from local groups, often women, in India, Nepal (Case Study 18), Bangladesh, Sri Lanka, Kenya, Sudan, Malawi, Zimbabwe and Peru. The following extracts are from ITDG publications concerning their operations in Kenya.

1 The most abundant building source, soil, has great advantages as a building material because it can be easily compressed, it absorbs heat well and transmits it slowly. Although it is vulnerable to erosion, soil can be 'stabilised'

Figure 19.36

Maasai House Improvement Project

Figure 19.37

Improved cooking stoves (*jiko*)

using lime, natural fibres or a small amount of cement. In Kenya, soil blocks stabilised with cement are replacing the more expensive concrete blocks or industrially produced bricks.

2 The Maasai are slowly being forced to live in permanent homes, but few can afford a new house. ITDG is helping them to find affordable ways to upgrade their existing houses (Figure 19.36). Women have told us that their primary concern is a leaky roof; they dread being woken up by their husbands and ordered to climb up in the middle of the night to smear more cow-dung over a crack in the roof. One answer is a thin layer of cement reinforced with chicken wire laid over the old flat mud roof. Plastering is a job the women can do themselves. By incorporating a gutter and a water-jar, women are spared another tiresome chore – collecting water from a spring or river possibly several kilometres away. By adding a small window and a chimney cowl, the inside of the house can be made lighter, less smoky and more healthy.

3 Kenyan women are almost entirely dependent upon wood as fuel for cooking. In the cities women must buy wood or charcoal (transported from rural areas) to use on their traditional stoves, while rural women spend long hours gathering the fuel they need. At the same time, Kenya's resources of wood are diminishing (page 543). An improved cooking stove (*jiko*) has been developed which drastically reduces the use of charcoal for urban households (Figure 19.37). It is based on traditional design and is made by local metal workers, from scrap metal, and incorporates a ceramic lining, produced by local potters. This development has improved the lives of many people:

• More than 40 000 households are now making sizeable savings in their fuel costs. In many homes the new stove pays for itself in about a month. Moreover, the improved stove reduces smoke and fumes in the kitchen and contributes to improved women's health.

• Large numbers of small-scale metal workers and potters are now engaged in manufacturing the *jiko* and have found new support from the government, which makes their business more stable.

Widespread use of the stove also has a potential impact on the environment, as any saving in urban fuel consumption helps relieve the pressure on rural wood resources.

	Average				
	1971–80	1981–90	1995	1998	1999
China	5.4	9.5	7.6	8.4	7.6
Hong Kong	6.3	7.1	5.3	−6.3	5.1
Indonesia	3.2	6.3	4.8	−5.4	−2.2
Japan	4.3	4.1	3.6	−1.4	2.1
Malaysia	4.0	5.1	8.6	−6.2	1.6
Philippines	1.6	1.1	1.6	−0.5	1.1
Singapore	7.2	7.0	6.1	−0.3	2.1
South Korea	6.8	10.1	8.1	−5.8	3.2
Taiwan	5.5	5.2	7.0	2.8	5.7
Thailand	4.0	7.6	8.4	−6.3	1.1

The economies of the Asian NICs continued to grow until the late 1990s when an economic crisis, beginning in Indonesia and the Philippines, spread throughout the whole region, and with effects that were to be felt world-wide. There were signs, however, by mid-1999, of a slow recovery (Figure 19.38).

Figure 19.38

Annual economic growth rate (%) in eastern Asia

Places 91   Malaysia: a newly industrialised country

Until the 1980s, Malaysia's economy was based upon primary products such as rubber and palm oil (Places 68, page 483), timber (Places 76, page 520), tin (Places 79, page 523) and oil (Figure 19.39a). The government at that time proclaimed its vision of Malaysia becoming a fully developed and industrialised nation by the year 2020. Since then the country has emerged as the leader of the second wave of Asian 'tiger economies', averaging – between 1990 and mid-1997 – an annual growth rate of 8 per cent. This allowed the World Bank to classify Malaysia as an 'upper middle income country', no longer a developing country. This has been achieved without high inflation or unmanageable foreign aid, mainly due to the policies of its prime minister, Dr Mahathir.

Malaysia's economic development was based upon its pivotal position as a gateway to ASEAN (Figure 21.28), it being a springboard to eastern Asia, its affordable land and liberal investment rules, and its encourage-ment, through tax incentives, for transnationals to locate there. The country's industrial strategy emphasised the development of

high-value goods for the domestic market and export (e.g. cars) and the encouragement of high-tech industries (e.g. electronics – Figure 19.39b). In 1985, the government founded the Proton car company, initially in conjunction with Mitsubishi (Figure 19.40) and in 1995, the Perodua company, in partnership with Daihatsu.

By the mid-1990s, industry was confined to specifically designed areas such as the new town of Shah Alam. This policy was good environmentally as only certain tracts of primary forest or farmland were used, but had the social disadvantage of concentrating jobs and development within a few core places (page 640). As firms newly locating in Shah Alam need not pay taxes for 10 years, then many of the world's better-known transnationals, together with Proton, located here. The town, with its wide dual-carriageways, is linked to three-lane motorways that lead to the adjacent present capital city (Kuala Lumpur), the chief port (Port Klang) and the old (Subang) and new (near Seremban) international airports.

Figure 19.39

Malaysia's changing exports, 1970 and 1998

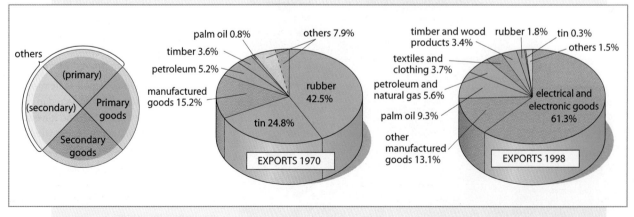

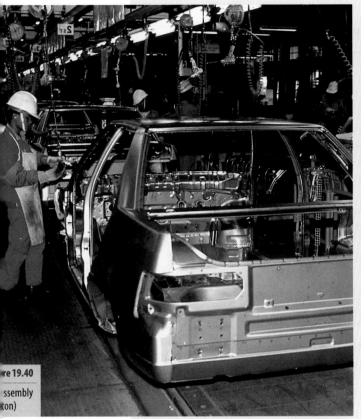

Figure 19.40

...ssembly
...ton)

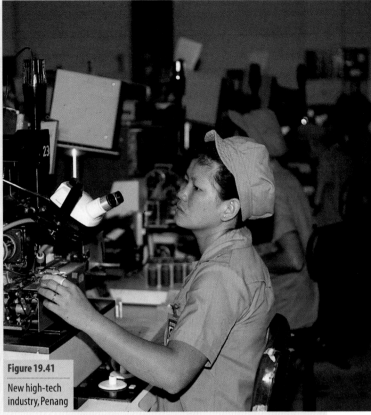

Figure 19.41

New high-tech industry, Penang

The government had also, during the early 1990s, invested less money in industries that required large workforces and more in those where the emphasis was on technology. Its 'Technology Action Plan' covered micro-electronics, biotechnology and information technology (Figure 19.41).

Malaysia's Second Industrial Master Plan, to operate between 1996 and 2005, focused on the manufacturing sector and R & D (research and development), together with the integration of support industries such as packaging, distribution and marketing. The Plan identified eight industrial groups with development potential:

1 electrical and electronics, including IT and multimedia

2 resource based, e.g. oleochemicals from palm oil, furniture from timber and gloves and condoms from rubber

3 agro-based and foods, including aquaculture, processed fruit and vegetables

4 chemicals, including petrochemicals and pharmaceuticals

5 disposable materials, mainly for use in the home

6 transport equipment, e.g. components for car and light aircraft

7 machinery such as machine tools and factory equipment

8 textiles and clothing, mainly 'high-value' goods.

By this time, Malaysia had embarked on a series of expensive 'prestige' projects, notably the construction of a new capital city at Putrajaya (planned for completion 2008), the Petronas Towers (the world's tallest building on its completion in 1998), a new deep-water ocean terminal at Westport (near Port Klang), and 'Linear City', a 2 km long, 10-storey high structure with shopping malls, restaurants, hotels, apartments and offices (to be built on stilts above the River Klang in Kuala Lumpur).

Then, in mid-1997, came a major recession among the Asian 'tigers'. Malaysia's currency, hit by a loss of confidence and speculation in foreign markets, was devalued. Several of the large-scale plans, notably Linear City, were delayed. The country also experienced a double labour crisis, with insufficient skilled people (partly due to a successful family planning programme and partly due to the rapid increase in high-tech industries), and a shortage of cheap, unskilled labour (prompting the need for overseas workers, mainly Filipinos, Indonesians and Bangladeshis – compare Places 44, page 369).

Recent industrial development in China

When the People's Republic of China was declared in 1949, the country's economy was still feudal, with virtually no agricultural or industrial development. There was a very low standard of living; most of the land and wealth was in the hands of a small minority; and life expectancy was just 32 years.

1949–76: the commune system State control was overseen by Mao Zedong. The government dictated where each factory had to be located, how many workers would be employed there, who would have jobs and what products would be made. The state controlled who was responsible for each stage in the manufacturing process, how much was to be produced, how much the wages would be, where the product would be sold and at what price. There were no individual entrepreneurs or decision-makers. The First Five-year Plan

(1953) concentrated on developing labour-intensive heavy industry and discouraged the production of consumer goods. China's huge coal reserves became the basis for an iron and steel industry from which ships, textile machinery, tractors and locomotives were produced. In 1958, Mao launched the 'Great Leap Forward' in an attempt to mobilise China's people and resources. In order to meet production targets, factory workers were organised, like their counterparts in agriculture, into self-sufficient communes. Mao's later attempts to create a classless society (the 'Cultural Revolution' of 1966–76) led to the virtual destruction of the country's economy.

1979 to the present: the responsibility system In 1979 the government, under Deng Xiaoping, began replacing the regimentation of farm production through

communes with the responsibility system (Places 68, page 468). This system, which incorporated capitalist ideas, turned China's peasants into tenant farmers, and was later applied to industry, with the result that total production and its quality improved. Encouragement was given to individual entrepreneurs to set up their own firms, especially in consumer goods. State-owned companies, still accounting for 70 per cent of the total firms, were allowed, once their production targets had been met, to sell surplus stock and to share the profits between their workers. Competition between firms was encouraged and an 'open-door' policy allowed overseas firms to invest and settle in the country (Pepsi-Cola at Shenzhen and Volkswagen in Shanghai).

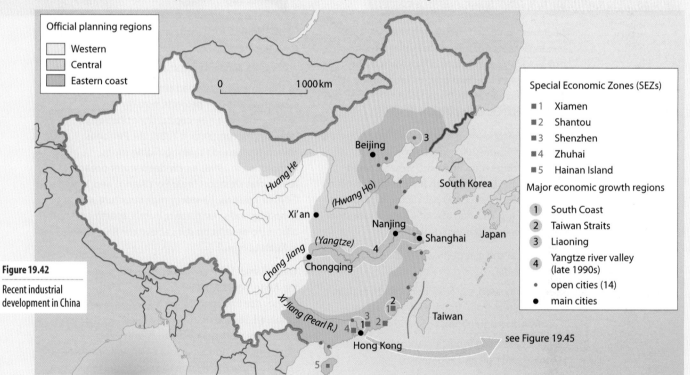

Figure 19.42

Recent industrial development in China

Opening up to the outside world

Since 1980 China has established five **Special Economic Zones** (SEZs) in Shenzhen, Zhuhai, Shantou, Xiamen, and

the whole of Hainan Island (Figure 19.42) so as to better effect reform and stimulate economic growth. Primarily geared towards exporting processed goods, the SEZs now 'integrate science and industry with trade, and benefit from preferential policies and

special economic managerial systems intended to facilitate exports. The SEZs also offer preferential conditions to foreign investors by granting them more favourable rates than in inland areas, and relaxing entry and exit procedures for business people. SEZs aim to attract foreign investment, to import advanced techniques, keep up to date with trends and activity in international markets, to expand export trade, to stimulate foreign exchange earnings, to facilitate participation in international economic and technological co-operation, and to provide a training ground for scientific and technological personnel specialising in international economics and trade' (*China 1998*, Beijing).

In 1984 China opened 14 coastal cities to overseas investment (Figure 19.42). These **open cities** have, according to *China 1998* (Beijing) 'the dual role of being "windows" opening to the outside world and "radiators" spreading economic development in building up an export-oriented economy. The economic and technological development zones established in these open cities

have become hot-spots for foreign investment'. In 1985, the state decided to expand the SEZs and open cities in one continuous coastal belt (Figure 19.42). Five years later, the Chinese opened several additional cities to overseas investment along the Yangtze River as far as Chongqing, with the Pudong New Zone of Shanghai as its 'dragon's head'. The development of Pudong must rank as the world's fastest-ever growth area, with huge commercial, industrial and residential zones and a new transport network (Places 102, page 619; and Figures 19.43 and 19.44). The Chinese see Pudong as the engine pulling Shanghai into position as a major international economic, financial and trade centre. Since 1992, the Chinese have opened 15 free trade zones, 32 state-level economic and technological zones, and 53 new high-tech industrial development zones.

Figure 19.43

Pudong's new skyline

Figure 19.44

Development in Pudong

Shenzhen

Since Shenzhen became an SEZ in 1980, its population has increased from 30 000 to 3.8 million (1997), its economic growth has risen by an average of 20 per cent a year, and it now has an income per capita of £420 compared with an average of £70 in rural China. Its rapid growth, apart from the financial incentives offered by being an SEZ, is due to its coastal location for trade, its proximity to the financial and commercial centre of Hong Kong (Figure 19.45), its plentiful supply of labour, which is adaptable but cheaper than other Asian NICs, and its low land values (rents are half those in Hong Kong).

Three main areas are being developed in Shenzhen: the Longan Grand Industrial Zone (110 km²), the High-Tech Industrial Village (11 km²), and the Futian Central Zone (5 km²). Shenzen has a broad industrial base (over 2000 products have been developed), with electronics, textiles, petrochemicals, machinery, building materials and food processing (Figure 19.46). Many of the earlier industries, including toys, textiles, tyres, chemicals and electronics, were those relocating from Hong Kong. These were followed by transnational organisations such as Sanyo, Hitachi, Matsushita (all Japanese), Siemens (German), IBM (American) and Great Wall (Chinese); in 1998 the *Shenzhen Economic Daily* expressed concern over the domination of foreign firms. In 1999 there were 181 R & D (research and development) organisations within the city, located in large and medium-sized enterprises. Many local scientific research and production joint-bodies have developed strong links with inland universities (compare Places 86, page 566). The annual output value of high-tech products make up one-third of Shenzhen's total industrial output and 90 per cent of its exports.

As the population continues to increase by almost 250 000 per year, many parts of the city resemble huge construction sites surrounded by modern buildings. Only from the tops of the numerous high-rise office and residential blocks can reminders be seen that, 20 years ago, this was primarily a fishing village in a farming area (Figures 19.47 and 19.48).

Figure 19.46

New software park
for Shenzhen

Figure 19.45

Shenzhen
Economic Zone

Software park starts construction

SHENZHEN – Work began yesterday on a 2-billion yuan (US$240 million) software industrial park in Shenzhen, China's vanguard city in its reform and opening-up drive. 'Cyber City Shenzhen Ltd, a Sino–US joint venture, aims at attracting international investment to help promote the development of China's information industry,' says Frank Chen, the park's general manager. The company has raised US$50 million worth of venture capital from the American market and is contacting big-name US companies including Microsoft for co-operation in several other information technology projects, said Chen. The new facility, located in the Shenzhen High-Tech Industrial Park, will have a total floor space of 500 000 square metres.

China Daily, 19 May 1999

Figure 19.47

Shenzhen – this was
all farmland in 1980

Figure 19.48

Today, Shenzhen is a
city of tower blocks

References

Barke, M. and O'Hare, G. (1991) *The Third World*, Oliver & Boyd.

IT, *Intermediate Technology*, IT Publications (ITDG).

Malaysia Official Year Book 1998 (1998) Ministry of Information.

Raw, M. (1993) *Manufacturing Industry: The Impact of Change*, Collins.

Schumacher, E. F. (1993) *Small is Beautiful*, Abacus.

Waugh, D. and Rowley, C.(2000) *Images of Change: China* CD Rom, Geopix.

Websites

Nissan in Europe: Operations and Activities page:
    http://www.nissan-europe.com/europa/eng/operation.htm

Swansea Bay City:
    http://sunset.swan.ac.uk/swantour/default.html

Intermediate Technology Development Group:
    http://www.itdg.org.pe/index.htm

Shenzhen Trade Development Bureau:
    http://sztdb.asiansources.com/FACTS/TOURISM.HTM

Statistics Singapore: Facts and Figures:
    http://singstat.gov.sg/FACT/fact.html

Statistics Bureau and Statistics Centre of Japan:
    http://www.stat.go.jp/l.htm

Merry Hill Centre:
    http://www.merryhill.co.uk/

J Sainsbury plc's Virtual Museum:
    http://www.j-sainsbury.co.uk/museum.htm

See also for more links:
    http://www.nelsonthornes.com/gaia

Q

1 a i What is 'manufacturing industry'? **(1 mark)**
  ii 'With the shift from an industrial to a post-industrial society it is sometimes unrealistic to try to draw clear boundaries between "manufacturing" and "services".' Explain the problems which led to this statement. **(2 marks)**
  iii The table below shows the proportion of the UK's population employed in different sectors of the economy.

	1851	1901	1951	1997
primary	24	12	6	4
secondary	48	54	49	25
tertiary	28	34	45	61

  Explain why the proportion of the population in secondary employment has fallen so sharply in recent years. **(6 marks)**
b The number of people employed in manufacturing has not fallen evenly across the country.
  i Name an area where a loss of manufacturing jobs has caused a serious local unemployment problem. **(1 marks)**
  ii Explain what caused the loss of manufacturing jobs in that area. **(3 marks)**
  iii Describe a strategy that has been used to create new employment opportunities in that area. Assess the success of the strategy. **(6 marks)**
c The gender structure of the workforce in the UK has changed rapidly since 1960. Describe and account for the changes. **(6 marks)**

2 a i What are 'high-tech industries'? **(2 marks)**
  ii It has been noted that firms involved in high-tech industries often have two quite distinct parts to their operations. These are:
  ■ research and development
  ■ mass manufacturing.
  Suggest why these two separate parts of the industry often locate in different types of location. **(3 marks)**
b i Name **one** area where a concentration of research and development centres for high tech industry has developed. **(1 mark)**
  ii Explain why the area you named in i is attractive to this industry. **(9 marks)**
c i Name **one** area where mass production for the high-tech industry has developed quite separately from research and development. **(1 mark)**
  ii Explain why the area you named in i is attractive to this industry. **(9 marks)**

AS

3 a In the Weber model of industrial location, what is the 'least cost location'? Explain why it is important. **(3 marks)**
b Show, with the aid of diagrams, how the least cost location in Weber's locational triangle may be:
  i near to the raw material
  ii near to the market
  iii midway between the raw material and the market. **(6 marks)**
c Study Figure 19.49, which shows an area with one raw material and one market. Assume two different situations:
  X – the raw material is pure
  Y – the raw material loses 50 per cent of its weight during manufacture.
  i Complete the table to show the total transport costs (in tonne km) for an industry located at each of the sites A–D. **(4 marks)**
  ii For each situation, X and Y, describe the least cost location. Give reasons for your answer. **(6 marks)**
d Describe **one** other model of industrial location which has been developed since Weber's model. Explain the advantages of the new model. **(6 marks)**

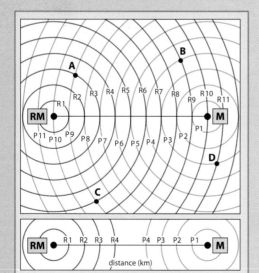

RM raw material

M market

— R 1 — isotims (for raw material)

— P 1 — isotims (for product)

Figure 19.49

Site	Total transport costs if the raw material is pure	Total transport costs if the raw material is gross and loses 50% of its weight during manufacture
A		
B		
C		
D		

A2

4 a Study the diagram below. It shows some of the factors that influence the location of manufacturing industry.

Raw materials

Power supply → Factory ← Labour

↓

Market

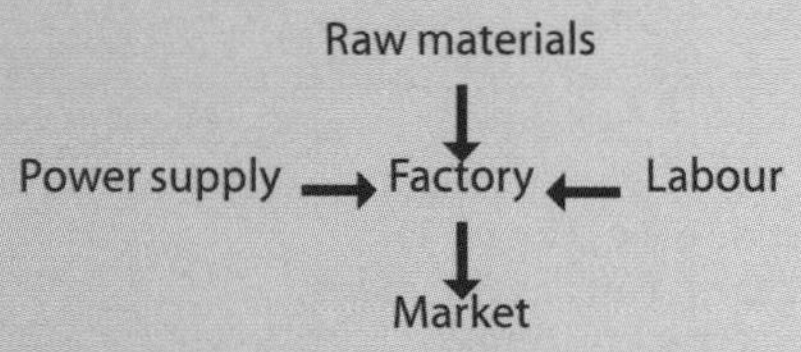

Give one example of an industry where the most important factor influencing its location is:
  i raw materials
  ii power supply
  iii labour supply
  iv access to market.
For each example you have given, explain why that factor is so important. **(12 marks)**

b i What is meant by:
  • footloose industry?
  • greenfield site?
  ii Name an example of a footloose manufacturing industry which has located on a greenfield site. Suggest why that site was a suitable location for that industry. **(5 marks)**

c i What is 'inward investment'? **(1 mark)**
  ii Choose an example of a factory that has been built by a foreign-based company investing in the UK. Suggest why that company chose to invest in the UK, and why that particular location was suitable for their investment. **(7 marks)**

5 a i Name a region in the UK which has suffered unemployment as a result of the decline of its traditional manufacturing industry. **(1 mark)**
  ii Explain why the traditional industry developed in that area and then declined. **(8 marks)**
  iii Describe the other social and economic problems that are found in that area as a result of the unemployment that followed the decline of the traditional industry. **(6 marks)**

b The government has developed several policies to try to attract new industry into regions which have suffered the loss of their traditional industry. Such policies have included:
  • Development Areas
  • Enterprise zones
  • direct attempts to attract foreign firms.
Choose any such government initiative, and describe how it has affected any **one** area. Try to assess how successful the initiative has been in attracting new manufacturing industry. **(10 marks)**

Industry	Region										Great Britain
	South East	East Anglia	South West	West Midlands	East Midlands	Yorkshire and Humber-side	North West	North	Wales	Scotland	
Shipbuilding	39.5	3.5	19.5	0	1.5	7.3	10.0	48.3	1.6	43.0	174.2
Textiles	19.6	2.6	8.5	16.6	85.4	65.0	64.2	11.0	8.5	41.7	323.1
All manufacturing	1683.5	186.0	395.7	800.7	533.4	578.9	799.8	339.4	238.2	502.0	6057.6

Figure 19.50

UK employment in shipbuilding, textiles and all manufacturing industries by region

6 Study the information in Figure 19.50.
  a Calculate the location quotient (LQ) for the following industries:
    i textiles in the North West
    ii textiles in East Anglia
    iii shipbuilding in Yorkshire and Humberside
    iv shipbuilding in Scotland. **(4 marks)**
  Use the formula:

$$\frac{\left(\dfrac{\text{number of people employed in industry A in area X}}{\text{number of people employed in all industries in area X}}\right)}{\left(\dfrac{\text{number of people employed in industry A nationally}}{\text{number of people employed in all industries nationally}}\right)}$$

    v Summarise what is shown by your results in i to iv. **(4 marks)**

b With reference to a named industrial region in the UK:
  i describe the problems that have been caused by a high concentration of employment in a small number of industries **(7 marks)**
  ii explain how one or more government initiatives have been used to try to broaden the base of employment in the area. **(10 marks)**

7 Changes in technology during the past 30 years have had a major effect on industrial location throughout the world. Describe **at least three** of these changes. Explain why they have taken place and how they have affected the location of industry. **(25 marks)**

8 a i Describe **three** differences between the formal and the informal economic sectors in economically less developed countries. **(6 marks)**

ii Explain why 'Jua Kali' workshops are very important in Kenya's economy. **(6 marks)**

iii 'If governments wish to encourage development that will benefit the poorer sections of the population, one of the most important actions they can take is to reduce the rules and regulations which hinder the development of the informal economy.'
Suggest why this is seen to be important. **(4 marks)**

b Discuss the advantages and disadvantages for less economically developed countries of investment by transnational companies. You should make specific references to one or more countries which you have studied. **(9 marks)**

9 a i In the 1980s four countries, Hong Kong, Singapore, South Korea and Taiwan, became known as the 'Tiger Economies.' Why? **(3 marks)**

ii Name one **other** southeast Asian country which is industrialising rapidly. **(1 marks)**

b Choose **either** one of the countries named in question a i, **or** the country named in a ii. For your chosen country:

i describe the traditional economy of the country before 1970 **(4 marks)**

ii explain how the government's policy towards industrial development has changed since 1970 **(5 marks)**

iii describe how industrial development has affected **one** named area of the country since 1970 **(8 marks)**

iv explain how the country was affected by the recession in southeast Asia in 1997. **(4 marks)**

10 a In Myrdal's model of industrial location he referred to 'cumulative causation' which is also sometimes called 'the multiplier effect'. What does this term mean? **(4 marks)**

b With reference to the industrialisation of:
either South Wales in the late 19th and early 20th centuries
or any other region with which you are familiar
explain how cumulative causation helped to cause the development of industry. **(10 marks)**

c Myrdal realised that as some areas industrialised this may cause a 'backwash effect' on other regions in the coountry. He named the areas that were effected 'the periphery'. Explain:

i the 'backwash effect' **(3 marks)**

ii the 'periphery' of a country or region **(3 marks)**

d Name an area which could be regarded as part of the economic periphery in a country or region which you have studied.
Describe the features that make this area part of the economic periphery **(5 marks)**

AS

A report earlier this year showed that there has been a net loss of 500 000 jobs in Britain's 20 biggest cities since the early 1980s compared with a 1.7m growth in employment in the rest of the country. The worst affected areas have been the core districts of the major conurbations – especially Clydeside, Greater Manchester and Merseyside, where the loss of manufacturing jobs amongst men has been especially severe.

In addition, Britain's economy looks a lot different now than it did 30 years ago, with a sharp decline in manufacturing jobs and a shift to the service sector. But whilst manufacturing has declined sharply in the cities it has survived quite well, or even prospered, in less urbanised parts of the country.

Just as in agriculture, the labour-to-land ratio in manufacturing is declining and large urban centres, where land is scarce or expensive to redevelop, are likely to shed labour more rapidly, even when manufacturing industry is increasing in the country as a whole.

The Guardian, 7 December 1999

Figure 19.51

11 Read the extract in Figure 19.51.

a i Explain why manufacturing employment has declined in the UK during the past 30 years. **(6 marks)**

ii Explain why this decline has been particularly severe in the central parts of the major conurbations and less severe in smaller towns. **(6 marks)**

b With reference to a named conurbation in the UK:

i Describe the problems that have been caused by the decline in manufacturing employment. **(6 marks)**

ii Examine the strategies employed by central government and/or local authorities to manage the consequences of the loss of traditional industrial employment. **(7 marks)**

12 a Describe the main features of Myrdal's theory of cumulative causation. **(6 marks)**

b Explain how cumulative causation can lead to the development of growth poles (or core regions) and peripheral regions in a country. **(6 marks)**

c With reference to a named peripheral region in a country outside the UK:

i explain what problems have been caused by its peripheral position **(6 marks)**

ii describe **one** scheme that has been tried in an attempt to overcome these problems, and discuss how successful the scheme has been. **(7 marks)**

13 Assess the importance of transnational corporations in the development of the global pattern of industrialisation in the late 20th and early 21st centuries. You should refer to their effect in both more and less economically developed countries. You should also refer to more traditional manufacturing and to modern, high-tech industry **(25 marks)**

A2

Chapter 20

Tourism

'In the Middle Ages people were tourists because of their religion whereas now they are tourists because tourism is their religion.'

Robert Runcie, former Archbishop of Canterbury

'Travel broadens the mind.'

Proverbs

What is meant by 'services'?

Services are normally divided into three groups:

1 **Consumer services** are those offering services directly to the consumer, i.e. the point of final demand. These include retailing (shops, high street banks and estate agents) and tourism and leisure services.

2 **Producer services** are intermediate services that help businesses of all kind carry out their activities (banking, law, transport, warehousing and consultants). (Some companies may have both consumer and producer elements in their operations, e.g. banking.)

3 **Public services** are a crucial element of any modern economy (e.g. electricity and water companies).

All societies, past and present, have had a service sector. There does appear, however, to be a strong positive correlation between the service sector and levels of development – i.e. the more

a country or region develops economically, the greater the number of its inhabitants employed in the service sector (Framework 19, page 635). In Britain, the highest proportion of employed people is in the service sector (see Figure 19.1).

This chapter focuses solely on leisure and tourism; retailing and commerce are studied in Chapter 15 and transport and trade in Chapter 21.

Leisure and tourism

Leisure and tourism is now the world's fastest-growing industry employing – according to one UN source – 7 per cent (or 1 in 15) of people in employment. Leisure can be defined as 'time at one's own disposal' and involves activities that we do and enjoy voluntarily. Whereas leisure can be enjoyed at home, tourism involves visiting places. The official UK definition is: 'a stay away from one's normal place of residence which includes at least one night but is for less than a year'. The UK travel and tourism industry consists of a wide variety of commercial and non-commercial organisations that interact to supply products and services to tourists. Even so it is often difficult to differentiate leisure and tourism from other forms of employment. For example, how can hotel owners in central London

Figure 20.1

Types and location of various leisure and tourist facilities

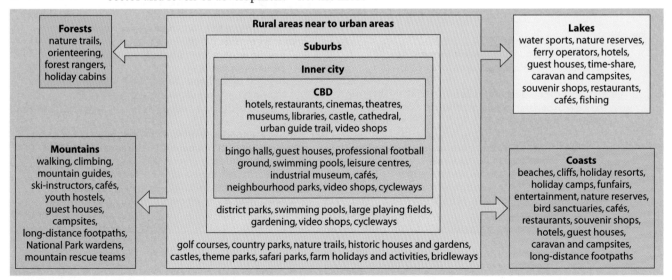

Forests
nature trails, orienteering, forest rangers, holiday cabins

Rural areas near to urban areas

Suburbs

Inner city

CBD
hotels, restaurants, cinemas, theatres, museums, libraries, castle, cathedral, urban guide trail, video shops

bingo halls, guest houses, professional football ground, swimming pools, leisure centres, industrial museum, cafés, neighbourhood parks, video shops, cycleways

district parks, swimming pools, large playing fields, gardening, video shops, cycleways

golf courses, country parks, nature trails, historic houses and gardens, castles, theme parks, safari parks, farm holidays and activities, bridleways

Lakes
water sports, nature reserves, ferry operators, hotels, guest houses, time-share, caravan and campsites, souvenir shops, restaurants, cafés, fishing

Mountains
walking, climbing, mountain guides, ski-instructors, cafés, youth hostels, guest houses, campsites, long-distance footpaths, National Park wardens, mountain rescue teams

Coasts
beaches, cliffs, holiday resorts, holiday camps, funfairs, entertainment, nature reserves, bird sanctuaries, cafés, restaurants, souvenir shops, hotels, guest houses, caravan and campsites, long-distance footpaths

differentiate between guests who are on holiday or on business, a fish and chip shop owner in Blackpool between tourists and residents, or farmers on a Greek or West Indian island between food sold to local people or to hotels?

People with limited leisure time tend to seek recreational amenities and activities that are near their homes. As the majority of British people live in towns and cities, then most amenities are located within or near to urban areas (Figures 20.1 and 20.2). People with more leisure time tend to travel further afield to scenic rural areas, especially those with added amenities (coasts, mountains and National Parks), to large urban areas (historical towns and cultural centres), and to places outside the UK.

As in other areas of geography, people have tried to classify aspects of tourism (Framework 7, page 167). Two such classifications are:

1 **Based on resources**
   a **primary resources** – scenic, climatic, ecological, historical and heritage
   b **secondary resources** – accommodation, catering, entertainment and infrastructure
2 **Based on type of tourism**
   a **domestic** – residents taking trips or holidays within their own country

   b **inbound** – non-resident visitors to a country
   c **outbound** – residents going outside their own country.

The growth in tourism

The Romans must rank amongst the earliest tourists, as many of their most wealthy families used to move to their country villas during the hot, dry summers. By the 18th and 19th centuries, many affluent British people were either visiting spa towns within England or making the 'Grand Tour' of Classical Europe, while the less well-off were beginning to popularise local seaside resorts. Today tourism has become part of everyday life and a major source of employment in many developed countries. Here, the rapid growth of the tourist industry in the last half-century can be linked to numerous factors such as greater affluence (wealth), increased mobility, improvements in accessibility and transport, more leisure time, product development and innovations, improvements in technology, changes in lifestyles and fashion, an increased awareness of other places and, more recently, the need for 'green' (sustainable) tourism (page 597). These factors are summarised in Figure 20.3.

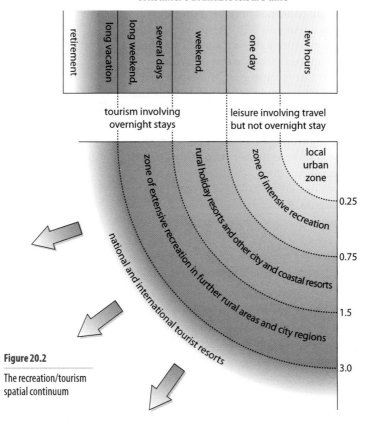

consumer's available leisure time

retirement | long vacation | long weekend | several days | weekend | one day | few hours

tourism involving overnight stays | leisure involving travel but not overnight stay

zone of extensive recreation in further rural areas and city regions
national and international tourist resorts
rural holiday resorts and other city and coastal resorts
zone of intensive recreation
local urban zone

0.25
0.75
1.5
3.0

Figure 20.2

The recreation/tourism spatial continuum

Greater affluence	– People in employment earn high salaries and their disposable income is much greater than it was several decades ago. – People in full-time employment also receive holiday with pay, allowing them to take more than one holiday a year and to travel further.	
Greater mobility	– The increase in car ownership has given people greater freedom to choose where and when they go for the day, or for a longer period. In 1951, only 1 UK family in 20 had a car. By 1999, 72 per cent had at least one car. – Chartered aircraft have reduced the costs of overseas travel.	
Improved accessibility and transport facilities	– Improvements in roads, especially motorways and urban by-passes, have reduced driving times between places and encourages people to travel more frequently and greater distances. – Improved and enlarged international airports (although many are still congested at peak periods). Reduced air fares. Package holidays.	
More leisure time	– Shorter working week (although the UK's is still the longest in the EU) and longer paid holidays (on average 3 weeks a year, compared with 1 week in the USA). – Flexitime, more people working from home, and more firms (especially retailing) employing part-time workers. – An ageing population, many of whom are still active	
Technological developments	– Jet aircraft, computerised reservation systems, use of the Internet.	
Product development and innovation	– Holiday camps, long-haul destinations, package tours.	
Changing lifestyles	– People are retiring early and are able to take advantage of their greater fitness. – People at work need longer/more frequent rest periods as pressure of work seems to increase. – Changing fashions, e.g. health resorts, fitness holidays, winter sun.	
Changing recreational activities	– Slight decline in the 'beach holiday' – partly due to the threat of skin cancer. – Increase in active holidays (skiing, water sports) and in self-catering. – Most rapid growth in mid-1990s was in 'cruise holidays'. – Importance of theme parks, e.g. Alton Towers, Thorp Park. – Large number of city breaks.	
Advertising and TV programmes	– Holiday programmes, magazines and brochures promote new and different places and activities.	
'Green' or sustainable tourism	– Need to benefit local economy, environment and people without spoiling the attractiveness and amenities of the places visited (ecotourism).	

Figure 20.3

Recent trends in tourism in the UK

Global tourism

By 1997, tourism receipts accounted for over 8 per cent of total world export of goods and almost 35 per cent of the total world exports of services. Of the total tourist receipts, 80 per cent were earned by countries in either Europe or North America (Figure 20.4). The travel account in developed countries showed a small surplus, with those regions for the first time earning more from tourist receipts than they spent overseas. Their previous trade deficit was mainly due to large sums of money spent by German and Japanese tourists. The travel account balance in the developing countries has been persistently in the surplus and continually widening. This is mainly because relatively few inhabitants of these regions can afford holidays in developed countries (Figure 20.5)

Figure 20.4

Growth in global tourism

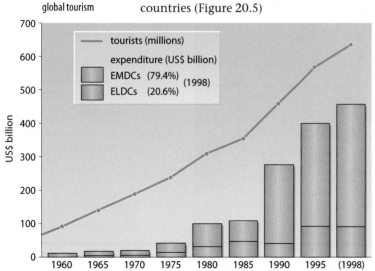

		Arrivals (thousands)	% world total
1	France	71.4	11.1
2	Spain	51.9	8.1
3	USA	46.9	7.3
4	Italy	35.8	5.6
5	China	27.0	4.2
6	UK	25.7	4.0
7	Mexico	20.4	3.2
	World	**642.4**	

		Earners (US$ million)	% world total
1	USA	71.116	16.2
2	Italy	30.427	6.9
3	France	29.700	6.8
4	Spain	29.585	6.7
5	UK	21.233	4.8
6	Germany	15.859	3.6
7	China	12.600	2.9
	World	**439.393**	

		Spenders (US$ million)	% world total
1	USA	51.220	13.5
2	Germany	46.200	12.2
3	Japan	33.041	8.7
4	UK	27.710	7.3
5	Italy	16.631	4.4
6	France	16.576	4.4
7	Canada	11.268	3.0
	World	**379.757**	

Figure 20.5

Leading tourist countries, 1999

UK tourism

Tourism in the UK grew from the development of spa towns to, in the 19th and early 20th centuries, the beginnings of mass tourism when many industrial workers and their families took a day, and later a few days, 'at the seaside' (Figure 20.6). Since the Second World War, the annual holiday has become part of most families' way of life with, increasingly, many of them spending more of their annual income on holidays and more travelling further afield. As a result, the industry has had to expand rapidly to meet the growing and diversifying needs of tourists. The scale of the UK travel and tourism industry is shown by certain indicators.

■ **Consumer spending** In 1998, UK residents spent £16 931 million overseas (£7280 in 1987) compared with overseas residents who spent £12 244 million in the UK (£6260 million in 1987) – a deficit on the travel account of £4687 million (a deficit of £1020 million in 1987). The World Tourist Organisation claimed that 25.7 million visited Britain in 1999, compared with 17.1 million in 1991. Recent figures show the average British family spends 15.6 per cent of its annual income on holidays.

■ **Number employed and type of job** Estimates suggest that over 2 million people are now employed directly in the tourist industry in Britain, with at least as many again employed indirectly. The wide range of jobs includes construction, hotels and catering, transport, travel agents, entertainment and tour guides.

■ **Number of tourists** In 1998, Britain received over 32 million visitors from overseas while, at the same time, 25.8 million UK residents took their holidays overseas (outgoing tourists). Spain still remained the main location for British tourists with 28 per cent going there, followed by France (18 per cent), Greece (8 per cent), North America (7 per cent), and Italy, Portugal, Africa and the Caribbean (each with 4 per cent). Of UK residents who stayed within Britain for their holidays (domestic tourists), 31 per cent visited the countryside, 64 per cent towns and cities and 5 per cent the coast. The West Country was the most popular region visited, followed by Wales and Scotland.

The travel and tourist industry is dynamic and has to change continually to meet consumer needs and perceptions. Its key features at present include the following:

■ It is predominantly private-sector led.

■ It is dominated by relatively few large-sized, often transnational, firms, e.g. tour operators (Lunn Poly, Going Places, Thomas Cook); hotel chains (Marriott, Sheridan, Holiday Inn), theme parks (Disney), and air operators (BA, Monarch). Despite this, the majority of enterprises are small and medium-sized.

■ It is vulnerable to external pressures such as currency fluctuations, government legislation, climatic change, together with war, terrorism and civil unrest.

■ There is extensive use of new technologies, especially in data handling, advance bookings and use of the Internet.

■ It is a complex structure consisting of a wide variety of interrelated commercial and non-commercial organisations (Figure 20.7).

■ It has both a positive and a negative impact on host communities (economic, social and cultural) and local environments (Figure 20.8).

Figure 20.6

Factors affecting the growth of the holiday industry in the UK

Factors	Specific examples	Example of area or resort
1 Transport and accessibility	• Early resorts (stage-coach), spa towns • Water transport (18th century) • Railways • Car and coach • Plane	Bath Margate Blackpool, Brighton Cornwall, Scottish Highlands Channel Islands
2 Scenery	• Sandy coasts • Coasts of outstanding beauty • Mountains, lakes and rivers	Margate, Blackpool Pembroke, Antrim Lake District, Snowdonia
3 Weather	• Hot, dry, sunny summers • Snow	Margate Aviemore
4 Accommodation	• Hotels and boarding house resorts • Holiday camps • Caravan parks and campsites	Margate, Blackpool Minehead, Pwllheli National and forest parks
5 Amenities	• Culture and historic (castles, cathedrals, birthplaces) • Active amenities (sailing, golf, water-skiing) • Passive amenities (shopping, cinemas) • Theme parks	York, Edinburgh Kielder, St Andrews Most resorts Alton Towers, Chessington
6 Ecotourism and sustainability	• Wildlife conservation areas • Heritage sites	SSSIs, nature reserves York

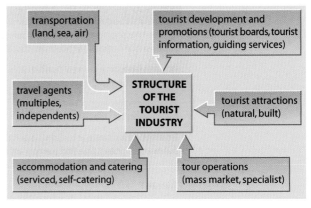

Figure 20.7

Structure of the tourist industry

Positive effects/benefits	Negative effects/problems
Economic	
Increases gross domestic product directly and indirectly via multiplier (see Myrdal, page 569).	May divert government expenditure from other needy areas of the economy.
Taxes on tourism increase government revenue.	Requires government expenditure on tourism.
Increased foreign exchange earnings.	Over-dependence on outside agencies and some external control on the economy.
Foreign investment.	Income reduced by 'leakages' or outflows.
Creates employment, including in unskilled occupations; labour-intensive.	Profits may go overseas.
Helps fund new infrastructure and facilities which local people can also use.	Overstretches infrastructure.
Stimulates and diversifies economic activity in other sectors – local craft revival, manufacturers, services and agriculture.	Spread effects limited and may therefore increase regional inequalities between tourist growth areas and less developed periphery (page 640).
May act as a seedbed for entrepreneurship, with spin-offs into other sectors.	Diverts labour and resources away from non-tourist regions and may (particularly) affect peripheral areas, leading to out-migration to tourist resort opportunities (Places 42, page 366).
Improves balance of payments through increased trade.	Labour unskilled and seasonal.
	Foreign personnel and firms dominate managerial and higher-paid posts, reducing opportunities for local people.
	Inflated prices for land, housing, food and clothes.
Social	
Cultural exchange stimulated with broadening of horizons and reduction of prejudices amongst tourist visitors and host population.	May cause polarisation between population in advancing tourist regions and less developed areas, creating a 'dual society'.
May enhance role and status of women in society, as opportunity for goals in tourism is created and outlook widened.	Increases rift between 'rich' and 'poor'.
Encourages education.	Breakdown of traditional family values creates material aspirations.
Encourages travel, mobility and social integration.	Breakdown of families due to stress between younger generation, who are affected by imported culture, and older members of household.
Improves services (electricity and health), transport (new roads, airports) and widens range of shops and leisure amenities.	Social pathology, including an increase in prostitution, drugs and petty crime.
	Increases health risk, e.g. AIDS.
Cultural	
May save aspects of indigenous culture due to tourist interest in them.	Impact of commercialisation may lead to pseudo-cultural activities to entertain tourists and, at extreme, may cause disappearance or dilution of indigenous culture.
Contact with other cultures may enrich domestic culture through new ideas and customs being introduced.	Mass tourism may create antagonism from host population who are concerned for traditional values, e.g. dress, religion.
Encourages contact and harmonious relations between people of different cultures.	Westernisation of culture, food (McDonalds) and drink (Coca-Cola).
Increases international understanding.	
Environment	
Improved landscaping and architectural standards in resort areas, including increased local funding for improvement of local housing, etc.	Poor building and infrastructure development – tourist complexes do not integrate with local architecture.
Promotes interest in monuments and historic buildings, and encourages funding to conserve and maintain them.	Destruction of natural environment and wildlife habitat – marine, coastal and inland.
May induce tighter environmental legislation to protect environment, i.e. landscape, heritage sites, wildlife.	Excessive pressure leads to air, land, noise, visual and water pollution, and breakdown in water supplies, etc.
Establishment of nature reserves and National Parks; growing tourist interest and awareness protects areas from economic and building encroachment.	Traffic congestion and pollution.
	Clearance of natural vegetation, loss of ecosystems.

Adapted from a World Tourist Organisation classification

Figure 20.8

Positive and negative
effects of mass tourism

Tourism and the environment

As the demands for recreation and tourism increase, so too will their impact on other socio-economic structures in society, tourist environments and wildlife habitats. Tourists will compete for space and resources with:

- local people living and working in the area, e.g. farmers, quarry workers, foresters, water and river authority employees (Figure 17.4)
- other visitors wishing to pursue different recreational activities, e.g. water skiers, windsurfers, anglers and bird watchers all visiting the same lake.

The development of recreation and tourist facilities creates pressure on specific places and environments in both urban and rural areas. Places with special interest or appeal that are very popular with visitors and which tend to become overcrowded at peak times are known as **honeypots**. Honeypots may include, in urban areas, concert halls (Albert Hall), museums (Madame Tussaud's), and historic buildings (Tower of London); and, in rural areas, places of attractive scenery (Lake District), theme parks (Alton Towers), and places of historic interest (Stonehenge). The problem of overcrowding within certain American National Parks (Yellowstone), together with congestion on access roads, has become so acute that permits are needed for entry and quotas are imposed on areas that are ecologically vulnerable (Case Study 17).

Sometimes planners encourage the development of honeypots, especially in British National Parks and African safari parks, to ensure that such sites have adequate visitor amenities (car parks, picnic areas, toilets, accommodation). It is now widely accepted that leisure amenities and tourist areas need to be carefully managed if the maximum number of people are to obtain the maximum amount of enjoyment and satisfaction (Figure 20.9).

Recently there has been a growth in **ecotourism** (Places 95, page 598), which aims at safeguarding both natural and built environments, being sustainable (Framework 16, page 499), and enabling local people to share in the economic and social benefits.

It is possible to identify three levels of recreation and tourism in rural areas.

1 High-intensity areas where recreation is the major concern (theme parks such as Alton Towers, honeypots as at Bowness on Windermere, and resorts such as Aviemore).
2 Average-intensity areas where there needs to be a balance between tourism and other land users, and between recreation and conservation (Peak District National Park, Places 92).
3 Low-intensity areas, usually of high scenic value, where conservation of the landscape and wildlife is given top priority (upland parts of Snowdonia and the Cairngorms – Places 94, page 595).

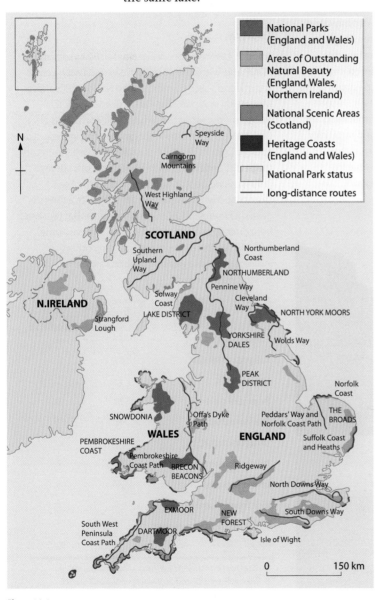

Figure 20.9

Protected areas in the UK

Following the passing of The National Parks and Access to the Countryside Act in 1949, the Peak District became the first National Park in 1951 (Figure 20.9). The Environment Act of 1995, which set up a National Park Authority to administer the affairs of each of the National Parks, defined the purposes of National Parks as:

- conserving and enhancing the natural beauty, wildlife and cultural heritage, and
- promoting opportunities for the understanding and enjoyment of their special qualities.

National Parks must also foster the economic and social well-being of local communities. They are also required to pursue a policy of sustainable development by which they must aim to improve the quality of people's lives without destroying the environment (Framework 16, page 499). Despite their often spectacular scenery, National Parks are not owned by the nation nor managed purely for their landscapes and wildlife. They are, rather, mainly farmed areas where many people live (38 000 within the Peak District) and work.

Figure 20.10

The White Peak: Lathkill Dale Nature Reserve

Figure 20.11

The Dark Peak: Kinder Scout

Land use

There are some 2700 farms in the Peak District National Park (PDNP), most of them under 40 ha. Some farms are owned by the National Trust and Water Companies and 60 per cent are thought to be run on a part-time basis (where the farmer has a second job). The Peak Park Authority manages 480 ha of woodland, much of it coniferous. There are 55 reservoirs of over 2 ha in the Park supplying the large urban areas on its fringes. Limestone and fluorspar are quarried.

Tourism

About 17 million people live within 100 km of the PDNP and, with over 30 million day visits each year, it is the world's second most visited National Park (after Mount Fuji in Japan). The Park is divided, scenically, into two – the attractive 'White Peak' consisting of Carboniferous limestone (page 196 and Figure 20.10) and the Dark Peak, which is a Millstone Grit moorland (page 201 and Figure 20.11). The result of a survey (1994) asking people why they visited the Peak District and which were their favourite attractions, is given in Figure 20.12. Estimates suggest that tourism here directly provides 500 full-time, 350 part-time and 100 seasonal jobs, as well as many others indirectly (people working in shops and other service industries).

	All reasons %	Main reason %
Scenery/landscape	61	39
Enjoyed earlier visit	37	9
Easy to get to	33	7
Peace and quiet	28	4
Outdoor activity	18	11
Event/attraction	16	10
As it is a National Park	15	3
Come every year	9	3
Visit friends/family	5	3
Own accommodation in area	4	2
Other	20	10

Most popular areas of the Peak District National Park are:
- **Bakewell**, with interesting buildings and a busy market.
- **Chatsworth**, home of the Duke of Devonshire.
- **Dovedale**, a spectacular limestone dale.
- **Hartington** village.
- **Goyt Valley** and its reservoirs.
- **Hope Valley** and the village of Castleton.
- **Upper Derwent** and the Ladybower and Derwent Reservoirs.

Figure 20.12

Why people visit the Peak District National Park

Conservation

National Parks were set up with the specific purpose of protecting areas of natural beauty in the countryside. Today, although facilities for suitable types of recreation (walking, climbing and fishing) are an important part of the National Parks, the aims of conservation have to take priority. Conservation in the PDNP involves more than just preventing damage and leaving things alone. It means:

- maintaining the best features of the landscape, e.g. well-managed moorland and listed buildings
- improving neglected features, e.g. rebuilding stone walls and replanting old woodlands
- managing developments so that damage is limited, e.g. new buildings and recreation activities.

Conservation of the landscape is not only the concern of the PDNP Authority. English Nature is responsible for Sites of Special Scientific Interest (SSSIs) which now cover 30 per cent of the PDNP. The Ministry of Agriculture, Food and Fisheries (MAFF) created the North Peak ESA in 1988 and the South Peak ESA in 1992 (page 496). The Countryside Stewardship Scheme (page 497), set up in 1996, encourages farmers, through grants, to enhance their land.

Conservation of villages (Figure 20.13) is also important, with the PDNP Authority controlling the erection of new buildings, limiting the range of building materials, and creating Conservation Areas in many villages that have buildings of historic or architectural interest.

In 1999, the PDNP Authority issued, for discussion, a new Management Plan. The Plan sets out a vision for the future and suggests ways it may be achieved. The vision is based on three main aims:

1 to conserve and enhance the special qualities of the National Park

2 to provide opportunities for visitors' enjoyment and understanding

3 to improve the quality of life for people who live, work and visit.

The Plan is based on two principles: sustained development, and understanding (Figure 20.14).

OBJECTIVE 1

Working together for an *environment* ...

where the qualities that people value – landscape, geology, clear earth, air and water, wilderness, tranquillity, remoteness, wildlife, biodiversity, archaeology, historic buildings, traditional farming, stone walls, small woodlands, open spaces, gardens, parks and villages – are all conserved and enhanced.

OBJECTIVE 2

Working together so that *people* ...

- can live in vibrant, innovative and sustainable communities
- can be involved in and take responsibility for caring for their environment and culture
- from every part of our society have access to the countryside and can experience, understand, enjoy and benefit from the culture and recreational opportunities that the National Park has to offer
- can make more use of improved public transport and rely less on their cars.

OBJECTIVE 3

Working together to generate an *economy* ...

- that is more diverse and based on sustainable development
- where communities meet the challenge of economic and social change
- where farming and business are helped to conserve the environment
- where visitors are welcomed and bring positive benefits
- where natural resources are used prudently and mineral working is restricted to essential national need or provides traditional building materials for the National Park.

OBJECTIVE 4

Working together to create *understanding* ...

- with access to information for all
- so that people can acquire and share the knowledge, skills and resources necessary for the sustainable management of the National Park and the wider countryside.

Figure 20.14

From the Discussion Document of the 1999 PDNP Management Plan

Figure 20.13

Enhancement Project in Eyam Square

The tourist resort/area life-cycle model

Despite some of the obvious disadvantages of tourism, the nightmare scenario for any tourist-dependent country, region or resort, is that people will find somewhere else to visit and to spend their money. New resorts develop; old resorts may become run-down; fashions change; places may receive a bad press; economic recessions occur; currency rates alter and new activities are designed. To survive, tourist places have to keep re-inventing themselves by, for example, including new attractions or changing their orientation to a wider or new client group. Places that fail, such as some older British seaside resorts and spa towns, begin to wither away. Places that manage to adapt, such as Blackpool, continue to be successful. On this basis, Briggs produced a useful life-cycle model (Framework 12, page 352) for tourist resorts (Figure 20.15); this may also be applied more widely to tourist regions (Places 93).

Figure 20.15

Tourist area/resort life-cycle model (*after* Briggs)

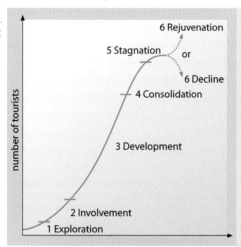

1 **Exploration:** small number of visitors attracted by natural beauty or cultural characteristics – numbers are limited and few tourist facilities exist, e.g. Chile.
2 **Involvement:** limited involvement by local residents to provide some facilities for tourists – recognisable tourist season and market areas begin to emerge, e.g. Guatemala.
3 **Development:** large numbers of tourists arrive, control passes to external organisations, and there is increased tension between local people and tourists, e.g. Florida.
4 **Consolidation:** tourism has become a major part of the local economy, although rates of visitor growth have started to level off and some older facilities are seen as second-rate, e.g. earlier Mediterranean coastal resorts.
5 **Stagnation:** peak numbers of tourists have been reached. The resort is no longer considered fashionable and turnover of business properties tends to be high, e.g. Costa del Sol (Places 93).
6 **Decline** or **rejuvenation:** attractiveness continues to decline, visitors are lost to other resorts, and the resort becomes more dependent on day visitors and week-end recreationalists from a limited geographical area – long-term decline will continue unless action is taken to rejuvenate the area and modernise as a tourist destination, e.g. Blackpool, British spa towns and older coastal resorts.

Places 93   The Spanish 'costas': the life-cycle of a tourist area

In the 1950s, Benidorm on the Costa Blanca was still a small fishing village (compare the Costa del Sol in Figure 20.16). During the 1960s, thanks to its sandy beaches, warm sea and hot, dry, sunny summers, it began to attract visitors from northern Europe. By the 1970s it had turned into a sprawling modern resort with high-rise hotels and all the amenities expected by mass tourism. By the 1980s it had reached Briggs' stage of consolidation, when the carrying capacity was reached. By the early 1990s it had begun to stagnate and to decline. Since then the Spanish government has tried to rejuvenate the area by encouraging the refurbishment of hotels, reducing VAT in luxury hotels and ensuring that both beaches and the sea have become cleaner (Spain has the most 'Blue Flag' beaches in the EU).

Figure 20.16

Life-cycle of a holiday area: tourists from the UK to the Costa del Sol, 1960s–1990s

	1960s	1970s	1980s	1990s
Tourists from UK to Spain	1960 = 0.4 million	1971 = 3.0 m	1984 = 6.2 m 1988 = 7.5 m	1990 = 7.0 m
State of, and changes in, tourism	very few tourists	rapid increase in tourism; government encouragement	carrying capacity reached; tourists outstrip resources, e.g. water supply and sewerage	decline (world recession); prices too high; cheaper upper-market hotels elsewhere; government intervention to rejuvenate tourism
Local employment	mainly in farming and fishing	construction workers; jobs in hotels, cafes, shops; decline in farming and fishing	mainly tourism: up to 70 per cent in some places	unemployment increases as tourism declines (30 per cent); farmers use water for irrigation
Holiday accommodation	limited accommodation; very few hotels and apartments; some holiday cottages and campsites	large hotels built (using breeze blocks and concrete); more apartment blocks and villas	more large hotels built, also apartments, time-share and luxury villas	older hotels looking dirty and run down; fall in house prices; only high-class hotels allowed to be built; government oversees the refurbishment of hotels
Infrastructure (amenities & activities)	limited access and few amenities; poor roads; limited streetlighting and electricity	some road improvements but congestion in towns; bars, discos, restaurants and shops added	E340 opened: 'the Highway of Death'; more congestion in towns; marinas and golf courses built	bars/cafés closing; Malaga by-pass and new air terminal opened; re-introduction of local foods and customs
Landscape & environment	clean, unspoilt beaches; warm sea with relatively little pollution; pleasant villages; quiet with little visual pollution	farmland built upon; wildlife frightened away; beaches and sea less clean	mountains hidden behind hotels; litter on beaches; polluted seas (sewage); crime (drugs, vandalism, muggings); noise from traffic and tourists	attempts to clean up beaches and sea (EU Blue Flag beaches); new public parks and gardens opened; nature reserves

At a conference in the late 1990s on 'sustainable mountain development', one speaker claimed: 'Mountains are suffering an unprecedented environmental crisis. Wherever you go in the world, you can find the same symptoms – landscapes wrecked by roads; forests cleared for, and slopes shredded by, skiing; vegetation worn away by walkers; litter left by tourists; and wildlife killed by acid rain'.

Places 94 The Cairngorms: a mountainous area under threat

The Cairngorms range includes four of Britain's five highest mountains. The arctic-alpine plateau is a fragile ecosystem which includes mosses, lichens and dwarf shrubs (page 333) and which provides an irreplaceable habitat for rare birds such as the golden eagle, ptarmigan, snow bunting and dotterel (Figure 20.17). On its slopes are the remnants of the ancient Caledonian pine forest that used to cover most of Scotland. Although the area contains three SSSIs (Places 92) and a proposed special protection area for birds, it is under threat by developers who wish to expand facilities for downhill skiing and for tourism.

The Cairngorm Chairlift Company, which has operated downhill skiing facilities on the northern flank of Cairn Gorm for 40 years, has twice tried to expand west into Lurcher's Gully. Twice the plans caused a furore and were rejected by the Scottish Office. In 1996, the company put forward a plan to build a 2 km funicular railway to within 150 m of the summit. The plan included a new chairlift, three new ski tows and four new ski runs. A 250-seater restaurant and interpretative centre would be built at the top of the funicular, at a height of 1100 m, tripling the capacity of the existing 'Ptarmigan' snack bar. A 250 m tunnel would mean that the funicular, which would allow visitors to reach the summit in three minutes, could operate during strong winds which, at present, force the chairlift to close on many days during the year. The funicular would quadruple the number of summer visitors (from 60 000 to 225 000). However, it was the prospect of all these extra feet tramping the fragile summit plateau during the short summer growing and nesting season that most alarmed conservationists. They fear plants will be crushed, birds disturbed and the landscape eroded. The Company attempted to overcome these objections by developing a 'visitor management plan' which would either limit access to the summit by charging people to walk under the supervision of a ranger, or operating a 'closed system' confining people to the restaurant, the ecological interpretative visitor centre and the indoor viewing area. Planning permission was eventually given, and the Company hoped to have both the railway and the Centre opened in the millennium year).

Figure 20.17

The Cairngorm arctic/ alpine environment

Other types of tourism

Heritage

According to the World Heritage Convention (WHC), created by UNESCO, 'Heritage is our legacy from the past, what we live with today, and what we pass on to future generations. Our cultural and natural heritage are both irreplaceable sources of life and inspiration.' **Cultural heritage** includes monuments, groups of buildings and sites such as the Pyramids, the Acropolis, the Taj Mahal (Figure 20.18a), Machu Picchu, Chichen Itza (Figure 20.23) and The Great Wall of China (Case Study 20). **Natural heritage** includes landscape and wildlife sites such as the Barrier Reef and Tanzania's Selous National Park. There are, at present, over 100 World Heritage Sites.

Theme parks and purpose-built resorts

Theme parks and purpose-built resorts have become centres of mass tourism in the last two or three decades. They include Disney World (Florida – Figure 20.18b), Disneyland (Paris), Legoland (Denmark), Seaworld (Queensland) and Alton Towers (England).

Wildlife

There has been a steady increase in the number of people wishing to see wildlife in its natural environment. The most popular is the African 'safari' in which tourists are driven around, usually in small minibuses with adjustable roofs to allow for easier viewing. Kenya, South Africa, Tanzania and Zimbabwe are all able to capitalise on their abundance of wildlife. Other tourists may go whale-watching (New Zealand), visit marine reserves, view threatened wildlife such as the giant panda (Case Study 20) and the mountain gorilla, or go to places with a unique ecosystem (Madagascar and the Galapagos Islands).

Wilderness holidays

These are popular in America: one or two people set off into largely uninhabited areas such as Alaska to 'live and compete with nature' (Figure 20.18d).

City breaks

Globally, more people take city breaks – often lasting just a few days – than any other type of holiday. In Britain, for example, 64 per cent of people taking a domestic holiday spend it in a town or city, the vast majority visiting London to take advantage of its cultural amenities (museums), entertainment (theatres), historic buildings (St Paul's Cathedral), shops (Oxford Street) and, during 2000, the Millennium Dome. In 1998, over 10 million UK residents visited London, compared with 1 million visiting Edinburgh, and between a one-third and one half a million visiting (in order) Glasgow, Birmingham, Oxford, Manchester, York, Cambridge and Bath.

Religious centres

Religious centres to which people make a pilgrimage include Mecca, The Vatican, Jerusalem, Salt Lake City and Varanasi (Benares).

Figure 20.18

Types of tourism

a Heritage: Taj Mahal

c Wildlife parks: Botswana

b Theme parks: Disney World

d Wilderness: Mt McKinley in Dynali National Park

Cruises

Cruising is the fastest-growing section of the tourist industry, with more, and larger, liners being built each year (Figure 20.19). Cruise holidays may follow a theme, such as whale-watching (Alaska), visiting historic/archaeological sites (the Mediterranean) or capital cities/large ports (the Baltic); or they can be an excuse for passengers to relax and enjoy the sun and life on board ship (the Caribbean). Cruise liners, like other types of tourism, have a mixed impact on the host port, especially if that port is located on a small island. While the scores of passengers create jobs for tour guides and shop assistants and generate income for bus companies, taxi drivers and local craft industries, they rarely spend large amounts of money because they eat and sleep on board ship, and their large numbers –up to 2500 on the latest super-cruise ships – swamp the local amenities and disrupt the residents' way of life.

Certain rivers are also popular for cruising (with the added advantage of having calmer water!). People sail along the Nile (to see ancient temples), the Mississippi (on paddle boats, Figure 20.20), the Amazon, the Rhine, and through the Yangtze Gorges (Figure 21.24). Canal holidays are a self-catering form of cruising.

Figure 20.19

Cruise liners in the Geiranger Fiord, Norway

Figure 20.20

A Mississippi river boat

Ecotourism

Ecotourism, sometimes known as 'green tourism', is a sustainable form of tourism (Framework 16, page 499) that is more appropriate to developing countries than the mass tourism associated with Florida and certain Mediterranean areas. Ecotourism includes:

- visiting places in order to appreciate the natural environment, ecosystems (page 295), scenery and wildlife, and to understand their culture
- creating economic opportunities (jobs) in an area while at the same time protecting natural resources (scenery and wildlife) and the local way of life.

Compared with mass tourists, ecotourists usually travel in small groups (low-impact/low-density tourism), share in specialist interests (bird-watching, photography), are more likely to behave responsibly and to merge and live with local communities, and to appreciate local cultures (rather than to stop, take a photo, buy a souvenir and then move on). They are likely to visit National Parks and game reserves where the landscape and wildlife which attracted them there in the first place is protected and managed. Places visited include Brazil (rainforests), the east coast of Belize and Mexico (coral reefs – Places 95), Nepal (mountains), Burundi (mountain gorillas) and the Arctic (polar bears).

Even so, ecotourists usually pay for most of their holiday in advance (spending little in the visited country), are not all environmentally educated or concerned, can cause local prices to rise, congregate at prime sites (honeypots), and may still cause conflict with local people. There is a real danger that tour operators, by adding 'eco' as a prefix, give certain holidays unwarranted respectability.

The Xcaret Eco-archaeological Park (Figure 20.21), on the Yucatan peninsula, won the Sunday Times Readers' Award in 1999 as the project which they considered was the most successful in protecting or improving the quality of the local environment. Xcaret is located (Figure 20.22) 72 km south of the mass tourist resort of Cancun and 270 km east of the former Mayan settlement, and now a Heritage City, of Chichen Itza (Figure 20.23). Xcaret is built on the ancient Mayan port of Pole (the Mayan civilisations in northern Yucatan lasted from about AD 900 to 1200). The five families who began the venture in 1990 had two main aims: to support Mexican science, and to encourage ecofriendly tourism. The latter was to be achieved not by preaching to visitors but by allowing them to relax in beautiful surroundings and to learn, almost by accident, the value of the ecosystem on show, and about the scientific work being done (compare Places 80, page 526).

Visitors to Xcaret are encouraged to arrive by bus or taxi, to avoid huge areas of car parking. At the entrance, they are asked to hand over any suntan oil (as this pollutes the sea-water) and are given a small bottle of ecofriendly (though less effective as a sunblock) lotion (it is only SPF 8). The inlet, home to thousands of multicoloured fish, is ideal for snorkelling, the warm, calm, crystalline waters for swimming, and the sandy beaches for relaxing. Two underground rivers, illuminated by sunlight streaming through natural holes and openings in the rock, allow tourists to explore the world's largest system of underground channels. Visitors, having been informed that touching coral can be enough to kill it, are taken to off-shore reefs in small boats. People are encouraged to swim with 30 bottlenose dolphins, albeit in a fairly enclosed area, following an educational introduction to their behaviour, habitat and influence on humans. There is a wild bird breeding centre (many are endangered species), a butterfly pavilion, botanical gardens (dedicated to preserving and reproducing species of local plants) and a coral reef aquarium. In the archaeological zone there is a museum with scale models of the principal Mayan cities (including that of Chichen Itza).

At night 'visitors are led to witness secret Mayan ceremonies in the depths of caverns lit only by candlelight. Guided by moonlight, one must walk through the jungle to the pulsating rhythms of a percussion group playing in front of a candlelit Mayan temple'. The show's finale, in the open air theatre, includes a famed folkloric ballet. Xcaret shows that you can combine care for the environment with having a good time.

Figure 20.21

Xcaret

©XCARET

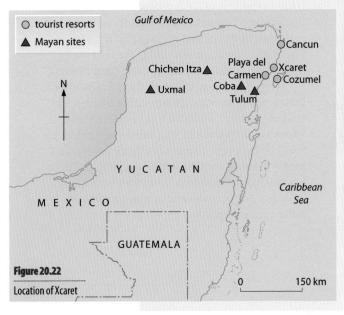

Figure 20.22

Location of Xcaret

Figure 20.23

Chichen Itza

The personal investigative study, or enquiry, is an important part of the examination assessment for AS and A2 Geography. It provides an opportunity for you to develop your individual interests in a particular part of the specification, to make use of fieldwork and to become an 'expert' on a small investigation. It is also an opportunity to develop and collect evidence of your Key Skills (see pages 101, 259 and 387).

Choosing your study

- Choose a topic in which you have personal interest. This will make it easier to study.
- Check in local papers for current issues which could prove a useful topic to investigate. Collect as much background information as you can before setting up your topic.
- Studies can involve combined fieldwork undertaken at field centres. However, your conclusions must be individual, even though the data collection may have been done as a group.
- Choosing a topic covering a human/ environmental theme may allow work in different sections or modules of the course to be linked to an investigation.
- Avoid a topic that will mean travelling long distances to collect data and to do fieldwork. This can be expensive in terms of both time and money, and it will not be easy to make return visits.
- Careful planning is essential – particularly the Schedule for your enquiry.

Collecting your data

- It is important to begin preliminary collection of ideas and materials as early as possible.
- Primary data is the basis of a study, and collecting the data has to be carefully planned, involving surveys, questionnaires, interviews, use of annotated photographs and map construction. Make sure that you choose appropriate dates and times for your fieldwork.
- Questionnaires need to be succinct and to the point – you need to know the types of answers that you require. You should make sure that you have a large sample in order to have well-founded results.
- Take as many photographs as possible of the study area. Carefully annotate and label them, and make sure they are relevant to your enquiry. Bear in mind that you will need to select only the most relevant photographs in your final report.
- If you are visiting an organisation or requesting information it is always useful to write a polite letter beforehand, outlining what you wish to find out and giving time for an answer. People are always busy, so be prepared to wait a few days before telephoning to make an appointment.
- Secondary data collection will mean visits to local libraries, researching newspapers for background, and using the Internet (see Framework 1, page 22). Old maps will show conditions at previous times. Keep a detailed record of all your sources.

Writing your report

- Plan the structure of your report before you start writing. You may find the following outline useful: Introduction – Aims – Data Collection – Data Analysis – Evaluation – Conclusion.
- Data that you collect will have to be analysed and displayed in maps and statistical form. Although bar charts and pie charts are clear and easy to display, try to use a variety of forms of presentation. Most exam boards require that you are able to use statistical methods effectively in your studies. This helps you to evaluate and then to explain the results of your investigation. Some of the most useful methods are Spearman's rank correlation and chi-squared (see Framework 19, page 635).
- Careful detailed analysis of your statistics and diagrams is vital – do not assume that the examiner or moderator will automatically understand what is set out.
- Thoughtful and detailed evaluation is a very important part of your study. You may have collected the opinions of a number of different groups in your investigations and you must set these out clearly and balance up the different values which may be apparent. Do not forget to include your own ideas.
- An extended conclusion will complete the study, drawing together the different opinions and values, weighing up the options and probably putting forward any alternative proposal you may consider to have value.

Remember...

- Presentation is important. Make your report look good – use a word processor if possible.
- Diagrams may be computer-generated but maps should be hand-drawn and not photocopied.
- Check that all maps, diagrams and photographs are labelled and annotated.
- Acknowledge any quotations, and draw up a clear bibliography of your references.
- Number all the pages and where necessary cross-reference diagrams and text.

Tourism in China

Before 1980 it was very difficult for the Chinese to travel freely around their own country, or for foreign tourists to gain access to China. Since 1980, and coinciding with the introduction of the Responsibility Scheme (Places 63 page 468, and Case Study 19), tourism has developed rapidly.

- Domestic tourism expanded as the Chinese were now allowed to earn money which meant, together with a new freedom of movement, they could afford to travel around their own country.

- Overseas visitors were encouraged through a more 'open door' policy. Numbers rose steadily to a peak in 1989 – the year of Tiananmen Square – after which numbers dipped dramatically. It took nearly a decade for the number of foreign tourists to recover from the events of that year. Figures published by the World Tourist Organisation (WTO) showed that in 1999, China received over 27 million overseas tourists (the Chinese themselves put the figure at 57 million) putting it fifth in the world rankings.

More recently, China has made a major effort to attract even more overseas tourists. Many new luxury hotels have been built, older hotels have been refurbished and upgraded, and an increasing choice of cheaper accommodation is being provided. The internal transport system has improved, despite the vast distances and large numbers of people. (Of the five state airlines that the author travelled with in 1999, all had new planes, used new or upgraded terminals, and not only flew to time, but took off early once everybody was on board).

Public transport is exceptionally congested (there is always a scrum to get onto a bus or train). Tour guides tend to be university students as they – unlike their more knowledgeable elders – have a reasonable grasp of English and Japanese. Shopping is easy (although language can be a problem), and food, especially in hotels, is usually of a high quality (though with the quantity provided one wonders how the Chinese remain so slim). Although foreigners are very safe walking and travelling alone (far more so than in the Western world), they are recommended to travel as an organised group, if only to avoid the language barrier. Naturally,

the problems of accommodation, travel and language increase with distance from the main cities. Most of the tours are arranged by the state-run China International Travel Service (CITS).

China is ideally suited to tourism. A leading Western travel writer on China, J. D. Brown, on asked why he had visited China so often in the last two decades, said, 'No other place on Earth quite possesses such an intensity of beauty and strangeness, no other nation has a longer or richer historical legacy, and no other nation is likely to be as important in the 21st century'. China has the potential to offer all types of holiday (Figure 20.25) although as yet the country has not developed ski-resorts (despite many areas having heavy winter snowfalls) nor mass-tourism coastal resorts (even though many parts of the coastline are spectacular). It is also noticeable that places visited by the Chinese are often very different from those that appear on the itinerary of foreigners (Figure 20.27).

Figure 20.24

Natural attractions

a Three Gorges on the Yangtze

b River Li near Guilin

c Mountains of Tibet

d Giant panda

	Domestic tourists		Overseas tourists	
A Natural attractions (Figure 20.24)				
– Rivers and lakes	Most lakes, waterfalls and rivers		Three Gorges on the Yangtze (Figure 20.24), River Li at Guilin (Figure 8.7)	
– Mountains	Many sacred mountains across China		Tibet	
– Wildlife/conservation	Giant panda (Chengdu) and Sea of Bamboo (Sichuan)		Giant panda (Chengdu), tropical rainforest of extreme south-west China	
B Human attractions (Figure 20.26)				
– Heritage/historic	Numerous Buddhist caves, paintings, stone statues and temples, Forbidden City and Summer Palace (Beijing), ancient city of Lijiang		Great Wall of China and Ming Tombs, Forbidden City and Summer Palace (Beijing), Terracotta Army (Xi'an), Silk Route	
– City breaks	Shenzhen (shopping – page 581), Suzhou (gardens)		Beijing, Shanghai and Shenzhou for shopping, Beijing Opera, Shanghai Acrobats	
– Religious centres	Buddhist caves of Dazu, Luoyang		Tibet	**Figure 20.25**
– Cruises	–		Middle Yangtze, River Li	Types of holiday in China

Figure 20.26

Human attractions

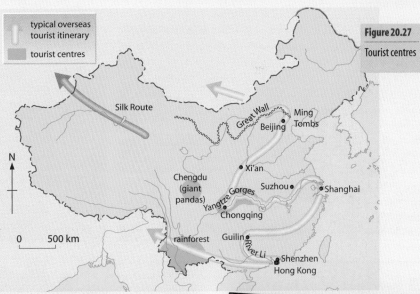

typical overseas tourist itinerary

tourist centres

Figure 20.27

Tourist centres

Silk Route

Great Wall

Beijing

Ming Tombs

Xi'an

Chengdu (giant pandas)

Suzhou

Shanghai

Yangtze Gorges

Chongqing

rainforest

Guilin

River Li

Shenzhen

Hong Kong

N

0   500 km

a The Great Wall

b Forbidden City, Beijing

c Terracotta Army (Xi'an)

d Beijing Opera

References

Carr, M. (1997) *New Patterns: Process and Change in Human Geography*, Thomas Nelson.

Waugh, D. (1998) *The New Wider World*, Thomas Nelson.

Waugh, D. and Rowley, C. (2000) *Images of Change: China*, CD Rom, Geopix.

Websites:

Yahoo!'s Parks page (for US parks):
http://parks.yahoo.com/

Travel Association of America Research Links:
http://www.tia.org/research/reslinks.asp#AirTravel

KENYA WEB's Tourism and National Park:
http://www.kenyaweb.com/tourism/parks/parks.html

The Peak District National Park:
http://www.peakdistrict.org/

Directory of National Parks Worldwide:
http://hum.amu.edu.pl/~zbzw/ph/pnp/swiat.htm

Estonian Ecotourism Web Site:
http://www.ee/ecotourism/index.html

See also for more links:
http://www.nelsonthornes.com/gaia

1 a What is tourism? **(1 mark)**
  b i Give four factors that have helped cause the growth of
     world tourism since 1960. **(4 marks)**
    ii For each of your answers in i, explain why this factor led
       to a growth of tourism. **(4 marks)**
  c With reference to a named resort or tourist area that you
    have studied, explain how the growth of tourism has
    brought both:
     i benefits, and
     ii problems to the people who live in the area. **(10 marks)**
  d If tourism starts to decline in an area it can cause serious
    economic problems. Name a tourist area where the
    industry has started to decline. Describe how the area
    has adapted to try to stop the decline. **(6 marks)**

2 Study Figure 20.28 a, b and c.
  a For each of the photographs:
     i Describe the attractions of the area that make it a
       suitable tourist destination. **(6 marks)**
    ii Suggest which sector of the holidays market this area
       will particularly appeal to. **(3 marks)**
   iii Suggest how tourism has brought advantages and
       disadvantages to the people of the area. **(6 marks)**
  b In many tourist areas the natural environment is a major
    attraction for tourists. Unfortunately the pressure of
    tourism threatens to destroy the natural environment.
    For a named tourist area, explain how management
    strategies have been, are being, or could be developed
    to allow tourism to continue without destroying
    the environment. **(10 marks)**

AS

3 Study Figure 20.28 a, b and c.
  a Describe the tourist attractions of each of the areas
    shown in the photographs. **(6 marks)**
  b Briggs's model of the life-cycle of a tourist resort
    shows the following stages:
    ■ exploration
    ■ involvement
    ■ development
    ■ consolidation
    ■ stagnation
    ■ rejuvenation or decline.
    Suggest, with reasons, which stage has been reached by
    each of the tourist areas shown in the photographs.
    (12 marks)
  c Name a tourist resort that has reached the later stages
    of the model, and explain what is being done to
    rejuvenate the tourist industry there. **(7 marks)**

4 a Describe the six stages of Briggs's model showing
    the life-cycle of a tourist resort or area. **(8 marks)**
  b Describe named holiday resorts or areas that illustrate
    each of the stages in the model. You may refer to areas
    as they are today, or as they have been in the past.
    (9 marks)
  c For a named holiday resort or area, explain how it
    is being rejuvenated, having reached stage 6 of
    the model. **(8 marks)**

Figure 20.28

a Spain, b Nepal, c Greece

5 Explain how tourism can bring both advantages and
  disadvantages to the people and environment in areas
  where it develops. Make reference to examples from
  around the world. **(25 marks)**

6 Can the development of tourism lead to sustainable
  development in poor, remote areas of the world? Discuss
  this with reference to examples that you have studied.
  (25 marks)

A2

7 a Each of the following factors is a major reason for the development of tourist resorts in the UK:
- scenery
- weather
- cultural and historic attractions
- active leisure amenities
- entertainment amenities
- conservation areas.

Choose **three** of the listed factors. For each of your chosen factors:
  i name a resort in the UK where that factor is important;
  ii describe the factor and explain why it is attractive to tourists. **(9 marks)**

b Explain how the following factors have helped the growth of the international travel and tourism industry:
  i improved transport technology
  ii the development of package holidays
  iii ecotourism. **(9 marks)**

c For a named example of a less economically developed country, explain how tourism is being used as a means of stimulating economic development. **(7 marks)**

8 a Refer to Figure 20.29. Name the most popular destinations for tourists from the UK in:
  i Europe
  ii regions outside Europe **(2 marks)**

b i With reference to holidays taken in Europe by residents of the UK, describe and account for the distribution of the main holiday destinations. **(5 marks)**
  ii The number of UK residents taking holidays in Europe in February is fairly small. Suggest, with reasons, how the distribution of holiday destinations is likely to be different from that shown on Figure 20.29. **(5 marks)**

c i Describe the changes shown on the graph in Figure 20.30. **(3 marks)**
  ii With reference to any one of the regions shown on this graph, account for the changes in the number of tourist arrivals since 1970. **(10 marks)**

9 a i What is 'ecotourism'? **(1 marks)**
  ii Name an example of a place in a less economically developed country where ecotourism has been developed. **(1 mark)**
  iii Describe the attractions for ecotourists of the area that you named in ii. **(4 marks)**
  iv Explain how ecotourism has brought specific benefits to the people and the environment in the area. **(6 marks)**

b With reference to the Cairngorms or another mountainous area in the UK that is being damaged by increased tourist pressure:
  i explain why the number of tourists has increased in recent years **(4 marks)**
  ii explain how the tourist pressure is damaging the environment **(4 marks)**
  iii describe one management strategy which aims to reduce the damage being done, and explain how the strategy is intended to work. **(5 marks)**

10 a With reference to Figure 20.29, describe and account for the distribution of destinations of holidaymakers from the UK. **(12 marks)**

b With reference to Figure 20.30, explain why the regions shown on this graph have shown such a dramatic change in their tourist arrivals since 1970. **(13 marks)**

11 The Peak District National Park Authority issued a Management Plan in 1999. The vision for the future outlined in the plan is based on three main aims:
- to conserve and enhance the National Park's special qualities
- to provide opportunities for enjoyment and understanding of the Park
- to improve the quality of life for people who live in, work in and visit the Park.

The Plan is based on two principles:
- sustained development
- understanding.

Referring *either* to the Peak District National Park *or* to any other tourist area that you have studied:
  i describe how conflicts can arise between different groups and individuals who use the land in the Park **(7 marks)**
  ii discuss how the aims and principles of the Management Plan for the Peak Park could help to manage and reduce land use conflict. You must make reference to specific conflicts in named places. **(18 marks)**

Figure 20.29

Foreign holidays taken by UK residents

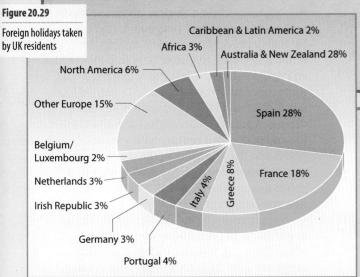

Figure 20.30

Increases in the number of tourist arrivals, 1950–96

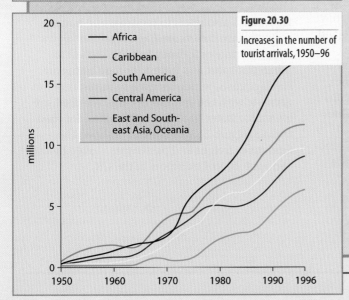

Chapter 21

Transport and interdependence

'I will build a motor car for the great multitude so low in price that no man will be unable to own one … and enjoy with his family the blessing of hours of pleasure in God's great open spaces.'

Attributed to Henry Ford, c.1900

There is not a single thing, however small, in the world that does not depend upon something that is higher – for everything is interdependent.'

Jewish text

Transport

The following references to the role of transport have been made at various stages in this text.

1 It may be viewed as an indicator of wealth and economic development, as measured by the number of cars per 1000 people. While the more developed countries have only 25 per cent of the world's population, they have 88 per cent of its rail traffic and 72 per cent of its cars and lorries.

2 Transport is essential in linking people, resources and activities and in enabling the exchange of goods (trade) and ideas (information).

3 Transport was considered to be a major factor in industrial location (Weber – page 557) and in determining agricultural land use (von Thünen – page 471). The relative decrease in transport costs to the average firm or farm has made this a less significant locational and land use factor since the 1950s (page 554).

4 In early economic/geographical theory, costs were thought to be proportional to distance (von Thünen's central market and Christaller's central place), especially on a flat plain where transport was equally easy and cheap in all directions. Later, costs were regarded as a function of a raw material's weight as well as the distance it had to be moved (Weber).

It is now accepted that, as transport costs comprise terminal costs plus haulage costs (Figure 21.1), cost per tonne/km declines with distance.

Characteristics of modern transport systems

Early tribespeople had to travel on foot or by animal. The first civilisations, after 6000 BC, used river transport (Egypt, Mesopotamia and China – Figure 14.1). By 4000 BC the wheel, arguably one of the most important inventions ever, was being used in ceremonial processions in Mesopotamia. Between then and the late 18th century came four significant technological advances: engineered roads (for military purposes, but also used for trade); the ocean-going sailing ship (for exploration and trade); the artificial pound lock (which enabled canals to traverse hills); and fixed track routes to enable heavier loads to be carried (colliery wagonways of Tyneside, originally horse-drawn). A greater technological advance came at the end of the 18th century with the use of steam power and the invention of the steam railway engine and steam ship. The first journey by a petrol-driven car took place in 1885 and the first powered flight in 1903.

A comparison of the characteristics of the major forms of present-day transport – canal, ocean shipping, rail, road, air and pipeline – is given in Figure 21.1. Each type has advantages and disadvantages over rival forms of transport. Figure 21.1 also refers to terminal and haulage costs. Terminal costs are fixed regardless of the length of time of journey and are highest for ocean transport and lowest for road transport. Haulage costs, which increase with distance but decrease with the amount of cargo handled, are lowest for water transport and highest for road and air (Figure 21.2). Figure 21.3 shows the amount of freight and passengers moved by the various forms of transport in the UK.

		Canals and rivers	Ocean transport and deep sea ports	Rail	Road	Air	Pipelines
Physical	**Weather**	Canals can freeze in winter. Drought/heavy rains make rivers unnavigable.	Storms, fog. Icebergs in North Atlantic.	Very cold (frozen points). Heavy snow (blocks line). Heavy rain can cause landslides.	Fog and ice both can cause accidents/pile-ups. Cross-winds for big lorries; snow blocks routes; sun can dazzle.	Fog, icing and snow: less since planes have had automatic pilots. Airports better if sheltered from wind and away from hills and areas of low cloud.	Not greatly affected.
	Relief	Width of channels. Need flat land or gentle gradients. Soft rock/soil for digging, problems with deltas. Rivers must be slow-flowing, have a constant discharge and have no rapids.	Harbours need to be deep, wide and sheltered. Tidal problems.	Cannot negotiate steep gradients so have to avoid hills. Estuaries can be obstacles. Flooding in valleys.	Avoids/takes detours around high land. Valleys may flood. May go around estuaries.	Large areas of flat land for runways, terminal buildings and warehousing. Firm foundations. Ideally, cheap farmland or land needing reclamation. Seas and mountains not a barrier.	Difficult to lay, then relief is not a problem.
Economic	**Speed/time**	Slowest form of transport. Long detours and possible delays at locks.	Slow form of transport, yet most economical.	Fast over medium-length distances.	Fast over short distances and on motorways. Urban delays.	Fastest over long distances, not over short ones due to delays getting to and passing through airports.	Very fast as continuous flow.
	Running or haulage costs (wages and fuel): increase with distance	Often family barge. Limited fuel use means the cheapest form of transport over lengthy journeys.	Expense (oil used as fuel) increases with distance.	Relatively cheap over medium-length journeys. Fuel and wages quite high.	Cheapest over shorter distances. Haulage costs rapidly increase with distance.	Very expensive, yet speed makes it competitive over very long distances.	Cheapest as no labour is involved (provided diameter is large).
	Terminal costs (loading and unloading costs and dues): no change with distance	Canals expensive to build and to maintain, unless natural waterways used.	Ports expensive. Harbour dues/taxes. Expensive to build specialised ships. Less since containers. Cheapest over long distances.	Building and maintenance of track/stations/signalling/rolling stock are very expensive.	Expensive building and maintenance costs, especially motorways. Car tax instead of dues, but roads built by community taxation therefore lower overheads.	Very expensive to build and maintain airports. High airport dues. Planes expensive to purchase and maintain.	Very expensive to build.
	Number of routes	Relatively few. Inflexible.	Relatively few ports, inflexible due to increased specialisation of ships. Links to hinterland.	Decreasing, as limited to linking main cities and power stations. Not very flexible. Recent increase in urban rail and new high-speed 'InterCity' routes.	Many and at different grades. Great flexibility, most in urban and industrial areas.	Often only a few: internal and international airports/routes. Not very flexible because of safety.	Limited to key routes. Inflexible and one-way flows.
	Goods and/or passengers carried	Heavy, bulky, non-perishable, low-value goods. Present-day tourists.	Heavy, bulky, non-perishable low-value goods. Cruise passengers.	InterCity passengers. Heavy, bulky (chemicals, coal) and rapid (mail) goods. Can carry several hundred passengers. Dependable and safe.	Many passengers. Perishable, smaller loads by lorry. Relatively few people carried by one bus or car.	Mainly passengers. Freight is light (mail), perishable (fruit) or high-value (watches).	Bulk liquid (oil, gas, slurry, liquid cement, water).
	Congestion	Very little.	Very little.	Little track congestion other than commuter trains.	Congestion in urban areas and at holiday times.	Usually little, other than at peak holiday times.	None.
	Convenience and comfort	Neither very convenient, unless for leisure/relaxation, nor very comfortable.	Not very convenient. Cruise liners very comfortable.	Not very, other than InterCity. Needs transport to and from station. Comfortable for passengers.	Door-to-door (except for some city centre destinations): most convenient and flexible. Safety is questionable; strain for drivers, but independent and private.	Country to country. Jet lag if more than three time zones crossed. Relaxing over shorter journeys; cramped, dehydrating and tiring over longer journeys.	Raw material or port to industry.
Environmental	**Environmental problems**	Some oil discharged, but relatively few problems.	Tankers discharging oil. Much land needed for ports, hard-standing and warehousing.	Noise and visual pollution limited to narrow belts. Noise decreases with welded rails, increases with high-speed trains. Electric trains cause less pollution.	A major cause of noise and air pollution. Effect on ozone layer, acid rain, and global warming (greenhouse effect). Uses up land, especially farmland. Structural damage caused by vibrations.	High noise levels. Some air pollution. Uses up much land for airports.	Few once 'buried'. Eyesore on surface.

Figure 21.1

Comparable characteristics of transport systems

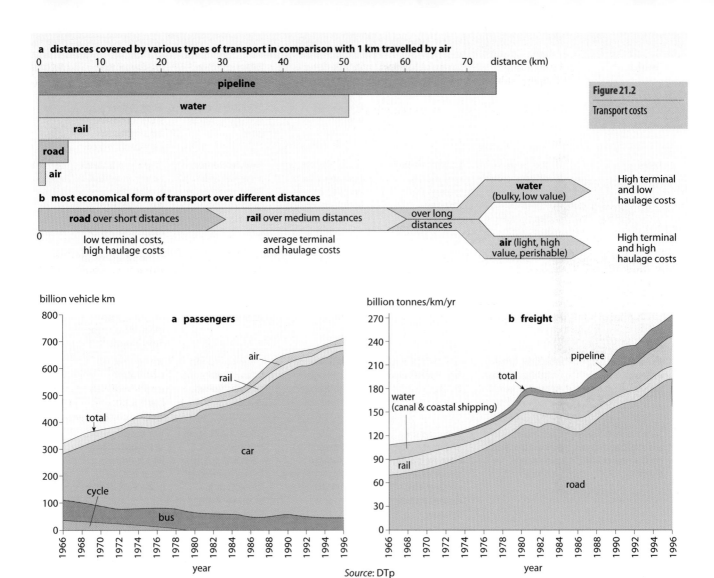

Source: DTp

Figure 21.2

Transport costs

a **distances covered by various types of transport in comparison with 1 km travelled by air**

distance (km)

pipeline
water
rail
road
air

b **most economical form of transport over different distances**

road over short distances | low terminal costs, high haulage costs

rail over medium distances | average terminal and haulage costs

over long distances

water (bulky, low value) | High terminal and low haulage costs

air (light, high value, perishable) | High terminal and high haulage costs

billion vehicle km

a passengers

air
rail
total
car
cycle
bus
year

billion tonnes/km/yr

b freight

pipeline
total
water (canal & coastal shipping)
rail
road
year

Figure 21.3

Passenger and freight traffic in the UK, 1966–96

Types of transport

Inland waterways

The major canal-building period in western Europe was in the late 18th and early 19th centuries. The virtual monopoly by the canals was shortlived due to competition, first from rail and later from road. However, since the 1970s the inland waterways have been rejuvenated. This began with the increase in road and rail charges caused by the rise in oil prices following the 1973 Middle East War, and an appreciation that, despite its slowness and inflexibility of routes, water transport remains the cheapest form for bulky, non-perishable goods. Huge 4400-tonne barges, propelled by push-barges, have been introduced by the EU to carry the equivalent of 110 railtrucks or 220 x 20-t lorries. To accommodate these 'super-barges', the EU governments have financed the widening and deepening of canals and the linking of the Mediterranean, North and Black Seas. Smaller canals, including those in Venice, Amsterdam and parts of England, are benefiting from increased tourism.

Elsewhere, sections of rivers have been made navigable by removing dangerous rocks and shoals, as on the Rhine and Yangtze (Places 96), or by trying to maintain river levels throughout the year, as on the Nile below the Aswan Dam (Places 73, page 490).

Ocean shipping

Many ports in western Europe and eastern North America developed either by trading with each other across the Atlantic or by importing raw materials from colonies and other developing countries and exporting manufactured goods in return. Recently, several major changes have occurred in British ports which have led to the growth of those in the south and east of England (with the exception of the port of London) and a corresponding decline elsewhere. The increase in ship size, especially purpose-built **bulk carriers** for oil and iron ore, has meant that wider and deeper estuaries are needed. The recession in world trade (page 625) has led to a general decrease in the number of ships and ports needed.

According to atlas maps, the Yangtze appears as the main artery from the coast to inland China. However, while it does indeed carry over 75 per cent of China's inland waterway traffic, only ships under 1500 tonnes can negotiate the Three Gorges section of the river at all times of the year (Figure 20.24a). This is because the Yangtze:

- has numerous sharp bends, narrow sections and shoals, rocks and rapids.
- can be too fast and dangerous during the annual flood, at which time the river levels rise by over 10 m (Figure 3.66)
- can be too shallow and dangerous during times of low water which occur early each year.

One of the advantages of the Three Gorges Dam, at present being built at Sandouping (Figure 3.65), will be to turn the river behind the dam into a lake 600 km in length. With the depth of the lake eventually to be 175 m higher than the present low-water level, this should ensure that ships of up to 10 000 tonnes can sail safely throughout the year upriver to the port of Chongqing, almost 2500 km from the sea. To negotiate the dam itself ships will either have to use the shiplock or the shiplift (Figure 18.24). The shiplock operates in both directions, with five 'steps-locks' allowing one ship, or the combined weight of several barges, of up to 10 000 tonnes to pass through at a time. The shiplift is a one-stage vertical lift of 175 m capable of carrying one 3000-tonne passenger or cargo boat. The completed scheme (after 2003) should reduce transport costs by 35 per cent. The main cargoes consist of coal and quarried stone (Figure 21.4).

Figure 21.4

Traffic on the Yangtze

A ship berthed at a quayside is not earning money. Two innovations have enabled the turn-around time (the time taken to unload and load cargo) to be shortened. The first was the development of **roll-on/roll-off** (Ro-Ro) methods whereby lorries loaded with freight are driven on board, so reducing the need for cranes. Rail Ro-Ro ferries have also been introduced (Dover–Dunkirk, Denmark–Sweden) in order to save time. (The first Ro-Ro ferries carried trains in the 1870s between the Jutland peninsula and the Danish islands of Fyn and Zealand). The second innovation was the introduction of **containerisation** (unitised traffic) in which goods are packed into containers of a specified size at the factory, taken by train or lorry to the container port, and easily and quickly loaded by specialist equipment (Figure 21.5). Containers have to be of internationally specified standard sizes (they are measured in TEUs, or 20-foot equivalents). Some of the larger container ships (containers are usually stacked on deck) can hold upwards of 4000 units. These technological advances have, however, decimated the dock labour force and reduced the number of ships required.

Felixstowe was Britain's ninth-largest port in 1960, handling only 1 per cent of the country's seaborne trade. By 1992 it was the largest container port, the second-largest in terms of foreign trade by value and in Ro-Ro units, and the fourth-largest for foreign trade by tonnage. Reasons for this growth include: the presence of the deep water necessary for handling large vessels, regardless of the state of the tide; its coastal site, so that no time is lost sailing up an estuary; its position on the major trade routes with the EU and Scandinavia (its main links are with the container ports of Rotterdam and Zeebrugge); the availability of Ro-Ro facilities and five container terminals; its site on marshy, poor-quality, cheap land, with space for future expansion; good road links with its hinterland; and good industrial relations (no restrictive industrial practices).

Transport and interdependence

On founding the modern port of Singapore in 1819, Sir Stamford Raffles decreed that it was open to all maritime nations. By 1998 over 800 shipping lines with links to over 600 ports had taken advantage of that decree and Singapore has been, since 1986, the world's busiest port in terms of shipping tonnage, and its major bunkering port. At any given time, over 800 ships are likely to be in port, with one arriving or weighing anchor every 6 minutes (140 922 vessels in 1998 compared with 81 000 in 1992). It takes less than 1 second to handle 1 tonne of cargo, and cargo is handled during every second of the year. Warehouses are automated and computerised. Vessels vary from modern supertankers and container ships to the more traditional bumboats and junks. In 1998 the port handled over 15 million TEUs of containers (Figure 21.5). To save docking time, harbour pilots now fly out to incoming vessels by helicopter.

Singapore is a **free port**, open to all countries, with seven free trade zones (six for seaborne cargo and one at nearby Changi International Airport). In these zones, goods can be made or assembled without payment of import or export duties, and profits can be sent back to the parent company without being taxed. Many high-tech firms assemble their products here and sell them at very competitive prices. (A visiting Japanese student was surprised to find his own country's brand names selling more cheaply in Singapore than in Japan.) Singapore is the world's largest manufacturer and exporter of disks. However, the port's largest money-earner is oil, a resource which the country does not possess. This is because Singapore imports crude oil from the Middle East, Indonesia and Malaysia, refines it within the free port (Figure 21.6) and then exports a range of oil products. It is the world's third-largest oil refining centre, with five major oil companies.

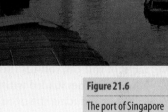

Figure 21.6

The port of Singapore

Figure 21.5

Containers at Singapore

Rail transport

Railways became the dominant form of transport in Britain during the Industrial Revolution. Although they are still the most convenient method of moving bulky goods and large numbers of passengers (Figure 21.1), the inflexibility of their routes and terminals has left them at a considerable disadvantage relative to road transport as cities have grown and industry has changed its location. Railways were constructed when cities were much smaller, and industry was sited near to the mainline station in what is now the inner city. Modern industry now tends to seek an out-of-town or small town location – sites that usually do not possess a rail terminal. The east-coast, mainline route between London and Edinburgh has now been electrified, the entire route being controlled from just seven modern signalling centres, and new high-speed electric locomotives introduced.

The Channel (Fixed Link) Tunnel project (Places 98), funded in Britain mainly by the private sector, will eventually allow passengers to travel between London and Paris in 2 hours 30 minutes.

The British Rail network was split up and privatised and deregulated (1994 and 1997) between 25 franchised train operating companies.

Rail transport has recently regained some trade (mainly passengers) in both North America and western Europe, partly because of congestion and other road problems and partly because of successful modernisation attempts. France (the TGV), Germany (Transrapid), Italy and Spain have all built new high-speed routes, but the track and rollingstock are not always compatible. The introduction of freightliners has enabled containers of internationally agreed standard size (page 607) to be carried by rail to specialist ports, which saves time and reduces labour costs and the risk of theft and damage.

Places 98 South-east England: the Channel Tunnel rail link

In January 1994, seven years after the 110 km, high-speed Channel Tunnel rail link between Kent and London was first proposed, the government finally confirmed, with one or two exceptions, the chosen route (Figure 21.7). The link is being built in two stages. The first, between Folkestone and North Kent, started in 1998 and is due to open in 2000–2003. Work on the second, between North Kent and St Pancras, will begin when the first section has been completed, and is scheduled for opening in 2007. Two proposed new stations, at Barking and Rainham, have been deleted from the original plan. The main reasons for the delays in the past have been economic and environmental. British Rail's initial favoured option, the cheapest, cut through some of the most attractive scenery and residential areas of Kent, as well as through several marginal parliamentary constituencies. Delays have also resulted from difficulties in raising private investment (compare Shanghai, Places 102 page 619).

Figure 21.7

The Channel Tunnel rail link and its potential environmental impact

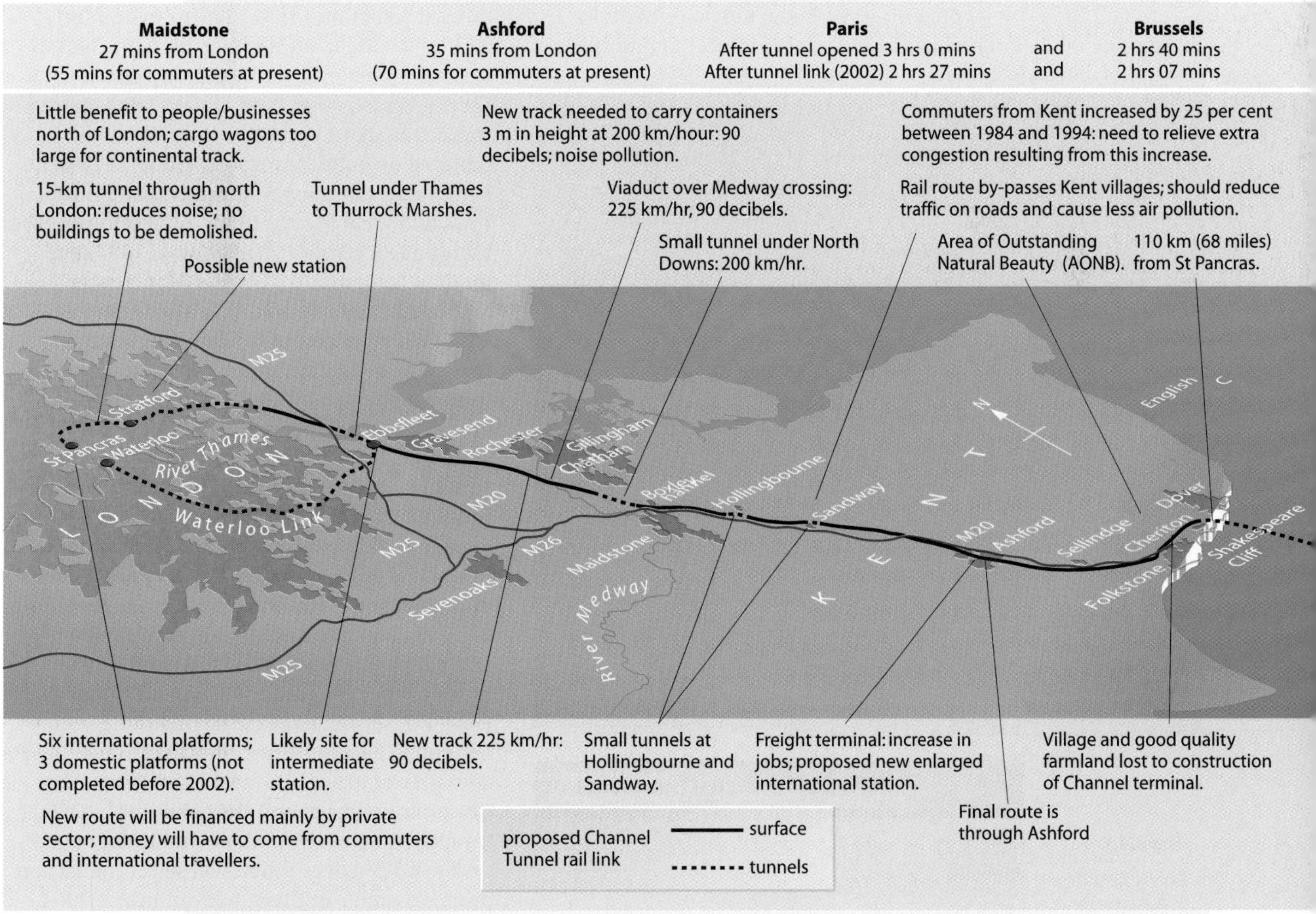

Maidstone
27 mins from London
(55 mins for commuters at present)
Ashford
35 mins from London
(70 mins for commuters at present)
Paris
After tunnel opened 3 hrs 0 mins
After tunnel link (2002) 2 hrs 27 mins
and
and
Brussels
2 hrs 40 mins
2 hrs 07 mins
Little benefit to people/businesses north of London; cargo wagons too large for continental track.
New track needed to carry containers 3 m in height at 200 km/hour: 90 decibels; noise pollution.
Commuters from Kent increased by 25 per cent between 1984 and 1994: need to relieve extra congestion resulting from this increase.
15-km tunnel through north London: reduces noise; no buildings to be demolished.
Tunnel under Thames to Thurrock Marshes.
Viaduct over Medway crossing: 225 km/hr, 90 decibels.
Rail route by-passes Kent villages; should reduce traffic on roads and cause less air pollution.
Possible new station
Small tunnel under North Downs: 200 km/hr.
Area of Outstanding Natural Beauty (AONB).
110 km (68 miles) from St Pancras.
M25
Stratford
St Pancras
Waterloo
River Thames
LONDON
Waterloo Link
Ebbsfleet
Gravesend
Rochester
Gillingham
Chatham
M20
M25
M26
Sevenoaks
Maidstone
River Medway
Hollingbourne
Sandway
KENT
N
M20
Ashford
Sellindge
Cheriton
Dover
Folkestone
Shakespeare Cliff
English C
Six international platforms; 3 domestic platforms (not completed before 2002).
Likely site for intermediate station.
New track 225 km/hr: 90 decibels.
Small tunnels at Hollingbourne and Sandway.
Freight terminal: increase in jobs; proposed new enlarged international station.
Village and good quality farmland lost to construction of Channel terminal.
New route will be financed mainly by private sector; money will have to come from commuters and international travellers.
proposed Channel Tunnel rail link
surface
tunnels
Final route is through Ashford

Road transport

The major advantage of road transport is its flexibility which allows it to operate from door-to-door, especially over short distances, and at the most competitive price (Figure 21.2). Yet while the building of motorways and urban roads seems to be essential if an area wishes to improve its accessibility and stimulate economic activity, the results of such construction often increase the amount of traffic and accentuate environmental problems (Figure 21.1). Older urban areas, built in a pre-automobile era, experience the modern problems of congestion, such as unplanned, narrow streets (unless redeveloped), car parking difficulties, delays caused by buses and delivery lorries, air pollution from exhausts, health problems (asthma), noise and visual pollution, roadworks (the repairing of old utility services), and old buildings being shaken by passing traffic. Even where redevelopment has occurred there is still heavy pollution; land is taken up by road widening; land values next to urban motorways are reduced; and the risk of accidents seems as great.

At present, seven out of ten households in Britain have a car, and the number of multiple car users is increasing (Figure 21.8). The 16.5 million licensed road vehicles of 1975 had risen to 26.3 million by 1996, a figure that is predicted to increase by over 50 per cent by 2026 (Figure 21.9) if the car continues to be a symbol of economic success for British people. It is this growth in car ownership that has enabled retail outlets to relocate in out-of-town, regional shopping centres (page 433) and industry to relocate in industrial, science and business parks (Places 86, page 566), and which has given individual families greater freedom of movement and choice in where they live, shop, work and spend their leisure time.

Although the length of Britain's motorways increased from 150 km in 1960 to 3226 km in 1996, this and the creation of outer ring roads to divert traffic from city centres and motorways to link large urban areas, have generated their own traffic (Figure 21.10). It is, for example, often quicker and easier for shoppers living on the edge of an urban area to travel by motorway to a regional shopping centre than it is to make the shorter journey to their local, possibly crowded, city centre. (Average traffic speeds in some urban areas at peak times are 17 km/hr in London, 27.8 in Sheffield, 28.9 in Nottingham, 34.3 in Greater Manchester, and 40.1 on Tyneside.) Several British cities have already implemented **rapid transit systems** by which the car is replaced by public transport. These, at present, involve the use of underground railways (the Tyne and Wear Metro, 1980), light railways (the London Docklands Light Railway, 1987) and modern 'supertrams' (Greater Manchester Metrolink, 1992; South Yorkshire Supertram, 1995). Of more recent rapid transit schemes, Croydon, and Birmingham to Wolverhampton, were due to open by 2000, as was Metrolink Phase 2 from Manchester to Salford Quays and Eccles. Otherwise only Nottingham, where construction was due to start in 1999, has received funding. By mid-1999, 16 local authorities had applied to take part in trials involving either new road tolling or workplace car park levies. Six West Midland metropolitan councils, including Birmingham and Wolverhampton, agreed to trial car park levies, while Leeds and Bristol opted to trial congestion tolling on roads. However, several large cities, including Liverpool, Newcastle and Sheffield, have decided not to apply, for fear of driving away inward investment and shoppers – as have many rural areas which lack viable public transport systems. The government will select the 10 best proposals and run them, but not before 2004.

Figure 21.8

Increase in car ownership, 1950–2011

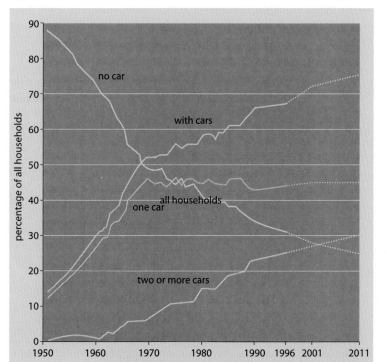

Figure 21.9

Percentage increase in traffic over the 1996 figures

	Car increase	All vehicles	Annual average over last 5 years
1996–2001	9	9	1.65
1996–2011	27	28	1.46
1996–2026	48	53	0.69

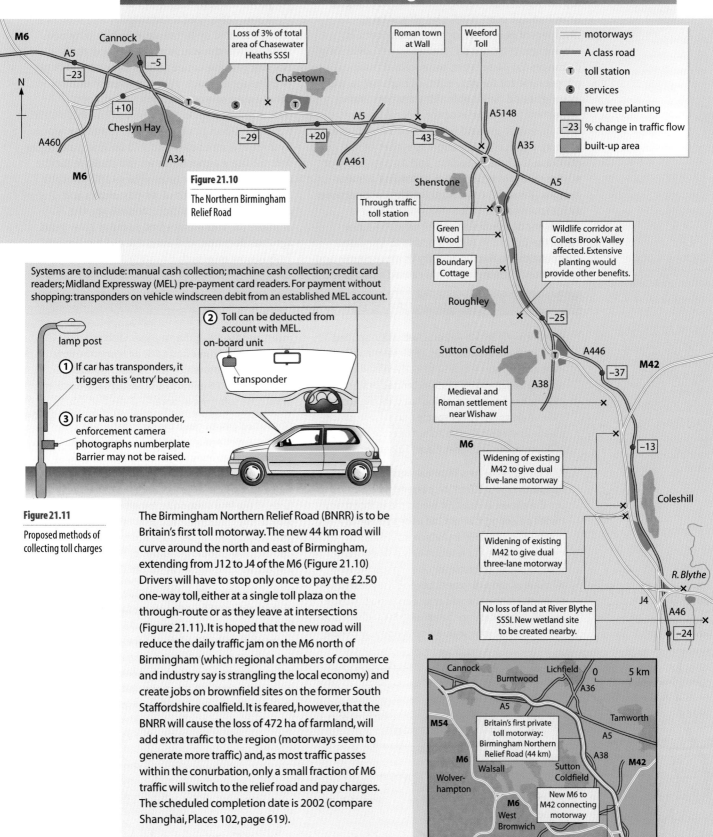

Figure 21.10

The Northern Birmingham Relief Road

Loss of 3% of total area of Chasewater Heaths SSSI

Roman town at Wall

Weeford Toll

motorways

A class road

T toll station

S services

new tree planting

−23 % change in traffic flow

built-up area

M6
Cannock
A5
−23
−5
+10
Cheslyn Hay
A460
A34
M6
Chasetown
−29
+20
A5
A461
−43
A5148
A35
A5
Shenstone

Through traffic toll station

Green Wood

Boundary Cottage

Roughley

Sutton Coldfield

Wildlife corridor at Collets Brook Valley affected. Extensive planting would provide other benefits.

−25

A446

M42

A38

−37

Medieval and Roman settlement near Wishaw

M6

−13

Coleshill

Widening of existing M42 to give dual five-lane motorway

Widening of existing M42 to give dual three-lane motorway

R. Blythe

J4

A46

No loss of land at River Blythe SSSI. New wetland site to be created nearby.

−24

a

Systems are to include: manual cash collection; machine cash collection; credit card readers; Midland Expressway (MEL) pre-payment card readers. For payment without shopping: transponders on vehicle windscreen debit from an established MEL account.

lamp post

① If car has transponders, it triggers this 'entry' beacon.

② Toll can be deducted from account with MEL.

on-board unit

transponder

③ If car has no transponder, enforcement camera photographs numberplate. Barrier may not be raised.

Figure 21.11

Proposed methods of collecting toll charges

The Birmingham Northern Relief Road (BNRR) is to be Britain's first toll motorway. The new 44 km road will curve around the north and east of Birmingham, extending from J12 to J4 of the M6 (Figure 21.10) Drivers will have to stop only once to pay the £2.50 one-way toll, either at a single toll plaza on the through-route or as they leave at intersections (Figure 21.11). It is hoped that the new road will reduce the daily traffic jam on the M6 north of Birmingham (which regional chambers of commerce and industry say is strangling the local economy) and create jobs on brownfield sites on the former South Staffordshire coalfield. It is feared, however, that the BNRR will cause the loss of 472 ha of farmland, will add extra traffic to the region (motorways seem to generate more traffic) and, as most traffic passes within the conurbation, only a small fraction of M6 traffic will switch to the relief road and pay charges. The scheduled completion date is 2002 (compare Shanghai, Places 102, page 619).

Cannock
Burntwood
Lichfield
0   5 km
A36
A5
M54
Tamworth
A5
M6
A38
M42
Walsall
Sutton Coldfield
Britain's first private toll motorway: Birmingham Northern Relief Road (44 km)
Wolver-hampton
M6
West Bromwich
A38 (M)
New M6 to M42 connecting motorway
M5
Birmingham
M6
b

Air transport

Air transport has the highest terminal charges, high haulage costs (fuel) and affects large numbers of people living near to airports. Its advantages (Figure 21.1) include speed over long distances for passengers, tourists and business people, and for freight, if it is of high value, light in weight and perishable. Air transport is also important to countries of considerable size (Brazil); where ground terrain is difficult (Sahara Desert); when crossing short stretches of sea (London to Dublin, Belfast, and Amsterdam); or where relief aid is necessary following a major human (Rwanda) or natural (earthquake) disaster or international incident (Kuwait).

Recent deregulation of airlines (USA since 1978, the EU in 1993) has led, due to increased competition, to the availability of more routes and lower fares. Although this has caused a decrease in the number of airlines, it has encouraged more people to travel by air, except during times of recession. Scheduled international and domestic traffic handled by British carriers has increased steadily over the years causing, during peak holiday periods, congestion at airports and competition for airspace. Whereas over 80 million people used one of London's four airports (Heathrow, Gatwick, Stansted and Luton) in 1996, with 54 million passing through Heathrow alone, by 2016 it is estimated that 173 million passengers will use these airports (hence the pressure for the fifth terminal at Heathrow (Places 100). Despite the worldwide growth of air-passengers and improvements in aviation technology, the demand for 1000-seater jets, popular a few years ago, has declined, and even the production of 747s is now minimal.

Places 100 London: Heathrow Airport

BAA (formerly British Airports Authority) applied, in 1994, for permission to build a fifth terminal at Heathrow Airport (Figure 21.12). The ensuing public enquiry began in May 1995 and ended in March 1999 (it had been expected to last until the end of 1996). A decision is not expected until 2001, which means a new terminal could not operate before 2006 (compare Shanghai, Places 102 page 619).

BAA says that the new terminal would be able to handle the extra 30 million passengers expected to use the airport annually; it would mean that many transit passengers would not have to travel to Paris, Amsterdam or Frankfurt (which would lose the UK income); the site at present is only a sludge farm, with no public amenities or open space; the new terminal would not extend beyond the present perimeter fence; it would create an extra 6000 jobs as well as maintaining the existing 8000 direct and 60 000 indirect jobs; it would be funded by private investment (costing the taxpayer nothing); and BAA has already invested £440 million in the new Heathrow Express which takes passengers to Paddington in 16 minutes and reduces the number of vehicles using the airport.

Opposition groups claim the terminal would be built on a wetland site within the green belt; it would cause an increase in noise, night flights, pollution from aviation fuel and a greater risk of a major accident; it will cover an area (280 ha) as large as the four existing terminals combined; it would only create unskilled jobs, and too many jobs would mean an increase in demand for housing; BAA should instead expand Stansted Airport, which is not yet at full capacity.

BAA claim there will not be a third runway or extra night flights, that noise levels are falling as technology improves (Figure 21.13), that extensions to the Heathrow Express and Piccadilly line will not cause extra road congestion; and there will be a freeze on car parking facilities (BAA employers get an 80 per cent ticket reduction if they travel to work by public transport).

Figure 21.12

The proposed Terminal 5 at Heathrow Airport

Figure 21.13

Reduction in noise levels over time at Heathrow Airport

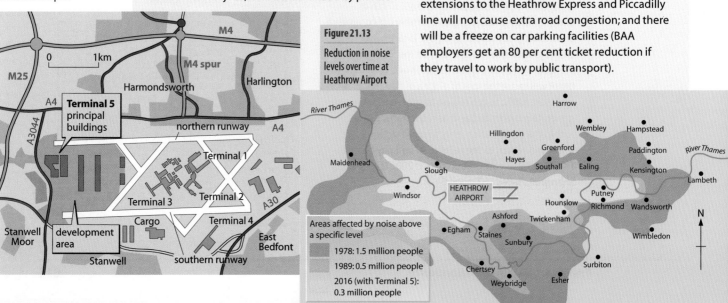

Areas affected by noise above a specific level

1978: 1.5 million people
1989: 0.5 million people
2016 (with Terminal 5): 0.3 million people

Transport and the environment

As the demand for greater mobility by an ever-increasing number of people, both in Britain and across the world, continues to grow, so too does the need for improved traffic systems and more efficient forms of transport. However, improvements that appear to allow people this greater mobility invariably seem to have adverse effects upon the environment. The unanswerable question is: 'How can we accommodate the ever-growing movement of people and goods without causing serious and irreversible damage to the environment?' In 1998, the DoE (Department of the Environment) produced a document called 'A New Deal for Transport'. It began by stating: 'We want a transport system that meets the needs of people and business at an affordable cost and produces better places in which to live and work. We want to cut congestion, improve our towns and cities and encourage vitality and diversity locally; helping to reduce the need to travel and avoid the urban sprawl that has lengthened our journeys and consumed precious countryside.' What the DoE considers this New Deal means for transport is shown in Figure 21.14. The DoE also spells out how it sees the New Deal for transport affecting **a** the motorist and **b** the public transport passenger (Figure 21.15).

The New Deal for transport means:

- cleaner air to breathe by tackling traffic fumes
- thriving town centres by cutting the stranglehold of traffic
- quality places to live where people are the priority
- increasing prosperity backed by a modern transport system
- reduced rural isolation by connecting people with services and increasing mobility
- easier and safer to walk and cycle
- revitalised towns and cities through better town planning.

Figure 21.14

The New Deal for transport in the UK

Figure 21.15

The New Deal for the motorist and the public transport passenger

A New Deal for the motorist	A New Deal for the public transport passenger
• improved management of the trunk road network to reduce delays, through e.g. Regional Traffic Control Centres in England	• more and better buses and trains, with staff trained in customer care
• investment focused on improving reliability of journeys	• a stronger voice for the passenger
• better-maintained roads – increased resources both locally and nationally	• better information, before and when travelling, including a national public transport information system by 2000
• updated Highway Agency's Road User's Charter to give more emphasis to customer service	• better interchanges and better connections
• more help for the motorist if their car breaks down on a motorway	• enhanced networks with simplified fares and better marketing, including more through-ticketing and travelcards
• reducing the disruption caused by utilities' street works	• more reliable buses through priority measures and reduced congestion
• improved road safety and safer cars	• cash boost for rural transport
• quality information for the driver – before and during journeys	• half-price or lower fares for elderly people on buses
• dealing with car crime	• improved personal security when travelling
• more secure car parks	• easy-access public transport – helping disabled and elderly people, and making it easier for everyone to use
• better information and protection when buying a used car	
• action on 'cowboy' wheelclampers	
• more fuel-efficient cars	
• less congestion on our roads and less pollution in our cars	

Sustainable transport

The Royal Commission on Environmental Pollution has produced two comprehensive reports on reducing transport's impact on the environment and proposed targets to drive the process. These reports focus the attention on reducing greenhouse gas emissions (Case Study 9B), improving local air quality and road safety, boosting rail freight and encouraging cycling (Figure 21.16).

Figure 21.16

The Royal Commission on Environmental Pollution

Current targets
- Greenhouse gases – legally binding target to reduce emissions to 12.5% below 1990 levels by the period 2008–2012, and a domestic aim to reduce CO_2 emissions by 25% by 2010.
- Air pollution – National Air Quality Strategy, encompasses health-based objectives for a range of air pollutants to be met by 2005.
- EU vehicle and fuel quality standards – to reduce toxic emissions and noise from new vehicles.
- Cycling – from a 1996 base, double cycling by 2002 doubling again by 2012 (from the National Cycling Strategy).
- Road safety – existing target for 2000, new target for 2010.

Targets for the future
- Freight on the railway – endorsement of the industry's targets for growth.
- EU vehicle standards – target to improve fuel efficiency and reduce CO_2 emissions by more than a third before 2010.
- Health – proposed targets in 'Our Healthier Nation' for reducing all accidents by a fifth by 2010 and reducing death rates from heart disease and strokes amongst people under 65 by a third by 2010.
- Green transport plans – for HQ/other key government buildings by 1999/2000.
- Walking – targets being prepared to reverse the decline in walking.
- Public transport – targets to encourage more use of public transport.
- Road traffic – assess impact of measures in this White Paper and consider national targets for the level of road traffic.

Transport routes and networks

A single route location

The shortest distance between two points on the Earth's surface is, due to the world being a sphere, a **great circle route**. This is why planes flying from London to airports adjacent to the Pacific Ocean (Los Angeles, Tokyo) fly over polar regions, and was one reason why the *Titanic* struck an iceberg well to the north of a straight-line route from Liverpool to New York. However, although over short distances the shortest route is approximately a straight line, in practice, direct routes have rarely been constructed, even by the Romans. Haggett (1969) suggested that deviations from a straight-line route were for either positive or negative reasons.

Positive deviations are when routes are diverted to gain more traffic. Figure 21.17a shows that although (assuming there are no physical obstacles) the straight-line route is cheapest to build, having the shortest distance, it is the most expensive to the user. In comparison, the least-cost route for the user (Figure 21.17b) is likely to be the longest and most expensive to build. The selected route will probably be a compromise between these two extremes (Figure 21.17c), provided that it gains acceptance from a third party – the environmental lobby.

Negative deviations result from routes being diverted to avoid physical obstacles such as a mountain range (Figure 21.18a) or a spur (Figure 21.18b).

Figure 21.17

Single route location: a positive deviation

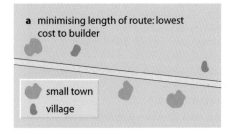

a minimising length of route: lowest cost to builder

small town

village

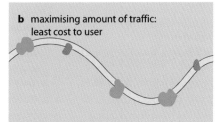

b maximising amount of traffic: least cost to user

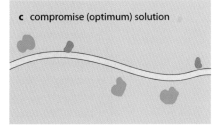

c compromise (optimum) solution

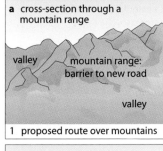

a cross-section through a mountain range

valley

mountain range: barrier to new road

valley

1 proposed route over mountains

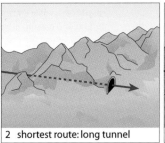

2 shortest route: long tunnel

3 longest route: no tunnel

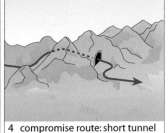

4 compromise route: short tunnel

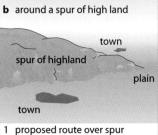

b around a spur of high land

town

spur of highland

plain

town

1 proposed route over spur

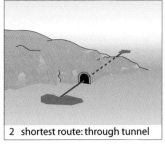

2 shortest route: through tunnel

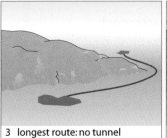

3 longest route: no tunnel

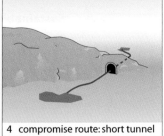

4 compromise route: short tunnel

Figure 21.18

Single route location: a negative deviation

Figure 21.19

Builder and
user costs

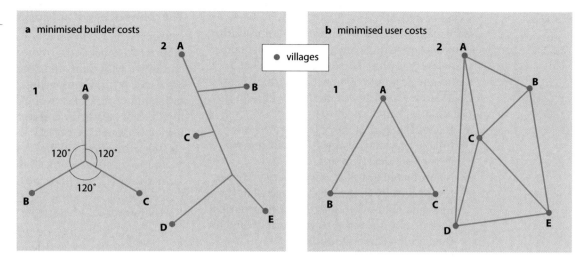

a minimised builder costs

b minimised user costs

• villages

Builder and user costs

The least-cost network to the **builder** is likely to be the one with the shortest overall length (Figure 21.19a). These are more likely to be found in areas with a sparse population and low traffic density (Australia); in highland areas where routes are funnelled along valley floors (Lake District); or where, as in the case of motorways, the construction cost per unit length is extremely high.

The least-cost network to the **user** is the one where the traveller can move as quickly and directly as possible between any point in the network (Figure 21.19b). These are likely to occur in conurbations where high population and traffic densities demand numerous alternative routes.

Transport networks

A **network** is defined as a group of places joined together by a number of routes to form a system. Network patterns vary greatly between and within countries, so it is necessary to adopt a universally acceptable method of description to allow comparison between different systems. Network analysis is achieved by showing the network as a series of straight lines. Drawing a **topological map** preserves the pattern of routes

and junctions but distorts directions and distances (like the London Underground map). The resultant map (Figure 21.20a) consists of vertices and edges. **Vertices** or **nodes** are points in the network. **Edges** or **links** form the direct lines of transport between these points.

Vertices (Figure 21.20a) may be a point of origin or destination (Swansea), a significant place *en route* (Cardiff), or a junction of two or more routes (Bristol). This division into vertices and edges helps to describe the **accessibility** of a place and the **efficiency** of the network. Figure 21.20b gives the accessibility matrix for points on the topological map. The aim is to follow a route between two places, passing through as few vertices as possible, since they can cause congestion and delay. The **Shimbel index** (which is 14 for Bath) is the total number of edges needed to connect any one place with all the other vertices. The place(s) with the lowest Shimbel index (Newport and Bristol with 11) has/have the highest **centrality** and is/are the most accessible. If the Shimbel index for all the vertices is plotted it is possible to draw an isopleth map to show the relative accessibility of places (isopleths for 12, 15, 18 and 21 could be added to Figure 21.20a).

Figure 21.20

A topological map and
accessibility matrix

a topological map

Gloucester

Swansea

Newport

Severn
Bridge

Cardiff

Swindon

Bristol

Bath

Taunton

b accessibility matrix

.	Bath	Bristol	Cardiff	Gloucester	Newport	Swansea	Swindon	Taunton	TOTAL
Bath	—	1	3	2	2	4	1	1	14
Bristol	1	—	2	1	1	3	2	1	11
Cardiff	3	2	—	2	1	1	3	3	15
Gloucester	2	1	2	—	1	3	1	2	12
Newport	2	1	1	1	—	2	2	2	11
Swansea	4	3	1	3	2	—	4	4	21
Swindon	1	2	3	1	2	4	—	2	15
Taunton	1	1	3	2	2	4	2	—	15

Shimbel
index

Types of network

Networks are shown by means of **planar graphs**, i.e. where all vertices and edges are in the same plane, so that while two edges may meet at a vertex they cannot cross over each other. Figure 21.21 shows six different methods of linking, or not linking, four vertices.

a In a **null graph**, none of the vertices is linked: they may divided by mountain barriers or political frontiers.

b A **connected graph** is where every vertex is linked directly to the adjacent vertex in the network, but only indirectly to other vertices (a cross-city bus route, or the Central Line in the London Underground).

c A **circuit**, or **closed path**, follows a 'circular' route linking up all the vertices (a milkround beginning and ending at a dairy, or the Circle Line in the London Underground). It is the shortest possible distance linking all points.

d In a **tree graph**, all the vertices are linked, but there are no circuits. Each edge, leading to a terminal point, acts as a branch line (motorways radiating from London).

e A **complete graph** is where each vertex is directly linked to every other vertex. It is more likely to be found in a developed country and equates with Christaller's $k = 3$ model, page 409.

f A **subgraph** is where part of the network is detached from the rest. This incomplete system may be associated with the pre-take-off stage of the Rostow model (Figure 22.13).

Figure 21.21

Types of network

The connectivity of networks

Connectivity is a means of **measuring the efficiency** of a network (the more vertices that are connected with each other, the more efficient the network) and **comparing**, quantitatively, different networks. Before looking at four methods of measuring connectivity, two basic principles should be noted.

■ The **minimum** number of edges, e (minimum), needed to link all the vertices in a network is 1 less than the total number of vertices (v). Therefore:

$$e_{(min)} = v - 1$$

e.g. in the tree and connected graphs with 4 vertices, $e = 3$.

■ The **maximum** number of edges, e (maximum), that can exist in a network without a duplication of routes (linkages) is found by subtracting 2 from the number of vertices and then multiplying by 3. Therefore:

$$e_{(max)} = 3(v - 2)$$

e.g. in a complete graph with 4 vertices,

$$e = 3(4 - 2) = 6.$$

Measuring connectivity

■ The **beta index** (β) is the easiest method of measuring connectivity, as it simply involves dividing the number of edges (e) in the network by the number of vertices (v).

Therefore: $\beta = \dfrac{e}{v}$

Three situations can result:

1 In simple networks, the beta index is less than 1.0,
e.g. in Figure 21.21a, $0 \div 4 = 0$ (null), and in Figures 21.21b and d, $3 \div 4 = 0.75$ (tree and connected).

2 In a network with one circuit, the beta index is 1.0.
e.g. in Figure 21.21c, $4 \div 4 = 1.0$ (circuit).

3 In a complete network, the beta index is greater than 1.0,
e.g. in Figure 21.21e, $6 \div 4 = 1.5$ (complete).

As can be seen, the higher the value of the beta index, the greater the degree of connectivity and the more efficient the system.

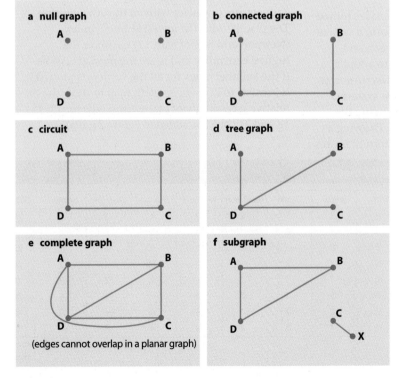

a null graph

b connected graph

c circuit

d tree graph

e complete graph

(edges cannot overlap in a planar graph)

f subgraph

- **The cyclomatic number** (c) refers to the number of circuits in a given network. The more edges there are in a network, the greater the number of circuits is likely to be. The formula is derived from the fact that a tree graph plus one edge gives a circuit. Therefore it can be assumed that the number of circuits = the number of edges (e) minus the number of edges needed to form a tree ($v - 1$) – i.e. the first of the two basic principles given above. Therefore:

$e - (v - 1)$, which is the same as
$c = e - v + 1$.

In a tree graph, where there are no circuits (Figure 21.21d):

$c = 3 - 4 + 1 = 0$.

In a complete graph, where there are three circuits (Figure 21.21e):

$c = 6 - 4 + 1 = 3$.

A disadvantage of this method is that networks with very different forms may have the same cyclomatic number.
- **The alpha index** (α) is very useful when comparing networks. It is obtained by expressing the number of actual circuits in a network (the cyclomatic number) as a percentage of the maximum possible number of circuits in that network.
This can be expressed by the formula:

$$\alpha = \frac{(e - v + 1)}{2v - 5} \times 100.$$

This can give a value between 0 and 100 per cent. A low alpha index means there is little or no connectivity, whereas one of 100 per cent indicates that the network is completely connected.
In this network:

$$\alpha = \frac{2}{8 - 5} \times 100.$$

Therefore:

$\alpha = 66\%$.

- **The gamma index** (γ) is considered less useful than the alpha index. It compares the actual number of edges in a network with the maximum possible number of edges in that network.

This is expressed as:

$$e_{(max)} = \frac{e}{3(v - 2)} \times 100.$$

Using the same two diagrams as for the cyclomatic number, the maximum links in the circuit will be:

$e_{(max)} = 3 (4 - 2) = 6$.

However, in the tree graph, where there are only three edges:

$$e_{(max)} = \frac{3}{3(4 - 2)} \times 100 = 50\%.$$

The gamma index ranges from a minimum value which varies, but is always above zero, and 100 per cent. This proves a weakness when comparing different transport networks or the same network over a period of time. It is of more value when comparing the efficiency of networks of different sizes.
- **The detour index** (DI) When using topological maps, the edges which link the vertices are drawn as straight lines. Yet it has been shown that in reality communications rarely follow the most direct route. The efficiency of a route can be measured by determining how far it deviates from the straight-line route. This is calculated by using the formula:

$$DI = \frac{\text{Shortest possible actual route distance}}{\text{Direct (straight-line) distance}}$$

The lowest possible index is 1.00 (some texts suggest this is given as a percentage). A detour index of 2.00 means that the actual route is twice as long as the most direct route.

Desire line maps In transport studies, a **desire line** is a straight line joining two places between which people, or goods, travel. On a desire-line map, one line is drawn for each separate journey or movement (Figure 21.22). The number and closeness of desire lines indicate the frequency and intensity of journeys. We have already seen, for example, that when shopping for convenience goods journeys are usually short in distance and frequent in occurrence, whereas journeys made for comparison goods are much longer and more infrequent.

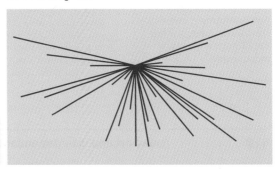

Figure 21.22
Desire lines

Interaction (gravity) models

As described on page 410, these models try to predict the degree of interaction, or movement, between pairs of places. They are based on two premises. First, as the sizes of both towns increase, so too will the amount of movement between them. Secondly, the further apart the two towns are, the less will be the movement between them (distance decay). The model predicts that movement will be greatest where two large cities are close together.

In transport terms, the gravity model may be used to estimate two features of a system:

- The **boundary** between two towns beyond which people find it easier (less distance, less time or less cost) to travel to one centre rather than the other (Reilly's breaking point, Figure 14.40).
- The **volume** of traffic (passengers, freight or, as initially suggested by Christaller, telephone calls) between various towns.

Geographers are interested not only in the direction of flow of people and goods, but also in predicting the volume of movement. The latter can be estimated by multiplying the population of two towns and dividing the product by the square of the distance between them.

This can be expressed by the formula:

$$\text{movement A to B} = \frac{p\text{A} \times p\text{B}}{d^2}$$

where: pA is the population of town A
   pB is the population of town B
   d is the distance between towns A and B.

A worked example is given in Figure 21.23. It shows that on the basis of the gravity model, more buses or trains (public transport) and better roads (private transport) are needed, for example, to cope with the volume of movement between Montrose and Arbroath than between Montrose and Forfar.

Figure 21.23

Interaction (gravity) model applied to an area of eastern Scotland

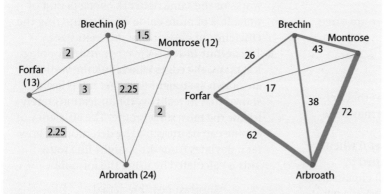

population (thousands): i.e. Brechin (8) = 8000
distance (tens of km): i.e. Brechin to Forfar [2] = 20 km

$$\text{Brechin to Forfar} = \frac{8 \times 13}{2^2} = 26$$

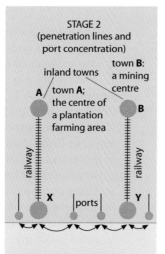
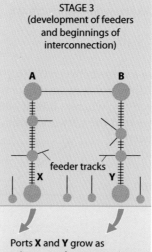
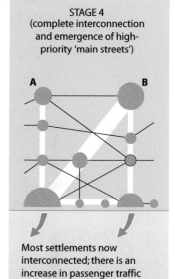

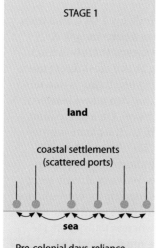

STAGE 1

land

coastal settlements (scattered ports)

sea

Pre-colonial days, reliance upon local sea-borne trade with tracks leading inland; port development limited by shallow seas, heavy surf, few natural inlets and deltas at river mouths.

STAGE 2
(penetration lines and port concentration)

town **B**: a mining centre

inland towns

town **A**; the centre of a plantation farming area

railway

X   ports   **Y**

railway

Early colonial development included routes inland, usually rail, to newly developed towns associated with the collection of a cash crop or the exploitation of a mineral: both of which were exported to the colonial power.

STAGE 3
(development of feeders and beginnings of interconnection)

A   B

feeder tracks

X   **Y**

Ports **X** and **Y** grow as primary products are exported; intermediate towns with short feeder routes grow up along the major inland routes (very few imports).

Figure 21.24

Model showing the evolution of a transport network in a developing part of the world (*after* Taaffe, Morrill and Gould)

STAGE 4
(complete interconnection and emergence of high-priority 'main streets')

A   B

Most settlements now interconnected; there is an increase in passenger traffic as opposed to earlier goods movements; major centres connected by all-weather roads, air and rail; major link is between the capital **B** and major port, **X**; some feeder tracks may be abandoned.

A model showing the evolution of a transport network in a developing country

In 1963 Taaffe, Morrill and Gould put forward a sequential model to show how a transport network might evolve in a developing part of the world. Their model, based on studies in Nigeria and Ghana, is summarised in Figure 21.24. The early stages reflected the needs of a controlling colonial power to:

1 link politically and militarily the centres of administration, which were on the coast, with inland centres
2 exploit mineral resources
3 develop plantations and increase agricultural production

4 carry freight in preference to passengers. The model has since been applied to other developing countries, especially in East Africa, South America, Africa and Malaysia. In many cases, only Stage 2 has been reached, mainly because following independence these countries had insufficient money to improve and modernise their networks beyond the main urban areas, except, perhaps, to a prestigious airport. To reach Stages 3 and 4 (Brazil and Malaysia), it appears that a country needs the development of a manufacturing industry and the emergence of an affluent sector in the population.

Places 101   Delhi: traffic

'In spite of all the controls, investment is already slanted in favour of the élite. To take one example, private transport is beyond the wildest dreams of most Indians, but the streets of Delhi are nevertheless clogged up with Japanese-designed cars and scooters which can compete in the international market. For the less affluent there are only decrepit, outdated and fuel-inefficient buses quite incapable of providing an efficient service even if the roads were cleared for them. There has

been no development of suburban railways worth the name, and not even any attempt to relieve the burden of the poor man's taxi-driver – the cycle rickshaw puller. He does not enjoy the fruits of modern aerodynamics, metallurgy or engineering. His vehicle hasn't changed in the twenty-five years I have known Delhi – it's still inordinately heavy, and doesn't even have such modern aids as gears.'

Mark Tully, *No Full Stops in India*, 1991

Places 102   Shanghai: traffic developments

Shanghai, with an estimated population of 25 million, is alleged to have, on its roads, 6 million cycles, 100 000 scooters, 30 000 taxis and numerous company cars (relatively few Chinese, as yet, own their own cars), lorries, buses and minibuses. By 1990, traffic was grid-locked most days and the city was virtually permanently shrouded in smog. Drastic

problems meant drastic solutions – solutions that would neither have been achieved nor tolerated in Britain. This is partly because, in China, there is neither a conservation lobby (and eco-warriors/protestors would not be tolerated) nor the means to hold public enquiries (compare Places 98, page 609 and Places 100, page 612) and partly because the abundance of cheap labour means that work on a project can continue 24 hours a day, 7 days a week.

The first attempted solution was the building of a 47 km elevated ring road (Figure 21.26), a project which took, from proposal to completion, less than three years. This was followed by a cross-city north–south motorway (1995) and an east–west motorway (1999). Each of these three-lane motorways, which were literally bulldozed straight through the city, has one carriageway built above the second (Figure 21.25). People living on the route of the new motorways were given the option of being moved to a new flat in the suburbs or being given money. While people were often pleased to be moved from old, crowded properties

Figure 21.25

Elevated motorways, Shanghai

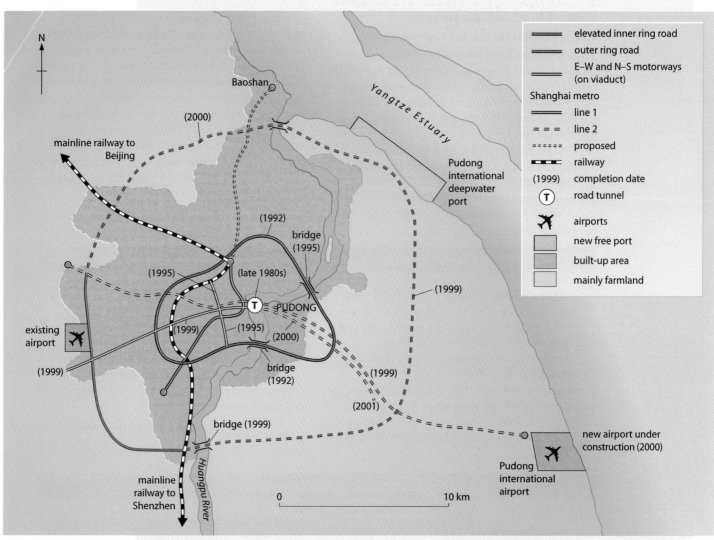

Figure 21.26

Recent transport changes in Shanghai

to new flats with modern amenities, the new properties were often on the edge of Shanghai many kilometres in distance and up to one hour in time from their place of work. Whereas the only method of crossing the Huangpu River in 1990 was by ferry, in 1999 there were three road bridges, two underground (metro) tunnels and one road tunnel. Shanghai's first underground (16 km in length) opened in the mid-1990s, its second (14 km) in 1999, and two more are planned by 2001. Many main roads within the city now have cycle tracks alongside them (Figure 21.27 – although many cyclists seem to ride wherever they like: often the wrong way down one-way streets). A new international airport was due to begin operating by 1999, a second ring road and a 100 m wide motorway to the new airport were both due to be completed by 2000, and a new international deepwater port in use by the early 21st century.

Drivers claimed (1999) that average travel speeds within the city had increased to 28 km/hr at peak times (17 km/hr in London), the journey to the old airport from the city centre had been reduced by 45 minutes, and atmospheric pollution had decreased considerably (although it still seems bad by British standards).

Figure 21.27

Cyclists

International trade

The development of international trade

Trading results from the uneven distribution of raw materials over the Earth's surface. Today, trade plays a major role in the economy of most countries, as none has large enough supplies of the full range of minerals, fuels and foods to make it self-sufficient.

While early peoples bartered in order to exchange goods, the first civilisations found that making transactions with money proved a method by which wealth could be accumulated. By 1500 BC, the Canaanites had learned to become 'middlemen' in the trade between Egypt, Asia Minor and Mycenae (Greece), while their descendants, the Phoenicians, traded five centuries later around and beyond the Mediterranean Sea. Until this time, trading was mainly the simple exchange of raw materials:

$$\text{Area A} \rightleftarrows \text{Area B}$$
$$\text{grew wheat} \qquad \text{reared animals}$$

Subsequently, certain communities proved to be more inventive than others and began to process some of these raw materials:

$$\begin{array}{c} \text{Country A grew} \\ \text{cotton and mined} \\ \text{iron ore} \end{array} \rightleftarrows \begin{array}{c} \text{Country B} \\ \text{manufactured textiles} \\ \text{and smelted iron} \end{array}$$

Such exchanges led to the emergence of European colonial powers who used the raw materials from their colonies to establish their own domestic industries. This saw the beginnings of modern international trade and the widening gap between those countries producing the raw materials, with the exception of oil, and those who made a much larger profit by manufacturing goods (Figure 21.31). Trade developed further as new and improved forms of transport were introduced, such as transcontinental railways, air traffic and refrigerated ships (Figure 21.1).

Since the 1960s there has been, especially among the industrialised nations of Europe and North America, a tendency to specialise in particular aspects of manufacturing. This creates greater benefits (the **law of comparative advantage**) than trying to compete with other countries that have equal, or better, opportunities. This can be seen in the modern car industry:

Country A:		Country B:		Country C:
brakes,	→	tyres,	→	clutches, oil
distributors	←	carburettors	←	pumps and
and glass		and headlights		pistons

The law of comparative advantage operated in 19th-century Britain. Britain could have produced more of its own food but, having a greater comparative advantage over other countries in manufacturing, opted to concentrate on industry where greater profits could be made.

Figure 21.28

Major global trading groups

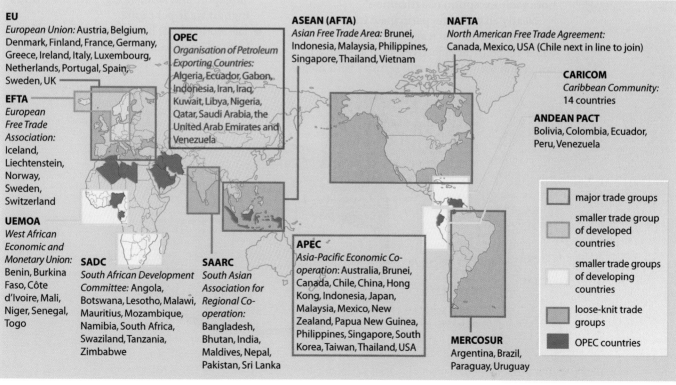

EU
European Union: Austria, Belgium, Denmark, Finland, France, Germany, Greece, Ireland, Italy, Luxembourg, Netherlands, Portugal, Spain, Sweden, UK

EFTA
European Free Trade Association: Iceland, Liechtenstein, Norway, Sweden, Switzerland

UEMOA
West African Economic and Monetary Union: Benin, Burkina Faso, Côte d'Ivoire, Mali, Niger, Senegal, Togo

SADC
South African Development Committee: Angola, Botswana, Lesotho, Malawi, Mauritius, Mozambique, Namibia, South Africa, Swaziland, Tanzania, Zimbabwe

SAARC
South Asian Association for Regional Co-operation: Bangladesh, Bhutan, India, Maldives, Nepal, Pakistan, Sri Lanka

OPEC
Organisation of Petroleum Exporting Countries: Algeria, Ecuador, Gabon, Indonesia, Iran, Iraq, Kuwait, Libya, Nigeria, Qatar, Saudi Arabia, the United Arab Emirates and Venezuela

ASEAN (AFTA)
Asian Free Trade Area: Brunei, Indonesia, Malaysia, Philippines, Singapore, Thailand, Vietnam

APEC
Asia-Pacific Economic Co-operation: Australia, Brunei, Canada, Chile, China, Hong Kong, Indonesia, Japan, Malaysia, Mexico, New Zealand, Papua New Guinea, Philippines, Singapore, South Korea, Taiwan, Thailand, USA

NAFTA
North American Free Trade Agreement: Canada, Mexico, USA (Chile next in line to join)

CARICOM
Caribbean Community: 14 countries

ANDEAN PACT
Bolivia, Colombia, Ecuador, Peru, Venezuela

MERCOSUR
Argentina, Brazil, Paraguay, Uruguay

- major trade groups
- smaller trade group of developed countries
- smaller trade groups of developing countries
- loose-knit trade groups
- OPEC countries

Transport and interdependence

Balance of trade and balance of payments

The balance of trade for a country is the difference between the income received from **visible exports** and the cost incurred in paying for **visible imports**. The balance of payments includes the balance of trade together with any **invisible** earnings or costs – i.e. from banking and insurance, tourism, remittances from migrant workers abroad, professional advice and air/sea transport. Countries that earn more from their exports than they pay for their imports, have a **trade surplus**. Those that spend more on their imports than they earn from their exports have a **trade deficit**.

Regional trade groups

The 1990s saw a substantial growth in the scope and number of regional trade agreements (RTAs). By 1999 there were more than 100 in force, with others under negotiation. The EU (Figure 21.28) was, until the formation of NAFTA, the world's largest single trading group, with 370 million people (NAFTA has 390 million). The EU has been, according to the World Bank, 'the main actor in the regionalism process: internally by removing numerous barriers to the free circulation of goods, services, persons, and capital under the Single Market, and externally by developing its network of Free Trade Agreements (FTAs). As a result of these agreements, the EU now trades duty- and quota-free with its immediate neighbours on most products, with the noticeable exception of agriculture'.

Nearly all countries participate in at least one agreement. Often, such agreements are aimed at wider geopolitical objectives, such as strengthening political relations and security with neighbouring countries. Some RTAs have led to structural reforms within member countries (the EU, page 493, and NAFTA) and others have helped newcomers integrate into the global economy (as with Vietnam's entry into ASEAN). The future seems to indicate the enlarging of existing RTAs, with the EU accepting former East European states and NAFTA embracing all – or most – of the Americas.

Direction of world trade

By the late 1990s:

- Most of the world's trade was between the advanced market economies (the EU, North America and Japan), although their share fell from 72 per cent in 1990 to 68 per cent in 1998 (Figure 21.29).
- The advanced market economies had relatively little trade with the economically less developed countries; they exported more to them in value, and less in weight, than they imported.
- There was relatively little trade between economically developing countries, many of which had low rates of economic growth and tended to export the same types of goods (usually raw materials).
- In more recent years, the older developed countries – the USA and those in western Europe – have faced increasing competition, initially from Japan and later from the newly industrialised countries (NICs, page 576). Figure 21.30 shows that two of the NICs, Singapore and Hong Kong, have the world's highest trade per capita.
- The world's trade has become dominated by large and powerful transnational corporations.

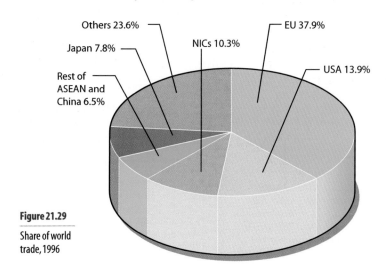

Figure 21.29

Share of world trade, 1996

Others 23.6%
Japan 7.8%
Rest of ASEAN and China 6.5%
NICs 10.3%
EU 37.9%
USA 13.9%

Figure 21.30

Trade per capita for selected countries, 1996

Trade per capita (US$)									
Developed countries		NICs		OPEC		Middle-income developing countries		Lower-income developing countries	
Germany	9815	Singapore	55 483	UAE	16 000	Mexico	1507	Kenya	110
Canada	9355	Hong Kong	45 761	Kuwait	12 009	Chile	1221	Sierra Leone	62
UK	6697	Taiwan	7713	Brunei	11 108	Argentina	877	Nepal	61
Australia	4993	Malaysia	4870	Venezuela	1 224	Peru	369	Bangladesh	54
Japan	4849	South Korea	3767	Indonesia	293	Brazil	343	India	50
USA	4611			Nigeria	197			Ethiopia	16

GATT and WTO

GATT (the General Agreement on Trade and Tariffs) was set up in 1948 with the aim of reducing tariffs (import duties), and to provide a forum for discussing problems of international trade.

The Uruguay Round of 1986 had the ambitious target of trying to replace the various protectionist blocs and self-interest groups with one large global free trade area. Although some 150 nations attended, in reality the representatives of developing countries were little more than spectators to the arguments between, and the decisions made by, the so-called **G7**. The G7 (Canada, France, Germany, Italy, Japan, the UK and the USA), which form the 'inner circle' of the **OECD** (Organisation for Economic Co-operation and Development), account for more than half of the world's trade. The two major areas of dispute were:

- between the EU and the USA over farm subsidies, and
- between Japan and the remaining six over tariffs (Japan at that time had a huge trade surplus yet discriminated against imports from other countries, while at the same time was free to export to those countries).

After several years of debate, and with the global economy in an unhealthy state, agreement to accept the GATT proposals was finally reached in April 1994. Tariffs were immediately reduced on many industrial products, a decision which was to benefit the NICs and, as increased competition lowered prices, consumers. However, farm subsidies and tariffs on agricultural goods were only to be reduced over a period of six years. This delay, resulting from the tactics of strong farming lobbies in the EU (page 493) and the USA, was to the detriment of the many developing countries that relied upon the export of agricultural products.

In 1995, GATT was replaced by the **World Trade Organisation (WTO)**. The WTO can serve the interests of developing countries in four ways:

- It facilitates trade reform.
- It provides a mechanism for settling disputes.
- It strengthens the credibility of trade reforms.
- It promotes transparent trade regimes that lower transaction costs.

These benefits explain the willingness of developing countries to join the WTO (in 1987, 65 developing countries were GATT members; in 1999, 110 developing countries were WTO members, accounting for almost 20 per cent of world exports).

The latest WTO trade Round took place in Seattle in the USA in December 1999 at a time, according to the World Bank, 'of slower trade growth and signs of increasing tensions in international trade relations. A successful Round would contribute to the restoration of market confidence, help create more competitive global markets for goods, services and technology, and reinforce the multilateral trading system. Sustaining and enhancing the growth of world trade is essential to a balanced and sustained recovery in the world economy.'

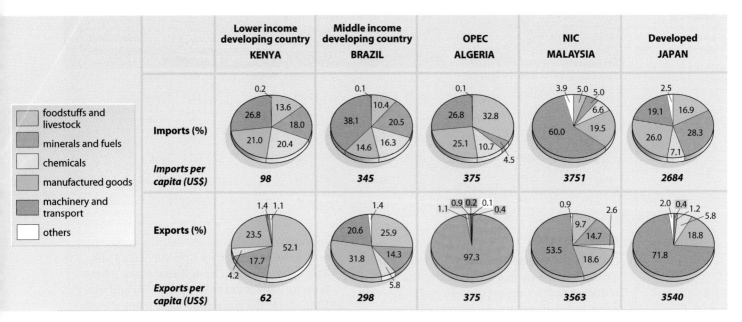

	Lower income developing country KENYA	Middle income developing country BRAZIL	OPEC ALGERIA	NIC MALAYSIA	Developed JAPAN
Imports (%)	0.2 / 13.6 / 18.0 / 20.4 / 21.0 / 26.8	0.1 / 10.4 / 20.5 / 16.3 / 14.6 / 38.1	0.1 / 32.8 / 10.7 / 4.5 / 25.1 / 26.8	3.9 / 5.0 / 5.0 / 6.6 / 19.5 / 60.0	2.5 / 16.9 / 28.3 / 7.1 / 26.0 / 19.1
Imports per capita (US$)	*98*	*345*	*375*	*3751*	*2684*
Exports (%)	1.4 / 1.1 / 52.1 / 17.7 / 4.2 / 23.5	1.4 / 25.9 / 14.3 / 5.8 / 31.8 / 20.6	1.1 / 0.9 / 0.2 / 0.1 / 0.4 / 97.3	0.9 / 2.6 / 14.7 / 18.6 / 53.5 / 9.7	2.0 / 0.4 / 1.2 / 5.8 / 18.8 / 71.8
Exports per capita (US$)	*62*	*298*	*375*	*3563*	*3540*

Legend:
- foodstuffs and livestock
- minerals and fuels
- chemicals
- manufactured goods
- machinery and transport
- others

Figure 21.31

Imports and exports of selected countries, 1996

Trade links between economically developed, economically developing and OPEC countries

Figure 21.31 gives an indication of the nature and value of exports of five contrasting economic groupings and shows the six customary categories into which the many items of world trade are manageably placed. The developed countries of the EU, the USA and Japan often have manufactured goods, machinery and transport equipment accounting for over half their exports. This has enabled these countries to accumulate the capital and technology needed to buy and to process requisite raw materials (fuels and minerals). In contrast, while most developing countries have some manufacturing, it is often primary processing or is operated by transnationals taking advantage of cheap labour rates (pages 573–574).

The world market in fuels (75 per cent of which is oil and natural gas) is dominated by the OPEC countries. Most oil is exported to the energy-short advanced market economies in Europe (excluding the UK and Norway) and to Japan. As developing countries rarely have the capital to afford these fuels, their economic development tends to be further retarded. The pattern of mineral exports is less obvious, with both developed (Australia and Canada) and developing (Jamaica, Zambia) countries being major exporters. It is again, however, the advanced market economies that are the chief importers.

Agricultural products often account for over three-quarters of a developing country's exports, yet an increasing number of African states are having to import cereals as their food production decreases (page 503). Developed countries have to import tropical goods but, on balance, export almost as much agricultural produce as they import. The USA, Canada and Australia, with their extensive agricultural systems, are net exporters accounting for over 75 per cent of the world's wheat (page 485), while others, like the Netherlands and Denmark (page 487), use their farmland intensively to obtain high yields. Most OPEC countries have to rely on imports as their natural environments rarely favour agriculture.

Developing countries have, for many years, made demands for a fairer trading system. One request is for higher and fixed prices for primary products as this might prevent a further widening of their trade gap. A second priority is better access to markets within developed countries. At present, if there is a world recession or if developing countries begin to increase their trade too much with industrialised countries, the latter often impose quotas to limit the number of goods imported or add tariffs so that the prices of imported goods increase and they become less competitive. While the developed countries thus protect their own jobs, they cause increased depression within developing countries. The knock-on effect can be that the developing countries, with a reduced income, will have less money to spend on goods manufactured in industrialised countries and the resultant trade recession will adversely affect both parties. Other demands have included changes in the international monetary system to eliminate fluctuations in currency exchange rates; to encourage industrialised countries to share their technology; to stop developed countries 'dumping' their unwanted and sometimes untested products cheaply; a reduction in interest rates; and an increase in aid free of economic, political and bureaucratic strings.

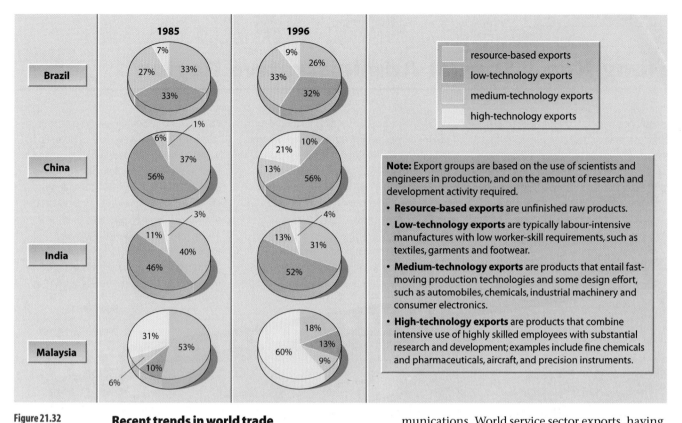

resource-based exports

low-technology exports

medium-technology exports

high-technology exports

Note: Export groups are based on the use of scientists and engineers in production, and on the amount of research and development activity required.

- **Resource-based exports** are unfinished raw products.
- **Low-technology exports** are typically labour-intensive manufactures with low worker-skill requirements, such as textiles, garments and footwear.
- **Medium-technology exports** are products that entail fast-moving production technologies and some design effort, such as automobiles, chemicals, industrial machinery and consumer electronics.
- **High-technology exports** are products that combine intensive use of highly skilled employees with substantial research and development; examples include fine chemicals and pharmaceuticals, aircraft, and precision instruments.

Figure 21.32

Change in the composition of exports for selected countries: 1985 and 1996

Recent trends in world trade

The past half-century has brought unprecedented prosperity and better living standards to many parts of the world, with world GNP 24 times greater in 1998 than it was in 1960. This achievement has been underpinned by the liberalisation and rapid expansion of trade. Between 1987 and 1997 world trade nearly doubled. Trade in services has grown at an even faster rate, aided by the technological revolution in computers and telecommunications. World service sector exports, having more than trebled since 1980, now amount to a quarter of world merchandise exports, with many of Asia's NICs (page 576 and Places 91, page 578) being the main beneficiaries (Figure 21.32).

Statistics released by the World Bank in 1999 show that during the period 1990 to 1998 it was the developing and newly industrialised economies that achieved the fastest rate of expansion (Figure 21.33), albeit sometimes from a low base. With several of the NICs in Asia having had an average annual increase of over 10 per cent, and with those in the Americas averaging 9 per cent, the developing and newly industrialised countries have been able to increase their share of world trade from 23 per cent to 27 per cent. This, however, still leaves them far behind the industrialised countries, even though the latter's share fell from 72 per cent to 68 per cent in the same period (countries that the World Bank categorises as being 'in transition' – i.e. the former centrally planned economies of the Communist bloc – remained constant at 5 per cent).

World trade growth in volume and value slowed sharply in 1998 as a result of the drop in demand in south and east Asia, the broader slowdown in global economic growth, and depressed commodity prices. Even so, to demonstrate the fickle nature of world trade, the *Financial Times* was able to print, as a headline in mid-1999, 'Chastened Asian tigers hit the road to recovery'.

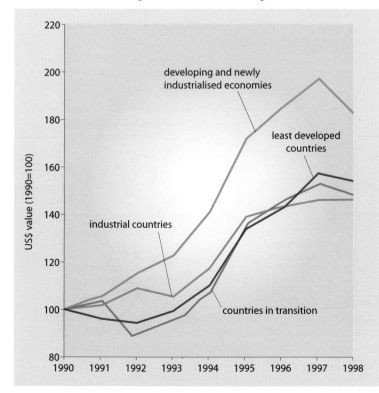

Figure 21.33

Differences in rates of trade expansion between different regions, 1990–98

Hong Kong: Special Administrative Region

Hong Kong originally grew as a result of its strategic trade route location and its large, deep, sheltered harbour. Growth continued partly as a result of increased industrialisation. Hong Kong was one of South-east Asia's four 'little tigers' (page 578), and trade with China in particular and with the remainder of the Pacific Rim in general has expanded rapidly.

Early transport was mainly restricted to water, due to the limited amount of flat land. As building on the steep hillsides proved difficult and hazardous (Case Study 2B), especially on Hong Kong Island, land has had to be repeatedly reclaimed from the sea for use by industry, housing and transport. Three forms of transport used at the beginning of this century are still in operation today (Figure 21.34). The Star Ferry transfers large numbers of people daily from Hong

Kong Island to Kowloon on the mainland; trams still link northern parts of Hong Kong Island (although land reclamation means their routes are no longer adjacent to the sea); and the Peak Tram funicular railway carries wealthy commuters and tourists to and from Victoria Peak (Figure 21.35). A fourth form of transport, the Kowloon Railway linking Hong Kong with the New Territories and the Chinese city of Canton (present-day Guangzhou), was opened in 1910.

Figure 21.34

Hong Kong's Star Ferry, funicular railway and tram

Figure 21.35

The development of transport in Hong Kong

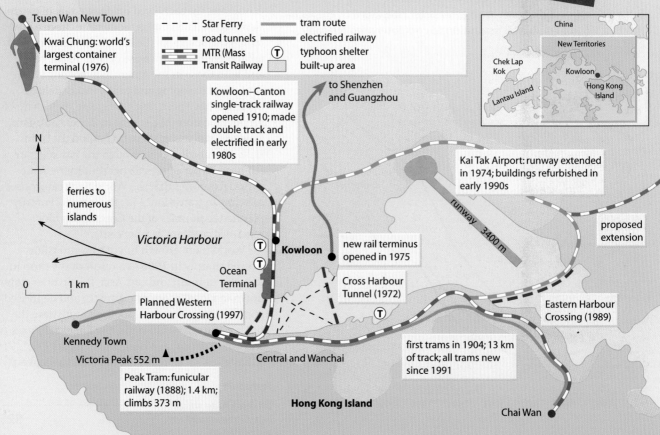

Present-day transport

Hong Kong is the world's largest container port handling over 14 700 TEUs in 1998 (9 million in 1993). Most containers are loaded and discharged at Kwai Chung (Figure 21.35) where, on land reclaimed from the sea, there will be 17 terminals by 2003. Hong Kong is also one of the busiest ports in terms of vessel arrivals (over 40 000 in 1998) and cargo and passenger throughput. The port handles over 90 per cent of the Special Administrative Region's trade by weight and, together with port-related industries, generates 15 per cent of its GNP.

Despite 90 per cent of journeys within the Special Administrative Region being made by public transport (30 per cent by rail), Hong Kong's roads have one of the highest vehicle densities in the world (268 vehicles/km of road in 1998). This high density, combined with difficult terrain and intensive building development, poses a constant problem to transport planning, construction and maintenance. The problem is accentuated by the large numbers of daily commuters between the mainland (Kowloon and various new towns) and Hong Kong Island and, since the reversion of the territory to China, 170 000 passengers daily between Hong Kong and Shenzhen (Case Study 19). A third road tunnel has been opened under Victoria Harbour, and a high-speed rail link is proposed between Hong Kong and Beijing. Over 2.3 million people daily use the Mass Transit Railway (MTR – Figure 21.35), making it one of the busiest underground railways in the world. The network of the early 1990s (three lines with 48 km of track and 38 stations) has been extended, with two new lines to the new international airport and the new town of Tung Chung, both on Lantau Island (Figure 21.38).

The biggest transport development has been the construction of the new international airport at Chek Lap Kok (Figures 21.36 and 21.38). It was built because the old airport at Kai Tak (Figure 21.35) had only one runway (which extended outwards into the sea); had reached full capacity (it was the world's fourth largest passenger airport); there was no space for further expansion; it had a dangerous approach over numerous high-rise flats; and it faced incoming typhoons (Places 30, page 237). The decision to relocate at Chek Lap Kok was finally taken in 1989 as part of a comprehensive strategy that was to include major road, rail and port developments in addition to the airport. The strategy, which after 1991 became known as the Airport Core Programme (ACP), consisted of 11 independent projects (Figures 21.36, 21.37 and 21.38). The airport itself was opened on 6 July 1998.

Figure 21.36

Hong Kong's new international airport at Chek Lap Kok

Figure 21.37

The Tsing Ma bridge

The Airport Core Programme (ACP)

New Airport
1 New Airport: at Chek Lap Kok, off Lantau Island

New Transport Links
2 Lantau Fixed Crossing: including a 1377 m suspension bridge
3 Airport Railway: 34 km local and express services
4 Western Harbour Crossing: the third cross-harbour tunnel
5 North Lantau Expressway: 12 km along the Lantau coast
6 Route 3: first section of a major highway network
7 West Kowloon Expressway: a highway on reclaimed land
8 Kwai Chung: enlarging major container port

New Land
9 West Kowloon Reclamation: 330 ha of new urban land
10 Central and Wanchai Reclamation: first section of 20 ha

New Town
11 Tung Chung: first phase of a new town serving the airport.

see Figure 21.38 on page 628

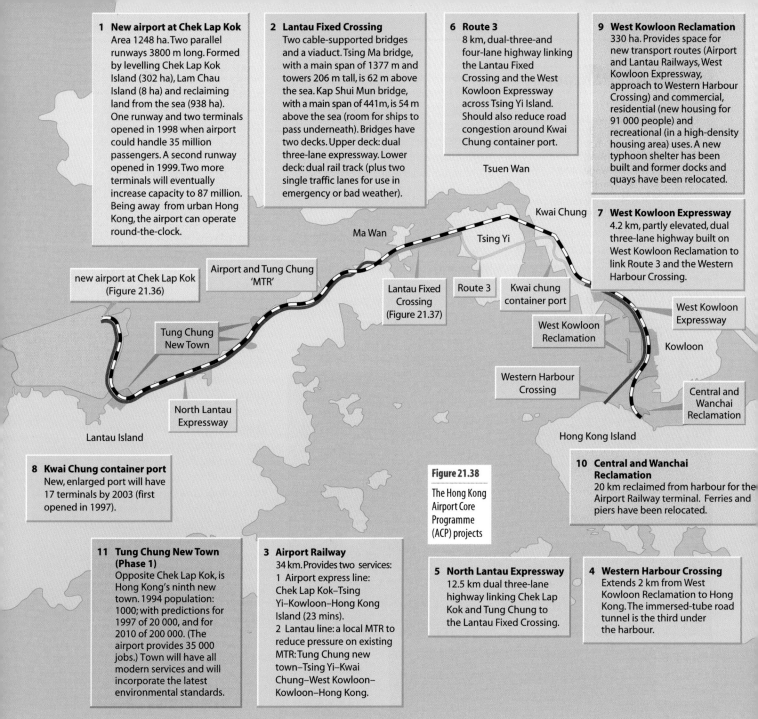

1 New airport at Chek Lap Kok
Area 1248 ha. Two parallel runways 3800 m long. Formed by levelling Chek Lap Kok Island (302 ha), Lam Chau Island (8 ha) and reclaiming land from the sea (938 ha). One runway and two terminals opened in 1998 when airport could handle 35 million passengers. A second runway opened in 1999. Two more terminals will eventually increase capacity to 87 million. Being away from urban Hong Kong, the airport can operate round-the-clock.

2 Lantau Fixed Crossing
Two cable-supported bridges and a viaduct. Tsing Ma bridge, with a main span of 1377 m and towers 206 m tall, is 62 m above the sea. Kap Shui Mun bridge, with a main span of 441m, is 54 m above the sea (room for ships to pass underneath). Bridges have two decks. Upper deck: dual three-lane expressway. Lower deck: dual rail track (plus two single traffic lanes for use in emergency or bad weather).

6 Route 3
8 km, dual-three-and four-lane highway linking the Lantau Fixed Crossing and the West Kowloon Expressway across Tsing Yi Island. Should also reduce road congestion around Kwai Chung container port.

9 West Kowloon Reclamation
330 ha. Provides space for new transport routes (Airport and Lantau Railways, West Kowloon Expressway, approach to Western Harbour Crossing) and commercial, residential (new housing for 91 000 people) and recreational (in a high-density housing area) uses. A new typhoon shelter has been built and former docks and quays have been relocated.

7 West Kowloon Expressway
4.2 km, partly elevated, dual three-lane highway built on West Kowloon Reclamation to link Route 3 and the Western Harbour Crossing.

8 Kwai Chung container port
New, enlarged port will have 17 terminals by 2003 (first opened in 1997).

10 Central and Wanchai Reclamation
20 km reclaimed from harbour for the Airport Railway terminal. Ferries and piers have been relocated.

Figure 21.38
The Hong Kong Airport Core Programme (ACP) projects

11 Tung Chung New Town (Phase 1)
Opposite Chek Lap Kok, is Hong Kong's ninth new town. 1994 population: 1000; with predictions for 1997 of 20 000, and for 2010 of 200 000. (The airport provides 35 000 jobs.) Town will have all modern services and will incorporate the latest environmental standards.

3 Airport Railway
34 km. Provides two services:
1 Airport express line: Chek Lap Kok–Tsing Yi–Kowloon–Hong Kong Island (23 mins).
2 Lantau line: a local MTR to reduce pressure on existing MTR: Tung Chung new town–Tsing Yi–Kwai Chung–West Kowloon–Kowloon–Hong Kong.

5 North Lantau Expressway
12.5 km dual three-lane highway linking Chek Lap Kok and Tung Chung to the Lantau Fixed Crossing.

4 Western Harbour Crossing
Extends 2 km from West Kowloon Reclamation to Hong Kong. The immersed-tube road tunnel is the third under the harbour.

Map labels: new airport at Chek Lap Kok (Figure 21.36) · Airport and Tung Chung 'MTR' · Tung Chung New Town · North Lantau Expressway · Lantau Island · Ma Wan · Lantau Fixed Crossing (Figure 21.37) · Route 3 · Tsing Yi · Kwai Chung · Tsuen Wan · Kwai chung container port · West Kowloon Reclamation · Western Harbour Crossing · West Kowloon Expressway · Kowloon · Central and Wanchai Reclamation · Hong Kong Island

References

Carr, M. (1997) *New Patterns: Process and Change in Human Geography*, Thomas Nelson.

China Business Handbook 1999, China Economic Review (1999).

Hong Kong 1998, Information Services Department (1999).

Hoyle, B. S. and Knowles, R. D. (eds) (1998) *Modern Transport Geography* (enlarged and revised 2nd edition), John Wiley.

Philip's Geographical Digest 1998–99 (1998) Heinemann-Philips.

Simon, D. (1996) *Transport and Development in the Third World*, Routledge.

Singapore, Ministry of Information and the Arts (1998) *Singapore 1998*, Singapore.

Tolley, R. S. and Turton, B. J. (1995) *Transport Systems, Policy and Planning: A Geographical Approach*, Longman.

Tully, M. (1992) *No Full Stops in India*, Penguin Books.

Waugh, D. (1998) *The New Wider World*, Thomas Nelson.

Websites
UK Department of the Environment, Transport and the Regions:
http://www.detr.gov.uk/

Port of Rotterdam:
http://www.port.rotterdam.nl/

Maritime and Port Authority of Singapore:
http://www.mpa.gov.sg/index.html

BAA:
http://www.heathrow.co.uk/BAA Home.htm

Stockley Park (a London business park with a progressive transport policy):
http://www.stockleypark.co.uk

World Trade Organisation:
http://www.wto.org/

See also for more links:
http://www.nelsonthornes.com/gaia

1 a Study Figure 21.39.
  i What is the cheapest way to transport 'one unit of volume' of goods for 100 km? **(1 mark)**
  ii Over what range of distances is barge transport cheaper than lorry or train for an equal volume of goods? **(1 mark)**
  iii Why does the rate of increase in transport costs per km increase more slowly as distance increases? **(5 marks)**
  b Compare the advantages and disadvantages (other than cost) of road and rail transport for the movement of:
  i passengers **(5 marks)**
  ii freight. **(5 marks)**
  c Assess the benefits and drawbacks of **two** current and/or proposed schemes to reduce the number of private cars in towns. **(8 marks)**

2 Study Figure 21.40.
  a Choose any two of the targets and for each:
  i suggest why this was considered to be an important target **(6 marks)**
  ii discuss one or more ways in which the government has attempted to meet that target since 1994. **(6 marks)**
  b Discuss the arguments **for** and **against**:
  i charging tolls on trunk roads, and using the money collected to pay for new roads **(6 marks)**
  ii charging motorists to drive into town centres and using the money raised to improve public transport in towns. **(7 marks)**

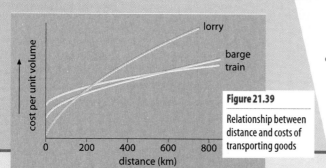

Figure 21.39

Relationship between distance and costs of transporting goods

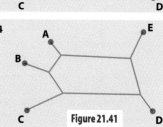

Figure 21.40

Targets

- Increase the proportion of journeys by public transport (measured in total passenger km) from 12 per cent in 1993 to 20 per cent in 2002 and 30 per cent in 2020.
- Raise the proportion of freight carried by rail from 6.5 per cent in 1993 to 10 per cent by 2000, and 20 per cent by 2010.
- Increase the fuel efficiency of new cars sold in Britain by 40 per cent by 2005.
- Cut the proportion of journeys in London made by cars from 50 per cent in 1993 to 35 per cent by 2020; corresponding reductions elsewhere.
- Raise urban bicycle use fourfold by 2000, from 2.5 per cent of all journeys to 10 per cent.
- Achieve full compliance by 2002 with World Health Organisation air-quality guidelines for pollution produced mainly by vehicles. Freeze emissions of carbon dioxide from road and rail at their 1990 level by 2000, and reduce them by 20 per cent by 2020.

Royal Commission's Report on Transport, October 1994

3 'Between 1990 and 1999 the developing and newly industrialised countries have increased their share of world trade from 23 per cent to 27 per cent.' With reference to both LEDCs and NICs, explain how this growth has been achieved. Discuss whether the growth in their share of world trade is likely to continue, and suggest how the more economically developed countries are likely to respond to these changes. **(25 marks)**

4 a In a transport network, what is:
  i a node?
  ii a positive deviation?
  iii a negative deviation?
  iv minimised builder costs?
  v minimised user costs? **(5 marks)**
  b Study Figure 21.41. The four diagrams show four different networks that might be built to connect a number of places. Describe the main advantages and disadvantages of each of the networks. **(6 marks)**
  c i Describe the model developed by Taaffe, Morrill and Gould to show how transport networks grow up in less developed countries. **(6 marks)**
  ii Choose a country or region in Asia, Africa or South America. Discuss the extent to which its transport network has developed along the lines suggested by the model. **(8 marks)**

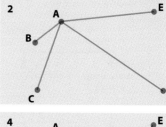

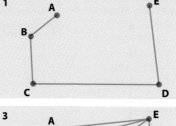

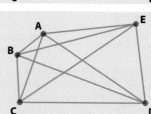

Figure 21.41

AS

A2

Chapter 22

World development

'Development is more than mere economics.'
MarkTully, *No Full Stops in India*, 1991

The concept of economic development

Frequent references have been made in earlier chapters to the inequalities in world development and prosperity. Gilbert, in his book *An Unequal World*, began by stating that:

'Few can deny that the world's wealth is highly concentrated. The populations of North America and Western Europe eat well, consume most of the world's fuel, drive most of the cars, live in generally well serviced homes and usually survive their full three score years and ten. By contrast, many people in Africa, Asia and Latin America are less fortunate. In most parts of these continents a majority of the population lack balanced diets, reliable drinking water, decent services and adequate incomes. Many cannot read or write, many are sick and malnourished, and too many children die before the age of five.'

Figure 22.1

Terms used in relation to world development

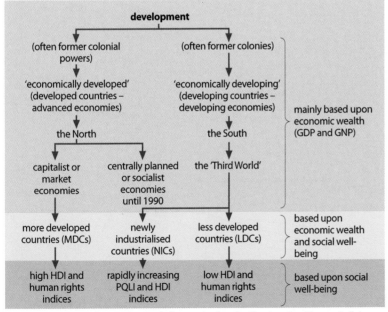

NB For consistency, the terms 'economically more developed' or 'developed', and 'economically less developed' or 'developing' (replacing `Third World' of the first edition), are mainly used in this book.

Definition of terms

Terms such as 'developed' and 'developing' have been used for several decades to indicate the economic conditions of a group of people or a country. By the 1980s, the term 'developing' had come to be regarded as a stigma and was replaced by the concept of the 'South' (**Brandt Report**, 1980) and, with increasing popularity, the 'Third World' (Figure 22.1). Later, with the growing realisation and appreciation that poverty is relative, not absolute, the terms 'more developed countries' (MDCs) or 'advanced economies, and 'less developed countries' (LDCs) or 'developing economies' were introduced. Since the publication of the first edition of this book (1990), world events have seen the collapse of the Communist bloc (the so-called, or assumed, 'Second World'), and its former members in Europe, now being collectively known as 'countries in transition' (Figure 21.33). At the same time, those nations once grouped together as belonging to the 'Third World' have themselves shown a widening spread of wealth and living standards – for example, there is a growing gap between the NICs (newly industrialised countries, page 576) and many of the countries in sub-Saharan Africa.

All of these definitions (summarised in Figure 22.1) were based upon, and overemphasised, economic growth. To those living in a Western, industrialised society, economic development tends to be synonymous with wealth, i.e. a country's material standard of living. This is measured as the **gross domestic product (GDP)** per capita and is obtained by dividing the monetary value of all the goods and services produced in a country by its total population. When trade figures for 'invisibles' (mostly financial services and deals) are included, the term **gross national product (GNP)** is used. It is possible to use either term – GDP is preferred by the EU, and GNP by the UN (our usual source of data) and the USA – as both aim to measure the wealth of a country and to show the differences in wealth between countries. GDP and GNP figures need to be treated cautiously due to problems with exchange rates, differences between countries in their methods of calculation, and difficulties in evaluating services.

Global inequalities in wealth are increasing. Anthony Giddens (first Reith Lecture, 1999) claimed: 'The share of the poorest fifth of the world's population in global income has dropped from 2.3 per cent to 1.4 per cent over the last 10 years, while the proportion taken by the richest fifth has risen from 70 per cent to 85 per cent. In sub-Saharan Africa, 20 countries have a lower income per head in real terms than they did two decades ago.'

Recently, an increasing number of definitions, often involving cultural development, social well-being and political rights, have been suggested as alternatives to those previously based solely upon economic criteria – i.e. they emphasise 'quality of life' in contrast to 'standard of living'. In the early 1990s, the UN introduced the term **Human Development Index (HDI)** – see page 632.

Development is not just the difference between the developed, rich and powerful countries and those that are less developed, poor and subordinate. Each country has areas of prosperity and poverty; contains people with different standards of living based on variations in race or tribe (Hutu and Tutsi in Rwanda), religion (Northern Ireland), language (Dutch-speaking Flemings and French-speaking Walloons in Belgium) or caste (India); and has inequalities between men and women.

Criteria for measuring development

1 Economic wealth

To many people living in developed countries, economic development has been associated with a growth in wealth based on GDP (or GNP). This implies that the GDP (or GNP) of a country has to increase if its standard of living and quality of life are to improve. An economic growth rate of 8 to 10 per cent, which is the highest, has been achieved in China and Ireland in recent years, and by several South-east Asian countries over the past decade or two (Figure 19.38). A rate of 1 per cent is considered disappointing.

Although GDP/GNP figures are easier to measure and to obtain than other development indicators such as social well-being, there are limitations to their use and validity. They are more accurate in countries that have many economic transactions and where goods, services and labour can be measured as they pass through a market place – hence the term 'market economies'. Where markets are less well developed, and trading is done informally or through bartering, and where much production takes place in the home for personal subsistence, GDP figures are less reliable. In the former centrally planned, socialist economies, with their relatively small role in international trade and with few services, GDP figures were difficult to calculate and interpret.

Figure 22.2

World GNP, 1997

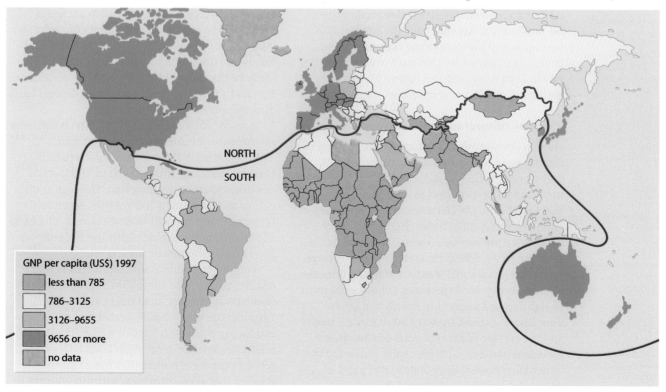

GNP per capita (US$) 1997

- less than 785
- 786–3125
- 3126–9655
- 9656 or more
- no data

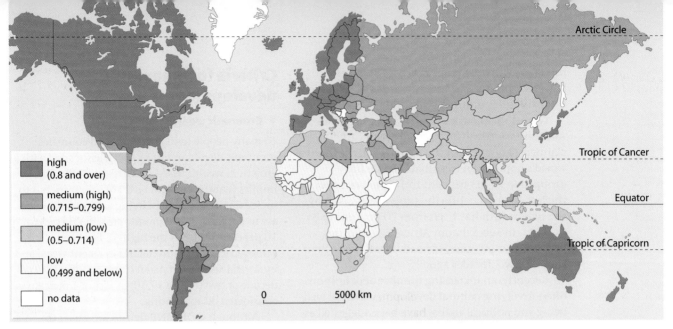

(0.8 and over)

medium (high)
(0.715–0.799)

medium (low)
(0.5–0.714)

low
(0.499 and below)

no data

0   5000 km

Arctic Circle

Tropic of Cancer

Equator

Tropic of Capricorn

Figure 22.3

The UN Human
Development Index
(1993) World
development

Comparison of GDP requires the use of a single currency, generally US dollars, but currency exchange rates fluctuate. The size and growth of GDP may prove to be poor long-term economic indicators and fail to take into consideration human and natural resources. GDP per capita is a crude average and hides extremes and uneven distribution of income between regions and across socio-economic groups, especially in less developed countries where there may be very few extremely wealthy people and a large majority living at subsistence level. Despite these limitations, GDP and GNP are still regarded as relatively good indicators of development and good measures for comparing differences between countries (Figure 22.2). Notice that it is the advanced economies and several of the oil-producing states that have the highest GDP per capita and the developing economies and countries in transition that have the lowest. The World Bank now produces figures for income inequality within some countries, e.g. Brazil.

2 Social, cultural and welfare criteria

Human development has changed the purpose of development to that of meeting human needs, and away from the old style of economic development based upon changes in a country's economy and wealth. The UN Development Programme's Human Development Index (HDI) gives every country a score between 0 and 1, based on its citizens' longevity, education and income. The three factors are given equal weight. Longevity is measured by average life expectancy at birth – the most straightforward measure of health and safety. Education is derived from the adult literacy rate and the average number of years of schooling. Income is based on GDP per capita converted to 'purchasing power parity dollars' (PPP) and is adjusted according to the law of diminishing

returns, i.e. what an actual income will buy in a country. The HDI value for a country shows the distance that it has already travelled towards the maximum possible value of 1, and also allows comparisons with other countries (Figure 22.3). The difference between the value achieved by a country and the maximum possible value shows the country's shortfall, i.e. how far the country has to go. Finding ways of reducing this shortfall is a major challenge for each country.

The 1998 HDI puts Canada top of the list, with a score of 0.99 (Japan was top in 1993 with 0.98). Sierra Leone was at the foot of the list of 174 countries with a score of 0.41 (replacing Guinea which had 0.45 in 1993). Countries with a score of over 0.9 correspond closely with the 'developed' countries of the 'North' while those with a score of under 0.25 equate closely with the economically least developed countries (Figures 22.2 and 22.3). Yet should the similarities between GDP and HDI really be that surprising? Longevity, a good education and a high purchasing power all depend fairly directly upon a country's wealth.

A major criticism of the HDI is that it contains no measure of human rights or freedom. Although the UNDP did produce a separate **Human Freedom Index (HFI)** in 1991, it has not done so since, arguing that 'freedom is difficult to measure and is too volatile, given military coups and the whims of dictators'. The issue of personal and political rights has become increasingly important since then.

Perhaps the main point about HDI is that it enables you to spot anomalies, e.g. countries that have a better (Canada, Sri Lanka and Tanzania) or worse (Saudi Arabia and other oil-producing countries) level of well-being than might be expected from their GNP. HDI can serve a purpose if it identifies where poverty is greatest (between countries, within a country or between groups of people in a country) or if it stimulates debate and action as to where aid, trade and debt alleviation needs to be focused.

3 Other criteria for measuring development

Further criteria have also been used to measure the quality of life as an indicator of levels of, or stages in, development. Several are linked to population as, in developing countries, birth rates are generally high, the natural increase is rapid, life expectancy is shorter and a high percentage of the population is aged under 15 (Figure 13.15). Higher death and infant mortality rates reflect the inadequacy of nutrition, health and medical care. In many developing countries, the prevalence of disease may result from an unbalanced diet, a lack of clean water and poor sanitation – a situation often aggravated by the limited numbers of doctors and hospital beds per person. The majority of people live in rural areas and are dependent upon farming, while in the country as a whole only a small percentage of the population is likely to find employment in manufacturing or service industries. Many jobs are at a subsistence level, in the informal sector (page 574) and the amount of energy consumed within the country is low (Figure 18.25). Economically less developed countries often import manufactured goods, energy supplies and sometimes even foodstuffs, especially grain. In return, they may export raw materials for processing in the developed world (Figure 21.31), accumulate a trade deficit and get increasingly into debt (page 643). High rates of illiteracy reflect a shortage of schools and trained teachers. The density of communication networks, circulation of news-papers and numbers of cars, telephones and television sets per household or per capita have also been used as indicators of development.

Social and economic development
Two often neglected factors in social and economic development (and given insufficient prominence here) are the role of women and the use of child labour. Regarding the key role of women, Figure 22.4 is a quotation from a talk by the head of the UN Population Division, Dr Nafis Sadik, at an International Conference in Delhi in 1993; Places 103 describes the lifestyle of a Kenyan woman who, like many other women across the world, provides a backbone for her family and local community. Places 104 discusses the problems of the use, and misuse, of child labour, an issue which has only recently been highlighted in the Western press.

'A cornerstone of new development thinking is the full integration of women into the mainstream of development and concern for progress in all aspects of their lives – health, education, employment, nutrition, legal and political rights. In traditional development thinking, investment in social development was seen as a luxury, a fruit of economic success. But we now know that the opposite is true: the basis of economic progress is a healthy, socially stable and slow-growing population. Instead of being the fruit of development, social programmes, especially those addressing the status of women, are its very foundation.'

Figure 22.4

Women and development

Places 103   Kenya: women and development

Marietta lives on a small *shamba* (farm) just outside Tsavo National Park in south-east Kenya (Figure 22.5). With her husband working 250 km away in Mombasa, and her nearest neighbour living 3 km away, Marietta is left alone to look after the farm and her seven children. Her day begins by sharpening the machete needed to collect the daily supply of dead wood (living trees are left for animal grazing), as this is her only source of energy (page 543), and by preparing a meal for the family. The eldest girls, before walking to school, collect water from the river 1 km away. Much of Marietta's day is spent collecting firewood and looking after her crops (maize, beans and sorghum). Although owning a few chickens and goats, Marietta's 'wealth' is her two cows which provide milk and are used to plough the hard ground. It is essential that these cows remain healthy for even if the vet, living over 50 km away, did call, Marietta would not be able to afford the bill. Helped by the Intermediate Technology Development Group (Places 90, page 577) Marietta has become a *wasaidizi* and has been given basic training in animal health care. Each week, she spends two mornings in her 'surgery' in the local village (a 4-hour round walk along a track where, just prior to the author's visit, a lion had killed a villager) and other days visiting other local farms. She earns a small commission from the sale of vaccines and medicine, but does not receive a salary.

Figure 22.5

Marietta at her 'surgery'

Elsewhere in Kenya, ITDG are helping local communities to develop skills. These community schemes, which are often run by women, include the improving of building materials and cooking stoves (page 577). In Kenya, as in many other developing countries, it is the Mariettas who form the backbone of the family, women's groups the backbone of the community, and women the backbone of the nation's development and, often, wealth. Yet this role of women as providers and generators of wealth is not matched in most societies by their status and influence. Women are often, and not just in developing countries:

- denied ownership of property (especially land), access to wealth, education and family planning (page 357), and equality in justice and employment

- kept subordinate by being granted lowly positions or given menial tasks and jobs which are often poorly rewarded and paid (agricultural work), and are heavy, tedious and time-consuming (collecting firewood and water) – the concept of double work within and outside of the home

- subject to violence, both physical and mental

- denied political influence which reduces their role in directing and influencing investment and aid patterns.

Places 104 The World: child labour

In developing countries about 260 million children between the ages of 2 and 14 work, at least 120 million of them full-time (61 per cent of children in Asia, 32 per cent in Africa and 7 per cent in Latin America). Around 70 per cent of all child labourers are unpaid family workers and fewer than 5 per cent are employed in export-related production (making clothes, training shoes and carpets). Most children who work in rural areas are involved in agricultural activities (Figure 13.66), while urban children tend to work in services and manufacturing. Although statistics suggest that more boys work than girls, this is only because boys tend to work in more visible types of employment (factories) while girls are likely to perform jobs as domestics (often involving longer hours than boys).

Not all child labour is harmful. Those living in a stable home environment can benefit from informal education and job training. Many working children are also studying and their wages can help their siblings attend school. However, some forms of employment, in particular prostitution and forced or bonded labour, involve working conditions that are hazardous to the children's health, both physical and mental. There are two significant correlations (Framework 19) regarding child labour – the number of working children decreases:

1 as the wealth of a country increases (numbers begin to fall rapidly when the GNP reaches around US$1200)

2 as educational enrolment rises and school quality improves.

Policies that reduce child labour have strong support on economic grounds, as young working children rarely develop the skills necessary to earn higher wages later in life and tend to have, when adult, low productivity rates. Several suggested approaches aimed at reducing child labour include:

- reducing poverty so that children's wages are no longer essential to the family's survival

- educating children, as the provision of primary school enrolment tends to reduce child labour; also, in rural areas, by fixing school terms so as to avoid times of peak agricultural activity

- providing support services to children, such as meals and night shelters (especially for those living on the streets – page 574)

- raising public awareness – of parents, employers and society

- introducing and, often more significantly, enforcing legislation – in June 1999 the International Labour Organisation (ILO) targeted the worst forms of child labour (slavery, prostitution, hazardous work, forced and bonded labour)

- imposing (very difficult) trade sanctions and consumer boycotts.

Adapted from the *World Development Report, 1999/2000*

Scattergraphs

It was suggested on pages 632–633 that there was a correlation between certain criteria and the level of development. 'Correlation' in this sense is used to describe the degree of association between two sets of data. This relationship may be shown graphically by means of a **scattergraph**. This involves the drawing of two axes: the horizontal or *x* axis and the vertical or *y* axis. Usually one variable to be plotted is dependent upon the second variable. It is conventional to plot the **independent variable** on the *x* axis and the **dependent variable** on the *y* axis.

Figure 22.6 shows two relationships, one from physical geography and one from human geography. In the physical example, rainfall is the independent variable, with runoff being dependent upon it. The human example shows GDP as the independent variable and energy consumption per capita to be dependent upon this measure of a country's wealth.

The data are plotted against the scales of both axes. The degree of correlation is estimated by the closeness of these points to a **best-fit** line. This line is usually drawn by eye and shows any trend in the pattern indicated by the location of the various points. One or two points, or **residuals**, may lie well beyond the best-fit line and, being anomalous, may be ignored at this stage. (Later it may be relevant to try to account for these **anomalies** or exceptions.)

The best-fit line may be drawn as a straight line (on an arithmetic scale) or as a smooth curve (on log or semi-log scales). If all the points fit the best-fit line exactly, there is a perfect correlation between the two variables. However, most points at best will lie close to and on either side of the drawn line. A **positive correlation** is where both variables increase – i.e. the best-fit line rises from the bottom left towards the top right (Figure 22.7a and b). A **negative correlation** occurs where the independent variable increases as the dependent variable decreases – i.e. the best-fit line falls from the top left to the bottom right (Figure 22.7d and e). In some instances, the arrangement of the points makes it impossible to draw in a line, in which case the inference is that there is no correlation between the two sets of data chosen (Figure 22.7c). In the event of one, or both, of the variables having a wide range of values, it may be advisable to use a logarithmic scale (Figures 3.22 and 18.25).

If the scattergraph shows the **possibility** of a correlation between the two variables, then an appropriate statistical test should be used to see if there is indeed a correlation, and to quantify the relationship.

Figure 22.6

Plotting the dependent and independent variables

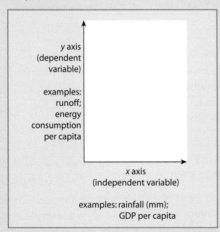

y axis (dependent variable)

examples: runoff; energy consumption per capita

x axis (independent variable)

examples: rainfall (mm); GDP per capita

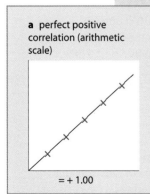

a perfect positive correlation (arithmetic scale)

= + 1.00

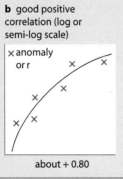

b good positive correlation (log or semi-log scale)

× anomaly or r

about + 0.80

c no correlation

0.00

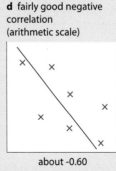

d fairly good negative correlation (arithmetic scale)

about -0.60

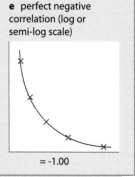

e perfect negative correlation (log or semi-log scale)

= -1.00

Figure 22.7

Types of correlation and their associated Spearman's rank coefficients

	GNP per capita		Energy consumption per capita				Birth rate			
	US$	Rank	kg oil-equivalent	Rank	d	d^2	per 1000	Rank	d	d^2
Switzerland	40 080	1	3622	6	5	25	11			
Norway	34 330	2	5284	2	0	0	11			
Japan	32 380	3	4058	4	1	1	10			
USA	29 340	4	8051	1	−3	9	15			
Germany	25 850	5	4267	3	−2	4	9			
UK	21 400	6	3992	5	−1	1	13			
Argentina	8970	7	1653	8	1	1	20			
Brazil	4570	8	1012	9	1	1	20			
Malaysia	3600	9	1950	7	−2	4	26			
Colombia	2600	10	799	11	1	1	21			
Egypt	1290	11	638	12	1	1	28			
China	750	12	902	10	−2	4	17			
India	430	13	476	13	0	0	25			
Kenya	330	14	466	14	0	0	32			
Sierra Leone	140	15	230	15	0	0	47			
						$\sum d^2 = 52$				$\sum d^2 =$

Spearman's rank correlation coefficient

This is a statistical measure to show the strength of a relationship between two variables. Figure 22.8 lists the GNP per capita for 15 selected countries. Fifteen is the minimum number needed in a sample for the Spearman's rank test to be valid.

The first stage is to see if there is any correlation between the GNP and the energy consumption per capita. This can be done using the following steps:

1 Rank both sets of data. This has already been done in Figure 22.8. Notice that the highest value is ranked first. Had there been two or three countries with the same value, they would have been given equal ranking, e.g. rank order: 1, 2, 3.5 , 3.5 (3.5 is the mean of 3 and 4) , 5, 7 , 7 , 7 (7 is the mean of 6, 7 and 8), 9, 10.

2 Calculate the difference, or d, between the two rankings. Note that it is possible to get negative answers.

3 Calculate d^2, to eliminate the negative values.

4 Add up ($\sum$) the d^2 values (in this example, the answer is 52).

5 You are now in a position to calculate the correlation coefficient, or r, by using the formula:

$$r = 1 - \frac{6\sum d^2}{n^3 - n}$$

where: d^2 is the sum of the squares of the differences in rank of the variables, and n is the number in the sample.

In our example it follows that:

$$r = 1 - \frac{6 \times 52}{3375 - 15}$$

$$= 1 - \frac{312}{3360}$$

= 1 − 0.093 (then do not forget the final subtraction)

= 0.91 (it is usual to give the answer correct to two decimal places).

In this example, there is a strong, positive correlation (remember, a perfect positive correlation is 1.00) between the GDP and energy consumption per capita.

Although the closer r is to +1 or −1 the stronger the likely correlation, there is a danger in jumping to quick conclusions. It is possible that the relationship described may have occurred by chance. The second stage is therefore to test the **significance** of the relationship. This is done by using the graph shown in Figure 22.9. Note that the correlation coefficient r is plotted on the y axis and the **degrees of freedom** on the x axis. Degrees of freedom are the number of pairs in the sample minus two.

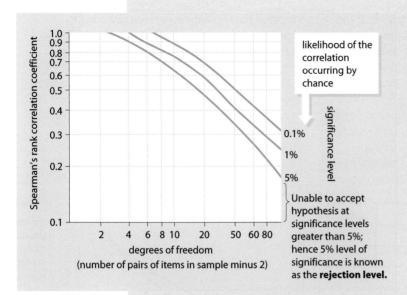

Figure 22.9

The significance of the Spearman's rank correlation coefficients and degrees of freedom

this distribution, then it might reasonably be expected that as area A covers 50 per cent of the total area, then half the villages would be located there. Similarly, areas B and C, each covering 20 per cent of the area, should both have 10 villages, leaving area D, with only 10 per cent of the area, with the remaining 5 villages. This means, as shown in Figure 22.11a, that we have two sets of data showing the **observed** (*O*) number and the **expected** (*E*) number of villages. In reality, however, Figure 22.10 shows that areas B and D have more villages than might be expected and A and C fewer than expected. It is tempting, therefore, to suggest that there could be a relationship between the observed and expected distributions and that this relationship is dependent upon the height of the land, whereas the difference may in fact be due entirely to chance factors. Chi-squared is used to estimate the probability that the differences are due to chance.

It is often best to begin with a null hypothesis, which in this case might be 'There is no significant relationship between the distribution of villages and the height of the land'. We can now use the formula for chi-squared, which is:

$$\chi^2 = \sum \frac{(O - E)^2}{E}$$

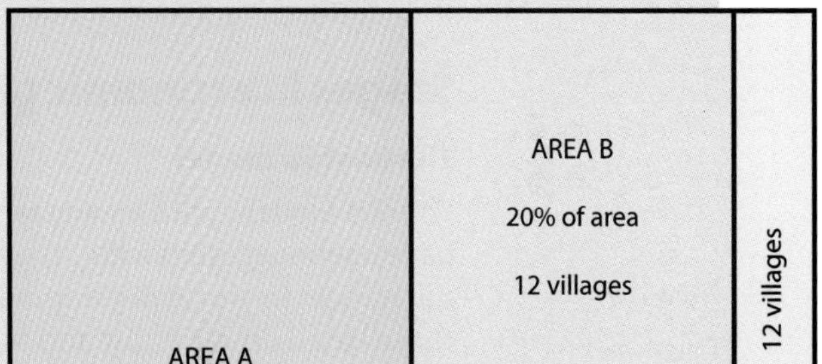

Using the correlation coefficient of GDP per capita and energy consumption per capita, which we have worked out to be 0.91, we can read off 0.91 on the vertical scale and 13 (i.e. 15 in the sample minus 2) on the horizontal. We can see that the reading lies just above the 0.1 per cent significance level curve. This means that we can say with 99.9 per cent confidence that the correlation has not occurred by chance. The graph also shows that if the correlation falls below the 5 per cent significance level curve then we can only say with less than 95 per cent confidence that the correlation has not occurred by chance. Below this point, the correlation or hypothesis is rejected in terms of statistical significance – i.e. there is too great a likelihood that the correlation has occurred by chance for it to be meaningful. Even if there is a significant correlation, the result does not prove that there is necessarily a *causal* relationship between variables. It cannot be assumed that a change in *A* causes a change in *B*. Further investigation is necessary to establish this.

Chi-squared

Whereas Spearman's rank seeks **associations** between *x* and *y* values, chi-squared looks for **differences** between groups (or areas). The symbol for chi-squared (*chi* is a Greek letter pronounced 'ky') is χ. Figure 22.10 shows the hypothetical distribution of villages over an area of land consisting of four contrasting categories of height, i.e. frequencies of 0–50 m, 51–100 m, 101–150 m and over 150 m (it could have been different types of soil, or rock type, etc.). Of the 50 villages located here, 20 are in area A, 12 in each of areas B and D, and 6 in area C. Had chance been the only factor affecting

Figure 22.10

Chi-squared: observed and expected villages

Figure 22.11b shows how to use the formula and, in this example, how we obtain a calculated value of chi-squared of 12.8. We can now, by using Figure 22.12, test for the significance of this value and determine the probability that the distribution was due to chance. Notice that, as in Spearman's rank (Figure 21.9), the horizontal axis is labelled 'degrees of freedom' (*df*). We read the degrees of freedom by subtracting 1 from the total number of distributions (areas), in this case 4 − 1 = 3. Using our two coordinates ($\chi^2 = 12.8$ and $df = 3$) we can obtain a location on the graph which is just above the 1 chance in 100 curve, i.e. our distribution is only likely to occur by chance once in every 100 situations. We can assume, therefore, that there is a possible connection between the distribution of villages and height of the land and so we can start looking for causes (had the location on the graph been below the 5 chances in 100 curve, then we could assume that there was no connection between village distribution and height of the land and therefore we need not spend time seeking reasons).

Figure 22.11

A worked chi-squared example

a

Area	A	B	C	D	Total
O (Observed)	20	12	6	12	50
E (Expected)	25	10	10	5	50

Using chi-squared

b

		A	B	C	D	
(i)	$(O - E)$	− 5	+2	−4	+7	
(ii)	$(O - E)^2$	25	4	16	49	
	$\dfrac{(O - E)^2}{E}$	1.0	0.4	1.6	9.8	
(iv)	$\Sigma \dfrac{(O-E)^2}{E}$ (sum of)	1.0 +	0.4 +	1.6 +	9.8 =	12.8

$\therefore \chi^2 = 12.8$

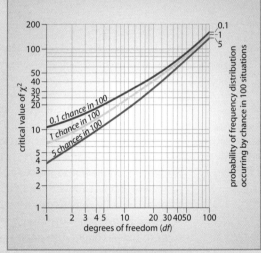

Figure 22.12

The significance of chi-squared and degrees of freedom

Stages in economic growth

The Rostow model

Figure 22.13

Rostow's model of economic growth

Various models, with a wide range of criteria, have been suggested when trying to account for differences in world development. These include those based on capitalist and Marxist systems as well as those more concerned with wealth, social and cultural differences. One of the first models to account for economic growth, and probably still the simplest, was that put forward by W. W. Rostow in 1960. Following a study of 15 countries, mainly in Europe, he suggested that all countries had the potential to break the cycle of poverty and to develop through five linear stages (Figure 22.13).

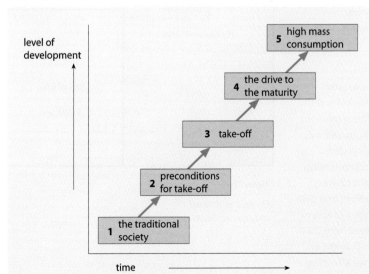

Approximate date of reaching a new stage of development

Country \ Stage	2	3	4	5
UK	1750	1820	1850	1940
USA	1800	1850	1920	1930
Japan	1880	1900	1930	1950
Venezuela	1920	1950	1970	—
India	1950	1980	—	—
Ethiopia	—	—	—	—

Figure 22.14

Changes in employment
structure based on
Rostow's model

	1 Primary	2 Secondary	3 Tertiary (services)
Stage 1	vast majority	very few	very few
Stage 2	vast majority	few	very few
Stage 3	declining	rapid growth	few
Stage 4	few	stable	growing rapidly
Stage 5	very few	declining	vast majority

Stage 1: Traditional society A subsistence economy based mainly on farming with very limited technology or capital to process raw materials or develop industries and services (Figure 22.14).

Stage 2: Preconditions for take-off A country often needs an injection of external help to move into this stage. Extractive industries develop. Agriculture is more commercialised and becomes mechanised. There are some technological improvements and a growth of infrastructure. The development of a transport system encourages trade. A single industry (often textiles) begins to dominate. Investment is about 5 per cent of GDP.

Stage 3: Take-off Manufacturing industries grow rapidly. Airports, roads and railways are built. Political and social adjustments are necessary to adapt to the new way of life. Growth is usually limited to one or two parts of the country (**growth poles** – page 569) and to one or two industries (**magnets**). Numbers in agriculture decline. Investment increases to 10–15 per cent of GDP, or capital is borrowed from wealthier nations.

Stage 4: The drive to maturity By now, growth should be self-sustaining. Economic growth spreads to all parts of the country and leads to an increase in the number and types of industry (the **multiplier effect**, page 569). More complex transport systems develop and manufacturing expands as technology improves. Some early industries may decline. There is rapid urbanisation.

Stage 5: The age of high mass consumption Rapid expansion of tertiary industries and welfare facilities. Employment in service industries grows but declines in manufacturing. Industry shifts to the production of durable consumer goods.

Criticisms of Rostow's model

Rostow's model, put forward in 1960, suffers the same criticisms as several other models, of being both outdated and oversimplified (Framework 12, page 352), although, as one critic concedes, 'the alternatives are just too difficult to explain and to apply'. You should be aware, however, of such valid criticisms:

- The model assumes, incorrectly, that all countries start off at the same level.

- While capital was needed to advance a country from its traditional society, often the injection of aid has been dwarfed by debt repayments which delayed, and has even prevented some countries (especially in Africa), from reaching the 'take-off' stage.
- The model underestimates the extent to which the development of some countries in the past was at the expense of others, e.g. through colonialism and imperialism.
- It predicts too short a timescale between the beginning of growth and the time when a country becomes self-sustaining. It over-emphasises the effect of the **learning curve**, i.e. the time taken for a country to develop diminishes as countries learn from others that are already developed. While the emergence of the NICs (page 576) seems to support Rostow's claim, the effect of the Asian financial crises (late 1990s) on them and the rest of the world, shows that he underestimated the effects of globalisation.
- The model has not seen a universal sequence and is, according to Barke and O'Hare amongst others, too Eurocentric.

Barke and O'Hare's model for West Africa

Barke and O'Hare (*The Third World*, 1984) claimed that although developed industrial countries may have moved through Rostow's five stages, it seems increasingly unlikely that countries that have yet to develop economically will follow the same pattern. This may be because capital alone is insufficient to promote take-off. Perhaps what is needed is a fundamental structural change in society which encourages people to save and invest and to develop an entrepreneurial, business class, as was the case in Hong Kong. Possibly the process which allows transition from traditional agriculture to advanced industry is a relict one, being applicable only to the early industrialised countries which had unlimited use of the world's resources and markets. Barke and O'Hare have suggested a four-stage model for industrial growth in developing countries, pointing out that elements from different stages often exist side by side, providing a 'dual economy'.

Stage 1: Traditional craft industries These were in existence before European colonisation, e.g. cloth weaving, iron working, wood carving and leather goods in northern Nigeria (Kano).

Stage 2: Colonialism and the processing of primary products Raw materials were initially exported in an unprocessed form (cocoa and palm oil) while the chief imports (textiles and machinery) came from the colonial power and, being cheaper, destroyed many local craft industries. Later, some processing took place, usually in ports or the primate city (page 405), if it reduced the weight for export (vegetable oils and sugar), if it was too bulky to import (cement), or if there was a large local market (textiles). To help obtain raw materials from their colonies, the European powers built ports (Accra and Lagos), but railways were only constructed if there were sufficient local resources to make them profitable (Figure 21.24). Education, along with the development of industrial and management skills, was neglected.

Stage 3: Import substitution During the Second World War and, later, following their independence, countries had to replace the import of textiles, furniture, hardware and simple machinery with their own manufactured goods. Production was in small units with limited capital and technology.

Stage 4: Manufacture of capital, goods and consumer durables As standards of living rose in several countries (notably in the NICs in Latin America and South-east Asia), there was an increased demand for heavier industry and 'Western'-style durable consumer goods. These industries, often because of the investment and skills needed, were developed by transnational companies wishing to take advantage of cheap labour, tax concessions and entry to a large local market (page 573). The American Valco company, for example, in the mid-1950s constructed a dam on Ghana's River Volta, a hydro-electric power station at Akosombo, and an aluminium smelter at Tema, in return for duty and tax exemptions on the import of bauxite and the export of aluminium, and the purchase of cheap electricity. Projects developed by transnationals are usually prestigious, of limited value to the country, and may be withdrawn (Volkswagen have stopped operating in Nigeria) should world sales drop. In other cases, where private capital was not forthcoming or where the dominance of transnational corporations was felt to be undesirable, as in China and India, large-scale industrial development was promoted through five-year national plans for economic development (Case Study 19).

Core–periphery model

Economic growth and development are rarely even. We have already seen how Myrdal (page 569) identified 'growth poles' which, he claimed, developed into core regions; how in the 19th century it was the coalfields that formed Britain's major industrial areas; and, since 1980,

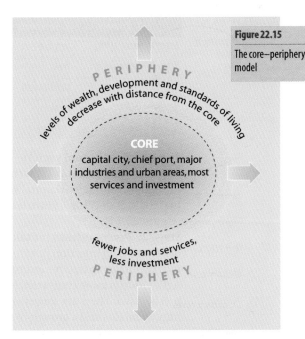

Figure 22.15

The core–periphery model

PERIPHERY

levels of wealth, development and standards of living decrease with distance from the core

CORE

capital city, chief port, major industries and urban areas, most services and investment

fewer jobs and services, less investment

PERIPHERY

how the artificially created Special Economic Zones became growth centres in China (Case Study 19). Economic activity, including the level of industrialisation and intensity of farming, decreases rapidly with distance from the core regions and towards the periphery – as shown in the '**core–periphery model**' (Figure 22.15).

The **core** forms the most prosperous and developed part of a country, or region. It is likely to contain the capital city (with its administration and financial functions), the chief port (if the country has a coastline) and the major urbanised and industrial areas. Usually, levels of wealth, economic activity and development decrease with distance from the core so that places towards the **periphery** become increasingly poorer.

As a country develops economically, one of two processes is likely to occur.

1 Economic activity in the core continues to grow as it attracts new industries and services (banking, insurance, government offices). As levels of capital and technology increase, the region will be able to afford schools, hospitals, shopping centres, good housing and a modern transport system. These 'pull' factors encourage rural in-migration (page 366). Meanwhile, in the periphery jobs will be relatively few, low-paid, unskilled and mainly in the primary sector, while services and government investment will be limited. These 'push' factors (page 366) force people to migrate towards the core. This process still seems to operate in the NICs and in many of the economically less developed countries (Kenya, Peru). Barke and O'Hare have suggested that 'just as it is possible to conceive of cores (MDCs) and

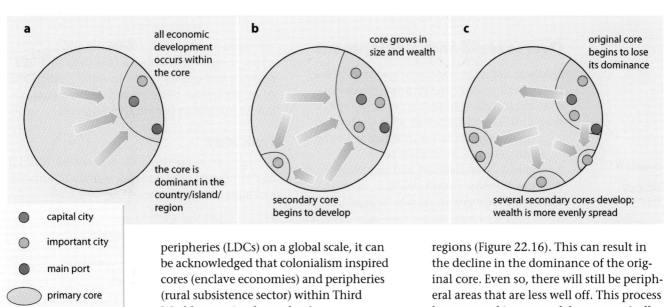

a all economic development occurs within the core

the core is dominant in the country/island/region

b core grows in size and wealth

secondary core begins to develop

c original core begins to lose its dominance

several secondary cores develop; wealth is more evenly spread

● capital city

○ important city

● main port

⬭ primary core

⬭ secondary cores

⬭ periphery

➜ migration of people

Figure 22.16

The hoped-for economic growth in a country

peripheries (LDCs) on a global scale, it can be acknowledged that colonialism inspired cores (enclave economies) and peripheries (rural subsistence sector) within Third World countries themselves'.

2 Industry and wealth begin to spread out more evenly. Initially, a second core region will develop followed by several secondary regions (Figure 22.16). This can result in the decline in the dominance of the original core. Even so, there will still be peripheral areas that are less well off. This process has occurred in many of the economically more developed countries, e.g. USA (Places 105) and Japan (Figure 19.20).

Places 105   The USA: core–periphery

By the late 19th century, the north-east of the USA had become one of the two major industrial regions in the world. Based on the local availability of coal, iron ore and limestone, the growth of a huge steel industry led the region, stretching from Chicago and Detroit on the Great Lakes to Boston, New York, Philadelphia and Baltimore on the Atlantic Coast, to become known as the 'manufacturing belt' (Figure 22.17a). The manufacturing belt, which was the country's core region for economic development, contained the capital city (Washington D.C.) as well as all the largest cities and ports (not just New York which was both the largest city *and* the largest port). The region was also the centre of commerce, trade, transport and retailing and was a major destination for immigrants from other parts of the world (mainly from Europe). Much of the remainder of the USA remained as a peripheral region.

During the mid-20th century, California emerged as a second core region (Figure 22.17b), with the growth of two large cities, the development of ports and an increase in in-migration, mainly from Mexico (Case Study 15) and other parts of the USA. The region attracted the aerospace industry (Second World War) together with electronics and high-tech industries (Silicon Valley south of San Francisco). The main attractions of California – notably its climate and access to an attractive environment – led, during the latter part of the 20th century, to the development of several other secondary core regions in the southern states (Figure 22.17c), in what has become known as the 'Sunbelt'. At the same time, the original core region began to lose much of its former dominance especially as the older industries declined in a region now referred to as either, economically, the 'Rustbelt' or, climatically, the 'Snowbelt' and as people move away from the region to the newer secondary 'cores'.

Figure 22.17

Core and periphery in the USA (see Figure 22.16 for key)

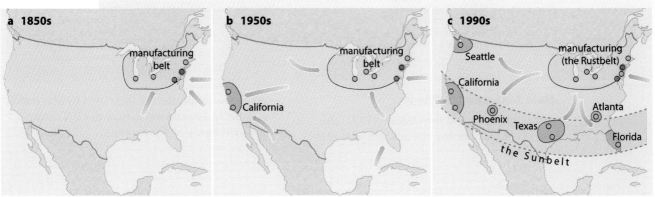

a 1850s — manufacturing belt

b 1950s — manufacturing belt, California

c 1990s — Seattle, California, manufacturing (the Rustbelt), Phoenix, Texas, Atlanta, Florida, the Sunbelt

Overseas aid and development

Overseas aid is the transfer of resources at non-commercial rates by one country (the donor), or an organisation, to another country (the recipient). The resource may be in the form of:

1 money, as grants or loans, which has to be repaid, even at low interest rates
2 goods, food, machinery and technology
3 know-how and people (teachers, nurses).

The basic aim in giving aid is to help poorer countries develop their economies and to improve services in order to raise their standard of living and quality of life. In reality, the giving of aid is far more complex and controversial as it does not always benefit the recipient.

Types of aid

Basically, there are two main types of aid: **official** and **voluntary**. The differences in their purposes and aims are summarised in Figure 22.18.

Figure 22.18

Official and voluntary aid

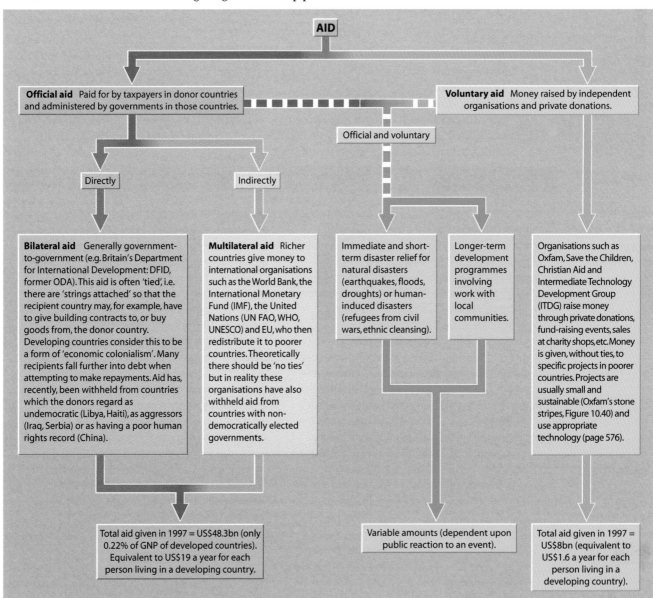

Main donors and recipients

Although Japan and the USA are (1997) the world's largest donors, the amount that each gives as a proportion of their own GNP is small (Figure 22.19) and well below the UN recommended figure of 0.7 per cent.

Figure 22.19

The major aid donor countries (1997)

Rank 1997	(1992)	Country	% GNP	US$bn	% change 1992–97
1	(3)	Denmark	0.97	1637	+3.7
2	(1)	Norway	0.86	1306	+0.5
3	(4)	Netherlands	0.81	2947	+1.4
4	(2)	Sweden	0.79	1731	−2.5
5	(6)	France	0.45	6107	−4.2
=6	(7)	Canada	0.34	2045	−4.0
=6	(8)	Switzerland	0.34	911	−3.1
8	(5)	Finland	0.33	379	−11.9
11	(10)	Germany	0.28	5857	−3.9
13	(14)	UK	0.26	3433	−0.3
18	(16)	Japan	0.22	9358	−5.8
20	(19)	USA	0.09	6878	−8.9

Rank 1997	(1990)	Country	% GNP	% change 1990–97
1	(38)	Rwanda	32.0	+18.4
2	(1)	Mozambique	29.6	−16.0
3	(15)	Mongolia	28.7	+11.6
4	(23)	Madagascar	24.3	+10.6
5	(5)	Mauritania	23.9	−1.9
6	(2)	Nicaragua	22.7	−16.3
7	(8)	Laos	19.5	−1.1
8	(9)	Mali	18.7	−2.3
9	(12)	Niger	18.6	+2.1
10	(17)	Zambia	16.7	−0.7
11	(57)	Burkina Faso	16.0	+6.9
12	(28)	Sierra Leone	15.9	+3.3

Figure 22.20

The major aid recipients (1997)

As for the recipients, while the two-thirds of the world's poorest countries (low-income economies) located in sub-Saharan Africa did receive 60 per cent of all overseas aid, there is no simple correlation between the level of poverty and the amount of aid received (Figure 22.20). Often aid is not given to countries with the greatest poverty and need, but to those with a strong political voice (Israel with its strong Jewish lobby in the USA), which have supported donor countries at times of war (Egypt and Turkey during the Gulf War), have a valuable natural resource (Kuwait's oil), have strong historic ties (Jamaica with Britain), or have strategic military locations (the Philippines).

The International Monetary Fund's (IMF) Structural Adjustment Programmes aim to help the poorest countries, many of which are in Africa, as well as the emerging economies, as with various South-east Asian countries in the wake of the late-1990s financial crisis. In contrast, the World Bank lends capital for specific projects.

Both agencies, however, have recently been through a period of self-examination about their priorities and support programmes. Meanwhile there was, in 1999, a high-profile campaign to cancel Third World debt for the year 2000.

While few people argue against emergency aid, except to say that it is often 'too little, too late', other forms of aid are more controversial. Some consider that no non-emergency aid should be granted, especially as it is usually given in the political, industrial and commercial interests of the donor country, without concern for the environment, and does little to improve the long-term quality of life in the recipient country. Aid tends to address the symptoms of poverty rather than the causes. Others feel that aid does make an important contribution to the economy of many poor countries and to the welfare of some of their poorest communities. Some of the arguments of the pro-aid and anti-aid groups are listed in Figure 22.21.

Figure 22.21

Arguments for and against the giving of aid

For

- Response to emergencies, both natural and human-induced.
- Helps in the development of raw materials and energy supplies.
- Encourages, and helps to implement, appropriate technology schemes.
- Provides work in new factories and reduces the need to import certain goods.
- Helps to increase yields of local crops (green revolution) to feed rapidly growing local populations.
- Provides primary health care, e.g. vaccines, immunisation schemes, nurses.
- Helps to educate people about, and to implement, family planning schemes.
- Grants to students to study in overseas countries.
- Can improve human rights.

Against

- Aid is a conscience-salver for the rich and former colonial powers.
- Better to use money on the poor living in the donor countries.
- An exploitation of physical and human resources.
- Used to exert political and economic pressure on poorer countries.
- Increases the recipient country's external debt.
- Often only goes to the rich and the urban dwellers in recipient countries, rather than to the real poor.
- Encourages corruption amongst officials in donor and recipient countries.
- Undermines local activities, e.g. farming.
- Does not encourage self-reliance of recipient countries.
- Often not given appropriate technology.

Health and development

Health is closely linked with development and yet often, until recently, it has often been neglected by geographers. Two of the HDI characteristics – life expectancy and infant mortality (page 632) – and several other measures of development – birth and death rates, calorie consumption, a balanced diet and the number of people per doctor or hospital bed (page 633) – are all health indicators. Health is more than just the absence of diagnosed physical disease. It is seen by the World Health Organisation (WHO) as a 'state of complete physical, mental and social well-being and not merely the absence of disease and infirmity'. It implies complex interactions between humans and their environments and, more particularly, between social and economic factors (Figure 22.22).

Bearing in mind Phillips's warning concerning the difficulties in trying to correlate economic development and health (Figure 22.23), there do appear to be marked differences in the major types of disease (Figure 22.24) and age at death (page 354) between developing and developed countries. Indeed, it has been suggested that as a country develops, it is likely to pass through several stages of **'epidemiological (or health) transition'** (Case Study 22). The *World Health Book* (1999) links the epidemiological transition to urbanisation – i.e. the concentration of people in urban areas increases exposure to communicable diseases and, among other things, adds new hazards such as pollution.

Figure 22.22

The complex inter-relationship between health and development

"It has long been acknowledged that the health status of the population of any place or country influences development. It can be a limiting factor, as generally poor individual health can lower work capacity and productivity; in aggregate in a population, this can severely restrict the growth of economies. On the other hand, economic development can make it possible to finance good environmental health, sanitation and public health campaigns — education, immunisation, screening and health promotion — and to provide broader-based social care for needy groups. General social development, particularly education and literacy, has almost invariably been associated with improved health status via improved nutrition, hygiene and reproductive health. Socio-economic development, particularly if equitably spread through the population — although this is rarely the case — also enables housing and related services to improve. The classical cycle of poverty can be broken by development.

However, it is notoriously difficult to provide generalisations about the relationship between economic development and a population's health status. We can cite examples in which correlations between GNP and life expectancy are not straightforward. There are many examples to show how economic development has contributed to improving quality of life and health status, via indicators such as increased life expectancy, falling infant, child and maternal mortality and enhanced access to services. By contrast, there are examples in which economic development, infrastructure expansion and agricultural intensification do not always coincide with improved human well-being. There is, in fact, a growing realisation that macroeconomic changes may not always filter down to benefit all of the population, and many perhaps soundly based policies in economic terms can have devastating human effects in increasing poverty and maldistribution of resources."

David Phillips and Yola Verhasselt

Figure 22.23

Differences in health care
a Cataract camp, Calcutta

b Intensive care unit, St Bartholomew's, London

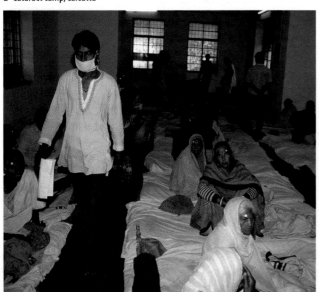

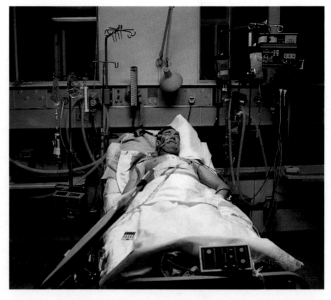

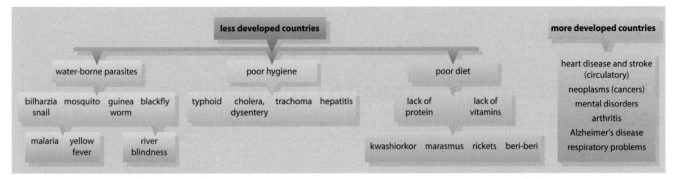

less developed countries			more developed countries
water-borne parasites	**poor hygiene**	**poor diet**	heart disease and stroke (circulatory)
bilharzia mosquito guinea blackfly	typhoid cholera, trachoma hepatitis	lack of lack of	neoplasms (cancers)
snail worm	dysentery	protein vitamins	mental disorders
malaria yellow river		kwashiorkor marasmus rickets beri-beri	arthritis
fever blindness			Alzheimer's disease
			respiratory problems

Figure 22.24

Differences in types of disease between less and more developed countries

AIDS and HIV

Acquired immune-deficiency syndrome (AIDS), first described in medical literature in 1981, presents one of the greatest threats to global public health. In 1999, the United Nations Population Division, which studies the demographic impact of AIDS in the world, stated the following:

- 7.9 million adult and child AIDS cases had occurred since the illness had first been recognised, with cases confirmed in 195 different countries.
- Over 33 million people were believed to be **HIV (human immunodeficiency virus)** affected, including 1.6 million born with the virus (both figures were estimates as only a small proportion of those infected are either aware of it or have been tested).
- The Division was paying special attention to 34 developing countries which are hardest hit, i.e. with an adult HIV prevalence of 2 per cent or more. Of these countries, 29 are in sub-Saharan Africa, 3 in Asia (India, Cambodia and Thailand) and 2 in Latin America (Brazil and Haiti). Of the 33 million people currently affected by HIV, 85 per cent (28 million) live in these 34 countries, and 91 per cent of AIDS deaths have been recorded there.

- Life expectancy at birth in the 29 hardest-hit African countries has fallen to 47 years – 7 years less than what could have been expected in the absence of AIDS (page 360). The demographic impact of AIDS is even more dramatic in the five worst-affected countries (Figure 22.25), where life expectancy has fallen by 10 years and could, by 2015, be 16 years less (47 years instead of 63 years).
- 2.6 million people were expected to die of AIDS in 1999 (2.3 million died in 1998).
- The three main means of HIV transmission are from the exchange of body fluids in sexual intercourse (with greater efficiency from male to female than vice versa), through infected blood (sharing needles/syringes and via contaminated blood trans-fusions), and parentally from mother to child during pregnancy or birth. The dominant forms of transmission vary worldwide (Figure 22.26).

Figure 22.26

Estimated cumulative global distribution of HIV infections, mid-1999

Figure 22.25

AIDS and the declining life expectancy in selected sub-Saharan countries

global total about 33 000 000

30 000

1 000 000

500 000

50 000   2 000 000 +

28 000 000 +

1 500 000 +

30 000

Pattern 1 countries:
Extensive spread occurred here in the late 1970s, predominantly amongst the homosexual, bisexual and intravenous drug-using community. Heterosexual spread is slowly increasing.

Pattern 2 countries:
Spread here also occurred in the late 1970s but predominantly by heterosexual transmission. Vertical transmission from mother to child and transmission via contaminated blood and blood products are also important routes.

Pattern 3 countries:
Here HIV infection was introduced later, probably in the 1980s by travellers and also by imported infected blood and blood products.
Source: WHO (1992).

The epidemiological (health) transition in South-east Asia

'It has long been recognised that societies pass through various patterns of **morbidity** (illness and disease) and mortality (causes of death) during the development process, even if not all the stages and sequences are identical in every case. In general, health improves, morbidity and mortality fall and come from different causes, and life expectancy increases; this comprises the **'epidemiological transition'** (after Omran, 1971). More recently the term **'health transition'** is being used, as it has a broader concept than epidemiological, i.e. it focuses on health rather than just on morbidity. These changes generally come with 'modernisation' and are indeed part and parcel of the process. They seem to occur at a different pace in varying countries and, in recent years, they are related to the application of modern medical techniques and technology as well as to changing standards of living, nutrition, housing and sanitation.'

David Phillips, *The Epidemiological Transition in Hong Kong*, 1988

The demographic transition model (Figure 13.10) suggests that fertility (birth rate) declines appreciably, probably irreversibly, when traditional, mainly agrarian societies are transformed by modernisation, industrialisation and bureaucratic urban-oriented societies. This rather straightforward and simplistic demographic transition assumes that, for example, a simple industrial-economic modernisation will occur in societies accompanied by changes in lifestyles, living conditions and health levels. Of greater interest to epidemiologists, health planners and medical geographers is that with 'modernisation' and increasing affluence and life expectancy comes a very different disease or ailment profile in most countries from that which previously existed in a 'traditional' state or developing country. Figure 22.27 has been adapted from Omran's epidemiological transition. Initially, three stages of the transition were envisaged:

1 the age of pestilence and famine which gradually merges into ...
2 the age of receding pandemics (worldwide diseases), giving way to ...
3 the age of degenerative and human-induced diseases.

More recent studies have suggested the emergence of ...

4 the age of delayed degenerative diseases and, associated with a lengthening of life, poorer health.

Omran suggested that there were three variations in the basic model:

1 The classical or 'Western' model which took place over a prolonged period (100 to 200 years).
2 The 'accelerated' model which occurred in Japan, after the Second World War, and more recently in Hong Kong, Singapore and other NICs in South-east Asia. This showed rapid declines in mortality and fertility.
3 The 'delayed' model which is common to many of today's less developed countries. It contains elements of morbidity and mortality from both degenerative and infectious diseases but, at the same time, lacks the marked reduction in fertility experienced in the 'Western' model.

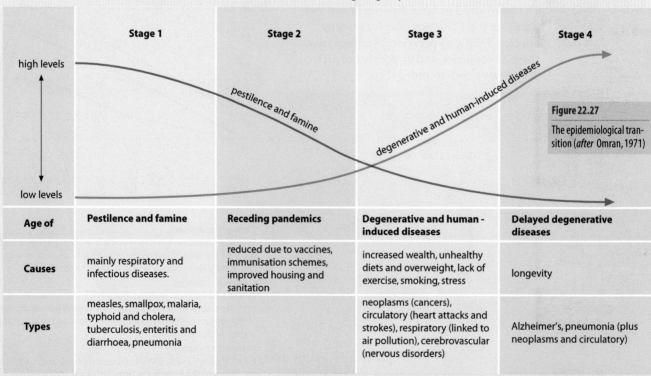

	Stage 1	Stage 2	Stage 3	Stage 4
Age of	**Pestilence and famine**	**Receding pandemics**	**Degenerative and human-induced diseases**	**Delayed degenerative diseases**
Causes	mainly respiratory and infectious diseases.	reduced due to vaccines, immunisation schemes, improved housing and sanitation	increased wealth, unhealthy diets and overweight, lack of exercise, smoking, stress	longevity
Types	measles, smallpox, malaria, typhoid and cholera, tuberculosis, enteritis and diarrhoea, pneumonia		neoplasms (cancers), circulatory (heart attacks and strokes), respiratory (linked to air pollution), cerebrovascular (nervous disorders)	Alzheimer's, pneumonia (plus neoplasms and circulatory)

Figure 22.27

The epidemiological transition (*after* Omran, 1971)

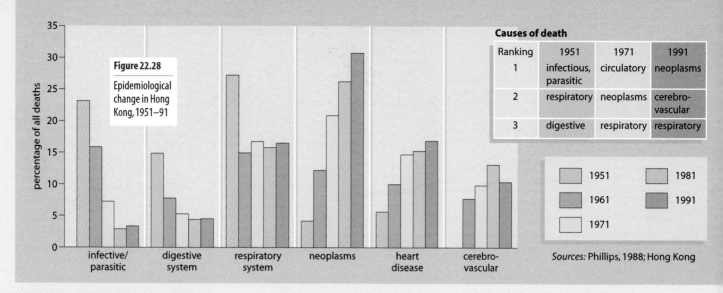

Figure 22.28

Epidemiological change in Hong Kong, 1951–91

Causes of death

Ranking	1951	1971	1991
1	infectious, parasitic	circulatory	neoplasms
2	respiratory	neoplasms	cerebro-vascular
3	digestive	respiratory	respiratory

Legend: 1951, 1961, 1971, 1981, 1991

Sources: Phillips, 1988; Hong Kong

Figure 22.28 shows the epidemiological change for Hong Kong between 1951 and 1991. The graph confirms three trends:

1 The rapid decline in infective/parasitic diseases (improved standard of living, housing conditions, immunisations) and digestive complaints (improved health care, better diet).

2 An initial drop in respiratory illnesses which has since been reversed (increased traffic emissions).

3 The rapid increase in deaths from 'Western' diseases, especially neoplasms (cancers) and heart disease.

Hong Kong, like Singapore, shows a remarkable closeness to Omran's accelerated model. Thus both countries may provide an epidemiological exemplar likely to be followed by other existing, and emerging, NICs in South-east Asia – Taiwan, South Korea, Thailand and Malaysia.

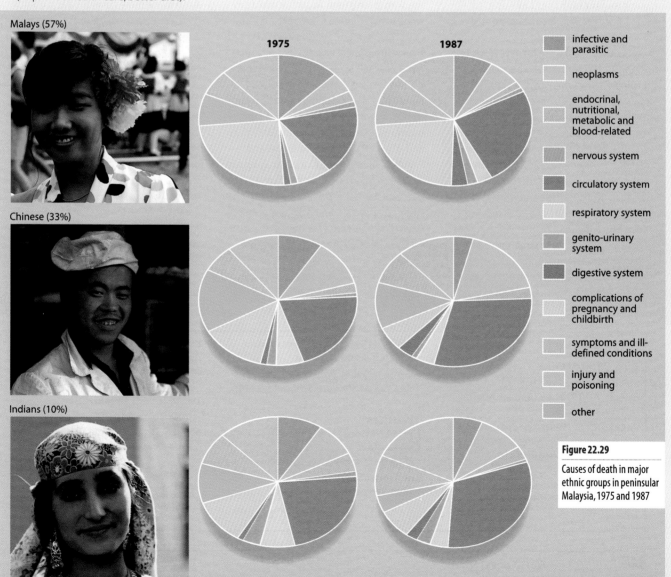

Malays (57%)

Chinese (33%)

Indians (10%)

1975   1987

Legend:
- infective and parasitic
- neoplasms
- endocrinal, nutritional, metabolic and blood-related
- nervous system
- circulatory system
- respiratory system
- genito-urinary system
- digestive system
- complications of pregnancy and childbirth
- symptoms and ill-defined conditions
- injury and poisoning
- other

Figure 22.29

Causes of death in major ethnic groups in peninsular Malaysia, 1975 and 1987

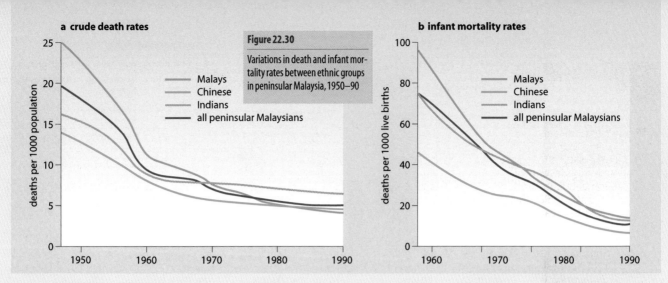

a crude death rates

deaths per 1000 population

- Malays
- Chinese
- Indians
- all peninsular Malaysians

1950 1960 1970 1980 1990

b infant mortality rates

deaths per 1000 live births

- Malays
- Chinese
- Indians
- all peninsular Malaysians

1960 1970 1980 1990

Figure 22.30

Variations in death and infant mortality rates between ethnic groups in peninsular Malaysia, 1950–90

However, Hong Kong and Singapore share three characteristics which are not all shared by other NICs in the region. They both:

1 have a fairly homogeneous ethnic population

2 form small, compact, island/city states with most of their inhabitants being urban-dwellers

3 have experienced rapid economic growth which they have used to improve their health service and standard of housing.

Some NICs in the region have a greater ethnic mix. In peninsular Malaysia, for example, there are three main ethnic groups – Malays, Chinese and Indians. While there has been a similar pattern of change in the causes of death for all three groups, the changes have affected them differently (Figure 22.29). Similarly, while the crude death and infant mortality rates for each group have declined, the rate of decline has differed noticeably between them (Figure 22.30). In other words, just as there are differences in epidemiological change between countries, so too there can be differences between ethnic groups within a country. These may be due to dif-

ferences in wealth, social status, religion, social customs and urbanisation.

The value of the epidemiological (health) transition

Perhaps the most important role the epidemiological transition can play is to provide a formal framework within which to set health and health-care strategies over the medium to long term. As such, it could provide a major stimulus to future health-care needs both within countries when directed by governments, or globally through international health agencies. It could help health planners in places where change is very rapid (NICs), is varied between social groups (rich and poor communities in developing countries) and ethnic groups (South Africa), and where health care is expensive and finance is limited (the UK). It could also point out the growing needs for care from causes like mental illness, especially in developing countries, and Alzheimer's disease, in more developed countries, which are both considerably underestimated in much health sector planning.

The epidemiological transition is relevant for manufacturers and suppliers of

medicine and health equipment, for researchers looking for new vaccines, and for governments trying to decide where best to allocate funds.

Finally, by identifying a fourth stage, that of the age of delayed degenerative diseases, the epidemiological transition draws attention to the world's ageing population (page 360). This stage, although at present confined to the more developed and wealthy 'Western' countries (Japan, the UK), suggests a lengthy old-age potentially dogged with chronic, but non-fatal, ailments. Old age, faced by an ever-increasing proportion of the population and whose health and social needs are often greater than those in younger age groups, may not be attractive unless public and family support are forthcoming. Although developing countries are further from this stage, nevertheless many are experiencing a rapid increase in longevity, resulting in more people needing care as they live longer. Due to the increasing numbers of the elderly in many developing countries (China – Case Study 13; India), and due to the absolute totals, it is necessary to start planning now for their future health and social care.

References

Barke, M. and O'Hare, G. (1991) *The Third World*, Oliver & Boyd.

Gilbert, A. (1992) *An Unequal World*, Thomas Nelson.

Phillips, D. R. (1988) *The Epidemiological Transition in Hong Kong*, Centre of Asian Studies, University of Hong Kong.

Phillips, D. R. and Verhasselt, Y. (eds) (1994) *Health and Development*, Routledge.

Tully, M. (1991) *No Full Stops in India*, Penguin Books.

Wilson, J. (1995) *Statistics in Geography for 'A' Level Students*, Schofield & Sims.

Websites

United Nations Development Programme:
 http://www.undp.org/

The World Bank Group:
 http://www.worldbank.org/

Canadian International Development Agency, Virtual Library on International Development:
 http://w3.acdi-cida.gc.ca/virtual.nsf

World Health Organization:
 http://www.who.int/

See also for more links:
 http://www.nelsonthornes.com/gaia

1 a i What is meant by 'gross domestic product' (GDP) per capita? **(2 marks)**

ii Why is this often chosen as a useful indicator of a country's level of development? **(2 marks)**

iii Sometimes the 'Human Development Index' (HDI) is used to indicate level of development, rather than using GDP/capita. What are the advantages of using the HDI? **(4 marks)**

b Study Figure 22.31.

i Name one cause of death which fell between 1968 and 1985. Suggest the reason for this fall. **(3 marks)**

ii Name one cause of death which rose during this period. Suggest the reason for this increase. **(3 marks)**

c A falling death rate is one good indicator of a country with an economy that is developing. Choose one of the following sets of statistics that can also be used to show development:

■ energy consumption/person
■ number of doctors/thousand people
■ level of education of females.

i Explain why it is a good indicator of the level of development of a country. **(4 marks)**

ii Why will an increase in the level of that indicator help to bring about further economic development? **(7 marks)**

2 a The Rostow model shows the economy of a country going through five stages:

■ traditional society
■ preconditions for take-off
■ take-off
■ the drive to maturity
■ high mass consumption.

i Briefly describe the characteristics of each stage. **(5 marks)**

ii Explain the importance of the role of capital in the model. **(3 marks)**

iii Explain why the problem of debt seems to be stopping some countries from moving through the stages of the Rostow model. **(5 marks)**

b In Myrdal's core–periphery model, why does population often move from the periphery towards the core? **(3 marks)**

c Name a country that shows evidence of having a core and a periphery. Explain how Myrdal's model helps you to understand the distribution of economic development in that country. **(9 marks)**

AS

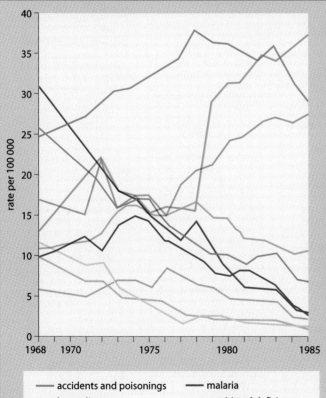

— accidents and poisonings

— heart diseases

— pneumonia

— diarrhoeal diseases

— tuberculosis of the respiratory system

— malignant neoplasms (mainly cancers)

— malaria

— nutritional deficiency diseases

— diseases of pregnancy and childbirth

— digestive diseases

Source: Institute for Population and Social Research, 1988

3 a Compare the Human Development Index (HDI) and gross domestic product (GDP) per capita as ways of measuring the level of development of different countries. **(10 marks)**

b Study Figure 22.31. Explain how economic development has helped to cause the changes shown on the graph. **(15 marks)**

4 a Explain the importance of capital investment in Rostow's model of industrial development. **(6 marks)**

b Barke and O'Hare developed a different model to help explain the way many African countries were developing. Explain the importance of transnational corporations in their model. **(5 marks)**

c Name a less economically developed country that has tried to develop its economy without relying on investment by transnational corporations.

i Explain how your chosen country has tried to promote industrial development. **(6 marks)**

ii Assess the extent to which the country has been successful in developing its economy. **(8 marks)**

5 With reference to **one** country where there are marked differences between the level of development in the core region and the periphery:

■ explain why the different levels of economic development have arisen
■ explain what the government is doing to try to reduce the differences between the core and the periphery. **(25 marks)**

Figure 22.31

Health transition in Thailand: crude death rates per 100 000 of population for major causes of death, 1968–85

A2

Index